T0136672

MINERALS OF NEVADA

MINERALS OF NEVADA

STEPHEN B. CASTOR
and GREGORY C. FERDOCK

NEVADA BUREAU OF MINES AND GEOLOGY

Special Publication 31
in association with the

UNIVERSITY OF NEVADA PRESS

Reno & Las Vegas

Nevada Bureau of Mines and Geology Special Publication 31

University of Nevada Press, Reno, Nevada 89557 USA

Copyright © 2004 by University of Nevada Press

Photographs by Stephen B. Castor, Jeff Scovil, Storm Sears, Jeffrey Weissman, and Sugar
White copyright © 2004 unless otherwise noted

Map of Mining Districts and significant mineral occurrences of Nevada by Nevada Bureau of
Mines and Geology

All rights reserved

Manufactured in the United States of America

Design by Barbara Jellow

Library of Congress Cataloging-in-Publication Data

Castor, Stephen B.

Minerals of Nevada / Stephen B. Castor and Gregory C. Ferdock.— 1st ed.

    p. cm.

Includes bibliographical references and index.

  ISBN 0-87417-540-2 (hardcover : alk. paper)

1. Minerals—Nevada. 2. Mines and mineral resources—Neavda. 3. Minerals—Collection
and preservation—Nevada. I. Ferdock, Gregory C., 1962– II. Title.

  QE375.5.N3 C37 2004

  549.9793—dc21   2003010951

The paper used in this book meets the requirements of American National Standard for
Information Sciences—Permanence of Paper for Printed Library Materials, ANSI Z39.48-1984.
Binding materials were selected for strength and durability.
First Printing

12 11 10 09 08 07 06 05 04

5 4 3 2 1

Frontispiece: Olivenite, 1.6-cm cluster, Majuba Hill Mine, Antelope district, Pershing County.
N. Prenn collection. Photograph by J. Scovil.

Publication of *Minerals of Nevada*
was made possible by funding from

AMERICAN PACIFIC CORPORATION

and the

BUREAU OF LAND MANAGEMENT

American Pacific Corporation, head-quartered in Las Vegas, Nevada, man-ufactures chemicals for use in solid rocket propellant, automotive airbag safety systems, and fire extinguishing sys-tems. The company is also active in real estate in Nevada and the production of environmental protection equipment.

American Pacific was organized in 1955. Fred Gibson Sr., the principal founder, came to Neva-da in 1929 and was actively involved in mining in the state and the western U.S. for the rest of his life. His sons Fred Gibson Jr. and John Gibson, both graduates of the Mackay School of Mines, have continued the tradition. In 1956, Fred Gib-son Jr. began a long career with the company, ris-ing through the ranks until he became CEO and president in 1985. John Gibson assumed leader-ship of the company with Fred's retirement in 1998 and has continued the American Pacific tra-dition of civic and philanthropic involvement in Las Vegas and throughout the state.

In the words of John Gibson,

Mining and the discovery, identification, and development of mineral resources have been fundamental to the economic prosperity of Nevada. American Pacific has, and will continue to be, active in the support of those industries related to mineral resources. It is an honor for American Pacific Corporation to assist in the publication of *Minerals of Nevada,* which we hope will be a valuable resource for the student, hobbyist, businessman, and citizen alike.

Within the state of Nevada, al-most forty-eight million acres of public land are administered by the Bureau of Land Management (BLM), an agency of the U.S. Department of the Interior. This land comprises 67 percent of the state. Nevada's public lands offer mining, live-stock grazing, wild horse viewing, scenic won-ders, and recreation and wilderness opportu-nities. Geologic, paleontologic, archaeologic, and cultural sites draw scientists from through-out the world for study. The BLM is responsible for inventorying, managing, and conserving Ne-vada's natural and cultural heritage for future generations.

In addition to helping with *Minerals of Nevada,* the BLM has assisted in the publication of the Nevada Bureau of Mines and Geology guide-books such as *Traveling America's Loneliest Road* and *Geologic Tours in the Las Vegas Area.*

Contents

Illustrations

Preface   STEPHEN B. CASTOR

Despite the importance of mineral wealth to Nevada's commerce and history, no prior work has been produced that deals comprehensively with occurrences of mineral species in the state. Publications that have dealt in part with the subject include Vincent Gianella's *Nevada's Common Minerals,* published in 1941 as a Nevada University Bulletin. Gianella listed 402 mineral species as occurring in the state, but included descriptions of occurrences for only the most common. In 1985, an issue of the journal *Mineralogical Record* was devoted to Nevada's minerals, mainly from the collector's standpoint. In 1999, an issue of the journal *Rocks and Minerals* focused on the minerals and fossils of Nevada.

This book consists primarily of a catalog of mineral occurrences in Nevada. Separate articles on some specific areas and types of mineral deposits in the state are also included, as is a small-scale map showing mining districts and sites that are important past or present mineral collection sites. The book is intended mainly as a reference source. It is designed to be used by collectors, mineral researchers, exploration geologists, prospectors, and interested laypersons.

Nevada is a wonderful state for mineral and gem collectors, in part because of the abundance and variety of mineral localities, and in part because it consists mostly of public land that is open to collecting. The mineral catalog in this book may be used as a guide to collecting sites; however, collecting is not possible at some of the sites listed for one reason or another, and permission to collect at a locality is not given or implied by its inclusion in the catalog.

Many sites where mineral specimens were found in the past are now inaccessible, such as those in caved-in mine workings or in workings where mining (or collecting) has completely removed all traces of the mineral. In other cases, active mining operations will preclude collection of specimens although commercial collectors may contract with a mine to collect and sell minerals, and the miners themselves may be a source of specimens. Deposits of valuable minerals, particularly gemstones, are often privately controlled by individuals or companies (by unpatented or patented mining claim or other ownership), and unauthorized collection at such sites is a criminal offense. At some privately held sites, such as the precious opal mines in Virgin Valley in northern Humboldt County and the crystalline quartz collecting area on Petersen Mountain in Washoe County, permission to collect may be paid for, and these payments may, in fact, be the main commercial activity at the site.

Although 87 percent of the land in Nevada is federally managed land and most of this is open to public access, there are many parcels, large and small, of private land. Here access for mineral collecting is at the option of the landowner, and readers are urged to contact such landholders prior to collecting. In addition, large tracts of land controlled by the U.S. Department of Defense and the U.S. Department of Energy have been withdrawn from public entry, because they are used for potentially dangerous or secret military activities. Such lands are fenced in or designated by signs along roads or other access, and some are actively patrolled; trespass upon them is considered a federal offense. The U.S. Bureau of Land Management, which has offices in Reno, Carson City, and Las Vegas, can provide maps of land ownership in Nevada, as well as information on unpatented mining claims in specific areas in the state.

In addition to access problems, legal or otherwise, there are safety problems related to collecting at some sites. Most of the mineral localities listed are inactive mines, and such sites may be hazardous for a number of reasons, including deep shafts, unstable workings, discarded explosives, and bad air. Over the years, a number of people have died as the result of accidents in old mine workings. Abandoned underground mines are particularly hazardous, but open-pit mines and even small open trenches can be dangerous. The Nevada Division of Minerals, with offices in Carson City and Las Vegas, can provide more information on hazards in abandoned mines.

Acknowledgments

Most of the work that went into this book was performed by Nevada Bureau of Mines and Geology (NBMG) personnel and by Mackay School of Mines students under contract to the NBMG. We acknowledge Jon Price, director of the NBMG, for his support during the years required for data collection and text preparation, Dick Meeuwig for helping with editing, and Kris Pizarro for her work on illustrations and maps.

In addition, Tom Lugaski of the W. M. Keck Museum, Mackay School of Mines, is thanked posthumously for providing access to specimens, particularly for photographic sessions. We are also grateful to Jeff Scovil, who took most of the photographs in this book, for providing excellent images of Nevada specimens, and to Sugar White for her fine photomicrographs. We also thank Jeffrey Weissman and Storm Sears for their mineral photographs.

The U.S. Department of the Interior Bureau of Land Management, which has provided partial funding for other NBMG publications, is acknowledged for similar help with *Minerals of Nevada*. In particular, we would like to thank Tom Leshendok and Jack Crowley for their support. The Nevada Division of Minerals, Walt Lombardo, and Steve Scott provided funding for photography. We are also grateful for contributions from the Nevada Mining Association, Mine Development Associates, and Neal and Camille Prenn.

This book has benefited tremendously from information given freely by mineral collectors. We thank Forrest Cureton, Bruce Hurley, Martin Jensen, Jack Kepper, Scott Kleine, Walt Lombardo, Jim McGlasson, Neil Prenn, Steve Pullman, Mark Rogers, Bruce and Jo Runner, Steve Scott, Dick Thomssen, and Bob Walstrom, who deserve recognition for their help. We thank Robert E. Jenkins for permitting us to use an unpublished manuscript that he prepared during his years in Nevada. We also thank Steve Pullman and Richard Bideaux for their reviews of the manuscript.

Others have also been helpful. Dave Tretbar assisted with the essay on Carlin-type deposits, particularly with unpublished data on the Getchell Mine. Don Hudson is acknowledged for sharing his knowledge of some Nevada mineral occurrences. Pete Vikre provided useful information on minerals in some classic mining areas.

Finally, we would like to acknowledge our wives—Valerie Castor for her support and patience during the evening, weekend, and vacation hours that the senior author devoted to *Minerals of Nevada,* and Christiane Ferdock for her support and for helpful reviews.

INTRODUCTION

PART I

The Importance of Minerals

STEPHEN B. CASTOR

Nevada has been blessed by far more than its share of mineral wealth, and minerals and their exploitation have long been a factor in mankind's interest in the mostly arid and hostile landscape that was to become Nevada. Prior to the presence of white men, Native American tribes such as the Anasazi were engaged in mining more than 1000 years ago. These early miners, who mined salt and turquoise in southern Nevada, actually engaged in underground mining practices. There is equivocal evidence that Mormon pioneers were mining in the Wahmonie district in 1846, and it is probable that hard-rock metal mining by Mormon settlers beginning in 1855 in the Goodsprings district near Las Vegas predated hard-rock mining in Virginia City along the Comstock Lode. The latter, however, is credited by many as the true beginning of the state's mining history.

Mineral wealth was the major factor in the settling of Nevada. Discoveries of metallic-mineral deposits led to foundation of the communities of Virginia City, Austin, Aurora, Unionville, and others during the time that Nevada was a territory. It has been suggested that when statehood was granted to the sparsely settled Nevada Territory in 1864, it was bestowed mainly on the basis of the wartime benefits to the Union that the mineral wealth of the Comstock Lode provided. This argument has been discredited because the Union would already have had access to mineral wealth in any territory under its dominion, but it is unlikely that statehood would have been granted in the absence of the population centers created by mining. Clearly, many mineral occurrences were recognized early in the state's history. In 1867, J. Ross Browne, reporting on mineral resources of states and territories west of the Rocky Mountains, listed more than 25 mineral commodities that he believed to be present in Nevada.

Although denigrated by many modern citizens as a blight on an innocent landscape, mining brought civilization and prosperity to an empty region in the case of Nevada. Initially unorganized and lawless (the boomtown of Pioche saw 40 murders go unpunished in its first two years), legal institutions were quickly established and some notable courthouses built in mining camps. Mining money brought roads and railroads, schools and churches, newspapers and writers, opera houses and divas, artists and artisans, and it gave hardworking (and lucky) individuals a chance to make something of themselves. On the other hand, mining and mining money brought politicians and other rabblerousers, gamblers, prostitutes, attorneys, speculators, and preachers as well as mine fires, labor unrest, and pollution. In short, mining provided all the trappings of civilization.

Nevada's mineral-based culture stretches over a 140-year history of intensive mineral exploration and mining, and the state is still fertile ground for prospecting. Silver, gold, copper, iron, mercury, and base-metal deposits were found throughout the state by early prospectors, and discoveries of important new metal deposits have continued to the present. Mineral discoveries over this period of time led to the founding or expansion of many of Nevada's towns, including Eureka, Tonopah, Beatty, Goldfield, Pioche, Ely, Yerington, Lovelock, Henderson, Winnemucca, and Elko.

Some of the best-known mining towns in Nevada had relatively short histories. Mining enterprises, by their very nature, tend to use up the essence of their livelihood—the ore. Virginia City, the best-known mining town in the state and possibly in the country, was only a boom-town for about 20 years (1860–80). The town has survived, but, today, even when filled with tourists on a summer weekend, it is just a shadow of its former glory.

Nevada's mineral endowment is varied. In addition to the metals noted above, the state is, or has been, a producer of barite, borate minerals, clay minerals, diatomite, dolomite, fluorite, gypsum, halite, limestone, lithium, magnesium minerals, manganese, molybdenum, perlite, and tungsten. Nevada has large unexploited deposits of zeolite minerals, garnet, and phosphate minerals, and is noted as a source of gemstones such as turquoise, variscite, and precious opals.

Some writers have promoted the importance of other enterprises, such as ranching and railroading, in the state's history. Although they have been historically significant to many and are commercially important to a few, these enterprises are of relatively minor importance to the state as a whole. Nevada's generally sparse flora support few livestock per acre, and railroads in large part cross the state without adding much in terms of population. It may be argued that even Reno, cited by some as the epitome of a "railroad town," owed more of its early commercial development to its proximity to the booming Comstock Lode than to its selection as a Central Pacific Railroad townsite. In addition to mining's early influence on the location of towns along transcontinental railways, some short-lived railroads were specifically created to haul the state's mineral wealth.

Many people rode the trains in Nevada, but most traveled through the state as quickly as possible. From the days of the covered wagon to the modern miracle of interstate highways, Nevada has mostly been considered a wasteland to be endured, a series of particularly boring wayside stops between the majesty of the Rockies and the lush glamour of California. In 1868, J. Ross Browne noted that "Frémont and other explorers . . . had regarded it [Nevada] as a sterile waste, and . . . sought only for routes or passes by which they could most expeditiously leave it." He stated further that "the desolation and sterility . . . of all the country lying between the Rocky Mountains and the Sierra Nevada had become so generally acknowledged, that the wish had been expressed that these ranges of mountains might come together, and this great region be obliterated from the surface of the earth."

Mining provided a reason to stay in Nevada, even if only for as long as metal prices and ore held out, and some mining-inspired communities have flourished. Several years ago, Elko was selected as the "best small town in the U.S.," mainly on the basis of its lively mining-based economy. Even Las Vegas, though clearly not a mining town today, benefited in its early days from exploitation of nearby deposits of silver, lead, gypsum, dolomite, limestone, and silica sand. Had Nevada's institutionalized gambling not evolved in the twentieth century, the state's economy would probably today be based primarily on exploitation of its mineral wealth.

As will be noted in the essay on mining history that follows this introduction, Nevada's past was dominated by a sequence of mining "booms" and a collection of "boomtowns." Nevada has passed through silver booms, copper booms, a barite boom, and is now in the throes of a gold boom. Silver and gold mining on the Comstock Lode during the nineteenth century led to major innovations in mining and ore processing to deal with ore from the "Big Bonanza." It has been said that Virginia City was the industrial center of the western United States during its heyday.

The modern gold boom has led to mining that is performed on a truly heroic scale, and Nevada currently produces more gold than any foreign nation except South Africa and Australia. Barrick Gold Corporation's huge trucks, once assisted by overhead electric lines, haul ore from the enormous Gold Strike open pit on the Carlin Trend. Newmont Mining Corporation's Twin Creeks Operation on the Getchell Trend includes a mine called the "Megapit." The Round Mountain Partnership produces nearly half a million ounces of gold per year from ore worth only about $8 per ton. Nevada's newest world-class gold mine, the Placer Dome Inc.'s Pipeline Mine—discovered in 1991 during drilling designed to disprove the presence of ore under millsite claims—was the largest gold producer in Nevada in 1999 at 1.3 million ounces.

Knowledge of the minerals in ore is often critical to the recovery of metal from it, and during early mining history the first people to make mineralogical determinations were often assayers, either in California or those who set up shop in active Nevada mining camps. During early exploitation of the Ophir discovery on the Comstock Lode, the miners recovered electrum, but according to Smith (1943) "they cursed and threw aside the heavy bluish sand because it clogged their rockers, not knowing that it was rich silver sulfide." Knowledge of the silver content of the "heavy bluish sand" did not come until the material was assayed in California.

Identification of the silver-bearing sulfides as "sulphuret of silver" (acanthite), stephanite, polybasite, and pyrargyrite came later. Modern miners are more likely than their predecessors to demand knowledge of the mineral composition of ore, and Nevada's minerals are far more actively studied today than during early mining. In addition to mining, academic study, environmental research, and the growth of the collectible mineral market spur current interest in minerals.

Geology and Minerals in Nevada

STEPHEN B. CASTOR

Compared with many other states, Nevada's geologic features are complex; they developed during a long and active history. Nevada has been the site of several major episodes of sedimentary rock deposition, igneous activity, and orogenic deformation (mountain building). This has contributed to the state's varied mineral wealth and to a wide array of collecting sites and mineral species.

Although small parts of the state belong to the Sierra Nevada, Colorado Plateau, and Columbia Plateau provinces, most of the state is in the Great Basin region of the Basin and Range physiographic province, a region characterized by a series of generally north-trending mountain ranges separated by alluvial basins. The mountain ranges are as much as 1500 m above the adjoining valleys, and are generally bounded by faults, many of which are still active. Consequently, Nevada is the third most seismically active state in the nation (behind California and Alaska). Most of these faults are "normal"; that is, they formed when the mountains went up and the valleys went down along inclined fault planes. However, some Nevada faults are "strike-slip" faults like the well-known San Andreas Fault in California, in which opposite sides of the fault move horizontally. The most apparent zone of strike-slip faults in the state is in an 80-km-wide swath called the Walker Lane that parallels the California border, deflecting the predominantly north-trending ranges. Rocks exposed in Nevada's mountains include metamorphic and granitic rocks that formed billions of years ago, as well as sedimentary and volcanic rocks that are only a few million years old or even younger.

THE PRECAMBRIAN

The oldest rocks in Nevada are Late Archean gneisses in the East Humboldt Range in the northeastern part of the state, which are considered to be more than 2.5 billion years old (Lush et al., 1988). Although isotopic evidence suggests that Early Proterozoic rocks between 2.0 and 2.3 billion years old may be present, particularly at depth, no rocks in Nevada have yielded dates in this age range. The next oldest rocks in Nevada are highly metamorphosed rocks as much as 1.8 billion years old that occur at, and south of, the latitude of Las Vegas. Igneous intrusions that are about 1.4 billion years old cut the metamorphic rocks in several areas in southern Nevada, and may include mafic intrusions in the Bunkerville district that have related copper, cobalt, nickel, and platinum group minerals. Proterozoic rocks also contain mineralogically interesting pegmatite occurrences in the Gold Butte district and in the Eldorado and Newberry Mountains near Searchlight.

A large gap in the state's geologic record, perhaps 500 million years, follows the Middle Proterozoic intrusive episode. In the Late Proterozoic, about 700 million years ago, southern and eastern Nevada were the sites of deposition of fine-grained sedimentary rocks (now weakly to moderately metamorphosed quartzites and siltstones) during the early stages of marine deposition in seas that lapped onto a mature, low-lying coastline. This type of marine deposition, called "miogeoclinal," marked the beginning of the Cordilleran Miogeocline, a shallow oceanic environment that was to endure in Nevada for half a billion years. Mineral deposits that formed during the Late Proterozoic are rare or nonexistent in Nevada; however, these rocks do host later mineralization, as in the Bare Mountain and Johnnie gold mining districts in southern Nevada.

THE PALEOZOIC ERA

During the Cambrian Period, at the beginning of the Paleozoic Era, fine clastic sedimentation was followed by carbonate rock deposition. Miogeoclinal carbonate deposition was to continue, with minor clastic sedimentary interludes, for the rest of the Paleozoic Era (about 250 million years) in the southern and eastern parts of Nevada. Some economic limestone, dolomite, and gypsum deposits were formed by sedimentary processes during this time in the miogeocline, but metallic-mineral deposition was not significant. However, rocks of this age in central and northern Nevada include deposits that are characterized by chert, fine clastic rocks, and local volcanic rocks. They contain many deposits of barite, including the Devonian-age Clipper, Greystone, and Argenta deposits near Battle Mountain and the Ordovician-age Rossi deposit near Carlin (Papke, 1984). The Rio Tinto Mine at Mountain City near Nevada's northern border produced high-grade copper, zinc, and silver ore during the 1930s and 1940s from a deposit considered to be of Ordovician age. In addition, interesting but uneconomic molybdenum mineralization took place during the Devonian in shale in the Fish Creek and Morey districts.

During the Antler orogeny in late Devonian and early Mississippian time, the deep-sea rocks were pushed eastward over miogeoclinal rocks in the central and northeastern parts of Nevada. This tectonic superposition took place along shallowly dipping thrust faults. The Roberts Mountains thrust, a regionally important structural feature identified by U.S. Geological Survey (USGS) mappers in the 1940s, apparently marks the lowest of such thrusts, although there are probably many lying above it; strata of the deep-sea assemblage are thought to have been "shuffled like a deck of cards" (Stewart, 1980) in the upper plate of this feature. Nickel-chromium-cobalt-platinum mineralization in Nye County was deposited in serpentinites thought to have been squeezed out of the earth's mantle and into the crust during the Antler orogeny.

Deformation similar to that which took place during the Antler orogeny also marked the end of the Paleozoic Era during the Sonoma orogeny along the Golconda thrust in approximately the same area in central and northeastern Nevada. Although both orogenies modified the depositional basins significantly, deposition of marine sediments continued in parts of Nevada during late Paleozoic and early Mesozoic times.

THE MESOZOIC ERA

The Mesozoic Era between about 245 and 65 million years ago, popularly known as the "age of the dinosaurs," was characterized in much of Nevada by erosion and nonmarine sedimentary deposition and by igneous activity, although deposition of deep-sea strata took place in the early Mesozoic well to the west of Nevada. These deep-sea rocks were added to the North American continent during later tectonism, as part of a process known as continental accretion. In southern and eastern Nevada, eastward thrusting of the Sevier orogeny took place in the latter half of the era.

Gradual retreat of the Paleozoic seas in the Triassic Period was followed by terrestrial sedimentary deposition in southern Nevada, giving rise to economic deposits of gypsum and eolian silica sand. Igneous activity related to subduction (descent of oceanic crust into the mantle under the continent) included widespread volcanism and local to batholithic-scale plutonism, which produced some important metallic-ore deposits. Magmatism during the Triassic Period may have been accompanied by metallization at a few sites in Nevada, including silver at Rochester, mineralogically significant silver and base-metal ores at Goodsprings, and minor metal deposits near Tonopah. By contrast, magmatic activity during the Jurassic Period produced more widespread metallic mineralization. Porphyry and skarn copper deposits in the Yerington area, which have been mined from 1865 to recently, have been related to the 169-million-year-old Yerington batholith. Base- and precious-metal deposits in the Pine Grove Hills south of Yerington are thought to be related to 186-million-year-old granodiorite. Porphyry copper prospects in the Gilbert and Crow Springs districts have been dated at 198 and 207 million years respectively, and the turquoise and base-metal mineralization in the Royston district is thought to be of the same age. Iron-ore deposits associated with the middle Jurassic Humboldt lopolith were mined in north-central Nevada; of these, the deposits in the Buena Vista Hills are probably best known. In addition, some minor tungsten deposits are considered to be of Jurassic age.

The Cretaceous Period was a time of locally intense intrusive activity throughout Nevada, and also of intense skarn, porphyry, and deep-vein mineralization. Ore deposition related to 110-million-year-old granitic porphyry in the rich Robinson district near Ely has been the most productive and is the best known. The Robinson district ranks highest in Nevada in the production of copper, which, until the current gold-mining boom, was the most important mineral product in the state in terms of total value (Table 1). The district also produced gold, silver, molybdenum, lead, zinc, manganese, and platinum group metals. Polymetallic skarn deposits in the Mount Hamilton district are associated with the 104- to 107-million-year-old Seligman stock. The Hall Mine in Nye County exploited molybdenum and copper mineralization related to a 77-million-year-old granitic intrusion. Tungsten skarns in the Mill City, Potosi, and Tem Piute districts, which ranked first, second, and third respectively in Nevada in terms of total tungsten produced, range in age from 72 to 93 million years old. Beryllium, tungsten, and fluorine mineralization related to 71- to 104-million-year-old granites occurs in a broad belt that extends from northeastern to southwestern Nevada and includes mineralogically interesting occurrences such as beryl-fluorite-scheelite veins in the Mount Wheeler area and tungsten mineralization in the Oreana district. Skarn mineralization related to 90-million-year-old intrusive activity near Hawthorne produced fine specimens of epidote at the Julie claims. Magnesium-rich skarns at Gabbs, where magnesite and brucite have been mined since 1935 and some rare copper-magnesium minerals have been found, are associated with granodiorite that is considered to be of Cretaceous age. In addition, pegmatites associated with Cretaceous granites contain topaz and amazonite along with some newly discovered unusual aluminum and lithium minerals at the Zapot claim in the Gillis Range in Mineral County, and rare earth minerals at Red Rock in Washoe County.

In northwestern Nevada, "suspect terranes" (large areas underlain by rocks that are thought to have been added to the continent during the Mesozoic by plate collisions) have been proposed by some geologists. These rocks include volcanic rocks and sedimentary rocks of deep-sea origin that locally contain commercial gypsum deposits, as at Empire in Washoe County.

THE CENOZOIC ERA

The First Wave of Tertiary Magmatism

Although Mesozoic mineralization contributed much to the mineral wealth of Nevada, hydrothermal activity related to Cenozoic magmatism and tectonism now appears to have been far more important to the origin of the state's precious metals. At present, Carlin-type gold deposits are the most important economic factor in the state's mining economy, producing more than half of the state's mineral wealth. At today's prices, the Lynn mining district, which includes several large Carlin-type deposits, is clearly the state's mineral powerhouse, having surpassed the Robinson district (Table 1). These deposits are also important to mineralogy, both as occurrences of newly discovered and rare minerals and as sources of collectible species. Many geologists involved in mining, exploration, and research on Carlin-type gold deposits now believe that most of this mineralization along the Carlin Trend, Getchell Trend, and in the Independence Mountains of Humboldt, Lander, Eureka, and Elko Counties took place in the Eocene 38 to 42 million years ago and that it was driven directly or indirectly by magmatism that took place at that time. For the most part, Carlin-type gold deposits are confined to an area in northeastern Nevada where Eocene magmatism is prevalent. However, some geologists ascribe Carlin-type mineralization in Nevada to magmatic or other processes that predate the Tertiary.

Though not so important economically, other ore deposition took place in northeastern Nevada during this period—mostly of skarn and porphyry types including porphyry molybde-

TABLE 1 Estimated value of production through 2001 from the top 30 Nevada mining districts

(on the basis of approximate 2002 commodity prices: Au $300/oz., Ag $5/oz., Cu $0.70/lb., Mn $0.54/lb., Zn $0.35/lb., Pb $0.20/lb, barite $40/t, lime $50/t, Li carbonate $2/lb, diatomite $100/t, magnesia $60/t, silica $20/t.)

Rank	District	County	$ Billion	Principal Commodities	Deposit Types	Largest Producer
1	Lynn	Eureka/Elko	8.7	Au	Carlin-type Au	Goldstrike Mine
2	Robinson	White Pine	5.0	Cu, Au, Ag	Porphyry Cu	Ruth Pit
3	Maggie Creek	Eureka	4.2	Au	Carlin-type Au	Gold Quarry Mine
4	Comstock (Virginia City)	Storey	3.6	Au, Ag	Volcanic-hosted Au-Ag	Consolidated Virginia Mine
5	Potosi (Getchell)	Humboldt	3.3	Au	Carlin-type Au	Twin Creeks Mine
6	Round Mountain	Nye	2.3	Au, Ag	Volcanic-hosted Au	Round Mountain Mine
7	Bullion	Lander	1.9	Au, barite	Carlin-type Au; barite	Cortez Joint Venture (Pipeline Mine)
8	McCoy	Lander	1.7	Au, Ag	Gold skarn, Carlin-type Au	McCoy/Cove Mine
9	Independence	Elko	1.6	Au	Carlin-type Au	Jerritt Canyon Mine
10	Goldfield	Esmeralda	1.4	Au, Ag, Cu	Volcanic-hosted Au	Goldfield Consolidated Mines
11	Tonopah	Nye/Esmeralda	1.3	Ag, Au	Volcanic-hosted Ag	Tonopah Mining Company Mines
12	Yerington	Lyon	1.2	Cu	Porphyry Cu	Yerington Mine
13	Eureka	Eureka	1.0	Au, Ag	Polymetallic replacement	Ruby Hill Mines
14	Rochester	Pershing	0.8	Ag, Au	Vein-type Ag	Rochester Mine
15	Silver Peak Marsh	Esmeralda	0.8	Li carbonate	Lithium brine	Chemetall Foote Corporation Operation
16	Bullfrog	Nye	0.7	Au, Ag	Volcanic-hosted Au	Bullfrog Mine
17	Apex	Clark	0.7	Lime	High-calcium limestone	Apex Quarry
18	Buffalo Mountain	Humboldt/Pershing	0.7	Au	Carlin-type Au	Lone Tree Mine
19	Pioche	Lincoln	0.6	Zn, Au, Ag, Pb, Mn	Replacement base metal	Combined Metal Mine
20	Battle Mountain	Lander/Humboldt	0.6	Cu, Au, Ag	Porphyry Cu; Carlin-type Au	Copper Canyon Mine; Marigold Mine
21	Paradise Peak	Nye	0.6	Au, Ag	Volcanic-hosted Au	Paradise Peak Mine
22	Awakening	Humboldt	0.6	Au, Ag	Volcanic-hosted Au	Sleeper Mine
23	Bald Mountain	White Pine	0.6	Au, Ag	Carlin-type Au	Bald Mountain Mines
24	Velvet	Pershing	0.5	Diatomite	Lacustrine diatomite	Celatom Mine
25	Rawhide	Mineral	0.4	Au, Ag	Volcanic-hosted Au	Denton-Rawhide Mine
26	Imlay	Pershing	0.4	Au, Ag	Vein-type Au	Florida Canyon Mine
27	Sulphur	Humboldt	0.4	Au, Ag	Volcanic-hosted Au	Hycroft (Crofoot/Lewis)
28	Gabbs	Nye	0.3	Magnesia	Mg skarn	Gabbs Magnesite-Brucite Mine
29	Reese River (Austin)	Lander	0.3	Ag, Au, Cu, Pb	Polymetallic vein	Manhattan Silver Mining Co. Mines
30	Moapa	Clark	0.3	Silica	Sandstone	Simplot Silica Mine

num at Mount Hope, silver-lead-zinc replacement mineralization at Cortez, silver-lead-zinc skarn in the Ward district, and porphyry copper and skarn gold deposition at Copper Canyon. These deposits are related to an early episode of Tertiary magmatism that took place in the Eocene between 42 and 35 million years ago in northeastern Nevada.

In contrast to the deep or "mesothermal" veins of California's Mother Lode, most vein-type precious-metal ore in Nevada formed in shallow or "epithermal" deposits. Most of these vein-type deposits are in Tertiary volcanic rocks, and where adequate age dating is available the mineralization may generally be shown to be of about the same age as the host volcanic rocks or related shallow intrusions. Except for silver-gold mineralization at Tuscarora, which at 38 million years old may be included in the first Tertiary magmatic episode described above, significant volcanic rock-hosted epithermal mineralization took place less than 30 million years ago.

The Second Wave of Tertiary Magmatism

The second wave of Tertiary magmatism, which took place between 34 and 17 million years ago in Oligocene and early Miocene time, was generally similar compositionally to the Eocene volcanism, but took place south and west of it in an arcuate 200-km-wide swath that extended from the northwest corner of the state through Pioche and Caliente on the southeast. Metallic mineralization of this age is generally restricted to the same swath and includes some important volcanic rock-hosted deposits, both in terms of economic geology and mineralogy. These include the fabulously rich gold ore in the Goldfield district, recently mined-out silver-gold ore at Paradise Peak, the huge Round Mountain gold deposit, the historic Tonopah silver-gold deposits, and gold-bearing veins in the Delamar district. In addition, porphyry-type, copper-tin mineralization at Majuba Hill and copper-molybdenum mineralization near Pyramid Lake were associated with 24- to 26-million-year-old intrusive activity. Although Carlin-type gold deposits in the Northumberland, Santa Fe, and Relief Canyon districts are thought to be of this age, they are poor relatives of their older cousins.

The Third Wave of Tertiary Magmatism

The third wave of Tertiary magmatism took place in middle to late Miocene times, 17 to 6 million years ago. It is mainly represented by "bimodal" volcanism consisting of mafic flows and rhyolitic ash-flow tuff sheets. Igneous rocks of this age occur in northern, western, and southern Nevada, but are conspicuously absent from a large area in the east-central part of the state. This period of magmatic activity is associated with important ore deposits that include the silver-gold bonanza ores of the Comstock Lode and shallow epithermal veins in the Bullfrog and Midas districts, at Aurora, at Manhattan, and at the Sleeper Mine.

Middle Miocene igneous activity also gave rise to the most important mercury deposits in the state in the McDermitt caldera, as well as largely unmined uranium deposits in the same area. The mercury mines have been the site of thorough mineralogic investigations and have yielded four new minerals. Fluorite veins and breccias in the Bare Mountain district, the most important fluorspar mining district in Nevada, were also deposited during this igneous episode. In addition, emplacement of topaz rhyolites in Nevada took place, producing no economic metal deposits but creating interesting occurrences of dark-red garnet near Ruth in the Robinson district and cassiterite and hematite in the Izenhood district.

Modern Cenozoic Magmatism

The most recent magmatic episode in Nevada, extending in age from latest Miocene (6 million years ago) to the present, has been relatively mild. It is represented mainly by andesitic to basaltic flows near the western margin of the state, although large-scale rhyolitic magmatism occurred nearby at the Long Valley caldera in California less than 1 million years ago. Mineralization associated with this relatively mild magmatic period has been correspondingly unspectacular, although economic precious-metal mineralization in the Silver Peak and Sulphur districts has been dated at about 5 and 2 million years respectively. Modern hot spring mineralization in the Steamboat Springs district near Reno is thought to be as old as 3 million years and is still taking place.

Tertiary Sedimentary Rocks

Tertiary sedimentary rocks in Nevada, ranging in age from Miocene to Pliocene, contain deposits of borate minerals, clay minerals, gypsum, limestone, diatomite, manganese, magnesite, halite, and precious opal that are being mined now or were exploited in the past. The highly profitable diatomite deposits in northwestern Nevada near Fernley and Lovelock, and the sepiolite, saponite, and montmorillonite clays mined from Ash Meadows in Nye County are of particular commercial importance. Halite deposits in Clark County, important as a source of salt to prehistoric Americans, are for the most part deeply buried and known only from drill samples or are under the waters of Lake Mead. At present, only one Tertiary gypsum deposit is mined in Nevada at the PABCO Mine, but several such deposits were exploited in the past in Clark County.

Holocene Deposits

Sedimentary deposits of Holocene age, which include modern playa deposits, are not generally considered as a source of collectible mineral specimens, although a large suite of evaporite minerals, some as nicely crystalline specimens, have come from the commercial brine field at Searles Lake in neighboring California. Nevada's evaporite deposits are generally unspectacular, although specimens of well-crystallized halite from the Humboldt Marsh in Churchill County are in the Keck Museum at the Mackay School of Mines. Several deposits of borate and other evaporate minerals were mined historically in Nevada from modern playas, and two are currently active. Lithium carbonate is produced from brines at Silver Peak in Clayton Valley, and a small amount of halite is harvested annually from a dry lake near Fallon.

The Mining and Mineralogical History of Nevada

STEPHEN B. CASTOR AND
JOSEPH V. TINGLEY

NATIVE AMERICAN MINING

Nevada's earliest miners were Native Americans. Archeologists who have studied early Amerind sites in Nevada have recorded the presence of various mineral commodities in addition to obsidian, chalcedony, and other stones used in chipped projectile points and tools. These include turquoise, salt, cinnabar, malachite (probably as coloring agents), quartz crystal, and soapstone.

Turquoise was especially prized by Native Americans, and at least two Nevada turquoise deposits show evidence of prehistoric mining. The best preserved of these prehistoric mines was discovered around 1890 in the Crescent Peak area west of Searchlight in Clark County. Here, around pits that exploited a steep turquoise vein, were found stone mining implements along with the remains of living quarters and a lapidary shop with rubbing and polishing stones. In the words of Morrissey (1968), "The aboriginal miners had followed this vein until the overhanging roof became a menace. . . . The usual way of extracting the ore was to build a fire against the face of the rock, then throw water on the hot stone, causing it to crack. Wedges were then driven into the cracks until the mass broke away." Tree-ring dating put the age of this mining operation at about A.D. 1300.

Other turquoise deposits in Nevada were known by Native Americans prior to their discovery by white miners. The Indian Blue Mine in the Toquima Range, Nye County, rediscovered in 1925, was determined to be the site of prehistoric mining on the basis of associated stone tools. A prehistoric shaft driven to a depth of as much as 5 meters was found to be so narrow that it was thought to have been worked by lowering a child into it headfirst (Morrissey, 1968).

Salt beds along the Virgin River in Clark County, now mostly beneath the waters of Lake Mead, were mined by the Anasazi people between about A.D. 750 and 1150. The Anasazi lived in an adobe pueblo near Overton called the "Lost City," which is also under water. Stone picks and ropes that were used by the Anasazi miners were found in the underground mines at this location. According to Horton (1964b), the salt was mined in a natural cave by chipping circular channels with stone hammers and prying out isolated slabs.

Cinnabar, the ore of mercury, was also mined by Native Americans. Early settlers reported its use as body paint in the Ione area, later the Union district, a moderately important mercury mining district (Bailey and Phoenix, 1944).

MEXICAN MINERS

Although Spanish and Mexican miners predated Anglo-Americans with early mining ventures in California and Arizona, this is not known to have been the case in Nevada. There has been some speculation that Spanish or Mexican miners operated in Nevada along the Colorado River in the Eldorado (Nelson) district south of Las Vegas prior to the discovery of the district in 1857 on the "Honest Miner claim." Writing in 1937, William O. Vanderburg of the U.S. Bureau of Mines noted: "This district is one of the oldest in the State and mining has been done there almost continuously for nearly 80 years. It was organized as the Colorado district in 1861. Old arrastras and prospect holes reported to have been found in this area in the sixties indicate that mining was carried on by Spanish adventurers probably several hundred years ago."

Mexican miners and mining technology were important factors in early hard-rock mining of Comstock Lode silver ore, and Mexican miners were also active in other early Nevada silver districts, including the Candelaria, San Antone, and Barcelona (Spanish Belt) districts.

MORMON MINERS

Although early-nineteenth-century American explorers such as Jedediah Smith, Peter Ogden, Kit Carson, Joseph Walker, and John C. Frémont spent time in Nevada, they produced no record of mineral discovery. The first American miners may have been Mormons, and settlements in Carson Valley and Las Vegas were originally founded as extensions of the land of Zion. On the basis of the discovery of a carved stone in the Wahmonie mining district on the Nevada Test Site in southern Nye County, Mormons may have engaged in mining in that area as early as 1846 (Quade and Tingley, 1984); however, this activity has not been verified by other evidence. Mormons en route to Carson Valley are thought to have made the first discovery of gold near Dayton at the mouth of Gold Canyon below the later Comstock Lode discovery, and this is relatively well documented. It is clear that Mormon miners, in search of lead for bullets, were the first hard-rock miners in the state, mining ore from the Goodsprings (Potosi) district near Las Vegas as early as 1855. According to H. J. Stewart (1913):

> A man by the name of Slade was made superintendent of the Potosi Mine in 1855, having been sent out by church authorities to supervise the lead mining. They made an attempt to smelt the ore at the mine, using pitch-pine for fuel, with no result save badly burned hands. They also tried cedar wood for that purpose, which was better, but still not successful. Not being satisfied with the results, they brought their ore down to the Las Vegas rancho. Dudd Leavitt and Isaac Grundy here built a furnace in a fireplace, using the chimney for making a draft. When the ore became too hot they devised a plan of placing an adobe brick in the furnace to even the temperature. In this crude way they succeeded in making a success of their smelting operations. They molded their lead in an old iron skillet, which gave the bars the appearance of miner's loaves of bread. In this manner they prepared and sent to Cedar City, Utah, ten thousand pounds of lead, which was put in charge of Bishop Smith and by him distributed.

THE COMSTOCK AND THE FIRST NEVADA BOOM, 1859–1900

Although the Mormon settlers and miners may have been first, their control over portions of Nevada was short-lived. Placer miners in Gold Canyon, a small tributary of the Carson River just south of what is now Virginia City, coexisted grudgingly with the Carson Valley Mormon settlement between 1852 and 1858, and in 1859 discovered the Comstock Lode, an event that was to change the territory quickly and irreversibly.

Grant Smith (1943), who grew up in Virginia City and became a mining attorney in San Francisco, recounted the history of the Comstock Lode. The tale of the Comstock is one of colorful characters and great riches, and the reader is referred to Smith's book for details. The Gold Canyon placer miners, about 100 men who worked their claims for part of each year, had processed most of the profitable ground by 1859 and a few of their number were searching for new placer ground when they discovered the lode ore. They were not geologists or mineralogists; if they had been, the lode potential in the area would have been recognized at an earlier date. Though sophisticated city dwellers such as Eliot Lord, who published a book on the Comstock mines and miners in 1883, had little appreciation for the early placer miners, Smith held a more idyllic view, writing in 1943: "Those men were free, their wants were few, their living assured. Time meant nothing to them—they slept, ate, and worked when they pleased. When there was water they worked hard; at other times they loafed. The hunting was good on the hills, the river teemed with trout. They owed the world nothing nor the world them."

Placer miners, among them James Finney, whose nickname "Old Virginny" was bequeathed on the town that grew up among the mines, discovered the southern part of the Comstock Lode at Gold Hill in 1859. Rich silver ore in the northern part of the lode was also discovered in 1859 at the Ophir diggings. According to Smith, the early miners, washing gold from dark sand from the upper, weathered part of the lode "cursed and threw aside the heavy bluish sand because it clogged their rockers, not knowing that it was rich silver sulphide." As the miners dug deeper, they encountered decomposed quartz-sulfide ore, and later in the year some of the rock was assayed at more that $3,800 per ton in California, and the "Washoe Rush" was on.

Annual production from the Comstock mines was small at first, ranging from $275,000 in 1859 to $6 million in 1862. In 1863, production jumped to more than $12 million, and stayed at about that level for 10 years, with the discovery and exhaustion of several bonanza orebodies and steady mining of lower grade ore. During this period, new developments in mining and ore dressing technology were developed to deal with Comstock ore. In 1873 the "Big Bonanza" orebody was discovered in the Consolidated Virginia Mine 1200 feet beneath the heart of Virginia City, and production increased to more than $20 million per year until 1877. During the 1880s and early 1890s, production continued, but at much lower levels, averaging a little more than $3 million annually. Thereafter production declined significantly, although the district was active into the 1980s. On the basis of recorded mineral production and modern metal values, the Comstock ranks fourth among Nevada mining districts in total value produced (Table 1).

Virginia City was a boomtown far longer than many Nevada mining communities. Nevertheless, after the turn of the century the district and the mill towns that served it along the Carson River were but shadows of their booming past. T. A. Rickard, a nationally known mining engineer who spent his childhood in Virginia City, poetically described the situation at the turn of the century in the abandoned Comstock district:

When I was there in 1901 the unlovely quiet of abandonment rested on the whole district. The Chinaman alone was superior to his surroundings . . . and his kitchen-garden formed a picture that was the very antithesis of the volcanic energies that once rioted at the mines of the Comstock. "Cold upon the dead volcano falls the gleam of dying day." The mines are idle, the sunlight falls on abandoned shaft-houses, the rain rusts the motionless machinery, but "the river still is winding, still is winding, past the gardens where the Mongol tends the cabbage and the leek."

The first technical report on the Comstock was written by a European geologist, Baron Ferdinand Von Richthofen, in 1865, and well-known U.S. Geological Survey (USGS) geologist Clarence King briefly described the geology of the area in 1870. G. F. Becker, another USGS geologist, reported more extensively on the geology and mineralogy of the district in 1882. Becker's work is noteworthy in that he used the polarizing petrographic microscope extensively to study rock types in the district, describing their mineral assemblages in some detail and correctly proposing Von Richthofen's "propylite" to be altered lava. Becker also studied minerals at Steamboat Springs, reporting on the deposition of sulfide minerals in the hot springs.

The basic mineralogy of Comstock ore appears to have been known by 1876, when Dan De Quille, a newspaperman and associate of Mark Twain, described ore in the Big Bonanza as two mainly drab types: "sulphuret ore" containing silver glance (acanthite) and ore containing chloride of silver (chlorargyrite). De Quille described other minerals in the ore as follows:

Throughout the mass of the ore in many places, however, the walls of the silver caverns glitter as though studded with diamonds. But it is not silver that glitters. It is the iron and copper pyrites that are everywhere mingled with the ore, which in many places are found in the form of regular and beautiful crystals that send out from their facets flashes of light that almost rival the fire and splendor of precious stones. There are also found in the mass of ore great nests of transparent and beautiful quartz crystals that are almost as brilliant as diamonds. Many of these crystals are three or four inches in length. Some of the nests of crystals are of a light-blue color, and then they may be classed among the precious stones, as they are amethysts. Some of these are almost as handsome as the precious amethyst. The miners always like to find these nests of crystals, as they indicate life and strength in the vein.

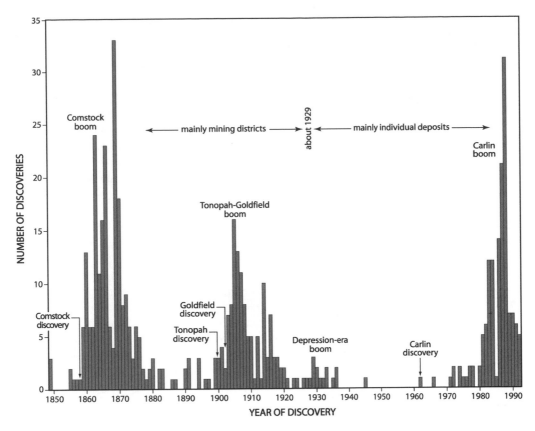

FIGURE 1 Discovery rate of Nevada mining districts and deposits, 1849–1992

De Quille also reported the presence of stephanite, free gold, and native silver, noting the latter to be found in places as "nests of pure, malleable silver in the shape of flattened wires that look as if they had been pulled in two and, in springing back after breaking, had coiled up against pieces of ore on which they are found. . . . Some of the smaller and finer wires, on being unrolled are found to be a foot or more in length." De Quille reported many "magnificent specimens for cabinets" were collected from the orebody.

Following the early success at the Comstock, intrepid prospectors quickly spread out across the Nevada Territory and new discoveries followed (Fig. 1). The distribution of these discoveries relative to the Comstock has been likened to "ripples spreading from the drop of a stone in a pool of water." In 1860, gold and silver veins were discovered in the Aurora (Esmeralda) district 80 miles southeast of Virginia City in what is now Mineral County. In the following year, the Buena Vista (Unionville) district was organized

about 100 miles northeast of Virginia City near the Humboldt River. In 1862, silver was discovered in the Reese River (Austin) district 125 miles east of the Comstock. This was followed in 1863 by the discovery of Pioche to the southeast, and in 1864 by the discovery of the Robinson district to the east, both about 250 miles from Virginia City. Major production from the latter districts was to come much later.

Eliot Lord recounted the spread of Nevada's mining camps as follows:

Yet the progress made in the development of the Comstock Lode mines during these years was not the only result of the discovery nor the most important one; for, incited by the rewards of the new industry and the hope of similar bonanzas, swarms of prospectors set out from this district as a centre to search for ledges in every direction. The bounding circle of their exploration constantly expanded, and newly organized districts extended like the rays of a star from its nucleus. When one of these outlying districts promised unusually rich returns, like the Reese River region in the spring of 1863, a rush followed which made it teem with prospectors for a few months, and then it in turn served as a base of supplies and was encircled by outlying camps. Sometimes the rush was rewarded by actual prizes, but commonly the movement and dispersion were rocket-like—a prolonged whir, a gleaming beacon, a moment of splendor as the centre of radiant stars, and then dwindling specks of glory succeeded by gloom and utter extinction. The Reese River district, or the Toiyabe range, was not such a treasure-trove as its explorers fancied—"a chain 200 miles in length, nearly every mile of which is rich as never hill or mountain was rich before;" still the rush toward it was not wholly unrewarded. From Austin, in the heart of the range, districts radiated like spokes of a wheel; southwest lay Ione, Union, and Mammoth; west, Augusta and Cold Spring; northwest, Ravenswood; northeast, Cortez; southeast, Jefferson and others; east, Mountain, Eureka, Diamond, and Gold Canon, chief among scores of others then notable, now nameless and forgotten.

Reports by early Nevada writers give some notion of the furor of early prospecting activity. Mark Twain traveled to the Nevada Territory in 1861 and wrote about the early days of the Comstock, Unionville, and Aurora areas in *Roughing It*. His description of searching for silver in the "Humboldt" (Unionville) area, written about 130 years ago, evokes the feeling of prospecting in Nevada even by modern exploration geologists.

> By and by I was smitten with the silver fever. "Prospecting parties" were leaving for the mountains every day, and discovering and taking possession of rich silver-bearing lodes and ledges of quartz. Plainly this was the road to fortune. . . . Knowledge of the nature of silver mining came soon enough. . . . We climbed the mountain sides, and clambered among sagebrush, rocks and snow till we were ready to drop with exhaustion, but found no silver—nor yet any gold. . . . Prospecting parties swarmed out of town with the first flush of dawn, and swarmed in again at nightfall laden with spoil—rocks. Nothing but rocks. Every man's pockets were full of them; the floor of his cabin was littered with them; they were disposed in labeled rows on his shelves.

By 1863 the discovery and founding of new mining districts became commonplace, and until 1873 new districts were discovered at the rate of nearly 18 per year. The "Rush to White Pine," a bonanza silver district that was to produce a small amount of silver relative to the Comstock, began in 1865. Although a lesser star than the Comstock, White Pine set something of a record for short duration; the boom was over by 1869. In three short seasons, however, it produced about $6 million, mostly from massive, shallow deposits of chlorargyrite (horn silver). USGS geologist Arnold Hague, in 1870, described White Pine as "probably the most remarkable occurrence of horn silver on record. In the early days of its development, channels or courses of ore were passed through that were almost solid hornsilver." Hague mentioned "boulders" of horn silver that weighed 2 and 3 tons, and the largest single mass discovered in the district is reported to have weighed 6 tons.

Other important districts discovered during the late 1860s were Yerington, Manhattan, Mountain City, Battle Mountain, and Bald Mountain. This period was a time of rapid change in Nevada: the Territory was granted statehood; the population of Virginia City grew to 25,000; the railroads came; and mining millionaires flourished, though they mostly lived in San Francisco. In 1873, silver was demonetized, leading to a slow decline in its price, and, although Comstock production increased temporarily, discoveries of new districts declined (Fig. 1). However, some mining districts that are notable for modern production were located in the years following 1873, including the Potosi (Getchell) and Maggie Creek gold mining districts. By 1880, the last of the Comstock bonanza ore had been mined, the price of silver had declined to about 80 percent of its value during the 1860s, and Nevada's economy entered a "bust" period that was to last until the discovery of Tonopah at the turn of the century.

During the 20 years of Nevada's first mining boom, the work of practitioners of the science of mineralogy, inexact and quaint though it seems to us now, was considered to be important to the fortunes of the mines. The first state mineralogist of Nevada, R. H. Stretch, described attempts to put together a Nevada collection in 1866 as follows:

> I regret that I have been able to procure but few specimens by donation or personally visiting the different mining districts of the state; but feeling a deep interest in the success of the office in the belief that it may be made really beneficial to the community, I have donated to the State the greater portion of my own private collection, and the cabinet in which it is contained, having no funds with which to purchase any other. . . . It is a shame that a State dependent on mining for its existence should have no intimation of the fact at its State Capital, and no public institution where reliable information relative to its resources can be easily obtained. The expense to each mining company of selecting and forwarding to Carson or Virginia a good series of ores from their mines, and samples of the country rock in which the veins occur, would be merely a nominal item (not more than $5); whereas if the State has to bear the entire aggregate expense, it would amount to quite a serious sum in the course of a year. Such a series should contain specimens showing every variety of mineral found in the lode, labeled to show the place in the mine from which they were taken; and would answer every purpose, if weighing from four ounces to one pound, though they need not by any means be limited to that size. . . . Such a collection cannot be got together in a day, and will be an imperative necessity to a mining school, whenever it may be organized in our State; and that such an institution will spring up in our midst there can scarcely be a doubt.

Stretch appealed to Nevada mining companies for exceptional ore specimens to be sent to the 1887 International Exposition in Paris. Apparently, such a collection was assembled and made it as far as San Francisco. It was ultimately returned to Nevada under the auspices of the second state mineralogist, and thereafter ostensibly made its way into the museum collection at the Mackay School of Mines (now the W. M. Keck Museum). In 1866, Stretch listed Nevada locations for 62 minerals:

Alum
Antimony, sulpuret (stibnite)
Arsenical antimony (stibarsen?)
Arsenolite
Blende or sulphuret of zinc (sphalerite)
Borate of lime (colemanite ± ulexite)
Bournonite
Brittle silver ore (stephanite)
Carbonate of soda (probably trona)
Cerussite
Chabazite
Chloride of silver (chlorargyrite)
Chrysocolla
Copper, carbonates of (Azurite, Malachite)
Copper, native
Copper, red oxyde of (cuprite)
Copper glance (chalcocite)
Copper pyrites (chalcopyrite)
Diallogite (rhodochrosite)
Dolomite
Embolite (bromargyrite?)
Epidote
Epsom salt (epsomite)
Feldspar (orthoclase)
Fluorspar
Galena or sulphuret of lead
Garnet
Gold
Graphite
Gypsum
Hematite, specular iron ore
Hornblende
Iodide of silver (iodargyrite)
Iron pyrites (pyrite)
Kaolin
Labradorite
Magnetite
Mercury
Mica (Biotite, Muscovite)
Mispickel (arsenopyrite)
Molybdate of lead (wulfenite)
Natrolite
Olivine
Polybasite
Proustite
Pryargyrite (pyrargyrite)
Pyrolusite
Pyromorphite
Pyrrhotine (pyrrhotite)
Quartz

Salt, chloride of sodium (Halite)
Saltpetre (nitre)
Scolecite
Silver, native
Silver glance (acanthite)
Spathic iron, carbonate of iron (siderite)
Sulphur
Tetrahedrite
Tourmaline
Tungstate of manganese (Hübnerite)
Tungstate of lime (scheelite)
Xanthocone (Xanthoconite)

Although precious-metal mining was clearly the driving force in the early history of Nevada, prospectors were not blind to other commodities. Salt was mined from Rhodes Marsh near Mina beginning in 1862, and from the Sand Springs Marsh near Fallon beginning in 1863. Borate minerals were also discovered in the Fallon area in the 1860s, but the earliest successful borate mining began in 1872 when the mineral ulexite was recovered from Columbus and Teels Marshes near Candelaria. Clearly, many mineral occurrences were recognized early in the state's history. In 1867, J. Ross Browne, reporting on mineral resources of states and territories west of the Rocky Mountains, described the variation of mineral resources as "Not only the precious, but also many of the useful metals as well as a large variety of mineral substances, are met with in the State of Nevada."

In 1868, the same Mr. Browne reported a catalog prepared by Charles Stetefeldt, assayer and metallurgist of Austin, of minerals that had been identified in the veins of eastern Nevada. The catalog included 33 species, mostly sulfide and other ore minerals in 15 mining districts: pyrargyrite, proustite, polybasite, stephanite, tetrahedrite, "pyrites of iron and copper" (pyrite and chalcopyrite), galena, "blende" (sphalerite), "horn silver" (chlorargyrite), native silver, "silver glance" (acanthite), "spathic iron" (siderite), quartz, "argentiferous sulphuret of copper" (?), "sulphuret of antimony" (stibnite), stetefeldtite (a new mineral named after Stetefeldt), stromeyerite, native gold, and "copper glance" (chalcocite). Browne also reported a list of "minerals of Reese River district, arranged according to Dana's system, by Eugene N. Riotte, M. E." The list included at least 60 mineral species.

Tungsten minerals were described early in Nevada's mining history. The first notice of tungsten in the state was probably a short article in the *Reese River Reveille* in 1865 that reported a new mineral, "tungstate of manganese," had been found by a Dr. Blatchly in the Enterprise and Erie ledges in the Mammoth district of Nye County. A few days later, the same paper carried a detailed mineralogical and chemical description of the new mineral by Riotte, who named the mineral "hubnerit" and noted that he had also found "the shelit, or tungstate of lime," at the same location.

Other deposits of hübnerite were discovered in the Osceola district in eastern White Pine County in 1889. It is reported that as early as 1885 prospectors from this district had sent samples of a heavy black mineral from white quartz veins to "the chemist of a prominent reduction works in California who reported same to be specular hematite" (Smith, 1902). It was not until four years later that a Denver laboratory correctly identified the black material as a tungsten mineral.

In 1890, R. A. F. Penrose Jr., while examining various manganese properties in Nevada, found tungsten in samples collected at Golconda in Humboldt County in a mixture of fine-grained manganese and iron oxide minerals, but did not specifically identify a tungsten-bearing mineral (Penrose, 1891).

Prospecting and mining activity waned in Nevada in the last 20 years of the nineteenth century, but it was not in complete eclipse, and the study of the state's mineral deposits continued. The Mineral Basin (Buena Vista) district, later an important source of iron, was discovered in 1880, and the Delamar gold mining district was discovered in 1891. Nickel and cobalt were mined from small, unusual deposits in the Table Mountain (Cottonwood Canyon) district between 1882 and 1908. In a "Catalogue of American Localities of Minerals," published in 1898 as part of his "System of Mineralogy," E. S. Dana listed localities for 54 Nevada mineral species, including the Comstock, Columbus (Candelaria), Tuscarora, Eureka, Reese River, Bristol, Belmont, and White Pine mining districts.

When the status of knowledge about minerals in Nevada prior to the turn of the century is considered, it must be remembered that most of the methods that mineralogists take for granted today, such as optical mineralogy, were not available or widely used. Goniometric measurements of crystal angles, specific gravity determinations using the Jolly balance, and elemental detection using blowpipe methods (Brush, 1875) were helpful, but in many cases insufficient for characterization and identification, particularly for fine-grained specimens. However, mineralogical science was to make substantial advances in the early twentieth century, during Nevada's second mining boom.

THE SECOND WAVE—
TONOPAH AND BEYOND, 1900–18

The discovery of silver-bearing veins at Tonopah at the beginning of the twentieth century began a second wave of intense exploration and mining in Nevada that was to last for about 20 years (Fig. 1). By the end of these two decades, almost all of the state's mining districts would be organized, although important deposits are still being located in them today.

Recorded total production from Tonopah between 1900 and 1961 was more than $150 million ($1.3 billion at today's metal prices; Table 1), making Tonopah the second most productive silver mining district in Nevada after the Comstock. Boom years at Tonopah (more than $4 million per year) were 1905–24. The story of the discovery of Tonopah ore by Smoky Valley rancher Jim Butler and his burro is well known. As told to Ben Edwards, a Tonopah business associate: "Driving two burros hitched to my buckboard I found one of my burros missing. I tracked him about three miles and found him sheltered from the wind alongside a rock cropping. I thought this dark rock might be quartz and broke off 30 to 40 pounds for assay."

An early report on Tonopah mineralogy by USGS geologist J. E. Spurr in 1903 listed 13 minerals including drusy crystals of argentite (acanthite), polybasite, and possible stephanite. A 1911 paper on crystalline cerargyrite (chlorargyrite), iodyrite (iodargyrite), and embolite (bromargyrite) at Tonopah by J. A. Burgess was a pioneer

work on the occurrence and origin of these silver halide minerals. This was followed by a University of California publication on Tonopah that listed 44 minerals and included excellent line drawings of crystal habits for some (Eakle, 1912). Bastin and Laney (1918) reported 35 hypogene, supergene, and gangue minerals from Tonopah mines, including the relatively rare minerals argyrodite and native selenium.

Mineralogic work performed during the second wave of discovery and mining in Nevada benefited from knowledge gained during the first wave. In addition, petrographic techniques were used widely following publication of the first edition of *Elements of Optical Mineralogy* (Winchell and Winchell, 1909). The effects of hydrothermal alteration were recognized, and petrographic examinations were used routinely by some USGS geologists, notably by F. L. Ransome, W. H. Emmons, S. F. Emmons, H. S. Gale, Henry Ferguson, and Adolf Knopf. The famous economic geologist Waldemar Lindgren, who worked with G. F. Becker at Steamboat Springs in 1885, published further observations on the site in 1906. Other well-known scientists of the time who worked on Nevada minerals were A. S. Eakle, A. N. Winchell, W. T. Schaller, and G. D. Louderback.

Following the discovery of Tonopah, a ripple effect of discoveries took place that was similar to those following the Comstock discovery. Gold was discovered at Round Mountain in 1901 and at Goldfield in 1902, 40 miles north and 30 miles south of Tonopah, respectively. After these discoveries, Nevada prospecting again became widespread. Other districts discovered during Nevada's second mining boom period include Divide (1901), Bullfrog (1904), Fairview (1905), Rawhide (1906), Lynn (1907), Jarbidge (1909), Awakening (1912), and McCoy (1914).

The Round Mountain district has about 10 million total ounces of gold in production and reserves, and is presently productive. It is known for rich veins of quartz, adularia, and electrum that have yielded exceptional electrum specimens.

The Goldfield district is one of Nevada's mineral treasure houses. It yielded very high-grade ore, one 47-ton shipment of which assayed 609 troy ounces of gold per ton (Rickard, 1940), and also has had a rich mineralogical history. Early research on Goldfield mineralogy was done by Frederick Ransome, who identified 53 mineral species in 1909, including goldfieldite, a new sulfide of copper, antimony, and tellurium. In addition, Ransome identified enargite and famatinite in some ore samples, and recognized the significance of alunite as an alteration mineral associated with ore, stating: "The recognition of alunite as a characteristic constituent of the Goldfield ores and the demonstration of its genetic relation to them establishes a new type—that of alunitic and kaolinitic gold-quartz veins—in the classification of epigenetic deposits based upon the kind of metasomatism effected in the wall rock by the ore-depositing solutions."

The richest Goldfield ore, called "jewelry ore" by some collectors, contained very fine-grained gold arranged in layers or shells along with black layers of other metallic minerals, mainly bismuthinite, around country rock clasts in siliceous matrix. This ore was sought after by the miners as a source of extra income, either as mineral specimens or for processing into bullion by bogus assay offices. In 1939, T. A. Rickard recited examples of such thievery:

> The mines in which such theft of ore has flourished most largely and successfully are those in which telluride minerals were mined, as at Cripple Creek, in Colorado, at Kalgoorlie, in Western Australia, and at Goldfield, in Nevada. . . . The larcenous miner today does not consider himself a thief; he thinks he's entitled to an occasional dividend by way of a little rich stuff; he might quote the Biblical reference which says "Muzzle not the ox that treadeth the corn." Therefore, the present term "high-grading" came into vogue in the mining districts of the western United States to cover the surreptitious removal of high grade ore. . . . The western American miner that regards a lethal bullet as the proper punishment for the theft of a horse would smile at stealing enough gold ore to buy a span of horses.

Reporting on the Goldfield situation specifically, Rickard (1940) noted:

> It was an orgy of theft. Regular harnesses were invented for carrying ore under the working

clothes of the miners. On reaching the surface, they would step heavily from the cage or bucket and stumble slowly on their way homeward, without being arrested, although the officials at the mine, as well as the bystanders, knew why they walked so clumsily. On one occasion, for instance, when the men were being hoisted to the surface at the Mohawk shaft, a man came from underground so loaded with concealed ore that, on stepping out of the bucket, he fell. The superintendent saw it and called out: "Someone help that son-of-a-gun."

In addition to ore in the newly organized districts, new deposits were found in old districts following the Tonopah and Goldfield discoveries. Zinc ore was recognized in the Goodsprings (Yellow Pine) district, site of Nevada's first recorded underground mining, and between 1905 and 1921 the district was the largest producer of the metal in the state. Later, the Goodsprings mines became known as an excellent source of mineral specimens, but it is not clear if their value was recognized in the early days, and many fine pieces were undoubtedly processed into metal. In 1905 gold was discovered in the Manhattan district, later an internationally known provider of superb specimens of antimony minerals. In 1912, large silver orebodies were found in the Rochester district, part of the Sacramento district that was originally formed in the 1860s, and a rush to the area ensued in 1913.

The Robinson district began producing copper in 1907 and quickly became the state's largest producer of the metal, as well as by-product gold, silver, zinc, and lead. The district, organized in 1868, "was first a silver camp for twenty years, then a gold camp for another 15 years; neither flourished" (Smith, 1976a). Potential for economic copper deposits in the district was first recognized in 1900 by two "dead-broke" young miners, Edwin Gray and Dave Bartley, from the Shasta copper country in California, who took a lease and option to buy on the Ruth and Kearsarge claims, which were thought to have gold and silver potential. Parsons (1933) provided the following account of their motivations:

> After the transaction was closed and the partners were alone, Gray asked Bartley what he thought of the ground. "If she's good for anything, it's copper" was the reply; and, so far as

the records disclose, that was the first time that anyone ever thought of Ely or the Robinson district, as it was called, as a possible copper area. There were some slight surface indications of copper, but it is probable that the fact that Bartley had just come from a copper district had something to do with his predilection for a metal that was much despised around Ely. In any event, his hunch was good and he spoke with the wisdom of an oracle.

Gray and Bartley, working on a grubstake from a store owner in Ely, laboriously drove underground workings to a depth of 200 feet at the Ruth mine, intersecting significant amounts of 2 percent copper ore. They received offers from local miners, but according to Parsons:

> continued to live on bacon, beans, and work . . . [and] instead of sickening of their task kept steadily at work until, at a depth of 300 feet, they encountered copper ore far richer than any that had been found before. They were a hospitable pair. In this first place, they were that way by inclination and habit, and in the second place, they had nothing to hide; in fact, they wanted everyone to know just how good their mine looked. Anybody who happened around was welcome to go into the mine and help himself to samples, wherever and however he pleased. A stranger, an old Comstocker who went by the name of Williams, was an occasional guest during a period of several weeks; he ate beans and swapped yarns with the boys and incidentally moiled a few samples here and there. He seemed to have no particular business, and Gray and Bartley rather missed his company when he left as unceremoniously as he had appeared on the scene two or three weeks before.

In 1902, a second stranger arrived at the Ruth mine who turned out to be somewhat more businesslike than Mr. Williams. The stranger, Mark L. Requa, a railroad man from Virginia City, purchased the property and organized the White Pine Copper Company, later the Nevada Consolidated Copper Company. The Robinson district was destined to be the most important of Nevada's mining districts for most of the twentieth century, and the total value of metals produced there was to surpass that of even the Com-

stock (Table 1) and to add substantially to the coffers of the Kennecott Copper Corporation, which absorbed Nevada Consolidated in 1933. Consolidated Coppermines Corporation was the other major producer in the district until it, too, was absorbed by Kennecott in 1958. Between 1909 and 1959, Nevada Consolidated/Kennecott was the largest producer of both copper and gold in Nevada, and Consolidated Coppermines was the second largest in most years.

Other Nevada districts that became important copper producers at about the same time as Robinson were the Yerington district, discovered in the 1860s and an important producer beginning in 1905, and the Santa Fe district, discovered in 1879 and mainly exploited after 1906. Large concentrators and smelters were built at McGill near Ely in 1908 and at Wabuska near Yerington in 1910, and the importance of mineralogy to mineral recovery inspired careful mineralogical studies of the ore. During such work, A. C. Spencer, who reported on the mineralogy of the Robinson district, and Adolf Knopf, who described the mineralogy of the Yerington district, both recognized the importance of supergene concentration of copper sulfide as proposed earlier by S. F. Emmons at Butte, Montana.

In a USGS Professional Paper on the Robinson district, Spencer and others (1917) noted:

> In copper deposits no primary ore has been developed. All the ores mined owe their value to chalcocite enrichment. . . . The cuprous sulfide, chalcocite (Cu_2S) carries approximately 80 per cent of copper. It is the most important ore mineral at Ely, as it is in many other districts where the copper deposits are of the disseminated type. In this district all occurrences of this mineral are of secondary origin. It is the characteristic mineral of the porphyry ores, in which it occurs to a certain extent in minute specks replacing small grains of pyrite or chalcopyrite, but principally in the form of films surrounding grains of these minerals. . . . Shallow deposits of copper ore, mostly worked out in former years, consisted in considerable part of massive chalcocite more or less altered to cuprite and to copper carbonates.

In addition, these authors, following Lindgren's earlier work at Clifton-Morenci in Arizona, described fluid inclusions associated with the Robinson porphyry deposit:

> In all the occurrences quartz must have formed in the presence of water or aqueous solutions, for in examples of normal porphyry, of metamorphosed porphyry, of jasperoid, and of veins may be found grains of the mineral that show cavities containing liquid and a vapor "vacuole." Some of these cavities contain also minute cubical crystals likely to be sodium chloride or potassium chloride.

> These fluid-containing cavities may be of irregular form, but many of them are "negative crystals," which show the bipyramidal form of quartz, and are of course oriented so as to correspond crystallographically with the host. They are so abundant as to be characteristic in the vein quartz and jasperoid quartz. In worked-over quartz in the groundmass of metamorphosed porphyry they can be made out in the study of many thin sections. In the original quartz of the groundmass, however, it is rather unusual to find cavities large enough to be resolved by even high powers of the microscope, though one suspects that almost omnipresent specks in such quartz are really cavities so minute that their nature is not ascertainable.

> If it is assumed that the quartz formed at such an elevated temperature that the cavities now partly filled were entirely full of liquid at the time the mineral crystallized, a consideration of the relative space now occupied by fluid and vapor indicates that the temperature of deposition was probably above 200 and below 400° C.

Mercury began to be mined in Nevada not long after the turn of the century. In 1906 discoveries were made in the Humboldt Range of Humboldt (now Pershing) County. Between this time and the start of World War I, many of Nevada's mercury districts were discovered. Numerous small retorts and furnaces, which still dot the Nevada landscape, were built during brief bursts of activity, but prospecting and production nearly ceased when the price was low. In 1907 deposits at Ione in the Union district were rediscovered and the Antelope Springs district was discovered in the same year; rediscovery activated the Barcelona district in 1908, and mercury was found in the Goldbanks district in 1912.

There is even a lost mine story that involves one of the mercury discoveries of this time. In the early summer of 1913, two cowpunchers chasing down a couple of strayed steers in the Pilot Mountains east of Mina recognized veinlets of cinnabar in the exposed face of an old prospect. After finding their lost stock and escorting them back to Mina, the cowhands now turned prospectors returned to claim what was later known as the Lost Steers Mine. As Knopf (1916c) told the story, this discovery was

> widely heralded as the rediscovery of the "lost Hawthorne quicksilver mine," named for Judge Hawthorne in whose honor it is said Hawthorne, the seat of Mineral County, is named. According to local report, Judge Hawthorne discovered in the seventies a rich quicksilver deposit which is believed to have been situated at the site of the recent discoveries. In returning from the mountains, so it is said, Hawthorne lost his bearings, and although he attempted annually to the end of his life to find the "quicksilver mine" he remained unsuccessful.

Knopf felt that this story was "highly improbable," but if true, deposits in the Pilot Mountains would take their place with those of the Barcelona and Steamboat Springs districts as the earliest mercury discoveries in the state.

In 1907 the first of the important contact metasomatic tungsten deposits discovered in the United States was found in the Toy district west of Brown's, a railroad siding near Lovelock, where scheelite was found associated with garnet and other dark silicate minerals in metamorphosed limestone. However, the first tungsten production from Nevada took place around the turn of the century from hübnerite-bearing veins in the Osceola district, and other hübnerite occurrences were discovered in 1907 at Round Mountain and in 1910 in the Eagle district. Scheelite-bearing veins were discovered in 1915 in the Shoshone and Silver Star districts, and the latter was to figure prominently as a source of tungsten during World War I.

Nevada also became a source of gemstones in the early part of the twentieth century. Turquoise was mined from several sites, including the Crescent district in Clark County, the Crow Springs district in Esmeralda County, and the Bullion district in Lander County, all prior to 1915. Fire opal was discovered at Virgin Valley in 1908 and intensely mined until 1913; however, this early opal mining ceased by about 1920.

In 1908, the Mackay School of Mines, named after Comstock luminary J. W. Mackay, began with the dedication of the original building that still stands on the University of Nevada campus in Reno. The building included a mining museum, now called the W. M. Keck Museum for a recent donor, to house mineral and ore specimens along with mining artifacts of early Nevada mines (Lugaski, 2000).

MINING AND MINERALOGY DURING WORLD WAR I

The onset of World War I in Europe in 1914 spurred metal production in Nevada, and, although precious metals were still important, the focus of mining in the state turned toward the discovery and extraction of the "war metals" copper, lead, zinc, tungsten, manganese, and antimony. The price of copper doubled between 1914 and 1917, and the value of the metal produced in Nevada exceeded that of gold or silver. Although quartz-scheelite veins at the Silver Dyke Mine in the Silver Star district near Mina were the most important tungsten source during the war, large tungsten skarn deposits that were to become much more important producers were found in the Mill City and Potosi districts in 1917. Production from these skarn deposits did not get underway in time to significantly contribute to the war effort.

World War I drastically changed this country's perception of its mineral resource endowment. Even before the U.S. entered the conflict, it became apparent that the loss of access to foreign supply was going to result in critical shortages of certain minerals. Soon after the declaration of war on April 6, 1917, concerned geologists and mining engineers organized a War Minerals Committee to investigate mineral supply needs for the war effort. Members were drawn from the American Institute of Mining Engineers, the U.S. Bureau of Mines (USBM), the USGS, the National Research Council, and the Association

of American State Geologists. Although this committee had no official status, it apparently had considerable influence in formulating mineral policy. It began its work by defining three classes of formerly imported minerals with which it would be concerned: minerals whose domestic production could be stimulated by ordinary commercial means; minerals whose domestic production would require some type of government support; and minerals that would have to continue to be imported because domestic sources were inadequate (Rabbitt, 1986). The last two of these classes evolved into what are now known as critical and strategic mineral commodities.

By early summer of 1917, the Division of Mineral Resources of the USGS embarked on a nationwide program of field investigations to aid in the discovery and development of minerals in short supply. One of the most important of these was manganese. In Nevada, Survey geologists J. B. Umpleby, E. S. Larsen, E. L. Jones Jr., and J. T. Pardee set out to examine manganese prospects across the state. Umpleby examined deposits in the Pioche district, Larsen described deposits near Golconda and Sodaville, Jones examined the newly discovered Three Kids property and other deposits in the Las Vegas district, while Pardee covered deposits in the Seigel and Nevada districts in White Pine County. Pardee and Jones (1920) collected much of this information and published it in a USGS bulletin, the first collection of Nevada manganese property descriptions to be published.

Tungsten was of equal importance and many contact metasomatic tungsten deposits had been discovered in the Great Basin region of Nevada, eastern California, and western Utah, prompting USGS geologist F. L. Hess to coin the name "tactite." Hess felt it too "irksome" to continually refer to the deposits as "contact metamorphosed rock" and did not like the miner's term "garnetite" as some did not contain much garnet. Hess's term tactite, derived from the Latin *tactus* (touching), seemed to be very appropriate for rocks formed by the contact metamorphism of limestone, dolomite, or other reactive rocks adjacent to intrusions (Hess, 1919). Hess, along with E. S. Larsen, undertook an extensive survey of intrusive-related tungsten occurrences in Nevada and other western states. Their work, published in 1920 as USGS Bulletin 725D, was the first descriptive study of contact metasomatic tungsten deposits. Eleven of the 23 tungsten districts described were in Nevada; the remainder were scattered in California, Utah, Oregon, Arizona, and New Mexico. Other field investigations carried out in Nevada as part of the USGS program included a survey of quicksilver deposits by F. L. Ransome, and examination of sulfur deposits by E. S. Larsen.

Mineralogical work performed during and after World War I mainly employed the same methods that were used so effectively during the second Nevada mining boom. The study of ore minerals was aided by publication of the first widely used book on petrology of opaque phases, *Microscopical Determination of the Opaque Minerals* (Murdoch, 1916).

NEVADA MINING AND MINERALOGY IN THE ROARING TWENTIES

With the end of World War I in late 1918, the federal government almost immediately withdrew from its wartime mineral investigation programs. In fact, with decreased demand, oversupply of some mineral commodities became a problem, and there were calls for price supports for some metals by industry. Strategic minerals did not receive much government attention throughout the 1920s and 1930s; supplies were adequate, prices were low, and the general consensus was that "mineral technology would overcome the increasing handicaps of nature, leaner ores, greater depths, and exhaustion, . . . and aside from the hazard of war, technology and its allies, exploration and transport, should be able to supply all the fuel and earth materials that the world can consume at prices not greatly different from those to which we became accustomed during the 1920s" (report of the National Resources Committee, 1937, quoted in Rabbitt, 1986).

After 1918 most metal prices dropped rapidly and Nevada's mining slowed, although silver mining flourished for several years because the silver price remained high due to the Pittman Act of 1923, which provided for purchase of silver by the U.S. Treasury. Mining in Nevada began a slow

recovery in 1924, with increases in base-metal production at Ely, Yerington, Pioche, and Goodsprings leading the way. Precious-metal mining, while less important, was active, but production from the Tonopah district slowed considerably during the late 1920s. Other commodities, such as arsenic (in demand for insecticide) and mercury (which reached the high price of $100 per flask in 1927) were actively prospected for and mined during the 1920s.

Nevada has mainly been known for metallic mineral production, but industrial mineral mining has also been historically important, and many of these deposits were initially discovered and mined in the 1920s. Mining of vein barite deposits, which began in Nevada in 1910, continued into the 1920s. Deposits of the borate mineral colemanite were discovered in the Muddy Mountains northeast of Las Vegas in the early 1920s and were considered to be of major importance at the time; however, mining of these deposits was discontinued before the end of the decade, because of competition from the huge Kramer borate deposit in California. Hoyt S. Gale, who studied saline deposits in Death Valley and Searles Lake, California, and later was to contribute substantially to the understanding of the Kramer deposit, described the colemanite deposit at the Callville (Anniversary) Mine in Nevada as follows:

> The colemanite in the Callville deposits is in a single large lenticular bed which at this place occurs as a regularly stratified member of the Tertiary succession. . . . It is possible that the original mineral may have been ulexite, as the colemanite occurs partly as nodular masses within the bed, and included in shaley bands or beds, but the present mineral of the deposit seems to be almost entirely colemanite.
>
> Colemanite still stands as the principal source of borax and boric acid in this country, its only competitor at the present time being the product manufactured from the brine of Searles Lake. Among the colemanite districts the largest, from the point of view of total reserves actually in sight, is undoubtedly the Furnace Creek field, when considered as a whole, but this is probably the only district that should be placed ahead of the Callville district in importance today.

Talc mining in the Palmetto and Sylvania districts of Esmeralda County began in 1925 and was to continue until the 1970s. Fluorite was mined for many years from the two largest producers—the Daisy Mine, Bare Mountain district, and the Baxter Mine, Broken Hills district—beginning in 1928. In the late 1920s, gypsum, limestone, and dolomite were mined in the Las Vegas area, and these commodities are still mined there today. Dumortierite, an aluminum silicate mineral used in the manufacture of spark plugs, was mined from the Champion Mine near Orcana in Pershing County for 20 years beginning in 1925.

The Gabbs brucite deposits were discovered in 1927; however, mining was not initiated until 1935. The earliest report on the geology and mineralogy of the brucite deposits was an unpublished 1929 report (now in the Nevada Bureau of Mines and Geology Information Office) by J. M. Carpenter identifying brucite, magnesite, and hydromagnesite at the property. The deposits, originally known as the Luning deposits, were described by Hill (1930):

> During 1927 a deposit of brucite (magnesium hydrate) of great purity was discovered on the west slope of Paradise Mountains, in Nye County, 30 miles north of Luning. The owner, H. E. Springer of Mina, states that 2,000,000 tons have been exposed, and that the deposit, which occurs on the contact of granite with dolomite, has not been bottomed. In Washington brucite occurs in considerable quantity along a fault vein in one of the large magnesite deposits but is described as "useless" for calcining purposes on account of the large amount of water held in combination. Tests made upon the Nevada brucite, however, are said to have demonstrated that the material can be used commercially. Transportation difficulties are not very serious, and, as the deposits are exceedingly large, this occurrence is likely to prove an important factor in the domestic situation.

The mineralogy of both the Oreana and Gabbs deposits was studied and described in the early 1930s by the well-known optical mineralogist P. F. Kerr, who was to be involved in the study of a wide variety of Nevada mineral deposits until the late 1960s. The Gabbs deposit is still being mined today (mainly for magnesite) and has figured prominently as a Nevada mineral locality.

Besides Gale and Kerr, other well-known scientists who worked on Nevada minerals during the 1920s were W. T. Schaller, Adolf Pabst, A. S. Rogers, and W. F. Foshag. In 1923, Edson Bastin of the USGS published some mineralogical information on the bonanza ore of the Comstock Lode, but by this time most of the original workings were inaccessible and good ore specimens were difficult to obtain. In 1928, an identification table for Nevada minerals was published by O. R. Grawe, Instructor in Geology and Mineralogy at the Mackay School of Mines; however, this publication did not contain specific information about mineral occurrences in the state. In 1929, the Bureau of Mines of the State of Nevada was created by the state legislature. In the following years this body, known today as the Nevada Bureau of Mines and Geology (NBMG), served as the state geological survey and published articles, often in collaboration with USGS scientists, on natural resources, geology, and mineralogy in the state.

In 1921, E. S. Larsen wrote *Microscopic Determination of Non-Opaque Minerals,* one of the first petrographic texts. Other advances in chemical and mineralogical sciences in the 1920s included the proliferation of graphs depicting the relationship between chemical, physical, and optical properties for many mineral series, and development of emission spectroscopy as an analytical tool. In the 1920s, the Mackay School of Mines began to publish reports on mineral studies in Nevada (for example, Grawe, 1928b). However, only a single Nevada type mineral, benjaminite, was identified during the decade (Shannon, 1925).

In the 1920s, the hobby of mineral collecting grew more widespread, and some Nevada mining districts became known as sources of fine mineral specimens. The first well-known Nevada mineral collector/dealer was H. G. Clinton, who lived in Manhattan and offered specimens of arsenic minerals and stibnite from that district (including "hair orpiment," later found to be the rare arsenic antimony sulfide mineral wakabayashilite). Clinton also provided specimens to mineralogists for research, including the type specimens for benjaminite. Another collector of note was J. D. O'Brien, mining engineer of Beatty; a re-creation of O'Brien's office and some of his specimens are on display in the W. M. Keck Museum.

The oxidized caps of the copper deposits in the Robinson district yielded fine specimens of native copper, cuprite, copper carbonate minerals, vivianite, and selenite, of which some fine examples that were donated by the Consolidated Coppermines Corporation may be found in the W. M. Keck Museum. On the basis of an unpublished and undated report by J. H. Courtright and Kenyon Richard in the NBMG Information Office, these samples were probably collected from the Richard Mine, where native copper, cuprite octahedrons over 1 cm in diameter, and selenite crystals to 25 cm in length were reported.

THE 1930S AND THE GREAT DEPRESSION— NEVADA MINING DOWN BUT NOT OUT

The Depression years of 1930–33 seriously affected Nevada mining, and the value of mineral production in 1933 was only one-sixth of that in 1929. This mining depression was nearly as severe as the post-Comstock, pre-Tonopah "bust" at the turn of the century, forcing the closure of silver mining operations at Tonopah and elsewhere in the state, drastic reductions in copper mining at Robinson, and cessation of a major new development at Goldfield. The only bright spot in Nevada mining during the Depression years was the development of the Rio Tinto Mine at Mountain City; however, significant production from this mine did not begin until 1935. The Rio Tinto deposit is unique in Nevada, consisting of partially oxidized strata-bound orebodies of massive sulfide ore grading nearly 10 percent copper with silver and gold credits. The mine yielded the highest grade of copper ore in the U.S. during much of the period that it operated. This rich orebody was discovered by S. F. Hunt, a prospector-geologist who located claims on an outcrop of gossan in 1919. According to Granger and others (1957), "Persistent work, under great difficulty and with inadequate financial resources, was rewarded in 1931 by the discovery of rich secondary copper ore beneath the gossan at a depth of 242 feet."

Post-Depression increases in the price of gold and silver in 1933 that were due to price-raising legislation stimulated precious-metal mining in Nevada, and in 1934 gold was the leading metal in terms of value. The nearly moribund Tonopah district was resurrected, and increased gold production from other districts, such as the Comstock and Manhattan, was reported. Copper production revived in 1935, and by 1937 was by far Nevada's most important commodity, with major production coming from the Robinson and Mountain City districts. Brucite mining was begun at Gabbs in 1935, and tungsten mining was reinvigorated in the Mill City district. Barite production increased in the 1930s, and the first extensive mining of barite in Nevada probably took place at a deposit that was to become part of the currently active Argenta Mine near Battle Mountain.

In the 1930s, the NBMG published reports on individual deposits and mining districts in Nevada, such as *Brucite Deposit, Paradise Range, Nevada* (Callaghan, 1933) and *The Tuscarora Mining District, Elko County, Nevada* (Nolan, 1936). Also in the 1930s, W. O. Vanderburg of the USBM began to publish reports on the mineral resources of Nevada on a county-by-county basis. This format was adopted by the NBMG following World War II.

P. F. Kerr, later a professor at Columbia University, worked extensively in Nevada during the 1930s, studying tungsten ore from the Mill City and Silver Star districts. In addition to microscopic techniques, X-ray diffraction mineral determinations began to be used at this time; Eugene Callaghan, in the NBMG report on the Gabbs deposits, acknowledged Kerr for mineral determinations using X-ray diffraction. In 1936, V. P. Gianella, Professor of Geology and Mineralogy at the Mackay School of Mines, published new work, including detailed microscopic information, on the Silver City district on the Comstock Lode.

In a 1930 American Mineralogist paper Charles Palache and David Modell of Harvard described the crystal habit of stibnite and orpiment in specimens from the White Caps Mine, Manhattan, that were provided by H. G. Clinton. This paper described orpiment crystals that "were in some cases slender needles of a clear golden color," probably a misidentification of waka-bayashilite, a mineral that was not to be named until 1972, from samples from Japan. Other mineralogists who studied Nevada minerals during the 1930s were W. F. Foshag, F. H. Pough, and W. T. Schaller.

Mineral collecting continued to grow as a hobby in the 1930s. In 1934 and 1935, John Melhase, retired Southern Pacific Railroad geologist, mineralogist, and collector from Berkeley, California, produced a series of articles in The Mineralogist (née The Oregon Mineralogist) that chronicled a 3000-mile mineral collecting "safari" through Nevada. Melhase visited many of Nevada's mining districts, including National, Austin, Robinson, Goldfield, Manhattan, Tonopah, and Yerington. His descriptions of occurrences and lists of mineral species at many sites are still useful, and, in fact, his list of minerals in the Tonopah district was used in the 1979 NBMG Bulletin on the Tonopah area. Melhase ended his series with these words:

> The writer spent much time during the years from 1916 to 1934 in traversing the territory covered by this 3000 mile safari and feels confident that just as many good minerals are obtainable now as in the halcyon days of old. Nevada presents a vast and open field for both the amateur and professional collector. . . . The collector is free to seek where he chooses; no anti-trespass signs glare from every fence-post or tree-trunk. Hospitality is the watchword of the people of Nevada, and none would be so thoughtless as to abuse their welcome by acts of discourtesy or vandalism. Fortunate indeed, is the collector who finds the opportunity to stow a few necessities in his car and head for an outing in Nevada, the Silver State, the mineral collector's Mecca.

WORLD WAR II AND THE CONCEPT OF STRATEGIC MINERALS

Although metal production dipped significantly in Nevada in 1938, the beginning of World War II in Europe in 1939 reversed this decline. Copper mining accelerated at Robinson and Mountain City and base-metal mining was revived in the Pioche district. Mercury and tungsten mining were also resurrected, with the development of new mercury mining operations in the Bottle Creek and Opalite districts and large tungsten orebodies in the Mill City district. Production of antimony and arsenic was renewed. Iron mining, insignificant in Nevada since the end of the First World War in 1918, was begun again in the Buena Vista Hills (Mineral Basin district). Manganese mining was initiated at several places during the later war years, notably at the Three Kids Mine near Henderson, following a cooperative USGS and USBM project to delineate ore reserves. Large-scale mining of brucite and magnesite at Gabbs was initiated for reduction in Henderson to metal needed for aircraft construction.

Even before the United States entered the fray, the short supply of certain mineral commodities, almost the same list of concern from World War I, became an important issue for both government and industry. In 1937, Nevada's Congressman Scrugham, along with USBM Director John W. Finch, called for special consideration of strategic and critical minerals. This apparently had little effect on Congress, however, and no special funding was provided for mineral studies. By 1939, there was no doubt that a European war was again to upset the U.S. mineral supply; facing this prospect, Congress finally considered legislation to provide support for strategic mineral investigations. The Strategic Materials Act, passed in 1939, contained a provision for strategic mineral studies, and the USGS and the USBM, with no increase in budget from Congress (Congressional support did not include money), diverted as much effort as possible to investigation of these commodities.

Following the entry of the United States into the war, production of metals came under the control of the federal government. The War Production Board was formed, which attempted to balance the need for military manpower with needs for workers to meet increased production targets in mining and other industries. Military deferments were issued for some miners, but the industry complained constantly about the lack of manpower and attempts to control mining from Washington. For the first time, women were widely employed in mining, particularly in extractive facilities. Miners and manufacturers were exhorted to produce more metals, household consumers to use less, and scrap recovery became important. In a 1942 article in the *Mining Journal,* H. O. King, chief of the Copper Branch of the War Production Board noted:

> There is just one reason why we must have nearly five times as much copper this year as our mines produced in 1938. That reason can be found by examining the material being used to fight this war. A medium tank takes almost a half ton of copper . . . one type of fighter plane uses over 800 pounds, and the big Flying Fortresses use over a ton and one-half . . . a front-line battleship uses two million pounds of copper. The copper that would normally go into electric refrigerators, which are not being made this year, will provide the copper and brass to complete 60 destroyers.
>
> Production is being increased by intensified effort in present mines, and by bringing into production the sub-marginal mines whose operation in peace-time had not been economic. The large producers (15 mines produce 98.5 per cent of our copper; 270 mines the rest) are working their properties as hard as the labor shortage will permit. Some mines are working 168 hours a week.

Gold mining also recovered during the early war years; the new Getchell Mine continued production in the Potosi district, and large-scale placer mines became important at Manhattan and Dayton. However, in late 1942 most gold mines were closed by order of the War Production Board, and most gold produced in Nevada in later war years was a by-product of copper mining. The Getchell Mining Company continued operation by exploiting arsenic and tungsten resources, and survived as the largest gold producer during the war by extracting gold as a by-product.

Under the impetus of wartime demand, Nevada deposits of mercury, tungsten, manganese, and tin were studied by USGS geologists, while the USBM personnel investigated alunite, copper, corundum, fluorite, iron, zinc, manganese, and tungsten properties across the state. Specific USGS projects included investigations by R. J. Roberts, later to figure in the discovery of the Carlin gold deposit, on mercury, tungsten, and manganese. In addition, tin deposits were studied at Majuba Hill by W. C. Smith and V. P. Gianella, and at Izenhood by C. Fries.

Studies carried out by the USBM were concerned with development of ore reserves at various properties, and their work usually involved detailed sampling and sometimes drilling. Mining engineer V. E. Kral, who later published reports on economic geology for the NBMG, investigated iron and fluorspar deposits. R. W. Geehan studied fluorite, iron, and zinc deposits. E. O. Binyon worked on corundum, manganese-nickel, and tungsten deposits. Other USBM strategic minerals projects in Nevada were carried out by G. H. Holmes, W. H. King, E. J. Matson, J. H. Soule, J. R. Thoenen, and R. R. Trengrove. Although work on projects was completed during the war years, most reports were published between 1946 and 1950 in the USBM Reports of Investigation series.

The mineralogy of Nevada's ore deposits became much better known during the war years due to research carried out by USGS, NBMG, USBM, and mineralogists from other institutions, and industry did their part to aid the war effort. The scientists who worked on Nevada's minerals in the 1940s included such luminaries as Clifford Frondel, Cornelius Hurlbut, P. F. Kerr, Charles Palache, F. H. Pough, and W. T. Schaller. Perhaps not as well known, Hatfield Goudey, mining geologist and mineral collector from Yerington, contributed articles on Nevada minerals to scientific and trade journals.

P. F. Kerr, building on his studies of Nevada tungsten deposits in the 1930s, described an unusual "hot springs" manganese-tungsten deposit near Golconda in 1940. Kerr proposed the varietal name "tungomelane" for tungsten-bearing psilomelane that he determined to be one of the ore constituents at the site (Kerr, 1940b). He eventually published a Geological Society of America Memoir on tungsten deposits in the United States in 1970 that contains descriptions of many Nevada deposits. Hurlbut published an article on supergene minerals in the Gabbs brucite-magnesite deposits in 1946, but Eugene Callaghan published most of the original geological and mineralogical work on the deposits in the 1930s, and the area containing the deposits was not mapped in detail until the 1950s. E. H. Bailey and D. A. Phoenix of the USGS, in the preface to their 1944 "Quicksilver Deposits in Nevada" publication for the NBMG, described their inventory project:

In 1940 quicksilver mining in Nevada became an industry with an annual output valued at nearly one million dollars. Shortly before this the U.S. Geological Survey began an extensive study of deposits of strategic metals which included quicksilver, a metal essential to the country in the time of war and not found in abundance in the United States. The writers have spent much of their time between November 1941 and December 1943 making detailed examinations of the quicksilver deposits in Nevada, and have visited nearly all of the more than 150 mines, prospects, and occurrences of quicksilver minerals described herein.

The most important Nevada mercury deposit during World War II was the Cordero Mine in the Opalite district. Discovered in 1924, the property was given little attention until 1941 when a furnace plant was built and the Cordero Mining Co. began production. In its first year, the mine became the largest mercury producer in Nevada. Additionally, this mine, together with the nearby McDermitt Mine developed later, became a mineralogical site of some repute owing to discovery of new mercury-bearing species.

The strategic minerals program did not end abruptly in 1945 as similar programs had ended following World War I. With the onset of the Cold War, the strategic minerals programs were continued. Two significant laws were enacted in 1946 that were to have important effects on the mineral industry; in July the Strategic and Critical Materials Stock-Piling Act was passed, and in August the Atomic Energy Commission was established. The stockpiling program effectively provided price supports for listed commodities. Miners in Nevada benefited from purchasing programs for tungsten, manganese, antimony, mercury, and to a smaller extent, uranium.

In the late 1940s, the Atomic Energy Commission encouraged private exploration for uranium, which had been sought by government geologists in secrecy during World War II. J. J. Johnson, the director of the commission's Division of Raw Materials was "convinced that the prospector, like the infantryman, is not outmoded. We still need the prospector to find mineral deposits. The geologist's technical knowledge is no substitute for the optimism and the persistence of the prospector" (Nininger, 1956). The intensity of the search was described by Nininger as involving "more than one hundred mining companies, including many of the largest companies, . . . and literally tens of thousands of prospectors—professional and amateur— . . . combing the hills" in "the gold rush of 1949."

MINING DURING THE COLD WAR, THE FABULOUS FIFTIES, AND THE SEXY SIXTIES

The end of World War II brought prosperity to the United States, and the years that followed were in general a period of escalating consumerism and use of natural resources. In general, Nevada's mining industry prospered and study of the state's minerals became more detailed.

Passage of the Defense Production Act of 1950 led government assistance to the mineral industry down a slightly different path than before. Under this act, the Defense Minerals Administration (DMA) was organized which, among other things, provided for assistance to the domestic mining industry in the form of technical advice and loans for the exploration of deposits of strategic minerals. In 1951, administrative changes placed all programs of the DMA under jurisdiction of the General Services Administrator who then delegated to the Department of the Interior the exploration assistance programs. Interior organized these programs into a new department, the Defense Minerals Exploration Administration (DMEA). In contrast to the earlier programs during which properties or entire districts were examined by USGS or USBM personnel and results of the work were made available to the general public through government publications, DMEA work was done at the request of individual property owners and the results of the work generally remained proprietary and in the hands of the property owner.

Along with the rest of the United States, Nevada was prospected for uranium during the

1950s, leading to new work on existing deposits and prospects and to new discoveries of radioactive mineral occurrences. Most publications on the geology and mineralogy of radioactive sites were issued by the Atomic Energy Commission (AEC) or by the USGS on behalf of the AEC. Publications that focused on uranium were released on Virgin Valley district opal by M. H. Staatz and H. L. Bauer in 1954, on oxidized base-metal ore in the Goodsprings district by P. B. Barton in 1956, and on the copper-tin ore at the Majuba Hill Mine by A. F. Trites and R. H. Thurston in 1958. Staatz and Bauer described an occurrence of uraniferous opal at Virgin Valley that was uneconomic as a uranium resource but was to be used for years as a source of brilliantly fluorescent opal for collectors.

Uranium deposits that were discovered in Nevada during this period included the Apex Mine near Austin in the Reese River district, a modest resource that was to be Nevada's largest uranium producer; the aptly named Stalin's Present claim in the Rye Patch district; and the Moonlight Mine in the Disaster district on the east side of the McDermitt caldera.

The Korean War stimulated mineral production between 1950 and 1953, and government supports for commodities continued into the 1960s. In the 1950s and 1960s the mining industry entered a period of expansion for many commodities, and modern mineral exploration was born. As a metallic commodity, copper was king during the 1950s and 1960s, particularly in the western United States, where new porphyry copper deposits were discovered and brought into production. In Nevada, production from the Robinson district increased due to expansion of open-pit mining beginning in the early 1950s by Kennecott Copper Corp. and Consolidated Coppermines Corp., and the Anaconda Copper Mining Company began production from its Weed Heights operation at Yerington in 1953. By 1953 copper production represented nearly half of the state's total mineral production value.

Tungsten, iron, mercury, lead, zinc, and manganese mining also fared well in the 1950s, but precious-metal mining was largely in eclipse. There were 32 tungsten-producing properties in Nevada in 1951, and the first substantial production of iron for shipment to Japan began in

that year. Several iron mines were active during the 1950s and 1960s, and descriptions of the state's iron deposits were published in a series of NBMG bulletins by R. G. Reeves, F. R. Shaw, and V. E. Kral. The Cordero Mine was the largest mercury producer in the United States for some years during the 1950s. In southern Nevada, the 1950s also saw renewed base-metal production from the Pioche and Bristol districts and production of manganese from the Las Vegas district at Henderson.

The end of the war in Korea in 1953 signaled another change in U.S. mineral policy. Price supports were phased out; the tungsten support price ended in 1957, manganese support was not renewed after 1961, and most other price supports ended by about 1964. The DMEA program ended in 1958, and its remaining programs were shifted to a new agency, the Office of Minerals Exploration (OME), still within the Department of the Interior. When the last of the old DMEA assistance contracts ended in 1962, the Department of the Interior estimated that the program had resulted in discoveries, nationwide, worth approximately $1 billion at present market prices for a cost of $32 million. OME continued to offer exploration assistance through about 1970. In 1961, gold and silver were added to the list of commodities eligible for assistance, and the last years of the program were dominated by gold and silver projects.

Although a nationwide depression in the late 1950s and the end of some government price supports largely brought Nevada's lead, zinc, and tungsten mining to an end, the early 1960s saw a rebound in mining led by copper and iron mining and aided by the renewal of gold mining at Getchell and the emergence of diatomite mining near Lovelock. Although copper accounted for more than 50 percent of the value of Nevada's mineral production in the mid-1960s and new copper mines were opened by the Duval Corp. in 1967, the first gold bar poured at the Carlin Mine in May of 1965 marked a turning point in the state's mining history.

NBMG publications on minerals and mineral deposits in the 1950s and 1960s included many county mineral-resource reports. As before, these consisted of text with 1 : 250,000-scale location maps for deposits (for example, Kral, 1951);

however, the report for Elko County (Granger and others, 1957) also included a geologic map at the same scale, and this format was to be followed until all Nevada counties were covered (Kleinhampl and Ziony, 1984, 1985). In addition, NBMG publications germane to mineral studies included *Antimony Deposits of Nevada* (Lawrence, 1963) and *Mineral and Water Resources of Nevada* (Nevada Bureau of Mines, 1964), which was published in collaboration with the USGS.

The 1950s and 1960s were marked by advances in chemical and mineral analytical techniques, including widespread use of atomic absorption spectrometry, X-ray fluorescence spectrometry, neutron activation analysis, X-ray diffraction methods, scanning electron microscopy, and electron microprobe analysis. Geophysical techniques also became more sophisticated, and were widely applied by explorationists. Mineralogical studies during the period included the description of nine Nevada type minerals and work by E. H. Bailey, G. T. Faust, R. C. Erd, A. S. Radtke, and S. A. Williams among others. Careful study of alteration minerals associated with porphyry copper deposits led to more-or-less worldwide acceptance of terminology for alteration assemblages. In 1968, the NBMG issued *Interpretation of Leached Outcrops,* a detailed work on oxidation mineralogy related to ore deposits, by Roland Blanchard, published posthumously.

In the 1950s and 1960s, mineral collecting became a popular hobby, and the terms "rockhound," "rock hammer," and "rock shop" became parts of the American lexicon. Mineral-collecting clubs, amateur mineralogists, and gemstone hunters grew common. Well-known mineral collectors and dealers who prospected Nevada during this period included Jack Parnau and Forrest Cureton. Turquoise and opal mining were revived somewhat during this period, with the latter mainly collected by amateurs for a fee. For many years during its operation, the copper mine at Copper Basin was an important source of turquoise; a local contractor in Battle Mountain had the contract to mine turquoise as copper mining exposed veins.

THE NEW GOLD DEPOSITS

When Newmont Mining Co. geochemist Alan Coope found high-grade gold in a trench cut on claims in the Lynn district of northern Eureka County in 1961, Nevada mining was poised for a change comparable to that inspired by the discovery of the Comstock Lode more than 100 years before. With subsequent sampling and drilling, this discovery became the Carlin gold orebody, 11 million tons of bulk-mineable ore averaging 0.32 ounces of gold per ton. Carlin, the first major gold discovery to be made in Nevada since gold mining had been curtailed at the start of World War II, started a frenzy of gold exploration, development, and mining that propelled Nevada into first place in gold production in the United States by 1980 and the country into second place worldwide in the 1990s.

Why was Carlin such a sensation? Carlin's gold was "invisible," as it could generally not be seen in the ore even with a high-powered microscope, and more importantly, it could not be detected by panning, the age-old prospecting technique depended upon by gold seekers everywhere. This property, with gold values up to 2 ounces gold per ton in float and surface outcrops (Coope, 1991), had been completely missed during many years of prospecting although a vein-type deposit with visible gold and associated placer deposits only a few miles to the northeast were mined as early as 1907.

Stranger yet, invisible gold was not unknown to western U.S. miners. At this point, it is worthwhile to backtrack a bit in mining and mineral history to show that "Carlin-type" gold was mined for 70 years prior to the discovery of the Carlin deposit.

Gold detectable only by assay was reported from the Mercur Mine in neighboring Utah in the early 1870s. This mine produced gold beginning in 1892, and was the first successful application of the cyanide gold recovery process in the United States. It is said that early Mercur prospectors considered bringing suit against an assayer who reported high gold values in rock that was obviously worthless on the basis of its failure to yield gold by panning.

The first production of gold from invisible gold ore in Nevada came in 1912 at the White Caps

Mine in the Manhattan district of Nye County. The ore was oxidized, very high in arsenic and antimony, and visible gold was completely lacking (Ferguson, 1924). Trouble with this ore quickly arose, however, when the oxidized material was mined out and deeper sulfide ores proved to be "refractory," a blanket term used for ore that was untreatable by available technology.

In 1922, another discovery was made at Gold Acres, Lander County, where miners reported "it is impossible to distinguish between ore and waste except by assay, and gold is present in such a state that it is impossible to obtain a single color by panning" (Vanderburg, 1939). Similar discoveries followed at Maggie Creek, Northumberland, Standard (Imlay district), and Getchell (Potosi district). The ore at each of these deposits contained gold detectable only by fire assay.

In 1925, a deposit of invisible gold was discovered along the Carlin Trend about 12 miles southeast of the future site of the Carlin Mine and staked as the Maggie claims. The property was evaluated by Consolidated Copper mines and an unknown company as a potential candidate for open-pit mining by steam shovels in 1935, and about 60 tons of ore were mined at the property in 1936. Although the Maggie claims were eventually to be absorbed in the giant Gold Quarry Mine in the 1980s, no further development of any significance was done on the property until 1963.

Getchell was found in 1934 by two prospectors grubstaked by Noble Getchell, a state senator from Battle Mountain who was active in mining ventures. Rather than seriously prospecting, the two had holed up for the winter in an area on the east side of the Osgood Range, northwest of Battle Mountain. With the coming of spring and needing something to show for Getchell's grubstake money, they collected samples from a nearby old pit dug by ranchers 20 or so years earlier on a bold outcrop of altered rock. The samples were fire assayed, found to contain gold, and several lode claims were staked. According to Hardy (1938), the ranchers had probably tested the rock by crushing and panning it, with negligible results, because the gold later found there was "invisible." Getchell obtained financial backing from Goldfield banker George Wingfield and other notables of the time, including

Bernard Baruch, Jack Dempsey, and Eleanor Roosevelt, to develop the property and construct a mill. Here is the special historical twist to the Getchell story: Getchell exposed Newmont Mining Co. to invisible gold long before their 1961 Carlin discovery provided the "type" name now applied to all of these deposits. Newmont joined Getchell's venture and provided financial and engineering help for the mill in return for a 17.5 percent interest in the property (Berger and Tingley, 1985). Similar to White Caps, Getchell was plagued with high arsenic content "refractory" ore, and Newmont personnel set out to solve this metallurgical problem. Plant construction at Getchell began in 1937, and gold production began in 1938. Newmont maintained its interest in the property for several years but sold after the company was unsuccessful in buying more ownership from Wingfield and Getchell.

Meanwhile, cyanide plants had been built at Gold Acres and Northumberland in 1936, and a cyanide plant was in operation at the Standard Mine by 1940. Through the last years of the 1930s and into the first years of the 1940s, four large "invisible gold" mines in Nevada were successfully producing gold that was not at all invisible in the form of bars of bullion. Still, as far as is known, no light of revelation illuminated these deposits as unique, and there appears to have been no rush to find other deposits of the future "Carlin type."

In 1936, W. O. Vanderburg, a mining engineer with the USBM, began a reconnaissance of mining districts throughout Nevada. When he completed his project in early 1939, Vanderburg had visited active mining operations and compiled data on prospects in Clark, Churchill, Eureka, Humboldt, Lander, Mineral, Nye, and Pershing Counties. These were the first countywide mine and mineral reports to be completed in Nevada, and they were published as individual Information Circulars by the USBM in 1936, 1937, 1938, 1939, and 1940. An astute observer, Vanderburg quickly grasped the significance of the "invisible gold." His 1939 report on Lander County closed with the following comments:

> In regard to future prospecting, a recent development in the northern part of the State has been the discovery and exploitation of a number of gold deposits in sedimentary rocks.

Among such discoveries may be mentioned the Getchell Mine in Humboldt County, the Standard Mine in Pershing County, and the Gold Acres Mine in Lander County. . . . In general, the sedimentary gold deposits do not possess easily recognizable surface indications, and consequently they were passed over in former years. Moreover, the gold in such deposits usually cannot be detected by panning, so that the early prospectors, who depended largely on panning, were misled. No doubt, deposits similar in character to those already found remain to be discovered in Lander County and other areas in the State.

Vanderburg's words did not seem to catch the imagination of the mining community. It may be unfair, however, to infer lack of vision; the time and technology were not right. Most of the "invisible gold" mines became plagued with recovery problems as soon as easily treatable oxidized ore became exhausted. Common to the unoxidized portions of these deposits were high concentrations of arsenic, antimony, and carbon, all of which seriously interfered with the cyanide process. Early attempts were made to roast these refractory ores in advance of cyanidation, but very high processing expenses soon outweighed the low mining costs obtainable in the large-tonnage, open-pit mines. Also, the war looming in Europe may have chilled thoughts of exploration for gold, invisible or otherwise. When war finally hit, Presidential Order L-208 declared gold mines to be nonessential to the war effort and ordered them closed to encourage miners to move to mines of copper, lead, zinc, and other strategic commodities. Even when the order was lifted at the end of World War II, Nevada's gold mines (including the ones with invisible gold) did not return. The Korean conflict, following close on the heels of World War II, kept the demand and price for strategic metals such as tungsten, mercury, and copper at levels that promoted search for deposits of these commodities in preference to gold. After post-war inflation and a government-controlled gold price were factored in, gold was not on most corporate exploration agendas through the 1940s and 1950s.

Then about 1960, the exploration climate began to change. The porphyry copper industry in the western United States had revolutionized mining and metallurgy with the development of new technology, and grades of copper ore that could be profitably mined progressively decreased as the scale of the operations grew. Newmont Mining Co. executives, possibly alone in their vision, saw the opportunity to apply the low-cost, bulk-mining methods used in porphyry copper operations to gold, but they needed a large, bulk-mineable gold deposit to work with.

A further link in the invisible gold story was provided by USGS geologist R. J. Roberts. Roberts came to Nevada about 1938. He, along with H. G. Ferguson and S. W. Muller during mapping in the Sonoma Range of Humboldt and Lander Counties, examined deposits of mercury, manganese, and tungsten all over Nevada for the strategic metals program. After a wartime stint in Washington, D.C., Roberts returned to Nevada to map the Antler Peak 15-minute Quadrangle in Lander County. About this same time, in the early 1940s, geologists mapping in Eureka County recognized a thrust fault of regional extent that they named the Roberts Mountains thrust for the mountains where they found it exposed (Merriam and Anderson, 1942). Roberts noted similar relationships at Antler Peak, saw evidence of major orogenic activity he believed to be responsible for the thrusting, and named this the Antler orogeny. He also noticed that in the small mining districts within the Antler Peak Quadrangle, the thrust fault and carbonate rocks immediately below the thrust appeared to be favorable spots for deposits of gold and other metals.

Roberts tied everything together in his publication "Alinement of Mining Districts in North-Central Nevada" (Roberts, 1960). This work linked the Antler orogeny to thrusting, described subsequent folding and doming of the thrust sheet, and proposed that mineral-bearing solutions—which liked the structural preparation of the folded, domed rocks as well as the chemical properties of the lower-plate carbonate rocks—found these sites favorable for formation of mineral deposits. Roberts pointed out that windows in the thrust sheet and mining districts in this part of north-central Nevada followed a northwest alignment, and that other areas along such favorable belts could be fruitful sites for prospecting. Roberts's introductory sentence of the 1960 paper is perhaps the best summary of his concept: "Few areas in the United States offer exploration targets like those in north-central Nevada. Here, a unique combination of geologic features permits us to make fair guesses as to where extensions of known districts may be found, and where we may look for new ones."

Now the two fields, economic geology and field geology, were merged in the person of John Livermore. John, a Newmont Mining Co. geologist, had long been interested in Nevada's gold potential. As a Stanford University student, he had worked at a summer job at the Standard Mine. He had read Vanderburg's predictions in the late 1940s and was deeply impressed by the exploration possibilities outlined therein (Coope, 1991). Livermore was working on a silver and base-metal exploration project at Eureka for Newmont in 1961 and, while in the area, attended a talk given by Ralph Roberts to a professional group in Ely. The two met, exchanged ideas, and Newmont was firmly set on the path that led to the Carlin discovery. Livermore had earlier been alerted to the Carlin area by Harry Bishop, manager at the Gold Acres Mine, who felt there was good potential for discovery of additional deposits of "fine gold" in northern Eureka County. Livermore, now joined by Alan Coope, began work on properties in the Lynn window, and, in the fall of 1961, claims were staked over the future site of the Carlin Mine.

The Carlin discovery set the stage for what was to follow, but still there was a hesitation before the modern gold rush got underway. With the help of the USGS Branch of Geochemical Exploration, the Cortez gold deposit was identified in 1966. The government still controlled the price of gold, however, and $35 per ounce did not provide much exploration incentive. In 1972, gold was released from government control and allowed to "float" or seek its own level on the world market. With the inevitable rise in price, exploration exploded, and discoveries soon followed. Freeport Gold Co. discovered deposits at Jerritt Canyon in the Independence Range of Elko County, Amselco Minerals set about developing discoveries at Alligator Ridge in White Pine County, and Newmont and others continued to find more deposits along the discovery mineral belt in Eureka and Elko Counties now known as the Carlin Trend.

Two new programs of the USGS began in the mid-1960s that had great impact on the mineral industry of Nevada. The first of these, the Heavy Metals Program, started in April 1966 as a joint effort by the USGS and USBM to stimulate domestic production of a group of metals (gold, silver, mercury, antimony, tin, bismuth, nickel, tantalum, and platinum-group metals) deemed by the government to be in short supply. Nevada, with a new multimillion-ton gold mine in production, was the right place at the right time to reap benefit from this new program.

The Carlin gold deposit focused attention on north-central Nevada as a potential new "gold belt." There were many questions to be answered about the nature of the new gold-deposit type, and the USGS set out to answer these questions. Areas in or near the "gold belt" selected for geologic mapping and geochemical sampling were the Carlin Mine area; the Bullion district (site of Gold Acres and Tenabo); the Cortez and Buckhorn districts; and the Independence, Adobe, Sheep Creek, and Tuscarora Ranges of Elko County. Other areas in Nevada also received attention. Geochemical and geological investigations were started in the Aurora district; in the Fallon to Manhattan mineral belt of Churchill, Mineral, and Nye Counties; in western Churchill County; and in the Comstock, Goldfield, Battle Mountain, and Robinson districts. Deep-drilling projects were undertaken at Iron Canyon in the Battle Mountain district, Tenabo in the Bullion district, and at Swales Mountain in Elko County.

Early mineralogical work on ore from the Carlin deposit by D. M. Hausen and P. F. Kerr in 1968 referred to the gold in ore from the Carlin Mine as "colloidal," perhaps in reference to its fine grain size (90 percent or more of the gold was estimated to be in particles of less than 2 microns). However, it turns out that the first study of "invisible" gold in a Carlin-type deposit in Nevada was that of Peter Joralemon on Getchell ore in 1951. Work by A. S. Radtke and F. W. Dickson in the 1970s and 1980s established a descriptive mineral, chemical, and geologic model for Carlin-type deposits, and reported on several new minerals, particularly thallium-bearing species, encountered during detailed mineralogical studies.

ENERGY-RELATED MINING
AND SILVER IN THE 1970S AND 1980S

During the 1970s and early 1980s short supplies of oil and ensuing high prices led to an "energy crunch" that stimulated exploration and production of some mineral commodities in Nevada. Barite, used heavily in oil well drilling, was an object of intense exploration and exploitation during a short-lived boom. In 1981, at the height of this boom, barite was mined at 30 sites in Nevada by 23 companies; however, by 1988 only five companies remained in the business, mining at six sites. In 1984 K. G. Papke published a Nevada Bureau of Mines and Geology report on 181 barite deposits and occurrences in the state, citing evidence for bedded barite mineralization during deposition of the Paleozoic sedimentary rocks that host it.

Uranium exploration came back into vogue during this period as well, and exploration geologists returned to areas discovered during the earlier uranium exploration period in the late 1940s and the 1950s. Areas of particular interest for uranium prospectors, easily recognized by their propensity to wave uranium-detecting scintillometers at every outcrop, during this new exploration were the Disaster and Opalite districts in the McDermitt caldera, the Tonopah district, and the Reese River (Austin) district. Interest in uranium potential in the McDermitt caldera led to intense study of its peralkaline rocks and the identification of large but uneconomic uranium deposits. In the late 1970s, a claim-staking frenzy in the caldera ensued, and public land in the 400-square-mile feature was completely covered by claims, creating a veritable forest of the perforated white PVC pipe in vogue as claim posts at the time. Discussions of the mineralogy of McDermitt caldera peralkaline rocks and ores may be found in Rytuba and Glanzman (1979) and Dayvault and others (1985).

Development of the North Slope oil field in Alaska and oil and gas developments elsewhere during the 1970s led to a period of unprecedented construction of oil and gas pipelines. The resulting need for high tensile-strength steel increased demand and prices for nonferrous steel additives, particularly molybdenum. Exploration for, and development of, "moly" deposits in the western U.S. included Nevada targets. The Hall molybdenum deposit in the San Antone district, known since the late 1930s, was put into production during 1981 and had a checkered history of operation by two major mining companies (Anaconda Corporation and Cyprus Minerals Company) until closure in 1991. Ore minerals of the deposit are molybdenite, ferrimolybdite, and powellite, and it is also known as a source of fine creedite specimens.

The Mount Hope deposit in the Mount Hope district and the Buckingham deposit in the Battle Mountain district, both in Eureka County and both intensively studied in the early 1980s (although Buckingham was discovered in the late 1960s), constitute large molybdenum resources that have never been mined. Catalogs of Nevada's molybdenum deposits and occurrences may be found in two NBMG reports by J. H. Schilling (1962a, 1979).

In the late 1970s and early 1980s, Nevada mining entered a short-lived revival of the glory of silver mining that many mining company executives and geologists would just as soon forget. Silver, at less than $5 per troy ounce in 1977, reached a high of nearly $50 in 1981 but declined rapidly to fall below $5 in 1982. At the same time, the price of gold skyrocketed to a high of about $800 per troy ounce, but did not drop as precipitously as silver, remaining above $300 until 1998. The late 1970s and early 1980s were years of intense prospecting in Nevada, with uranium, barium, silver, and gold all commanding high interest, and even molybdenum and tungsten getting into the act. Many prospectors and exploration geologists look back on this period with fondness. It was also a period of intense mineralogical study; in 1978 and 1979 six Nevada type minerals were described. Mineralogists active during this period were W. S. Wise, S. A. Williams, F. W. Dickson, and A. S. Radtke.

The silver price was driven up by the Hunt brothers of Dallas, Texas, using borrowed money in collusion with Saudi princes. Many Nevada silver mines were operated for a short period in the early 1980s, but quickly closed in the mid-1980s when the price tumbled. However, several silver mines were successfully operated, including the Candelaria, Taylor, and Rochester Mines. The latter, which actually produced its first silver in 1986 when the price hovered around $6, is still operating in 2003.

THE 1980S AND 1990S— GOLD MINING ON A HEROIC SCALE

Building on gold-mining methods perfected at Carlin during the 1960s and 1970s, and following exploration methods developed during that time, Nevada became one of the most active mining areas in the world in the 1980s and 1990s (Fig. 1). Gold replaced copper as king. This was fortunate for the miners and geologists in Nevada, because copper, tungsten, lead, and zinc were essentially done for as commodities by 1980 (although the Duval operation at Copper Canyon managed to hang on for a while), and the demise of molybdenum, mercury, iron, and barite were soon to come. Gold became the only game in town—but what a game it turned out to be. Annual gold production in the state rose steadily from 250,000 troy ounces in 1980 to 9 million in 1999. This made Nevada, if it were to be considered a nation (and sometimes Nevadans seem to have that attitude), the third largest gold-producing country in the world.

The star of the Nevada gold show is a feature called the Carlin Trend, a linear belt about 50 miles long and 3 miles wide with gold production and reserves that are estimated at over 100 million troy ounces. This amount of gold is approximately equal to the known gold production, both lode and placer, of the gold-producing area along the western slope of California's Sierra Nevada, an area of about 200 miles by 40 miles, and only slightly less than eastern Canada's Abatibi gold-mining zone, also a considerably larger area.

Significant mines in the Carlin Trend include the Carlin Mine, whose discovery in 1961 was described above; the Goldstrike property (first mined in 1980); the Dee Mine (first ore mined in 1984); the Gold Quarry Mine (first pour in 1985); the Genesis deposit (first mined in 1986); and the Meikle Mine (first pour in 1996). However, there are many smaller deposits in the Carlin Trend, which includes the Bootstrap, Lynn, Maggie

Creek, Carlin, and Railroad mining districts. Some of these deposits are mined out, some are currently being mined, and some are unmined. They have names like Screamer, Bobcat, Beast, Turf, Pete, Tusc, Mac, and Rain. There are so many gold deposits packed between the Meikle and Carlin Mines, a distance of about 8 miles, and so many high-grade ore discoveries have been found at depth, that at a sufficiently low cut-off grade, gold ore might be considered continuous over this distance.

With so many closely spaced mines, some of which are very large, the Carlin Trend has become almost surrealistic real estate. In the words of Seabrook (1989):

> To someone whose notion of a gold mine involves little ore carts pushed by men with lights on their hats, it is perhaps disenchanting to see these pits: the work is essentially an earth-moving operation, and the miners are really just truck drivers. But in their own way the pits are fantastic. Standing in them insults one's sense of scale. The axles of the hundred-ton haul trucks are ten feet from the ground, and the shovels on the Demag 285s are eleven yards wide. Everything is cartoonishly large. A typical pit descends in concentric circles, in a sequence of tiers: tier, twenty-foot drop, tier, to the bottom. Some pits are shaped like coliseums; others, more like ancient Greek theatres. Sometimes the job is simply to tear a mountain apart. Nearby there is always a new mountain, of waste rock, going up.

As has been the case during Nevada's earlier mining booms, serendipity and the vagaries of fate have played a hand in many of the new discoveries. As reported in Seabrook (1989), Witt "Dee" DeLaMare related the discovery of one such deposit in 1980 that was later named after him:

> I came to this big ol' outcrop, and it had that winy, purplish color you like to see, so I whacked off a hunk of it. But I was thinking to myself, It's too obvious. If that's the ore body sticking its snout up, somebody would have found it by now. So I hunted around some more. I came to another outcrop, same kind of rock—that altered limestone—but much smaller and less impressive. Kind of tucked

away. So I took a sample of that, too. I took maybe eight samples in all—not that many—and then I hopped back in the truck and left. Later, when I got the assay results back from the lab, it was like I figured—that big, handsome-looking outcrop was just about dead, but that little outcrop kicked like a mule.

Although Nevada mining came to be dominated almost completely by international companies in the 1980s and 1990s, the independent prospector remained an important fixture. Many of the original discoveries, such as at Alligator Ridge, were made by independent prospectors, who have not always bought in completely to the hegemony of the big mining companies. In the words of A. Wallace, also reported by Seabrook (1989), in reference to a disgruntled millionaire prospector and a lawsuit over royalty payments on a deposit along the Carlin Trend:

> [T]here's a whole shopful of other things that have been bothering Phil over the years. He thinks we ought to be exploring for barite. He thinks we're not looking hard enough for more gold on his property. He never thought we knew how to look for gold in the first place. You get that all the time out here—guys who just know they got gold on their claims, and they know right where it is, and they aren't going to take any contradiction from anybody, and especially not from some pointy-headed geologist. They think it's their land, so they know about it. Well, it is their land, sort of. I think the mining law is a great law in principle, but when you've got guys tying up big chunks of land, not letting you explore on it unless you do it their way, that's not doing anybody any good.

In addition to discoveries in the Carlin Trend, sedimentary rock–hosted disseminated gold deposits were found in many parts of Nevada, leading to proposals of other trends, most notably the Battle Mountain and Getchell Trends to the west of the Carlin Trend. Although it is not clear that all of these deposits should be called Carlin-type deposits, geological, mineralogical, and geochemical studies performed during the 1980s and 1990s led many geologists to the conclusion that the deposits should be put in a single class. Openings of new mines in such

deposits included Pinson (1980), Jerritt Canyon (1981), Alligator Ridge (1981), Northumberland (1982), Sterling (1984), Bald Mountain (1986), Florida Canyon (1987), Horse Canyon (1988), Marigold (1989), Rabbit Creek (1990), Lone Tree (1991), Pipeline (1997), and Ruby Hill (1998). In the Potosi district, the Getchell Mine was revived and the nearby Chimney Creek deposit began production in 1987, the Rabbit Creek Mine poured its first gold bar in 1990 and was combined with Chimney Creek to form Twin Creeks in 1993, and mining at the Turquoise Ridge deposit began in 1997.

Carlin-type deposits were not the only important discoveries during this period; volcanic rock–hosted gold and silver deposits began to produce gold in a big way in the 1980s. The Paradise Peak Mine poured its first gold bar in 1986. Modern open-pit mining at Round Mountain Mine, which began in 1977, produced more than 100,000 ounces of gold for the first time in 1987. Production also began at Sleeper (1986), Bullfrog (1989), Rawhide (1990), Rosebud (1997), and Midas (1999). Other types of precious-metal deposits added to the state's gold production beginning in the 1980s. These included the Fortitude (first pour in 1986), McCoy (1986), and Cove (1989) Mines.

In the 1980s and 1990s, computerized structural analyses and the ion microprobe were added to the arsenal of mineralogical technology. Electron microprobe and scanning electron microscope research on minerals was widespread. The accuracy of chemical analyses was enhanced by inductively coupled plasma emission and mass spectrometry methods. In addition, computer-generated modeling of orebodies and the hydrothermal processes that formed them have added to the knowledge of mineral deposits.

Because of the impact on extractive processes, a large amount of mineralogical study was done on Nevada's precious-metal deposits. Studies of Carlin-type deposits, building on the earlier work of P. Joralemon, D. M. Hausen, P. F. Kerr, A. S. Radtke, F. W. Dickson, and others, generally focused on the occurrence of gold. In 1985, Radtke proposed that the gold was in organic matter; however, studies by Hausen, B. M. Bakken, G. B. Arehart, and others showed that the primary occurrence of gold was mainly in the form of very tiny (micron-sized or smaller) native-metal particles and also possibly in atom-sized substitutions in sulfide minerals such as pyrite, marcasite, and arsenopyrite. In addition, new phases continued to be identified in Carlin-type deposits, including sulfides, native metals, and a host of secondary phases (see Ferdock, this volume).

Mineralogical work on volcanic-hosted gold deposits, though generally not as rewarding as that on Carlin-type deposits, was also productive. Mineral assemblages of high-sulfidation versus low-sulfidation deposits were established by P. Heald and others. The Nevada type mineral uytenbogaardtite was identified in ore from the Comstock Lode by M. D. Barton and others, and later in ore from Bullfrog. Electrum textures in vein-type deposits were classified by J. A. Saunders and others.

Mineral collecting in Nevada entered a new stage, with the state becoming more widely known as a source of museum-quality specimens. Carlin-type deposits turned out to be very rich in mineral species, and as at Manhattan and Getchell during earlier mining periods, localities such as the Meikle, Murray, and Twin Creeks Mines began to yield cabinet-quality specimens, particularly of barite, stibnite, realgar, orpiment, and calcite. New collectors and dealers arrived on the scene, and through their efforts began to make Nevada mineral localities known worldwide, including H. Gordon, M. Jensen, C. and J. Jones, N. Prenn, W. Lombardo, S. Pullman, R. Thomssen, and S. Kleine, many of whom were helpful during the preparation of this book.

PRESENT AND FUTURE MINING

Mining in Nevada today is very different in some ways than during its colorful past. Although independent prospectors are still combing the same deserts and mountain ranges that Mark Twain and his contemporaries wandered, they are much more likely to be in a sport utility vehicle than leading a mule. Like the family farm, the effectiveness of the independent miner is limited, and, as has been true for many years, the playing field is dominated by large companies. Costs and delays of permitting have driven many

small miners out of the game, and even the large corporate players may eventually have to give up and move overseas. Mineral commodity prices, which have generally trended downward in terms of real value in recent years due to foreign competition, have also not provided much incentive.

Although metal use is at an all-time high, there has been tremendous public pressure for the mining industry to clean up its act, and mining companies have responded, though slowly and grudgingly, ever mindful of the bottom line, to this pressure. Water pollution, air pollution, dust pollution, noise pollution, destruction of wildlife habitat, and a host of other problems are closely monitored by watchful governments, the media, and the public. Mining operations, by the very nature of their business, are environmentally disruptive, because they must move rock and dirt. In addition, most mining operations are located in relatively pristine areas and, by contrast, are messy and objectionable to some people. Because of their visibility and the amount of waste produced, open-pit mining methods, once regarded as the economic salvation of large-scale modern mining, have come increasingly under attack. Underground mining is now in favor because of its small "footprint" and the increasing depth of new deposits. However, underground mining is more expensive and dangerous than open-pit mining, and miner injuries and fatalities are likely to increase, leading to further outcry by the same public that consumes mineral products in ever increasing amounts.

Despite all these negative factors, mining is relatively healthy in Nevada. The state probably provides the kindest regulatory and public environment for the industry in the United States, and those of us who are involved in mining can only hope that it continues to be viable here. The industry is clearly still important to Nevada economically. At present, about 60 significant mining operations in Nevada produce about $3.3 billion worth of mineral products and employ more than 14,000 workers directly with a total payroll in excess of $700 million.

The mineral collecting business is, to some extent, a small subset of the mining industry. Mineral collectors and dealers rely increasingly on mining operations for specimens. In some cases, small mining operations are completely dependent on mineral sales. Examples include opal miners at Virgin Valley, quartz crystal miners on Petersen Mountain, and gold miners in the Winnemucca area. In other cases, collectors and dealers depend on large operating mines for specimens such as barite from the Meikle Mine, orpiment from the Twin Creeks Mine, stibnite from the Murray Mine, and galkhaite from the Getchell Mine. In addition, most amateur Nevada mineral collecting is done at abandoned mine sites.

Nevada's minerals have been an important part of her culture since the first Native Americans mined turquoise. Since that time access for both miners and collectors has been relatively easy in Nevada because of relatively minor population pressures and the large amount of public land. Let's hope it stays that way.

Nevada Type Localities

STEPHEN B. CASTOR

Thirty-seven minerals that are accepted as species by the International Mineral Association (IMA) were first discovered in Nevada (Table 2). DeMouthe (1985) describes the type localities for most of these minerals.

The first Nevada mineral, hübnerite (sometimes written huebnerite), was described in 1865 from veins in the Mammoth district, Nye County, by mining engineer Eugene Riotte, who documented its occurrence at the first tungsten discovery in the state. Initial publication of this occurrence has been ascribed to Palache and others (1951), perhaps because Riotte's description was in the *Reese River Reveille,* a mining camp newspaper published in Austin, rather than in a technical journal. However, Riotte's description gave physical, chemical, and crystallographic data that were sufficient to establish species validity at the time, and it is clear that he was its discoverer, although his determinations of crystal class and density for the mineral are now known to be incorrect.

Hübnerite ($MnWO_4$) forms a solid-solution series with the mineral ferberite ($FeWO_4$), which was described from Spain in 1863. Specimens of intermediate composition are wolframite, a more ancient mineral name that has been discredited as a species. Hübnerite-bearing veins have been found elsewhere in Nevada and it is a tungsten ore mineral, but such ore has accounted for less than 0.5 percent of the state's tungsten production.

In addition to hübnerite, Mr. Riotte first named the mineral stetefeldtite, silver antimony oxide hydroxide, in ore from the Belmont district in Nye County in 1867, though some researchers believe that this material is actually a mixture of minerals (Anthony et al., 1995). Another nineteenth-century–Nevada type mineral was metastibnite (Sb_2S_3), first reported in sinter at Steamboat Hot Springs near Reno in Washoe County by George Becker in 1888.

New minerals in Nevada have come from many different sites that represent a wide variety of occurrence types; however, a few sites are notable for the number of minerals for which they are type localities. As befits Nevada's present glory in gold mining, and the uniqueness of Carlin-type gold deposits, eight new minerals have come from such deposits and five are from the namesake for the deposit type, the Carlin Mine in Eureka County. Most of these minerals are thallium-bearing sulfide minerals, but they include frankdicksonite (BaF_2), a barium analog to fluorite. Other Carlin-type gold deposits are type localities; getchellite ($SbAsS_3$) was first identified at the Getchell Mine, and goldquarryite ($CuCdAl[F,PO_4] \cdot H_2O$) and Nevadaite ($[Cu_2VAl]Al_8[PO_4]_8F_8[OH]_2[H_2O]_{20}[H_2O]_{-1.5}$) were first identified at the Gold Quarry Mine.

Nevada mercury mines have also been prolific Nevada type localities. The Red Bird Mine near Lovelock, which may be the site of overlapping mercury and base-metal hydrothermal systems, is the type locality for the lead antimony sulfide mineral robinsonite, and the co-type locality for dadsonite and schuetteite. The McDermitt Mine in northern Humboldt County, where mercury was extracted from high-level epithermal ore hosted by volcaniclastic sedimentary rocks in the McDermitt caldera, is the type locality for kenhsuite and radtkeite, both mercury sulf-halide minerals. The Cordero Mine, located only a half-mile from the McDermitt Mine, is the type locality for a third mineral from the caldera, the mercury sulf-chloride corderoite.

The historical Goldfield district, and in particular the Mohawk Mine, has two type locality species to its credit, goldfieldite and mackayite, both tellurium-bearing minerals. A third Gold-field species, blakeite, has apparently recently been discredited as a mixture of tellurite and goethite (DeMouthe, 1985); although it was listed as a valid species as recently as 1995 (Fleischer and Mandarino, 1995), it has apparently fallen into disfavor (Mandarino, 1999).

Two other Nevada localities are multiple type mineral sites. The Majuba Hill Mine, an important collecting locality for a variety of arsenate minerals, is also the type locality for two minerals, parnauite and goudeyite, both copper arsenate species that were named after mineral collectors active in Nevada. The Van-Nav-San claims are the type locality for schoderite and metaschoderite, the latter being a dehydration product of the former.

In addition to blakeite, eight other Nevada species that have been proposed in the literature were either never recognized or invalidated following recognition. These include belmontite, a silicate of lead from Belmont that was never recognized; esmeraldite, a discredited species from an unspecified locality in Esmeralda County; and vegasite, a discredited hydrated sulfate from Goodsprings. Two phosphate phases from the Manhattan district have been denied species status; clinobarrandite was discredited and trainite never recognized. The Gabbs district claims two discredited species, cuproartinite and cuprohydromagnesite, in addition to the valid species callaghanite.

Several researchers have been notably productive in terms of Nevada type localities. Arthur Radtke, USGS geologist who contributed significantly to original research on Carlin-type gold deposits, has been the most prolific Nevada type locality producer and is senior author or coauthor of seven papers that introduced Nevada type minerals. Frank Dickson, a Radtke co-worker and researcher of western U.S. mercury deposits, contributed to five papers on Nevada type localities. Sidney Williams, noted Arizona mineralogist and collector, has been responsible for publications on minerals from four Nevada type localities. Other well-known mineralogists and geologists who have contributed Nevada type localities include Frederick Ransome, Edgar Bailey, Clifford Frondel, Frederick Pough, Eugene Foord, Charles Palache, and Richard Erd.

Advances in the study of Nevada minerals, as elsewhere in the world, has partly been a function of technological advances. Early Nevada mineralogists relied on physical property measurements such as specific gravity, compositional determinations by wet chemical methods, and crystallographic determinations by contact or optical goniometer measurements. Blowpipe testing was widely used, but yielded only qualitative results. Determinations on small particles and very small specimens were difficult at best, and mineral assemblages for many historic mining camps are poorly known. For instance, the list of reported minerals from the Comstock Lode only amounts to about 40 species (Stretch, 1867; Smith, 1985; Vikre, 1989), a relatively small number for such a productive and historically rich district. In the words of Smith (1985):

> Mineral specimens from the Comstock are very scarce today, even in museums. In 1913 when Edson Bastin went to Virginia City to procure some material for a microscopic study of the ores, specimens from the shallow workings and bonanza ore specimens were difficult to obtain (Bastin, 1922). No doubt there are several reasons for this scarcity of specimens. The isolated location of the Comstock in the early days, the lack of appreciation of specimen material by the mostly uneducated people working the mines, lack of tradition for specimen collecting and selling, the devastating fires, particularly the one in October of 1875 when much of the town was destroyed, and the fact that attractive specimens in the Comstock mines were probably not abundant. The bonanza ores were usually dull black mixtures of acanthite, polybasite, stephanite, and tarnished silver that were only rarely crystallized. It was certainly not something that a miner would want decorating his hearth. And even in the case of attractive specimens, most material that he high-graded was probably turned into bullion as soon as he needed money.

Unfortunately, collections of ore specimens for other historical Nevada mining camps are similarly incomplete, and it is possible that we will never have a full accounting of all species for many districts even when modern methods are used in the study of presently available specimens.

The history of Nevada type mineral localities

TABLE 2   Type localities in Nevada

Mineral	Chemical formula	Initial Citation	Locality
Aurorite	$(Mn,Ag,Ca)Mn_3O_7 \cdot 3H_2O$	Radtke et al., 1967	Aurora Mine, White Pine County
Benjaminite	$(Bi,Ag,Cu,Pb)_5S_6$	Shannon, 1925	Outlaw prospect, Nye County
Callaghanite	$Cu_2Mg_2CO_3(OH)_6 \cdot 2H_2O$	Beck and Burns, 1954	Gabbs Magnesite-Brucite Mine, Nye County
Carlinite	Tl_2S	Radtke and Dickson, 1975	Carlin Gold Mine, Eureka County
Christite	$TlHgAsS_3$	Radtke et al., 1977	Carlin Gold Mine, Eureka County
Corderoite	$Hg_3S_2Cl_2$	Foord et al., 1974	Cordero Mine, Humboldt County
Curetonite	$Ba_4TiAl_4(PO_4)_4(O,OH)_6$	Williams, 1979	Redhouse barite prospect, Humboldt County
Dadsonite*	$Pb_{21}Sb_{23}S_{55}Cl$	Jambor, 1969	Red Bird Mine, Pershing Co.
Ellisite	Tl_3AsS_3	Dickson et al., 1979	Carlin Gold Mine, Eureka County
Elyite	$CuPb_4SO_4(OH)_8$	Williams, 1972	Silver King Mine, White Pine County
Faustite	$(Zn,Cu)Al_6(PO_4)_4$	Erd et al., 1953	Copper King Mine, Eureka County
Frankdicksonite	BaF_2	Radtke and Brown, 1974	Carlin Gold Mine, Eureka County
Getchellite	$SbAsS_3$	Weissberg, 1965	Getchell Gold Mine, Humboldt County
Goldfieldite	$Cu_{12}(Te,Sb,As)_4S_{13}$	Ransome, 1909	Goldfield, Esmeralda County
Goldquarryite	$CuCdAl(F,PO_4) \cdot H_2O$	Roberts et al., 2003	Gold Quarry Mine, Eureka County
Goudeyite	$Cu_6(Al,Y)(AsO_4)_3(OH)_6 \cdot 3H_2O$	Wise, 1978	Majuba Hill Mine, Pershing County
Heyite	$Pb_5Fe_2O_4(VO_4)_2$	Williams, 1973	Betty Jo claim, White Pine County
Hübnerite	$MnWO_4$	Riotte, 1865	Erie and Enterprise Mines, Nye County
Huntite	$CaMg_3(CO_3)_4$	Faust, 1953	Currant Creek, White Pine County

Mineral	Chemical formula	Initial Citation	Locality
Kenhsuite	$\acute{\text{Y}}$-$Hg_3S_2Cl_2$	McCormack, 1998	McDermitt Mine, Humboldt County
Mackayite	$FeTe_2O_5(OH)$	Frondel and Pough, 1944	Mohawk Mine, Esmeralda County
Metaschoderite	$Al(PO_4,VO_4)\cdot3H_2O$	Hausen, 1962	Van-Nav-San claims, Eureka County
Metastibnite	Sb_2S_3	Becker, 1888	Steamboat Springs, Washoe County
Mopungite	$NaSb(OH)_6$	Williams, 1985	Lake prospect, Mopung Hills, Churchill County
Natrojarosite	$NaFe_3(SO_4)_2(OH)_6$	Palache, 1951	Soda Springs Valley, Esmeralda County
Nevadaite	$(Cu_2VAl)Al_8(PO_4)_8F_8(OH)_2$ $(H_2O)_{20}(H_2O)_{-1.5}$	F. Hawthorne, w.c., 2003	Gold Quarry Mine, Eureka County
Parnauite	$Cu_6(AsO_4)_2(SO_4)(OH)_{10}\cdot7H_2O$	Wise, 1978	Majuba Hill Mine, Pershing County
Pottsite	$PbBiH(VO_4)_2\cdot2H_2O$	Williams, 1988	Linka Mine, Lander County
Radtkeite	Hg_2S_2ClI	McCormack et al., 1991	McDermitt Mine, Humboldt County
Retgersite*	$NiSO_4\cdot6H_2O$	Frondel and Palache, 1949	Nickel Mine, Churchill County
Robinsonite	$Pb_4Sb_6S_{13}$	Berry and others, 1952	Red Bird Mine, Pershing County
Schoderite	$Al2(PO_4)(VO_4)\cdot8H_2O$	Hausen, 1962	Van-Nav-San claims, Eureka County
Schuetteite*	$Hg_3(SO_4)O_2$	Bailey et al., 1959	Silver Cloud Mine, Elko County, and others
Simmonsite	Na_2LiAlF_6	Foord et al., 1999	Zapot claims, Mineral County
Stetefeldtite	$Ag_2Sb_2(O,OH)_7$	Riotte, 1867	Belmont district, Nye County
Uytenbogaardtite*	Ag_3AuS_2	Barton and others, 1978	Comstock Lode, Storey County
Weissbergite	$TlSbS_2$	Dickson and Radtke, 1978	Carlin Mine, Eureka County

*Nevada sites are co-type localities.

is illustrative of the influence of modern analytical methods on the study of mineralogy. Between 1860 and 1880, which included most of the first wave of Nevada prospecting and mining, two Nevada type localities were reported, both by the versatile mining engineer Eugene Riotte. These species, hübnerite and stetefeldtite, are considered valid today, and their early discovery is a tribute to the skill of Mr. Riotte because the three following 20-year periods yielded only one valid Nevada type locality each.

During the late nineteenth century, the polarizing microscope, still indispensable in mineralogical and petrographic studies, came into widespread use. Work on the changes caused by hydrothermal alteration in the Comstock district by Becker (1882) is a classic example of the early use of this technology. Becker (1888) was also responsible for the discovery of metastibnite at Steamboat Springs, the state's third type mineral locality.

By 1940, Nevada had only five mineral type localities to its credit, including the three noted above. In addition, goldfieldite, a tellurium-bearing sulfide, was described by a well-known USGS geologist during the heyday of the Goldfield district (Ransome, 1909a); and benjaminite, a bismuth-bearing sulfosalt, was described by Shannon (1925). The validity of both of the latter minerals has been disputed, but both are currently listed as species (Mandarino, 1999).

Between 1940 and 1960, eight new minerals were described from Nevada. This sudden increase corresponds with widespread use of X-ray diffraction (XRD) analytical technology following the 1941 founding of the Joint Committee on Chemical Analysis by X-ray Diffraction Methods, an organization that catalogued and published XRD data for minerals and other substances (reconstituted as the Joint Committee on Powder Diffraction Standards, or JCPDS, in 1969). Nevada type localities described during this period included a wide variety of minerals: carbonates, sulfates, a phosphate, a sulfide, and an oxide. One of these minerals, mackayite from Goldfield (Frondel and Pough, 1944), was named after John Mackay, who had no connection with Goldfield but was arguably the most important figure in Nevada's mining history.

In the 1960s, microbeam technology (microscopic examination and analysis using an electron beam) revolutionized the study of minerals by enabling chemical analysis of tiny (about 1/1000 mm) mineral grains. The first commercial microprobe was developed in France in 1958 and the first commercial scanning electron microscope in England in 1965. With these additional tools, and the recognition and large-scale exploitation in Nevada of a new class of ore deposits called Carlin-type gold deposits, new mineral discovery became relatively commonplace. In all, 17 Nevada mineral type localities were described between 1960 and 1980, including 6 discovered in Carlin-type deposits by USGS scientists.

At this writing, only seven new minerals have been discovered in Nevada since 1980. This decrease in production of type localities in the state is problematic; one would expect that the combination of improving technology and the most profitable mining boom the state has ever known would have accelerated discovery. The answer may lie with a decrease of scientific interest in the search for new minerals—some earth scientists believe that the discovery and description of new mineral species is no longer an important objective and the training of new mineralogists is not a high priority in modern earth science education. The increase in the number of mineral species worldwide over the past 20 years is about the same or slightly less than the increase during the previous 20. Another possibility is that the geologists and mineralogists most likely to be contributing new mineral species from Nevada have been too busy looking for gold ore.

MINERAL DEPOSITS

PART II

Carlin-Type Gold Deposits

GREGORY C. FERDOCK

Carlin-type gold deposits are a relatively newly identified type of mineral deposit. F. W. Dickson and A. S. Radtke first used the term in the 1970s to refer to deposits similar to the one at the Carlin Mine, which began operation in 1965 in the Lynn district of Eureka County. Carlin-type gold deposits have also been referred to as "sedimentary rock–hosted gold deposits," because they mainly occur in sedimentary rocks, predominantly those that contain carbonate. An important characteristic of these deposits is the presence of "invisible gold," gold that cannot be detected in the native form, even using an electron microscope. Although the Carlin Mine became the model for these deposits, several mines exploited similar deposits prior to 1940 in Nevada, including White Caps (Manhattan district), Getchell (Potosi district), Standard (Imlay district), and Gold Acres (Bullion district). In addition, mining of such deposits at Mercur, Utah, actually took place before the end of the nineteeth century.

Carlin-type deposits occur throughout the state of Nevada but most are concentrated in clusters along four major "trends": Carlin, Battle Mountain-Eureka, Getchell, and the Independence Mountains (Fig. 2). Total production and reserves at these deposits, which include at least 75 individual deposits, total more than 150 million ounces of gold.

Although individual Carlin-type deposits share some general characteristics, particularly in chemistry, mineralogy, and general geologic setting, there is considerable variation in detail. Because of the variation of sizes, grades, structural complexity, areal distribution, and details in host rock lithology and mineral assemblages, it has been reported that "when you've seen one Carlin-type deposit, you've seen one."

Since the initial description of Carlin-type deposit models, new discoveries have been made, such as recognition of the variable nature of the orebodies and the importance of gold-bearing iron and arsenic sulfide minerals. In addition, similar mineralization has been found in igneous and metamorphic rocks as well as in noncarbonate sedimentary rocks including chert and orthoquartzite, although the most common host lithology remains clastic-rich, calcite-bearing marine sediment, generally silty limestone, as at Carlin. On the basis of this new knowledge, the term "sedimentary rock–hosted" should probably be abandoned. If a single term is desired for these deposits, which do share chemical and mineralogical attributes that are generally universal, "Carlin-type" is probably as good a name as any.

The mineralogy of Carlin systems is, at first glance, staggering. Close to 300 mineral species have been identified in the various deposits (nearly 160 at the Goldstrike Mine alone) ranging from the mundane to very exotic and extremely rare. Nine new mineral species have been described from these deposits, and others are pending. The mineral assemblages vary from deposit to deposit, depending on host rock composition and variations in the chemistry of the ore fluids. However, general mineralogy is similar from deposit to deposit, as is paragenesis (the sequence of mineral deposition).

F. W. Dickson and A. S. Radtke conducted previous work on the mineralogy of these systems in the 1960s and 1970s at the Carlin Mine. Their work led to a descriptive mineralogical and chemical model as well as the discovery of several new mineral species that are typical of Carlin-type deposits. Further studies have been conducted by Kuehn (1989) and Bakken (1990) at the

FIGURE 2   Location of Carlin-type gold deposits in the western United States. The trends are indicated by fine dashed lines. Modified from Hofstra et al., 1999.

Deposits cited in the text: AR = Alligator Ridge, C = Carlin Mine, Ge = Getchell, GP = Gold Pick Pit, GQ = Gold Quarry, GS = Goldstrike Mine (Betze-Post-Screemer-others), M = Murray, MC = Mercur, Me = Meikle, N = Northumberland, S = Sterling, TC = Twin Creeks, WC = White Caps.

Carlin Mine; Ilchick (1990) and Hitchborn and others (1996) at the Bald Mountain Mines; Arehart and others (1992, 1993), Peters and others (1998), and Ferdock and others (1997) at the Goldstrike Mine; Hofstra (1994) and Phinisey and others (1996) at the Independence Mountains Mines; Bloomstein and others (1993) and Panhorst (1996) at the Lone Tree Mine; and Tretbar and others (in press) at the Getchell Mine among others at various deposits around Nevada. Although careful mineralogical work has been done at a number of deposits, most of the work has focused on ore minerals and otherwise only the most common phases are reported. Comprehensive descriptive work on complete mineral assemblages, including identification and listing of secondary and collectable minerals, has not been performed on many of these deposits.

THE MINERALS

Mineral assemblages at most Carlin-type deposits are complex, containing dozens, or in some cases more than a hundred, mineral species. This mineral complexity frequently results in problems with gold recovery or difficulties with waste rock due to acid generation by sulfide-rich rock. Concentrations of sulfides, carbonate, carbon, phosphates, and various metals such as nickel, copper, and zinc are often encountered in places in orebodies, forcing mine operators to blend ore to achieve maximum return from mills and leach facilities. To provide for effective gold recovery and waste rock management, it is imperative that the mineral makeup of deposits be well defined prior to mining.

The complexity of mineralization is most often related to multiple generations and uneven distributions of primary mineral species and their alteration products. Paragenesis is often confused by overprinting of multiple generations of the same or similar mineral species. Multiple stages of iron and base-metal sulfides, quartz, and clays are found in the same general area as high-grade ore. Furthermore, in many deposits, the paragenesis was complicated by overlapping hydrothermal events, tectonic and hydrothermal brecciation, and dissolution and collapse of host rocks.

Although the hydrothermal processes are poorly understood, it is possible that a single hydrothermal event can produce the variety of primary hydrothermal minerals (sulfides, sulfosalts, carbonates, and silicates) found in Carlin-type deposits. As ore-bearing fluid evolves, various physiochemical parameters are met and systematic deposition of minerals takes place (Woitsekhowskiya and Peters, 1998).

Of the array of minerals found in Carlin-type deposits, only two generally have economic value: native gold and gold-bearing pyrite. Relatively small amounts of realgar, orpiment, stibnite, barite, cinnabar, turquoise, and variscite have been mined economically.

Minerals found in Carlin-type deposits generally fall into one of six genetic categories, from earliest to latest: I) original sedimentary and diagenetic minerals; II) igneous and metamorphic minerals in Paleozoic to Cenozoic mafic to silicic intrusive and volcanic rocks and accompanying metamorphic halos; III) early pre-gold hydrothermal minerals with clay and chemically simple sulfides; IV) gold-bearing sulfides; V) waning gold deposition with formation of As-Sb-Zn-(Tl) sulfides and sulfosalts along with Ba-(Ca) sulfate (including the bulk of the complex sulfosalts and rare halides); VI) post-gold secondary mineralization related to hypogene and supergene processes (Table 3). Few species form in more than one of these categories with the exceptions of pyrite, some base-metal sulfides (e.g. sphalerite), quartz, calcite, dolomite, illite, and possibly kaolinite, graphite, and gypsum.

I. Sedimentary minerals. Only 19 species have been found to be of unequivocal sedimentary origin. Quartz, calcite, dolomite, illite, and pyrite are most common. Lesser amounts of barite (rarely strontium-bearing) and apatite occur in carbonate host rocks, but are much more common in adjacent clastic sedimentary rocks. Sphalerite formed as a diagenetic mineral in siliceous rocks in the Independence Range (Hofstra, 1994) and was reputedly of diagenetic origin at the Meikle Mine (Emsbo and others, 1997). Other species such as zircon occur in trace amounts and are of detrital origin.

II. Igneous/metamorphic. Minerals of metamorphic and igneous origin account for 65 species, including some that are not confined to this genetic group, such as quartz, calcite, and pyrite. Significant amounts of dolomite, base-metal sulfides, and zeolites fall into this group. Minor non-sulfide-bearing gold mineralization in metamorphic and igneous rocks was probably related to the emplacement of intrusive bodies, commonly of Mesozoic age. Bismuth and tellurium minerals are associated with these rocks.

III. Early-stage pre-gold hydrothermal mineralization. This was the first of the multiple stages of Carlin-type mineralization. At least 32 species are delineated in this group, mostly "simple" sulfides. Arsenopyrite was almost exclusively deposited during this stage. Rare tungsten, nickel, lead, and copper sulfides and pyrrhotite are often found as micron-sized inclusions in pyrite. Bismuth-bearing compounds were also deposited during this stage, often above and distal to the main disseminated gold mineralization. Dolomite replaced calcite, and illite and muscovite (or sericite) were the dominant silicates formed. Some apatite may have also formed during this stage.

IV. Gold-bearing sulfide deposition. Gold-bearing sulfide and associated compounds that formed during the distinctive main-stage gold phase are few, seven in all. Gold was principally deposited in submicroscopic form in iron sulfides, mainly in pyrite, and to a lesser degree in marcasite,

TABLE 3 Minerals occurring in Carlin-type deposits

Mineral	Depositional Phase	Mineral	Depositional Phase
Arsenates		Gold	[II,1][IV,15+]
Arseniosiderite	[VI,1]	Graphite	[II,1][III(?),3][V(?),3]
Carminite	[VI,1]	Selenium	[VI,1]
Chalcophyllite	[VI,1]	Silver	[V,3]
Conichalcite	[VI,4]	Sulfur	[VI,3]
Dussertite	[VI,2]	Tellurium	[V(?),1]
Gaitite	[VI,1]	*Halides*	
Guérinite	[V,1]	Chlorargyrite	[VI,6]
Haidingerite	[VI,2]	Corderoite	[VI,1]
Hörnesite	[VI,1]	Fluorite	[II(?),1][V,4]
Mansfieldite	[VI,2]	Frankdicksonite*	[V,1]
Mimetite	[VI,1]	Halite	[V,1]
Olivenite	[VI,2]	*Molybdates*	
Pharmacolite	[VI,2]	Ferrimolybdite	[VI,1]
Pharmacosiderite	[VI,3]	*Oxides and Hydroxides*	
Picropharmacolite	[VI,1]	Anatase	[I,3][III(?),2]
Pitticite	[VI,3]	Arsenolite	[VI,4]
Rauenthalite	[VI,1]	Avicennite	[VI,1]
Rösslerite	[VI,1]	Bismite	[VI,1]
Sarmientite	[VI,1]	Cassiterite	[II,1]
Scorodite	[VI,9]	Cervantite	[VI,4]
Symplesite	[VI,2]	Chalcophanite	[VI,2]
Weilite	[VI,1]	Chromite	[II,1]
Antimonates		Claudetite	[VI,1]
Bindheimite	[VI,3]	Cuprite	[VI,5]
Arsenides		Gibbsite	[VI,1]
Löllingite	[IV,1]	Goethite	[VI,8+]
Carbonates		Hematite	[VI,7+]
Ankerite	[V(?),3]	Ilmenite	[II,1]
Aragonite	[VI,6]	Ilsemannite	[VI,4]
Aurichalcite	[VI,1]	Lepidocrocite	[VI,1]
Azurite	[VI,8]	Magnetite	[I,1][II,1]
Bismutite	[III,1]	Minium	[VI,1]
Calcite	[I,20+][II,2]	Pyrolusite	[VI,3]
	[III,10+][VI,1]	Romanèchite	[VI,1]
Carbonate-cyanotrichite	[V,20+][VI,10+]	Rutile	[I,4][II,3]
Cerussite	[VI,4]	Stibiconite	[VI,3]
Dolomite	[I,20+][II,1][III,10+]	Thorianite	[I,1]
Malachite	[VI,7]	Tripuhyite	[VI,2]
Elements		Uraninite	[I(?),2]
Arsenic	[IIIorV,2]	Valentinite	[VI,2]
Arsenolamprite	[IIIorV,1]	*Phosphates*	
Bismuth	[III,2]	Augelite	[VI,1]
Cadmium	[III(?),1]	Autunite	[VI,1]
Copper	[VI,6]	Burangaite	[VI,1]
Electrum	[V,1]	Cacoxenite	[VI,2]

Mineral	Depositional Phase	Mineral	Depositional Phase
Carbonate-fluorapatite	[I,1][III,3][V,4][VI,5]	*Silicates*	
Carbonate-hydroxylapatite	[VI,2]	*Nesosilicates*	
		Almandine	[II,1]
Crandallite	[VI,5]	Andalusite	[II,1]
Cyrilovite	[VI,1]	Andradite	[II,1]
Dufrénite	[VI,1]	Grossular	[II,2]
Englishite	[VI,2]	Titanite	[II,1]
Faustite*	[VI,2]	Zircon	[I,4][II,1]
Fluellite	[VI,1]	*Sorosilicates*	
Fluorapatite	[II,1][VI,2]	Epidote	[II,2]
Goldquarryite	[VI, 1]	Vesuvianite	[II,1]
Hinsdalite	[VI,1]	Zoisite	[II,1]
Huréaulite	[II,1]	*Cyclosilicates*	
Hydroxylapatite	[VI,1]	Cordierite	[II,1]
Jahnsite-(CaMnMg)	[VI,1]	Tourmaline	[I,2]
Kingite	[VI,1]	*Inosilicates*	
Kingsmountite	[VI,1]	Actinolite	[II,3]
Leucophosphite	[VI,3]	Augite	[II,1]
Libethenite	[VI,1]	Diopside	[II,1]
Lipscombite	[VI,1]	Hedenbergite	[II,1]
Meta-autunite	[VI,1]	Magnesiohornblende	[II,1]
Metatorbernite	[VI,2]	Pargasite	[II,1]
Metavariscite	[VI,1]	Tremolite	[II,2]
Meurigite*	[VI,2]	*Phyllosilicates*	
Mitridatite	[VI,1]	Aliettite	[II,1]
Monazite-(Ce)	[I(?),1][II,1]	Allophane	[VI,1]
Monazite-(La)	[II,1]	Biotite	[I(?),1][II,3]
Montgomeryite	[VI,2]	Chamosite	[III(?),1]
Nevadaite*	[VI,1]	Chrysocolla	[VI,4]
Phosphosiderite	[VI,1]	Clinochlore	[II,2]
Planerite	[VI,1]	Dickite	[V,2]
Plumbogummite	[VI,1]	Fraipontite	[III,2]
Pseudomalachite	[VI,1]	Halloysite	[V,1][VI,3]
Richellite	[VI,1]	Hisingerite	[III,1]
Sincosite	[VI,1]	Hydroxyapophyllite	[II(?),1]
Strengite	[VI,4]	Illite	[I,5+][III,5+]
Tinticite	[VI,1]	Kaolinite	[V,8+][VI,6]
Torbernite	[VI,1]	Lizardite	[II,1]
Turquoise	[VI,7]	Montmorillonite	[III,2][VI,1]
Variscite	[VI,9]	Muscovite	[II,1][III,2+]
Vivianite	[VI,2]	Natroapophyllite	[II(?),1]
Wavellite	[VI,3]	Nontronite	[III(?),1]
Whitlockite	[VI,2]	Phlogopite	[III,1]
Xenotime-(Y)	[I,1][II,2]	Prehnite	[II(?),1]
		Roscoelite	[II(?),1]
Selenates and Selinides		Wollastonite	[II,3]
Mandarinoite	[VI,1]		
Tiemannite	[V,1][VI,1]		

(continued next page)

TABLE 3   Carlin-type deposits, *continued*

Mineral	Depositional Phase	Mineral	Depositional Phase
Tectosilicates		*Sulfides*	
Albite	[II,2]	Acanthite	[V,6][VI,1]
Andesine	[II,1]	Aktashite	[V,1]
Anorthite	[II,1]	Andorite	[V(?),1]
Anorthoclase	[I,1][II,1]	Arsenopyrite	[II,1][III,13][IV,1]
Fluorapophyllite	[V(?),1]	Balkanite	[V,1]
Gismondine	[VI,1]	Berthierite	[V,1]
Labradorite	[II,1]	Bismuthinite	[III,3]
Laumontite	[II(?),1]	Bornite	[III,3][V,1]
Marialite	[II,1]	Boulangerite	[II,1][V,1]
Microcline	[II,1]	Bournonite	[II,1][V,2]
Mordenite	[II,1]	Carlinite*	[V,2]
Natrolite	[VI,1]	Chalcocite	[VI,5]
Opal	[VI,2+]	Chalcopyrite	[II,3][III,11][IV,2]
Orthoclase	[I,2][V,5]	Chalcostibite	[V,1]
Quartz	[1,20+][II,2+][V,20+]	Christite*	[V,2]
Stilbite	[II,1]	Cinnabar	[II(?),1][V,11]
		Covellite	[VI,3]
Sulfates		Digenite	[III(?),1]
Alunite	[VI,6]	Ellisite*	[V,1]
Anglesite	[VI,3]	Freibergite	[II,1][V(?),2]
Anhydrite	[VI,1]	Galena	[II,1][III,1][V,7]
Antlerite	[VI,1]	Galkhaite	[V,4]
Barite	[I,13][V,20+]	Gersdorffite	[III,3]
Beudantite	[VI,3]	Getchellite*	[V,4]
Brochantite	[VI,1]	Gortdrumite	[V,1]
Celestite	[VI,1]	Gratonite	[III(?),1]
Chaidamuite	[VI,1]	Greenockite	[VI,2]
Chalcanthite	[VI,2]	Gruzdevite	[V,1]
Chalcophyllite	[VI,1]	Hypercinnabar	[V,1]
Cyanotrichite	[VI,1]	Jordanite	[V,2]
Epsomite	[VI,3]	Jordisite	[III,2]
Gunningite	[VI,1]	Kermesite	[VI,5]
Gypsum	[V,2][VI,9]	Laffittite	[V,1]
Halotrichite	[VI,1]	Lorandite	[V,3]
Hexahydrite	[VI,1]	Marcasite	[III,8+][IV,2+][VI,2]
Hydronium jarosite	[VI,2]	Meneghinite	[II,1]
Jarosite	[VI,8]	Metacinnabar	[V,2]
Kieserite	[VI,1]	Metastibnite	[V,5]
Melanterite	[VI,3]	Miargyrite	[V,1]
Morenosite	[VI,1]	Millerite	[III,1][V,1]
Natroalunite	[VI,2]	Molybdenite	[II,3]
Natrojarosite	[VI,2]	Orpiment	[V,15+]
Rozenite	[VI,2]	Pararealgar	[VI,2+]
Schuetteite	[VI,1]	Pentlandite	[III,2]
Starkeyite	[VI,1]	Polhemusite	[V(?),1]
Szomolnokite	[VI,2]	Polybasite	[V,2]
		Pyrargyrite	[V,2]

Mineral	Depositional Phase	Mineral	Depositional Phase
Pyrite	[I,2+][II,2]	*Tungstates*	
	[III,15+][VI,2]	Hübnerite	[II,2]
Pyrite (arsenian)	[IV,20+]	Scheelite	[II,2]
Pyrrhotite	[II,1][III,3]		
Realgar	[V,15]	*Vanadates and Vanadium Oxysalts*	
Siegenite	[V,1]	Descloizite	[VI,1]
Sphalerite	[I,1][II,1][III,3][V,11]	Fervanite	[VI,1]
Spionkopite	[VI,1]	Häggite	[VI,1]
Stephanite	[V,1]	Hewittite	[VI,1]
Stibnite	[V,18]	Heyite	[VI,1]
Tennantite	[V(?),2]	Hummerite	[VI,1]
Tetrahedrite	[II,1][V,12]	Kazakhstanite	[VI,1]
Tungstenite	[III,1]	Metatyuyamunite	[VI,1]
Tvalchrelidzeite	[V,1]	Pascoite	[VI,1]
Violarite	[V,1]	Schubnelite	[VI,1]
Wakabayashilite	[V,2]	Turanite	[VI,2]
Weissbergite*	[V,1]	Tyuyamunite	[VI,1]
Wurtzite	[III,1]	Vanadinite	[VI,1]
		Vésigniéite	[VI,2]
Tellurides		Volborthite	[VI,1]
Calaverite	[IV,1]		
Coloradoite	[V,2]		
Tellurobismuthite	[V(?),1]		

Minerals in Carlin-type deposits. Roman numerals in brackets show depositional phase (I-VI). Arabic numerals report number of occurrences, + indicates species observed at more localities than reported in current literature, ? denotes uncertain paragenesis. Multiple listings indicate episodic deposition of species. * indicates type locality is a Carlin-type deposit. Italics indicate non-species or group name.

arsenopyrite, and chalcopyrite. Some researchers believe that it was deposited on an atomic scale in the crystal lattice of these species, rather than as particles of native metal. Native gold is very rare in these systems; it is often very late in, and in places postdates, the main gold-depositional stage. Calaverite is the only other reported gold-bearing mineral species found thus far and has been reported from only one occurrence.

V. Waning gold deposition. During this stage nearly 70 different species of complex sulfosalts, halides, and some sulfates and carbonates formed. Although minor amounts of gold were still being deposited in minerals such as galkhaite, realgar, and orpiment, gold deposition had largely ceased by this point. Rare native silver and tellurium formed during this stage. Quartz replacement and flooding was most pro-nounced, mainly occurring during the middle of this phase. Large crystals of realgar, orpiment, calcite, barite, and stibnite formed. Crystals of galkhaite, fluorite, stilbite, quartz, kaolinite, millerite, dolomite, and ankerite formed in open cavities, mainly in silicified rock but also in large carbonate-hosted caverns, hollowed out by preceding stages of acidic hydrothermal solutions. Oil and deactivated carbon concentrated around the periphery of many deposits, and oil concentrations formed within the gold deposits themselves (e.g., Alligator Ridge Mines).

VI. Post-gold secondary mineralization. By far the most numerous group of minerals found in Carlin-type deposits are secondary alteration products of sulfide and sulfosalts along with subsidiary carbonates, phosphates, and silicates. More than 150 species are ascribed to this group of minerals, which formed during post–gold stage, low-temperature hydrothermal and supergene deposition. Many of the species in this group are phosphates, some of which were only previously described from pegmatites. Sulfates and oxide/hydroxide minerals are abundant. Other minerals formed during this depositional period include arsenates, carbonates, and vanadates. The native elements sulfur and selenium were also deposited during this stage.

The abundance of secondary phosphates, and locally of vanadates, in rock overlying gold ore in some deposits suggests that primary phosphorus and vanadium minerals originally deposited by sedimentary or later hydrothermal activity were broken down and redeposited as new phases in fractures and cavities during oxidation. Such redistribution may form secondary phases mixed with sulfides, lending the appearance of cogenesis and adding to confusion in interpretation of ore genesis.

SPECIMENS

With nearly 300 species available in Carlin-type deposits, one would expect to garner a few for the mineral cabinet. Unfortunately, specimens suitable for the "macro" collector are seldom found. The exceptions have been outstanding, many claiming the title of "world's finest," or at the very least competitive with other deposits.

The world-class yellow-to-amber-colored barite crystals from the Meikle Mine (Bootstrap district, Elko County) are considered by many to be the finest among this class (Jensen, 1999). These crystals, which have glorious proportions and appeal, were first discovered in 1993 during the sinking of the Meikle shaft and quickly became world classics. The Meikle Mine has also produced what are likely the best calcite specimens to come from Nevada; water-clear "dog-tooth" scalenohedrons in excess of 25 cm in length have been recovered.

Fine specimens of golden-yellow barite have also been produced from the Northumberland Mine in Nye County (Kokoninos and Prenn, 1985). Here crystals were found lining pockets in silicified rock, many of which could be entered by a collector. In addition, pale-blue barite crystals have come from mines in the Independence

district and from the Marigold Mine, Buffalo Mountain district.

For nearly half a century, world-class specimens of realgar in crystals to 10 cm or longer and orpiment in crystals to 1 cm or more have come from the Getchell Mine in the Potosi district of Humboldt County. Some of the finest have been recently collected. In addition, the mine is known for deep-red crystals of galkhaite to 5 mm or more, as well as specimens of rare minerals such as laffittite and getchellite (Stolburg and Dunning, 1985).

Superb orpiment specimens with lustrous orange-brown crystals to 3 cm also come from a new occurrence at the Twin Creeks Mine (Cook, 2000), which is located near the Getchell Mine. These pockets were first discovered in 1997, with systematic excavation to preserve these crystals undertaken in 1999. Fine orpiment has also been produced from the Gold Pick pit at the Gold Bar Mine in the Roberts Mountains of Eureka County.

Internationally known stibnite in crystals as much as 10 cm long were found at the White Caps Mine, Manhattan district, in Nye County beginning in the 1920s. Melhase (1935d) described the specimens as "superb clusters of radial splendent crystals, sometimes reticulated or perhaps with individual crystals capped with a crystal of 'nail-head' calcite." The world's finest wakabayashilite specimens, originally sold as "hair orpiment," were produced from this classic locality in the 1930s and 1940s. Other rare arsenic minerals were also collected from this mine, including arsenolite, haidingerite, claudetite, and pharmacolite (Gibbs, 1985).

World-class stibnite in individual crystals reported to be as long as 35 cm was recently recovered from the Murray Mine in the Independence Mountains, Elko County. These specimens are often found perched on aesthetic groups of barite crystals coated with lustrous drusy quartz (Cook, 1999).

A number of rare and beautiful phosphate minerals, many as the best examples yet discovered, have been recovered from the oxidation zones of several mines, including the Gold Quarry (Jensen and others, 1995) and Goldstrike Mines in Eureka County; the Lone Tree Mine, Humboldt County; and the Rain Mine, Elko County.

Large clusters of aragonite needles, some with individual crystals to 7 cm in length, have been recovered from the Sterling Mine in the Bare Mountain district of Nye County, the most southerly of the Carlin-type deposits in Nevada.

CONCLUSIONS

Detailed mineral studies have been published for few Carlin-type deposits. However, the understanding of the genesis and distribution of minerals in the deposits is developing. Several stages of mineralization may be observed at each deposit, beginning with sedimentary deposition, in some cases overprinted by the effects of igneous intrusion, followed by deposition of minerals related to gold mineralization, and lastly overprinting by supergene mineralization. With almost 300 species from many diverse deposits, Carlin-type gold systems are a major source not only of gold but also of rare and beautiful mineral species.

Precious-Metal Deposits in Volcanic Rocks

STEPHEN B. CASTOR AND
RICHARD W. THOMSSEN

Although Nevada's modern precious-metal production is mainly from Carlin-type, sedimentary rock–hosted gold deposits, historic production of gold and silver was largely from deposits in Tertiary volcanic rocks, and the state's most famous camps—the Comstock, Tonopah, and Goldfield districts—were founded on precious-metal wealth in vein-type deposits in volcanic rocks. In addition, large modern precious-metal mines, such as at Round Mountain, Paradise Peak, Bullfrog, Sleeper, Midas, and Rawhide, exploit such deposits or have done so in the recent past.

Many volcanic rock–hosted deposits are in large veins or consist of networks of small closely spaced veins. The veins may be divided into two types: "fissure" veins, which have definite boundaries with host rock and can be thought of as the result of deposition from fluids in open fractures or fissures; and replacement veins, which form by the replacement of host rock by fluids along fractures or other weak zones and generally have gradational boundaries with host rock. Fissure veins are commonly finely banded, showing the effects of successive deposition of vein minerals in layers in open fractures, as at Bullfrog, Midas, and Sleeper. They also show other features suggestive of open-space deposition, such as comb texture and the presence of vugs that contain crystals of vein minerals. In contrast, replacement veins generally do not show evidence of significant open-space filling except where crystals partially fill cavities produced by dissolution. Well-described replacement veins occur in the Comstock, Tonopah, and Tuscarora districts.

Some of Nevada's epithermal precious-metal deposits yielded very high-grade ore, generally referred to as "bonanza" ore. Notable among these deposits were the silver-rich bonanza ore-bodies of the Comstock Lode and the gold-rich "jewelry ore" from Goldfield. Modern-day mining has exploited high-grade veins at Sleeper and Round Mountain, although the latter mainly produces gold and silver from low-grade disseminated ore. Enrichment by supergene processes produced some extremely high-grade silver ore, generally by the formation of rich silver halide deposits (mostly chlorargyrite, or horn silver) as in the Treasure Hill area of the White Pine district where masses as large as 6 tons were mined. Supergene enrichment of silver ore by the deposition of sulfide minerals such as acanthite has been a factor in some deposits.

In addition to veins, volcanic rock–hosted deposits may be in the form of breccia, as at Paradise Peak, or occur as disseminated and generally low-grade deposits in which hydrothermal fluids pervasively affected large volumes of relatively porous volcanic rock, as at Round Mountain. Many deposits contain ore in more than one form; for instance, at the Sleeper Mine high-grade fissure veins and low-grade breccia deposits are both present, and historic production at Round Mountain was from high-grade veins, but modern production is mostly from disseminated low-grade ore.

In contrast to the deep or "mesothermal" veins of California's Mother Lode, most vein-type precious-metal ore in volcanic rocks in Nevada formed in a shallow or "epithermal" environment. Epithermal precious-metal mineralization mainly took place in Nevada between 38 and 6 million years ago and was associated with volcanic and shallow intrusive magmatic activity of the same age. While there is no strict definition, most economic geologists would agree that mineralization that took place within 1000 m of the surface is epithermal. In some cases, precious-metal mineralization took place at very shallow depths, and surficial hot-spring deposits are associated with ore as in the Sulphur district.

Two important types of epithermal precious-metal deposits have been defined: the high-sulfidation and low-sulfidation types (Bonham, 1988). High-sulfidation epithermal precious-metal deposits, which are also called quartz-alunite or acid sulfate deposits, are characterized by the presence of enargite-group sulfosalt minerals and aluminum-rich alteration minerals such as alunite, pyrophyllite, and diaspore. They range from gold-dominated deposits such as Goldfield to silver-dominated deposits such as Paradise Peak, and are thought to have been formed by processes that were directly related to magmatism on the basis of isotopic and fluid inclusion evidence. Primary tellurium, bismuth, and tin minerals are commonly present and carbonate minerals are rare or absent. Oxidized portions of such deposits may contain colorful sulfate and arsenate minerals, such as at the Burrus Mine in the Pyramid district. High-sulfidation veins are typically replacement veins, which in many districts are referred to as ledges and do not show finely banded vein structure.

High-sulfidation deposits in Nevada are restricted to a relatively narrow zone in the western part of the state that extends from the Mount Lassen area in California; through the Pyramid, Wedekind, and Castle Peak districts near Reno; the Borealis and Paradise Peak districts near Hawthorne; to a cluster of districts in the Goldfield area that includes the rich mines of Goldfield itself, along with the Cactus Springs, Gold Crater, and other districts on the Nellis Air Force Range. Mineralization in these deposits took place between about 22.5 million years ago (Pyramid district) and 6 million years ago (Castle Peak district) and was associated with subduction-related igneous activity (ancestral Cascade volcanism).

Low-sulfidation epithermal deposits are thought to be indirectly related to magmatism, although this is hotly debated among economic geologists. Magmatism may only have been involved as the source of heat during the circulation of groundwater, and some believe that the ore metals were extracted from surrounding rocks by the circulating fluids. Other geologists think that these metals were derived from magmatic fluids that were subsequently diluted by groundwater. Low-sulfidation deposits, also referred to as quartz-adularia deposits, are typified by the presence of the potassium feldspar variety adularia and by a lack of enargite-group sulfosalt minerals. Tellurium, bismuth, and tin minerals are also generally rare. Carbonate minerals (or quartz pseudomorphs after them) are as common in low-sulfidation type deposits as they are rare in high-sulfidation deposits. Some low-sulfidation veins contain significant occurrences of electrum—sometimes as spectacular specimens. Many low-sulfidation vein deposits are geochemically clean or lacking in heavy metals such as arsenic and antimony; the Bullfrog and Olinghouse veins are of this type. The Midas low-sulfidation veins have similarly low arsenic and antimony, but are unusually rich in selenium. The famous silver deposits in the Comstock and Tonopah districts are low-sulfidation type as shown by the presence of adularia. In these deposits, which also produced significant amounts of gold, the ore contains important amounts of silver-bearing sulfide minerals such as acanthite and polybasite. Base metals and elements such as arsenic and antimony may be locally common.

Low-sulfidation type vein deposits are more widely distributed in Nevada than high-sulfidation types. They include the Jarbidge veins in the northeastern corner of the state, the Hog Ranch Mine in the northwest, the huge Round Mountain system in the center of the state, and the Bullfrog deposit to the south. The Tuscarora veins are notable because they are relatively ancient (about 38 million years old); most low-sulfidation deposits were formed between 26 and 10 million years ago.

Although widespread and abundant, Nevada's volcanic rock–hosted precious-metal deposits have been relatively unimportant as collecting localities, except as a source of gold and electrum specimens. With a few exceptions, the deposits are not known for well-developed zones of the colorful secondary species that are sought after by collectors probably because many of the deposits are formed by low-sulfidation mineralization without significant amounts of elements such as arsenic, zinc, copper, phosphorus, car-

TABLE 4 Minerals reported from the Goldfield district
(includes nearby Diamondfield district)

Primary sulfides, oxides, and native elements	Telluride minerals	Alteration and gangue minerals	Secondary (oxidation) minerals
Bismuthinite	Calaverite	Alunite	Alunogen
Bornite	Hessite	Anatase	Anglesite
Cassiterite	Krennerite	Anhydrite	Azurite (?)
Chalcopyrite	Petzite	Barite	Bismite
Chalcostibite	Sylvanite	Clinoptilolite	Bismoclite
Cinnabar		Diaspore	Chlorargyrite
Colusite		Gypsum	Copiapite
Emplectite		Kaolinite	Copper
Enargite		Mordenite	Coquimbite
Famatinite		Natroalunite	Emmonsite
Galena		Opal	Goethite
Gold		Pyrophyllite	Halotrichite
Goldfieldite*		Quartz	Hematite
Kuramite		Rutile	Iodargyrite
Luzonite		Tobermorite	Jarosite
Marcasite			Mackayite*
Molybdenite			Malachite
Polybasite (?)			Melanterite
Proustite (?)			Römerite
Pyrite			Rozenite
Sphalerite			Sonoraite
Stephanite			Tellurite
Stibnite			Turquoise
Sulfur			Variscite
Tellurium			
Tennantite			
Tetrahedrite			
Wurtzite			

*Nevada type minerals

bon, and sulfur that are necessary to produce colorful minerals.

The high-sulfidation deposits, with relatively high contents of metals such as copper, arsenic, bismuth, and tellurium, have more interesting and complex mineral assemblages. Most notable among these is the Goldfield district, from which more than 70 mineral species have been reported (Table 4), including two minerals for which the district is the type locality: the complex tellurium-bearing sulfide goldfieldite; and the secondary tellurium mineral mackayite. A third Goldfield-type mineral, blakeite, is not considered to be valid by the IMA (Mandarino, 1999), although it is shown as a valid species by others (for example, Nickel and Nichols, 1991). Three enargite-group minerals have been identified at Goldfield: enargite, luzonite, and famatinite. In addition, Goldfield mines have yielded specimens with complex assemblages of primary sulfide and telluride minerals, as well as some rare oxidation products such as the tellurite phases emmonsite and sonoraite, in addition to mackayite. Many of these minerals were identified and described during the early days in the district (Ransome, 1909a), but subsequent work has also been fruitful (for example, Tolman and Ambrose, 1934; Frondel and Pough, 1944). Other high-sulfidation districts, such as the Pyramid and Cactus Spring districts, have produced interesting specimens (e.g., richelsdorfite).

Nevada's low-sulfidation precious-metal districts have produced fine specimens of native gold and electrum. The *Glossary of Geology* (Bates and Jackson, 1987) defined electrum as a naturally occurring alloy with more than 20 percent silver (presumably by weight); however, *Fleisher's Glossary of Mineral Species* (Mandarino, 1999) does not list electrum, relegating the alloy to either the gold or silver species. If one accepts a division at 50 atomic percent, then any mixture with more than 35 weight percent silver should be considered as the mineral silver (because of the significantly higher atomic weight of gold). Unfortunately, many of the low-sulfidation deposits in Nevada contain native metal that ranges between 60 and 70 weight percent gold (fineness of 600 to 700), making the nonspecies electrum an excellent descriptive term.

The historic Comstock and National districts produced excellent examples of electrum. In recent years, fine specimens have come from the Sleeper Mine, and the Round Mountain, Seven Troughs, Ten Mile, and Olinghouse districts. At many of these sites, the electrum has dendritic form, suggesting disequilibrium conditions, with the formation of more crystalline dendrites at higher temperatures (Saunders and others, 1996). In addition to electrum, the Comstock Lode yielded fine native-silver specimens during its early years, but good examples are rarely seen.

Low-sulfidation precious-metal deposits, despite their abundance and relatively long history in the state, have not been prolific as mineral type localities. The Comstock Lode was a co-type locality for the rare gold-silver sulfide uytenbogaardtite, a relatively recent discovery (Barton and others, 1978), and discoveries of this mineral in low-sulfidation deposits in the Bullfrog and Tolicha districts have followed. In addition to uytenbogaardtite and specimens of electrum and wire silver, crystalline specimens of silver-bearing sulfides such as stephanite, polybasite, acanthite after argentite, and some fine specimens of amethyst came from the Comstock district. Pale amethystine quartz occurs in many low-sulfidation precious-metal deposits; examples include the Hog Ranch Mine, and veins in the Bullfrog, Gold Circle (Midas), and Olinghouse districts.

Other specimen localities include the Rosebud Mine, which has produced fine crystallized specimens of barite, marcasite, and miargyrite; the Tonopah-Divide Mine, known for crystals of powellite and ferrimolybdite; and the Tonopah district, which has yielded relatively large crystals of pyrargyrite. Recently, the Olinghouse Mine has provided excellent specimens of wire and crystalline electrum in association with crystalline quartz, rare scheelite crystals, and several zeolite species. In addition, the banded low-sulfidation veins of the Gold Circle (Midas) district are notable for the occurrence of the precious-metal selenide minerals naumannite, aguilarite, and the rare mineral fischesserite.

The formation of secondary minerals in low-sulfidation precious-metal deposits is largely a function of the oxidation of local concentrations of pyrite, and jarosite and goethite are the dominant species. The secondary phosphate, cacoxenite, has been found in oxidized rock at Rawhide and in silicified breccia at Talapoosa.

In summary, the volcanic rock–hosted precious-metal mines of Nevada have produced relatively few new mineral species and have not been important sources of collectible minerals except for gold and electrum. Nevertheless, such deposits have been important economically since the discovery of the Comstock Lode in the 1860s, and intense study of their minerals will likely continue.

Porphyry and Contact Metasomatic Deposits

STEPHEN B. CASTOR

Porphyry-type and contact metasomatic metal deposits are both related to the comparatively high-level intrusion of igneous rocks. In some mining districts the two types commonly occur side by side, and in some instances the distinction between them is obscure. In some places ore that is clearly hosted by skarn has been lumped with porphyry ore, and such districts are referred to simply as porphyry districts, ignoring the importance of contact metasomatic processes in ore formation.

PORPHYRY-TYPE DEPOSITS

Porphyry-type ore deposits are mesothermal (medium depth and temperature) metal deposits that form in intrusive rocks that have porphyritic texture. In general, the intrusions have granite to granodiorite compositions. Porphyry-type ore minerals are commonly disseminated, occurring in grains or fine veinlets that are scattered through altered intrusive rock. Alteration assemblages associated with porphyry deposits, in particular porphyry-copper deposits, were first defined by studies in the western United States, particularly in Arizona, Utah, and Nevada. The definition of propylitic, argillic, and potassic alteration types, which are now used internationally, was based on that work. In many cases, porphyry orebodies and associated altered rocks are in the form of regular shells in the upper and outer parts of the intrusions; however, in other examples ore occurs in veins or along other structural features.

Nevada's porphyry deposits lie in a belt of deposits in western South and North America that are thought to be related to subduction of the Pacific Plate that has been taking place since the Jurassic Period. While not as important economically or mineralogically as the numerous large porphyry-copper deposits of Arizona, Nevada's porphyry deposits have nonetheless been productive, both as a source of metals and minerals. These deposits include copper, molybdenum, tin, and gold porphyries.

The porphyry-copper deposits of the Robinson (Ely) district have been the most productive as the source of 80 percent of the more than 4 billion pounds of copper that came from the district, although significant amounts also came from skarn as noted below. The Robinson deposits also produced large amounts of by-product silver, gold, and molybdenum, along with platinum-group metal credits. Most of the copper that was produced in the Yerington district came from the large open-pit porphyry deposit exploited by the Anaconda Copper Mining Company, which alone produced about 1.5 billion pounds of the metal.

Mineral assemblages of large and commonly very profitable porphyry-copper deposits, such as at Robinson and Yerington, are generally not complex or exciting; except, of course, to economic geologists and mining companies, who find them endlessly fascinating. Only 78 mineral species (Table 5) have been identified in the large Robinson district; and some of these are unrelated to the ore deposits, such as the almandine, cristobalite, and tridymite that occur in post-ore rhyolite. Many of these minerals were described in the early work by Spencer and others (1917), but species have continued to be identified by more recent research. At Robinson and Yerington, vast amounts of rock contain primary ore assemblages that consist mainly of disseminated grains of chalcopyrite with lesser bornite and chalcocite that are associated with pyrite and alteration minerals such as biotite, potash

TABLE 5 Minerals of the Robinson mining district

Actinolite	Covellite	Kaolinite	Saponite
Almandine	Cristobalite	Langite	Scapolite
Alunite	Cuprite	Magnetite	Scheelite
Andalusite	Delafossite	Malachite	Smithsonite
Andradite	Devilline	Melanterite	Sphalerite
Anglesite	Diopside	Molybdenite	Sylvite
Anorthoclase	Epidote	Montmorillonite	Thaumasite
Aragonite	Epsomite	Muscovite	Titanite
Aurichalcite	Ferberite	Natroalunite	Tremolite
Azurite	Ferrocolumbite	Nontronite	Tridymite
Biotite	Fluorite	Orthoclase	Turquoise
Bornite	Galena	Phlogopite	Vanadinite
Brochantite	Gypsum	Plumbojarosite	Vivianite
Cerussite	Halite	Pyrite	Willemite
Chalcanthite	Halloysite	Pyrrhotite	Wollastonite
Chalcocite	Hematite	Quartz	Wulfenite
Chalcopyrite	Hemimorphite	Rhodochrosite	Zircon
Chlorargyrite	Hisingerite	Rosasite	Zunyite
Chrysocolla	Hydrozincite	Rutile	
Copper	Ilmenite	Sanidine	

feldspar, and muscovite (as the variety sericite). Secondary sulfides consist mainly of chalcocite with a little covellite and digenite. More complex and colorful mineral assemblages occur in oxidized zones of porphyry-copper deposits, with copper carbonates, silicates, phosphates, and sulfates as common minerals. At Yerington, rock that contained abundant chrysocolla was important as ore.

Other porphyry copper-deposits have been discovered in Nevada, but most have been too small or too low-grade to mine. However, one such deposit, the Majuba Hill Mine in the Antelope district of Pershing County, from which 2.8 million pounds of copper and 21,000 pounds of tin were produced (MacKenzie and Bookstrom, 1976), has been one of the state's best-known mineral localities, with a list of 93 mineral species identified (Table 6). It is the type locality for two Nevada type minerals, goudeyite and parnauite. At Majuba Hill, a series of high-level rhyolite porphyry intrusions is associated with ore containing chalcopyrite and cassiterite, and the deposit has been likened to porphyry-tin deposits in Bolivia.

Porphyry-type metal deposits in Nevada include several porphyry-molybdenum deposits. The Hall Molybdenum Mine in the San Antone district was the only Nevada porphyry-molybdenum deposit to have been mined, producing more than 5 million pounds of molybdenum during the 1980s. As in other such deposits, ore at the Hall Mine was found to consist of a network (stockwork) of small molybdenite-bearing quartz veins in altered granite porphyry. However, the Hall deposit is also known for the presence of ferrimolybdite and powellite, which are relatively uncommon in such deposits. In addition, the Hall Molybdenum Mine has produced Nevada's best specimens of creedite. Other porphyry-molybdenum deposits were identified by exploration geologists in the 1960s and 1980s, including the Cucomungo Spring deposit in the Tule Canyon district, the Buckingham deposit in the Battle Mountain district, and the Mount Hope deposit in the Mount Hope district. None of these deposits were mined, and two are essentially blind, with molybdenum-rich specimens restricted to drill samples.

The Cove Mine in the McCoy district may be an example of a third type of porphyry deposit;

TABLE 6 Mineral species from the Majuba Hill Mine

Acanthite	Clinoclase	Iodargyrite	Pyrolusite
Agardite-(Y)	Clinozoisite	Jarosite	Pyrrohotite
Anatase	Conichalcite	Kaolinite	Quartz
Arsenolite	Connellite	Langite	Rhabdophane-(Ce)
Arsenopyrite	Copper	Laumontite	Rosasite
Arthurite	Cornetite	Lavendulan	Schorl
Atacamite	Cornubite	Libethenite	Scorodite
Azurite	Cornwallite	Lindackerite	Silver
Bindheimite	Covellite	Luetheite	Spangolite
Bismuth	Cubanite	Malachite	Sphalerite
Bismuthinite	Cuprite	Melanterite	Stannite
Bornite	Cyanotrichite	Metatorbernite	Strashimirite
Brochantite	Delafossite	Metazeunerite	Tenorite
Brookite	Devilline	Mixite	Thenardite
Cassiterite	Digenite	Molybdenite	Thorite
Chalcanthite	Enargite	Olivenite	Torbernite
Chalcocite	Epsomite	Orthoclase	Tyrolite
Chalcomenite	Fergusonite-(Y)	Parnauite*	Wroewolfeite
Chalcophyllite	Fluorite	Pharmacolite	Xenotime-(Y)
Chalcopyrite	Galena	Pharmacosiderite	Zeunerite
Chamosite	Goethite	Pitticite	Zircon
Chenevixite	Gold	Posnjakite	
Chlorargyrite	Goudeyite*	Pseudomalachite	
Chrysocolla	Gypsum	Pyrite	

*Nevada type minerals

it has been referred to as a porphyry-gold deposit by some geologists. The ore, which occurs in granitic intrusive rock, is associated with a variety of sulfide minerals, including an assemblage of rare tin-bearing phases including canfieldite, chatkalite, and kësterite. Other porphyry-gold and porphyry-gold-copper deposits have been identified in Nevada, but none have had significant production.

With the exception of the Majuba Hill Mine, porphyry-type deposits in Nevada are not prolific providers of sought-after mineral specimens. The Robinson district produced excellent specimens of oxide copper minerals such as cuprite and azurite; however, most such material came from skarn deposits as at the Richard Mine noted below. Likewise, specimen localities in the Yerington district were mainly in skarn deposits, although Anaconda's open pit yielded some attractive specimens of chrysocolla, malachite, libethenite, and cornetite.

The Majuba Hill Mine has been known as a source of rare and colorful minerals since the 1940s. Mostly, these collectible minerals are copper-bearing arsenates such as clinoclase, olivenite, and chalcophyllite. Specimens of the iron arsenates scorodite and pharmacosiderite and the copper-uranium phosphate torbernite from this locality are also well known. Details on collectible minerals at the Majuba Hill Mine may be found in Jensen (1985).

CONTACT-METASOMATIC DEPOSITS

Mineral deposits formed as a result of concentration during alteration of rocks adjacent to or near igneous intrusions have yielded significant amounts of metallic commodities in Nevada in the past, but are presently of relatively minor importance. Commodities have included copper, lead, zinc, tungsten, iron, magnesite, silver, and gold. Contact-metasomatism, or replacement of preexisting minerals in the host rock by minerals that contain chemical components introduced during igneous intrusion, is the primary mechanism of ore formation, and a variety of gangue minerals are also formed during the process. As perceived by most geologists, the process is a variant of the more inclusive process of contact-metamorphism, which does not always involve the introduction of chemical components. The introduction of chemical components involves movement in fluids, generally outward from the intruding magma, but in some cases, inward from the intruded rocks. Intrusive rocks associated with contact metasomatic deposits are typically of quartz diorite to granitic composition; deposits associated with more mafic intrusions are rare, probably because of their lack of hydrous components. Host rocks are generally reactive lithologies such as limestone, dolomite, and mafic igneous rocks.

The term "skarn" is mostly used to refer to contact metamorphic rocks formed by the metasomatism of carbonate-bearing rocks such as limestone, dolomite, or calcareous clastic rocks into lime-bearing silicate minerals by the introduction or remobilization of silica and other elements. The rock name "tactite" is synonymous with skarn, but has fallen into disfavor. Intrusive rocks modified by assimilation of country rock or introduction of components from intruded rocks during contact-metamorphism have also been referred to as skarn by some geologists, generally as "endoskarn," in contrast to "exoskarn," which refers to skarn in pre-intrusive rocks. Skarn may also be subdivided into early "prograde" skarn that is mostly composed of anhydrous phases such as garnet- and pyroxene-group minerals, and later "retrograde" skarn characterized by hydrous phases such as chlorite- and amphibole-group minerals. In most cases, ore minerals are closely associated with retrograde skarn, and in many cases with late-stage veins.

Some contact metamorphic deposits are closely associated with porphyry-type deposits, particularly porphyry-copper deposits. In such deposits, the distinction between endoskarn and porphyry alteration is blurred, a problem that some geologists have solved by referring to the deposits as either skarn deposits or porphyry deposits, despite the fact that both types of mineralization are present. Hence, copper orebodies in the Robinson district are thought of by many simply as porphyry-copper deposits, although large amounts of ore were clearly mined from skarn. In Nevada's second most productive copper district, most of the copper came from a porphyry deposit at the Yerington Mine, and it is easy to overlook the early production of tens of millions of pounds of copper in the Yerington district from deposits associated with garnet-rich skarn at the Bluestone, Mason Pass, Douglas Hill, and Ludwig Mines.

Mineral assemblages of contact metasomatic deposits range from simple to complex, depending on the composition of intruded and intrusive rocks. Silica is generally the most abundantly introduced component; in the case of skarn, this combines with calcium from the host rocks to form calc-silicate minerals such as diopside, wollastonite, and grossular. Introduction or remobilization of iron results in the formation of hedenbergitic pyroxene and andradite. Addition or remobilization of alkali metals leads to reactions that form feldspar and mica minerals, fluorine produces minerals such as fluorite and fluorapatite, and boron enters into axinite- and tourmaline-group minerals.

The list of ore minerals in contact metamorphic deposits is varied. Common iron ore minerals in these deposits include magnetite, hematite, pyrite, pyrrhotite, and arsenopyrite. Common base-metal sulfides include chalcopyrite, sphalerite, galena, bornite, and sulfosalt minerals such as tetrahedrite and bournonite. In tungsten ore, ore phases include scheelite, wolframite, and powellite, often in association with sulfide minerals such as molybdenite and the base-metal sulfides noted above.

In addition to the important copper deposits at Robinson and Yerington discussed above, copper-bearing skarns were mined in the Copper Canyon and Copper Basin areas of the Battle Mountain district, in several mines of the Santa Fe district, and in a few mines in the Spruce Mountain district. In addition, base-metal skarn deposits in Nevada include zinc-lead-copper-silver deposits in the Ward and White Pine districts.

Rare minerals found in copper and base-metal skarns include elyite, a Nevada type mineral found in small amounts as a secondary phase in the Ward district. In addition, rare primary

TABLE 7   Minerals from the Ward district

Acanthite	Chalcocyanite	Hawleyite	Pyrite
Altaite	Chalcopyrite	Hessite	Pyrrhotite
Anglesite	Chlorargyrite	Joséite-β	Quartz
Aurichalcite	Chrysocolla	Kawazulite	Serpierite
Barite	Covellite	Langite	Silver
Brochantite	Devilline	Linarite	Smithsonite
Calcite	Elyite*	Malachite	Sphalerite
Caledonite	Fluorite	Molybdenite	Stephanite
Cerussite	Galena	Nontronite	Tetrahedrite
Chalcocite	Greenockite	Paratellurite	

*Nevada type minerals

phases such as kawazulite and joséite-β have also been identified in specimens from the Ward district, which has yielded a variety of unusual primary and secondary mineral species (Table 7). Magnesio-axinite, a rare skarn mineral identified in a museum specimen from a locality identified only as "Luning," was probably found in contact metasomatic rock prospected for copper in the Santa Fe or Fitting districts. At the Killie Mine, in the Spruce Mountain district, an extensive suite of secondary minerals, including the rare secondary lead and copper phases bayldonite and vauquelinite, have been identified in specimens from a base- and precious-metal deposit in skarn (Table 8).

Collectible specimens have also come from base-metal skarn deposits, including classic specimens of cuprite with copper, malachite, and azurite from the Richard and Alpha Mines in the Kimberly area of the Robinson district. Yerington district copper skarns are known for fine specimens of brochantite, libethenite, and other secondary copper minerals. The Chalk Mountain district has produced fine micromount specimens of a variety of secondary minerals from skarn-hosted deposits (Table 9).

Nevada contact metamorphic mineral deposits include tungsten skarn deposits that were important sources of the metal during both world wars and into the 1970s. Nevada tungsten skarn deposits were among the first to be studied in detail (for example, Hess and Larsen, 1920; Kerr, 1934). Descriptions of most of the tungsten skarn deposits, as well as other types of tungsten deposits, may be found in Stager and Tingley (1988).

Tungsten skarn deposits in the Mill City district were Nevada's most productive, yielding about 22,000 tons of WO_3 (equivalent to more than 17,000 tons of tungsten metal). Some of the Mill City ore was high grade; during World War I the Stank Mine produced more than 11,000 tons of ore grading 3.1 percent WO_3. Other important tungsten skarn districts in Nevada were the Potosi, Tem Piute, Leonard, Lodi, and Nightingale districts. In addition to scheelite and other tungsten-bearing phases, minerals in tungsten skarn include calcite, garnet (generally andradite- and grossular-rich), pyroxene (mostly diopside and hedenbergite), amphibole (tremolite and actinolite), vesuvianite, wollastonite, and epidote-group minerals (epidote, zoisite, and clinozoisite).

Despite the abundance of tungsten skarn deposits in Nevada, collectible minerals are few. Noteworthy specimens of garnet and clinozoisite with quartz crystals have come from the Nightingale district (Crowley, 2000). The tungsten mines of the Leonard district, near Rawhide, have produced epidote and quartz crystals, along with specimens that contain the rare secondary minerals ferritungstite and cuprotungstite. The Linka Mine, a moderately productive tungsten skarn deposit, has yielded specimens of the Nevada type mineral pottsite, a very rare but homely species, as well as crystalline epidote and garnet.

Several skarn-hosted iron ore deposits have been found in Nevada, including the Minnesota Mine in the Buckskin district, which produced as much as 5 million tons of iron ore, mostly in the 1950s to 1970s, for shipment to steel mills in

TABLE 8 Primary and secondary minerals from the Killie Mine, Spruce Mountain district

Anglesite	Chalcocite	Hematite	Pyromorphite
Arsenopyrite	Chalcopyrite	Hemimorphite	Rosasite
Azurite	Chrysocolla	Jarosite	Smithsonite
Bayldonite	Corkite	Malachite	Sphalerite
Beudantite	Covellite	Matlockite	Tenorite
Bornite	Fornacite	Mimetite	Vauquelinite
Carminite	Galena	Plumbojarosite	Wulfenite
Cerussite	Goethite	Pyrite	

TABLE 9 Minerals from the Chalk Mountain district

Anglesite	Fluorite	Mimetite
Aurichalcite	Galena	Molybdenite
Cerussite	Goethite	Plumbojarosite
Chalcopyrite	Hemimorphite	Pyrite
Chlorargyrite	Hydromagnesite	Pyromorphite
Chrysocolla	Jarosite	Scheelite
Descloizite	Leadhillite	Vanadinite
Epidote	Mcguinnessite	Wulfenite

Japan. Less productive and unmined iron skarns include deposits in the Yerington, Carson, and McCoy districts. Iron deposits that are arguably of contact metasomatic origin, such as at the Buena Vista, Segerstrom-Heizer, and Thomas Mines, occur in intermediate to mafic Triassic volcanic rocks and adjacent Jurassic diorite in the Mineral Basin district, which probably produced more than 4 million tons of iron ore. Descriptions of these deposits may be found in Reeves and Kral (1955). Contact metasomatic minerals at these deposits include scapolite, amphibole, and chlorite. The ore consists mainly of magnetite with minor amounts of other iron minerals, and it typically contains apatite, in places as crystals to more than 2.5 cm in diameter. At one locality, the iron-vanadium spinel and Nevada type mineral coulsonite was identified. Similar iron deposits in Mesozoic volcanic rock were mined in the Jackson Mountains district.

Recent exploration for gold in northern Nevada disclosed important contact metasomatic precious-metal deposits, including those at the Fortitude and McCoy Mines. The Fortitude gold skarn is related to nearby copper-bearing skarn of the West deposit in the Copper Canyon area of the Battle Mountain district. The ore occurs in prograde hedenbergite-andradite skarn and is associated with a retrograde assemblage that includes actinolite, chlorite, and prehnite. The McCoy deposit, located in the district of the same name, is in diopside-andradite skarn with retrograde alteration consisting of chlorite, amphibole, and other minerals. Brooks and others (1991) provided a detailed mineralogical description for the McCoy deposit.

A few deposits in altered limestone in eastern Nevada that contain beryllium, fluorine, and tungsten have also been described as skarns (Barton and Trim, 1991). The best known is the Wheeler Peak Mine in the Washington district, where beryl, phenakite, bertrandite, scheelite, and pyrite occur with fluorite and muscovite in veins and masses restricted to a single limestone bed. At the Bisoni (McCulloughs Butte) deposit in the Fish Creek district, beryl, sphalerite, scheelite, and molybdenite occur in a large low-grade fluorite deposit that was investigated by several major mining companies. In addition, occurrences of beryllium-rich vesuvianite in skarn have been reported in Elko and White Pine Counties. Except for minor shipments of tungsten ore from the Wheeler Peak Mine, these mineralogically interesting deposits have not been productive.

The highly productive magnesite and brucite deposits at Gabbs are associated with minerals but are considered hydrothermal (Schilling, 1968a). These deposits occur in dolomite, have minor amounts of associated tungsten skarn, and appear to be related to granodiorite or granite intrusions. They are discussed further in the section on industrial minerals.

Some geologists believe that the mineralogically complex base-metal, silver, gold, and platinum ores from the Goodsprings district should be classified as skarns; however, the evidence for this is inconclusive and most consider the Goodsprings deposits to be replacement deposits, somewhat akin to those at Pioche. Detailed information on Goodsprings mineral deposits will be found in the chapter by J. C. Kepper in this volume.

Mercury Deposits

STEPHEN B. CASTOR AND
JOSEPH V. TINGLEY

Mercury minerals have been mined for many centuries. Cinnabar, the chief ore mineral, was widely used as a coloring agent, and the metal was used for concentration of precious metals by amalgamation. Cinnabar was collected from near Ione by Native Americans, and as many as 200,000 flasks (at 76 pounds per flask), mostly from the New Almaden and other mines in California, are thought to have been used for amalgamation in the Comstock district prior to the turn of the century. More modern uses in munitions and medicines expanded demand during the two world wars, spurring modern mercury exploration and mining in Nevada.

Mercury deposits were mined in Nevada from the 1870s to 1990. Environmental concerns about the use of mercury arose in the 1970s, ultimately curtailing most uses of the metal. Even with rising awareness of the potential health risks of mercury, one market remained strong in the 1970s. At the Pershing Mine, Antelope district, Pershing County, cinnabar coarsely crystalline enough to be mined as "crystal" commanded a price of hundreds of dollars a pound when shipped to the Orient to be used for medicinal purposes. However, this was a local phenomenon. Because of decreased markets worldwide, the mercury price dropped to a level too low to support domestic mining, and, over the next few years, all the primary mercury mines in the United States closed. Closure of the McDermitt Mine in 1990 marked the end of primary mercury mining in Nevada and the United States as well.

Nevada, however, remains the number one producer of newly mined mercury in the U.S., a position it maintains solely because of by-product mercury collected at a number of large gold-mining operations. Production figures for by-product mercury are not released by most of the mining companies involved, so it is not possible to compare annual by-product production with Nevada's record production years in the 1970s and 1980s.

The mineral source of mercury from the gold mines is also not clear. Mercury is recovered along with gold and silver by the cyanide process, either in conventional milling or in heap leaching operations. The mercury in the ore must, therefore, be in a form that is soluble in cyanide. Bailey and others (unpublished manuscript, NBMG files) believed that mercury in oxidized ore at the Carlin Mine occurred as mercury sulfate (schuetteite?). Radtke (1985) identified mercury in several forms in the Carlin ores: (1) in or adsorbed onto iron oxides formed by oxidation of pyrite; (2) as sparse, remnant grains of cinnabar coated by secondary schuetteite; (3) as a secondary mineral containing mercury, arsenic, and oxygen; and (4) in a phase containing mercury, iron, and oxygen. Radtke (1985) attributed mercury in unoxidized gold ores at Carlin mainly to thin films coating pyrite grains and to mercury-organic and mercury-gold-organic compounds.

It is generally agreed that most economic concentrations of mercury minerals are formed at low temperatures in relatively shallow epithermal deposits. Some formed at very shallow levels, as shown by the discovery of cinnabar in sinter deposits in the Steamboat Springs district by Becker (1888) and in active vents at the Senator fumaroles in Churchill County. In addition, it was proposed that cinnabar formed in fumarolic vents at the McDermitt Mine (Roper, 1976); however, subsequent work showed that mercury min-

TABLE 10 Minerals from the Cordero and McDermitt Mines,
Opalite district

Alunite	Fluorite	Mordenite
Barite	Goethite	Mosesite
Buddingtonite	Gypsum	Opal
Calcité	Hematite	Pyrite
Calomel	Heulandite	Quartz
Chapmanite	Jarosite	Radtkeite*
Cinnabar	Kaolinite	Realgar
Clinoptilolite	Kenhsuite*	Stibnite
Copiapite	Kleinite	Terlinguaite
Corderoite*	Marcasite	Tridymite
Cristobalite	Melanterite	Tripuhyite
Eglestonite	Mercury	
Erionite	Montmorillonite	
Fibroferrite	Montroydite	

*Nevada type minerals

erals in this deposit were probably deposited below the surface (McCormack, 1986). Although most mercury deposits are shallow, it is clear that cinnabar can also be deposited in deep deposits, because it was found at a depth of 1,800 m in the active geothermal system in Calistoga, California. In Nevada, most significant mercury deposits occur in the western half of the state, and most of the mercury has come from deposits in, or associated with, Tertiary volcanic or shallow intrusive rocks. However, Mesozoic sedimentary rocks contain significant mercury deposits in the Antelope Springs and Pilot Mountains districts.

Bailey and Phoenix (1944) described more than 150 mercury mines and prospects and occurrences of mercury minerals in Nevada. With such a large number of occurrences, it is not surprising that mercury deposits in Nevada have yielded a large number of Nevada type minerals, six in all.

By far the most productive mercury mines in Nevada were in the Opalite district in the peralkaline volcanic rocks and related sediments of the McDermitt caldera in northern Humboldt County. This district, site of the highly productive Cordero and McDermitt Mines, was also the site of several other less productive properties, such as the Opalite Mine, that are directly north of the state line in Oregon. The Cordero and McDermitt Mines, located about 1 km from each other, together produced over 400,000 flasks

(about 14 million kg) of mercury. They are well known as mineral localities, with three Nevada type species between them as part of an extensive assemblage of mercury phases, including several mercury-bearing halides. All of the type minerals are sulf-halides of mercury, and one, corderoite, is not solely of mineralogical interest, because it was also an important ore mineral in the deposits. Table 10 lists mineral phases identified in specimens from the two mines, including primary and diagenetic minerals in the volcanic rocks that host the mercury deposits, and alteration, gangue, and ore minerals in the deposits.

A second Nevada mercury mine that is well known in mineralogical circles is the Red Bird Mine in the Antelope Springs district of Pershing County. This mine produced about 3,300 flasks of mercury from a pipe-like orebody in limestone that also contains antimony, lead, and zinc minerals. It is a co-type locality for two minerals, schuetteite and dadsonite, and the type locality for robinsonite. The main ore mineral is cinnabar, but some mercury was reportedly recovered from metacinnabar, native mercury, and mercury-bearing bindheimite.

Other important Nevada mercury-mining areas include the Union (Nye County), Pilot Mountains, Fish Lake Valley, Bottle Creek, Castle Peak, Ivanhoe, and Goldbanks districts, each of which had more than 1,500 flasks of production. Mineralogy of mercury deposits in most of these districts is relatively simple; cinnabar is the only significant ore phase in most deposits, and associated minerals are commonplace species such as quartz and calcite. However, the Nevada type mineral schuetteite was identified as a post-mining phase in specimens from the Fish Lake Valley, Ivanhoe, and Goldbanks districts. Metacinnabar, a rare dimorph of cinnabar, has been reported from the Ivanhoe and Pilot Mountains districts. Native mercury, also generally rare, has been reported from the Castle Peak, Goldbanks, and Union districts.

Considering the number of mercury mines and prospects in Nevada, the production of showy mineral specimens has been minor. The Cahill Mine in the Poverty Peak district of Humboldt County is an exception. Here cinnabar crystals of more than 2.5 cm occur in calcite veins, and a variety of associated minerals are present, including metacinnabar, eglestonite, terlinguaite, calomel, and native mercury. Other good cinnabar specimens from Nevada include brilliant crystals from the Red Bird Mine, tiny crystals in fluorite and fine disseminations in opal (myrickite) from the Bare Mountain district, and tiny crystals from the Castle Peak district. Mercury halide phases from the McDermitt Mine, including corderoite, radtkeite, and kenhsuite, are microcrystalline and generally colorful when first encountered, but darken on exposure to light. Kleinite, the oxychloride phase from the same locality, which also deteriorates in bright light, has been found in yellow crystals as much as 2 mm long.

Nevada's Industrial Minerals

KEITH G. PAPKE AND
STEPHEN B. CASTOR

Nevada is well endowed with industrial minerals and rocks; it probably has a greater variety of these materials than any other western state. The 1999 production of industrial minerals in Nevada was valued at nearly 400 million dollars. The materials produced were barite, cement, clays, diatomite, gypsum, limestone and dolomite, lithium, magnesite, perlite, pumice and volcanic cinder, salt, sand and gravel, silica, decorative and crushed stone, and zeolite. In addition, the state has had significant past production of andalusite, borax, feldspar, graphite, pyrophyllite, sodium carbonate and sulfate, sulfur, and talc. Nevada is the sole domestic producer of lithium and magnesite, the largest producer of barite, the second largest of diatomite, and the sixth largest of gypsum. About 430 industrial mineral deposits (excluding sand, gravel, and crushed stone) with present or past production, or with potential for future production, have been identified in the state; these include 22 commodities.

Many industrial minerals and rocks are prized because of their unique physical properties. For example, they include the hardest mineral (diamond) and the softest (talc), the whitest (magnesite) and the blackest (graphite), one of the heaviest (barite) and the lightest (pumice), and one of the toughest (jade) and one of the most brittle (sulfur). Industrial minerals are almost everywhere in our lives: gypsum in house walls, talc as a filler in cloth, pumice as an abrasive in toothpaste, clay in nondrip paint, kaolin as a coating on paper, diatomite in the cat box, and rare earths in the TV screen. We can't get along without industrial minerals.

Some early production and uses of industrial minerals in Nevada are noteworthy. One of these was the production of common salt (halite). This started in 1862 when camels were used to transport the surface salt from Rhodes Marsh in Mineral County to the Comstock district, where it was used to process the silver ores. Closer sources of salt were soon discovered, and the transportation by camels was discontinued. Another example is the borate industry, which essentially got its start in 1872 when a young prospector, searching for gold or silver, discovered borax at Teels Marsh in Mineral County. For the next 10 years, most of the world's borax came from surface crusts here and at the nearby Rhodes, Columbus, and Fish Lake Marshes. The prospector, Francis Marion "Borax" Smith, organized the Pacific Coast Borax Company, and it dominated the borax industry for many years. Prior to the completion of the Carson and Colorado Railroad through this area, all the borax was hauled more than 130 miles by teams to the existing railroad at Wadsworth. Nevada has had no production of borax since 1930, but a plant in southern Nye County processes borate mined in Death Valley.

Seven modes of origin have been recognized for the Nevada industrial minerals: those formed by hydrothermal activity, by sedimentary deposition, as evaporites, as igneous rocks, by alteration by groundwater, as pegmatite, and as brines. Fifty percent of the Nevada deposits were formed by the first two processes.

Industrial minerals and rocks are widely distributed throughout the state. Because of their unique modes of origin, however, some types are restricted in geographic occurrence. Thus, almost all the diatomite deposits are in the western one-third of the state, where Pliocene freshwater lakes were abundant, whereas almost all the bedded barite deposits are in the north-central and northeastern part of the state, where deep-sea

TABLE 11 Minerals from the magnesite and brucite mining area near Gabbs

Aluminohydrocalcite	Chondrodite	Huntite*	Periclase
Antigorite	Chromite	Hydromagnesite	Pokrovskite
Artinite	Clinochlore	Magnesite	Rutile
Brucite	Clinochrysotile	Mcguinnessite	Scheelite
Callaghanite*	Dolomite	Nakauriite	Talc
Chalcopyrite	Forsterite	Orthochrysotile	Tremolite
Chlorite	Grossular	Parachrysotile	Vesuvianite

*Nevada type minerals

rocks that represent the geological environment necessary for the deposition of bedded barite occur.

Although most of Nevada's industrial mineral deposits are uninteresting to mineral collectors and mineralogists, a few have been important sources of mineral specimens. The Gabbs magnesite and brucite deposits, which have been mined since the 1930s, are probably the most interesting in this respect.

The Gabbs deposits are large irregular masses that occur in dolomite adjacent to granodiorite. In addition to the economic minerals, the deposits contain a suite of mainly magnesium-bearing minerals, some quite rare. The suite includes two Nevada type minerals, callaghanite and huntite, although the type locality for the latter is a magnesite deposit in the Currant district (Table 11). In addition, the discredited species cuproartinite and cuprohydromagnesite, both of which are now considered nakauriite, were first described from Gabbs.

The Gabbs mineral assemblage also includes minerals such as scheelite, grossular, and vesuvianite, which were found in the Betty O'Neal Mine, a small tungsten skarn deposit in the magnesite and brucite mining area. The magnesium deposits, which may be of contact metasomatic origin, have also been described as hydrothermal deposits (Schilling, 1968a).

Two interesting Nevada industrial mineral localities are barite mines: the Redhouse barite prospect, Potosi district, type locality of cure-tonite, and the Little Britches Mine, Golconda district, a locality for the rare barium silicate mineral cymrite. In general, economic Nevada barite deposits are of the bedded type and seldom yield handsome crystalline barite. However, gold mines, such as in the Lynn and Northumberland districts, have yielded world-class barite specimens from veins.

Nevada fluorspar deposits are also relatively uninteresting as a source of specimens, although the Boulder Hill Mine in the Wellington district has yielded crystals to 6 cm. Nevada's gypsum mines have not been a source of fine crystalline specimens, but clear selenite cleavage blocks as much as 3 feet in diameter have been taken from the Empire Gypsum Mine near Gerlach. In addition, old mica mines and prospects in the Ruby Mountains of Elko County and the Eldorado and Virgin Mountains of Clark County have yielded specimens of beryl, chrysoberyl, and other pegmatite minerals.

Handsome sulfur specimens have come from a historical mine in the Alum district near Silver Spring, and from a silica pit in the Steamboat district. The former deposit was also once a source of unique kalinite "rams-horn" specimens. Another mineralogically interesting deposit is the Champion Mine near Oreana, the only known commercial deposit of the mineral dumortierite in the world; it was mined for material to make spark plugs from the 1920s to the 1940s.

Gemstones of Nevada

CHRISTOPHER ROSE AND
GREGORY C. FERDOCK

A gemstone is defined as "any mineral, rock, or other natural material that, when cut and polished, has sufficient beauty and durability for use as a personal adornment or other ornament" (Bates and Jackson, 1987). Nevada contains many gemstone occurrences; the vast majority fall into the semi-precious category, such as agate, jasper, and petrified wood. Precious-gem material is rare in the state; however, Nevada is noted as one of the world's premier producers of precious opal and fine turquoise. In addition, the state has recently become the source of North America's largest gem topaz crystals and some of the most desirable colors of amethyst known in the industry. Total value of Nevada's gemstone production is not known, but it is likely in excess of $100 million at current prices, mostly from production of precious opal and turquoise.

BERYL

Beryl is an important gemstone; some varieties, such as emerald and red beryl, rank with diamond and sapphire in value. Beryl has many varietal names depending upon its color and source. The best-known gem varieties, aquamarine and emerald, have both been reported from Nevada. Common beryl is dull white, gray, or gray-blue and opaque; such material has been used as a source for beryllium metal. Beryl is a relatively common mineral, found mainly in pegmatite and metamorphic rocks, and it has been reported from several localities in Nevada (see the mineral catalog in this volume).

Beryl was first described from Nevada in 1878 by Hoffman, who described sparse dull-bluish-ash crystals to 2 cm, 16 km north-northwest of Silver Peak. This locality is probably that noted by Olson and Hinrichs (1960) on Mineral Ridge in Esmeralda County. Occurrences of gem-quality and near-gem-quality beryl are listed below.

Virgin Mountains, Clark County

At the Taglo Mine, beryl occurs in pale-bluish-green to sea-green crystals as much as 2 cm across in zoned pegmatite dikes and sills; it is associated with chrysoberyl. At another site just west of the Arizona state line in the Virgin Mountains, about 10 km south of Bunkerville, beryl occurs in light-green to nearly colorless crystals as much as 1.5 cm in diameter with mica and possible phenakite in pegmatite dikes in Precambrian rocks.

Ruby Mountains, Elko County

Two geologists discovered gem-quality beryl in the upper reaches of Lamoille Canyon in 1988 during examination of pegmatite dikes that radiate from granitic rocks into surrounding high-grade metamorphic rocks. Here, transparent sky-blue aquamarine was found along with common bluish-gray opaque beryl. The largest gem-quality crystals found were approximately 2 and 3 cm in largest dimension. Several other small, imperfect aquamarine crystals were also recovered from the area.

Another occurrence of small aquamarine crystals is on the eastern side of the Ruby Mountains in the area of Dawley Canyon, where numerous beryl-bearing pegmatites have been prospected for beryllium. A few azure-blue crystals were recovered from underground workings, particularly those located closest to Ruby Valley; one was reportedly translucent.

Teels Marsh, Esmeralda County

Light-blue beryl crystals have been reported from quartz-wolframite-scheelite veins in a hill west of Teels Marsh 6 to 13 km southwest of Marietta.

Bisoni Prospect, Eureka County

Small yellow crystals of beryl occur with white beryl and other minerals at a prospect 24 kilometers southwest of Eureka, on the flank of Ditto Peak in the Fish Creek Range.

Oreana Mine, Pershing County

Green outer zones on pale beryl crystals were found in a pegmatite in limestone at the Oreana Mine (Sinkankas, 1997) in the Rye Patch district, but this material was reportedly "neither flawless nor transparent" (Ball, 1939). The green color was said to be rich enough to be considered emerald, and the material is possibly suitable for small cabochons (Sinkankas, 1997).

Wheeler Peak Mine, White Pine County

Blue beryl crystals to 5 cm have been reported with phenakite and other minerals from this deposit in the Snake Range.

CHRYSOBERYL

Chrysoberyl is a relatively uncommon beryllium mineral found in pegmatites and associated with beryl. Gem-quality chrysoberyl commands a high price because of its rarity.

Taglo Mine, Bunkerville District, Clark County

Chrysoberyl occurs as yellow to apple-green plates as much as 1.9 cm long with beryl crystals in pegmatite bodies at the Taglo Mine in the Virgin Mountains. This material is not high-quality gemstone, but has been used for decorative stone, some being carved into objets d'art.

CORUNDUM (SAPPHIRE)

Next to diamond, corundum is the second-hardest mineral. It is most common in high-grade metamorphic rock or in placers from such sources. It is rare in Nevada; blue corundum has been reported from several sites, but none are known to have yielded gem-quality stones.

Buckskin Mine, Douglas County

Corundum occurs as blue crystals and crystal aggregates with andalusite, pyrophyllite, diaspore, and rutile at the Buckskin Mine and the Blue Metal (Blue Danube) deposit in the Buckskin Range, at the north end of the Smith Valley.

Dover and Green Talc Mines, Mineral County

Corundum is found as dark-blue crystals at the Dover and Green talc mines, Fitting district, in the Gillis Range northeast of Thorne. It is most common in veins and usually granular, and occurs with andalusite, pyrophyllite, and other minerals.

DUMORTIERITE

Dumortierite is generally found in metamorphic rocks. The colors of this mineral make it desirable for stone carvings and cabochons. Material from Nevada is known for its blue and pink colors as well as the quantity of the deposits.

Oreana, Pershing County

Dumortierite occurs at the Champion Mine on the west flank of the Humboldt Range, in Humboldt (Gypsy) Queen Canyon, 10 km east of Oreana. It occurs as euhedral blue crystals, lavender or pink masses, and as isolated pink crystals or fibrous veins with andalusite, sericite, quartz, and coarse muscovite in metamorphosed, schistose tuff. This material was originally mined from a large open pit and several smaller workings for use in spark plugs. The size of the deposit is remarkable and production from the site was large. The mineral also occurs as cobalt-blue to deep-pink crystalline masses and lavender to pink veins in schist at Lincoln Hill, about 10 km northeast of Oreana.

GARNET

Garnet is a group name covering nearly a dozen different species; most are used for gemstones when found in sufficient quality. Garnet comes in red, orange, green, and brown colors, and is prized as faceted stones in fine jewelry. Nevada contains large amounts of garnet-bearing rock formed by contact and regional metamorphism, but most of this garnet is of mineralogical interest only. No deposit of garnet has been commercially mined in the state for gemstones, but two deposits are known to have produced at least a few stones used in jewelry.

Ruby Mountains, Elko County

Grossular in gemmy-orange crystals of the variety hessonite occurs as dodecahedrons to 4 cm in metamorphosed limestone in the central Ruby Mountains. These crystals contain zones of glassy orange that would facet into exquisite stones. Unfortunately, few stones have been recovered and some of this material is in a wilderness area and thus unlikely to be mined.

Adelaide, Humboldt County

Iridescent brown to honey-colored andradite occurs in contact metamorphic rock 6 km east of Adelaide in the Gold Run district (Ingerson and Barksdale, 1943). It is reportedly similar to gem garnet from Mexico, but its suitability for gemstones is not known (Sinkankas, 1997).

Garnet Hill, White Pine County

Almandine-spessartine crystals are found in cavities in rhyolite in the Garnet Fields recreation site on Garnet Hill near the town of Ruth. The crystals are mirror-lustrous, deep-red to black dodecahedrons as much as 3 cm in diameter. Material that can be faceted is rare due to the dark color of the stones, but some small deep-sherry-colored and dark-red stones have been cut. In addition, some detached crystals have been mounted to form naturally faceted jewelry.

JADE

Jade is technically either the mineral jadeite or nephrite, a compact variety of actinolite. Gemgrade jadeite is rare in the Americas, but nephrite is widespread, particularly in serpentinized ultramafic rock, and has been used extensively in the gem trade for cabochons or carvings. Two occurrences have been reported in Nevada, but details are sketchy.

Black Jade, Esmeralda County

Black nephrite jade, similar in appearance to "Edwards Black" from Wyoming, has been sporadically mined from serpentine in the Silver Peak Range near Fish Lake Valley. This may be the same locality reported near Tonopah by Austin (1991).

FELDSPAR

Several species of the feldspar mineral group have been used as gems, including the well-known gemstone varieties amazonite and moonstone. Nevada is listed as the source of transparent stones of labradorite, oligoclase, and anorthite, but the localities are not known (Gray, 1999). The occurrence noted below is a relatively new source of gem feldspar.

Zapot Pegmatite, Mineral County

Since its discovery in 1984 by Jerry Gray of Reno, blue amazonite (microcline) has come from a pegmatite occurrence about 25 km northeast of Hawthorne. The amazonite, which is recovered by Harvey Gordon of Reno, is reportedly "far more blue in hue than any hitherto mined in North America" and has been compared with that from Colorado (Sinkankas, 1997). The deposit is also known for the production of topaz and smoky quartz crystals, as well as the presence of rare alumino-fluoride minerals including the new species simmonsite.

OPAL

Opal is a widespread low-temperature mineral that is generally confined to surface or near-surface deposits. It includes several varieties. "Common opal" is generally white to brown and is often found in petrified wood and hydrothermal veins. Other varieties are named for color and other features, including precious opal, which is the most sought after for its brilliant play of colors.

Nevada contains some of the richest precious opal beds in the world and has produced individual specimens valued in excess of $1 million. Most of the precious opal has been found as wood replacement in lake-bed deposits of Virgin Valley in Humboldt County.

Other opal has been mined for carvings, cabochons, and decorative stone, mostly from numerous petrified wood deposits in Tertiary sedimentary and volcanic rocks throughout the state. The Nevada localities are too numerous to describe on a site-by-site basis, and the localities included are those with most recent or prolific production. Information on other sites may be found in this volume's mineral catalog and in the references cited therein.

Virgin Valley, Humboldt County

The Virgin Valley opal beds in northwestern Humboldt County, which are the most famous gemstone locality in Nevada, have been a favorite with jewelry makers since their initial discovery in 1905 by cowboys. Virgin Valley opal includes precious opal that occurs as a replacement of Miocene trees, principally conifers and rarely deciduous trees, but ginkgo has also been identified (Sinkankas, 1997). After the trees died, they accumulated in a lake where they were subsequently covered with sediment, including ash from nearby volcanic eruptions. Logs up to 6.5 meters long have been found that are replaced by opal and chalcedony.

The opal-bearing zones are covered by overburden up to 35 meters thick, which is removed by bulldozers and excavators. Prior to modern excavation, opal was recovered by underground mining; many of the old tunnels are still open and are quite extensive.

Virgin Valley opal has unusually high water content, up to 13 percent. Because of this, the opal tends to crack upon dehydration following excavation, and many stones have been lost due to this process. As a result, much of the opal is immediately placed in water to prevent deterioration. In many cases, outer portions of larger pieces desiccate and crumble, leaving behind small stable pieces. Cores of large limbs or logs may remain stable without treatment.

Virgin Valley opal is considered by most experts to be the brightest precious opal available, and it is also considered to include the world's finest black opal. Flashes of green, red, violet, blue, and yellow are spectacular in this material. The largest and most famous specimen is the Hodson opal, a 3.2 kg fiery mass of transparent gem opal.

As a result of its fame, nearly every acre of ground in the valley is staked, and there are literally hundreds of mining claims. But as with many mining districts and claims, only a few have produced commercial quantities of opal.

BONANZA OPAL MINE

Consisting of six patented mining claims, this deposit is on the northern side of Virgin Valley and is one of the original mines. The Hodson family owned it for many years until it was sold to a group who leases digging rights to the property. Rockhounds are allowed to work the deposit as long as they turn over a percentage of their recovery to the company, which also mines the deposit commercially with heavy equipment. Originally developed by underground methods, the mine is now an open pit. It has consistently produced large limb casts of precious opal.

RAINBOW RIDGE OPAL MINE

Owned by the Hodson family since the 1940s, this is one of the oldest and most consistently productive mines in the valley. The property consists of four patented claims that were originally developed as an underground mine and are now being worked as an open pit. Commercial mining is ongoing and rockhounds are allowed to comb through the dumps for a fee. The famous Hodson opal came from this locality.

ROYAL PEACOCK OPAL MINE

Located in the southwestern corner of the Virgin Valley, this property consists of eight patented claims. The property includes a number of open pits that are not currently being commercially mined. For a fee, rockhounds are allowed to dig in the opal-bearing clays exposed in the pits. The Royal Peacock Mine is one of the most prolific producers of opal in the valley. It has likely produced the largest quantity and average limb-cast size of any of the mines in the district and probably contains the largest reserves of opal in the valley with deposits extending over most of the length of the property.

JOY LU 55 OPAL MINE

This deposit is located on the southern side of the Virgin Valley between the Rainbow Ridge and Royal Peacock Mines. It was commercially worked in two open pits in the 1970s, but is now inactive. Lake-bed sediments at this site contain abundant opalized wood, and seams in this opalized wood are filled with late-stage common and precious opal. Petrified pine cones are common. This deposit probably contains more common opal-replaced petrified wood than any other property in the valley.

BECKY-MUCKET CLAIMS

Located on the southern side of Virgin Valley, this deposit is small but contains abundant twig and limb casts replaced by precious opal. These occur in clay beds containing more rounded gravel than other mines. This mine borders Virgin Valley wetlands and is very close to the creek; as a result the locality has been withdrawn from mineral entry and is not claimable.

SWORDFISH MINE

Located in the southeastern part of the Virgin Valley near the Rainbow Ridge Mine, this is an old mine that was rediscovered in the late 1990s by David Church. It is now being worked by hand and is producing limb casts replaced by precious opal.

LEMON MERINGUE MINE

This mine is located just north of the CCC campground in Virgin Valley. The opal was discovered in 1996, has been worked by hand every season since discovery, and is marketed as "Nevada Golden Opal" that is principally sold on the Asian market. The deposit contains large pods of transparent to translucent "jelly" opal in a rhyolite flow. The pods range in size from several centimeters to 2 meters long, are linearly oriented with respect to each other, and seem to occur within a particular horizon of the flow.

Little Joe and Hoss Opal Mines, Humboldt County

These mines are in the Calico Hills near the Black Rock Desert north of Gerlach. The opal was discovered by Ray Duffield in the 1960s and is now on two patented claims. The host rock is vesicular basalt and many of the vesicles are filled with clear, yellow, orange, and red opal that commonly has purple, blue, green, red, orange, and yellow fire. Much of the opal is transparent, and faceted opals from this site are valuable even if they lack fire. Unfortunately, a significant portion of this opal crazes after it is mined and develops cracks, sometimes falling apart. Crazing can cause the opal to become opaque, particularly immediately after mining.

Royal Rainbow Mine, Humboldt County

Located on Willow Creek several kilometers north of the Little Joe Mine on the eastern side of the Calico Hills north of Gerlach, the geology and mineralization at this site are similar to that of the Little Joe. It is operated as a campground and fee-digging site for rockhounds.

Firestone Opal Mine, Humboldt County

This deposit, located in the Santa Rosa Mountains north of Winnemucca, was worked in the past by a small open pit. At one time, it was a popular fee-digging area and campground and produced valuable opals. The precious opal, which has a base color of white, yellow, orange, red, and black, is opaque to transparent with plays of red, blue, green, and yellow fire. It occurs sparsely in vesicles in basalt, but opal-filled vesicles contain a high percentage of precious opal.

Webber Opal Mine, Lincoln County

This is a new deposit, located north of the Mormon Mountains and north of Moapa Valley, discovered by the Webber family of Beatty, Nevada. The opal is a blue-based precious variety formed in gas cavities and fractures in an extremely tough, highly siliceous, dense brown rhyolite flow. White common opal on the surface becomes translucent, blue-based precious opal at depth. Several open pits explore the deposit, and have intersected large masses of precious opal, including a single specimen weighing almost 20 kg. The enclosing rock is of sufficient toughness to warrant blasting, which causes some fracturing of the opal. After mining, the opal must be sawn from this tough matrix. The opal quality reportedly improves with depth.

Starfire Opal Mine, Nye County

This deposit, 21 km north of Gabbs, was originally mined by Tiffany's of New York early in the century. Later it was developed as a fee dig, and during the 1970s was again worked commercially. The property is now inactive. Surficial opal is opaque and white, but about 1 meter below the surface, it has a rich blue color, some with a play of color in reds, greens, and blues. The opal occurs in veins and breccia filling within generally clay-altered rhyolite in an area of over 20 acres.

QUARTZ

Quartz is the most abundant mineral in the Earth's crust and, consequently, is the most common mineral used in the gemstone industry. Many varieties of quartz occur in Nevada; a short list of material available in the state includes agate (some amethystine), bloodstone, bogwood, carnelian, chalcedony, chert, crystal, geodes, jaspagate, jasper, opalite, and petrified wood. Large amounts of semi-precious quartz, such as agate, chalcedony, and crystal, have been mined or just "picked up" from locations in all of the counties of Nevada. The principal use of this material is decorative, as in the case of petrified wood, or in carvings, cabochons, and covers for triplets and doublets of less durable stones. It is used more rarely in faceted stones, particularly the clear amethyst and citrine from Petersen Mountain.

The following discussion presents a variety of mines and deposits currently or recently producing for the gem trade. A complete list is beyond the scope of this volume. Several general rockhound guides describe localities and types of material available, including *Gem Trails of Nevada* by J. R. Mitchell, 1991, and *Nevada-Utah Gem Atlas* by R. N. Johnson, 1978. Other localities can be found in J. Sinkankas's *Gemstones of North America* volumes and G. F. Kunz's *Gems and Precious Stones of North America* (1890).

Amethyst Sage Mine, Humboldt County

Located south of Denio, this area has produced large quantities of gray to deep-purple agate with black dendrites that is popular among custom jewelers. It is owned by a Washington state company, which has operated the site since the late 1990s. Agate from this site is hosted by andesite and tends to be gray near the surface, with a purple tint that increases with depth. Gray exteriors are common on larger pieces of agate, coating attractive dendritic purple interiors. Material from this locality is used for making cabochons and for carving.

Mount Airy Blue Chalcedony Mine, Lander County

Located in the Reese River Valley near Austin, this deposit was discovered in 1996 by Chris Rose and Larry Everitt and has been worked continuously since that time. The chalcedony occurs in "thunder eggs" that range from 7.5 to 90 cm in diameter and occur in well-defined layers in clay-altered ash-flow tuff. The productive zone can be traced for more than 1.6 km, and blue chalcedony, quartz, and amethyst crystals are found throughout this zone.

Blue chalcedony has recently become very fashionable in the gem and jewelry business and is currently one of the most sought-after gemstones on the market. This material is cut in cabochons or carved and mounted in jewelry. According to the Gemological Institute of America, the blue color is a result of iron in the quartz.

Petersen Peak, Washoe County

This property is being worked by two mines, individually owned and operated by separate owners, on the northern end of Petersen Peak, north of Reno near Highway 395. Mining is by hand, augmented by backhoes and small bulldozers.

This locality produces world-class, museum-quality smoky quartz, citrine, and amethyst scepter crystals to 30 cm long from pockets in quartz vein stockwork in weathered granodiorite. The locality also produces faceting-grade citrine, amethyst, and smoky quartz. Very large amethyst and ametrine (mixed amethyst and citrine) gems have been cut, and the rough commands prices as high as $700 per inch. The amethyst includes "grape juice"-colored stones, which are very desirable. The site is often open for fee digging, and many weekend collectors have come away from the site with world-class material.

TOPAZ

Topaz is a widespread mineral in high-temperature quartz veins, in cavities in rhyolite and granite, and in some granite pegmatite. When found in pegmatite, it is associated with microcline, quartz, and in some cases tourmaline and beryl. In Nevada, topaz is usually found as accessory crystals in rhyolites and pegmatites up to, but usually much smaller than, 2 cm. Crystals found at the Zapot claims are an exception.

Zapot Claims, Mineral County

The Zapot property is located in the Gillis Range near Hawthorne. It has been worked commercially for the past 10 years, and at the time of this writing was still in production. Most of the commercial production comes from two open cuts and has focused on matrix mineral specimens primarily of smoky quartz and amazonite.

The deposit includes the zoned Zapot pegmatite and other masses of pegmatite in granitic rock. Crystal-lined cavities contain faceting-grade light-blue and light-green-blue topaz that, upon irradiation, darkens to a more gemmy, darker blue. Crystals to 20 cm and weighing several kilograms have recently been found at the site; these are the largest gem topaz crystals yet found in North America, though they are small compared to non-gem crystals found in Virginia. The topaz is found with crystals and masses of amazonite and smoky quartz.

TURQUOISE, VARISCITE, FAUSTITE, AND CHALCOSIDERITE

Turquoise, variscite, faustite, and chalcosiderite form a series of hydrated phosphates of copper, aluminum, iron, and zinc. Variscite is dominated by aluminum, turquoise contains aluminum and copper, chalcosiderite is the iron-bearing analog of turquoise, and faustite is a zinc-bearing relative. All of these minerals are secondary in origin, forming near the surface through the action of infiltrating water on aluminous sedimentary and igneous rocks and using metal scavenged from nearby metalliferous deposits.

Nevada has produced turquoise and turquoise-like minerals since prehistoric times, and became internationally known for turquoise in the late nineteenth century. From the 1930s to 1950s, Nevada led the United States in turquoise production, producing between 3500 and 5000 kgs per year (except during the immediate post-war years) from as many as 38 simultaneously operating properties. Turquoise is probably second only to precious opal in total gemstone dollar value for the state.

Until recently, faustite and chalcosiderite were considered rare and of secondary importance to the gem trade. Work on odd-colored turquoise has revealed that much of this material is not turquoise but contains the other two species. All of these minerals have roughly the same hardness and take a good polish when found in compact masses. Most deposits in Nevada are found as veins and nuggets, although microcrystals of turquoise are found at the Silver Coin Mine in Humboldt County, and crystalline variscite has been found at the Goldstrike and Rain Mines in the Carlin Trend.

Turquoise group minerals have been found in all counties in Nevada save Carson City. The following is a discussion of currently producing or important past-producing properties. The reader may refer to this volume's mineral catalog for a more comprehensive list, and details on many deposits may be found in Morrissey (1968).

Simmons Turquoise Mine, Clark County

Located about 19 km west of Searchlight on the southwestern flanks of Crescent Peak in Clark County, this deposit was discovered in 1889 by George Simmons, who found turquoise in the float, as well as in abandoned Native American encampments and mines. Historical workings include open cuts and underground workings, and total production is believed to have been in excess of $1 million.

Turquoise occurs as veins and nodules within shear zones and fractures in sheared, argillized granite that contains abundant quartz stockwork veins. Some of the turquoise appears to have been associated with the quartz veining. Turquoise from this mine is generally a bright-blue color, some with black spider-web patterns.

Candelaria Variscite Mine, Esmeralda County

This deposit, in the Candelaria Hills just north of Columbus Salt Marsh, was first worked around 1905 by the Los Angeles Gem Company in six open pits. Variscite and chalcosiderite occur as veins, paper thin to 5 cm thick, and as nuggets to 0.5 kg in a 15- to 30-m-thick layer of pale decomposed shale that is overlain by brown marl that contains little gem material. Most of the variscite and chalcosiderite is moderately to intensely spider-webbed with black matrix that adds to its beauty. Colors range from white to emerald green. Some of the variscite veins are gray to black but grade into a green color over a short distance. Nuggets are often difficult to identify, because they are coated with white material and are best located by shape. Nearly all nuggets are spider-webbed and are more colorful at depth. Many of these nuggets have been marketed as "variquoise."

Lone Mountain Turquoise Mine, Esmeralda County

Lee Hand discovered this deposit, which is about 1.6 km east of Paymaster Canyon southwest of Tonopah, in 1920. It was developed as an underground mine with more than 500 m of lateral workings, to a depth of 85 m. In the 1970s, the turquoise was mined from an open pit. The turquoise occurs in argillized shear zones as hard, high-quality, solid-blue, and spider-webbed nodules in shear zones in strongly folded and faulted calcareous shale.

Royal Blue Turquoise Mine, Esmeralda County

This deposit is on the Nye-Esmeralda county line about 38 km northwest of Tonopah. Two prospectors named Workman and Davis discovered it in 1902. In 1907 the property was purchased by the Himalaya Mining Company and became one of the most productive turquoise mines in the United States. For a period of time, the mine produced more than 600 kg of turquoise per month. Total production from the mine has exceeded $5 million in cut stones.

The turquoise is found as seams to 13 cm thick, lenses, breccia fillings, and nodules in altered porphyry. It ranges in color from light blue to dark sky blue, much of it showing a greenish tint. The deeper-colored material is high in silica and very hard.

Number 8 Turquoise Mine, Eureka County

Now located in the Blue Star open-pit gold mine in the Lynn district, the Number 8 Mine was worked by underground and open-pit methods since the 1920s. In 1954, a nugget weighing 70 kg was found. Until the development of the gold mine, the Number 8 Mine produced over 5 metric tons of turquoise valued at $1.4 million.

The host rock is variably altered monzonite and variously decalcified and silicified black limestone of the Popovich Formation. Much of the turquoise recovered prior to 1970 was as nuggets, some from placers. Veins and spider-webbed varieties were encountered in open pits, and in 1990 a solid 30-cm-wide vein of hard turquoise varying from light to dark blue and greenish blue was encountered. A single 25 kg boulder of spider-webbed turquoise was recovered at this time, but much of the turquoise uncovered in 1990 was sent to the waste dump.

Carico Lake Turquoise-Faustite Mine, Lander County

Located in the Carico Lake Valley north of Austin, this mine has been in production since at least the 1970s and produces from what may be the world's largest open pit excavated solely for the production of turquoise and faustite. Total production is reported at about 350,000 kg.

Turquoise and faustite occur primarily as nuggets in shear zones and along bedding planes, and much is spider-webbed. The turquoise ranges from pale blue to dark blue, and the faustite is a bright fluorescent-appearing yellow-green color. In the 1970s, faustite was held in low regard, but by the 1990s it became widely popular and valuable in the gem trade. Presently, this mine is one of the world's largest producers of faustite.

Ackerman Canyon Variscite Mine, Lander County

This deposit, located on the mountain west of Ackerman Canyon, north of Highway 50, was first worked in the 1970s by open pit and is now worked a few weeks each year. The variscite occurs in veins from 6 to 75 mm thick in black chert and gray shale. The veins are irregular but numerous, and occasionally nodules are found. The color of the variscite ranges from light to dark green, the best being a translucent emerald green resembling the highest quality Burmese jadeite. High silica content produces very hard material. To date, very little of the spider-webbed variety has been found here.

Apache Variscite Mine, Lander County

This mine is located on the western slope of the Toiyabe Range several kilometers north of Austin. It was discovered by Lee Britton of Fallon in the 1990s and has been continuously worked by him since. Much of the material was initially marketed as "Apache Turquoise," but was subsequently identified as variscite. This deposit contains unusually thick veins (to about 9 cm) of light-green variscite mottled with black inclusions, allowing it to be easily identified on the market. Zones of higher quality, darker-green variscite as nuggets have been uncovered at depth. This property is one of the world's largest variscite producers, having produced thousands of pounds of rough since its discovery.

Blue Gem Turquoise Mine, Lander County

This mine is located about 1.2 km north-northeast of Copper Basin, south of Battle Mountain in Lander County. It was discovered in 1934 by Duke Goff, and was developed by underground workings and later by open pit above these workings. High-quality turquoise occurs in veins up to 2 cm thick, closely associated with pyritic quartz veins in limonite-stained shear zones that cut argillized quartz monzonite. Total output from the mine is believed to have been over

$1 million in rough turquoise. It is now part of a large open-pit, gold-copper mine.

Damele Mine, Lander County

This mine, located east of Austin and just north of Hickson Summit, is currently active. The deposit is in commercial production from an open pit and has thus far produced an estimated $200,000 to $300,000 of gem materials. It is likely unique in the world for the production of the rare mineral faustite, along with variscite and chalcosiderite, for use in the gemstone industry.

Faustite is similar to turquoise except it contains zinc instead of copper as its primary anion. At this deposit, it is a bright, neon yellow-green color. Nuggets of faustite, variscite, and chalcosiderite are found in dark carbonaceous phosphate-rich shale. Faustite and variscite also occur as vein breccias, forming a much-desired spider-webbed variety. The variscite and chalcosiderite are bright emerald green in strong contrast to the black enclosing rock.

Faustite is very popular and in heavy demand by the gem industry, commanding $500 per pound for rough and $10 per carat for cut stones. Current demand exceeds supply because it has become a fashionable, but extremely rare, gemstone.

Fox Turquoise Mine, Lander County

This mine is located 2.5 km southwest of the mouth of Cortez Canyon. Native Americans knew the deposit; in 1914 Charles Schmidtlein developed it into a mine. Turquoise occurs along bedding planes in folded and faulted clay-altered chert and was mined from several open pits and numerous smaller cuts. Through the 1960s, this mine had produced more turquoise than all other turquoise mines in the state combined. For a time, more than 1000 kg per month were produced, with an estimated 250,000 kg total production. Much of the material produced was only of fair quality and required treatment for jewelry manufacture.

Godber Turquoise Mine, Lander County

The Godber deposit, which is 8 km north of Hickson Summit, east of Austin, was discovered in 1932 by Bob Burton and Joe Potts and was developed by open-pit and underground methods. Turquoise occurs as nuggets and veins in argillized gray, light-brownish-gray, and maroon shale. It is high-quality, medium- to dark-blue turquoise, and much of it is spider-webbed. Estimated total production exceeds $500,000.

Lander Ranch Chalcosiderite Mine, Lander County

Located several miles north of Tenabo near Lander Ranch, this mine is owned by Eddy Mauzy of Battle Mountain. This mine consists of open pits and produces high-quality, spider-webbed chalcosiderite consisting of rounded white masses circled by black lines. It is used in contemporary and Native American jewelry.

Super X Turquoise Mine, Lander County

Ted and Harold Johnson discovered this site in 1941 about 4 km northwest of Tenabo. It was originally developed by a 15-m shaft and later by open cuts. The mine has produced over $200,000 in high-grade, spider-webbed turquoise that occurs as veins and nuggets in a gouge zone in folded, sheared, and brecciated chert.

Taubert Turquoise Mine, Lyon County

The Taubert Mine near Mason Pass west of Yerington consists of several open pits. Turquoise is found as veins and nuggets in highly fractured zones in intensely clay-altered granite porphyry. The veins are commonly as much as 2 cm wide, and the nuggets are most frequently found in clay-altered gouge. The turquoise is hard, bright, and translucent, and ranges in color from deep blue to green with limonite dendrites.

Pilot Peak Turquoise Mine, Mineral County

The Pilot Peak Turquoise Mine is located on the eastern side of the Pilot Mountains near Troy Springs. It consists of numerous open pits and several tunnels. It was heavily worked in the 1970s, and sporadically in the 1980s and 1990s. Production is unknown, but is believed to be large. The turquoise occurs as veins and nodules in altered and fractured quartzite and quartz monzonite. A wide variety of blue and green colors and matrix patterns are found at this site.

Turquoise Bonanza Mine, Mineral County

Located on the eastern side of the Pilot Mountains about 8 km northeast of Pilot Peak, this mine was discovered in 1908, developed by Carl House in 1943, and subsequently sold to Ted Johnson. Johnson and others did considerable work on the deposit until the 1970s. The deposit, which has been developed by numerous open pits and short tunnels, is still producing. Total estimated production is around $400,000.

The turquoise occurs in veinlets, as breccia filling, and less commonly as nuggets in altered, fractured quartzite. Veinlets from 3 to 50 mm thick are recovered from the tough enclosing rock using explosives. A wide range of colors has been observed, including almost all shades of blue, green, and bluish-green, and the turquoise takes a good polish. Matrix patterns are varied and beautiful, including spider-webbing. Many veins are encrusted with pale to dark limonite that penetrates the turquoise, producing attractive color contrasts and patterns. There are more than 20 deposits that mostly occur in three linear zones, each containing a distinct variety of turquoise.

Easter Blue Turquoise Mine, Nye County

Located about 13 km northwest of the Royston mining district near Tonopah, this deposit was discovered in 1907 by Lew Cirac. The deposits are worked from open pits and underground workings. Total production is over 7000 kg of rough turquoise. The turquoise occurs as relatively thin veinlets and lesser amounts of nuggets along altered shear zones in a fine-grained white quartzite. Material from this deposit is of excellent quality; it is very hard, deep blue, and has beautiful matrix patterns. Blue turquoise predominates over green at this locality.

ZOISITE

The mineral zoisite is of interest to the gem trade as the pink variety thulite if it contains sufficient quartz to allow polishing. In stones without sufficient quartz, the fibrous mineral falls apart during cutting. Most thulite found in Nevada occurs in contact metamorphic rock, but small veins and pods can also be found in weakly altered granitic rocks. The following localities are noted for production of material that can be cut.

Four Clovers Mine, Douglas County

This deposit is located east of Gardnerville in the Pine Nut Mountains. David Smith of Castro Valley, California, discovered it in the 1950s. The material mined from this locality is a mixture of pink thulite and green diopside and is marketed as "Lapis Nevada." It occurs in a contact metamorphic deposit located between limestone and a granitic intrusion, and is associated with garnet, quartz, and a scapolite species. It is used for cabochons, beads, and carvings. The pink and green colors give the stone a "flower-garden" appearance, and as such it is particularly well suited for carvings and has been exported to Asian countries in large quantities. More than 110,000 kg of the stone have been produced from several open pits.

Ludwig, Lyon County

Zoisite occurs as small masses at the contact between granite and limestone in the Singatse Range. A small amount of cut-stone production came from this area.

Ryan Canyon, Mineral County

Massive thulite has been mined for the gem trade from a deposit located 8 km southeast of Thorne near Luning.

DESIGNER GEMS

This class of gemstone is limited to facetable materials that are too soft for general wear or are so rare that only a few stones have been cut. "Gem-grade" barite from the Meikle Mine in Elko County has been faceted into golden gems as large as 45 carats, and calcite from the same locality has also been cut. Creedite from the Hall Mine in Nye County has been fashioned into wonderful small purple gems. Vesuvianite from the Yerington district in Lyon County forms beautiful, but opaque, emerald-green stones, and realgar from the Getchell Mine in Humboldt County has also been cut. Unusual faceted stones have been made from clear quartz with red-orange spessartine inclusions from the Newberry Range near Searchlight in Clark County. In addition, cinnabar from near Lovelock in Pershing County has been faceted (Gray, 1999).

DECORATIVE STONES

Decorative stones are rocks that are suitable in hardness and appearance to be shaped into attractive objects. They consist of one or more minerals. Materials representative of this class of stone are wonderstone, obsidian, and some types of colored rock, principally metamorphic rock. Decorative material may be cut into cabochons or used for functional or artistic stone objects. Some notable Nevada occurrences of this type of material are listed below.

Wonderstone

Wonderstone is decorative rhyolite marked by the presence of Liesegang bands, colorful rings or bands of iron oxide. Several Nevada occurrences have gained recognition for the variety of colors and designs found in this material.

CHURCHILL COUNTY
Colorful swirled and banded rhyolite is found 22 km east of Fallon and north of Salt Wells on Wonderstone Mountain. Much of the material is porous and does not polish well. The site is open to rock collectors and has numerous small pits and prospects.

ELKO COUNTY
Large quantities of wonderstone occur near the old silver mines at Tuscarora.

HUMBOLDT COUNTY
Intricately patterned and banded rhyolite, locally referred to as "McDermitt Wonderstone," occurs in the hills northwest of McDermitt. The finest of this material takes a dull polish and

exhibits "pictures" as well as swirls and bands. Numerous small pits, some of which are under claim, mark this site.

MINERAL COUNTY

Some of the earliest commercial mining of wonderstone took place just north of Tonopah.

WHITE PINE COUNTY

Wonderstone is found in quantity at Little Antelope Summit on U.S. Route 50, 64 kilometers west of Ely.

Obsidian, "Apache Tears," and "Moonstone"

This material, which is volcanic glass, has been gathered since prehistory for use as decorative stones, jewelry, and tools. Most obsidian is black, but brown, green, red, and yellow varieties are known. "Apache tears" are small "droplets" of obsidian commonly found in volcanic tuff and perlite, or scattered about near where they were eroded from such rocks. "Moonstone" referred to here is obsidian that displays a play of color, or chatoyance, when moved or viewed from different angles. This material should not be confused with the feldspar variety of moonstone, which is also noted for the presence of chatoyance.

Nevada has numerous obsidian localities, particularly in the western part of the state. Most of these deposits are of interest to rockhounds, but obsidian has been mined commercially, particularly for the colorful and banded varieties. A few notable sites are listed below.

ESMERALDA COUNTY

"Apache tears" are common in the Fish Lake Valley southwest of Coaldale Junction. Here jet black to banded mahogany and brown types are found. The Goldfield Gem Claim is a fee-digging operation about 8 km west-northwest of Goldfield where "Apache tears" and numerous varieties of cryptocrystalline quartz can be mined and purchased by the pound.

NYE COUNTY

Abundant black "Apache tears" can be recovered by searching the light-colored ground near Crow Springs northwest of Tonopah. In addition, a vast field of gem-quality "Apache tears" is 13 km south of Scottys Junction on Route 95; most of this material has slight chatoyance.

WASHOE COUNTY

Obsidian and "Apache tears" are found in quantity between Route 140 and southwest to Cedarville, California. Here large quantities of material have been and continue to be collected by rockhounds and lapidary enthusiasts. Local and rare occurrences of "moonstone" have been reported from this area.

Marble, Travertine, Tufa, and Magnesite

Marble, onyx, and tufa have been used for decorative stone. They are mostly composed of the carbonate calcite, although dolomite occurs in some marble. Marble is a metamorphic rock that can range in color from pure white to shades of red, blue, green, brown, or black. The best-known Nevada locality, only mined for a short time, is near the abandoned site of Carrara south of Beatty in Nye County. Travertine, or onyx, is generally white to brown banded material. It forms in open fissures and caves where calcium carbonate is precipitated in layers from groundwater; it also forms in surficial spring deposits. Impurities give the precipitate its color. The Lehman Caves of Great Basin National Park in White Pine County contain travertine formations, but no collecting is allowed. Tufa is most frequently encountered in Nevada along ancient shorelines of Pleistocene lakes, where it may be found in fantastic shapes. The best-known tufa occurrences are along Pyramid Lake in Washoe County; however, collecting from these deposits is prohibited.

Magnesite, also a carbonate mineral, has been used for carving and other ornamental purposes, and has recently been marketed as "ivoryite." Sinkankas (1997) reported that the Currant Creek magnesite deposits, along the Nye/White Pine county line, produce nodules of "exceptional, very fine-grained and uniform-textured magnesite of carving grade."

Nevada Meteorites

DAVID A. DAVIS

Only four meteorite discoveries have been confirmed in Nevada. As of 1985, the surrounding states of California, Arizona, and Utah reported 26, 23, and 9 confirmed meteorite discoveries respectively (Mustoe, 1999). A fifth reported Nevada meteorite is apparently an Arizona transplant, and a possible sixth one is lost. Location and other data for these meteorites are in Table 12. More Nevada meteorites have probably been found but not reported, especially with the advent of prospecting with metal detectors, and many more likely are waiting to be found. This chapter will describe reported finds and possible areas and methods for finding more.

METEOROIDS, METEORS, AND METEORITES

It is beyond the scope of this chapter to describe every aspect of meteorites and how to positively identify them. Details can be found in books such as *Rocks from Space* (Norton, 1998), and most of the general information in this chapter is from that source.

The terms "meteoroids," "meteors," and "meteorites" are many times mistakenly interchanged. A meteoroid is the rock still in outer space; larger space rocks are referred to as asteroids. If the rocky material is mixed with frozen water, carbon dioxide, and other volatile material, the object is referred to as a comet. The vast majority of these meteoroids are dust or sand sized and travel at cosmic speeds exceeding 26 miles per second. A meteor is the flash of light seen as a meteoroid enters the Earth's atmosphere and burns up. Friction with even the rarified air of the upper atmosphere causes these particles to be heated to 1250°C and to burn up in a few seconds at altitudes of 60 miles. Larger particles may pass lower into the atmosphere with a more brilliant show, commonly referred to as a fireball that may be seen across several states. A meteorite is a meteoroid that makes it to the ground.

Contrary to popular belief, small meteorites do not create craters when they hit the ground. The Earth's atmosphere is quite effective in retarding the speed of incoming meteoroids. Although many variables are involved, a large portion of the meteoroid's mass is burned or broken off during passage through the atmosphere. A baseball-sized meteoroid will likely not survive passage to the surface. A meteoroid of up to a ton will likely be slowed to terminal velocity (the speed obtained as if it were thrown out of an airplane) and leave a depression that is not much larger than the surviving meteorite. Larger meteoroids will retain a portion of their cosmic velocity. Those with masses much greater than 10 tons are generally destroyed upon impact. The kinetic energy of a falling meteoroid is converted to heat energy upon impact, and for large ones, the kinetic energy is great enough to cause enough heat to vaporize the meteorite and surrounding rock, resulting in an explosion and crater formation. Also contrary to popular belief, small meteorites are not hot when they hit the ground. Space is very cold, and meteorites only spend a few seconds passing through the atmosphere. They tend to be relatively poor conductors of heat, and despite the high heat on their surface from friction and even the loss of most of their original mass, they can still remain cold internally. Meteorites recovered immediately after fall are usually only warm to the touch and are commonly still cold.

METEORITE COMPOSITION

Meteorites are broadly grouped into three basic types: irons, stones, and stony-irons. The nickel iron content for each type averages 98 percent, 23 percent, and 50 percent respectively. About 94 percent of recovered meteorites from witnessed falls are stones, which indicates that they are the most common type. Irons comprise 5.8 percent and stony-irons 0.2 percent of witnessed falls. However, because irons are heavy and metallic and just "look different" from surrounding rocks, and stones superficially look like terrestrial rocks, irons are the most commonly found meteorite type. Three of Nevada's confirmed meteorites are irons and one is a stone (Table 12).

Stony meteorites are divided into chondrites and achondrites. These classifications can be further subdivided, but it is beyond the scope of this chapter to summarize them all. The difference between chondrites and achondrites is that most chondrites have chondrules, and most achondrites do not. Chondrules are small spherical inclusions, usually of olivine or pyroxene, up to but rarely more than a millimeter across in the finer-grained matrix of the meteorite.

The largest classification of chondrites is the ordinary chondrites, which make up about 85 percent of all observed falls. The petrographic composition of ordinary chondrites ranges from 40 to 50 percent olivine, 15 to 25 percent pyroxene, 3 to 23 percent nickel-iron metal (5 to 25 percent of which is nickel), 5 to 15 percent plagioclase, and minor amounts of troilite. Other classes of chondrites, such as E-chondrites, R-chondrites, and carbonaceous chondrites, are based on chemical and petrographic differences from the ordinary chondrites and may contain magnetite, enstatite, pentlandite, pyrrhotite, serpentine, and very rarely quartz. Achondrites, which usually lack chondrules and generally resemble terrestrial mafic plutonic and basaltic rocks, are divided into at least six classes based upon petrography and chemistry. A subgroup referred to as the rare SNC achondrites are probably Martian in origin. Irons consist largely of an alloy of iron and nickel. Nickel concentrations range from 5 to 50 percent but are usually between 7 and 13 percent, forming the two minerals kamacite (4.5 to 6.5 percent nickel) and

taenite (more than 20 percent nickel). Irons have been classified structurally as hexahedrites, octahedrites, and ataxites. A more recent chemical classification using Roman numerals I through IV and letter designations based on the ratio of nickel to iridium is also employed. Hexahedrites and ataxites are rare, and the bulk of iron meteorites are octahedrites. All three Nevada iron meteorites are octahedrites.

Hexahedrites are low-nickel meteorites of pure kamacite. Upon impact, the crystal lattice of kamacite slips and forms twinning planes. When a hexahedrite is etched with acid, these planes form a pattern of several sets of closely spaced thin lines referred to as Neumann lines. Ataxites contain greater than 13 percent nickel and generally consist of a fine-grained mixture of kamacite and taenite, showing no patterns upon etching with acid.

Octahedrites make up about 85 percent of iron meteorites. These contain 7 to 13 percent nickel, and, because there is too much nickel to form pure kamacite and not enough to form pure taenite, octahedrites consist of an intergrowth of these two crystals. Both kamacite and taenite have cubic crystal structures, but kamacite is body-centered and taenite is face-centered, resulting in internal structural differences. As the nickel-iron mixture cools to form the octahedrite, taenite forms first. With continued cooling, solid diffusion of the iron and nickel atoms occurs, resulting in the growth of kamacite crystals in the shape of elongated plates or lamellae that truncate the corners of the taenite crystals. Ultimately, the kamacite and taenite form hybrid crystals in the shape of eight-sided or octagonal bipyramids, hence the name octahedrite. Cutting and etching of octahedrites show the lamellae forming a crosshatched pattern called Widmanstätten figures. Octahedrites are further classed according to the thickness of the lamellae, which roughly corresponds to nickel content.

Iron meteorites commonly contain inclusions of other minerals. Troilite is always present in varying amounts in the form of spherical nodules. Schreibersite is also always present, commonly as a shell around troilite nodules and sometimes as long needlelike inclusions next to kamacite lamellae. Cohenite forms with schreibersite but is only present when the nickel con-

TABLE 12 Nevada meteorite data

Minerals: Cohenite - Coh, Daubréelite - Dau, Graphite - Gr, Kamacite - Kam, Plessite - Ple, Schreibersite - Sch, Taenite - Tae, Troilite - Tro

Meteorite Name	Discovery Date	Discoverer	Mass	Dimensions (cm)	Type	Structural Class	%Ni	Minerals	Location (Lat/Long)	County
Quinn Canyon	1908	Unnamed prospector	1450 kg	110 × 90 × 50	Iron	Medium Octahedrite	8.9	Tae, Kam, Sch, Tro, Ple	38°10′N, 115°45′W	Nye
Las Vegas	1930	J. P. King	3045 g	12 × 8 × 8	Iron	Coarse Octahedrite	7	Tae, Kam, Sch, Tro, Coh, Gr	36°25′N, 114°45′W	Clark
C. C. Boak	1935	L. J. Murphy	24 kg	24 × 20	Iron	Unknown	n.d.	Unknown	Amargosa Desert near Rhyolite	Nye
Quartz Mountain	1935	J. T. Waldis	4832 g	15 × 12 × 7	Iron	Medium Octahedrite	7.9	Tae, Kam, Sch, Tro, Ple, Dau	37°12′N, 116°42′W	Nye
Hot Springs	1995	F. Keiper	4565 g	20 × 15 × 8	Iron	Medium Octahedrite	7.8	Tae, Kam, Tro	39°40′N, 118°58′W	Churchill
Majuba	1999	H. McCormick	330 g	8 × 5 × 5	Stone	Chondrite	n.d.	Unknown	40°38′N, 118°25′W	Pershing

tent is less than 7 percent. Carbon can also be present in nodules and rarely occurs as small, impure diamonds. Some iron meteorites contain silicate mineral inclusions. Stony-irons are rare and resemble brecciated volcanic rocks. They are generally classified into pallasites and mesosiderites. Pallasites consist of angular fragments and grains of olivine in a matrix of nickel-iron metal. The nickel-iron to olivine ratio may range from almost all nickel-iron to 1:2, but it is generally about 1:1. The nickel-iron has an octahedrite structure and composition and will display Widmanstätten patterns. Troilite and schreibersite inclusions are usually present.

Mesosiderites are polymict breccias with a roughly 1:1 ratio of nickel-iron and silicates. The metal averages 7 to 10 percent nickel and has an octahedrite structure and composition but is rarely in concentrations large enough to display Widmanstätten patterns. The silicates resemble pieces of achondrites and consist of plagioclase and calcium pyroxene with some hypersthene, bronzite, and rounded masses of olivine.

METEORITE APPEARANCE

Meteorites range in size and mass from dust to an iron meteorite measuring 2.8 x 2.8 x 1.1 m and weighing about 55 metric tons in Namibia, Africa. Most meteorites are pebble- to cobble- and rarely boulder-sized objects that are found by ordinary people looking for interesting rocks.

"Fresh" meteorites still have a fusion crust. As a meteoroid passes through the atmosphere and heats up, material melts and is ablated. Upon cooling, the remaining surface material forms a fusion crust up to several millimeters thick. Crusts on stony meteorites may appear glassy, but on irons they are commonly bluish-black and look like freshly welded steel. The crust may show striations, flow structures, and cracks. Fusion crusts are delicate and rapidly weather away. Old falls will likely lack a crust and have a weathered surface like most terrestrial rocks. Iron meteorites rapidly rust, and old falls may have a thick coating of rust or be completely oxidized.

Meteorites come in an assortment of shapes but tend to be rounded by passage through the atmosphere. A meteoroid may have rough edges if it broke up near the end of its passage or shat-tered upon impact. Meteoroids commonly tumble or spin, but if they otherwise remain oriented in one direction, they may end up with a rounded conical or pyramidal shape. Knobs of troilite or other nonresistant minerals will preferentially ablate upon passage through the atmosphere, leaving pits and holes. Shallow pits resembling the depressions that one's thumb would leave in clay are referred to as regmaglypts.

Dark, heavy, metallic-looking terrestrial rocks, slag, and fragments of bombs and iron tools are often mistaken for meteorites. The most commonly mistaken rocks contain abundant magnetite or hematite, especially if they are nearly black. A couple of easy tests can be applied in the field. Iron meteorites are commonly, though not always, attracted to magnets like magnetite. Hematite is not attracted to a magnet. Magnetite leaves a black streak, and hematite leaves a reddish-brown streak. Stony meteorites usually leave no streak, and iron ones may leave a faint metallic streak if they still have a crust. A magnet, streak plate, and hand lens are useful when looking for meteorites.

WHERE TO FIND
METEORITES IN NEVADA

Meteorites are relatively common though very few are ever recovered. It is estimated that meteorites weighing more than a gram fall at an annual rate of eight per square mile, and several hundred meteorites weighing more than a ton strike the Earth annually, though most fall in the ocean (Mustoe, 1999).

As noted above, meteorites are generally found because they contrast with the surrounding rocks and soil. Irons are most conspicuous because they are dark, heavy, and metallic. The metal in a meteorite can activate a metal detector. Nevada is mostly desert covered with sagebrush, sparse grasses, and other low-growing shrubs, which makes it easy to look for interesting rocks. Unfortunately, Nevada is also blessed with many dark rocks such as basalt and andesite, and magnetite is common in many areas. Most exposed rocks also develop a thin, dark-reddish-brown to almost black layer of rock varnish. In addition, 150 years of mining in some areas and 50 years of bombing in other areas have left slag and bits and pieces of metallic mining and other equipment and shrapnel that eventually weather and may superficially resemble iron meteorites.

When looking for meteorites, look for dark rocks that tend to be more rounded than surrounding rocks. Meteorites may also exhibit regmaglypts, which can be more distinct than normal irregularities in the surfaces of most terrestrial rocks. However, gas bubbles in dark volcanic rocks may be mistaken for regmaglypts. A fresh fusion crust may stand out as darker and glassier than the usual rock varnish. Meteorites do not contain quartz, which is common in many terrestrial rocks.

Possibly the best place to look for meteorites is on Nevada's playas. The centers of playas consist of light-colored fine sediments and alkali. Rocks are generally sparse and meteorites should be easy to spot. However, old falls probably would not survive the occasional playa flooding.

NEVADA METEORITES

Quinn Canyon Meteorite

In late August 1908, in some low andesitic foothills near Goat Ranch Springs on the western side of the Quinn Canyon Range, a borax prospector discovered an iron-nickel meteorite half buried in the soil in the Twin Springs district. The prospector cut off a few pieces of the meteorite with a cold chisel and returned to Tonopah, about 90 miles to the west (Jenney, 1909a). He sold his interest to Mr. Eugene Howell, a bank cashier, and left the area (Tonopah Miner, 1908).

Professor Walter P. Jenney, a geologist and mining engineer, was hired to find the meteorite and bring it back to Tonopah (Jenney, 1909a). The meteorite was found on a claim staked under the name of Mrs. Eugene Howell (Tonopah Daily Bonanza, 1908) after two automobile trips covering 430 miles. Later, three men with a freight wagon, a team of six horses, a derrick, and chain pulleys were sent out to retrieve the meteorite. The round trip took eight days. The meteorite was wrapped in sacking and brought to Tonopah, and with a crowd looking on, it was placed in the offices of the Tonopah Banking Corporation (Tonopah Miner, 1908; Jenney, 1909a).

At first, the meteorite was thought to be from a great fireball seen passing eastward from San Francisco to beyond Candelaria and Belmont on February 18, 1894 (Jenney, 1909b). However, later examination of the surface weathering indicated that the meteorite had lain on the ground for much longer than 14 years (Buchwald, 1975). Professor Jenney had two samples sent to Rochester and New York for analysis, and it was determined the meteorite was 5 to 10 percent nickel and over 90 percent iron (Tonopah Sun, 1908; Jenney, 1909a). Later, the meteorite was etched to show the Widmanstätten patterns.

Several institutions, including the Smithsonian Institution, made offers to purchase the meteorite. Ms. Jeanne Wier of the Nevada Historical Society attempted to raise funds to purchase it and keep it in the state but failed. It was purchased for $1 a pound by the Field Museum of Natural History in Chicago in 1909 and studied and described in detail by O. C. Farrington in 1910. The Quinn Canyon meteorite is presently on indefinite loan to the Fleischmann Planetarium at the University of Nevada, Reno.

Data for the Quinn Canyon meteorite are summarized in Table 12. The meteorite is broadly cone shaped with numerous regmaglypts that are generally 4 to 7 cm in diameter. The portion of the meteorite that was originally buried has undergone several millimeters of surface loss by corrosion resulting in subduing of the regmaglypts and the formation of numerous pits 5 to 15 mm in diameter and locally a surface Widmanstätten pattern. The ablation of troilite nodules has also resulted in numerous cylindrical holes 5 to 20 mm in diameter and up to 50 mm deep. The fusion crust is also still present in numerous places and the surface temperature may have reached 1400°C during passage through the atmosphere. The kamacite lamellae in the Widmanstätten pattern are about 1.1 ± 0.15 mm wide (Buchwald, 1975).

Las Vegas Meteorite

The Las Vegas meteorite was reportedly found in 1930 by J. P. King in a wash about four miles east of the Comstock Service Station on U.S. Highway 91 about 30 miles northeast of Las Vegas. In May 1939, Cliff Whitmore of Las Vegas sold the meteorite to the U.S. National Museum at the Smithsonian Institution. The meteorite has hammer and chisel marks and was said to have been heated by a blowtorch in an attempt to cut it. Signs of artificial heating, however, were not conclusively found.

The Las Vegas meteorite contains vesicular troilite nodules 5 to 15 mm in diameter. Some of the troilite forms slag-like cakes on the surface, indicating shock melting, with 1- to 30-mm-long veins of the once-melted material injected into the underlying taenite and kamacite. Associated schreibersite appears to have been partially melted, and cohenite and graphite have been melted and recrystallized into spherules 10 to 100 microns in diameter. The kamacite lamellae in the Widmanstätten pattern are about 1.9 ± 0.4 mm wide (Buchwald, 1975).

The Las Vegas meteorite is petrographically, structurally, and chemically similar to, and contains shock-melted features almost identical to, the Canyon Diablo meteorite of Meteor Crater in Arizona, which is 375 km east-southeast from where the Las Vegas meteorite was found. It is considered unlikely that two meteorites would have fallen with almost identical characteristics, or that such a large piece of the Canyon Diablo meteorite would have been thrown so far. Therefore, the original location story is suspect, and the Las Vegas meteorite is probably a transported piece of Canyon Diablo material (Buchwald, 1975).

C. C. Boak's Meteorite

Sometime before early 1935, Mr. L. J. Murphy was prospecting in the Amargosa Valley near Rhyolite and discovered what seems to have been an iron-nickel meteorite. Mr. C. C. Boak of Tonopah came into possession of this specimen and took it to Dr. J. A. Fulton, Director of the Mackay School of Mines, who also thought it was a meteorite and one of the most interesting he had yet seen. However, no tests or analyses were reported to have been conducted to confirm this. Mr. Boak was uncertain what he would eventually do with his prize and added it to his collection for the time being (Dake, 1935). That appears to be the last that was ever heard of this meteorite.

Mr. Boak was a 50-year resident of Tonopah, an officer in several mining companies, a 10-term Nevada state assemblyman, and an avid mineral collector and dealer. He died in 1954 (*Tonopah Times-Bonanza*, 1954), and the bulk of his mineral collection was donated to the Nevada State Museum. The Boak collection at this museum was searched, but unfortunately the meteorite was not found (D. Nylen, oral communication, 2000).

It is possible the rock was not a meteorite and was later identified as such. It is also possible Mr. Boak or his heirs sold or gave it away or kept it as a family heirloom. The location of its discovery was never reported, though any mining claims Mr. Murphy may have staked near Rhyolite could be clues.

Quartz Mountain Meteorite

On April 25, 1935, a group of men at the Manitouac mining property about 8 km southeast of Quartz Mountain were removing surface material to prepare for driving a tunnel. One of the group, John T. Waldis, discovered a meteorite lying on bedrock under 60 cm of surface detritus. The meteorite was analyzed and confirmed as such by the Nevada State Analytical Laboratory at the University of Nevada, Reno. Dr. V. P. Gianella of the Mackay School of Mines published a preliminary report in 1936. Mr. Waldis and Mr. Leslie Green donated the meteorite to the Mackay School of Mines Museum (Gianella, 1936c; Buchwald, 1975), and it is presently on display at the Fleischmann Planetarium at the University of Nevada, Reno. Unfortunately, the discovery site is now in a restricted area of the Nellis Air Force Range.

The Quartz Mountain meteorite is roughly cone shaped with one relatively flat smooth side and another convex side with numerous regmaglypts generally 15 to 20 mm in diameter. The surface has undergone only 15 to 20 mm of terrestrial weathering, and although most of the fusion crust is gone, the exterior likely retains its ablation-sculptured surface. The meteorite contains a deep 7-cm-long fissure that is partially filled with troilite. It is uncertain if the fissure formed during atmospheric deceleration or was a feature of the original meteoroid. Troilite also forms scattered nodules 0.2 to 1 mm in diameter, and the meteorite consists of about 10 percent daubréelite. Kamacite lamellae in the Widmanstätten pattern are about 1.1 ± 0.2 mm wide (Buchwald, 1975).

Hot Springs Meteorite

On March 8, 1995, Mr. Fred Keiper was riding his dirt bike near the southeastern flank of the Hot Springs Mountains. He stopped to rest and noticed an odd, dark rock eroding out of whitish rock on the side of a moderately steep canyon about 25 m deep. He brought it to the Fleischmann Planetarium where it was identified as an iron meteorite. It was sent to the Institute of Geophysics and Planetary Physics at the University of California, Los Angeles, for study. The Fleischmann Planetarium has since acquired the meteorite, where it is on display. It is sometimes referred to as the Eagle Rock meteorite (Fleischmann Planetarium, unpublished data).

A detailed list of minerals in the Hot Springs meteorite was not available to the authors at the time of this writing; however, visual examination indicates the presence of kamacite, taenite, and troilite. The Hot Springs Meteorite is rounded and appears to have no regmaglypts. It has a well-preserved fusion crust and is thought to represent a fall of just a few years to several decades old (Grossman, 1997; Fleischmann Planetarium, unpublished data).

Interestingly, on May 25, 1911, a series of five bright fireballs with accompanying sizzling sounds and explosions passed roughly west to east over Doyle, California, and Fort Sage Mountain. It was thought then to have hit Tule Peak about 30 km east of Doyle (*Nevada State Journal*, 1911; *Reno Evening Gazette*, 1911). If the path had actually been somewhat east-southeast, the fireballs would still have passed over Doyle, Fort Sage Mountain, and Tule Peak, and would have headed for the Hot Springs Mountains about 100 km from Doyle.

Majuba Meteorite

In the spring of 1999, Mr. Harold McCormick and a friend were prospecting with metal detectors for gold in the Majuba Placers just west of Rye Patch Reservoir. The metal detector was set to detect only gold and to screen out "hot rocks" (that is, rocks containing other metals or tin cans). Mr. McCormick had already found a very small gold nugget when his detector surprisingly buzzed on an odd-looking "hot rock." Mr. McCormick brought the rock to the Nevada Bureau of Mines and Geology. A visiting geologist who had worked one time as a meteoriticist identified it as a stony meteorite, and the rock was taken to Fleischmann Planetarium. From there it was sent to the University of Arizona at Tucson, Arizona, for study. A later search of the locality site produced no more fragments. The meteorite had been found in the berm along the edge of a gravel road and had obviously been moved, though likely no more than a few hundred meters, and it most likely fell in the area. The site is on a pediment gently sloping eastward from the Majuba Mountains about 1 km to the west. The Fleischmann Planetarium has since acquired the meteorite, where it is on display (Fleischmann Planetarium, unpublished data).

A complete report on the Majuba meteorite was not available to the author at the time of this writing. However, the meteorite is rounded with a chip missing from one end. It is heavily weathered, though it still has a partial fusion crust. An initial XRD scan made at the NBMG showed olivine and troilite. It contains small flecks of iron-nickel metal as well as chondrules. It contains an area that polished a little differently than the rest of the meteorite. This may be a clast, which suggests that the meteorite is part of a breccia (Fleischmann Planetarium, unpublished data).

The Goodsprings (Yellow Pine) Mining District, Clark County

JOHN C. KEPPER

"If the old Nevada theory still holds good that a mining camp is no account unless it is hard to get at, then Yellow Pine is possessed of claims to distinction, for it is very difficult to access," wrote J. V. Kealy of the Goodsprings (Yellow Pine) district in the *Scientific Press* of September 19, 1893. Further, he said "the climate is something awful. Nothing but sharp, shiny sand and brilliant sunshine, the thermometer ranges from 90 to 120, water four cents a gallon and everything to eat, drink or use packed 50 miles through sand hub deep makes an undesirable combination. But the gold is there and lots of it." Although gold production at the Keystone Mine lived up to Kealy's claim, the district is best known today for its production of zinc and lead.

The Goodsprings district is in the southern Spring Mountain Range immediately west of Goodsprings, Clark County, about 30 miles south of Las Vegas (Fig. 3). The district is some 30 miles long and 10 miles wide, and contains over 70 abandoned mines in addition to a number of prospect pits, short exploration adits, and shafts. In the early part of the twentieth century, it was one of the nation's major producers of zinc. For detailed descriptions of the orebodies and maps of the mines, the reader should consult Hewett (1931) and Albritton and others (1954).

The Goodsprings district was organized in 1882 as the Yellow Pine district and was named after The Yellow Pine Mining Company, which, in 1868, controlled a number of silver and copper claims in an area a few miles west of Goodsprings known as Porphyry Gulch. The earliest mining was done by Mormons at the Potosi Mine at the northern end of the district in 1856. But the galena ore contained too much zinc, making the lead flaky and unusable for bullets, and the operation was abandoned the following year. Mining

efforts shifted to the area due west of Goodsprings in the 1880s, and copper and gold dominated production. The principal gold producer was the Keystone Mine where wire gold occurred in clay-altered porphyry. To process this ore, a milling and cyanide operation was set up in the town of Sandy, just west of the Spring Mountains.

For many years, gray-white material accompanying the lead ore was considered waste rock and discarded on mine dumps. Fortunately, T. C. Brown, an engineer from Socorro, New Mexico, visited Goodsprings in 1906 and recognized this material as similar to zinc ore mined at Magdalena, New Mexico. The discarded rock was found to be hydrozincite and smithsonite, which together with hemimorphite made up zinc ore at most of the mines. Hydrozincite was later found to be the dominant ore mineral in the district. From 1906 to 1911 the milling process involved hand sorting to separate galena from zinc ore. In the *Los Angeles Mining Review* (September 14, 1910), this milling was described as "sacking the galena product." Galena was the source of both lead and silver from the district. After 1911, the Yellow Pine Mining Company constructed a mill in Goodsprings to separate the denser lead minerals from the zinc minerals, and zinc concentrate was really tailings of this separation. The Shenandoah mill at Sandy and the Milford Mine mill were subsequently set up to increase production for the district. Because some ore from the Potosi Mine was pure zinc carbonate, the owners constructed a kiln and the ore was calcined. This raised the percentage of zinc in the concentrate and reduced its weight, lowering freight costs. Unfortunately, the presence of hemimorphite in many of the ore zones at the Potosi and other mines produced poor results, and the practice of calcining ore was not used

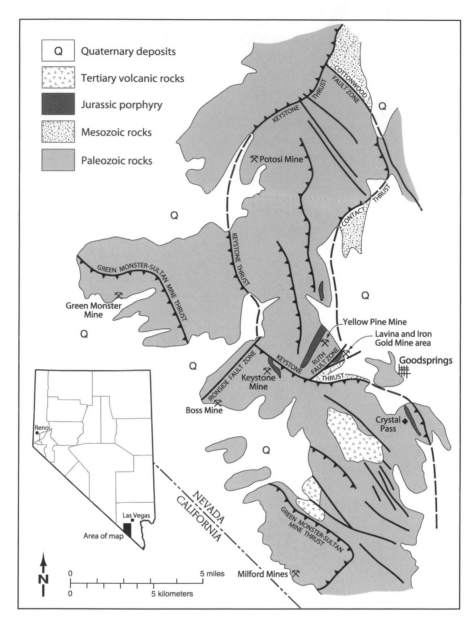

FIGURE 3 Generalized geologic map of the Goodsprings district

again. The separation of zinc and lead minerals from the carbonate host in low-grade ore was a persistent problem for the district that was never fully resolved. In addition to copper, gold, silver, lead, and zinc production, small amounts of cobalt, molybdenum, vanadium, and platinum/palladium ores were mined in the early 1900s.

Platinum- and palladium-bearing ores discovered at the Boss Mine in 1914 were of great inter-est, because carbonate rocks do not typically host these metals. Excitement over this discovery led to the laying out of the town of Platina on the flats just west of the Boss Mine (Paher, 1970). By 1916, Platina had a general store, a post office, and seven houses and numerous tents to accommodate the miners. However, the boom was over by 1918 when the owners found that the platinum and palladium could not be economically extracted, and Platina quickly became a ghost town.

Goodsprings concentrates were hauled by wagon to railheads some 75 miles away and shipped to Missouri or Salt Lake City for smelting and refining. The completion of the San Pedro, Los Angeles and Salt Lake Railroad to Jean, Nevada, in 1906 significantly reduced the high cost of shipping the ore to the smelters. Although ore grades were much higher than in lead-zinc districts in the Midwest and eastern United States, transportation costs were a big factor in slowing development of the Goodsprings district. The Yellow Pine Mining Company built a narrow-gauge railroad to Jean, further reducing shipping costs. Jean was the main railhead, but it was less costly for the mines in the northern part of the district to haul ore by wagon or later by truck to Arden, and for those in the south to Roach just over the California state line.

Access to the railroad was the beginning of a long history of zinc and lead production that peaked during World War I, but continued through several small mining cycles until 1952. At least five mines were operating in the 1940s to produce zinc for the strategic metal stockpile at Jean, Nevada.

GEOLOGY

The geology of the Goodsprings district is complex, and the timing of the various events recorded in the rocks is still being deciphered. Much of the following information on the geology is drawn from Hewett (1931), Albritton and others (1954), Carr (1983, 1987), and Kepper (2000). Stratified rocks in the Goodsprings district are dominantly Paleozoic limestone and dolomite, along with Mesozoic sandstone and shale with subordinate limestone and gypsum. The Cretaceous Lavina Wash sequence is a succession of sandstone, conglomerate, and volcanic tuff overlying folded Triassic redbeds in the upper plate of the Bird Spring thrust in the Lavina Wash area (Carr, 1983, 1987). Flat-lying Tertiary ash-flow tuffs occur at Table Mountain in the southern part of the district. Farther south, at Devil Peak, is a Miocene rhyolite plug surrounded by a narrow band of volcanoclastic sediments. Intrusive rocks are concentrated in the central part of the district where a series of dikes and sills of granitic porphyry are exposed (Fig. 3). The 780-foot-thick Yellow Pine sill in Porphyry Gulch is perhaps the largest of the exposed intrusions. These intrusions are early Jurassic (180 to 190 million years old; Garside and others, 1993). Hewett (1931) described minor intrusions of basalt in the central part of the district and an alignment of lamprophyre dikes (Singer, Puelz, and Star Mines) south of Columbia Pass. The ages of these mafic intrusions are unknown, but they indicate that the central part of the Goodsprings district was the focus of multiple episodes of igneous activity.

The northwesterly striking, high-angle Cottonwood fault that passes through Mountain Spring Pass marks the district's northern boundary (Fig. 3). South of this fault is a sequence of four major northerly striking and westerly dipping thrust faults. From bottom to top, these are the Bird Spring thrust along the base of the Spring Mountains, the Contact thrust at the base of distinctive steep limestone and dolomite cliffs, the Keystone thrust in the central part of the range, and the Green Monster/Sultan thrust. This structure is complicated by secondary thrusts, tear faults, and folds in each thrust sheet.

Hewett (1931) and Albritton and others (1954) proposed that the thrusts and associated tear faults functioned as conduits for the intrusions, and that thrusting preceded the emplacement of the porphyry. If this is the case, at least some of the thrusting occurred as early as Late Triassic. A second episode of thrusting probably took place in the Late Cretaceous, shearing and truncating the porphyry dike at the Keystone Mine.

AGE OF MINERALIZATION

Detailed mine mapping shows that both the intrusions and faults acted as permeability barriers to ore-bearing fluids. The orebodies appear to be post-intrusion and, therefore, post–Early Jurassic, possibly related to hydrothermal activity following the intrusion of the porphyry. The dikes and the sills may represent the higher portions of a deeply buried porphyry system. An additional reason for associating the ores with the emplacement of the porphyry is the zoning of metals and the distribution of the intrusions.

Virtually all outcrops of porphyry, as well as most of the gold, silver, copper, cobalt, vanadium, and platinum/palladium ore, occur in the central part of the district. Only minor amounts of these metals occur outside this area where ore fluids appear to have been guided by major thrust faults.

Additional support for a pre–Late Cretaceous age for the mineralization comes from copper-bearing garnet-diopside skarn at the Iron Gold Mine. This skarn and porphyry at the nearby Lavina Mine are incorporated in a fault slice within the upper plate of the Bird Spring thrust. Bird Spring thrusting pre-dated movement on the Keystone thrust (Carr, 1983), suggesting that copper mineralization took place prior to the Late Cretaceous.

AGE OF OXIDATION

The depth and extent of the oxide zone in the Goodsprings district indicates that it formed above a deep water table during a period of sufficient infiltration of meteoric water to thoroughly oxidize the primary sulfides. This deep oxidation may have taken place prior to the deposition of Tertiary volcanic and conglomerate units that unconformably overlie ore-bearing structures in the district. Latitic and andesitic volcanic rocks at Table Mountain, southwest of Goodsprings, rest on an erosion surface that truncates mineralized faults. Outcrops of cobble and boulder conglomerate, mapped as "Early Alluvium" by Hewett (1931), may be equivalent to Miocene quartzite-bearing fanglomerates and associated volcanic rocks described in the Kingston Range to the west of the district in California (McMakin and Prave, 1991). Subsequent erosion removed most of this Tertiary cover. The oxide zone in the Goodsprings district either developed during the formation of this Tertiary unconformity or formed during a period when the southern Spring Mountains were exhumed and Tertiary deposits eroded.

STRUCTURAL AND STRATIGRAPHIC CONTROLS OF ORE

Combinations of structure and stratigraphy appear to have controlled the movement of mineralizing fluids and the concentration of ore in the Goodsprings district. Mineralization in the porphyry is uncommon and most of the ore is hosted by Devonian and Mississippian carbonates, which include dolomite and limestone. Dolomite in the Cambrian Bonanza King Formation, which rarely contains ore, is stratigraphic in the sense that individual units can be mapped throughout the district as well as regionally. It probably formed by diagenetic dolomitization during the Cambrian and not during Mesozoic or later events. In contrast, dolomite is erratic in distribution in overlying Paleozoic units, such as the Sultan Limestone and the Monte Cristo Formation. Dolomite in the orebodies is typically light colored and coarsely grained, commonly occurring in bedded breccia zones or as bands of bedding-parallel white dolomite (zebra striping). According to Hewett (1931) and Albritton and others (1954), crystallization of this coarse dolomite preceded deposition of primary sulfide ore. Typically, ore occurs in the dolomite in preference to the adjacent limestone beds. Pre-ore dolomite in the Cambrian units may have been less permeable than the structurally prepared zones in the overlying limestone units. Thrusts and the associated secondary structures formed the plumbing system for the ore fluids, and dolomitizing fluids preceded the ore solutions, moving through the plumbing system and concentrating in the fractured limestone beds.

In detail, permeability differences played a significant role in the distribution of orebodies at the mines. Permeability barriers included porphyry sills and dikes, basal shales and sandstones of the Bird Spring Formation, and clay gouge developed along some fault zones. The Yellow Pine and Prairie Flower orebodies were localized in the Yellow Pine Member of the Monte Cristo Formation beneath either the Yellow Pine sill or the basal shales of the overlying Bird Spring Formation. The Como Dike, in the northern portion of the Yellow Pine Mine, functioned as a lateral barrier to the ore fluids. At the Potosi Mine, ore occurs against clay gouge along a thrust or, locally, beneath basal shales of the Bird Spring Formation. Albritton and others (1954) demonstrated that strike-slip movement along sinuous, high-angle faults created zones of intense fracturing and the localization of ore at the Yellow

Pine Mine. Similarly, strike-slip movement along sinuous northeast striking faults parallel to the Ironside tear fault zone in the upper plate of the Keystone thrust contributed to fracturing and the localization of orebodies at the Boss Mine and other nearby mines. The Ironside tear fault may be made up of older faults reactivated during thrusting and is not inconsistent with a pre–Late Cretaceous age for mineralization. At the Boss Mine (Fig. 3) intersecting northeast- and northwest-striking faults probably provided conduits for hydrothermal fluids.

Orebodies in the Milford Mine #3 illustrate how the mechanical properties of individual rock units influenced permeability. Milford #3 refers to three adits located at the western end of the ridge on which the Milford Mine (Milford #1 here) is plotted by Hewett (1931). A high angle, southern-dipping fault brings thinly bedded, cherty dolomites in the Anchor Member of the Monte Cristo Formation on the northern side up against massive dolomite of the overlying Bullion Member. The dip of the fault decreases to an angle close to the 30° dip of the beds where it crosses into the Anchor on the footwall and the thinly bedded dolomite is highly fractured. Most of the ore is in the Anchor with minor amounts in the massive Bullion.

To date the only significant mineralization found in the granitic porphyry is at the Keystone Mine (Fig. 3) where clay-altered porphyry contains gold. The porphyry was emplaced as a dike in the Banded Mountain Member of the Bonanza King Formation, just above the Keystone thrust. The dike is truncated at its northern end by the thrust. Minor quartz veinlets containing proustite and pyrargyrite were reported to occur in the Lavina Mine porphyry (Hewett, 1931). At Crystal Pass, a highly silicified porphyry dike and adjacent recrystallized dolomite contain abundant secondary silver minerals such as chlorargyrite.

PRIMARY ORE MINERALOGY

Because most of the deposits in the Goodsprings district are thoroughly oxidized, the nature of the primary minerals is poorly known. The primary copper mineral is chalcopyrite, typically rimmed by secondary chalcocite, and these were probably the chief copper ore minerals. Hewett (1931) reported small quantities of bornite and noted additional sulfide minerals including tennantite, proustite, stibnite, and cinnabar. Limonite after pyrite is common throughout the district, although the amount of pyrite in the primary assemblage varied considerably within and between orebodies.

Sphalerite and galena are primary sulfides at the lead-zinc mines. Sphalerite from the Potosi Mine (Fig. 3) is typically a medium to dark-brown color and occurs as rounded crystals to 1/2 inch that consist of tetrahedrons modified by dodecahedral faces. These crystals line white calcite-filled open-space cavities in breccia. Sphalerite also occurs as crystalline masses and disseminations in gray dolomite. Similar sphalerite occurs on mine dumps at Milford #1 and at the Green Monster Mine. Masses of galena, some as crude cubic crystals, are enclosed in sphalerite and hydrozincite ore at the Potosi Mine. Sphalerite veinlets that cut galena indicate that the latter formed first. Small amounts of finely crystalline pyrite occur in the carbonate matrix surrounding the galena and sphalerite. Takahashi (1960) reported greenockite associated with the sphalerite at the Potosi Mine.

Jedwab (written communication, 1999) and Jedwab and others (1999), who have been working on oxygenated platinum group metals from localities in Africa, Europe, and South America, are conducting studies on similar material from Goodsprings. Grains of individual mineral species as small as 35 angstroms in samples of bituminous material from the Boss, Azurite, Oro Amigo, Blue Jay, and Kirby Mines have been examined using SEM and EMP methods. The bitumen presumably came from carbonaceous limestone host rock and migrated into the ore zone during hydrothermal activity.

Bitumen from the Boss Mine contains elevated amounts of palladium, platinum, gold, and mercury. Grains of potarite and gold (including mercurian gold), which are considered as primary minerals, occur in the bitumen and as inclusions in quartz. Sulfide grains identified in the bitumen include pyrite, galena, arsenopyrite, bornite, covellite, pyrrhotite, acanthite, bismuthinite, and cinnabar. Associated secondary minerals include chalcocite, cinnabar, florencite (REE ratios not determined), plumbojarosite, argentojarosite, hydronium jarosite, arseniosiderite, and unidentified bismuth oxychloride species. Knopf (1915b) reported anatase crystals in the platinum- and palladium-bearing siliceous material, and this has been confirmed. P. Hlava (written communication, 1999) reported chromium-bearing hematite as inclusions in smoky quartz crystals in vugs in silicified ore from the Boss.

Bituminous material from the Azurite Mine contains an unusual suite of minerals including auriferous tocornalite, eucairite, and either moschellandsbergite or schachnerite (native alloys of silver and mercury). These minerals occur with native silver and iodargyrite, but platinum-group minerals, cinnabar, or plumbojarosite were not found.

HOST ROCK MINERALS

Granitic porphyry in the Goodsprings district is of mineralogical interest, because it is the source of well-formed orthoclase crystals to 5 cm long. Orthoclase weathers easily out of the rock, and crystals are collected by picking through the soil above exposures. Drugman (1938) and Gilmour (1961) described single and twinned crystals from the Crystal Pass area southeast of Goodsprings (Fig. 3). Carlsbad twins are the most abundant, but Baveno and Manebach twins are present. Plate 1 in Drugman's paper contains excellent illustrations of all of these, including various combinations of twins. The porphyry exposures at the Yellow Pine Mine and near the Lavina Mine are also sources of fine orthoclase crystals.

Small vugs lined with sharp, millimeter-size, golden-brown to grayish-brown andradite occur in the garnet-diopside skarn at the Iron Gold Mine. The garnets are dodecahedrons, in some cases modified by trapezohedrons. Small octahedrons of magnetite are present in the skarn at the Iron Gold Mine and in a recrystallized dolomite near the Monarch Mine south of Crystal Pass.

SECONDARY MINERALS

Weathering of a primary sulfide orebody by reaction with downward percolating oxygenated water results in vertical zoning of supergene secondary minerals. In many deposits, the zoning consists of gossan at the surface followed downward by a leached zone and an oxide zone. A sulfide-enriched zone may occur at and just below the water table. In the Goodsprings district the two upper zones may have largely been removed by erosion; however, the oxide zone is extensive and was the source of most metal production. There is no evidence of an enriched zone of secondary sulfides. Host rocks for the primary ore were carbonates that may create an alkaline environment unfavorable for the formation of an enriched zone of the chief secondary copper sulfide chalcocite. Copper was also immobilized in the oxide zone in the form of secondary carbonates and oxides, reducing the availability of copper to form a basal sulfide-enriched zone. Because of relatively low solubility, much of the galena remained in the orebodies surrounded by secondary minerals, and the lead that was released during weathering formed secondary minerals in the oxide zone. Sphalerite is relatively soluble, and zinc was either removed by groundwater or trapped in the oxide zone as one of the zinc-bearing carbonate minerals.

Oxide zone minerals typically occur in cellular openings in yellow-brown, dark-brown, or iridescent black goethite. At Milford #1 a shiny, dark-brown material composed of hisingerite (hydrous iron silicate) forms irregularly shaped shells and botryoidal crusts lining these openings.

Minerals of Copper-Bearing Oxide Ore

With the exception of the Blue Jay Mine, copper minerals in the oxide zone, although common, are not spectacular in the Goodsprings district. The dominant oxide subzone copper minerals are tenorite, malachite, and chrysocolla. Tenorite occurs as black to reddish-black, lustrous material typically rimmed with, or laced with veinlets of, malachite. Malachite appears as clusters of stubby, prismatic to acicular, dark-green crystals lining openings in boxwork goethite. The most interesting occurrence is at the Blue Jay Mine where malachite pseudomorphs after 1- to 5-cm azurite crystals are found. Sky-blue to greenish-blue chrysocolla is abundant at a number of mines in disseminated or veinlet form. Chrysocolla pseudomorphs after prismatic malachite occur at the Boss Mine. Dioptase occurs as millimeter-size, greenish-blue, prismatic crystals at both the Blue Jay and Green Monster Mines. Other oxide zone copper minerals identified in the district include aurichalcite, rosasite, brochantite, libethenite, nissonite, conichalcite, gilalite, and turquoise. The aurichalcite is present as radiating sprays of acicular, light-blue crystals, and is present at both copper mines and lead-zinc mines. Aurichalcite masses to 5 pounds were taken from the Yellow Pine Mine (Hewett, 1931). Rosasite forms bluish-green globular arrays of crystals, principally in the lead-zinc mines. Nissonite, a new mineral for Nevada, was recognized in 1996 at the Boss Mine, where this rare copper phosphate forms diamond-shaped plates of light-blue crystals. The first reported Nevada occurrence of gilalite, a very rare copper silicate, was also at the Boss Mine as green globules. Adams (1998) identified green globular arrays of needlelike crystals of conichalcite at the Singer Mine.

One of the most intriguing mineral occurrences noted by Hewett (1931) was a report of linarite crystals to 4 inches (10 cm) long from the Yellow Pine Mine (Fig. 3). The whereabouts of such material is presently unknown, but this may have been a world-class example of the species. A specimen of linarite from the Yellow Pine Mine is in the W. M. Keck Museum at Mackay School of Mines (Von Bargen, 1999). Linarite and caledonite were also reported at the Root Mine, but have not been observed there during recent collecting. Knopf (1915b) reported brochantite from the Boss Mine.

Cobalt and bismuth minerals occur in the oxide zone at copper mines in the Goodsprings district. Cobalt is present in the black, botryoidal mineral heterogenite. It is also the source of the pink color in some of the dolomite host rock at the copper mines. Heterogenite-3R has recently been reported from the High Line Mine (*Excalibur Newsletter,* May 2000). Bismuth carbonates (bismutite and beyerite) were reported from SEM/EDS studies of Boss Mine specimens (P. Hlava, written communication, 1998).

The jarosite mineral group is abundantly represented at Goodsprings. Jarosite, plumbojarosite, argentojarosite, hydronium jarosite, and beaverite have all been identified from mines in the central part of the district (Knopf, 1915a;

Hewett, 1931; Jedwab and others, 1999). These minerals are indistinguishable without X-ray or EDS analysis, because they typically occur as nondescript yellow-orange powdery to granular material. However, jarosite does occur as yellow-gray, light-tan, golden-brown, and dark-reddish-brown flattened rhombs at the Kirby Mine. Jedwab (written communication, 1999) noted the presence of Au, Ag, Hg, and Pb in Goodsprings jarosite, and reported that plumbojarosite from the Boss Mine contains zones with concentrations of Bi, Fe, Cu, Ca, Se, and Au. Jarosite minerals are best developed in the oxide zone at copper mines and probably reflect a greater abundance of pyrite and chalcopyrite there.

Gypsum does not appear to be common in oxidized copper ore at Goodsprings. This may result from intense silicification often associated with the primary copper mineralization that would, in turn, have reduced the availability of calcium during subsequent oxidation. However, Jedwab (written communication, 1999) noted tiny granules of native iridium in gypsum from the Boss Mine. Barite occurs as minor vein fillings at a few of the copper mines. Hewett (1931) reported masses of white alunite from the Kirby Mine.

Minerals of Lead-Zinc Oxide Ore

Oxidation of the lead-zinc orebodies at Goodsprings resulted in a wide spectrum of secondary minerals. The two primary minerals, sphalerite and galena, responded quite differently to oxidation. Sphalerite is much more easily oxidized than galena, and, in most of the mines, masses of galena are surrounded by secondary zinc minerals. The chief secondary zinc mineral is the hydrated zinc carbonate, hydrozincite. Although normal hydrozincite is colorless to white, it may be colored by various trace elements (largely iron or manganese) leading to shades of brown, pink, and purple. At the Milford #1 Mine, white to brown hydrozincite forms thin, highly contorted laminae in the host dolomite. The pattern is similar to rhythmic Liesegang banding found in iron-bearing weathered sandstone. The pattern is attributed here to the interdiffusion of zinc from oxidizing fluids and carbonate from the host rock source at a millimeter scale. Clear, flattened, paper-thin crystals of hydrozincite that line the interior of cavities in massive hydrozincite are not uncommon (Kepper, 1995). The hydrozincite has a striking bright-blue fluorescence.

Two other secondary zinc ores of importance are hemimorphite and smithsonite. Smithsonite is widespread in the district, but is subordinate in quantity to hydrozincite. At the Milford #1 Mine smithsonite occurs as white, botryoidal masses of fine radiating needles, often as the initial lining in a vug or cavity in the host dolomite. Rhombs of white smithsonite, attached to thin acicular prisms of hemimorphite and millimeter-size scalenohedrons of pale-pink smithsonite, have been identified optically from samples collected at the Yellow Pine Mine. Takahashi (1960) determined that hydrozincite orebodies were typical at shallow levels and smithsonite at lower levels. Hewett (1931) made a similar observation at the Yellow Pine Mine, one of the largest and deepest of the zinc producers. Ore specimens show smithsonite, followed by hydrozincite, lining the walls of vugs and of bedding-parallel openings. Central portions of these openings often contain clear crystals of hemimorphite, generally the last of the zinc ore minerals to form.

At the Milford #1 and #3 Mines, hemimorphite shows great variation in habit (Kepper, 1995). The earliest hemimorphite replaces host dolomite, forming sheets of flattened sheaves of platy, clear crystals along the walls of openings in breccia or vugs in dolomite. Later hemimorphite forms single and small clusters of water-clear, stubby, prismatic crystals, either perched on hemimorphite sheaves or as solitary crystals on goethite and hisingerite in cellular openings. Some of these solitary hemimorphite crystals have hopper structure resulting from rapid growth of the crystal faces in the [001] zone relative to the faces on the termination. The latest hemimorphite occurs as small acicular prisms perched on the earlier stubby ones. In the breccia zone at the 100-foot level in Milford #1 Mine, where hemimorphite is associated with plattnerite, the hemimorphite is strongly etched and partially replaced by hydrozincite. This suggests high pH and strongly oxidizing conditions. Willemite, another zinc silicate, is rare but was recognized by Takahashi (1960) and later by Adams (1998) at the Singer Mine. Willemite also occurs as crystal clusters in goethite at the 100-foot level of the Milford #3 Mine. It has distinctive hexagonal prismatic form and yellow-green fluorescence.

Oxidation of sulfides resulted in high sulfate concentrations in oxide-zone water, which combined with calcium from the host rocks to form gypsum as an early sulfate mineral at some lead-zinc mines. At the Potosi Mine, polished slabs of primary ore contain small amounts of gypsum along cleavages and fractures in sphalcrite (Takahashi, 1960). Specimens show veinlets of clear gypsum cutting both sulfide and host rock and, in one such specimen, the veinlet walls are lined with gypsum and the interior of the vein contains hydrozincite. Masses of white gypsum crystals mixed with hematite occur on the dump at the Milford #1 Mine.

Anglesite is present in limited amounts at a few of the mines. It is typically gray to grayish-brown, resinous-to-vitreous material that rims galena. Excellent microcrystals of anglesite are found at the Potosi and Contact Mines. These are generally water clear, locally slightly wine-colored crystals that line vugs or occur along cleavages in galena.

Much of the original anglesite was replaced by cerussite, the most common secondary lead mineral in the district. It forms gray to nearly black resinous to vitreous rims on, and nodular replacements of, galena. Clusters of clear to white or slightly yellow, well-formed, stubby to elongated prismatic cerussite crystals occur in the vicinity of galena at many of the mines. Hematite often forms a dark-reddish-brown to black coating on these crystals. Elbow and rare cyclic twins of cerussite occur at the Milford #1 and Yellow Pine Mines. Paragenetic studies on Milford Mine samples show that cerussite is replaced by mimetite and pyromorphite or by wulfenite. Bladed, dark-gray cerussite crystals to a centimeter long occur on the dumps at the Yellow Pine Mine.

Pyromorphite and mimetite occur in small amounts at most of the lead-zinc mines. White calcian pyromorphite is common throughout the district, ranging from stubby hexagonal prisms to very thin needles. Yellow and green pyromorphite typically appears in prismatic forms, but steep, tapering dipyramidal forms occur at the Milford and Singer Mines. Adams (1998) described white to yellow calcian pyromorphite consisting of barrel-shaped crystals showing subparallel growths with multiple terminations from the Singer Mine. He also reported the occurrence of light-green balls composed of acicular crystals of conichalcite from the same mine. SEM images by P. Hlava of yellow pyromorphite crusts from the Milford #1 Mine show dipyramidal pyromorphite crystals with small, cauliflower-shaped lumps of mimetite attached to crystal faces. Images of bright-orange crusts associated with cerussite rims on galena contain zoned hexagonal prisms consisting of an arsenate-dominated (mimetite) core and phosphate-dominated (pyromorphite) exterior. Hewett (1931) reported excellent prismatic, green and yellow pyromorphite crystals from the Root Mine; however, EMP analysis by J. McCormack indicates that this material is mimetite with elevated concentrations of phosphate. McCormack also noted the possible occurrence of heliophyllite, a rare oxychloride of lead and arsenic, in the yellow pyromorphite crusts from the Milford #1 Mine. Goethite after mimetite or pyromorphite in the form of thin, hollow prismatic shells occurs at several mines.

Excellent vanadinite specimens have come from the Ruth Mine. The mineral occurs as sharp, brownish-red to golden-brown hexagonal prisms with pinacoidal terminations on a matrix of finely crystalline light-brown to yellow chervetite or on black mottramite. Some of the vanadinite crystals are zoned along the c-axis with alternating brownish-red and golden-brown zones. Vanadates are also present as the minerals descloizite and mottramite. Descloizite occurs as brownish-black flattened dipyramidal crystals and mottramite as yellow-green, olive-green, and black crusts. J. Jedwab (written communication, 1999) noted that some Boss Mine mottramite contains bismuth. Lustrous dark-green microcrystals of mottramite are present at the Whale Mine.

The chromate fornacite occurs as yellowish-green crystals at the Yellow Pine Mine (S. White, oral communication, 1999) and as wedge-shaped plates, both as single crystals and as twins, at the Singer Mine (Adams, 1998). Fornacite forms wedge- to chisel-shaped, olive-green crystals associated with chlorargyrite, pyromorphite, and wulfenite at a silver prospect near Crystal Pass and as thin crusts of crystals on dolomite at the Alice Mine. I have collected very finely crystalline plates and small masses of a grayish-pink zinc-chromium mineral perched on orange wulfenite from the dump at the Milford #3 Mine. It was subsequently found in situ on the 100-foot level at the same mine and in minor amounts on the 100-foot level at Milford #1 Mine. The X-ray diffraction pattern (A. Roberts, Canadian Geological Survey, Ottawa) indicates a zinc-chromium mineral belonging to the hydrotalcite group, structurally very close to reevesite.

Wulfenite is one of the most interesting minerals found in the Goodsprings district. It has been identified at the Milford #1 and #3 (Kepper, 1995), Mobile, Whale, Yellow Pine, Prairie Flower, Ruth, Green Monster, and Potosi Mines. Wulfenite from the Mobile and Whale Mine is yellowish gray, and is quite platy. The largest wulfenite crystals from the district probably came from the Mobile Mine as crystal plates to more than a centimeter. At the Ruth Mine, bright-reddish-orange, tabular to stubby, pyramidal wulfenite associated with mottramite and vanadinite was once found in abundance. Unusual, acicular, yellow-gray to pale-brown wulfenite associated with hemimorphite occurs in specimens from the Potosi Mine.

Wulfenite crystals from the Milford Mines (Fig. 3) are normally in the 2 millimeter or less range, but single crystals to over a centimeter do occur. The upper adit at the Milford #3 Mine has produced wulfenite that consists of grayish-yellow, paper-thin, square plates, often in the range of 5 to 15 mm. Of all the Goodsprings occurrences, the Milford #1 Mine wulfenite shows by far the greater variety in form and color (Kepper, 1998). In the 100-foot-level stope at this mine, wulfenite occurs as translucent, lemon-yellow, paper-thin plates with familiar stop-sign morphology on dark-brown shells of hisingerite lining cavities in cellular goethite. Small, clear hemimorphite crystals are attached to the edges of some crystals. Orange, square to rectangular, tabular wulfenite also appears in some specimens. Rarely, zoning occurs in these crystals around a distinct red-orange center. Some crystals are clearly twins because of the presence of a pronounced re-entrant structure developed along the composition plane (Hurlbut, 1955). In the same stope, barrel-shaped, steep pyramidal,

TABLE 13   Minerals from the Goodsprings district

Acanthite	Chlorargyrite	Hisingerite	Plattnerite
Alunite	Chromite	Hydronium jarosite	Plumbojarosite
Anatase	Chrysocolla	Hydrozincite	Potarite
Andradite	Cinnabar	Ilmenite	Proustite
Anglesite	Conichalcite	Iodargyrite	Pyrite
Annabergite	Copper (native)	Iridium	Pyromorphite
Aragonite	Covellite	Jarosite	Pyrrhotite
Argentojarosite	Cuprite	Kaolinite	Quartz
Arseniosiderite	Descloizite	Kasolite	Rosasite
Arsenopyrite	Digenite	Libethenite	Rutile
Aurichalcite	Dioptase	Linarite	Scheelite
Azurite	Dolomite	Litharge	Scorodite
Barite	Dumontite	Magnetite	Sepiolite
Beaverite	Epsomite	Malachite	Siderite
Beudantite	Eucairite*	Massicot	Silver
Beyerite*	*Florencite*	Mimetite	Smithsonite
Bismuth	Fluorite	Minium	Sphalerite
Bismuthinite	Fornacite	Mirabilite	Stibiconite
Bismutite	Galena	Molybdenite	Stibnite
Boltwoodite	Gilalite*	Moschellandsbergite	Tennantite
Bornite	Goethite	or Schachnerite*	Tenorite
Brochantite	Gold	Mottramite	Tocornalite*
Calcite	Greenockite	Natrojarosite	Turquoise
Caledonite	Gypsum	Nissonite*	Uranophane
Carnotite	Heliophyllite*	Olivenite	Vanadinite
Cerussite	Hematite	Orthoclase	Vauquelinite
Chalcocite	Hemihedrite	Osmiridium	Willemite
Chalcopyrite	Hemimorphite	Pentlandite	Wulfenite
Chervetite*	Heterogenite	Platinum	Xenotime-(Y)

*Mineral species new to Nevada

tan-to-pale-orange wulfenite crystals occur on colorless hemimorphite that has been partially replaced by hydrozincite. Black needles of plattnerite occur as overgrowths on the etched surface of these pyramidal crystals. The barrel-shaped crystals are deeply etched along the crystallographic c-axis and in some cases are actually shells of wulfenite with the interior leached away. Translucent honey-brown, and rarely yellow-green, wulfenite displaying first- and second-order pyramidal forms in combination with pedions line cellular openings in goethite in a few specimens.

The 200-foot-level stope at the Milford #1 Mine contains an abundance of grayish-yellow, platy, layered wulfenite associated with plattnerite along a silicified fault zone. This wulfenite shows multiple cycles of precipitation and dissolution. The hemimorphic character (that is, different crystal forms at opposite ends of the crystal) of this wulfenite is indicated by arrays of small pyramidal crystals in parallel growth along one of the flat surfaces (positive pedion) and square etch figures on the opposite surface (negative pedion). A few crystals show small pyramidal growths on the opposite sides and represent twinned crystals (Hurlbut, 1955).

At this locality, plattnerite occurs as radiating sprays of fine black needles attached directly to plates of wulfenite or as inclusions in some of the wulfenite layers. These black needles are also found perched on calcareous fine silt- to clay-size sediment coatings on wulfenite crystals, and sediment coatings also appear to occur within the layered structure of individual wulfenite crystals. These sediment coatings suggest that the water table may have periodically risen to allow fine sediment to accumulate. Growth and dissolution (etching) of the wulfenite oscillated, perhaps with the position of the water table. Takahashi's work suggests that wulfenite may not be stable under the highly oxidizing conditions favorable to the formation of plattnerite, and etching cycles may equate to times when plattnerite was formed.

Secondary Carbonate Minerals

Secondary aragonite, calcite, and dolomite are well represented throughout the district. Aragonite occurs as pale-brown to white, finely crystalline, banded material associated with post-ore solution-precipitation processes. This banded aragonite serves as the base for centimeter-long, clear to white bladed prisms of aragonite. The aragonite has a yellow-green fluorescence. Excellent examples occur at the Potosi, Oro Amigo, Prairie Flower, and Milford Mines.

Calcite occurs as clear to white rhombohedrons perched on earlier secondary copper, lead, or zinc minerals. These rhombs are combinations of positive and negative rhombohedrons, are occasionally sharp, but are often rounded in appearance with step-growth patterns on each face. At the Milford #3 Mine the calcite is zoned with white cores and a clear exterior. Flattened rhombs or nail-head calcite to several centimeters also occur in the district. Scalenohedral forms are rare, but have also been recognized. Dolomite appears as sharp, clear rhombohedrons perched on the edges of acicular hemimorphite crystals at Milford #3 Mine.

Uranium Minerals

Unpublished reports prepared for the Atomic Energy Commission (AEC) in the 1950s, during the nationwide search for uranium resources, include descriptions of several mines in the Goodsprings district. Some of these reports are stamped with the acronym *SOM*, the code word for uranium. This is an artifact of the days when the uranium search was under the aegis of the Manhattan Project and anything related to development of the atomic bomb was secret (Chenoweth, 1998). Kasolite and dumontite were collected in one of the adits at the Green Monster Mine (Fig. 3) during this work (AEC unpublished reports; Albritton and others, 1954). Excalibur Minerals (Weissman and Nikischer, 1999) pictured a specimen of platy, yellow dumontite on brown kasolite from the Green Monster Mine. Millimeter-size bright-yellow sprays of radiating acicular crystals of boltwoodite were found at the Green Monster by Dick Thomssen (personal communication, 2002) and bright-yellow, platy kasolite crystals nestled within clusters of aurichalcite occur in a few dump samples from the same mine.

CONCLUSIONS

The Goodsprings district has provided mineral researchers and collectors with an extensive list of mineral species (Table 13). The potential for the discovery of mineral species new to Nevada or of entirely new species is very high for the Goodsprings district, particularly given the long list of trace metals available throughout the district. Most of the mines, or at least the mine dumps, are accessible; however, the district has not attracted large numbers of collectors because crystals of most mineral species tend to be in the millimeter or less size range. Mineral identification requires initial microscope work, followed by X-ray, SEM, and EMP analysis to make many mineral determinations. Nonetheless, I am sure that serious collectors will continue to be rewarded and that the list of mineral species from the Goodsprings district will lengthen over time.

MINERAL COLLECTING

PART III

A Personal Nevada Collecting History

FORREST CURETON

I started collecting in Nevada in 1951 and was introduced to collecting in the state by old-time collectors Jack Parnau and Norman Pendleton. My first trip to Nevada was in the spring of 1951 to the Getchell Mine in Humboldt County, and I continued to collect there for the next 30 years. On the first trip, it was possible to collect orpiment crystals to 5 cm in the open pit. During the following years I was able to collect and distribute many thousands of crystallized realgar specimens, and some of these were specimens to 45 cm with 5-cm realgar crystals. In 1972, I rediscovered getchellite and provided thousands of samples of that rare mineral. During the 1960s and 1970s, galkhaite was available in some quantity but in very small crystals; now superb crystals to 1 cm or more have been found.

Other early collecting was in the Yerington area. Fine specimens of spangolite, libethenite, cornetite, and other phosphates were readily available at various prospects and small mines. East of Yerington, on the eastern side of Walker Pass, I collected fine green vesuvianite in sharp crystals to 5 cm in diameter.

Another of my early collecting sites was the Nightingale district, Pershing County. Here, in the mountains approximately 11 miles west of Toulon, was a rather small deposit of clinozoisite crystals to 1 x 6 cm with crystallized quartz. The clinozoisite is pale greenish gray and relatively transparent. Hundreds of specimens to 15 cm in size were collected from this site.

I also collected, on a fairly regular basis, in the Goodsprings area in the 1950s and 1960s. I was able to go into the Yellow Pine Mine while it was still being worked on a small scale and found an abundance of very nice heterogenite, aurichalcite, smithsonite, hydrozincite, and rosasite specimens. In the Yellow Pine Mine, I visited the level where Hewett (1931) reported fine linarite crystals as much as 4 inches long. However, I was unable to find linarite anywhere on that level, only azurite. I never saw a sample of the linarite reported from this locality until 1999, when I was introduced to one with crystals to 1.5 cm in the W. M. Keck Museum at the Mackay School of Mines.

The Alice Mine, which is near the Yellow Pine Mine, provided fine white crystals of orthoclase by the thousands. It was also very easy to obtain matrix specimens of high quality there. The Green Monster Mine produced good specimens of the very rare mineral dumontite with kasolite, and in the 1950s these were readily available. The Boss Mine seemed to have an endless supply of sharp crystallized descloizite (variety cuprodescloizite) and crystallized plumbojarosite as well as other uncommon and unknown minerals.

The Azurite Mine on the southern slope of the hills southwest of Goodsprings provided exceptional malachite balls to 3 cm on matrix, as well as some azurite. Throughout the Goodsprings district, fine specimens of many minerals were found in oxidized ore. A few nice specimens of sulfide minerals such as sphalerite and galena were present, but these were rare.

Another of my early and long-term collecting sites was at the Myler Mine at Majuba Hill, Pershing County. Martin Jensen and Scott Kleine have described many of the minerals found there. As an addition to their observations, I would like to point out the two Nevada type minerals from Majuba Hill, goudeyite and parnauite, were found during my days as a collector and dealer. Also, in my early days I remember that we collected very good spangolite crystals by breaking boulders on the dump.

I was fortunate to visit the White Caps Mine at

Manhattan in Nye County during the 1950s when small-scale mining was under way. I was allowed to mine in any accessible areas and found some very high-quality specimens. Prisms of stibnite intermixed with cinnabar were plentiful, as were specimens of stibnite with realgar. I was able to collect some good crystallized stibnite, but the best of my specimens of that type came from the local store where a fine 30-cm crystal group was being used as a doorstop. Good specimens were still so common around town that they held no value. I also found hair orpiment (now known to be the species wakabayashilite) on the dumps with little effort. Realgar and orpiment crystals were also collectible on the dumps but were not of the quantity or quality of the Getchell material so they were of little interest.

The phenakite crystals at the Walker Mountain (Mount Wheeler) Mine should also be noted. They are glassy, clear to white or tan, and well formed to 1 cm, and were plentiful in the mine during the 1950s. By this I mean that you could collect twenty to thirty 1.5- to 8-cm specimens in 4 to 6 hours of work, but I remember that the drive was hard.

West of Imlay in Pershing County, superb clear quartz crystals to 8 cm and groups to 12 cm were found with sharp inclusions of stibnite crystals and of stibiconite after stibnite crystals. I believe the best of these were collected in the 1960s, but smaller specimens can still be found.

The Steamboat Springs sulfur deposit is about 500 m southwest of the junction of Highway 395 and the Mount Rose Highway in Washoe County. Many thousands of fine single sulfur crystals to 5 cm and crystal group specimens were collected in the 1960s and 1970s. The largest specimen I collected was approximately 45 × 75 cm with crystals to 3 cm.

The Black Diamond Mine east of Eureka in Eureka County had a number of watercourses or caves that were intersected at various levels (mostly between the 600- and 900-foot levels). Many of these provided exceptional specimens of aragonite as beautiful clear to white flos-ferri, but none were sharply crystallized. The following is an aside to show how many miners treated us in those days (in the early 1960s). The Black Diamond was shut down but was in operating

condition. This was my first trip to that mine, and I had never met anyone there. The watchman, who lived there, said it was much too far to walk into the caves so he loaned us a fully charged electric engine with a flatcar for all our gear and told us how to get to our destination. He told us to come to the house whenever we came out so he would know we were okay. On the way out, I got going too fast and derailed the flatcar. We had a terrible time getting it back on track.

I visited various mines and prospects at Pioche and Eureka only a few times, mostly with Jack Parnau who knew the area well. We collected fine specimens of mostly crystalline, pure atacamite, and also azurite with malachite. We usually stopped by these areas on our way to Utah.

Over the years, the Red Bird Mine, northeast of Lovelock in Pershing County, has provided a fair number of very nice crystallized cinnabar specimens, at times with crystallized stibnite. In addition, it is a type locality for the rare minerals robinsonite and dadsonite. Other small mines in the area also produced nice crystallized cinnabar.

The exceptional opal deposits in Virgin Valley, Humboldt County, must be Nevada's most important mineral specimen location when measuring value. These deposits have, for many years, produced some of the world's finest fire opal for gemstones as well as specimens. The gem opal in this area is primarily in the form of petrified wood and makes exceptional specimens. We have a 2-cm faceted fire opal from Virgin Valley in our collection. There is also a ledge of highly fluorescent common opal that has produced tons of student-teaching material.

The relatively new Zapot Mine, in Mineral County, produces gem and specimen pegmatite minerals, and is also the type locality for the new mineral simmonsite. This location should produce for years to come and is being operated solely for the specimen trade.

The shores of Lake Lahontan provide an endless array of petrified wood limbs and chips. The colors are not bright but some of the wood detail is exceptional.

The Gold Quarry Mine in Eureka County, while closed to general collecting as an operating gold mine, has produced an exceptional array of rare minerals. Martin Jensen has written an outstanding article on the locality (Jensen and others, 1995).

There are many other Nevada localities worthy of note, a few of which stand out in my mind. I have collected svanbergite crystals, pyrophyllite, andalusite, and diaspore from the Dover Mine, Mineral County; fine ram's horns of kalinite to 15 cm with sulfur from the Alum Mine near Silver Peak, Esmeralda County; naumannite in rich crystalline pieces to 6 cm from Seven Troughs in Pershing County; perhamite, morinite (cuprian), and turquoise, along with unknown species at the Gold Coin Mine, Humboldt County; owyheeite, pyrargyrite, and other minerals from the Morey district of Nye County; aikinite, krupkaite, and ilsemannite from Cucomungo Springs, Esmeralda County; bindheimite, stibnite, and stibiconite from the Arabia district, Pershing County; and calcite (variety thinolite), gypsum, and anhydrite from Pyramid Lake in Washoe County.

In summary, the state contains a plethora of mineral localities that have yielded excellent mineral specimens. I believe it is possible for a collector of any level to develop a fine collection of Nevada minerals.

Nevada from a Mineral Collector's Standpoint

MARTIN C. JENSEN

Nevada is a state that has generally been ignored by serious mineralogists and collectors since its statehood in 1864. Although there was a state mineralogist early in the state's history, his budget was limited, his knowledge slight, and his successes few. This was indeed unfortunate, for it was during these early times (1860–1900) that many of the rich and great ore districts were discovered and exploited. Mines in the Comstock, Tuscarora, Reese River, and Eureka districts were in full production, in high-grade ores, and innumerable exquisite specimens were undoubtedly lost or wasted. Later, during the revival of the Nevada mining boom at Tonopah and Goldfield (1900–1920), wonderful specimens were again exposed but tragically neglected.

During the 1930s, mineral specimen awareness increased due to the exploits of John Melhase and his publications in *The Mineralogist,* a former periodical of the Oregon Gem Society. Melhase toured Nevada widely and was responsible for noting a large number of specimen localities and for generating mineralogical interest in the state. Shortly thereafter, production of stibnite specimens from the now-famous White Caps Mine at Manhattan, and realgar and orpiment from the Getchell Mine, began to finally "put Nevada on the map" for collectors. A small occurrence of crystallized cinnabar at the Snowdrift adit of the Cahill Mine in Humboldt County was essentially the only other noteworthy contribution of the state to mineral collecting through the 1970s. However, with increasing worldwide recognition of mineralogy and its importance since that time, Nevada's significance as an area of interest to mineral collectors has become much better known.

Today, references on mineral specimen localities in Nevada are easily available, and the collector now has a wide range of enjoyable localities, both historic and current, that he or she may choose to visit. The following discussion presents a brief review of some of the more significant and important occurrences that should be on the Nevada mineral collector's list.

Surprisingly, the most spectacular and significant mineral specimens ever discovered in Nevada were exposed only recently at the Meikle underground gold mine in Elko County. Here, magnificent pockets and cavities of barite crystals of beautiful orange and yellow color, and of very large sizes, were found. Although the majority of these occurrences were quickly mined away or backfilled, many world-class crystals were saved. This incredible locality was described, and the natural beauty of its specimens shown, in a paper by the author (Jensen, 1999).

The White Caps Mine in Manhattan in Nye County probably ranks second in importance in terms of specimen beauty and number of pieces saved. This underground gold mine encountered well-crystallized stibnite in pockets between the 335- and 565-foot levels (Ferguson, 1924; Gibbs, 1985). Crystals comparable to the now readily available Romanian stibnite were encountered, but were not plentiful. Choice pieces were made available and offered for sale by H. G. Clinton, who collected during the 1920s and 1930s, and by Forrest Cureton, who collected during mining operations in the 1950s. Interesting associated minerals included small orpiment and realgar crystals perched on calcite, and also the world's finest wakabayashilite crystals. Some of the workings are still accessible today, although only small examples of selected species remain to be found.

Another locality that ranks with the White Caps Mine in terms of mineral specimen production is the well-known Getchell open-pit (and now underground) gold mine in Humboldt County. The widespread reputation of this locality as a source of realgar crystals, orpiment, getchellite (type locality), and fine galkhaite crystals is misleading. Good specimens of the above species have been found here, but the actual quantity of these pieces has remained quite small. The chief reason for this is the fact that few good pockets of specimens have been exposed. Stolburg and Dunning (1985) discussed the mineralogy of this deposit and provided general information. Recent mining has been confined to underground ore zones, where locally good realgar, calcite, and galkhaite crystals have been encountered.

Another enjoyable and interesting locality is the small Majuba Hill Mine, a former copper, tin, and uranium producer in Pershing County. This underground mine has produced a large number of rare copper and iron arsenate species, some in very attractive examples. In a paper by Jensen (1985), species such as clinoclase, olivenite, arthurite, pharmacosiderite, scorodite, chalcophyllite, torbernite, and zeunerite, all in very fine crystals, were discussed. Even though considerable field collecting has taken place here over the years, good specimens continue to be found.

For those who enjoy quartz crystals and pegmatites, granodiorite outcrops in the area surrounding Incline Village in Washoe County have produced some of the most surprising specimens. Dark-smoky-brown, well-formed single crystals to more than 65 kg have been collected here, along with a relatively small number of associated species such as microcline, albite, titanite, and schorl. These specimens are the largest and finest examples of quartz crystals found within the state. Jensen (1993a) described these deposits and their mineralogy, and an example of a large smoky quartz crystal from the area may be seen in the W. M. Keck Museum at the University of Nevada, Reno.

A favorite for many years to collectors is the spessartine/almandine garnet occurrence at Garnet Hill, near Ely, in White Pine County. The specimens are relatively small, but quite attractive. The garnet occurs as dark-red-brown to black, lustrous crystals perched inside small lithophysal (gas) cavities in pink rhyolite. Abundant specimens have been collected here over the years, and a brief paper by Hollabaugh and Purcell (1987) provided an account of the deposit.

There are perhaps a dozen or so additional known specimen localities of primary interest across the state, ranging from quartz epimorphs after barite to extensive assemblages of secondary phosphate minerals; unfortunately, these have not provided a large enough number of pieces to the collecting market to become well known. Current large-scale gold mining activities, predominantly in the northeastern portion of the state, continue to encounter occurrences of well-crystallized minerals and will likely provide future important discoveries. By reading through the present volume, however, the collector will be able to locate and select occurrences of specialized interest that should fill the needs of his/her collection. Nevada is a state with a very wide diversity of mineral species, and access for collectors is generally very good. One need only have the time, physical ability, and perseverance to assemble a collection of which any student of mineralogy could be proud.

Mineral Collectibles from Nevada

SCOTT KLEINE

Nevada is becoming known as a premier state for connoisseur-quality mineral specimens, and I believe that this recognition is justified. A series of recent discoveries, including the world-class barite at the Meikle Mine, huge stibnite prisms from the Murray Mine, and superb gemmy orpiments from the Twin Creeks Mine have put Nevada minerals in the collector's spotlight. I feel this is only the beginning. There are a variety of other Nevada localities where exceptionally fine pieces have been produced, but have been overlooked by many serious collectors. It is my goal to change this by increasing awareness of some of Nevada's better specimen-producing sites. Listed below are 35 of my personal favorites. The following locality and specimen descriptions have been compiled from personal observations, direct communication with other field collectors, or from photographs. Many of these sites have been previously described. It is not my intent to rehash published information but to provide updated information and additional locality history. Note that no permission is given or implied for collecting at or gaining entry to any of these sites, many of which are either privately held or extremely dangerous, even life-threatening, to enter.

Barite Outcrop, Tybo District, Nye County

A deposit of quartz epimorphs after prismatic, sometimes doubly terminated, barite crystals reaching 15 cm in length was found in a hillside outcrop near the town of Warm Springs. The deposit was discovered in the mid-1990s and quickly exhausted. Most specimens were found as loose shells formed around single crystals, with only sucrosic drusy quartz casts where these

large barite crystals used to be. Rare pieces consist of beautiful groups to 30 cm on matrix. Unreplaced grayish-yellow, blocky barite crystals on and off matrix, in sizes up to 8 cm, were also discovered (L. Griffin, oral communication, 1996).

Burrus Mine, Pyramid District, Washoe County

The long-dormant Burrus underground mine, near the northern end of the Pah-Rah Mountain Range, once tapped rich copper and silver ore. In the main stope, 20 m deep and only 1.5 m wide, what may be the world's best richelsdorfite was discovered in 1991. The Canadian Geological Institute identified this mineral. During collecting trips from 1991 to 1995, I found neon-blue, gemmy, tabular-hexagonal crystals to 1 mm of this very rare secondary copper arsenate; however, the mineral was most often found as pulverulent vug coatings formed by oxidation of enargite. I also found fine examples of chalcophyllite with the largest crystals, in a flat-lying gemmy spray, measuring an impressive 1 cm. Some chalcophyllite was found with 1- to 3-mm cyanotrichite crystals. Specimens of golden pyrite octahedrons to 5 mm with terminated, black enargite crystals up to 1 cm growing atop them were also collected.

Candelaria Mine, Candelaria District, Mineral County

The Candelaria open-pit mine on the southern edge of Mineral County was mined in olden days, mostly for silver and high-quality turquoise. Modern-day mining, which ended in 1997, produced large amounts of silver. In a very unusual occurrence, the secondary nickel arsenate species annabergite was discovered in fine, sharp, lustrous emerald-green plates to 5 mm (M. Jensen, oral communication, 1993).

Chalk Mountain Mine, Chalk Mountain District, Churchill County

At Chalk Mountain, an island of rock on the southern edge of Dixie Valley, is an oxidized base-metal deposit that was explored by a deep underground mine. The best-known mineral specimens from here have 1- to 5-mm-long, acicular, freestanding crystal sprays of arsenic-bearing vanadinite (endlichite) on thin crusts of crystalline orange descloizite. Lesser-known specimens from this locality are wulfenite crystals to 1.2 cm that nearly rival specimens from the Red Cloud Mine of Arizona in quality. A few specimens of fine leadhillite crystals to 2 cm have also been collected here (L. Griffin, oral communication, 1996).

Comstock Lode, Comstock (Virginia City) District, Storey County

This, the most famous and historic of all districts in Nevada, actually produced very few silver ore specimens, as the miners did not keep much for posterity. The best of what was kept must certainly be the superb, near Kongsberg-quality, wire silver specimens. These are extremely rare, and it is unclear how many were saved. The best is a 7-cm piece pictured in Bancroft (1984); other fine specimens are shown in Witters (1999). Acanthite, polybasite, pyrargyrite, and stephanite are other collectible silver-bearing species from here, all very rarely seen in collections. Modern-day field collecting on mine dumps in Virginia City yields white quartz crystals, as well as rare, prismatic (sometimes boldly phantomed) *amethyst* crystals to 6 cm in singles and groups. All of the underground mines are very dangerous and should not be entered. In the late 1990s, two people died after being overcome by bad underground air.

Dauntless Mine, Robinson District, White Pine County

Near the Garnet Hill Recreation Area, a small adit opens up into a series of drifts and winzes where an oxidized base-metal deposit was explored. Aurichalcite, hemimorphite, and smithsonite are found in abundance throughout the mine as smears, fibers, and small crystals in a matrix of massive orange limonite. In 1994, I found a hemimorphite pocket measuring 10 x 15 x 20 cm that was densely lined with thick, well-terminated, snow-white, lustrous crystals to 2 cm. I also collected aesthetic specimens of aurichalcite associated with hemimorphite. This is likely the best Nevada occurrence of hemimorphite and one of the better North American localities for the species.

Dee Mine, Bootstrap District, Elko County

In 1999, in this active underground gold mine on the northern Carlin Trend, a 2-m pocket was found to be loaded with many fine barite crystals to 5 cm. A collecting recovery specialist was called in to harvest specimens of sharp, lustrous, gemmy, prismatic crystals with a brown-gold color on matrix plates to 1 m (S. Werschky, oral communication, 1999).

Douglas Hill Mine, Yerington District, Lyon County

The long-abandoned Douglas Hill underground mine recovered high-grade copper ore from a skarn deposit. In one stope, many collectors have found superb specimens of brochantite. The ones I found have two distinctly different forms: acicular crystals in 0.2- to 3-cm balls; and 0.1- to 1.2-cm, single, flat-lying crystals. Both varieties occur on solid garnetite that is coated with either pseudomalachite or opaline chrysocolla. In 1995 or 1996 the stope collapsed, probably sealing off the site forever to collectors.

East Gold Pick Pit, Gold Bar District, Eureka County

At an elevation of about 3000 meters in the Roberts Mountains is the small East Gold Pick pit. In 1993, during a Mackay School of Mines field trip, I collected thick calcite-filled voids lined with terminated orpiment crystals. During the trip I talked with a geologist who showed me a 10 x 5 x 3 cm crystal. I do not know of another locality in the world that has produced such large crystals of this species. Specimens from this locality are distinctive, because they have rough, dulled faces and pleasing orange color, and many have twisted prismatic forms.

Garnet Hill Recreation Area, White Pine County

The federal government has set aside an area a few miles west of Ely for recreational use and public mineral collecting. The area is underlain by flow-banded, pink rhyolite. In certain places, abundant open vesicles can be found, and some of these contain vapor-phase almandine/spessartine garnet of exceptionally sharp crystal form and stunning luster. These garnet crystals, which are almost black to maroon-red, average 0.5 to 1 cm in size. Rarely do they exceed 2 cm and are extremely attractive and desirable specimens. In most cases, only a single garnet crystal occurs in each quartz-lined vug; less commonly, six or more crystals may be present. Associated minerals in these vesicles include sanidine, stibiotantalite, and tridymite.

Getchell Mine, Potosi District, Humboldt County

This is one of Nevada's classic mineral localities, best known for the country's finest specimens of realgar. Over the last few decades, fine realgar crystals in calcite, as well as specimens of orpiment and tiny crystals of the rare species galkhaite, came from the open-pit gold mine. In 1965, the type species getchellite was first described from here. Recently, mining activity went underground, digging deeper into this prolific deposit.

During 1997, at a depth of a little more than 300 meters, an incredible find of blood-red galkhaite crystals to 6 mm was made. The fine color and richness of this pocket is unsurpassed by any other locality in the world. Some of these specimens show associations with barite, calcite, realgar, stibnite, and stilbite. Other galkhaite finds have since been made at deeper levels, with

crystals to 1 cm, but these are dark red to metallic silver in color with little or no transparency. Studies of these crystals at the Mackay School of Mines have attributed the color difference to relatively high thallium and low mercury contents when compared to the blood-red variety.

As underground mining continued, local pockets of superb realgar were discovered. The best was found in 1997 and consisted of a 1-m void lined with stunning red, freestanding crystals to an amazing 7 cm, associated with lustrous, white, highly modified poker-chip calcites to 5 cm. A few deep pumpkin-orange orpiment crystals to 1.5 cm of high luster and gemminess were also found in one pocket (D. Tretbar, unpublished data).

Gold Quarry Mine, Maggie Creek District, Eureka County

The Gold Quarry Mine is about 40 km northwest of Elko in the Carlin Trend. This very large open-pit gold mine is owned and operated by Newmont Mining Corporation and is the company's flagship operation. Between 1990 and 1993, world-class specimens of the following species were recovered: fluellite as clear, lustrous crystals to 1 cm; hewettite as superb freestanding crystal sprays to 5 cm; and kazakhstanite as drusy crystals to 0.5 mm (M. Jensen, oral communication, 1993). While working for Newmont in 1994, I collected a few superb specimens of volborthite in lath-like crystal groups to 1.2 cm, some with malachite and azurite, and a piece with a patch of sharp, idiomorphic 1-mm volborthite crystals.

Goldstrike Mine, Lynn District, Eureka County

Located about 30 km northwest of Elko, this gigantic open-pit gold mine is the largest in Nevada. As of 1999, Barrick Gold Corporation was mining more than 500,000 tons of ore and waste from it every day. To date, around 150 mineral species have been identified from this pit by G. Ferdock (unpublished data), who reported the following examples: barite as brilliant golden crystals to 2 cm in association with 1- to 3-mm stibnite sprays, carbonate-fluorapatite crystals to 0.5 cm; greenockite as prismatic crystals to 3 mm; jarosite as sharp brilliant crystals to 0.5 cm; millerite in crystal sprays to 0.5 cm; orpiment crystals to 1 cm; realgar crystals to 3 cm; stibnite as superb crystals to 6 cm; whitlockite as lustrous rhombs to 3 mm; and variscite as gemmy emerald-green blocky crystals to 2 mm. Other unusual finds include fine examples of marcasite and sphalerite as pseudomorphs after millerite.

Hall Mine, San Antone District, Nye County

This mine is a cluster of medium to large open pits, where mining of molybdenite took place until 1991. Near the bottom of one of the pits, the nation's best specimens of lavender-colored creedite were discovered in the mid-1990s. The finest examples had sharp, gemmy, deeply colored crystals to 2 cm long occurring as very aesthetic radial groups and singles on plates to 30 cm (M. Jensen, oral communication, 1994).

Incline Village Pegmatites, Washoe County

Randomly scattered in the mountains east of Incline Village are rare bubblelike, simple pegmatite pockets up to 2 m in diameter that have produced immense black to smoky quartz crystals of fine luster and form. Prismatic crystals to more than 140 kg and 1 m in length are among the largest from these finds. In addition, specimens of perthitic feldspars to 20 cm, needlelike schorl, and rare hydroxylapatite crystals to 1 cm were also recovered (M. Jensen, oral communication, 1993).

Julie Claims, Pamlico District, Mineral County

In the early 1980s, Mike Smith found a superb locality for epidote crystals while working on his master's thesis and named the property after his wife. Mike collected sharp, deep-green, mirror-lustrous crystals to 17 cm in length. The specimens rate among the finer epidote specimens found in the United States. In the early 1990s, Harvey Gordon acquired the claims and began a brief but rewarding recovery project with heavy equipment. Fine crystals were collected, some with very aesthetic associations of prismatic quartz. The best specimens measured as much as 10 cm long and 5 cm thick.

Ken Snyder Mine, Gold Circle (Midas) District, Elko County

The Ken Snyder underground mine is working a very unusual gold-silver deposit. One of the primary ore minerals is the relatively rare species naummanite (Ag_2Se). However, very few crystals of this species have thus far been discovered. Another silver- and selenium-bearing species, aguilarite, has been recovered in specimens to 5 cm, with 1- to 12-mm intergrown crystals, most displaying a semi-melted look. Non-crystalline nagyágite, aguilarite, naumannite, and native gold have been found together in the form of rich ore bands in white quartz veins (M. Hunerlach, oral communication, 1999).

Lone Tree Mine, Buffalo Mountain District, Humboldt County

This mine is near Interstate 80 between Battle Mountain and Winnemucca. In 1995, soon after it was opened, specimens were found with 0.25- to 0.75-cm crystallized balls of satinlike, orange, translucent to transparent rhodochrosite perched very aesthetically on and with 1-cm, tabular, white barite crystals on a matrix of black manganese oxide. Crystals of white tabular barite to 2 cm, green pseudocubic pharmacosiderite to 3 mm, and glassy jarosite crystals have been collected since (J. Leising, oral communication, 1996).

Majuba Hill Mine, Antelope District, Pershing County

One of my favorite collecting localities is the Majuba Hill Mine located about 65 km northeast of Lovelock. It has had a long and illustrious history of recovery of superb specimens of several arsenate species. The Copper stope, a large chamber about 60 m long, 30 m wide, and 2 to 20 m high, yielded a total of 75 species, including arthurite, goudeyite, parnauite, and strashimirite as world-class specimens. Other minerals, such as agardite-(Y), chalcophyllite, clinoclase, pharmacosiderite, olivenite, scorodite, and zeunerite occur in specimens that rival those found anywhere in the United States. Some colorful associations have been found in only this stope, including arthurite with pharmacosiderite, scorodite and chalcophyllite, chalcophyllite on adularia, zeunerite with olivenite, and strashimirite with olivenite. During my many collecting trips to this mine, the maximum crystal sizes I have measured for the more important arsenate species include the following: agardite-(Y) as acicular sprays to 3 mm, arthurite in crystallized balls to 5 mm, chalcophyllite as superb crystals to 1 cm, clinoclase in crystallized balls to 2 cm, goudeyite as acicular sprays to 2 mm, olivenite as prismatic crystals to 3 cm, parnauite in crystallized balls to 4 mm, pharmacosiderite as pseudocubes to 4 mm, scorodite in crystallized balls to 4 mm, acicular strashimirite in balls to 1 cm, and zeunerite as tabular crystals to 8 mm. In addition, native copper and native silver have also been found in this stope.

In the Tin stope of the mine, the world's best formed crystals of clinoclase occur as 0.1- to 0.9-cm prismatic, well-terminated, sharp, mirror-lustrous, gemmy crystals in highly aesthetic, divergent sprays, some on cornubite-coated matrix and others with pale-green olivenite fibers. The Middle adit, which is directly below the Copper stope, has produced a variety of rare, secondary copper sulfate species: brochantite as green crystals to 2 mm, chalcanthite as single crystals to 1 cm and stalactitic growths to 10 cm, connellite in acicular sprays to 0.5 mm, langite as superb crystals to 1 mm, and spangolite as barrel-shaped crystals to 3 mm.

At the time of this publication, the Majuba Hill Mine is continually becoming more dangerous. Improper and unsafe collecting, as well as vandalism, has left the entrance to the Copper stope and the Middle adit unstable. The wooden supports under the Copper stope are being deformed by excess strain. In time, this stope and all inside of it will be sealed off forever by collapse. It is good to know, though, that because of decades of past field collecting at this mine, a fine supply of excellent specimens will remain available for generations to come.

Meikle Mine, Bootstrap District, Elko County

This active underground gold mine on the Carlin Trend is likely the finest golden barite locality in the world. Starting at the 1075-foot level, gigantic pockets, some measuring tens of meters in any direction, were found to be completely lined with amazingly sharp, gemmy, mirror-lustrous barite crystals to 10 cm, with tabular-blocky form and exquisite, deep-gold to cognac colors. Other cavernous voids, such as the so-called George Bush pocket, contained striking (though less gemmy and sharp) spear-shaped barite crystals to the incredible size of 20 cm, dramatically perched on mounds of white to tan-yellow rhombic calcite. From 1995 to 1998, top-quality pockets containing barite crystals to more than 15 cm were recovered in groups to 40 cm across. Even more intriguing were eyewitness reports by miners of perfect crystals to more than 50 cm in unrecoverable portions of these pockets that were seen briefly before being destroyed by the gold mining process. Sad as this may be, many fabulous, world-class specimens were recovered by miners and professional specimen recovery projects before the pockets were lost.

Besides these incredible barites, fine specimens of lustrous, sometimes gemmy, scalenohedral calcite crystals, some twinned along the c-axis and others modified with rhombic faces, were found as crystals to over 50 cm. In a few pockets, smaller, lustrous calcite crystals occurred in association with gemmy, golden barites. Other pockets were found to contain calcite crystals with either an unusual green color, or noticeable, tightly stacked, gray carbon phantoms. More notable finds were that of pyrite-coated barite crystals to 7 cm on plates to 20 cm and superb millerite sprays to 4 cm perched on solution-fluted arsenian pyrite with rare stalactite forms.

Mount Wheeler Mine, Lincoln District, White Pine County

Underground workings of the Mount Wheeler Mine, which explored a sizable beryllium deposit, extend more than 8 km into the west side of the rugged Snake Range at an elevation of more than 2500 m. During a visit in 1994, I found mud-filled pockets containing aquamarine as opaque light-blue, curving crystals to 4 cm. I also collected clear to white crystals of phenakite with good luster, resembling stubby quartz crystals, up to 1.5 cm in size. Associated minerals include punky, boxwork, purple fluorite, and quartz. There are rumors of a superb scheelite crystal that was found here as well.

Murray Mine, Independence Mountains District, Elko County

The Murray Mine is now known as the finest stibnite locality in North America and one of the top localities in the world. In 1997, very large stibiconite-after-stibnite pseudomorphs were found in lengths surpassing 25 cm. It was surmised that as the mine went deeper, unaltered stibnite specimens might be found. Then, in 1998, it happened. Pockets of unknown size were discovered in blasted muck in the underground mine, and miners began finding huge, metallic-silver stibnite crystals to 29 cm in length and 5 cm in width. Along with these immense crystals, large blocky barite crystals to 5 cm were found in later vugs. Most of the barite crystals are coated with a layer of drusy white sucrosic quartz, and some occur in association with freestanding stibnite blades. Specimen recovery specialists have retrieved more fine pieces since the initial find (C. Jones, oral communication, 1999).

Northumberland Mine, Northumberland District, Nye County

This mine is another classic Nevada mineral locality. Though closed some time ago, during its active period barite prisms of superb quality poured out. Many fine specimens with white to slightly gray crystals to more than 10 cm were collected. A few superb specimens with gemmy, golden, lustrous crystals to 7 cm, some doubly terminated, were also recovered. This is certainly Nevada's best locality for prismatic barite, and aesthetic specimens from it grace museums and private collections around the world. A spectacular plate of these barites, weighing over 50 kg, can be seen at the W. M. Keck Museum on the University of Nevada, Reno, campus.

Olinghouse Open-Pit Mines, Olinghouse District, Washoe County

The Olinghouse district is about 50 km east of Reno near the town of Wadsworth. In the past, gold was found here in rich deposits that were extensively mined. Much of this gold was crystalline. In 1995, Alta Gold Company began large-scale, open-pit mining. Immediately, fine crystallized gold specimens displaying a wide range of morphologies were found, including wires to 8.4 cm long, garlands to 7 cm, spongy masses of more than 100 grams, octahedrons to 1 mm, dodecahedrons to 6 mm, and leaves to 5 cm. Many of these forms occur in very aesthetic combinations. A diverse assemblage of minerals is associated with the gold: adularia, calcite, cerussite, chalcopyrite, chrysocolla, clinozoisite, epidote, epistilbite, galena, goethite after pyrite, heulandite, laumontite, pyrite, quartz (white to clear and amethystine), scheelite, and scolecite. Of particular interest is a single specimen of wire gold and lustrous, white scheelite crystals to 7 mm, a combination that may be unique in the world. A specimen recovery project that I participated in has dispersed over 1100 specimens of Olinghouse crystallized gold.

Petersen Mountain, Hallelujah Junction, Washoe County

About 40 km north of Reno, east of U.S. Highway 395, is a favorite locality of many northern Nevada field collectors where organized collecting tours are run during the summer months. Jon Johnson and Foster Hallman claimed this noteworthy and prolific locality for commercial specimen mining in 1976. As of 1999, many of the world's finest and largest examples of amethyst-capped sceptered quartz have been found in huge cavities, the largest more than 5 m long and 0.5 m wide, that occur along faults in decomposed granite. These pockets are commonly packed with gemmy, etched smoky quartz prisms and finger-slicing, razor-sharp crystal shards. More unusual cavities contain scepters and "turkey-heads" (large, complex, overgrown scepter-heads with no exposed stem). The largest single turkey-head crystal so far found weighed an amazing 10 kg, and the finest specimen, in terms of color and clarity, 6.5 kg.

Quartz scepters are the most desirable specimens from this locality. Ranging from 2 to 40 cm long, some have dramatic overgrowths and are perched on gemmy, etched smoky stems. The scepters display one or more distinctive head colors: amethyst (reddish purple, purple, and smoky purple), smoky, citrine, and clear to white or milky. The majority are translucent to opaque milky white. In extraordinary specimens, the crystals are on matrix and have gemmy, smoky-colored, or amethystine heads. The finest piece found thus far, discovered by lucky diggers Aaron and Jade Wieting, has a superb scepter measuring approximately 18 cm long, with a 4-cm-wide shaft and a 10-cm-wide purple head, growing from a 25-cm, smoky crystal studded plate. At the top of the list of rarities are a mere handful of ultra-desirable "dumb-bells"—doubly terminated, etched, smoky quartz crystals with scepters, some amethystine, at both ends of the prism.

Other Petersen Mountain quartz oddities include crystals with green chlorite phantoms and those with liquid-filled chambers, some displaying moving vapor bubbles more than 3 cm long. Smoky-colored, Japan-law twin specimens to 4 cm have also been discovered, but less than a half-dozen have thus far been seen. World-class "ametrine"—gemmy interlayered amethyst and citrine—and the nation's largest faceted stone of flawless, fine amethyst (approximately 300 carats) have also been recovered.

Red Bird Mine, Antelope Springs District, Pershing County

The Red Bird underground mine, located near the southern edge of Black Ridge, was a source of mercury until 1969. At this mine, fine-twinned cinnabar crystals to more than 2 cm have been found encased in calcite (M. Jensen, oral communication, 1993). While collecting at this mine, I found mirror-brilliant, 2- to 4-mm, perfect penetration-twinned cinnabar crystals lining drusy, quartz-lined vugs a few centimeters in diameter. Large, sharp, undamaged cinnabar crystals from this locality are difficult to acquire, especially on matrix, and are prized. This mine is also a type locality for dadsonite, robinsonite, and schuetteite.

Regent Mine, Rawhide District, Mineral County

The Regent Mine consists of underground workings located along the western edge of a volcanic rock–hosted gold deposit being mined by Kennecott Rawhide Mining Company. Here, in long-abandoned workings, unusual and aesthetic specimens of included barite have been found. The inclusions, of cinnabar, metacinnabar, orpiment, realgar, and stibnite, are amazingly intricate and occur in lustrous, clear, tabular barite crystals to 3.5 cm. In 1996, I collected 9 flats of these fine specimens from a pocket 1 m long and 10 cm wide.

Round Mountain, Round Mountain District, Nye County

The Round Mountain open-pit gold mine, located in the northwestern part of Nye County near State Route 376, is on the site of historic vein and placer gold production. In recent years, it produced disseminated gold of little interest to collectors from ore that was too low grade to be of interest to the old-timers. In about 1994, the new workings intersected a high-grade zone that produced hundreds of kilograms of free gold, mostly as semi-crystalline masses. The largest of these masses weighed an impressive 3 kg, probably the largest piece of gold ever preserved from Nevada. Also found were well-crystallized gold leaf specimens to 7 cm, some with wirelike edges, as well as specimens with sharp trigonal faces on broad leaves.

Steamboat Springs, Steamboat Springs District, Washoe County

Steamboat Springs, an active geothermal area, is located at the southern edge of the city of Reno. Along with its reputation in mineralogical circles as the type locality for metastibnite, the area also has the distinction of being one of few active hydrothermal systems with associated anomalous gold and silver in North America. It has also produced thousands of specimens of crystallized sulfur, with the best among the finest ever found in the United States. Very aesthetic, hoppered, elongate octahedrons to 5 cm have been recovered on matrix, along with loose crystals to more than 7 cm (F. Cureton, oral communication, 1999).

Sulfur Pit, Crescent Valley, Eureka County

This locality, discovered by Greg Ferdock and Brian Bond, is one of the only known sources for crystallized sideronatrite and metasideronatrite in the world. Aesthetic, mostly flat-lying, golden, radial crystal sprays to 2.5 cm occur in massive, crystalline sulfur, along with a few freestanding crystals to 1 cm. Also found were sulfur specimens showing sharp, hoppered octahedrons to 1 cm (G. Ferdock, unpublished data).

Twin Creeks Mine, Potosi District, Elko County

Twin Creeks is a Newmont Mining Corporation open-pit mine about 50 km northeast of Winnemucca that exploits a Carlin-type gold deposit. A few hundred feet from the original surface, zones containing pockets lined with stunning, world-class orpiment specimens were found in early 1998. Brian Lees retrieved deeply colored, mirror-lustrous crystals to 5 cm in masses to 12 cm before the specimen-bearing zones were mined through. Out of this find, about 40 specimens were recovered with attractive barite crystals to 5 cm perched upright on well-crystallized orpiments. This colorful association is highly prized by collectors.

Victoria Mine, Dolly Varden District, White Pine County

The Victoria Mine, an underground mine located approximately 125 km north of Ely in a now-abandoned district, explored copper and bismuth mineralization in skarn. Though the mine is now completely reclaimed, superb bismuthinite crystals were recovered from it while it was open. The best of these specimens includes a crystal cluster approximately 5 cm long, terminated and freestanding, and a 5 cm free-standing crystal associated with a mirror lustrous 3 cm pyrite cube, on druse quartz coated matrix. Most other specimens have similarly large crystals, but all were found locked in a matrix of bournonite and other copper-bearing sulfides (M. Houhoulis, oral communication, 2002).

Virgin Valley Opal Fields, Virgin Valley District, Humboldt County

In Virgin Valley, near the northwestern corner of Humboldt County, about 20 km from the Oregon border, is a world-class occurrence of precious opal, particularly black fire opal. It is also known as the source of the world's best specimens of petrified wood with fire opal. Fine pieces have been recovered of intense white and black fire opal, some with fire flecks over 3 cm, replacing twigs, branches, roots, and in rare cases log sections, of ancient conifer, deciduous, and gingko trees (Sinkankas, 1997). Single specimens weighing up to 13.6 kg have been found with 30 percent of the wood replaced by precious opal. Some of the most exciting and spectacular specimens are the so-called "fire cones," pine cones up to 8 cm long that have been replaced by spectacular white, or more rarely black, fire opal. Only a single cone is found every several years by lucky fee-diggers or the owners of the properties. Fee digging can be done during the spring through fall months at both the Royal Peacock and Rainbow Ridge Mines.

White Caps Mine, Manhattan District, Nye County

Near the town of Manhattan is a cluster of underground mines that were more-or-less continuously active from the time gold was discovered there in 1905 until modern mining ceased in 1991. At the White Caps Mine, one of the largest workings, miners explored deep into the hillside. In 1921, "hair orpiment" was reported to occur with calcite and realgar. It was collected by the curious, and specimens were sent to various museums, via H. G. Clinton. Then in the 1970s, this same material was found in Japan, discovered to be a new species, and named wakabayashilite. Not long after, specimens from White Caps were found to contain the same mineral species but were significantly better examples. Today, specimens of White Caps wakabayashilite are highly prized by rare species collectors from Nevada and around the world. One of the finest specimens known, in the O'Brien collection at the W. M. Keck Museum, measures approximately 15 cm across and is composed of crystalline white calcite richly covered with thousands of acicular, golden wakabayashilite fibers several centimeters in length. Only a dozen or so other examples, miniature-sized and up, possessing freestanding crystals of this species, are known to exist.

In addition to wakabayashilite, fine specimens of stibnite also came from the White Caps Mine. Superb crystals more than 12 cm in length were discovered in cavities near the ore zones. Special care must be taken when adding specimens of this species, and the arsenolite from this mine, to one's collection. In 1958, a fire was started in the main shaft of the mine. The heat from this fire crystallized new stibnite and many of the known examples of arsenolite from this locality. Afterwards, these specimens were collected and dispersed onto the open market as "natural." The mine-fire stibnite can be easily identified by the melted appearance of terminal crystal faces.

Zapot Claims, Fitting District, Mineral County

The Zapot claims, owned and operated by Harvey Gordon, a Reno-area mineral dealer, have sporadically produced limited amounts of specimen-grade amazonite feldspar, smoky quartz, and topaz. In 1999, Harvey found a pocket containing some of the finest natural blue topaz ever found in the United States. Sharp, lustrous, blue-green crystals to an incredible 15 cm were recovered, on and off matrix, along with two larger yellow-brown matrix crystals measuring 21 and 24 cm. Many of the specimens were damaged by post-mineralization faulting but are nonetheless superb. Large, single, gemmy smoky quartz crystals to 20 cm and cabinet-sized plates of smoky quartz and blue-green amazonite have also been found.

CONCLUSIONS

I believe that the above list is just the tip of the mineralogical iceberg for Nevada. As time goes on, continued commercial mining and diligent field collecting will add to it and contribute to the knowledge, diversity, and collectibility of Nevada's minerals.

MINERAL CATALOG

PART IV

Nevada Mineral Catalog

GREGORY C. FERDOCK,
STEPHEN B. CASTOR,
FREDERICK J. BREIT,
DAPHNE D. LAPOINTE,
LIANG-CHI HSU,
and JOHN H. SCHILLING

The catalog of Nevada mineral occurrences is organized along the lines of similar references for some other states (for example, *Mineralogy of Arizona* and earlier editions of *Minerals of California*). Mineral species are listed in alphabetical order, and occurrences of these species are listed by county in alphabetical order. Nonspecies (mineral group or variety) names are italicized, and occurrences are listed for some of these (e.g., *apatite*) where data are not sufficient to establish species.

Locations are given as precisely as available information permits. Location data include specific mines or prospects where available, but in some cases only general information is known. If available, legal locations are given for some occurrences that are not reported as specific mines or prospects. In addition, legal locations are given for many of the most notable mineral occurrences.

Sources of information that were searched for the catalog include academic journals such as *The American Mineralogist* and *Economic Geology,* collectors' journals including the *Mineralogical Record* and *Rocks and Minerals,* and trade journals such as *The Lapidary Journal.* Government publications, particularly those of the Nevada Bureau of Mines and Geology (NBMG) and the U.S. Geological Survey (USGS), were important sources. Publications of the Geological Society of Nevada were also researched.

In addition to published references, unpublished lists of Nevada specimens in museums throughout the world were helpful. Museums that provided useful information, and the abbreviations that denote them in the catalog, are shown in Table 14. Examination of specimens in the W. M. Keck Museum of the Mackay School of Mines, University of Nevada, Reno, were particularly helpful.

Other sources of unpublished data used in the catalog included the authors' own files. Such data is referenced as follows: GCF = G. C. Ferdock; JHS = John H. Schilling; LCH = L. C. Hsu; SBC = S. B. Castor.

Mineral identification files from the Nevada Bureau of Mines and Geology Analytical Laboratory, as collected by the late John Schilling, were also used. Personal communications from mineral collectors, mineral dealers, mining professionals, and earth science researchers were also helpful. Mineral collectors and dealers who have been particularly helpful include Jack Crowley, Forrest Cureton, Bruce Hurley, Martin Jensen, Jack Kepper, Scott Kleine, Walt Lombardo, Jim McGlasson, Neil Prenn, Steve Pullman, Mark Rogers, Bruce Runner, Steve Scott, Dick Thomssen, Bob Walstrom, and Sugar White. References from such sources are listed as either o.c. (oral communication) or w.c. (written communication). An unpublished manuscript, "Mineralogy of Nevada," prepared in 1981 by collector Robert E. Jenkins was particularly useful.

Mineral occurrence sites described in the catalog may be specific mines or prospects, but some are generalized (for example, unremarkable occurrences of a common mineral such as galena that is found widely in workings throughout a mining district). Other occurrences are essentially unmarked outcrops, areas where specimens occur loose on the surface (for example, *agate* in the Carson Sink), or occurrences where

TABLE 14   Museums that provided information on Nevada mineral specimens

Abbreviation in catalog	Museum or Institution Name	Location
AM	Australian Museum	6 College Street, Sydney NSW 2000, Australia
AMNH	American Museum of Natural History	Central Park West at 79th Street, New York, New York
ASDM	Arizona-Sonora Desert Museum	2021 N. Kinney Road, Tucson, Arizona
BMC	Bryn Mawr College Collections	336 Rockland Road, Wayne, Pennsylvania
CGM	Colburn Gem and Mineral Museum	P.O. Box 1617, Asheville, North Carolina
CIT	California Institute of Technology	1201 E. California Blvd., Pasadena, California
CSM	The Geology Museum, Colorado School of Mines	1500 Illinois Street, Golden, Colorado
EMP	École des Mines de Paris	60, boulevard Saint-Michel, Paris, France
FUMT	Freiberg University of Mining and Technology	Brennhausgasse 14, Freiberg, Germany
HMM	Harvard Mineralogical Museum	24 Oxford Street, Cambridge, Massachussets
HUB	Humboldt-Universitat zu Berlin Museum fur Naturkunde	Invalidenstrasse 43, Berlin, Germany
LANH	Natural History Museum of Los Angeles County	900 Exposition Boulevard, Los Angeles, California
MM	The Morris Museum	6 Normandy Heights Rd., Morristown, New Jersey
MV	Museum of Victoria	222 Exhibition Street, Melbourne, Australia
NMBM	New Mexico Bureau of Mines and Mineral Resources	801 Leroy Place, Socorro, New Mexico
NYS	New York State Museum	3140 Cultural Education Center, New York
RMO	Royal Museum of Ontario	100 Queen's Park, Toronto, Ontario, Canada
SDSM	South Dakota School of Mines and Technology	Rapid City, South Dakota
SM	The Springfield Museums	220 State Street, Springfield, Massachussets
TUB	Technische Universitat Berlin	Ernst-Reuter-Platz 1, Berlin, Germany
UAMM	University of Arizona Mineral Museum	Flandrau Science Center, Tucson, Arizona
UCB	Department of Geology and Geophysics	University of California, Berkeley, California
UMNH	Utah Museum of Natural History, University of Utah	1390 East Presidents Circle, Salt Lake City, Utah
WMK	W. M. Keck Museum, Mackay School of Mines	University of Nevada, Reno, Nevada

specimens may be found by digging into dry lakes (such as ulexite at the Columbus Salt Marsh, Esmeralda County). In addition, some sites are listed by the formal rock formation or specific rock type in which a mineral occurs; in these cases the mineral may occur in small amounts over a wide area (blue chatoyant sanidine that occurs in the tuff of Chimney Spring in the Reno area is an example).

In general, mineral identifications for the catalog have not been verified by instrumental or chemical techniques. The reader may use the references provided to determine identification methods in some cases; however, methods are not reported in most references. In many cases, mineral identifications have been made by geologists and mineralogists in the field; such identifications are generally accurate, particularly for common minerals. In some cases, particularly for oral or written communications on unusual minerals, identification methods are noted in the catalog as X-ray diffraction (XRD), scanning electron microscope with energy dispersive analysis (SEM/EDX), or electron microprobe analysis (EMP). Where given, chemical analyses are generally reported in weight percent (wt. percent), but some are given in atomic percent (at. percent). In addition, mineral amounts are reported as volume percent (vol. percent) in some cases.

Measurements are generally given as originally cited in the literature. Therefore, both English units (e.g., inches and feet) and metric units (e.g., centimeters and meters) are used in the catalog. For the most part, longer distance measurements arc given in miles, rather than kilometers.

ACANTHITE

Ag_2S

Acanthite is the most important silver mineral in Nevada, occurring in most of the state's epithermal precious-metal deposits. In older literature it was generally referred to as *argentite,* a polymorph that is only stable at high temperatures. Prior to 1900 it was commonly called "silver glance" or "silver sulphurette."

CHURCHILL COUNTY As dark metallic bands, streaks, and masses in quartz veins at the Fairview Silver and Nevada Hills Mines in the Fairview district. (Greenan, 1914; Schrader, 1947; Quade and Tingley, 1987)

With chlorargyrite, gold, and other minerals, unspecified locality, Holy Cross district. (Schrader, 1947)

Reported with chlorargyrite and wire silver at the Kinney Mine, Sand Springs district. (Schrader, 1947)

In quartz-*adularia* veins with pyrite, chalcopyrite, galena, sphalerite, and other silver sulfides at the Nevada-Wonder Mine, Horgan and Flynn Lease, Queen No. 11 Lode, Jackpot Mine, Bald Eagle Mine, and the Gold King Lode, Wonder district. (Gianella, 1941; Schilling, 1979; WMK)

CLARK COUNTY Reported at the Boss Mine in the Goodsprings district. (Jedwab, 1998b)

Minor with native gold in comb quartz at the Lucy Grey Mine in the Sunset district. (Jenkins, 1981)

DOUGLAS COUNTY With stephanite in vein quartz at the Veta Grande Mine in the Gardnerville district. (Overton, 1947)

ELKO COUNTY With cerussite, galena, copper oxides and carbonates, iron oxides, and nontronite in a quartz vein at the Hickey Mine in the Contact district. (Schrader, 1935b)

Reported as *argentite* with pyrite and tetrahedrite in banded quartz veins in andesite at the Cornucopia, Panther, and Leopard Mines, Cornucopia district. (Emmons, 1910; Lawrence, 1963; LaPointe et al., 1991)

With gold-bearing pyrite in a vein, Rex Mine, Gold Circle district; also reported from the St. Paul Mine. (LaPointe et al., 1991; Emmons, 1910)

With gold in quartz-*adularia* veins at the Bourne, Pick and Shovel, Starlight, Van Alder, and Winner Mines in the Jarbidge district. (Schrader, 1912)

At various mines in the Mountain City district. (Gianella, 1941)

Important in ore from many mines in the Tuscarora district. (Emmons, 1910; Nolan, 1936)

ESMERALDA COUNTY As sooty black coatings in quartz-breccia veins at the Tonopah-Divide Mine in the Divide district. (Knopf, 1921a; Schilling, 1979)

Reported with pyrargyrite and chlorargyrite at the Mammoth Mine in the Gilbert district. (Ferguson, 1928)

Identified as *argentite* in a specimen from an unspecified locality in the Montezuma district. (AMNH)

Reported at the Sixteen-to-One, Mohawk, Nivloc, and other mines in the Silver Peak district. (Keith, 1977)

EUREKA COUNTY At various mines including the Tenabo Mine, Cortez district. (Emmons, 1910; WMK)

As grains associated with a purple efflorescence composed of kaolinite, wurtzite, barite, and cadmium; also as grains and micron-sized crystals in ore at the Goldstrike Mine in the Lynn district. (AMNH)

Reported with other silver minerals in replacement bodies and breccia at the Grant Mine in the

Mineral Hill district. (Emmons, 1910; Schilling, 1979; AMNH)

Reported at the Zenoli Mine in the Safford district. (Vanderburg, 1938b)

HUMBOLDT COUNTY In gold-silver veins with abundant *electrum* and other minerals at the Sleeper Mine in the Awakening district. (Wood, 1988; Saunders, 1994)

Rare at the Lone Tree Mine in the Buffalo Mountain district. (R. Braginton, o.c., 1995)

A minor component with pyrite at the Moonlight Uranium Mine in the Disaster district. (Dayvault et al., 1985; Castor et al., 1996)

Reported in silver ore at the National Mine in the National district. (WMK)

Reported with quartz at the Paradise and Wild Goose Mines in the Paradise Valley district. (WMK)

Identified by XRD at the Getchell Mine in the Potosi district. (Dunning, 1987)

Reported in a specimen with chlorargyrite at an unspecified locality in the Sulphur district. (WMK)

Reported in a specimen with argentiferous pyrite at the 1910 Property in the Winnemucca district. (WMK)

LANDER COUNTY With chlorargyrite on the Gold Butte claims in the Battle Mountain district. (Vanderburg, 1939)

Reported at the Garrison Mine in the Cortez district. (S. Pullman, w.c., 1999)

Common with tetrahedrite and other minerals in ore at the Betty O'Neal and other mines in the Lewis district. (Vanderburg, 1939; Gilluly and Gates, 1965)

Dull black euhedral crystals to 1 mm at the McCoy Mine in the McCoy district. (M. Jensen, w.c., 1999)

Reported as *argentite* with argentiferous manganese at the Manhattan, New York Canyon, North Star, Reeds Tunnel, Roosevelt, and other mines, Reese River district. (Ross, 1925, 1953; NYS; WMK)

LINCOLN COUNTY As minute inclusions in galena at the Groom Mine in the Groom district. (Humphrey, 1945)

Reported in a specimen from the Silver King Mine in the Silver King district. (WMK)

An important ore mineral with galena, cerussite, sphalerite, and *wolframite* at the Comet

Mine in the Comet district. (M. Jensen, w.c., 1999)

LYON COUNTY Reported in ore from unspecified localities in the Como district. (Gianella, 1941)

A minor mineral in silver-gold ore in the Silver City district. (Gianella, 1936a)

Questionably identified at the West Willys claim in the Washington district. (Garside, 1973)

With gold, pyrite, and chalcopyrite in quartz veins and as a secondary mineral in altered porphyry at the Wheeler Mine in the Wilson district. (Dircksen, 1975; Princehouse and Dilles, 1996)

LYON/MINERAL COUNTY Reported as *argentite* with other minerals in quartz-calcite veins in the Benway district 10 miles north of Schurz on the south end of Painted Mesa. (Schrader, 1947)

MINERAL COUNTY Acanthite, locally selenium-bearing, occurs in quartz-*adularia* veins with *electrum* and naumannite at the Juniata vein system in the Aurora district. (Hill, 1915; Osborne, 1985, 1991)

Reported at the Broken Hills, Baxter's Fluorite, and Sundown Mines in the Broken Hills district. (Schrader, 1947; WMK)

In relatively unoxidized ore at the Candelaria Silver Mine in the Candelaria district. (Chavez and Purusotam, 1988)

With chlorargyrite and native silver at the Mabel Mine in the Garfield district. (Ponsler, 1977)

Present in ore at the Lucky Boy Mine in the Lucky Boy district. (Hill, 1915)

In ore at the Gold Pen, Nevada Rand, and Lone Star Mines in the Rand district. (Schrader, 1947)

In rich ore from the Rawhide district. (Schrader, 1947)

Reported in a specimen with chalcopyrite at an unspecified locality in the Silver Star district. (WMK)

NYE COUNTY With chlorargyrite and native gold at an unspecified locality in the Antelope Springs district. (Kral, 1951)

With pyrargyrite and stibnite in veins at the Eaton prospect in the Arrowhead district. (Lawrence, 1963)

Reported in a specimen with stephanite at the Belmont Mine in the Belmont district. (WMK)

In veins with pyrite, chalcopyrite, galena, and hessite at the Fairday Mine in the Cactus Springs district. (Tingley et al., 1998a)

Reported in quartz veins at a locality in Stone Cabin Valley in the Clifford district. (Ferguson, 1916)

In veins with base-metal minerals at the Iron Rail group and Orizaba Mine in the Cloverdale district. (Kral, 1951; Kleinhampl and Ziony, 1984; Hamilton, 1993)

With chlorargyrite and native silver at the Silverton claims in the Currant district. (Kral, 1951)

Probably present at the Return Mine in the Ellsworth district. (Kral, 1951)

Tentatively identified at the Prussian vein in the Jefferson Canyon district. (Kral, 1951)

As small grains with brochantite in quartz veins in the Hasbrouck Shaft in the Lodi district. (GCF)

Reported at various mines in the Morey district. (S. Pullman, w.c., 1999)

In tiny veinlets cutting gray, fine-grained quartz in gold ore at an unspecified locality in the Northumberland district. (Jenkins, 1981)

Reported as *argentite* in various forms with antimony and base-metal minerals in vuggy quartz veins in silicified Tertiary rhyolite at the Antimonial Mine in the Reveille district. (Kral, 1951; Lawrence, 1963)

In considerable quantities with arsenopyrite and silver at the Blue Horse and Hillside Mines and the Catlin Claim in the Silverbow district. (Tingley et al., 1998a)

In veins with pyrite, barite, galena, and *electrum* at the Life Preserver Mine in the Tolicha district. (Tingley et al., 1998a)

An important mineral occurring as both a primary and supergene mineral, either massive or well crystallized in vugs, in silver-gold ore in the Tonopah district with "ruby silver" (pyrargyrite), polybasite, and stephanite at various mines. (Spurr, 1905; Bastin and Laney, 1918; Melhase, 1935c; Nolan, 1935; CSM; NMBM; WMK)

Reported from several properties in the Bullfrog district; particularly noted with *electrum* and uytenbogaardtite in high-grade gold ore at the Bullfrog and Original Bullfrog Mines. (Jenkins, 1981; Castor and Weiss, 1992; Castor and Sjoberg, 1993)

PERSHING COUNTY As grains to 1 mm with arsenopyrite in the Tin Stope of the Majuba Hill Mine in the Antelope district. (M. Jensen, w.c., 1999)

As greenish stains and sooty masses in quartz veins with pyrargyrite at the Arizona Mine in the Buena Vista district. (Cameron, 1939; WMK)

Reported at the Gilbert, Sunshine, and Twilight Mines in the Mill City district. (WMK)

The principal source of silver as bluish-black specks in patches in quartz veins and bodies in the Rochester district. (Knopf, 1924)

Reported in the Rosebud district; specifically noted as arborescent crystals to 1 cm in vugs and as platy crystals in honeycomb-like clusters at the Hecla Rosebud Mine. (Johnson, 1977; M. Jensen, w.c., 1999; GCF)

With native gold, pyrargyrite, naumannite, and *electrum* in epithermal quartz veins in the Seven Troughs district. (Ransome, 1909b)

In quartz-*tourmaline* veins at the Bonanza King Mine in the Spring Valley district. (Campbell, 1939)

STOREY COUNTY The main silver ore mineral in most mines in the Comstock district. Abundant in bonanza ore as massive dull black mixtures with polybasite and stephanite. Based on museum specimens, silver minerals in ore from the Comstock Lode were mainly acanthite and *electrum* associated with quartz, calcite, sphalerite, galena, chalcopyrite, and pyrite. In ore from shallow levels, acanthite was reported to be partly replaced by native silver in quartz-lined vugs and also to occur after galena. Crystals to 2.5 cm on silicified matrix have been noted in specimens from the Comstock district. (Becker, 1882; Bastin, 1922; Smith, 1985; WMK; F. Cureton, w.c., 1999)

WASHOE COUNTY In quartz-calcite veins with pyrite, sphalerite, galena, and enargite at the Golden Fleece, Standard Metals, and Fravel Mines in the Peavine district. (Hill, 1915; Bonham and Papke, 1969)

With pyrite and other sulfide minerals at the Burrus and Ruth veins in the Pyramid district. (SBC)

Reported in silver ore at the Wedekind Mine in the Wedekind district. (WMK)

WHITE PINE COUNTY Reported at the Star and Imperial Mines in the Cherry Creek district. (Schrader, 1931b)

In ore at the Paymaster and other mines in the Ward district. (Hill, 1916)

Along fault zones in the silver belt in the White Pine district. (Humphrey, 1960)

ACTINOLITE

$\square Ca_2(Mg,Fe^{2+})_5Si_8O_{22}(OH)_2$

A very common amphibole-group mineral in metamorphic rocks of both low-grade regional and contact origin; also widespread in hydrothermally altered rock. *Uralite* is amphibole replacing *pyroxene,* generally with composition similar to actinolite; *nephrite* is a type of jade composed of finely crystallized actinolite.

ELKO COUNTY In skarn and adjacent granodiorite with diopside, wollastonite, *axinite,* and other calc-silicate minerals in the Contact district. (Bailey, 1903; Schilling, 1979)

Plentiful in contact-metamorphic zones in the Dolly Varden district. (Melhase, 1934d)

Reported in a specimen from the Lone Wolf Mine in the Lone Mountain district. (AMNH)

ESMERALDA COUNTY Reported as *uralite* in a specimen from near Oasis Divide in the Silver Peak district. (CSM)

Black *nephrite* has been mined from a locality in the Silver Peak Range. (C. Rose, o.c., 2000)

EUREKA COUNTY As medium green, acicular to fibrous crystals to 1.5 cm, in calcite with pyrite in a vein in diorite at the Goldstrike Mine in the Lynn district; also an accessory mineral in skarn. (GCF)

HUMBOLDT COUNTY Reported as *uralite* in gray fibrous masses to 1 cm with *garnet,* quartz, and hematite in skarn from unnamed prospects on hills south of Gold Run Creek in the Gold Run district. (B. Hurley, w.c., 1999)

LANDER COUNTY As a constituent of gold-bearing skarn at the West orebody, Copper Canyon Mine, and Fortitude Mine in the Battle Mountain district. (Roberts and Arnold, 1965; Doebrich, 1993)

LYON COUNTY In skarn at the Mason Valley Mine in the Yerington district. (JHS)

MINERAL COUNTY Reported with *axinite,* vesuvianite, and other skarn minerals from an unspecified locality near Luning. May be the same locality reported by Melhase (1935e) in the Fitting district about 7 miles north-northwest of Luning. (Dunn et al., 1980; S. Pullman, w.c., 1999)

As deep green material with large epidote crystals at the Julie Claim in the Pamlico district. (Smith and Benham, 1985)

Reported in a specimen from an unspecified locality in the Rawhide district. (WMK)

NYE COUNTY Part of the gangue assemblage with *chlorite, sericite,* phlogopite, calcite, and dolomite at the Phelps Stokes Mine in the Ellsworth district. (Reeves et al., 1958)

As perfect, acicular, green-black crystals to 3 cm in calcite with titanite and diopside at the Great Basin (Victory) Mine in the Lodi district. (M. Jensen, w.c., 1999)

PERSHING COUNTY A major constituent of skarn in the Mill City district. (FUMT; HUB; S. Kleine, o.c., 1999)

Part of the gangue in iron ore deposits in the Mineral Basin district in the Buena Vista Hills. (Reeves and Kral, 1955)

STOREY COUNTY A widespread alteration mineral in the Comstock district. (D. Hudson, o.c., 2000)

WASHOE COUNTY A product of propylitic alteration of Tertiary rocks with calcite, epidote, clinozoisite, *chlorite,* and other minerals in the Peavine and Wedekind districts. (Hudson, 1977)

WHITE PINE COUNTY An alteration mineral in skarn and porphyry in the Robinson district. (James, 1976)

ADAMITE

$Zn_2(AsO_4)(OH)$

Adamite, a product of oxidation of arsenic-rich zinc ore, commonly occurs with conichalcite, duftite, and various iron arsenates. The copper-bearing variety *cuproadamite* is not recognized as a separate species.

CLARK COUNTY Reported in specimens from the Goodsprings district. (R. Thomssen, w.c., 1999)

ESMERALDA COUNTY As microcrystals associated with *agardite-(Nd)* at the Big 3 Mine in the Railroad Springs district. (M. Jensen, w.c., 1999)

Green *cuproadamite* crystals to 0.5 mm occur with arseniosiderite on matrix at the Weepah open-pit mine in the Weepah district. (F. Cureton, w.c., 1999; M. Jensen, w.c., 1999)

HUMBOLDT COUNTY *Cuproadamite* occurs as 0.5 mm green to white crystals in sprays to 1 mm with duftite in the Silver Coin Mine in the Iron Point district. (Thomssen, 1998; M. Jensen, w.c., 1999; GCF)

LANDER COUNTY Reported in specimens from

the Snowstorm Mine in the North Battle Mountain district. (R. Thomssen, w.c., 1999; J. McGlasson, w.c., 2001)

LINCOLN COUNTY Reported as pale-yellow-green blocky crystals of *cuproadamite* to 2 mm in specimens from the Silver King Mine in the Silver King district. (R. Thomssen, w.c., 1999; GCF)

MINERAL COUNTY The first noted occurrence of adamite in the United States was as small white crystal aggregates lining vugs in smithsonite in oxidized lead-zinc ore from the Bell district; specifically reported with pyromorphite and is locally copper-bearing at the Simon and Calavada Mines. (Knopf, 1921b; M. Jensen, w.c., 1999; S. White, w.c., 1999)

Reported at an unspecified locality in the Silver Star (Mina) district. (Heyl and Bozion, 1962)

NYE COUNTY Well-crystallized *cuproadamite* occurs in single prisms and aggregates to 1 cm with mimetite in the Lodi district; adamite also occurs in pale-blue to yellowish-green colors at the San Rafael Mine, Sec. 30, T14N, R36E. (ASDM; Jenkins, 1981; R. Thomssen, w.c., 1999)

PERSHING COUNTY Reported in a specimen from the Jersey Valley Mine in the Jersey Valley district. (R. Thomssen, w.c., 1999)

AEGIRINE

$NaFe^{3+}Si_2O_6$

A *pyroxene* group mineral that occurs in alkaline igneous rocks, carbonatites, and metamorphic rocks.

DOUGLAS COUNTY In aplite-pegmatite dikes at the Kingsbury Grade Prospects in the Genoa district. (Garside, 1973)

HUMBOLDT COUNTY As a groundmass mineral in peralkaline volcanic rocks of the McDermitt caldera. (Conrad, 1984)

LINCOLN COUNTY Reported in a specimen from near Alamo. (R. Thomssen, w.c., 1999)

NYE COUNTY As a groundmass and vapor-phase mineral in peralkaline rocks of the Silent Canyon volcanic center. (Noble et al., 1968)

AENIGMATITE

$Na_2Fe_5^{2+}TiSi_6O_{20}$

A primary constituent of sodium-rich alkaline igneous rocks.

HUMBOLDT COUNTY A constituent of peralkaline ash-flow tuffs of the McDermitt caldera with alkali *feldspar,* quartz, arfvedsonite, and clinopyroxene. (Wallace et al., 1980; Conrad, 1984)

AGARDITE

$(Y,La,Ca)Cu_6(AsO_4)_3(OH)_6 \cdot 3H_2O$(generalized)

A mineral found in oxidized portions of arsenic- and copper-bearing hydrothermal deposits. *Agardite* has been subdivided into two species on the basis of the dominant rare earth element (lanthanum or yttrium), and some have proposed species with other elements as dominant components. Species have not been identified at the following localities.

ESMERALDA COUNTY Reported at the Broken Toe Mine in the Candelaria district, and as *agardite-(Nd)* at the Big 3 Mine in the Railroad Springs district. (R. Thomssen, w.c., 1999)

At the best Nevada occurrence as pale-blue to green fibrous crystals to 1 mm at the Big 3 Mine in the Railroad Springs district; reported as *agardite-(Nd).* (LANH; M. Jensen, w.c., 1999)

HUMBOLDT COUNTY Reported as *agardite-(Ca)* in pale-green 0.5-mm fibers with heterogenite and adamite, and as *agardite-(Pb)* in 0.5-mm fibers associated with aurichalcite, hemimorphite, and chlorargyrite at the Silver Coin Mine in the Iron Point district. (M. Jensen, w.c., 1999)

LINCOLN COUNTY Reported at the Silver King Mine in the Silver King district. (R. Thomssen, w.c., 1999)

MINERAL COUNTY Reported as *agardite-(Nd)* in bright blue-green microcrystalline tufts to 3 mm at the Simon Mine in the Bell district. (B. and J. Runner, o.c., 2000)

NYE COUNTY At the San Rafael Mine in the Lodi district as green botryoids on matrix and coating wulfenite crystals. (S. Pullman, w.c., 1999; GCF)

PERSHING COUNTY As *agardite-(Ce)* in pale-green acicular microsprays on quartz crystals at the Majuba Hill Mine, Tin stope, in the Antelope district. (M. Jensen, w.c., 1999)

AGARDITE-(Y)

$(Y,Ca)Cu_6(AsO_4)_3(OH)_6 \cdot 3H_2O$

See description for *agardite.*

ESMERALDA COUNTY Reported as pale-blue to green sprays to 2.5 mm with *agardite-(Nd)* at the Big 3 Mine in the Railroad Springs district. (S. White, w.c., 2000)

MINERAL COUNTY Reported as blue sprays to 0.5 mm with conichalcite at the Simon Mine in the Bell district. (S. White, w.c., 2000)

NYE COUNTY Reported at the San Rafael Mine in the Lodi district. (R. Thomssen, w.c., 1999)

PERSHING COUNTY As deep green velvety crusts to 1 mm thick with cornwallite and olivenite from the Copper stope, Majuba Hill Mine, in the Antelope district. (Jensen, 1993b)

AGUILARITE

Ag_4SeS

Aguilarite is a relatively rare mineral that occurs in epithermal silver deposits containing selenium.

ELKO COUNTY A primary silver ore mineral in banded quartz veins with naumannite and *electrum* at the Ken Snyder Mine in the Gold Circle district. (Casteel and Bernard, 1999)

HUMBOLDT COUNTY A minor mineral in veins with *electrum* at the Sleeper Mine in the Awakening district. (Wood, 1988)

A trace mineral in unoxidized ore with gold, *electrum,* galena, naumannite, sphalerite, chalcopyrite, eucairite, and selenium at the Crofoot-Lewis Mine in the Sulphur district. (Ebert et al., 1996)

LANDER COUNTY Identified by EMP with other sulfides at the Dean Mine in the Lewis district. (J. McGlasson, w.c., 2001)

MINERAL COUNTY Tentatively identified at the Rawhide Mine in the Rawhide district. (Black, 1991)

NYE COUNTY As disseminated grains in trace quantities with chalcopyrite, pyrite, and pyrrhotite at the Blue Horse and Hillside Mines and Catlin claim in the Silverbow district. (Tingley et al., 1998a)

Reported as crystals to 5 mm in the Tonopah district. (Von Bargen, 1999)

STOREY COUNTY In dark-gray sectile masses, intimately mixed with other minerals, replacing base-metal sulfides and calcite, and replaced by acanthite, *electrum,* and stephanite at an unspecified locality in the Comstock district. (Coats, 1936a)

AIKINITE

$PbCuBiS_3$

Aikinite, a widespread but seldom abundant mineral, generally occurs with gold and other bismuth minerals in hydrothermal deposits that contain bismuth.

ESMERALDA COUNTY As 3.2-cm crystals in white quartz from an unspecified locality in the Buena Vista district. (TUB; Carnegie Museum Auction Catalogue, August 1998)

As coarse, steel gray prisms to 5 cm in white quartz with coatings of blue-gray bismutite in the Sylvania district near Cucomungo Spring; also reported with white bismite pseudomorphs at an unnamed prospect. (MV; F. Cureton, w.c., 1999; Jenkins, 1981)

NYE COUNTY As minute crystals embedded in quartz or rarely as free-standing crystals in vugs at the Outlaw Mine in the Round Mountain district; also as irregular masses to 5 cm and as crude crystals and intergrowths with benjaminite and other sulfosalts. (Harris and Chen, 1975; Foord et al., 1988; Dunning et al., 1991)

AKAGANÉITE

$\beta\text{-}Fe^{3+}(O,OH,Cl)$

Akaganéite, an oxidation product of iron sulfide minerals, has been identified in hydrothermal deposits, soils, sea-floor nodules, and meteorites. It is possible that the mineral is a widespread iron oxide phase that has been characterized in many places as *limonite.*

ELKO COUNTY With dussertite in a specimen from an unspecified locality in the Wells district. (Jenkins, 1981; EMP; AMNH)

HUMBOLDT COUNTY Reported from Black Rock. (Anthony et al., 1997)

PERSHING COUNTY Reported as lustrous to earthy brown masses on quartz at an unspecified locality in the Sierra district. (EMP; Jenkins, 1981; AMNH)

AKTASHITE

$Cu_6Hg_3As_4S_{12}$

A rare sulfide mineral of hydrothermal origin.

EUREKA COUNTY Tentatively identified as a microgranular phase in the ore zone at the Goldstrike Mine in the Lynn district. (GCF)

HUMBOLDT COUNTY Reported in a specimen from the Getchell Mine in the Potosi district. (EMP)

ALABANDITE

$Mn^{2+}S$

Alabandite is a moderately uncommon mineral in epithermal sulfide vein deposits; it has also been identified in meteorites.

WHITE PINE COUNTY As grains to 5 mm in the St. Anthony vein with rhodonite, rhodochrosite, quartz, calcite, pyrite, and galena at the Siegel Mine in the Muncy Creek district. (Pardee and Jones, 1920; Hewett and Rove, 1930)

ALBITE

$NaAlSi_3O_8$

Albite, the sodium-rich end member of the *plagioclase feldspar* group, includes the varieties *oligoclase* and *andesine*. It is an extremely abundant and common mineral in many rock types, and occurs in large crystals in granite pegmatite; only a few Nevada occurrences are listed below.

CHURCHILL COUNTY *Andesine* occurs as a primary mineral in the wall rock in the Fairview district. (Schrader, 1947; Saunders, 1977)

CLARK COUNTY As white to blue crystals to 2 cm in miarolitic cavities and pegmatites in the Eldorado and Newberry Mountains; also in perthite and as small crystals in granite. (S. Scott, w.c., 1999)

ELKO COUNTY Reported in a specimen, probably of pegmatite, from the west flank of the Ruby Mountains on the north side of Gilbert Canyon in the Gilbert Canyon district. (WMK)

As a primary mineral in dike rock at the Jerritt Canyon Mines in the Independence Mountains district. (Hofstra, 1994)

With quartz, beryl, and other pegmatite minerals in Dawley Canyon in the Valley View dis-trict; also reported as *oligoclase*. (Olson and Hinrichs, 1960)

EUREKA COUNTY As crystals of *andesine* to 4 mm in diorite at the Goldstrike Mine in the Lynn district. (Walck, 1990; GCF)

As *andesine* as phenocrysts in andesite at the Barth Iron Mine in the Safford district. (Shawe et al., 1962)

MINERAL COUNTY Cream-colored crystals to 1 cm in pegmatite at the Zapot claims in the Fitting district. (M. Jensen, w.c., 1999; Foord et al., 1996)

Reported at the Silver Dyke Mine in the Silver Star district. (S. Pullman, w.c., 1999)

PERSHING COUNTY *Oligoclase* occurs as a major constituent of tungsten-bearing pegmatite with beryl, muscovite, and native antimony at the Oreana (Little) Tungsten Mine in the Rye Patch district. (Olson and Hinrichs, 1960; F. Cureton, w.c., 1999)

STOREY COUNTY An alteration and vein mineral in the Comstock district. (Vikre, 1989)

WASHOE COUNTY An alteration mineral at the Olinghouse Mine in the Olinghouse district. (Jones, 1998)

Part of the propylitic alteration assemblage and a constituent of the volcanic host rocks at the Nevada Dominion Mine in the Pyramid district. (B. Park-Li, o.c., 1996)

In several *allanite*-bearing pegmatite bodies at the Red Rock prospect in the Stateline Peak district; also in pink perthite with quartz and minor muscovite at a *feldspar* mine. (JHS; Bonham and Papke, 1969)

Reportedly Nevada's best occurrence as lustrous euhedral cream-colored crystals to 3 cm with microcline and smoky quartz in pegmatites at Incline Village. (Jensen, 1993a)

ALEKSITE

$PbBi_2Te_2S_2$

A rare mineral of hydrothermal origin.

NYE COUNTY Reported as black grains to 1 mm in rhyolite in a specimen from Keystone, probably in reference to Keystone Canyon; also reported from Tybo in the Tybo district. (Anthony et al., 1990; HUB; M. Jensen, w.c., 1999)

ALIETTITE

$[Mg_3Si_4O_{10}(OH)_2] \cdot [(Ca_{0.5},Na)_{0.33}(Al,Fe,Mg)_{2-3}(Al,Si)_4O_{10}(OH)_2 \cdot nH_2O]$ (Anthony et al., 1995)

A clay mineral of the *smectite* group that occurs in altered dolomite and *serpentine*.

EUREKA COUNTY With lizardite as white, vitreous or resinous to plasterlike fracture coatings along the interface between hornfels and unmetamorphosed carbonate at the Goldstrike Mine in the Lynn district. (GCF)

ALLANITE

$(Ce,Y,Ca)_2(Al, Fe^{2+},Fe^{3+})_3(SiO_4)_3(OH)$ (generalized)

A relatively common epidote-group accessory mineral in many igneous and metamorphic rocks; it also occurs as a detrital mineral in placers and sedimentary rocks and as coarse crystals in pegmatite. *Allanite* is divided into two species on the basis of specific rare earth content, which has not been determined for the following occurrences.

CLARK COUNTY As fine-grained crystals with xenotime, zircon, and *monazite* in aplite-pegmatite dikes in Precambrian rocks in the Crescent district. (Garside, 1973)

In Precambrian rapakivi granite and pegmatite in the Gold Butte district. (Volborth, 1962b; Garside, 1973)

In granitic rock in the Eldorado and Newberry Mountains. (S. Scott, w.c., 1999)

DOUGLAS COUNTY Reported at the Hunch and Lucky Strike claims in the Genoa district. (Garside, 1973)

ELKO COUNTY Along a contact between granite and limestone at an unspecified locality in the Contact district. (Garside, 1973)

HUMBOLDT COUNTY Reported in a specimen from an unspecified locality in the Pueblo district. (WMK)

LANDER COUNTY In pre-Tertiary sedimentary rock as a detrital mineral at the West orebody at Copper Canyon in the Battle Mountain district. (Theodore and Blake, 1978)

NYE COUNTY A common accessory mineral in the tuff of Wilson's Camp in the Antelope Springs and Wilson districts. (Ekren et al., 1971)

An accessory mineral in rhyodacite making up part of the Roller Coaster laccolith on the Nellis Air Force Range. (Ekren et al., 1971)

As an accessory mineral in the Fraction Tuff in central Nye County. (Ekren et al., 1971)

An accessory in tuff at Yucca Mountain, particularly in the Crater Flat Group and older units. (Broxton et al., 1989)

PERSHING COUNTY With uraninite at the Long Tungsten Mine in the Wild Horse district. (Garside, 1973)

WHITE PINE COUNTY An accessory mineral as euhedral crystals to 1.5 mm in quartz monzonite near the area of beryllium mineralization at the Mount Wheeler Mine in the Lincoln district. (Lee and Bastron, 1962)

ALLANITE-(Ce)

$(Ce,Ca,Y)_2(Al,Fe^{2+},Fe^{3+})_3(SiO_4)_3(OH)$

Allanite species dominated by light rare earth elements (see above for further information).

WASHOE COUNTY As a major constituent (5–30 percent) of small, irregularly shaped, radioactive, aplitic pegmatite bodies in euhedral to subhedral crystals to 6 cm long and 1 cm wide at the Red Rock Prospect in the Stateline Peak district, NE1/4 SW1/4 Sec. 27, T22N, R18E; dominated by light rare earth elements on the basis of SEM/EDX analysis. (Volborth, 1962a; UMNH; Tingley et al., 1986)

ALLOPHANE

$Al_2O_3 \cdot 1.3\text{-}2.0(SiO_2) \cdot 2.5\text{-}3.0(H_2O)$

Allophane is a widespread but seldom recognized mineral that occurs in volcanic ash as a weathering product, in chalk and coal beds, in hydrothermally altered igneous rock, and in oxidized portions of hydrothermal ore deposits.

EUREKA COUNTY In colorless to sky-blue porcelaneous crusts to 1 mm thick at the Gold Quarry Mine in the Maggie Creek district. (M. Jensen, w.c., 1999)

Reported in a specimen from near Palisade in the Safford district. (WMK)

STOREY COUNTY As an alteration product of *hornblende* in volcanic rocks in the Comstock district. (Whitebread, 1976)

WHITE PINE COUNTY In jasperoid at an unspecified locality in the Cherry Creek district. (Lovering, 1972)

ALMANDINE

$Fe_3^{2+}Al_2(SiO_4)_3$

A *garnet*-group mineral in series with pyrope and spessartine. The most common *garnet* species; typically in metamorphic rocks, but also in granite and as a detrital mineral.

CLARK COUNTY In the Virgin Mountains as pods to 15 cm across with beryl, chrysoberyl, *tourmaline,* and magnetite in pegmatite; also as porphyroblasts to 1 cm in Precambrian schist and gneiss in the Bunkerville district. (Beal, 1965)

With cordierite, sillimanite, and other minerals in gneiss in the Gold Butte district. (Volborth, 1962b)

ELKO COUNTY *Garnet* with a composition of 65 percent almandine, 18 percent andradite, and 17 percent pyrope occurs in schist at the Errington-Thiel Mica Mine in the Valley View district. (Olson and Hinrichs, 1960)

Reportedly abundant at various locations, particularly in the northern part of the Ruby Range. (Melhase, 1934c)

HUMBOLDT COUNTY As crystals to several centimeters in a contact zone several meters in thickness at the Alpine Mine in the Potosi district. (S. Sears, w.c., 1999)

MINERAL COUNTY Reported as *grandite* (calcium-bearing almandine) in skarn at the Julic claim in the Pamlico district; crystals are commonly replaced by epidote, quartz, and calcite. (Smith and Benham, 1985)

WHITE PINE COUNTY As lustrous dark-red to black crystals (composition 69.3 percent almandine and 28.5 percent spessartine) to 4 cm in vesicles with drusy quartz and as smaller phenocrysts in rhyolite at Garnet Hill in the Robinson district 5 miles northwest of Ely; prior to 1934, large quantities were found in washes, but these were exhausted by commercial collectors. (Melhase, 1934d; Pabst, 1938; Staatz and Carr, 1964; Hollabaugh and Purcell, 1987)

With staurolite in schist and as placer concentrations of poorly formed crystals to 7 mm in the Mount Moriah district in Hampton Creek Canyon, Mount Moriah area, Snake Range. (Hose et al., 1976)

ALTAITE

PbTe

Altaite occurs in some hydrothermal precious metal deposits where it is generally associated with other telluride minerals and a variety of sulfides.

NYE COUNTY With *electrum,* hessite, acanthite, and other sulfide minerals in gold ore at the Bullfrog Mine in the Bullfrog district. (Eng et al., 1996)

WHITE PINE COUNTY Intergrown with hessite in lead-silver ore at the Silver King Mine in the Ward district. (Everett, 1964)

ALUMINOHYDROCALCITE

$CaAl_2(CO_3)_2(OH)_4 \cdot 3H_2O$

Aluminohydrocalcite is a rare secondary mineral that occurs in some contact deposits in limestone, or in rocks which have been affected by carbonatization.

NYE COUNTY Reported as a supergene mineral in a magnesite-brucite-*serpentine*-sulfide contact zone in an abandoned brucite pit at the Gabbs Magnesite-Brucite Mine in the Gabbs district. (Oswald and Crook, 1979)

ALUNITE

$K_2Al_6(SO_4)_4(OH)_{12}$

Alunite is a common mineral in hydrothermal ore deposits where it may be of hypogene or supergene origin. Hypogene alunite is a characteristic mineral in some high sulfidation-type precious-metal deposits, such as at Goldfield.

CLARK COUNTY Abundant with pyrite in altered rock in the Alunite district near Railroad Pass. (Tingley, 1992)

In the Goodsprings district as earthy white lenses made up entirely of minute, perfect rhombohedrons at the Kirby Mine; also locally in chalky masses replacing shale near the top of the Goodsprings Dolomite. (Hewett, 1931)

DOUGLAS COUNTY Reported in a specimen from an unknown locality in the Buckskin district. (WMK)

ELKO COUNTY With kaolinite in white, earthy veins and massive pods to 12 feet across along joints and faults, most commonly at depth in the deposit, but also as veins and breccia fillings to 3 inches in the discovery jasperoid at the Rain Mine in the Carlin district. (D. Heitt, o.c., 1991)

In two forms, one fine-grained and claylike, the other coarse-grained (grains to 0.4 mm) in veinlets with pyrite and silica, at the Hollister gold deposit in the Ivanhoe district. (Deng, 1991)

ESMERALDA COUNTY Abundant and widespread as massive white alteration product at numerous prospects in the north part of the Cuprite district. (Melhase, 1935b; Swaze et al., 1993; J. Crowley, w.c., 1999)

Identified in a specimen from the Diamondfield district. (JHS)

Widespread in the Goldfield district as soft, white, or slightly pink material that resembles clay, and as pale-pink crystalline aggregates to 5 mm in diameter after *feldspar* phenocrysts and minute grains replacing groundmass *feldspar* in altered volcanic rock. Also reported in quartz veins and as breccia fillings, where it may occur with gold. Most commonly associated with quartz, pyrite, kaolinite, and hematite, and may be mixed with natroalunite. (Ransome, 1907, 1909a; Keith et al., 1979)

EUREKA COUNTY White, pale-blue, and pink veins and masses to 4 cm wide in unoxidized, sulfide-rich rock from as much as 1,200 feet below the surface at the Goldstrike Mine in the Lynn district; also with stibnite and antimony oxides in silicified breccia, and as yellow coatings with jarosite in oxidized rock. (GCF)

As massive, cream-white, finely crystalline veinlets and lustrous, equant, trigonal crystals lining vugs in silicified barite with carbonate-fluorapatite at the Gold Quarry Mine in the Maggie Creek district. (Arehart et al., 1992; Jensen et al., 1995)

HUMBOLDT COUNTY As a fine-grained replacement mineral and rarely as crystals to 100 microns with kaolinite and after *adularia* at the Lone Tree Mine in the Buffalo Mountain district. (Panhorst, 1996)

Reported at the Silver Coin Mine in the Iron Point district. (Thomssen, 1998)

Locally as very fine-grained, white, chalky material, commonly in balls to 10 cm in diameter, in the McDermitt Mine in the Opalite district; also as white masses at shallow levels in the Cordero Mine. (McCormack, 1986; Fisk, 1968)

Massive in veins and as an alteration mineral in the Sulphur district in the foothills south of Sulphur. (Clark, 1918; Ebert et al., 1996; J. Crowley, w.c., 1999)

LINCOLN COUNTY Reported from an unspecified locality in the vicinity of Caliente. (AMNH)

MINERAL COUNTY As a common mineral in the veins at numerous gold mines in the Rawhide district. (M. Jensen, w.c., 1999)

In the Aurora district as pink rosettes to 2 cm in diameter in silicified rock and in micaceous form with quartz, as part of a high sulfidation alteration assemblage in the vicinity of Brawley Peaks. (Bill Silberman, o.c., 1992)

As crystals to 1 cm long in ledges south of Garfield Flat in the Marietta district, northern Excelsior Mountains. (Hudson, 1983)

NYE COUNTY Reported from an unspecified locality in the Antelope Springs district. (JHS)

Fine-grained alunite occurs at the Sterling, Telluride, and Mother Lode Mines in the Bare Mountain district. (Castor and Weiss, 1992)

One of the principal alteration minerals with quartz and pyrite in the Cactus Springs district; specifically noted at the Cactus Nevada Mine. (Tingley et al., 1998a)

A primary alteration product in the Gold Crater district. (Tingley et al., 1998a)

In gold-rich rock from quartz ledges at the Gold Reed Mine in the Kawich district. (Tingley et al., 1998a)

Probably of supergene origin and often associated with jarosite as thin fracture coatings, replacing sanidine phenocrysts in tuff, and as veinlets to 5 cm wide in hydrothermal breccias and stockwork zones in the Round Mountain district on Round Mountain and Stebbins Hill. (Tingley and Berger, 1985)

Part of the alteration assemblage with kaolinite, illite, and quartz at the Franz Hammel, Golden Chariot, and other mines in the Jamestown district. (Tingley et al., 1998a)

PERSHING COUNTY With *electrum* and fine-grained quartz in veins at an unspecified locality in the Scossa district. (Vanderburg, 1936b)

STOREY COUNTY Common in the Comstock district as an alteration mineral with no clear spatial relation to lodes and faults; however, it is a major constituent of the Occidental lode. Also found along Geiger Grade to the northwest of the district. (Vikre, 1989; Nicols, 1991; D. Hudson, w.c., 1999)

WASHOE COUNTY In hydrothermal breccia in semi-radiating clusters or as clear intergrown laths to 4 mm in the Peavine and Wedekind districts; associated minerals include quartz, diaspore, hematite, and pyrite. (Hudson, 1977)

As white veins and fracture fillings with chalcanthite at the Nevada Dominion Mine in the Pyramid district. (Park-Li, o.c., 1996)

Reported in the Silica Pit in the Steamboat Springs district. (Wilson and Thomssen, 1985)

Just below the water table replacing silicate minerals around Washoe Hot Springs. (Dickson and Tunnell, 1968)

WHITE PINE COUNTY An oxidation product along quartz veins with jarosite and barite at the Vantage gold deposit on Alligator Ridge in the Bald Mountain district. (Ilchick, 1984)

Green and massive in Ruth Pit in the Robinson district. (UMNH)

ALUNOGEN

$Al_2(SO_4)_3 \cdot 17H_2O$

Alunogen is a moderately common secondary mineral that occurs as efflorescences or coatings where aluminum-bearing species have been altered by acidic solutions.

ESMERALDA COUNTY Reported in a specimen from an unknown locality in the Goldfield district. (WMK)

MINERAL COUNTY Questionably identified in oxidized ore in the Rawhide district. (Schrader, 1947)

NYE COUNTY On mine walls with apjohnite and melanterite at the Urania Mine in the Cactus Peak district. (J. Price, o.c., 1999)

PERSHING COUNTY Reported in a specimen from an unknown locality in the Gerlach area. (WMK)

STOREY COUNTY Capillary, silky alunogen is reported in a specimen from the Comstock district. (UCB)

WASHOE COUNTY With chalcanthite as fibrous post-mining efflorescent growths to 1 cm in the Burrus Mine in the Pyramid district; also as fibrous efflorescences to 5 cm long in the Nevada Dominion Mine. (Jensen, 1994; GCF)

As whitish, glassy crystals to 2 mm at Steamboat Hot Springs in the Steamboat Springs district. (Wilson and Thomssen, 1985)

AMMONIOALUNITE

$(NH_4)_2Al_6(SO_4)_4(OH)_{12}$

A very rare mineral found at active and fossil hot springs; only found at one locality outside Nevada.

ELKO COUNTY Identified in a fossil hot spring deposit in the Ivanhoe district with opal, quartz, and alunite; also reported with buddingtonite in the district. (Altaner et al., 1988; Krohn et al., 1993)

WASHOE COUNTY As a white mass with natrojarosite in an unnamed prospect in the Nightingale district; identified by XRD. (GCF)

ANALCIME

$Na[AlSi_2O_6] \cdot H_2O$

Analcime is a common mineral, occurring in amygdules in mafic igneous rocks, in some pegmatites, and as a product of diagenetic and/or hydrothermal alteration of volcanic tuffs and volcaniclastic sediments. Occurrences of the latter type are particularly common in Nevada.

ESMERALDA COUNTY Reported in the Coaldale district 4 miles west of Blair Junction on the Tonopah-Goldfield Railroad. (UCB)

As clear, white, and bluish-gray trapezohedrons to 2 cm with calcite, fluorite, and quartz in veins and vugs in tuff at the Sharon claims in the Silver Peak district. (L. Thompson, o.c., 1999)

HUMBOLDT COUNTY Closely associated with hectorite in the interior portions of the McDermitt caldera complex in the Opalite district. (Rytuba and Glanzman, 1979)

MINERAL COUNTY Identified in a specimen from the Teels Marsh district. (JHS)

NYE COUNTY With *clinoptilolite* in ash-flow tuff in the Currant district near Stone Cabin. (Scott, 1965)

As crystals to 2 mm with quartz, calcite, and fluorite in veins in drill core from Yucca Mountain. (Carlos et al., 1995; Castor et al., 1999b)

PERSHING COUNTY Identified in a specimen from an unspecified locality in the Mineral Basin district in the Buena Vista Hills. (JHS)

ANAPAITE

$Ca_2Fe^{2+}(PO_4)_2 \cdot 4H_2O$

A rare mineral that occurs in sedimentary iron deposits and in pegmatite.

NYE COUNTY As finely crystalline, botryoidal, pale-green material from an unspecified locality. (S. Pullman, w.c., 1999)

ANATASE

TiO_2

Anatase, a low-temperature polymorph of rutile, is a common accessory mineral in some igneous metamorphic rocks; it also occurs in granite pegmatite, as a product of hydrothermal alteration, and as a detrital or authigenic mineral in sedimentary rocks.

CHURCHILL COUNTY As small crystals with rutile in albitite masses adjacent to quartz veins in the Corral Canyon district; also reported as masses to 5 cm in gold-bearing quartz veins with pyrite, calcite, *sericite,* titanite, and rare sphalerite in granite. (Ferguson, 1939; Vanderburg, 1940)

Reported as *octahedrite* in a specimen from the Dixie Valley Mine in the Dixie Valley district. (WMK)

CLARK COUNTY As inclusions in cavernous gray quartz with rutile at some copper and silver mines in the Goodsprings district. (Hewett, 1931)

ELKO COUNTY Reported at the Rain Mine in the Carlin district. (D. Heitt, o.c., 1991)

As very small (<50 microns) grains of rodlike clusters disseminated in host rock at the Hollister gold deposit in the Ivanhoe district; may form up to 3 percent of the rock by volume. (Deng, 1991)

Reported in a specimen from the Ruby Mountains. (R. Thomssen, w. c., 1999)

ESMERALDA COUNTY A minor constituent of altered rocks in the Goldfield district. (Ashley and Albers, 1975)

EUREKA COUNTY As tiny crystals and blebs at the Goldstrike Mine in the Lynn district. (Berry, 1992)

In very small, rounded and partially corroded crystals in breccia with later phosphate, uranium, and mercury minerals at the Gold Quarry Mine in the Maggie Creek district. (Jensen et al., 1995)

LANDER COUNTY As euhedral black lustrous crystals to 0.1 mm on muscovite at the McCoy Mine (underground) in the McCoy district. (M. Jensen, w.c., 1999)

As deep blue crystals to 0.5 mm associated with galena and sphalerite crystals on quartz crystals at the Driscoll Mine near the Battle Mountain district. (M. Jensen, w.c., 1999)

NYE COUNTY Reported as *octahedrite* in a specimen of *sericite* schist with talc from an unspecified locality in the Bellehelen district. (WMK)

PERSHING COUNTY As rare disseminated crystals in altered rhyolite at the Majuba Hill Mine in the Antelope district. (Jensen, 1985)

STOREY COUNTY A minor constituent with alunite of some altered volcanic rocks in the Virginia City area in the Comstock district. (Whitebread, 1976)

ANDALUSITE

Al_2SiO_5

Andalusite, a common constituent of regionally and contact-metamorphosed aluminous rocks, has also been identified in granitic rock and granite pegmatite. *Chiastolite* is a variety containing cross-shaped zones of dark inclusions that is found in regionally metamorphosed rock.

CHURCHILL COUNTY *Chiastolite* occurs in schist at the Michigan prospect in the Mountain Wells district, La Plata Canyon, Stillwater Range. (R. Walstrom, w.c., 1999)

In black metamorphic rocks as small white *chiastolite* crystals in road cuts along the south side of U.S. Highway 50, 28 miles east of Fallon, on the west slope of the Sand Springs Range. (Horton et al., 1962)

CLARK COUNTY Reported in a specimen from an unknown locality in the Muddy Mountains district near Overton. (WMK)

DOUGLAS COUNTY With corundum, pyrophyllite, and rutile in shear zones in altered andesite at the Blue Metal (Blue Danube) property in the Buckskin district. (Binyon, 1946a; M. Jensen, w.c., 1999)

ELKO COUNTY Reported to replace *garnet* in a specimen from the west flank of the Ruby Mountains on the north side of Gilbert Canyon in the Gilbert Canyon district. (WMK)

Reported from an unspecified locality in the northern Pinon Range. (Emmons, 1910)

With muscovite, quartz, beryl, and many other minerals in pegmatite dikes in Dawley Canyon in the Valley View district; also reported as a pink variety with sillimanite and other minerals in pegmatite 1/4 mile north of the Errington-Thiel Mine. (Olson and Hinrichs, 1960)

HUMBOLDT COUNTY With biotite and cordierite in contact-metamorphosed pelitic rocks adjacent to granodiorite stocks in the Potosi district in the Osgood Mountains. (Hotz and Willden, 1964)

In a specimen of hornfels with cordierite and *feldspar* from the Bloody Run Hills in the Sherman district. (WMK)

LINCOLN COUNTY Found with sillimanite in hornfels on Blind Mountain in the Highland Range. (Westgate and Knopf, 1932)

MINERAL COUNTY In the Fitting district in metamorphosed rocks of the Excelsior Formation as fine disseminations and scattered blebs; also noted as coarse blue to lavender crystals to 10 cm with pyrophyllite, halloysite, diaspore, and other minerals at the Dover (Donnelly Andalusite), Bismark, and Green Talc Mines. (Vanderburg, 1937a; Klinger, 1952; Ross, 1961; Tatlock, 1964; HMM)

A constituent of hornfels at the Bataan Mine in the Garfield district. (Ponsler, 1977)

As minute crystals or aggregates with *sericite* at the Deep and Kenjay prospects in the Pamlico district. (JHS)

NYE COUNTY In hornfels adjacent to granodiorite at the Victory Tungsten Mine in the Lodi district. (Humphrey and Wyatt, 1958)

PERSHING COUNTY In disseminated grains in irregular lenses in schistose rock that predated, and were partly replaced by, dumortierite at the Champion Mine in the Sacramento district. (Knopf, 1924; Jones and Grawe, 1928; Kerr and Jenney, 1935; Gianella, 1941)

With dumortierite at Lincoln Hill in the Rochester district. (Grawe, 1928b; WMK; Tatlock, 1964)

With *sericite* and pyrophillite at the Pinite Mine in the Spring Valley district. (Kerr, 1940a; Johnson, 1977)

As *chiastolite* crystals to 20 cm in schist at the Auld Lang Syne Mine in the Blue Wing Mountains. (D. Hudson, w.c., 1999)

WASHOE COUNTY A constituent of the rocks of the Nightingale sequence in the Fox Range. (Bonham and Papke, 1969)

Abundant in the metamorphic Peavine sequence, particularly in the Hungry Valley area. (Bonham and Papke, 1969)

WHITE PINE COUNTY *Chiastolite* was found as crystals in fragments of black shale enclosed in obsidian 500 yards due north of the Ruth Shaft in the Robinson district. (Melhase, 1934d)

ANDORITE

$PbAgSb_3S_6$

Andorite is a moderately rare sulfosalt in some hydrothermal deposits of the precious metals. It is generally associated with other sulfides of lead and silver, and in some cases with tin minerals.

ESMERALDA COUNTY Reported at an unnamed prospect in the Sylvania district near Cucomungo Spring. (LANH)

LANDER COUNTY Identified in a specimen from the Cortez Mine in the Cortez district. (JHS)

MINERAL COUNTY A single complex crystal was found in a quartz vug in a specimen with owyheeite and sphalerite at the Broken Hills Mine in the Broken Hills district. (Jenkins, 1981)

NYE COUNTY As small crystals and masses in silver-tin veins with stephanite and pyrargyrite at the Morey and Keyser Mines in the Morey district; some crystals are partly or completely replaced by owyheeite, diaphorite, or fizélyite. (Shannon, 1922; Kral, 1951; Williams, 1968; S. Pullman, w.c., 1999)

WHITE PINE COUNTY Reported as tiny, prismatic crystals in complex sulfosalt ore in the Taylor district; also as crude crystals with miargyrite, tetrahedrite, and pyrostilpnite at the Patriot Mine. (Williams, 1965; Jenkins, 1981)

ANDRADITE

$Ca_3Fe_2^{3+}(SiO_4)_3$

Andradite, which forms series with grossular and schorlomite, is a widespread *garnet*-group mineral that is most common in contact-metamorphosed rocks, but also occurs in schist, serpentinite, and some igneous rocks.

CLARK COUNTY Tentatively identified as microgranular masses in magnesian rocks of the Moenkopi Formation at the Lavina Mine in the Goodsprings district. (Hewett, 1931)

ELKO COUNTY Identified using density and refractive index in skarn in the Railroad district. (Ketner and Smith, 1963)

Reported in a specimen from the Aljo prospect in the Merrimac district. (R. Thomssen, w.c., 1999)

ESMERALDA COUNTY *Garnet* with a composition very near the andradite end member occurs at a pyrometasomatic deposit at an unspecified locality in the Weepah district in Paymaster Canyon. (Gulbrandsen and Gielow, 1960)

EUREKA COUNTY As dark-red-brown anhedral crystals in the calc-silicate mineral assemblage at the Goldstrike Mine in the Lynn district. (Berry, 1992)

LANDER COUNTY As crystals to 15 mm in skarn at the West orebody in Copper Canyon in the Battle Mountain district. (Theodore and Blake, 1978)

As small transparent brown crystals in rhyolite in the Izenhood district at Izenhood Ranch. (Fries, 1942)

LINCOLN COUNTY In large amounts at the Manhattan Mine in the Pioche district. (Westgate and Knopf, 1932)

Zoned andradite-grossular garnet, 60–95 percent andradite, occurs in skarn at the Lincoln Tungsten Mine in the Tem Piute district. (Buseck, 1967; JHS)

LYON COUNTY As overgrowths and in crosscutting veins in skarn at the Douglas Hill and Mason Valley Mines in the Yerington district. (Harris, 1980; LANH)

Massive reddish-brown andradite is a constituent of iron ore along a limestone-granite contact in the Red Mountain district about 12 miles northeast of Dayton on the Lyon/Storey county line. (Harder, 1909)

MINERAL COUNTY Reported with *axinite* and other skarn minerals in the Fitting district in the Gillis Range, 7 miles north of Luning. (S. Pullman, w.c., 1999)

As subhedral to euhedral crystals in contact-metamorphosed conglomerate at the Silver Gulch and Birdsong Mines in the Marietta district. (Jenkins, 1981)

Reported at an unspecified locality in the Mountain View district near Schurz. (LANH)

A constituent of fresh skarn at the Nevada Scheelite Mine in the Leonard district. (Warner et al., 1959)

PERSHING COUNTY At the MGL Mine in the Nightingale district as orange-brown crystals to 3 inches; also at the Alpine Mine as lustrous brown, euhedral crystals to 2 inches. XRD indicates a grossular component. (Crowley, 2000; GCF)

With grossular in skarn in the Panther Canyon area on the west flank of the Humboldt Range in the Rye Patch district; both *garnets* have a small proportion of the spessartine molecule. (Vitaliano, 1944)

WASHOE COUNTY In skarn zones in the Peavine district on Peavine Mountain. (Jenkins, 1981)

WHITE PINE COUNTY Reported in the Robinson district as red-brown grains with polychromatic twinning; most *garnet* in contact-metamorphosed rock in the district is andradite-rich. (Spencer et al., 1917; James, 1976)

ANGLESITE

$PbSO_4$

Anglesite is not as common as the carbonate cerussite as an oxidation product of primary lead minerals; nonetheless, it is widespread near the surface in many base-metal deposits in Nevada. Notable crystals have come from the Eureka and Goodsprings districts.

CHURCHILL COUNTY In oxidized lead-silver veins as microcrystals and masses to 1 cm in galena with wulfenite, cerussite, chlorargyrite, plumbojarosite, and vanadinite at the Chalk Mountain Mine in the Chalk Mountain district. (Vanderburg, 1940; Schilling, 1962a; Schilling, 1962b; Jensen, 1990b)

Reported in a specimen from along a pipeline road on the west side of Fireball Ridge in the Truckee district. (R. Thomssen, w.c., 1999)

CLARK COUNTY Widespread in the mines of the Goodsprings district; reported as excellent crystals from the Yellow Pine and Prairie Flower Mines. (Hewett, 1931; Melhase, 1934e; Sinkankas, 1967; Jedwab, 1998b; LANH; S. Pullman, w.c., 1999)

DOUGLAS COUNTY With cerussite, galena, and other minerals at the Iron Pot prospect in the Red Canyon district. (Lawrence, 1963)

ELKO COUNTY With cerussite and trace amounts of vanadinite at an unknown locality in the Contact district. (WMK)

Silver-bearing anglesite and cerussite occur with galena, chalcopyrite, pyrite, chrysocolla, malachite, and rare wulfenite and bismuthinite in replacement bodies along faults in limestone at the Keystone Mine in the Dolly Varden district. (Hill, 1916; Melhase, 1934d; Schilling, 1979)

Reported in a specimen from the Grey Eagle Mine in the Railroad district. (WMK)

In the Spruce Mountain district at the Keystone Mine (Banner Hill claim) with wulfenite, cerussite, galena, and other minerals in a fracture zone in limestone; also at the Killie Mine as concentric rims around galena and as small, yellow, subhedral crystals in cavities in gossan. (Hill, 1916; Schilling, 1979; Schrader, 1931b; Dunning and Cooper, 1987)

ESMERALDA COUNTY Tentatively identified with cerussite and pyromorphite at the Nevada Goldfield and Centennial Mines in the Goldfield district. (Ransome et al., 1907)

A minor mineral with cerussite at an unspecified locality in the Klondyke district. (Chipp, 1969)

With cerussite, chlorargyrite, smithsonite, and hemimorphite at the Sylvania Mine in the Sylvania district. (Schilling, 1979)

Reported at the Weepah and 3 Metals Mines in the Weepah district. (S. Pullman, w.c., 1999; R. Thomssen, w.c., 1999)

Reported at the Rock Hill Mines in the Rock Hill district. (S. White, w.c., 1999)

EUREKA COUNTY In large lead-silver-gold orebodies replacing limestone with cerussite, plumbojarosite, mimetite, galena, and other minerals at the Richmond-Eureka (Ruby Hill) Mine in the Eureka district; also reported at the Gasperi prospect. (Nolan, 1962; WMK; S. Pullman, w.c., 1999)

Reported with cerussite and pyromorphite in oxidized lead-silver ore in the Mineral Hill district. (Emmons, 1910)

HUMBOLDT COUNTY A minor mineral in the oxidation assemblage at the Lone Tree Mine in the Buffalo Mountain district. (R. Braginton, o.c., 1995)

As colorless plates at the Silver Coin Mine in the Iron Point district. (Thomssen, 1998; J. McGlasses, w.c., 2001)

A minor mineral with galena, chalcopyrite, covellite, and sphalerite at the Silver Hill Mine in the Dutch Flat district. (Willden, 1964)

LANDER COUNTY With other lead minerals in a number of mines in the Battle Mountain district. (Roberts and Arnold, 1965)

With pyromorphite, cerussite, and oxidized zinc minerals at an unspecified locality in the Cortez district. (Gianella, 1941)

As a replacement of galena at the Betty O'Neal Mine in the Lewis district. (J. McGlasson, w.c., 2001)

LINCOLN COUNTY Reported on various claims in the Atlanta district. (Hill, 1916)

Reported with numerous secondary copper and lead minerals at the Jackrabbit Mine in the Bristol district. (Melhase, 1934d)

With cerussite and residual galena in oxidized lead-silver ore at the Groom Mine in the Groom district. (Humphrey, 1945; Tingley et al., 1998a)

With cerussite and malachite at the Rosario Arcurio Mine in the Pahranagat district. (Tschanz and Pampeyan, 1970)

A common mineral in oxidized lead-silver deposits of the Pioche district; specifically reported in the Raymond & Ely vein. (Westgate and Knopf, 1932; WMK)

LYON COUNTY With cerussite, wulfenite, and linarite at the Old Soldier Mine in the Churchill district. (Jenkins, 1981)

Crystallized anglesite occurs with aurichalcite, caledonite, and other minerals at the Jackpot Mine in the Wilson district. (Jenkins, 1981)

MINERAL COUNTY With *limonite* at the Potosi and Candelaria Mines in the Candelaria district. (Page, 1959; Moeller, 1988)

Reported with chrysocolla, malachite, jarosite, native silver, and chlorargyrite at the Mabel Mine in the Garfield district. (Couch and Carpenter, 1943; Ross, 1961)

Reported at the Lucky Boy Mine in the Lucky Boy district. (Hill, 1915)

In a specimen from the Greek Hills Mine in the Pamlico district. (WMK)

With cerussite, hemimorphite, wulfenite, and other minerals at an unspecified locality in the Santa Fe district. (Jenkins, 1981)

Reported in the Broken Hills and Aurora districts. (Schrader, 1947; Jenkins, 1981)

NYE COUNTY In a quartz vein with pyrite,

galena, cerussite, acanthite, sphalerite, and malachite in the Iron Rail group in the Cloverdale district; also with cerussite and galena in the Orizaba Mine. (Hamilton, 1993; Kral, 1951)

Minor with cerussite and wulfenite in gossan at an unspecified locality in the Gabbs district. (Kral, 1951)

In a specimen of lead ore from an unknown locality in the Johnnie district. (WMK)

Excellent crystals of small size occur with cerussite, wulfenite, and mimetite at the San Rafael Mine in the Lodi district. (Kral, 1951)

Tentatively identified with cerussite and chlorargyrite at prospects in the Reveille district. (Kral, 1951)

At the Outlaw Mine in the Round Mountain district as minute crystals in coatings on aikinite with sulfur, corkite, and koechlinite. (Dunning and Cooper, 1991)

With cerussite and silver-bearing galena at the Florence Mine in the San Antone district. (Kral, 1951)

With cerussite in cracks in a galena-sphalerite-pyrite replacement orebody in shale at the Haller property in the Troy district. (Hill, 1916)

With galena and cerussite at the Two G Mine in the Tybo district. (Jenkins, 1981; WMK)

In ore from the Union and Washington districts. (Kral, 1951)

STOREY COUNTY Reported with wulfenite, pyromorphite, and other minerals in oxidized ore from the Comstock district; also noted in drill-hole cuttings with quartz. (Koschmann and Bergendahl, 1968; Schilling, 1990)

WASHOE COUNTY Reported at the Leadville Mine in the Leadville district. (Gianella, 1941)

As yellow-orange crystals to 2 mm with galena and gold at the Olinghouse Mine in the Olinghouse district. (GCF)

With sphalerite, galena, pyrite, and barite at the Burrus and Ruth Mines and Sure Fire prospect in the Pyramid district. (SBC)

With cerussite, silver halide minerals, and native gold in oxidized ore in the Wedekind district. (Gianella, 1941; Overton, 1947)

WHITE PINE COUNTY Identified with cerussite and chlorargyrite at an unspecified locality in the Cherry Creek district. (Hose et al., 1976; JHS)

With galena, cerussite, and other secondary lead minerals at the Cuba Mine in the Granite district. (Melhase, 1934d)

With cerussite and galena at an unspecified locality in the Hunter district. (Melhase, 1934d)

Reported with other secondary lead and zinc minerals at the Grand Deposit Mine in the Muncy Creek district. (Hose et al., 1976)

In oxidized lead-zinc deposits with galena in the Robinson district. (Spencer et al., 1917; WMK)

Reported with cerussite, bindheimite, galena, and a large variety of other minerals at the Monitor Mine in the Taylor district. (Lovering and Heyl, 1974; WMK)

As boxworks after galena with cerussite, linarite, caledonite, elyite, and other minerals at the Silver King Mine in the Ward district. (Jenkins, 1981)

With cerussite, silver-bearing galena, and oxidized copper and zinc minerals in the lead belt near Mount Hamilton in the White Pine district. (Humphrey, 1960; Hose et al., 1976)

ANHYDRITE
$CaSO_4$

Anhydrite, a common mineral in marine gypsum deposits and in salt domes, is considered by some to have formed by the dehydration of gypsum and is common in some Nevada gypsum deposits. It is also found in some sulfide-bearing hydrothermal deposits.

CARSON CITY Reported in a specimen from the Carson City area, but probably from the Mound House district in adjacent Lyon County (see below). (AMNH)

CLARK COUNTY As lenticular beds or irregular masses in sharp contact with gypsum at the Blue Diamond Gypsum Mine, a major Nevada producer in operation since 1925. Also in masses as large as 100 tons in gypsum beds in underground mines in NE 1/4 Sec. 7, T22S, R60E in the Arden district. (Papke, 1987)

ESMERALDA COUNTY Identified in drill core at the Goldfield Deep Mines (2100-foot level) in the Goldfield district. (WMK)

Formed by drying gypsum crystals at surface temperatures at the Clayton Valley playa in the Silver Peak Marsh district. (JHS)

EUREKA COUNTY As microcrystalline grains in blue efflorescent material with szomolnokite and gunningite at the Goldstrike Mine in the Lynn district. (GCF)

LYON COUNTY Light-gray anhydrite occurs in the deeper parts of a gypsum mass that has been mined since 1914 at the Adams Mine in the Mound House district. (Papke, 1987; Tingley et al., 1998b)

As a pure white crystalline mass below massive beds of gypsum at the Ludwig Mine in the Yerington district. (Jones, 1912; Rogers, 1912a; Melhase, 1935e; UCB)

NYE COUNTY Reported in a specimen from an unknown locality in the Gabbs district. (NMBM)

PERSHING COUNTY Light-gray finely crystalline anhydrite occurs at depths in excess of 85 feet below the surface in large gypsum orebodies that have been mined since 1922 at the Empire Gypsum Mine in the Gerlach district. (Papke, 1987)

Identified in a specimen from an unknown locality in the Hooker district; may be from the Empire Mine in the Gerlach district. (JHS)

Identified in a specimen that is probably from a gypsum deposit in Muttlebury Canyon in the Muttlebury district. (JHS; Papke, 1987)

ANKERITE

$Ca(Fe^{2+},Mg,Mn)(CO_3)_2$

A carbonate mineral related to dolomite that occurs as a gangue mineral in base-metal veins, as a low temperature metasomatic mineral in sedimentary rocks, and in some iron-rich high-grade metamorphic rocks.

ELKO COUNTY Partially replaces dolomite and found as veinlets at the Rossi Mine in the Bootstrap district. (Snyder, 1989)

Part of the host rock assemblage at the Jerritt Canyon Mines in the Independence Mountains district. (Hofstra, 1994)

EUREKA COUNTY In calcite veins and in calc-silicate rock at the Goldstrike Mine in the Lynn district. (Berry, 1992)

HUMBOLDT COUNTY Reported at the Silver Coin Mine in the Iron Point district. (Thomssen, 1998)

LANDER COUNTY Reported as a gangue mineral in ore from the Reese River district. (Stewart et al., 1977a)

NYE COUNTY Reported from the Currant (Gold Point) district. (J. McGlasson, w.c., 2001)

As lustrous brown rhombohedral crystals to 2 mm near Warms Springs in the Tybo district. (GCF)

PERSHING COUNTY In a specimen from the Montgomery and Pershing Mine area in the Antelope Springs district. (JHS)

With cinnabar and dolomite at the Paymaster prospect in the Black Knob district. (Bailey and Phoenix, 1944)

STOREY COUNTY Identified in a specimen from the Castle Peak Mine in the Castle Peak district. (JHS)

ANNABERGITE

$Ni_3(AsO_4)_2 \cdot 8H_2O$

Annabergite is a product of surface oxidation of primary nickel arsenides such as nickeline.

CHURCHILL COUNTY As apple-green masses and euhedral crystals to 0.5 mm in partially oxidized Ni- and Co-bearing veins with morenosite, retgersite, *garnierite,* nickeline, millerite, marcasite, quartz, and dickite at the Lovelock and Nickel Mines in the Table Mountain district. (Ransome, 1909b; Ferguson, 1939; Frondel and Palache, 1949; M. Jensen, w.c., 1999)

CLARK COUNTY As bright green masses at the Yellow Pine Mine in the Goodsprings district; also reported from the Copper Chief and Price's Navajo Mines. (Hewett, 1931; WMK)

LANDER COUNTY Reported in a specimen from an unknown locality in the Battle Mountain district. (WMK)

MINERAL COUNTY Reportedly Nevada's best occurrence, with euhedral lustrous crystals to 5 mm associated with conichalcite and olivenite at the Johnson and Candelaria open-pit mines in the Candelaria district. (UCB; F. Cureton, w.c., 1999; M. Jensen, w.c., 1999)

ANNITE

$KFe_3^{2+}AlSi_3O_{10}(OH)_2$

Annite is an iron-rich member of the mica group that is related to biotite and occurs in magnesium-poor igneous and metamorphic rocks.

LINCOLN COUNTY Reported from rhyolite domes in the Kane Springs Wash volcanic center. (Novak, 1985)

ANORTHITE

$CaAl_2Si_2O_8$

Anorthite is the calcium end member of the *plagioclase feldspar* group. It is most common in mafic igneous rocks, but also occurs in skarn, high-grade regionally metamorphosed rocks, and in meteorites. Includes the intermediate varieties *labradorite* and *bytownite,* which are considered species by some mineralogists.

CLARK COUNTY Reported as *labradorite* in a specimen from Boulder Canyon. (WMK)

EUREKA COUNTY As gray chatoyant *labradorite* grains to 8 mm in diorite at the Goldstrike Mine in the Lynn district. (GCF)

As *labradorite* phenocrysts in andesite that hosts iron ore at the Barth Iron Mine in the Safford district. (Shawe et al., 1962)

LANDER COUNTY Part of the skarn assemblage at the West orebody in Copper Canyon in the Battle Mountain district. (Theodore and Blake, 1978)

NYE COUNTY At Lunar Craters as xenocrysts in basalt flows and at Easy Chair Crater. (Bergman, 1982)

PERSHING COUNTY Reported as *labradorite* in a specimen from 3 miles east of Placerites Camp. (UCB)

ANORTHOCLASE

$(Na,K)AlSi_3O_8$

A species of the alkali *feldspar* group that is intermediate between sanidine and albite. Generally occurs in sodium-rich igneous rocks, particularly rhyolite.

EUREKA COUNTY Reported as intergrowths with albite at the Goldstrike Mine in the Lynn district. (Walck, 1989)

LINCOLN COUNTY In large phenocrysts that comprise 40 percent of a trachyte flow near the mouth of Boulder Canyon, Kane Springs Wash volcanic center; associated with augite, pigeonite, ilmenite, and titanomagnetite; also occurs in lesser amounts in other trachyte flows and in syenite of the same volcanic center. (Novak, 1985)

NYE COUNTY In the Lunar Craters volcanic field as xenocrysts in basalt flows, and as loose, clear to cloudy white cleaved crystals to 9 cm at Easy Chair Crater. (Bergman, 1982)

WASHOE COUNTY As phenocrysts, along with sanidine, *plagioclase,* and biotite, in the widespread Nine Hill Tuff. (Deino, 1985)

WHITE PINE COUNTY Identified by optical properties in a specimen from Garnet Hill in the Robinson district. (R. Thomssen, w.c., 1999)

ANTIGORITE

$(Mg,Fe^{2+})_3Si_2O_5(OH)_4$

Antigorite, a *serpentine*-group mineral, is widespread in small quantities. It occurs in serpentinite, in some contact zones in magnesian limestones, and as an alteration mineral in various igneous rocks.

NYE COUNTY A constituent of contact rocks as a replacement of forsterite at the Gabbs Magnesite-Brucite Mine in the Gabbs district. (Cleveland, 1963; Schilling, 1968a)

STOREY COUNTY With possible magnesite as an alteration product of *pyroxene* phenocrysts in meta-andesite in the Clark district in the northern Virginia Range near Clark Station, 36 miles east of Reno. (Rose, 1969)

ANTIMONY

Sb

Native antimony is relatively uncommon and generally found in hydrothermal veins. The pegmatite occurrence listed below (Oreana Mine, Pershing County) may be unique.

ELKO COUNTY Reported from an unspecified locality in the Charleston district. (LANH)

LINCOLN COUNTY Reported in a specimen from an unknown locality in the Pioche district. (SM)

NYE COUNTY Reported in specimens from an unspecified locality in the Jett district in Wall Canyon. (WMK; LANH)

PERSHING COUNTY Rare as small irregular pods and tiny prismatic crystals in a quartz-*oligoclase*-albite-phlogopite pegmatite with scheelite, fluorite, beryl, and related antimony oxides at the Oreana (Little) Tungsten Mine in the Rye Patch district. (Olson and Hinrichs, 1960; Lawrence, 1963)

ANTLERITE

$Cu_3^{2+}(SO_4)(OH)_4$

Antlerite is an uncommon or seldom recognized mineral in the oxidized zones of some copper deposits. It is usually found with malachite, chrysocolla, brochantite, and other minerals.

ESMERALDA COUNTY As green crystals at the 3 Metals Mine in the Weepah district. (M. Jensen, w.c., 1999)

EUREKA COUNTY Light-green crusts to vitreous filiform crystals and coatings as an efflorescence near sulfide veins in diorite at the Goldstrike Mine in the Lynn district. (GCF)

LANDER COUNTY Reported in a specimen from the Estella vein, Betty O'Neal Mine in the Lewis district; also reported from the Lucky Rocks claims. (R. Thomssen, w.c., 1999; J. McGlasson, w.c., 2001)

LYON COUNTY Reported in a specimen from an unknown locality in the Yerington district. (HMM)

MINERAL COUNTY As crystalline masses and coatings with malachite, chalcanthite, brochantite, and cuprite at the Northern Light Mine in the Mountain View district. (Palache, 1939; Ross, 1961)

PERSHING COUNTY Green crystals to 1 mm at the Big Mike Copper Mine in the Tobin and Sonoma Range district, Sec. 23, T31N, R39E. (M. Jensen, w.c., 1999)

APATITE

$(Ba,Ce,K,Na,Pb,Y,Ca,Sr)_5[(As,P,Si,V)O_4]_3$
(F,Cl,OH) (general)

A mineral group name. *Apatite* is a common accessory mineral in almost all igneous rocks. It is also found in granite pegmatite, metamorphic rocks, sedimentary rocks, veins, and carbonatite. Species have not been identified at the following localities, which are a partial listing of occurrences in Nevada.

CHURCHILL COUNTY In varying amounts in veins at the Black Joe prospect in the Copper Kettle district. (Shawe et al., 1962)

Reported at an unspecified locality in the Dixie Valley district. (WMK)

As an accessory mineral in volcanic rocks in the Fairview district. (Schrader, 1947)

ELKO COUNTY Reported as large crystals with zircon in monzonite at an unspecified locality in the Dolly Varden district. (Melhase, 1934d)

Part of the detrital assemblage in Paleozoic sedimentary rocks at the Jerritt Canyon Mines in the Independence Mountains district. (Hofstra, 1994)

In pegmatite dikes with quartz, beryl, and other minerals in Dawley Canyon in the Valley View district. (Olson and Hinrichs, 1960)

EUREKA COUNTY One of the principal gangue minerals in iron ore at the Modarelli Mine in the Modarelli-Frenchie Creek district. (Shawe et al., 1962)

LYON COUNTY Reported with quartz, wollastonite, and epidote at the Bluestone and other mines in the Yerington district. (Melhase, 1935e)

NYE COUNTY As phosphatic nodules in mudstone, chert, and barite-rich conglomerate in east Northumberland Canyon in the Northumberland district. (Shawe et al., 1969)

PERSHING COUNTY Reported at the Heizer Iron Mine and Emery-Fisk prospect in the Mineral Basin district; with magnetite in pods up to 2 feet wide at the latter. (Shawe et al., 1962; UCB)

In small amounts in scheelite-bearing skarn in the Nightingale district. (Smith and Guild, 1944)

APHTHITALITE

$(K,Na)_3Na(SO_4)_2$

A soluble sulfate mineral, also known as *glaserite,* that occurs in altered volcanic rocks and as an evaporite mineral in saline lakes.

ESMERALDA COUNTY Precipitated from lithium-rich brine in solar evaporation ponds in the Silver Peak Marsh district. (JHS)

APJOHNITE

$Mn^{2+}Al_2(SO_4)_4 \cdot 22H_2O$

Manganese *alum;* generally occurs in efflorescent deposits in caves and mine workings.

NYE COUNTY As silky, pale-yellow fibers to 3 cm at the Twentieth Century Mine in the Cactus Springs district; also reported with alunogen and melanterite from the Urania Mine. (M. Jensen, w.c., 1999; J. Price, o.c., 1999)

WASHOE COUNTY As a white to pale-yellow

efflorescence with hexahedrite and pickeringite at the Wedekind Mine in the Wedekind district. (GCF)

ARAGONITE

CaCO₃

Much less common than its polymorph calcite, aragonite is metastable at normal temperatures and pressures, reverting to calcite with time. It forms as a primary precipitate in sediments; is common in glaucophane schist; and also occurs in hot springs, in caves, and as a secondary mineral in ore deposits.

CHURCHILL COUNTY Forms tan, hoppered, spear-shaped crystals and crystal clusters to 5 cm long in vugs in altered limestone at the Coppereid prospect in the Copper Kettle district. (GCF)

Reported as a secondary mineral at the Fairview Mine in the Fairview district. (Schrader, 1947)

As crystals in vuggy veins at the Green prospect in the Lake district. (R. Walstrom, w.c., 1999; R. Thomssen, w.c., 1999)

Reported in a specimen from the Nickel Mine in the Table Mountain district. (R. Thomssen, w.c., 1999)

CLARK COUNTY Reported as fine crystals in caves at the Prairie Flower, Mobile, and Shenandoah Mines in the Goodsprings district. (Hewett, 1931)

Reported as crystals to 1 cm in miarolitic cavities and pegmatites in the Eldorado and Newberry Mountains. (S. Scott, w.c., 1999)

ELKO COUNTY Found in cracks along with calcite at an unspecified locality in the Contact district. (Bailey, 1903)

ESMERALDA COUNTY Reported at the Weepah Mine in the Weepah district. (S. Pullman, w.c., 1999)

EUREKA COUNTY In fine crystals and stalactitic groups with calcite at the Black Diamond and Richmond Mines in the Eureka district. (BMC; UCB; F. Cureton, w.c., 1999)

As white to clear, needlelike crystals to 5 mm and massive white veins in carbonaceous limestone at the Goldstrike Mine in the Lynn district. (GCF)

As colorless to white, acicular crystals from the Knob Hill area at the Gold Quarry Mine in the Maggie Creek district. (Jensen et al., 1995)

HUMBOLDT COUNTY Reported at the Silver Coin Mine in the Iron Point district. (Thomssen, 1998)

LANDER COUNTY Reported in a specimen from the Reese River district. (WMK)

LINCOLN COUNTY Exceptional examples of white crystal clusters to 15 cm at the Bristol Silver Mine in the Bristol district, Secs. 30 and 31, T3N, R66E; also reported in unusual forms such as balls, branches, and flos ferri. (M. Jensen, w.c., 1999; WMK; AMNH; UCB)

LYON COUNTY Reported in a specimen from the Jackpot Mine in the Wilson district. (LANH)

MINERAL COUNTY As crystals to 4 cm replaced by calcite at the Northern Light Mine in the Mountain View district; many specimens were collected in the 1950s or 1960s. (M. Jensen, w.c., 1999)

NYE COUNTY Reported in a specimen with calcite from the San Rafael Mine in the Lodi district. (LANH)

As radiating aggregates at the Northumberland Mine in the Northumberland district. (Kokinos and Prenn, 1985)

Reported in a specimen from an unspecified locality in the Tonopah district. (AMNH)

PERSHING COUNTY The lead-bearing variety *tarnowitzite* was reported with stibnite and cinnabar at the Red Bird Mine in the Antelope Springs district. (Bailey and Phoenix, 1944)

With fluorite, calcite, and *chalcedony* at the Valerie Fluorspar Mine in the Imlay district. (Johnson, 1977)

Reported in a specimen from the Champion Mine in the Sacramento district. (LANH)

WASHOE COUNTY As radiating groups of euhedral, transparent, colorless crystals to 1 mm in vugs with malachite, azurite, cornwallite, goethite, and chrysocolla at the Burrus Mine in the Pyramid district. (Jensen, 1994)

Deposited in steam well casings at Washoe Hot Springs. (Dickson and Tunnell, 1968)

Reported as "coralloidal" aragonite along the borders of Winnemucca Lake. (Melhase, 1934a)

WHITE PINE COUNTY Reported in a specimen from Goshute Cave, on the east side of the Cherry Creek Range about 1.5 miles north of Goshute Creek, Sec. 1, T25N, R63E. (EMP)

In groups of white to pale-green, radiating acicular crystals in mammillary crusts at the

Robust Mine in the Robinson district. (Spencer et al., 1917)

With calcite in stalactites at Lehman Caves. (WMK)

ARFVEDSONITE
$NaNa_2(Fe_4^{2+}Fe^{3+})Si_8O_{22}(OH)_2$

A sodium-rich member of the amphibolite group, arfvedsonite mostly occurs in alkaline igneous rocks.

HUMBOLDT COUNTY Fluorine-bearing arfvedsonite is found as a constituent of pantellerite tuff with alkali *feldspar,* quartz, aenigmatite, and *clinopyroxene* in the McDermitt caldera. (Conrad, 1984; Rytuba and McKee, 1984)

LINCOLN COUNTY In peralkaline rocks of the Kane Springs Wash caldera. (Noble, 1968)

NYE COUNTY As a groundmass and vapor-phase mineral in peralkaline rocks of the Silent Canyon volcanic center. (Noble et al., 1968)

ARGENTOJAROSITE
$Ag_2Fe_6^{3+}(SO_4)_4(OH)_{12}$

Argentojarosite is a member of the jarosite group and a product of the oxidation of precious-metal deposits under acid conditions.

CHURCHILL COUNTY In gold ore at the Last Hope and Cripple Queen Mines in the Holy Cross district. (Peters et al., 1996)

CLARK COUNTY As micron-sized grains with potarite, gold, and other minerals at the Boss Mine in the Goodsprings district. (Jedwab et al., 1999)

ELKO COUNTY Identified with acanthite after pyrite in silicified breccia at the Dexter Mine in the Tuscarora district. (Henry et al., 1999)

LINCOLN COUNTY Reported in a specimen from the Bristol Mine in the Bristol district. (AMNH)

Tentatively identified as the ore mineral at the King Midas Mine in the Ely Springs district. (Tschanz and Pampeyan, 1970)

MINERAL COUNTY Reported at the Rawhide Mine in the Rawhide district. (Black, 1991)

WASHOE COUNTY In silicified ledges in the Peavine district. (D. Hudson, w.c., 1999)

ARGYRODITE
Ag_8GeS_6

Argyrodite is a rare mineral; it is most common in high-temperature tin-silver deposits in Bolivia, but also occurs in low-temperature hydrothermal deposits. The locality listed below is one of only two occurrences reported in the United States.

NYE COUNTY Tentatively identified in minor amounts with polybasite and pyrargyrite in polished sections of high-grade ore from various mines in the main part of the Tonopah district. (Bastin and Laney, 1918)

ARSENDESCLOIZITE
$PbZn(AsO_4)(OH)$

A very rare mineral found with other secondary minerals in the Tsumeb copper deposit, Africa.

HUMBOLDT COUNTY Pale-yellow pulverulent coatings on altered matrix at the Silver Coin Mine in the Iron Point district. (Thomssen, 1998; HUB)

PERSHING COUNTY As orange crystals to 1 mm associated with smithsonite at the Red Bird Mine in the Antelope Springs district; possibly the world's largest crystals for this species. (M. Jensen, w.c., 1999)

ARSENIC
As

Arsenic is a relatively rare native metal that occurs in association with sulfide minerals in hydrothermal deposits, mostly of nickel, cobalt, and silver.

DOUGLAS COUNTY Reported in a specimen from an unspecified locality at the south end of the Pine Nut Range. (WMK)

ELKO COUNTY With lorandite, cinnabar, realgar, orpiment, arsenopyrite, tetrahedrite, marcasite, antimonial sphalerite, and an unidentified sulfide in a veinlet cutting an andesite dike at one of the Jerritt Canyon Mines in the Independence Mountains district. (Hofstra, 1994; Phinisey et al., 1996)

EUREKA COUNTY As nearly spherical grains to 30 microns in argillaceous siltstone and in barite associated with gold, stibnite, galena, sphalerite, jordanite, and tennantite at the Carlin Mine in the Lynn district. (Hausen and Kerr, 1968)

HUMBOLDT COUNTY Reported after arsenopyrite at the National Mine in the National district. (Lindgren, 1915)

PERSHING COUNTY Massive with traces of bismuth at an unspecified locality near Rochester. (Schrader et al., 1917)

STOREY COUNTY Reported at the Ophir Mine in the Comstock district. (Palache et al., 1944)

WASHOE COUNTY Reported in considerable quantity in a prospect a few miles south of Pyramid Lake; this material may actually be arsenopyrite. (Schrader et al., 1917; WMK)

ARSENIOSIDERITE

$Ca_2Fe_3^{3+}(AsO_4)_3O_2 \cdot 3H_2O$

Arseniosiderite is an uncommon product of surface oxidation of ore containing arsenic sulfides.

CHURCHILL COUNTY Reported in a specimen from the Fireball Mine in the Truckee district. (R. Thomssen, w.c., 1999)

CLARK COUNTY Reported from the Boss Mine in the Goodsprings district. (Kepper, this volume)

ELKO COUNTY As brown resinous masses with dussertite at the Rain Mine in the Carlin district. (GCF)

ESMERALDA COUNTY Reported at the Weepah Mine in the Weepah district. (S. Pullman, w.c., 1999)

HUMBOLDT COUNTY As rust-colored spheroids with pharmacosiderite, jarosite, barite, and dufrénite as fracture coatings in oxidized rock at the Lone Tree Mine, Sequoia Pit, in the Buffalo Mountain district. (GCF)

LANDER COUNTY In specimens from the Wilson-Independence Mine in the Battle Mountain district. (R. Thomssen, w.c., 1999)

LINCOLN COUNTY As brown micaceous material partly replacing scorodite after arsenopyrite in veins at the Advance and Old Democrat veins in the Chief district. (Callaghan, 1936)

Reported in a specimen from the Silver King Mine in the Silver King district. (R. Thomssen, w.c., 1999)

MINERAL COUNTY With barite crystals in a prospect in the Fitting district west of Rhyolite Pass and north of Luning. (M. Jensen, w.c., 1999)

NYE COUNTY As red-brown micaceous material

in masses to several millimeters with beudantite and scorodite in cavities in crustified quartz at the San Rafael Mine in the Lodi district. (Jenkins, 1981)

ARSENOLAMPRITE

As

A dimorph with arsenic; occurs with sulfides and carbonate minerals.

EUREKA COUNTY A copper-bearing variety (1 to 2 wt. percent Cu) was identified by XRD in silver-gray, sectile grains to 1 mm in silicified collapse breccias in the Screamer deposit with quartz, calcite, realgar, pyrite, and kaolinite at the Goldstrike Mine in the Lynn district. (D. Brosnahan, o.c., 1997)

ARSENOLITE

As_2O_3

Arsenolite is a relatively uncommon soluble mineral that occurs as a secondary mineral in deposits containing arsenic sulfides; also formed by mine fires. The Manhattan occurrence listed below has produced some exceptional crystals that are among the world's best.

CHURCHILL COUNTY Identified by XRD in a sample with annabergite and erythrite from the Lovelock Mine in the Table Mountain district. (M. Coolbaugh, o.c., 2001)

ESMERALDA COUNTY Reported as fluffy, white material with arsenopyrite and scorodite at the Mickspot Mine in the Gilbert district. (Lawrence, 1963)

EUREKA COUNTY Reported in a specimen with malachite at an unspecified locality in the Eureka district. (UAMM)

An oxidation-zone mineral at the Carlin Mine in the Lynn district. (Hausen and Kerr, 1968)

HUMBOLDT COUNTY Minute octahedrons of arsenolite occur in a *smectite*like mineral on massive realgar, usually associated with pharmacolite and guérinite at the Getchell Mine in the Potosi district. (Dunning, 1988)

Reported at the Columbia Mine in the Varyville district. (Vanderburg, 1938a)

NYE COUNTY As white, yellow, and reddish octahedrons in gold-arsenic-antimony ore in the White Caps Mine in the Manhattan district. Also formed under oxidizing conditions during the mine fire of 1958 as clear, yellowish-red (from included minerals) octahedral crystals to 2 cm on charred mine timbers and on the walls of the main shaft. (Ferguson, 1924; Melhase, 1935d; Palache et al., 1944; Gibbs, 1985; AMNH)

PERSHING COUNTY Reported in a specimen from the Majuba Hill Mine in the Antelope district. (AMNH)

STOREY COUNTY Reported to coat native arsenic at the Ophir Mine in the Comstock district. (Palache et al., 1944)

ARSENOPYRITE

FeAsS

Arsenopyrite is the most abundant and common arsenic mineral, mostly occurring in hydrothermal deposits. It occurs at many localities in Nevada, a few of which have produced fine crystals.

CARSON CITY As lenticular bodies in schist at the Rafetto property in the Voltaire district. (Overton, 1947)

CHURCHILL COUNTY With stibnite, pyrite, and sphalerite in quartz veins in the Antimony King Mine; and with marcasite, pyrite, and several antimony species in the Hoyt Mine in the Bernice district. (Lawrence, 1963)

Reported in gold-silver ore at the Puzzle claim in the Sand Springs district. (WMK)

Reported in arsenic ore at an unspecified locality in the White Cloud (Coppereid) district. (WMK)

CLARK COUNTY Reported in bitumen from the Boss Mine in the Goodsprings district. (Kepper, this volume)

DOUGLAS COUNTY With stibnite, galena, jamesonite, and pyrite at the Iron Pot Mine in the Red Canyon district. (Lawrence, 1963)

In a specimen from the Go Getter Mine in the Wellington district. (WMK)

As silvery crystals to 3 mm in quartz veins with massive scorodite in bulldozer cuts 1 mile east of Holbrook Junction. (M. Jensen, w.c., 1999)

Reported in a specimen from the south end of the Pine Nut Range. (WMK)

ELKO COUNTY As diamond-shaped crystals to 0.3 mm in gold ore at the Rossi Mine in the Bootstrap district. (Snyder, 1989)

Reported in quartz veins in the Burner and Island Mountain districts. (Cornwall, 1964)

In small veins at the Rain Mine in the Carlin district. (D. Heitt, o.c., 1991)

Common in quartz veins with stibnite, sphalerite, pyrite, chalcopyrite, and tetrahedrite at the Graham, Prunty, and Rescue Mines in the Charleston district. (Lawrence, 1963)

With pyrite and galena in gold-bearing quartz veins in quartzite at the Lucky Boy and Lucky Girl Mines in the Edgemont district. (Emmons, 1910; Cornwall, 1964)

With pyrite in a brecciated quartz vein at the Link Mine in the Gold Circle district. (LaPointe et al., 1991)

With tetrahedrite, stibnite, and pyrargyrite at the Buckeye and Ohio Mine in the Good Hope district; also in quartz veins at the Good Hope Mine. (Emmons, 1910; Lawrence, 1963)

As euhedral diamond-shaped crystals to 20 microns at the Jerritt Canyon Mine and Big Springs Mines in the Independence Mountains district. (Youngerman, 1992; Hofstra, 1994)

As tiny grains with marcasite at the Hollister deposit in the Ivanhoe district. (Deng, 1991)

In minor amounts with pyrite, sphalerite, and cinnabar in a Carlin-type gold deposit at the Kinsley Mountain deposit in the Kinsley district. (Monroe, 1991)

With gold and silver at the Nelson Mine in the Mountain City district. (Emmons, 1910)

As small grains with marcasite and yellow antimony oxide in quartz at the Red Cow, Rock Creek, and other prospects in the Rock Creek district. (Lawrence, 1963)

Massive arsenopyrite occurs sparingly with pyrite in unoxidized portions of replacement veins in limestone at the Killie Mine in the Spruce Mountain district. (Dunning and Cooper, 1987)

In unoxidized ore in the Tuscarora district; noted as pseudomorphs after stibnite. (Emmons, 1910; Melhase, 1934c; WMK)

ESMERALDA COUNTY With scorodite at the Mickspot Mine in the Gilbert district. (Lawrence, 1963)

Reported as gold-bearing in a specimen from an unspecified locality in the Silver Peak district. (WMK)

Reported at an unspecified locality in the Weepah district. (WMK)

EUREKA COUNTY Reported in gold ore with calcite, realgar, orpiment, stibnite, quartz, and pyrite at an unspecified locality in Mill Canyon in the Cortez district. (WMK; JHS)

In unoxidized ore with galena, pyrite, and sphalerite in the Eureka district; specifically reported from the Fad Shaft. (Nolan, 1962; Schilling, 1979; M. Jensen, w.c., 1999)

In gold ore with pyrite at the Goldstrike Mine in the Lynn district. (Ferdock et al., 1997)

In siliceous barite breccia as minute grains with pyrite, sphalerite, tetrahedrite, galena, and greenockite at the Gold Quarry Mine in the Maggie Creek district; also occurs as slightly corroded crystals to 1 mm in matrix and as freestanding individuals in quartz-lined vugs. (Jensen et al., 1995)

HUMBOLDT COUNTY A minor constituent, as fine particles, of the sulfide assemblage at the Lone Tree Mine in the Buffalo Mountain district. (Panhorst, 1996)

A minor mineral in radioactive rock with pyrite, zircon, and powellite at the Moonlight Uranium Mine and Horse Creek prospects in the Disaster district. (Castor et al., 1996)

As euhedral crystals rimming stibnite crystals in quartz vugs in the Bell vein on Buckskin Mountain in the National district; also found in the National Mine and in the Birthday vein on Radiator and Round Hills. (Vikre, 1985; Lindgren, 1915)

Reported as gold-bearing crystals to 500 microns and as inclusions in pyrite with quartz, arsenian pyrite, and rare native gold at the Twin Creeks Mine in the Potosi district. (Hardy, 1940; Simon et al., 1999)

Common with stibnite in quartz veins at the Snowdrift Mine in the Red Butte district. (Lawrence, 1963)

With tetrahedrite in veins in the Ten Mile (Winnemucca) district; also reported with sphalerite, quartz, and galena at the Pansy-Lee Mine. (Lawrence, 1963; WMK)

Abundant with stibnite in gold-bearing quartz veins at the Juanita claim group in the Varyville district. (Vanderburg, 1938a; Lawrence, 1963)

LANDER COUNTY As crystals to 1.2 cm at the Copper Canyon Mine in the Battle Mountain district; also found in gold-bearing skarn with pyrrhotite at the Fortitude Gold Mine and the

principal sulfide at the Irish Rose Mine. (Roberts and Arnold, 1965; Schilling, 1979; Doebrich, 1993; M. Jensen, w.c., 1999)

In early veins with chalcopyrite and other sulfides at the Lone Tree Mine in the Buffalo Mountain district; also noted as rims on pyrite. (Bloomstein et al., 1993)

As euhedral crystals to 3 mm with crystals of boulangerite, bournonite, and tetrahedrite at the Little Gem Mine in the Cortez (Tenabo) district. (M. Jensen, w.c., 1999)

Reported at the Kattenhorn and Hilltop Mines, Hilltop district. (Vanderburg, 1939; J. McGlasson, w.c., 2001)

Reported in primary ore in the Lewis district; specifically noted at the Dean Mine, Lewis district. (Vanderburg, 1939; R. Thomssen, w.c., 1999)

Euhedral crystals to 1 mm with pyrite, silver, galena, and muscovite at the McCoy-Cove Mine in the McCoy district. (R. Walstrom, w.c., 1999; M. Jensen, w.c., 1999)

Reported in primary ore from the Reese River district. (Ross, 1953)

LINCOLN COUNTY An important primary mineral in ore, now more or less replaced by scorodite, beudantite, arseniosiderite, and other minerals at the Advance and Old Democrat veins in the Chief district. (Callaghan, 1936)

Reported with tennantite at the Prince Mine in the Pioche district. (Westgate and Knopf, 1932)

MINERAL COUNTY As simple bipyramidal crystals to 1 mm in quartz vugs with owyheeite and tetrahedrite at the Broken Hills Mine in the Broken Hills district. (Jenkins, 1981)

Reported at the Potosi and other mines in the Candelaria district. (Knopf, 1923)

Found in silver-lead ore in the Bell district. (Knopf, 1921b)

NYE COUNTY As small crystals with pyrite at the Sterling Mine in the Bare Mountain district. (J. Marr, o.c., 1996)

Very abundant in primary ore at the San Rafael Mine in the Lodi district, but generally oxidized to scorodite, arseniosiderite, and other minerals. (ASDM; Jenkins, 1981)

With realgar, orpiment, scorodite, and other arsenic minerals at the White Caps Mine in the Manhattan district. (Ferguson, 1924; Melhase, 1935d; Gibbs, 1985)

As small euhedral crystals in quartz veins and

adjacent wall rocks at an unspecified locality in the Morey district. (Williams, 1968)

As small grains and crystals in rhyolite dikes at the Northumberland Mine in the Northumberland district. (Kokinos and Prenn, 1985)

With acanthite and silver at the Blue Horse and Hillside Mines and Catlin Claim in the Silverbow district. (Tingley et al., 1998a)

Reported in the Last Chance vein at the 500-foot level in the Tonopah district; also as minute crystals on quartz in a specimen from the Tonopah Merger Mines. (Bastin and Laney, 1918; WMK)

With galena and pyrite at the Two G, Rescue, and other mines in the Tybo district. (Ferguson, 1933)

With sphalerite, galena, and pyrite in a banded quartz, calcite, and siderite vein at the St. Louis and Richmond workings in the Washington district. (Hill, 1915)

With pyrite, galena, sphalerite, and the related oxidation products of these minerals in a vein in limestone at the Combined Metals Reduction Company claim in the Willow Creek district. (Kral, 1951)

PERSHING COUNTY In the Majuba Hill Mine in the Antelope district with pyrrhotite and copper-bearing sulfides; also in a 20-cm-thick vein as crystals to 1 cm at the Arsenic prospect and reported at the Antelope Mine. (Smith and Gianella, 1942; Trites and Thurston, 1958; Stevens, 1971; Mackenzie and Bookstrom, 1976; Jensen, 1985; M. Jensen, w.c., 1999)

Reported at the Montezuma Mine in the Arabia district. (Knopf, 1918a; Lawrence, 1963)

In minor amounts in scheelite-bearing skarn in the Nightingale district. (Smith and Guild, 1944)

Reported from unspecified localities in the Sacramento and Wild Horse districts. (Vanderburg, 1936b)

Reported at the Auld Lang Syne Mine in the Sierra district. (Schrader et al., 1917)

Reported in quartz veins in the Buena Vista, Haystack, Kennedy, Rochester, and Star districts. (Cameron, 1939; Vanderburg, 1936b; Johnson, 1977; Knopf, 1924)

STOREY COUNTY Reported in ore from the Comstock Lode and on the west side of Mount Davidson in the Comstock district. (Bastin, 1922; WMK)

WASHOE COUNTY With other sulfide minerals at the Union (Commonwealth), Rocky Hill, and Denver Mines in the Galena district. (Bonham and Papke, 1969; Geehan, 1950)

Reported in ore with other sulfide minerals in the Pyramid district. (Wallace, 1980)

In cavities in sinter with stibnite, cinnabar, and chalcopyrite at Steamboat Hot Springs. (Lawrence, 1963)

WHITE PINE COUNTY As disseminated grains less than 80 microns across at the Vantage deposit on Alligator Ridge in the Bald Mountain district. (Ilchick, 1990)

ARTHURITE

$Cu^{2+}Fe_2^{3+}(AsO_4,PO_4,SO_4)_2(O,OH)_2 \cdot 4H_2O$

Arthurite is one of a suite of arsenate minerals that originate by oxidation of copper sulfides, pyrite, and arsenopyrite. The Nevada locality listed is only the second reported occurrence in the world for this rare mineral and a source of many world-class specimens.

PERSHING COUNTY As apple-green crystals to 1 mm and radiating groups to 3 mm with pharmacosiderite and scorodite in the Copper stope at the Majuba Hill Mine in the Antelope district; also as yellow-green fibers, less than 0.5 mm long, in cavities in rhyolite. (MacKenzie and Bookstrom, 1976; Jensen, 1985)

ARTINITE

$Mg_2(CO_3)(OH)_2 \cdot 3H_2O$

Artinite is an uncommon mineral derived from alteration of material containing magnesite, dolomite, brucite, or other magnesium minerals. There are only a few localities for the mineral in the United States.

NYE COUNTY As delicate, radiating aggregates of acicular white crystals to 1 cm in brucite and hydromagnesite veins that cut brucite ore at the Premier Chemicals Mine (Gabbs Magnesite-Brucite Mine) in the Gabbs district; also with hydromagnesite as a weathering product of brucite in a 10- to 15-foot blanket over brucite bodies. (Melhase, 1935c; Hurlbut, 1946; Schilling, 1968a, 1968b; Dunn et al., 1980)

ATACAMITE

$Cu_2^{2+}Cl(OH)_3$

Atacamite is derived from the oxidation of copper sulfides or other copper oxysalts in arid environments.

CHURCHILL COUNTY As green crystals to 0.5 mm with gypsum at the Lovelock Mine in the Table Mountain district. (M. Jensen, w.c., 1999; S. Pullman, w.c., 1999)

ESMERALDA COUNTY Reported in specimens from the Broken Toe and Rock Hill Mines in the Rock Hill district. (R. Thomssen, w.c., 1999)

Reported 4-1/2 miles north of Coaldale Junction. (R. Thomssen, w.c., 1999)

HUMBOLDT COUNTY As green prisms at the Silver Coin Mine in the Iron Point district. (Thomssen, 1998; J. McGlasson, w.c., 2001)

LANDER COUNTY Reported in a specimen from the Turquoise Mine in the Battle Mountain district. (LANH)

LINCOLN COUNTY With malachite, cuprite, tenorite, and chrysocolla at the Ida May Mine in the Bristol district. (CSM; WMK)

Reported at an unspecified locality in the Pioche district. (S. Pullman, w.c., 1999)

LYON COUNTY In deep-green prismatic crystals to 1.0 mm with paratacamite and other minerals at the Old (Lost) Soldier Mine in the Churchill district. (Jenkins, 1981)

Reported in a specimen from the Yerington district from Ludwig; also noted at the Douglas Hill Mine. (LANH; R. Thomssen, w.c., 1999)

PERSHING COUNTY As green stains at the Majuba Hill Mine in the Antelope district. (Jensen, 1993b)

AUGELITE

$Al_2(PO_4)(OH)_3$

A rare phosphate mineral that is generally found in granite pegmatite. The Nevada occurrence may be unique.

EUREKA COUNTY As pale-blue seams to 1 cm thick and as rarer single crystals from stockwork veinlets in silicified siltstone at the Gold Quarry Mine in the Maggie Creek district. (Jensen et al., 1995)

AUGITE

$(Ca,Na)(Mg,Fe,Al,Ti)(Si,Al)_2O_6$

A common rock-forming mineral of the *pyroxene* group that generally occurs in mafic igneous rocks. There are undoubtedly many more occurrences in Nevada than are listed below.

CHURCHILL COUNTY As a primary mineral in wall rock in the Fairview district. (Schrader, 1947)

EUREKA COUNTY In microscopic grains along a hematite, galena, stibnite, and calcite vein at the Goldstrike Mine in the Lynn district. (GCF)

As phenocrysts in andesite at the Barth Iron Mine in the Safford district. (Shawe et al., 1962)

LYON COUNTY In andesite at the Crest of Hill, Foreman Shaft, and Haywood Mines in the Silver City district. (WMK)

AURICHALCITE

$(Zn,Cu^{2+})_5(CO_3)_2(OH)_6$

Aurichalcite, a relatively uncommon mineral in the oxidation zones of base-metal deposits, occurs with hemimorphite, smithsonite, cerussite, and other minerals.

CHURCHILL COUNTY Reported at the Corey No. 2 claim in the Chalk Mountain district. (R. Walstrom, w.c., 1999)

In well-developed crystals at the Westgate Mine in the Westgate district. (R. Walstrom, w.c., 1999)

CLARK COUNTY With malachite, smithsonite, and other minerals at the Azure Ridge (Bonelli) Mine in the Gold Butte district, which has produced some superb specimens. (Longwell et al., 1965; LANH)

As plumose aggregates of blue-green crystals with galena, cerussite, hemimorphite, chrysocolla, hydrozincite, malachite, and wulfenite at the Smithsonite Mine in the Goodsprings district; also reported in pale-blue masses at the Yellow Pine Mine, and noted at other mines. (Hewett, 1931; Melhase, 1934e; F. Cureton, w.c., 1999)

DOUGLAS COUNTY Reported with hydrozincite, brochantite, and hemimorphite at an unspecified locality in the Wellington district. (NMBM)

ELKO COUNTY As light-blue acicular crystals to 7 mm in sprays and mats in fractures in the Lucin district on the Utah-Nevada state line; this locality may be in Utah. (Heyl and Bozion, 1962; GCF)

ESMERALDA COUNTY Reported at the General Thomas and Alpine Mines in the Lone Mountain district. (Heyl and Bozion, 1962; Phariss, 1974)

Reported in specimens from the Broken Toe Mine in the Rock Hill district. (R. Thomssen, w.c., 1999)

Reported in a specimen from the 3 Metals Mine in the Weepah district. (R. Thomssen, w.c., 1999)

EUREKA COUNTY Reported at the Mill Canyon and Tenabo area in the Cortez district; also reported at the Garrison Mine. (Gianella, 1941; S. Pullman, w.c., 1999)

HUMBOLDT COUNTY As light-blue-green acicular crystals to 3 mm with fornacite, barite, and chlorargyrite at the Silver Coin Mine in the Iron Point district. (Thomssen, 1998; GCF)

LINCOLN COUNTY Reported at the Bristol Silver Mine in the Bristol district. (Gianella, 1941)

LYON COUNTY As blue-green crystal aggregates at the Jackpot Mine in the Wilson district. (Heyl and Bozion, 1962)

MINERAL COUNTY Reported in a specimen from the Simon Mine in the Bell district. (R. Thomssen, w.c., 1999)

In oxidized material at the Potosi Mine in the Candelaria district. (Jenkins, 1981)

Reported at the Silver Star Mine in the Silver Star district. (Heyl and Bozion, 1962)

NYE COUNTY As bright blue-green plumose aggregates in ore at the Downieville Mine in the Gabbs district. (Heyl and Bozion, 1962; M. Jensen, w.c., 1999)

PERSHING COUNTY As blue spherules and masses associated with small smithsonite crystals at the Nevada Superior Mine in the Antelope district. (M. Jensen, w.c., 1999)

WHITE PINE COUNTY As silky, blue-green, translucent microcrystalline fibers with white aragonite in the Muncy Creek district; also noted as exceptional specimen material with hydrozincite, hemimorphite, smithsonite and other minerals at the Grand Deposit Mine. (Gianella, 1941; Hose et al., 1976; UMNH; WMK)

In the Liberty Mine in the Robinson district as blue-green fibers to 5 mm perpendicular to the walls of veinlets; also reported from the Dauntless Mine. (Jenkins, 1981; WMK; S. Kleine, o.c., 1999)

Reported in a specimen from north of Muncy Creek in the Seigel district. (WMK)

In small crystals to 1 mm at the Paymaster decline in the Ward district. (Heyl and Bozion, 1962; M. Jensen, w.c., 1999)

AURORITE
$(Mn^{2+},Ag,Ca)Mn_3^{4+}O_7 \cdot 3H_2O$

Aurorite, a Nevada type mineral, is an uncommon silver-bearing phase that occurs in veins with other manganese oxides.

NYE COUNTY With cryptomelane and hollandite in drill core from Yucca Mountain. (Carlos et al., 1995)

STOREY COUNTY Identified by AMNH in a dull gray mixture with other manganese oxides in white quartz in a specimen from the Potosi shaft in the Comstock district. (F. Cureton, o.c., 2000)

WHITE PINE COUNTY The type locality as irregular masses and platy or scaly grains to 8 microns in black manganese-bearing calcite with cryptomelane, todorokite, other manganese oxides, and native silver at the North Aurora and Ward Beecher Mines in the White Pine district, Sec. 30, T16N, R58E. (Radtke et al., 1967)

AUSTINITE
$CaZn(AsO_4)(OH)$

Austinite is a rare secondary mineral found in oxidized portions of base-metal deposits, commonly in association with other arsenate minerals.

CHURCHILL COUNTY A copper-bearing variety is found as green crystal sprays to 1 mm with cobalt-bearing köttigite, heterogenite, and calcite at an unnamed prospect near the southwest end of the Clan Alpine Mountains on the east side of Stingaree Valley. (M. Jensen, w.c., 1999)

Reported as copper-bearing in light-green clusters to 1.5 mm with hemimorphite and copper-bearing adamite from a prospect 1/4 mile south of the Chalk Mountain Mine in the Chalk Mountain district. (R. Thomssen, w.c., 1999)

ESMERALDA COUNTY Reported as microcrystalline aggregates of a copper-bearing variety at the Weepah Mine open pit in the Weepah district. (S. Pullman, w.c., 1999; M. Jensen, w.c., 1999)

In specimens from the Big 3 Mine in the Railroad Springs district. (R. Thomssen, w.c., 1999)

HUMBOLDT COUNTY Rare as green copper-bearing microcrystals at the Silver Coin Mine in the Iron Point district, Secs. 1 and 2, T35N, R40E. (Thomssen, 1998; M. Jensen, w.c., 1999; J. McGlasson, w.c., 2001)

LANDER COUNTY Reported from the Gray Eagle Mine in the Shoshone Range. (R. Thomssen, w.c., 1999)

LINCOLN COUNTY Reported in a specimen from the Silver King Mine in the Silver King district. (R. Thomssen, w.c., 1999)

MINERAL COUNTY As green crystals to 0.3 mm at the Simon Mine in the Bell district. (S. White, w.c., 2000)

NYE COUNTY As greenish-white needles and groups to 0.4 mm with adamite in cavities in quartz at the San Rafael Mine in the Lodi district. (Jenkins, 1981)

Reported in a specimen from the Gila Mine in the Reveille district. (R. Thomssen, w.c., 1999)

WASHOE COUNTY Reported in a specimen from Tick Canyon in the Dogskin Mountain district. (R. Thomssen, w.c., 1999)

AUTUNITE

$Ca(UO_2)_2(PO_4)_2 \cdot 10\text{-}12H_2O$

The most commonly reported secondary uranium mineral. Unless found in relatively moist environments (such as in poorly ventilated underground workings), most autunite is actually the lower hydrate meta-autunite, from which it is generally not distinguished in the literature.

ELKO COUNTY Reported from many radioactive prospects in the Mountain City district; specifically noted as yellow, tabular crystals with torbernite and metatorbernite in an altered fracture zone in quartz *monzonite* at the Autunite and October group of claims. (Schilling, 1962a, 1962b; Garside, 1973)

Reported in a narrow shear zone between a quartz vein and a parallel pegmatite dike; also noted as sparse flakes in pegmatite dikes in the Valley View district. (Olson and Hinrichs, 1960)

HUMBOLDT COUNTY Reported with quartz, pyrite, fluorite, and torbernite in a fault zone at the Moonlight Uranium Mine in the Disaster district. (Sharp, 1955; Garside, 1973; Dayvault et al., 1985)

As crystals on matrix with opal near the Virgin Valley Ranch in the Virgin Valley district; also noted as fracture coatings in opalized wood and in tuffaceous sedimentary rocks. (Castor et al., 1996; NYS; WMK)

LANDER COUNTY As yellow-green translucent crystals with other secondary uranium minerals, uraninite, gypsum, and sulfide minerals at the Apex (Rundberg) Uranium Mine in the Reese River district; reportedly Nevada's best crystals to 3 mm. (Garside, 1973; M. Jensen, o.c., 2000; GCF)

LYON COUNTY Reported at the Clyde Garrett property in the Mound House district in the foothills of the Virginia Range near the border between Carson City and Lyon County. (WMK)

Reported as encrustations and colliform masses with torbernite, phosphuranylite, and secondary copper minerals along fractures in granite at the Flyboy claims in the Cambridge Hills. (Garside, 1973)

Reported at the Quartz Mine claims, Halloween Mine, and Silver Pick property in the Washington district. (Garside, 1973)

MINERAL COUNTY With uranophane at the Black Horse uranium prospect in the Fitting district. (Garside, 1973)

NYE COUNTY As sparse crystals in calcite veins with opal, fluorite, and cinnabar at the War Cloud prospect in the Jackson district. (Hamilton, 1993)

Rare along cracks with torbernite at the Rainbow and Hazel prospects in the Northumberland district. (Garside, 1973)

Reported at several properties in the Round Mountain district; specifically described as small yellow to apple-green flakes to 2 mm and earthy coatings on fractures at the Henebergh Tunnel. (Kral, 1951; Garside, 1973; Lovering, 1954)

As very rare, small, disseminated flakes northwest of Tonopah. (JHS)

PERSHING COUNTY As scattered flakes at the Dart Mine in the Nightingale district. (Garside, 1973)

WASHOE COUNTY Reported in uranium prospects in or adjacent to Tertiary ash-flow tuff at the Sunnyside occurrence in the Dogskin Mountain district. (Bonham and Papke, 1969; Castor et al., 1996)

With opal, hematite, uraninite, and other secondary uranium minerals in tuff at the

DeLongchamps and Red Bluff Mines in the Pyramid district; also noted at other uranium mines and prospects. (Brooks, 1956; Castor et al., 1996; WMK)

With *gummite* and possible uranophane in fractures and siliceous veinlets in rhyolitic tuff at the Buckhorn Mine in the Stateline Peak district. (Hetland, 1955; Castor et al., 1996)

AVICENNITE

Tl_2O_3

Avicennite is an extremely rare, three-locality mineral. It is a product of the oxidation of thallium-bearing sulfides or sulfosalts, themselves very rare.

EUREKA COUNTY In polycrystalline, porous, dark-gray to black grains to 0.5 mm in silicified limestone, and in quartz veinlets in oxidized gold ore in and adjacent to fault zones, at the Carlin Mine in the Lynn district; also as coatings on carlinite from relatively unoxidized ore. (Radtke et al., 1978)

AXINITE

$(Ca,Fe^{2+},Mg,Mn^{2+})_3Al_2BSi_4O_{15}(OH)$(general)

Group mineral name; species are ferro-axinite, manganaxinite, magnesio-axinite, and tinzenite. *Axinite* group minerals generally occur in contact-metasomatic zones in carbonate rocks in association with such minerals as prehnite, *garnet,* and actinolite. Most of the following listings, for which species have not been determined, are probably for ferro-axinite.

CHURCHILL COUNTY Reported in a contact zone in calcareous shales at an unspecified locality in the White Cloud (Coppereid) district. (Ransome, 1909b)

ELKO COUNTY Massive, yellowish-gray *axinite* is a major constituent of gangue in contact-metamorphic copper deposits with actinolite, diopside, epidote, *garnet,* and other minerals in the Contact district; specifically reported at the Brooklyn Mine. (Schrader, 1912; Melhase, 1934c; Schilling, 1979)

Questionably reported at an unspecified locality in the Elk Mountain district. (Schrader, 1912)

LYON COUNTY Reported to occur in skarn in the Ludwig area in the Yerington district. (Gianella, 1941)

MINERAL COUNTY As small tan to brown crystals in skarn with epidote and other minerals at the Julie claim in the Pamlico district. (Smith and Benham, 1985)

PERSHING COUNTY In the A. McGuinness collection as a single brown crystal 6 inches long from an unspecified locality in the Nightingale district. (S. Pullman, o.c., 2000)

Associated with dumortierite ore at the Champion Mine in the Sacramento district; probably ferro-axinite. (M. Jensen, w.c., 1999)

AZURITE

$Cu_3^{2+}(CO_3)_2(OH)_2$

Azurite is a widely distributed secondary mineral in oxidized zones of copper deposits; it is generally associated with malachite but less common.

CHURCHILL COUNTY Common as an oxide mineral in the Coppereid area in the Copper Kettle district. (M. Jensen, w.c., 1999)

As blue coatings with clinotyrolite and cobalt and nickel minerals at the Lovelock Mine in the Table Mountain district; also reported in a specimen from the Nickel Mine. (Ransome, 1909b; R. Thomssen, w.c., 1999)

CLARK COUNTY With malachite, brochantite, and other secondary copper minerals at the Key West Mine in the Bunkerville district. (Beal, 1965)

Reported in a specimen from the Ironside Mine in the Goodsprings district. (LANH)

Reported in a specimen from the Duplex Mine in the Searchlight district. (WMK)

Reported in a specimen from the Murphy Mine in the Sunset district. (WMK)

ELKO COUNTY As blue stains and coatings with copper-rich quartz-calcite veins at an unspecified locality in the Contact district. (Purington, 1903; Bailey, 1903)

Reported as crystals to 4 mm in vugs in silicified limestone at the Victoria Mine in the Dolly Varden district. (Melhase, 1934d; GCF)

Common as blue stains, crusts, and small masses in shale with malachite at the Rio Tinto Mine in the Mountain City district. (Coats and Stephens, 1968)

Minor as thin coatings and crystals associated with malachite and cerussite along fractures in oxidized ore at the Killie Mine in the Spruce Mountain district. (Dunning and Cooper, 1987)

Reported as fine crystals at Bull Run. (Hoffman, 1878)

ESMERALDA COUNTY As earthy blue coatings along cracks in oxidized quartz veins at the Bullfrog-George prospect in the Gold Point district. (Schilling, 1979)

Reported in a specimen from the Ludwig Mine in the Goldfield district; probably from the Yerington district, Lyon County. (WMK)

Identified in a specimen from the General Thomas Mine in the Lone Mountain district. (JHS)

Reported in specimens from the Gold Crown and Pandora Mines in the Montezuma district. (JHS; WMK)

Reported in specimens from the Cuprite and Klondyke districts. (WMK; JHS)

EUREKA COUNTY Reported in a specimen from the Eureka Consolidated Mine in the Eureka district. (WMK)

In oxidized base-metal veins in diorite at the Goldstrike Mine in the Lynn district. (GCF)

As bladed crystals to 6 mm with malachite, brochantite, cyanotrichite, and cuprite in the Gold Quarry Mine in the Maggie Creek district; also noted as druses and powdery blue coatings on joint surfaces. Also reported as masses to 5 cm thick with malachite at the Nevada Star (Good Hope) Mine, and as veins and fracture coatings at the Copper King Mine. (Rota, 1991; Jensen et al., 1995; GCF)

Reported as crystals with malachite at an unspecified locality in the Mineral Hill district. (Hoffman, 1878; M. Jensen, w.c., 1999)

HUMBOLDT COUNTY As blue coatings on fractures in the deeper portions of the Silver Coin Mine in the Iron Point district where it occurs in the vicinity of lipscombite. (GCF)

Reported in specimen from the west slope of Delong Peak at an altitude of 7,000 feet in the Jackson Mountains district. (WMK)

As blue stains and coatings on quartz veins at the Desert View prospect in the Leonard Creek district. (Schilling, 1979)

With malachite, tetrahedrite, and galena from the Prodigal Mine in the Red Butte district. (WMK)

LANDER COUNTY In oxidized ore with other copper minerals at the Copper Canyon, Copper Basin, and Little Grant Mines in the Battle Mountain district. (Schilling, 1979; WMK)

After tetrahedrite at the Dry Canyon Mine in the Big Creek district. (Lawrence, 1963)

Reported in a specimen from an unknown locality in the Hilltop district. (WMK)

In quartz veins at the Betty O'Neal and Highland Chief Mines in the Lewis district; also with tetrahedrite and owyheeite in the Eagle vein. (Vanderburg, 1939; J. McGlasson, w.c., 2001)

LINCOLN COUNTY With other secondary copper minerals at the Jackrabbit Mine in the Bristol district. (Melhase, 1934d)

Reported in a specimen with chrysocolla, malachite, and *psilomelane* at an unspecified locality in the Pioche district. (WMK)

LYON COUNTY Reported with other copper minerals at the Ludwig, Yerington, and Mason Valley Mines in the Yerington district. (Melhase, 1935e; LANH; WMK)

MINERAL COUNTY In crystals to 2 mm at the Simon Mine in the Bell district. (S. White, w.c., 2000)

As blue stains and small crystals at the Candelaria Mine in the Candelaria district. (M. Jensen, w.c., 1999)

With malachite in veins at the Lucky Boy Mine in the Lucky Boy district. (Lawrence, 1963; WMK)

As sparkly blue crystals to 2 mm with malachite and antlerite at the Northern Light Mine in the Mountain View district. (M. Jensen, w.c., 1999; S. Pullman, w.c., 1999)

Reported from an unspecified locality in the Pilot Mountains district. (Foshag, 1927)

With malachite, chrysocolla, and cuprite in fissures in limestone at the Wallstreet and Turk Mines in the Santa Fe district. (Ross, 1961; WMK)

NYE COUNTY Reported at the Dollar and Last Chance Mines in the Jett district. (Lawrence, 1963)

Reported as a rare mineral at the White Caps Mine in the Manhattan district. (Ferguson, 1924)

Coating cracks with malachite at the Northumberland Mine in the Northumberland district. (Kokinos and Prenn, 1985)

With malachite, cerussite, and galena in a gangue of quartz and gypsum at the Antimonial Mine in the Reveille district. (Lawrence, 1963)

As light-blue masses with malachite from an unspecified locality in the Tonopah district. (Melhase, 1935c)

With malachite in bonelike quartz in the Troy district. (Kral, 1951)

As dark-blue crystals to 1 mm in limestone at Grantsville in the Union district. (R. Thomssen, w.c., 1999; S. White, w.c., 2000)

With malachite and chrysocolla at unnamed prospects in the Wagner district. (Tingley et al., 1998a)

Reported at unspecified localities in the Bare Mountain and Oak Springs districts. (Melhase, 1935b)

PERSHING COUNTY As small single crystals or groups in the middle adit with malachite and a host of rare copper arsenate minerals at the Majuba Hill Mine in the Antelope district. (Smith and Gianella, 1942; MacKenzie and Bookstrom, 1976; Jensen, 1985)

Identified in specimens from the Buena Vista and Iron Hat districts. (JHS)

As minor staining at an unspecified locality in the Goldbanks district. (Dreyer, 1940)

WASHOE COUNTY In stringers with malachite in quartz diorite in the Freds Mountain district, Sec. 24, T22N, R19E. (Bonham and Papke, 1969)

With malachite in small prospects in the hills rimming Hungry Valley in the McClellan district, NE 1/4 Sec. 10, T21N, R20E. (Bonham and Papke, 1969)

Rare as clusters to 1 mm of small crystals with malachite or brochantite at the Burrus Mine in the Pyramid district. (Overton, 1947; Jensen, 1994)

In a quartz vein with malachite and tenorite at the Antelope Mine in the Stateline Peak district. (Bonham and Papke, 1969)

With other secondary copper minerals in the Peavine and Wedekind districts. (Hudson, 1977)

WHITE PINE COUNTY In a vein with quartz, malachite, chalcopyrite, chalcocite, *limonite,* and gold at the Bald Mountain Mine (Top Pit) in the Bald Mountain district. (SBC)

Some of Nevada's finest specimens are from the Robinson district as dark-blue crystals to 5 mm with cuprite, *limonite,* and malachite from the Alpha and Richard Mines; also as blue crystalline crusts to 1 mm from the Los Angeles pit. (UCB; WMK; M. Jensen, w.c., 1999)

Reported in specimens from the Cherry Creek and Taylor districts. (JHS; R. Thomssen, w.c., 1999)

BALKANITE

$Cu_9Ag_5HgS_8$

A rare sulfide mineral reported from three localities worldwide, balkanite was first identified in a high-grade, stratiform, base-metal deposit.

NYE COUNTY Reported from an unspecified locality in the Manhattan district. (Anthony et al., 1990; Gaines et al., 1997; HUB, RMO)

BARITE

$BaSO_4$

Barite is a widespread and common mineral that occurs in stratiform deposits in deep-sea sedimentary rocks and in veins. See Papke (1984) for descriptions of many of the barite occurrences in Nevada, which include some of the largest commercial deposits in the United States. High-quality crystalline barite specimens have come from some of Nevada's Carlin-type gold deposits, including the Meikle and Northumberland Mines.

CARSON CITY A vein of white barite is in a small prospect at the Strong Mine in the Carson City district. (Papke, 1984)

Reported as *heavy spar* at an unspecified locality in the Delaware district. (WMK)

CHURCHILL COUNTY Reported in veins and as a replacement mineral in wall rock at the Fairview Silver Mine in the Fairview district. (Schrader, 1947)

With cinnabar in calcite veins at the Cinnabar Hill Mine in the Holy Cross district. (Vanderburg, 1940)

In small tabular crystals with stibiconite and mopungite at the Green prospect in the Lake district; also in well-developed tabular crystals in silicified rocks at an unspecified locality in the central Mopung Hills. (R. Walstrom, w.c., 1999)

As white tabular crystals to 5 mm with acanthite at the Summit Queen Mine in the Sand Springs district. (M. Jensen, w.c., 1999)

Reported from the Nickel Mine in the Table Mountain district. (R. Thomssen, w.c., 1999)

After gypsum in a specimen from an unspecified locality near Fallon. (WMK)

As colorless tabular crystals to 1 cm in a mine near Jessup. (M. Jensen, w.c., 1999)

CLARK COUNTY With manganese oxide, quartz, and gypsum in irregular masses in sedimentary rock at the Lagarto Mine in the Alunite district; also as coarse-grained white to pink veins at the Searway prospect. (Papke, 1984)

Reported at the Boss Mine in the Goodsprings district. (Jedwab, 1998b)

As nodules in layers with petrified wood in the Chinle Formation at an unspecified locality in the Muddy Mountains. (WMK)

ELKO COUNTY Reported as crystals at an unspecified locality in Bull Run Canyon in the Aura district. (WMK)

As magnificent specimens of gemmy, yellow to orange, tabular crystals to 20 cm lining the walls of cavernous pockets along with gemmy calcite in the gold orebody in the upper levels of the Meikle Mine in the Lynn district; even larger crystals were encountered, reportedly in places in masses to several meters, but most were obliterated by rapidly advancing mining and backfilling. (GCF; NMBM; Jensen, 1999)

At the Rossi Mine in the Bootstrap district, one of Nevada's largest barite producers, in Ordovician chert and fine-grained sedimentary rocks as bedded barite, as crystals and rosettes to 1 cm, and in white microcrystalline veins; also mined from a bedded Ordovician deposit at the Queen Lode claims, and with stibnite and calcite in quartz veins at the Bootstrap Mine. (Papke, 1984; Lawrence, 1963)

In the Carlin district at the Rain Mine as white crystals to 2 cm, as veins and fracture coatings, as large masses of euhedral crystals in red clay, and in massive bodies along the Rain fault; also reported as replacement bodies to 8 feet thick at the Evans Mine, and in silicified rocks in the SMZ and Gnome deposits. (D. Heitt, o.c., 1991; LaPointe et al., 1991)

As stratiform bodies to 25 m thick in Devonian dolomite at the Judy Mine in the Cave Creek district; also in zones about 4.5 m thick at the B & P claims. (Papke, 1984)

Reported with cerussite, *jasper,* and quartz at the Hice prospect in the Contact district. (Schrader, 1935b; LaPointe et al., 1991; WMK)

As blue to bluish-gray doubly terminated crystals to 10 cm at the Divide Mine in the Divide district. (GCF)

In gold-silver veins with calcite, *adularia,* quartz, fluorite, and pyrite at the Ken Snyder Mine in the Gold Circle district. (Casteel and Bernard, 1999)

As bedded deposits in Ordovician rocks at the Hidden Hills, Snow Canyon, and Taylor Canyon deposits in the Independence Mountains district. Also found as blue-gray crystals to 6 cm with quartz and stibnite in veins and vugs at the Jerritt Canyon, Enfield-Bell, and Murray gold mines; as white, pseudocubic crystals to 5 cm with antimony minerals at the Sage Hen prospect; as colorless to gray tabular crystals to 5 cm in jasperoid at the Pie Creek prospect; as masses and crystals in quartz veins at the Burns Basin Mine and nearby prospects; and as transparent blue crystals to 1.5 cm in fractures in a road cut on the north side of Nevada Highway 226 near the west end of Taylor Canyon. (Papke, 1984; Hawkins, 1982; Jones and Jones, 1999; M. Jensen, w.c., 1999; J. Crowley, w.c., 1999; GCF)

In brown *jasper* veins with iron oxide at the Jay claims in the Larrabee district. (Papke, 1984)

Eleven properties with bedded barite deposits in Ordovician rocks are in the Snake Mountains district, with the Snoose Creek, Jungle, Consolation-Boies, Loomis Creek, and Stormy Creek Mines having had significant production. (Papke, 1984)

ESMERALDA COUNTY As veins in Triassic shale at the Red Bank claim in the Candelaria district, Sec. 3, T3N, R35E. (Papke, 1984)

In cinnabar deposits at an unspecified locality in the Fish Lake Valley district. (Bailey and Phoenix, 1944)

In the Goldfield district reported from the McGinnity shaft, Mohawk Mine, and from the Sandstorm Mine; also noted as colorless crystals to 1 cm with pyrite and stibnite inclusions from a large shaft 2 miles west of Blackcap Mountain. (S. Pullman, w.c., 1999; R. Thomssen, w.c., 1999; M. Jensen, w.c., 1999)

In small pods with *limonite,* quartz, cerussite, galena, malachite, and wulfenite at the McNamara Mine in the Palmetto district. (Schilling, 1979)

In veins in siltstone, limestone, and hornfels at the American Barium Mine, Weepah district. (Papke, 1984)

EUREKA COUNTY Bedded in the Ordovician Vinini Formation at the Sansinena Mine in the Beowawe district. (Papke, 1984)

Reported as stringers, veins, and irregular pods related to gold mineralization in the Carlin Mine in the Lynn district; also in the Goldstrike Mine as transparent, honey- to tan-colored crystals to 4 cm in fractures and vugs in silicified limestone, commonly with sprays of stibnite to 2 cm. (Papke, 1984; GCF)

At the Gold Quarry Mine in the Maggie Creek district as coarsely crystalline vein barite, locally in rounded crystals colored bright yellow by included tyuyamunite, and at one location as spherical patches to 5 mm of white, cottonlike fibers; also reported in a 10-foot-wide vein at the Maggie Creek Mine, as white veins to several feet wide at the Nevada Star (Good Hope) Mine, and as beds in the Ordovician Vinini Formation at the Queen Anne Mine. (Jensen et al., 1995; GCF; Rota, 1991; Vanderburg, 1938b; Papke, 1984)

HUMBOLDT COUNTY Clear, bluish gray, doubly terminated crystals up to 10 cm have been collected in the Marigold Mine in the Battle Mountain district; also reported from the Copper Canyon Mine. (SBC; Schilling, 1979)

Reported as lathlike crystals to 1 cm with quartz in vugs and fractures at the Lone Tree

Mine in the Buffalo Mountain district; also noted as white to colorless zoned tabular crystals to 2 cm with rhodochrosite. (Panhorst, 1996; M. Jensen, w.c., 1999)

Bedded with shale and limestone in the Cambrian Preble Formation at the Little Britches Mine in the Golconda district. (Papke, 1984)

As blocky colorless crystals to 3 mm with pyrite and carbonate minerals at the Silver Coin Mine in the Iron Point district. (Thomssen, 1998; J. McGlasson, w.c., 2001)

As bladed crystals to 6 mm, perched on and containing cinnabar and marcasite inclusions, at the Cordero Mine in the Opalite district. (Fisk, 1968)

With *adularia* in veins cutting bedded barite at the type locality for curetonite at the Redhouse (Hogshead Canyon) barite prospect in the Potosi district. (DeMouthe, 1985; M. Jensen, w.c., 1999)

As tiny patches in opal near the Virgin Valley Ranch in the Virgin Valley district. (Castor et al., 1996)

LANDER COUNTY Bedded in the Devonian Slaven Chert at ten properties in the Argenta district, including the Argenta, Beacon, Shelton, and H and L Mines. The Argenta Mine, with more than 2 million tons of barite mined, has been one of Nevada's largest producers. (Papke, 1984; SBC)

Bedded in the Devonian Slaven Chert at five properties in the Bateman Canyon district, including the Pleasant View, Slaven Canyon, and Barse Mines. (Papke, 1984)

Bedded barite in the Devonian Slaven Chert and the Ordovician Valmy Formation occurs at 14 properties in the Bullion district, including the Greystone, Clipper, Bateman Canyon, Bradshaw, and White Rock Mines. The Greystone Mine has been the most productive barite property in Nevada, at over 3 million tons. Bedded barite at the Clipper Mine, also a large producer, includes a bed rich in brachiopod fossils. (Papke, 1984; SBC)

Reported in a specimen from the Pittsburg Red Top Mine in the Hilltop district. (WMK)

In crystals to 2 cm at the Dean Mine, Lewis district; also noted as colorless crystals to 1.5 cm at an unspecified locality at the northern end of the Shoshone Range. (R. Thomssen, w.c., 1999; J. McGlasson, w.c., 2001; M. Jensen, w.c., 1999)

Bedded in the Devonian Slaven Chert and the Ordovician Valmy Formation at three properties

in the Mountain Springs district, including the Mountain Springs Mine with more than 1 million tons of production. (Papke, 1984)

Bedded in the Devonian Slaven Chert at the Rimrock and Cutler Mines in the North Battle Mountain district. (Papke, 1984)

With cinnabar at the Warm Springs prospect in the Warm Springs district. (Bailey and Phoenix, 1944)

LINCOLN COUNTY As crystals to 3 cm in dolomite adjacent to a fault and as crystalline masses in ore at the Atlanta Mine and Solo Joker prospect in the Atlanta district. (LaBerge, 1996; M. Jensen, w.c., 1999)

As a network of tabular crystals in a quartz vein at the Gold Chief Mine in the Chief district. (Callaghan, 1936; WMK)

Reported at an unspecified locality in the Pioche district. (UCB)

LYON COUNTY As colorless to pale blue tabular crystals to 5 cm, some containing realgar inclusions at the Boulder Hill fluorite prospect in the Wellington district; also noted locally as clear or zoned pseudorhombohedral crystals to 2 cm, and associated with abundant fluorite crystals. (M. Jensen, w.c., 1999; GCF; S. Sears, w.c., 1999)

MINERAL COUNTY As white to light-gray veins at the Gravity Mine in the Buckley district. (JHS)

Bedded in the Ordovician Palmetto Formation at four properties in the Candelaria district, including the productive Noquez and Little Summit (Giroux) Mines. (Papke, 1984)

Coarse, white, and nearly pure in a vein to 2.5 m wide and 900 m long in metavolcanic rocks that was mined in the 1920s and 1930s at the Eagleville Barite Mine in the Eagleville district. (Papke, 1984)

In quartz veins at the Kinkead and Lucky Boy Mines in the Lucky Boy district. (Lawrence, 1963; WMK)

As coarse white material in a vein to 3.5 m wide and 100 m long in Mesozoic andesite at the Crystal Mine in the Pamlico district, one of the earliest barite mines in Nevada. (Papke, 1984)

At the Regent Mine in the Rawhide district as lustrous euhedral tabular crystals to 2.5 cm with inclusions of orpiment, realgar, stibnite, cinnabar, and metacinnabar; also noted with rare inclusions of antimony-arsenic sulfide, possibly wakabayashilite. (M. Jensen, w.c., 1999; GCF)

As white tabular crystals to 6 cm with cinnabar inclusions, associated with cornwallite, dussertite, and barium-pharmacosiderite in an unnamed prospect near Rhyolite Pass in the Fitting district. (M. Jensen, w.c., 1999)

Reported from unspecified localities in the Pilot Mountains and Silver Star districts. (Foshag, 1927; WMK)

NYE COUNTY As a major constituent of mercury-bearing quartz veins at the Mariposa Canyon prospect and Senator Mine in the Belmont district. (Bailey and Phoenix, 1944)

As white masses and crystals to 5 cm at the King Solomon Mine in the Danville district. (Lawrence, 1963)

Bedded with chert, limestone, and argillite of early Paleozoic age at the Jumbo Mine in the Ellendale district. (Papke, 1984)

As colorless tabular crystals to 1 cm on stibnite and calomel at the Paradise Peak Mine open pit in the Fairplay district. (M. Jensen, w.c., 1999)

In silicified rock in the Gold Crater district. (Tingley et al., 1998a)

Reported at the White Caps Mine in the Manhattan district. (Gibbs, 1985)

Twelve properties in the Northumberland district contain bedded barite with chert, claystone, and shale, mostly of Devonian age but also in Ordovician rocks. The largest producers were the All Minerals pits and the IMCO Mine, but the only property operating after 1986 was the P & S Mine, a producer of small amounts of white bedded barite. In addition, the Northumberland Gold Mine yielded fine specimens of greenish-brown, golden-brown, gray-blue, and colorless crystals to 12 cm from pockets in jasperoid; and small water-clear tabular crystals were reported from the bedded deposit at the P & S Mine. (Shawe et al., 1969; Papke, 1984; SBC; Kokinos and Prenn, 1985; R. Walstrom, w.c., 1999)

In veins with sulfide and *electrum* at the Life Preserver Mine in the Tolicha district. (Tingley et al., 1998a)

As crystals in quartz veins and coating fractures at the Keystone Shaft and other mines in the Tonopah district. (Spurr, 1905; Melhase, 1935c; Nolan, 1935; CSM)

In flesh-colored crystals to 5 mm at the Hall Mine in the San Antone district. (M. Jensen, w.c., 1999)

Rare in gold placers in Manhattan Gulch in the Manhattan district. (Vanderburg, 1936a)

As euhedral thick lustrous tabular crystals to 8 cm with quartz and quartz epimorphs after barite in an outcrop near Warm Springs in the Tybo district; also as colorless crystals to 5 mm with carbonate-fluorapatite in unnamed turquoise prospects west of Warm Springs. (M. Jensen, w.c., 1999; S. Kleine, o.c., 2000)

PERSHING COUNTY As colorless tabular crystals to 1.5 cm and as fans to 5 mm of white, bladed crystals on fractures at the Pershing Quicksilver Mine in the Antelope Springs district; also reported as crystals to 2 mm with cinnabar and *adularia* on quartz at the Red Bird Mine. (M. Jensen, w.c., 1999; GCF)

In a vein at the Land Rover prospect in the Iron Hat district. (Papke, 1984)

With cinnabar and stibnite at the Polkinghorne Mine in the Mount Tobin district. (Lawrence, 1963)

As tabular crystals to 2 mm with antlerite at the Big Mike Copper Mine in the Tobin and Sonoma Range district. (M. Jensen, w.c., 1999)

As colorless to pale-purple, sharp, tabular crystals to 10 cm, some doubly terminated, with pyrite and marcasite at the Hecla Rosebud Mine in the Rosebud district. (M. Jensen, w.c., 1999; GCF)

In colorless druses at the Willard Mine in the Willard district. (Jensen and Leising, 2001)

STOREY COUNTY With cinnabar at the Castle Peak Mine in the Castle Peak district. (Bailey and Phoenix, 1944)

WASHOE COUNTY In tabular crystals with quartz and calcite in geodes weathering out of volcanic rock in Leadville Canyon in the Leadville district, Sec. 35, T37N, R23E. (R. Walstrom, w.c., 1999)

Common as milky-white, tabular crystals with quartz, enargite, and other vein minerals at the Burrus and other mines in the Pyramid district; also rarely in vugs with chalcopyrite and barium-pharmacosiderite and in quartz-lined cavities with enargite and pyrite crystals. (Overton, 1947; Jensen, 1994)

WHITE PINE COUNTY As veins and disseminations at the Vantage deposit on Alligator Ridge in the Bald Mountain district. (Ilchick, 1984, 1990)

In veins in jasperoid at the Dixie Lee prospect in the Butte Valley district. (Papke, 1984)

As lustrous pinkish zoned crystals to 1.5 cm at the Nighthawk Ridge Mine in the White Pine district. (M. Jensen, w.c., 1999)

In crystals to 1 cm at the Piermont Mine in the Piermont district. (M. Jensen, w.c., 1999)

As transparent yellow crystals to 5 cm at the Ward Mine (underground) in the Ward district. (M. Jensen, w.c., 1999)

BARIUM-PHARMACOSIDERITE

$BaFe_8^{3+}(AsO_4)_6(OH)_8 \cdot 14H_2O$

A relatively rare secondary mineral that occurs in hydrothermal deposits.

MINERAL COUNTY As pseudocubic microcrystals with cornwallite and barite in an unnamed prospect west of Rhyolite Pass in the Fitting district. (M. Jensen, w.c., 1999)

WASHOE COUNTY As yellow to pale-green cubes to 1 mm with scorodite and olivenite at the Burrus Mine in the Pyramid district; rarely on enargite or drusy quartz. (Jensen, 1994)

BARNESITE

$(Na,Ca)_2V_6^{5+}O_{16} \cdot 3H_2O$

Barnesite is a rare mineral derived from oxidation of primary vanadium-bearing minerals or organic substances.

EUREKA COUNTY As greenish, earthy masses with bokite on organic shale at the Gibellini vanadium deposit in the Gibellini district. (Jenkins, 1981)

BARRERITE

$Na_2[Al_2Si_7O_{18}] \cdot 6H_2O$

Barrerite is a very rare zeolite group mineral found in Italy with *heulandite* on fractures in weathered volcanic rock.

PERSHING COUNTY An alteration mineral in rhyolite with *clinoptilolite* and quartz at NE 1/4 SW 1/4 SW 1/4 Sec. 24, T30N, R28E, and other localities in the Seven Troughs district; originally identified as *Na-stellerite* by XRD. (D. Hudson, w.c., 1999)

BARTONITE

$K_3Fe_{10}S_{14}$

Extremely rare; reported elsewhere only from a single occurrence in an alkalic diatreme in California.

NYE COUNTY Reported as inclusions in kaersutite in basalt at the Lunar Craters. (Bergman, 1982)

BARYSILITE

$Pb_8Mn(Si_2O_7)_3$

Barysilite is a very rare mineral found in iron ore in Sweden and in zinc ore from Franklin, New Jersey.

NYE COUNTY Reported as yellow silicate of lead with stetefeldtite in a specimen donated to the Royal Natural History Museum of Vienna by Küstel in 1873. The specimen is said to be from Old Belmont, southeastern Toquima Range, and is probably from the Belmont district. (EMP; JHS)

BASTNÄSITE-(Ce)

$(Ce,La)(CO_3)F$

Bastnäsite-(Ce) is a rather uncommon mineral that occurs in some rare earth–bearing pegmatites and certain carbonatite bodies, as at Mountain Pass, California.

CLARK COUNTY Tentatively identified with *apatite* and *monazite* in the Crescent district. (Garside, 1973)

BAYLDONITE

$PbCu_3(AsO_4)_2(OH)_2 \cdot H_2O$

Bayldonite is a rare oxide zone mineral in base-metal deposits.

ELKO COUNTY In irregular pale- to apple-green granular masses mixed with iron oxides and local carminite after cerussite in the oxidized zone at the Killie Mine in the Spruce Mountain district, Sec. 14, T31N, R63E; possibly the mineral described as lead oxychloride by Schrader (1931b). (Dunning and Cooper, 1987)

ESMERALDA COUNTY As green botryoids and crystals with brochantite, hemimorphite, and mimetite at the 3 Metals Mine in the Weepah district. (B. Runner, o.c., 2000; GCF)

HUMBOLDT COUNTY In green crusts at the Silver Coin Mine in the Iron Point district, also reported as finely granular crystals with pyromorphite and *apatite*. (Thomssen, 1998; J. McGlasson, w.c., 2001)

MINERAL COUNTY As green botryoids to 4 mm with olivenite at the Simon Mine in the Bell district. (B. Runner, o.c., 2000; GCF)

NYE COUNTY As compact green botryoids to 0.3 mm with wulfenite, mimetite, and *agardite* at the San Rafael Mine in the Lodi district. (R. Thomssen, w.c., 1999; S. White, w.c., 2000; GCF)

BEAVERITE

$Pb(Cu^{2+},Fe^{3+},Al)_6(SO_4)_4(OH)_{12}$

Beaverite is an uncommon member of the jarosite group that is formed by surface oxidation under arid conditions, generally in association with plumbojarosite.

CLARK COUNTY Common in gold-platinum-palladium ore with plumbojarosite, bismutite, and quartz in a vertical pipe, Boss and Copperside Mines, and in a prospect southeast of the center of Sec. 25, T24S, R59E, in the Goodsprings district. (Hewett, 1931; Melhase, 1934e; Mount, 1964)

LINCOLN COUNTY Microcrystalline on siderite as pseudomorphs after gypsum crystals at an unspecified locality in the Pioche district; also reported as crystals to 1.5 mm from the Bristol Silver Mine, Bristol district. (F. Cureton, w.c., 1999; R. Thomssen, w.c., 1999; J. McGlasson, w.c., 2001)

MINERAL COUNTY Reported in a specimen from the Dispozitch Mine in the Silver Star district. (HMM)

NYE COUNTY Reported to be moderately common as coatings of minute, canary-yellow, hexagonal scales on quartz with jarosite, malachite, azurite, chrysocolla, and stetefeldtite at the Transylvania Mine in the Belmont district. (Jenkins, 1981)

BEIDELLITE

$(Na,Ca_{0.5})_{0.3}Al_2(Si,Al)_4O_{10}(OH)_2 \cdot nH_2O$

A clay mineral of the *smectite* group that commonly occurs as an alteration product in hydrothermal mineral deposits, particularly in porphyry deposits; it also occurs as a constituent of bentonite and in soil.

MINERAL COUNTY In white to buff or pink masses in pegmatite at the Zapot claims in the Fitting district. (Foord et al., 1999b)

BEMENTITE

$Mn_8^{2+}Si_6O_{15}(OH)_{10}$

Bementite is a widespread mineral in manganese deposits, occurring with manganese oxides, rhodochrosite, and rhodonite; it may be of primary or secondary origin.

HUMBOLDT COUNTY Reported at an unspecified locality near Golconda, probably at the tungsten-bearing hot-spring manganese deposits described by Kerr (1940) in the Golconda district. (MV; MM)

PERSHING COUNTY In small amounts in massive braunite-*chalcedony* ore with rhodonite, piemontite, and other minerals at the Black Diablo Mine in the Black Diablo district. (Ferguson et al., 1951; Johnson, 1977)

Questionably reported with manganese oxide in cavernous *travertine* at Sou Hot Spring. (Hewett et al., 1963)

BENJAMINITE

$(Ag,Cu)_3(Bi,Pb)_7S_{12}$

The Nevada type mineral benjaminite has had a history of discreditation and reacceptance. The mineral occurs with other sulfides in hydrothermal deposits, generally with other bismuth minerals and molybdenite.

NYE COUNTY Type locality as irregular masses to 5 cm in fractures in milky quartz veins with fluorite, muscovite, chalcopyrite, covellite, pyrite, and molybdenite at the Outlaw Mine in the Round Mountain district, NE 1/4 NE 1/4 Sec. 15, T9N, R44E; it is intergrown with aikinite and gustavite, and can only be distinguished from other sulfosalts with difficulty using microprobe and single crystal X-ray analyses. (Shannon, 1925; Melhase, 1935d; Nuffield, 1953; Harris and Chen, 1975; Nuffield, 1975; Shawe, 1977; Dunning et al., 1991; DeMouthe, 1985; UAMM)

BERAUNITE

$Fe^{2+}Fe_5^{3+}(PO_4)_4(OH)_5 \cdot 4H_2O$

An uncommon mineral that occurs in iron ore deposits and as an alteration product of primary phosphate minerals in pegmatite.

HUMBOLDT COUNTY Reported at the Silver Coin Mine in the Iron Point district. (Thomssen, 1998)

BERRYITE

$Pb_3(Ag,Cu)_5Bi_7S_{16}$

A rare mineral that generally occurs in quartz veins with other sulfide minerals.

NYE COUNTY Identified in museum specimens from the Outlaw Mine in the Round Mountain district with aikinite, benjaminite, and other sulfosalts. (Nuffield, 1953; Karup-Møeller, 1972; Harris and Chen, 1976; Dunning et al., 1991)

BERTHIERITE

$FeSb_2S_4$

Berthierite is a mineral that occurs in some low-temperature hydrothermal antimony veins.

EUREKA COUNTY As red-brown coatings on quartz, barite, and stibnite on the 4720-foot level at the Goldstrike Mine in the Lynn district. (GCF)

HUMBOLDT COUNTY With pyrite, tetrahedrite, and fibrous stibnite at the National Mine in the National district. (Vikre, 1985)

BERTRANDITE

$Be_4Si_2O_7(OH)_2$

Bertrandite is a relatively uncommon mineral that generally occurs in beryllium-bearing pegmatites, but is also known from hydrothermal deposits.

WHITE PINE COUNTY Identified with phenakite, beryl, fluorite, and scheelite in veins and altered limestone at the Mount Wheeler Mine in the Lincoln district. (Cohenour, 1963; Hose et al., 1976)

BERYL

$Be_3Al_2Si_6O_{18}$

Beryl is a widespread mineral and was formerly the world's foremost source of metallic beryl-

lium. It is generally found in pegmatites, occasionally in some contact-metamorphic zones, and rarely in hydrothermal deposits or as a vapor-phase mineral in rhyolite.

CLARK COUNTY As pale-bluish-green to sea-green crystals to 3/4 inch with muscovite, chrysoberyl, and possible phenakite in zoned pegmatite dikes and sills that cut Precambrian schist and gneiss at the Taglo (Santa Cruz) Mine in the Bunkerville district; also as light-green to nearly colorless crystals to 4 inches at the Virgin Mountain chrysoberyl property. (Beal, 1965)

Opaque green beryl occurs with quartz, muscovite, *garnet,* fluorite, and other minerals in a pegmatite at the Feldspar Mine in the Crescent district. (Olson and Hinrichs, 1960)

With various radioactive minerals in pegmatites in the Gold Butte district. (Olson and Hinrichs, 1960)

As crystals to 3 cm in miarolitic cavities and pegmatites in the Eldorado and Newberry Mountains; specifically reported from the Sunrise and Moonbeam claims in the Searchlight district. (S. Scott, w.c., 1999; Olson and Hinrichs, 1960)

ELKO COUNTY In small amounts with lepidolite in a pegmatite dike in the Corral Creek district head of Corral Creek, Ruby Mountains. (Olson and Hinrichs, 1960)

In a prospect as blue-green to honey-colored crystals with minor *columbite-tantalite, garnet,* and uraninite(?) in the Gilbert Canyon district, SE 1/4 Sec. 9, T29N, R57E. (Olson and Hinrichs, 1960)

In pegmatite dikes and sills with quartz, *columbite-tantalite,* uraninite, and *garnet* at the Harrison Pass prospect in the Harrison Pass district; also reported in skarn at the Campbell and Sliper Mines. (Olson and Hinrichs, 1960)

Reported in a specimen from an unspecified locality in the Ruby Valley district. (WMK)

With muscovite in pegmatite dikes that cut Precambrian metamorphic rocks and Cretaceous granite at many sites in the Valley View (Dawley Canyon) district as white, green, and blue crystals to 3 inches in diameter; for example, reported at the Hankins Canyon beryl occurrence as white to azure blue crystals to 0.5 inch in diameter. Also reported as small white to blue-green crystals at the Errington-Thiel Mica Mine, and as gray to white crystals to 1.5 inches long in schist. (Olson and Hinrichs, 1960; Smith, 1976; LaPointe et al., 1991)

In pegmatite at the Robinson Creek beryl prospect, Sec. 19, T33N, R60E. (LaPointe et al., 1991)

Sky-blue, gray-blue, and greenish-gray crystals to 4 cm in diameter and 10 cm long, some gemmy, occur in pegmatite near the head of Lamoille Canyon in the Dollar Lakes and Island Lake area. (GCF)

ESMERALDA COUNTY In pegmatite at an unspecified locality in the Sylvania Mountains. (Olson and Hinrichs, 1960)

EUREKA COUNTY Reportedly abundant as white and yellow needles with fluorite, sphalerite, molybdenite, and scheelite in Ordovician quartzite, dolomite, and limestone at the Bisoni (McCulloughs Butte) prospect in the Fish Creek district; also reported with fluorite at the Reese and Berry prospects. (Papke, 1979; Holmes, 1963)

LANDER COUNTY As green crystals to 1 inch long with calcite, *feldspar,* scheelite, and quartz in a vein to 4 feet thick on Lynch Creek in the Birch Creek district; also as small colorless to green masses and disseminations in skarn at the T-Bone tungsten property. (Hall, 1962; Stewart et al., 1977a)

MINERAL COUNTY Light-blue beryl was reported with *bismutosphaerite* (bismutite) in a *wolframite*-scheelite vein at the Pine Crow tungsten prospect in the Marietta district. (Olson and Hinrichs, 1960)

NYE COUNTY As small green crystals with purple fluorite and local scheelite in greisen veins and replacement bodies in schist and slate at the Hiatt prospect in the Manhattan district. (Papke, 1979; Stager and Tingley, 1988)

PERSHING COUNTY Colorless, in pegmatite pods and veinlets with scheelite, quartz, fluorite, and dark-blue *tourmaline* at the Lakeview Mine in the Imlay district. (Johnson, 1977)

In pegmatites cutting the Prida Formation in Limerick Canyon in the Rochester district. (Johnson, 1977)

As crystals to an inch wide and 4 inches long with scheelite, fluorite, and native antimony in quartz-rich pegmatite bodies at the Oreana (Little) Tungsten Mine in the Rye Patch district; some small crystals are semitransparent dark-green, approaching emerald quality. (Olson and Hinrichs, 1960; Johnson, 1977; JHS)

In trace amounts in quartz veins at the Hamilton Beryl Mine in the Sacramento district. (Johnson, 1977)

WHITE PINE COUNTY Reported in a specimen from an unspecified locality in the Bald Mountain district. (WMK)

Blue crystals to 5 cm occur with *allanite, monazite,* phenakite, bertrandite, fluorite, and scheelite in limestone at the Wheeler Peak (Mount Wheeler) Mine and other properties in the Lincoln (Mount Washington) district; beryl is also reported with quartz, muscovite, pyrite, and *wolframite* in veins in the same area. (Cohenour, 1963; Griffiths, 1964; M. Jensen, w.c., 1999; Barton and Trim, 1991)

BERYLLONITE

NaBePO$_4$

Beryllonite is a rare mineral that generally occurs in beryllium-bearing pegmatite. The Nevada occurrence listed is unusual in that the mineral is apparently a product of the oxidation of a hydrothermal deposit.

ELKO COUNTY Beryllonite occurs in minor amounts with duftite in oxidized portions of replacement deposits of probable hydrothermal origin at the Standing Elk Mine (Aladdin Mine) in the Railroad district. (Ketner and Smith, 1963; Griffiths, 1964; Smith, 1976)

BERZELIANITE

Cu$_2$Se

Berzelianite is a rare mineral found in a few vein-type deposits, generally in association with other selenides. The occurrences below are the only ones known in the United States.

ELKO COUNTY Reported with other selenide minerals and *electrum* in veins in the Gold Circle (Midas) district. (P. Goldstrand, o.c., 2000)

MINERAL COUNTY As a steel-gray mineral disseminated in quartz in the Durant vein, Chihuahua stope, in the Aurora district; identified by X-ray methods. (Berry and Thompson, 1962; AMNH; S. Pullman, w.c., 1999)

BETAFITE

$(Ca,Na,U)_2(Ti,Nb,Ta)_2O_6(OH)$

A radioactive oxide mineral that generally occurs in granite pegmatite and rarely in carbonatite.

MINERAL COUNTY Bismuth-rich betafite is intergrown with plumbopyrochlore in brown to black masses in the Zapot pegmatite in the Fitting district. (Foord et al., 1999b)

BEUDANTITE

$PbFe_3^{3+}(AsO_4)(SO_4)(OH)_6$

Beudantite is a relatively rare secondary mineral that occurs where arsenic-rich hydrothermal deposits have undergone oxidation under arid conditions.

CLARK COUNTY In red-brown crystals to 0.5 mm at the Addison Mine in the Goodsprings district. (M. Jensen, w.c., 1999)

ELKO COUNTY As crystals to 1 mm with yellow arsenates at the Standing Elk (Aladdin) Mine in the Railroad district. (J. Crowley, w.c., 1999)

As complex intergrowths of minute, pseudo-cubic crystals lining cavities in iron-rich gossan with jarosite and plumbojarosite at the Killie Mine in the Spruce Mountain district. (Dunning and Cooper, 1987)

ESMERALDA COUNTY Reported in a specimen from the Weepah pit in the Weepah district. (R. Thomssen, w.c., 1999)

Reported from the Rock Hill Mines in the Rock Hill district. (R. Thomssen, w.c., 1999)

EUREKA COUNTY Questionably identified with plumbojarosite in oxidized ore in the Eureka district. (Nolan, 1962)

HUMBOLDT COUNTY Reported at the Silver Coin Mine in the Iron Point district. (Thomssen, 1998)

In crystals to 0.5 mm with whitlockite at the Twin Creeks Mine in the Potosi district. (M. Jensen, w.c., 1999)

LANDER COUNTY Reported in a specimen from the Gray Eagle Mine in the Shoshone Range. (R. Thomssen, w.c., 1999)

LINCOLN COUNTY As olive-green masses and veinlets in a pit 180 feet northwest of the Old Democrat shaft in the Chief (Caliente) district. (Callaghan, 1936)

MINERAL COUNTY As golden-brown microcrystals at the Simon Mine in the Bell district; also as red and green coatings with carminite. (S. White, w.c., 2000; GCF)

NYE COUNTY As blocky golden-brown crystals to 1.5 mm with arseniosiderite and scorodite in cavities in iron-stained, crustified quartz at the San Rafael Mine in the Lodi district; also as yellow-green crusts with bayldonite, jarosite, and strashimirite. (Jenkins, 1981; B. Runner, 2000; GCF)

PERSHING COUNTY In yellow-brown crusts with carbonate-fluorapatite at a prospect near Imlay Canyon. (M. Jensen, w.c., 1999)

BEYERITE

$(Ca,Pb)Bi_2(CO_3)_2O_2$

A secondary mineral that typically occurs in pegmatites and forms by the alteration of bismutite or other primary bismuth minerals.

CLARK COUNTY Identified by SEM/EDX in rock from the Boss Mine in the Goodsprings district by P. Hlava. (J. Kepper, this volume)

ESMERALDA COUNTY Tentatively identified as tiny, lustrous tan to beige plates after large aikinite crystals with perite, wulfenite, mimetite, and bismutite in a specimen from a prospect on the north side of Cucomungo Wash in the Tule Canyon district. (R. Thomssen, w.c., 1999)

BINDHEIMITE

$Pb_2Sb_2O_6(O,OH)$

Bindheimite, a relatively common mineral in oxidized hydrothermal deposits that contain primary lead and antimony minerals, is frequently derived from jamesonite and occurs with antimony oxides, cerussite, and anglesite.

CHURCHILL COUNTY In clasts with stibnite at the Green prospect in the Lake district. (Lawrence, 1963)

ESMERALDA COUNTY Reported in quartz veins with other minerals a mile north of the site of Columbus. (R. Walstrom, w.c., 2000)

Reported at the Carrie Mine in the Gilbert district. (Ferguson, 1928)

EUREKA COUNTY One of many minerals in oxidized ore in the Eureka district. (Nolan, 1962)

As earthy yellow coatings and masses with coarse galena and minor cerussite at the Gold

Quarry Mine in the Maggie Creek district. (Jensen et al., 1995)

HUMBOLDT COUNTY Reported at the Silver Coin Mine in the Iron Point district. (Thomssen, 1998)

LANDER COUNTY A secondary mineral in sulfide-bearing quartz veins in the Goat Ridge window, T29N, R45E. (Prihar et al., 1996)

LINCOLN COUNTY Tentatively identified with cerussite at the Eagle Rock claims in the Patterson district. (Schrader, 1931a)

Reported as yellow masses with tripuhyite on quartz at an unspecified locality in the Pioche district. (Jenkins, 1981)

MINERAL COUNTY With chlorargyrite, cerussite, and other minerals at the Broken Hills Mine in the Broken Hills district. (Schrader, 1947)

With, and as a common oxidation product of, jamesonite and galena at the New Potosi (Potosi) and other mines in the Candelaria district. (Knopf, 1923; Page, 1959; Ross, 1961; AMNH)

Silver-bearing bindheimite was reported as disseminated grains and as light-brown or bright yellow-orange masses in streaks and lenses at the Wamsley Mine in the Pamlico district; also reported as yellow to orange, powdery to vitreous specks, pods, and veinlets after fibrous jamesonite in quartz veins at the Lowman Mine. (Ross, 1961; AMNH; Lawrence, 1963)

In mercury ore with cinnabar and stibnite at the Drew Mine in the Pilot Mountains district. (Foshag, 1927)

With brochantite, linarite, minium, and other minerals at the Dispozitch Mine in the Silver Star district. (Ross, 1961)

NYE COUNTY A common yellow oxidation product of owyheeite and jamesonite at the various mines in the Morey district. (Williams, 1968)

PERSHING COUNTY Reported in a specimen from Majuba Hill in the Antelope district. (AMNH)

A mercury-bearing variety is common with cinnabar and stibnite at the Red Bird Mine in the Antelope Springs district. (Bailey and Phoenix, 1944)

Silver-bearing bindheimite was reportedly a major source of metals in bonanza silver ore at the Aztec, Montezuma, Jersey, West Springs, and other mines in the Arabia district where it was found as yellow to yellowish-brown, earthy masses after jamesonite with plumbojarosite,

arsenopyrite, cerussite, and gypsum in milky quartz veins. (Knopf, 1918a; Lawrence, 1963; CSM; FUMT; WMK)

In silver veins in the Rochester district. (Knopf, 1924)

With other antimony oxides as oxidation halos around jamesonite at the Green Antimony Mine in the Wild Horse district. (Lawrence, 1963; Johnson, 1977)

WHITE PINE COUNTY Reported in oxidized silver-lead ore in the Taylor district. (Lovering and Heyl, 1974)

BIOTITE

$$K(Mg,Fe^{2+})_3(Al,Fe^{3+})Si_3O_{10}(OH,F)_2$$

Biotite, the most common mica group mineral, is widespread and abundant, occurring as a primary mineral in most rock types and as a relatively high-temperature hydrothermal mineral in association with porphyry and other types of metal deposits. Only a few of the many Nevada occurrences are noted below.

CHURCHILL COUNTY As a primary and alteration mineral in wall rock and in veins at the Fairview Mine in the Fairview district. (Schrader, 1947)

CLARK COUNTY As crystals to 8 cm in miarolitic cavities and pegmatites in the Eldorado and Newberry Mountains. (S. Scott, w.c., 1999)

In a quartz-*feldspar*-biotite body at the Old Dad prospect, Gold Butte district. (Garside, 1973)

ELKO COUNTY Plentiful in contact-metamorphic zones in the Dolly Varden district. (Melhase, 1934d)

With quartz, beryl, and other minerals in pegmatite dikes in Dawley Canyon in the Valley View district. (Olson and Hinrichs, 1960)

EUREKA COUNTY Common as black to red-brown crystals and clusters to 2 cm in diorite and associated skarn at the Goldstrike Mine in the Lynn district. (GCF)

As phenocrysts in andesite at the Barth Iron Mine in the Safford district. (Shawe et al., 1962)

LANDER COUNTY Part of the skarn and intrusive mineral assemblages at the West orebody in the Battle Mountain district. (Theodore and Blake, 1978)

LYON COUNTY Comprises as much as 50 percent of altered rock with rutile and titanite at the Wheeler Mine in the Wilson district. (Jackson, 1996; Princehouse and Dilles, 1996)

WHITE PINE COUNTY Brown biotite is an important alteration mineral in the porphyry at the Kimberly and Tripp pits in the Robinson district. (Spencer et al., 1917; WMK)

BIRNESSITE

$Na_4Mn_{14}O_{27} \cdot 9H_2O$

Birnessite is a rarely identified mineral formed by the oxidation of primary manganese minerals and an important component of rock varnish; it probably occurs at many more sites in Nevada than the one listed below.

WHITE PINE COUNTY Identified with cryptomelane, aurorite, and other minerals at the North Aurora Mine in the White Pine district. (Radtke et al., 1967)

BISMITE

Bi_2O_3

Bismite is a moderately rare mineral formed where primary bismuth minerals have undergone oxidation under arid conditions. The associated carbonate, bismutite, is much more widespread.

ESMERALDA COUNTY At the January, Combination, and Sandstorm Mines in the Goldfield district as minute pearly scales with brilliant luster and of silvery whiteness, generally with *limonite* and commonly in rich gold ore on the walls of cavities, as frostlike films, and as spongy aggregates with quartz; in some cases in quartz lining prismatic hollow casts after bismuthinite. (Ransome, 1909a)

EUREKA COUNTY With bismuthinite in the rich gold placers in the Lynn district. (Lincoln, 1923; Mount, 1964)

MINERAL COUNTY Reported with native bismuth near Candelaria in the Candelaria district. (Schrader et al., 1917)

Reported with bismutite and wulfenite in a specimen from Rawhide. (WMK)

BISMOCLITE

BiOCl

Bismoclite is an uncommon mineral that occurs in the oxidation zones of deposits containing primary bismuth minerals. Crystalline occurrences such as at Goldfield are very rare.

ESMERALDA COUNTY Similar to and mistaken for bismite as single, silvery white pearly scales or frostlike films on the walls of cavities in limonitic rock, commonly with iodargyrite and associated with rich ore at the January, Combination, Grizzly Bear, and Sandstorm Mines in the Goldfield district; also reported as minute, rounded, rhombohedral crystals. (Schaller and Ransome, 1910; Schaller, 1941; LANH; S. Pullman, w.c., 1999)

NYE COUNTY Reported at the Omar (Omer) Marie Mine in the Manhattan district. (WMK)

An alteration product of aikinite as abundant, thin, square plates at the Outlaw Mine in the Round Mountain district; a close match in chemistry with bismoclite from Goldfield. (Dunning et al., 1991)

BISMUTH

Bi

Native bismuth is a moderately common mineral that occurs in hydrothermal veins, skarns, and granite pegmatites.

CLARK COUNTY Reported with bismutite at the Boss Mine in the Goodsprings district. (Schrader et al., 1917)

ELKO COUNTY Near the center of lenslike tungsten orebodies in contact-metamorphic rock with quartz, pyrite, phlogopite, *pyroxene,* and bismuthinite at the Star Mine in the Harrison Pass district. (Hill, 1916; Mount, 1964)

With scheelite, bismuthinite, and other sulfide minerals at the Valley View Mine in the Valley View district. (Stager and Tingley, 1988; Berger and Oscarson, 1998)

ESMERALDA COUNTY With bismutite at the Hecla Mine in the Candelaria district. (M. Jensen, w.c., 1999)

EUREKA COUNTY As a minor mineral in pods and irregular replacement bodies of complex Au-Ag-Pb ore in brecciated dolomite at the Lord Byron and Kelly Mines in the Eureka district. (Nolan, 1962; Mount, 1964)

HUMBOLDT COUNTY Reported in a specimen from the Golconda district. (WMK)

Tentatively reported from quartz veins on Auto Hill in the National district. (Lindgren, 1915; Melhase, 1934c)

LINCOLN COUNTY With galenobismutite, cosalite, bismuthinite, and galena at the Free Tunnel Mine in the Tem Piute district. (Buseck, 1967)

MINERAL COUNTY With bismite at an unspecified locality near Candelaria in the Candelaria district. (Schrader et al., 1917)

Reported in a specimen from Gold Range in the Silver Star district. (WMK)

PERSHING COUNTY With molybdenite and bismuthinite at the Majuba Hill Mine in the Antelope district. (Jensen, 1993b)

Reported in trace amounts with native arsenic at an unspecified locality near Rochester in the Rochester district. (Schrader et al., 1917)

WHITE PINE COUNTY Reported at the Gold King Mine in the Bald Mountain district. (Schrader et al., 1917)

Reported from 6 miles south of the Joy Post Office in the Ruby Hill district. (Schrader et al., 1917)

BISMUTHINITE

Bi_2S_3

Bismuthinite is a widespread mineral in mesothermal veins that contain tin and copper and in tungsten skarn deposits; it also occurs in some epithermal veins in association with native gold, such as at Goldfield.

CLARK COUNTY Reported with bismuth, copper minerals, cinnabar, pyrite, and numerous base-metal and bismuth-bearing selenides, arsenides, and iodides at the Boss Mine in the Goodsprings district. (Jedwab, 1998a)

ELKO COUNTY Nevada's and possibly the world's best crystals came from the Victoria Mine, Dolly Varden district, Sec. 5, T28N, R66E, as dull gray, radiating sprays and crystal clusters to 7 cm in clay-altered rock just above the water table. Also reported from the same site with pyrite, chalcopyrite, calcite, and quartz in skarn in limestone adjacent to a quartz monzonite intrusion. At the Butte claims in the same district with pyrite, gold, and copper minerals in quartz veins up to 2 feet thick in quartz monzonite

porphyry. (Atkinson et al., 1982; M. Jensen, w.c., 1999; GCF; Hill, 1916; Melhase, 1934d; Schilling, 1979)

With native bismuth in skarn at the Star Tungsten Mine in the Harrison Pass district. (Cornwall, 1964)

With native bismuth and pyrite in a gangue of quartz, rutile, phlogopite, and *pyroxene* at the Valley View Mine in the Valley View district. (Stager and Tingley, 1988)

ESMERALDA COUNTY As slender, lead-gray, prismatic crystals associated and commonly intergrown with famatinite and gold in cryptocrystalline quartz in breccia and vein ore at the Combination, January, Jumbo Extension, Mohawk, and Sandstorm Mines in the Goldfield district; other associated phases are goldfieldite, marcasite, pyrite, and telluride minerals. (Ransome et al., 1907; Ransome, 1909a; Lindgren, 1933; HMM)

Questionably identified in lead-silver ore from the Montezuma district 6 miles west of Goldfield. (Cornwall, 1964)

Reported in a specimen from the Elkton Lode in the Silver Peak district. (WMK)

EUREKA COUNTY Questionably identified in gold-silver-lead ore at the Lord Byron and Kelly Mines in the Eureka district. (Mount, 1964)

As a minor microgranular constituent of ore at the Goldstrike Mine in the Lynn district; reported elsewhere in the district with bismite and gold in veinlets of quartz and silicified Tertiary rhyolite, and with gold in placers. (GCF; Lincoln, 1923; Vanderburg, 1936a)

HUMBOLDT COUNTY Present in tungsten ore in the Potosi district; specifically reported at the Riley Mine. (Cornwall, 1964; Hotz and Willden, 1964; Schilling, 1979)

LANDER COUNTY In gold ore in skarn at the Fortitude Mine in the Battle Mountain district. (M. Jensen, w.c., 1999)

With heyrovskýite, bismuth, galena, and other minerals in skarn at a prospect near Austin in the Reese River district. (Jenkins, 1981)

LINCOLN COUNTY Questionably identified at the Jackrabbit Mine in the Bristol district. (Mount, 1964; Tschanz and Pampeyan, 1970)

In tungsten ore at the Free Tunnel, Lincoln, and Schofield Mines in the Tem Piute district. (Cornwall, 1964; Buseck, 1967)

MINERAL COUNTY Reported in bismuth ore at the Hecla Mine in the Candelaria district. (WMK)

Reported with pyrite at the Nevada Scheelite Mine in the Leonard district. (Cornwall, 1964; F. Cureton, w.c., 1999)

Reported as bismuth sulfide with bismutite at the Pine Crow tungsten prospect in the Silver Star district. (Cornwall, 1964)

Reported at the Garnet claim group in the Pilot Mountains district. (Cornwall, 1964)

NYE COUNTY With benjaminite and chalcopyrite at the Outlaw Mine in the Round Mountain district. (Harris and Chen, 1975)

Reported in a specimen from the Tonopah district. (UAMM)

As laths to 2 cm surrounded by bismutite in skarn at the Great Basin (Victory) Mine in the Lodi district. (M. Jensen, w.c., 1999)

PERSHING COUNTY As minute grains in drill core with native bismuth and molybdenite at the Majuba Hill Mine in the Antelope district. (Jensen, 1993b)

WASHOE COUNTY Intergrown with enargite at the Burrus Mine in the Pyramid district. (Mike Lear, o.c., 1998)

WHITE PINE COUNTY With cubes of purple fluorite, hübnerite, and triplite at the Red Hills Mine in the Eagle district. (Hill, 1916; Melhase, 1934d)

As silvery laths to 2 cm with bismutite in quartz at the Silver Bell Mine in the Shoshone (Minerva) district. (M. Jensen, w.c., 1999)

BISMUTITE

$Bi_2(CO_3)O_2$

Bismutite is a typical product of the oxidation of primary bismuth minerals, and may be more common in Nevada than indicated by the localities listed below.

CHURCHILL COUNTY As green masses associated with clinobisvanite, vanadinite, and pottsite at an unnamed locality northeast of Chalk Mountain in the Clan Alpine Mountains. (R. Walstrom, w.c., 1999)

CLARK COUNTY With beaverite, plumbojarosite, and quartz as greenish-yellow powder in cavities and on fracture surfaces in gold-platinum-palladium ore at the Boss Mine in the Goodsprings district. (Hewett, 1931; Cooper, 1962; Mount, 1964)

ESMERALDA COUNTY With native bismuth at the Hecla Mine in the Candelaria district. (M. Jensen, w.c., 1999)

Reported after aikinite in quartz at a prospect near Cucomungo Spring in the Sylvania Mountains in the Sylvania or Tule Canyon district. (Jenkins, 1981)

EUREKA COUNTY Reported as nearly pure masses with gold and bismuthinite in placer deposits in the Lynn district. (Schrader et al., 1917; Vanderburg, 1936a)

LANDER COUNTY With scheelite and molybdenite(?) in tungsten deposits in the Spencer Hot Springs district; specifically reported at the Linka Mine with junoite, pottsite, and other minerals. (McKee, 1976; Williams, 1988)

MINERAL COUNTY Bluish or greenish bismutite was questionably identified in ore at the Blue Bell No. 2 Mine in the Eagleville district. (Schrader, 1947)

Reported with bismite and wulfenite in a specimen from Rawhide in the Rawhide district. (WMK)

Reported as *bismutosphaerite* at the Pine Crow tungsten prospect in the Silver Star district. (Olson and Hinrichs, 1960)

NYE COUNTY As yellowish-white crusts coating quartz near altered aikinite at the Outlaw Mine in the Round Mountain district. (Dunning et al., 1991)

As greenish to yellowish waxy rinds to 2 mm on bismuthinite crystals at the Great Basin (Victory) Mine in the Lodi district. (M. Jensen, w.c., 1999)

WHITE PINE COUNTY Reported in quartz with bismuthinite at the Silver Bell Mine in the Shoshone (Minerva) district. (M. Jensen, w.c., 1999)

BIXBYITE

$(Mn^{3+},Fe^{3+})_2O_3$

A manganese oxide phase that is generally found as a vapor-phase mineral in rhyolite or in metamorphosed manganese ore.

MINERAL COUNTY Reported as micron-size grains at the Candelaria Mine in the Candelaria district. (Chavez and Purusotam, 1988)

BOKITE

$(Al,Fe^{3+})_7(V^{5+},V^{4+},Fe^{3+})_{40}O_{100} \cdot 37H_2O$

Bokite is a rare secondary product of the alteration of primary vanadium-rich materials.

EUREKA COUNTY As red or brown to black earthy masses with metahewettite, barnesite, huemulite, and minyulite in organic-rich shale at the Gibellini vanadium deposit and Van-Nav-San claim in the Gibellini district. (AM; Jenkins, 1981; R. Walstrom, w.c., 1999)

BOLÉITE

$Pb_{26}Ag_{10}Cu_{24}^{2+}Cl_{62}(OH)_{48} \cdot 3H_2O$

A rare secondary mineral in oxidized base-metal deposits.

ESMERALDA COUNTY As blue microcrystals in a quartz vein at the Rock Hill Mines in the Rock Hill district. (R. Thomssen, w.c., 1999; S. White, w.c., 1999; S. Pullman, w.c., 1999)

BOLTWOODITE

$HK(UO_2)SiO_4 \cdot 1/2H_2O$

An uncommon secondary mineral that is generally found in sandstone-type uranium deposits and in pegmatites.

CLARK COUNTY As pale-greenish-yellow radiating groups of microcrystals at the Green Monster Mine in the Goodsprings district; also reported as yellow-orange pulverulent masses. (S. White, w.c., 1999; S. Pullman, w.c., 1999; GCF)

BONATTITE

$Cu^{2+}SO_4 \cdot 3H_2O$

Bonattite is a rare mineral that occurs in fumarolic deposits with other sulfates such as copiapite and coquimbite.

WASHOE COUNTY Pale-blue-white to blue-gray botryoidal crusts and crystalline masses with melanterite and coquimbite in sinter deposits at Steamboat Hot Springs in the Steamboat Springs district. (Wilson and Thomssen, 1985)

BOOTHITE

$CuSO_4 \cdot 7H_2O$

Boothite, a member of the melanterite group, is a secondary copper mineral that is similar to chalcanthite.

LANDER COUNTY Reported in a specimen from the county without further notation. (MV)

BORAX

$Na_2B_4O_5(OH)_4 \cdot 8H_2O$

Borax is a relatively common evaporite mineral in playa deposits in the western United States and elsewhere in the world.

CHURCHILL COUNTY Reported from Soda and Little Soda Lakes (now flooded) in the Soda Lake district. (Gianella, 1941; J. Crowley, w.c., 1999)

ESMERALDA COUNTY Reported in a specimen from the Columbus Marsh district. (WMK)

Reported with ulexite and tincalconite in marsh deposits in the Fish Lake Marsh district. (Gianella, 1941)

MINERAL COUNTY With ulexite in playa deposits near the center of Rhodes Marsh in the Rhodes Marsh district where halite precipitates. (Melhase, 1935e; Gianella, 1941; Papke, 1976; WMK; UCB)

As a crust on the surface of Teels Marsh in the Teels Marsh district, the world's largest producer of borax prior to borate mineral discoveries in California in the late nineteenth century; more than 8,000 tons of refined borax were produced from this locality prior to 1883. (Gianella, 1941; Ross, 1961; Papke, 1976)

BORNITE

Cu_5FeS_4

Bornite is a common mineral in many of the world's copper deposits, and in some cases it is an important ore mineral. It occurs as both a primary and a secondary sulfide and is found in porphyry-type, vein, and skarn deposits.

CARSON CITY Reported in a specimen from the North Carson Mines in the Carson City district. (CSM)

With other copper minerals in a specimen from the Brunswick Mine in the Delaware district. (WMK)

CHURCHILL COUNTY Reported from the Webber shaft of the Fairview Silver Mine in the Fairview district. (Schrader, 1947)

CLARK COUNTY With chalcopyrite and secondary copper minerals at the Sundog Mine in the Crescent district. (Longwell et al., 1965)

Reported at the Lincoln Mine in the Gold Butte district. (Hill, 1916)

Thin films of Se-bearing bornite were reported on chalcopyrite at several copper mines in the Goodsprings district; identified with other copper minerals, cinnabar, potarite, and gold at the Boss Mine. (Hewett, 1931; Jedwab, 1998a)

DOUGLAS COUNTY With chalcopyrite in a specimen of copper ore in the Genoa district. (WMK)

ELK COUNTY With chalcopyrite and other copper minerals in skarn in the Contact district; specific localities include the Allen, Boston, and Ivy Wilson Mines. Also widespread but rare in quartz-calcite veins, as at the Boston Mine. (Purington, 1903; Bailey, 1903; Schrader, 1912, 1935b; Schilling, 1979; LaPointe et al., 1991)

Reported at the Butte claim group in the Dolly Varden district. (Hill, 1916)

Reported at the Kinsley Consolidated claim group in the Kinsley district. (Hill, 1916)

With chalcopyrite, chalcocite, and pyrite in a quartz and calcite vein in limestone at the Eldorado Mine in the Lime Mountain district. (Emmons, 1910)

Common after pyrite and chalcopyrite immediately below the water table at the Rio Tinto Mine in the Mountain City district. (Gianella, 1941; Coats and Stephens, 1968)

With pyrite, chalcopyrite, possible tetrahedrite, and secondary minerals in veins at the Standing Elk (Aladdin) Mine and other mines in the Railroad district. (Emmons, 1910; Ketner and Smith, 1963)

With wulfenite and base-metal minerals in oxidized ore and in primary ore at the Banner Hill and Killie Mines in the Spruce Mountain district. (Hill, 1916; Schrader, 1931b; Schilling, 1979)

Reported with chalcopyrite and pyrite at an unspecified locality in the Tuscarora district. (Melhase, 1934c; WMK)

ESMERALDA COUNTY Identified in a specimen from the Goldfield district. (JHS)

Partly replaced by digenite and covellite in some ore from the Klondyke district. (Chipp, 1969)

EUREKA COUNTY Occurs sparingly in unoxidized, silicified siltstone and with bladed barite and stibiconite at the Gold Quarry Mine in the Maggie Creek district. (Jensen et al., 1995)

HUMBOLDT COUNTY A minor mineral at the Lone Tree Mine in the Buffalo Mountain district. (R. Braginton, o.c., 1995)

Reported in a specimen from the west slope of Delong Peak in the Jackson Mountains district. (WMK)

In a specimen with malachite from the Felex Mine in the Potosi district. (WMK)

LANDER COUNTY With other copper minerals at the Copper Canyon and Copper Basin Mines in the Battle Mountain district. (JHS)

With tetrahedrite and chalcocite at an unspecified locality in the Jersey district. (Johnson, 1977)

With uraninite and other uranium minerals at the Apex (Rundberg) Uranium Mine in the Reese River district. (Garside, 1973)

LINCOLN COUNTY Reported as a minor mineral at the Delamar Mine in the Delamar district. (Callaghan, 1937)

Disseminated in quartz with other sulfides at the Pennsylvania Mine in the Pennsylvania district. (Tschanz and Pampeyan, 1970)

With chalcopyrite at an unnamed prospect on Blind Mountain near Pioche. (Westgate and Knopf, 1932)

LYON COUNTY With pyrite and chalcopyrite in the Yerington district; an important copper ore mineral. (Knopf, 1918b)

MINERAL COUNTY With chalcopyrite in a specimen from an unnamed locality in the Santa Fe district. (WMK)

With molybdenite, chalcocite, and other minerals in narrow quartz veins at the Silver Dyke Mine, lower adit, in the Silver Star district. (Garside, 1979; WMK)

NYE COUNTY With covellite after chalcopyrite on the 1000-foot level at the Tonopah-Belmont Mine in the Tonopah district. (Bastin and Laney, 1918)

In minor amounts with other sulfides and antimony oxide in quartz veins at the Murphy and Teichert Mines in the Twin River district. (Hague, 1870; Kral, 1951; Lawrence, 1963)

Identified in a specimen from an unnamed locality in the Union district. (JHS)

In quartz veins with chalcopyrite and copper oxides at an unnamed prospect on Limestone Ridge in the northern Belted Range on the Nellis Air Force Range. (Tingley et al., 1998a)

With other sulfides and copper carbonates in an orebody in limestone at the Dresser Mine in the Willow Creek district. (Kleinhampl and Ziony, 1984)

PERSHING COUNTY Rarely with chalcopyrite in copper ore at the Majuba Hill Mine in the Antelope district. (Jensen, 1985)

An ore mineral with chalcopyrite and digenite in a massive sulfide deposit at the Big Mike Mine in the Tobin and Sonoma Range district. (Johnson, 1977)

STOREY COUNTY Rare with quartz in drill-hole cuttings at the Flowery lode in the Comstock district. (Schilling, 1990)

WASHOE COUNTY Rare with chalcopyrite and pyrite in gold-bearing quartz veins in the Olinghouse district. (Overton, 1947)

In quartz veins with schorl, chalcocite, pyrite, chalcopyrite, gold, and secondary copper minerals at the Red Metal Mine and unnamed prospects in the Copperfield area on the north and west slopes of Peavine Mountain in the Peavine district. (Hill, 1916; Overton, 1947; Hudson, 1977; WMK)

Reported as a minor mineral in the Pyramid district. (Wallace, 1980)

WHITE PINE COUNTY Reported in some vein deposits in the Cherry Creek district. (Hose et al., 1976; JHS)

Reported in specimens from the Robinson district. (Gianella, 1941; JHS)

With other sulfides in veins that cut skarn in the Centennial zone near Mount Hamilton in the White Pine district. (Myers et al., 1991)

BOTRYOGEN

$MgFe^{3+}(SO_4)_2(OH) \cdot 7H_2O$

Botryogen is an uncommon mineral that is slightly soluble in water. It occurs in some mines as efflorescences resulting from the oxidation of pyrite, typically in arid climates.

NYE COUNTY As brick-red to orange-brown spherules composed of acicular crystals to 0.1

mm with rozenite, fibroferrite, copiapite, and other minerals in oxidized silver ore containing pyrite at the Upper Magnolia workings of the Morey Mine in the Morey district. (Jenkins, 1981)

BOULANGERITE

$Pb_5Sb_4S_{11}$

Boulangerite is widespread in small amounts in low- to moderate-temperature hydrothermal deposits, particularly in lead-silver ore that contains antimony. One of the so-called "feather ores" of yesterday's miners, modern techniques show that it is a common sulfosalt mineral.

ELKO COUNTY In trace amounts at the Meikle Mine in the Lynn district. (Jensen, 1999)

EUREKA COUNTY Identified in a specimen from the Cortez Mine in the Cortez district. (JHS)

As rims on galena in veins with other sulfides in diorite at the Goldstrike Mine in the Lynn district. (GCF)

LANDER COUNTY A probable supergene mineral with cerussite and iron oxides at the Little Giant and Copper Canyon Mines in the Battle Mountain district. (Roberts and Arnold, 1965)

As acicular, metallic-gray crystals to 1 cm with pyrite at the Little Gem (Cracker Jack) Mine in the Cortez (Tenabo) district. (M. Jensen, w.c., 1999; S. Pullman, w.c., 1999; Excalibur Mineral Co., 2001c)

Silver-bearing boulangerite occurs as small blades in coarse calcite crystals in the Mud Spring area in the Bullion district, northern Shoshone Range. (Gilluly and Gates, 1965)

Reported from the Dean Mine in the Shoshone Range. (R. Thomssen, w.c., 1999)

MINERAL COUNTY As stout prisms to 5 mm in vugs in quartz with jamesonite and pyrargyrite at the Broken Hills Mine in the Broken Hills district. (Jenkins, 1981)

Identified by L. Larson in polished sections from the Candelaria Mine open pit in the Candelaria district. (UCB; M. Jensen, w.c., 1999)

NYE COUNTY As coarsely crystalline masses intergrown with jamesonite and zinkenite in the Magnolia and Keyser tunnel levels of the Morey Mine in the Morey district; also noted in quartz-rhodochrosite veins in the same district. (Williams, 1968; Jenkins, 1981)

PERSHING COUNTY Intergrown with robinsonite at the Red Bird Mine in the Antelope Springs district. (Berry et al., 1952)

Identified in a specimen from an unknown locality in the Buena Vista district. (JHS)

Reported from the Rye Patch (Echo) district. (Ford, 1932)

With bournonite and gold in quartz veins at the Marigold claims in Buena Vista Canyon in the Unionville district. (Cameron, 1939)

BOURNONITE

$PbCuSbS_3$

Bournonite, which forms a series with seligmannite, is widespread in hydrothermal base- and precious-metal deposits. Because it is easily confused with tetrahedrite in hand specimen, it may be more common in Nevada than the following listing indicates.

EUREKA COUNTY Identified in a specimen from an unknown locality in the Cortez district. (JHS)

As brownish-gray anhedral crystals in quartz veins with pyrite, galena, freibergite, chalcopyrite, and arsenopyrite in diorite at the Goldstrike Mine in the Lynn district. (GCF)

HUMBOLDT COUNTY A minor mineral at the Lone Tree Mine in the Buffalo Mountain district. (R. Braginton, o.c., 1995)

LANDER COUNTY In a specimen with malachite and azurite from an unknown locality in the Battle Mountain district. (UCB)

Reported at the Little Gem Mine in the Bullion district. (S. Pullman, w.c., 1999)

With tetrahedrite and galena at the Betty O'Neal Mine in the Lewis district; also reported in the Eagle vein as microcrystals with quartz and other sulfosalt minerals. (Gilluly and Gates, 1965; M. Jensen, w.c., 1999)

Reported from an unspecified locality in the Reese River district. (Palache et al., 1944)

MINERAL COUNTY As silvery, twinned, euhedral crystals to 1 cm with tetrahedrite, arsenopyrite, boulangerite, and rattlesnakes at the Candelaria Mine open pit in the Candelaria district. (M. Jensen, w.c., 1999)

NYE COUNTY A late mineral in quartz-rhodochrosite veins in the Morey district. (Williams, 1968)

PERSHING COUNTY Reported in a specimen from near Lovelock in the Antelope district. (WMK)

Identified in a sample from an unspecified locality in the Buena Vista district. (JHS)

Common in ore from the Santa Clara Mine, but sparse in other veins in the Star district. (Cameron, 1939)

STOREY COUNTY Tentatively reported at the Ophir Mine in the Comstock district. (HUB)

WHITE PINE COUNTY As excellent single and twinned crystals to 8 mm with enargite, andorite, and other minerals in vugs in quartz at the Monitor Mine in the Taylor district. (Williams, 1965)

Reported with seligmannite in a specimen from the Mount Hamilton area in the White Pine district; noted with other sulfides in veins in the Centennial zone. (WMK; Myers et al., 1991)

BRANNERITE

$(U,Ca,Y,Ce)(Ti,Fe)_2O_6$

A primary complex oxide mineral in pegmatite, vein, paleoplacer, and modern placer deposits.

ELKO COUNTY As grains to 10 microns with pyrite, *apatite,* and hydrothermal silica at the Meikle Mine in the Lynn district. (Armstrong et al., 1997)

HUMBOLDT COUNTY Tentatively identified in radioactive silicified breccia in a prospect on the east flank of Buff Peak in the Bottle Creek district. (Castor et al., 1996)

BRAUNITE

$Mn^{2+}Mn_6^{3+}SiO_{12}$

A widespread mineral in weathered and metamorphosed manganese deposits.

HUMBOLDT COUNTY The principal manganese mineral in a silica-rich bed in Paleozoic sedimentary rocks at the Black Diamond claim in the Black Diablo district. (Johnson, 1977)

LANDER COUNTY Reported with *psilomelane* and pyrolusite at the Black Rock Mine in the Buffalo Valley district. (Stewart et al., 1977a)

LINCOLN COUNTY Reported with dufrénoysite and other minerals at the Prince Mine in the Pioche district. (Melhase, 1934d)

Reported with pyrolusite and *psilomelane*

at the Culverwell Mine in the Viola district. (Tschanz and Pampeyan, 1970)

PERSHING AND HUMBOLDT COUNTIES With bementite, rhodonite, and piemontite intergrown with *chalcedony* at the Black Diablo Mine in the Black Diablo district. (Ferguson et al., 1951; Willden, 1964)

WHITE PINE COUNTY Reported from the Central Pit shaft of the Vietti Mine in the Nevada district. (Pardee and Jones, 1920)

BRITHOLITE-(Y)

$(Y,Ca)_5(SiO_4,PO_4)_3(OH,F)$

A rare mineral that is found in pegmatite; the tentative Nevada occurrence in rhyolite is unusual.

HUMBOLDT COUNTY Tentatively reported as a 10-micron inclusion in zircon with thorite in topaz rhyolite on Buff Peak in the Bottle Creek district; identified by SEM/EDX (Castor and Henry, 2000; SBC)

BROCHANTITE

$Cu_4^{2+}(SO_4)(OH)_6$

Brochantite, which may be more common than reported below because of its similarity to malachite, is a secondary mineral in many copper deposits, particularly those in arid climates. Good specimens have come from the Yerington and Luning-Mina areas of Nevada.

CHURCHILL COUNTY Reported in nickel-cobalt ore at the Nickel and Lovelock Mines in the Table Mountain district. (Ferguson, 1939; M. Jensen, w.c., 1999; R. Thomssen, w.c., 1999)

CLARK COUNTY With other secondary copper minerals at the Key West Mine in the Bunkerville district. (Beal, 1965)

Reported after azurite at the Boss Mine in the Goodsprings district. (Hewett, 1931; S. Pullman, w.c., 1999)

Reported at the Quartette Mine in the Searchlight district. (Callaghan, 1939)

ELKO COUNTY With malachite, azurite, and chrysocolla on fractures, in veinlets, and in a fault zone in dolomitic limestone at the Copper Creek prospect in the Larrabee district. (Bentz et al., 1983)

ESMERALDA COUNTY Reported from an unspec-ified locality in the Candelaria district. (S. Pullman, w.c., 1999)

As good crystals with many associated minerals at the 3 Metals Mine in the Weepah district. (B. and J. Runner, w.c., 1998)

As dark-green crystals to 5 mm on quartz at the Broken Toe Mine in the Rock Hill district. (M. Jensen, w.c., 1999; R. Thomssen, w.c., 1999; S. White, w.c., 2000)

Reported from the Monte Cristo Range near Coaldale Junction. (R. Thomssen, w.c., 1999)

EUREKA COUNTY As pulverulent patches and flattened green crystals to 0.5 mm with malachite, azurite, and cyanotrichite at the Gold Quarry Mine in the Maggie Creek district. (Jensen et al., 1995)

HUMBOLDT COUNTY In green acicular crystals to 1 mm with malachite, mixite, fluorite, and creedite at the Silver Coin Mine in the Iron Point district. (Thomssen, 1998; M. Jensen, w.c., 1999; S. White, w.c., 1999; J. McGlasson, w.c., 2001)

Reported in a specimen with cuprite, malachite, and *limonite* from Barren Mountain. (UCB)

LANDER COUNTY Reported in a specimen from the Pedro Mine in the Battle Mountain district. (LANH)

As crystals with cuprite in limestone at the Bronco Mine. (R. Walstrom, w.c., 1999)

LINCOLN COUNTY Reported in a specimen from the Bristol Mine in the Bristol district. (WMK)

LYON COUNTY As deep green crystals with linarite at the Old (Lost) Soldier Mine in the Churchill district. (Jenkins, 1981)

As deep green acicular crystals at the Jackpot Mine in the Wilson district. (Jenkins, 1981)

In superb specimens of euhedral green crystals to 5 mm and radiating acicular clusters to 1 cm with pseudomalachite and chrysocolla in skarn at the Douglas Hill Mine in the Yerington district; also reported from the Bluestone, Empire-Nevada (Weed Heights), and other mines. (Knopf, 1918b; Melhase, 1935e; Jenkins, 1981; M. Jensen, w.c., 1999)

In a specimen from the Flyboy Mine in the Cambridge Hills. (S. White, w.c., 1999)

MINERAL COUNTY As well-developed crystals with green sprays of libethenite at an unnamed prospect on Copper Mountain in the Buckley district at the northern end of the Gillis Range. (R. Walstrom, w.c., 1999)

With linarite after tetrahedrite at the Lucky Boy Mine in the Lucky Boy district. (Ross, 1961)

With antlerite and malachite at the Northern Light Mine in the Mountain View district. (Ross, 1961)

Reported with chrysocolla, malachite, and linarite at the Dispozitch and Silver Gulch Mines in the Silver Star district. (Ross, 1961)

With azurite, malachite, chrysocolla, tenorite, and other minerals at the Bataan and Bluelight Mines in the Garfield district. (Ponsler, 1977)

With malachite, linarite, and chrysocolla in mercury mines in the Pilot Mountains district; specifically reported at the Dunlap Mine. (Foshag, 1927; M. Jensen, w.c., 1999)

As bottle-green crystals to 4 mm with malachite, chrysocolla, and other minerals at the New York Canyon Mine in the Santa Fe district. (Jenkins, 1981; CSM)

NYE COUNTY Sparse acicular crystals with linarite and caledonite in vugs in quartz at the Transylvania Mine in the Belmont district. (Jenkins, 1981)

As grayish-green crystals and aggregates to 3 mm with malachite on white quartz from the Illinois Mine in the Lodi district; also noted with linarite at the San Rafael Mine and with acanthite at the Hasbrouck shaft. (Jenkins, 1981; GCF)

Reported in prospects near Oak Springs in the Oak Spring district. (Melhase, 1935b; JHS)

As crude, deep green crystals with turanite, volborthite, and chrysocolla in a zone of contact-metamorphism in black slate at an unnamed prospect south of Manhattan in the Manhattan district. (Jenkins, 1981)

PERSHING COUNTY As small, dark-green crystals to 1 mm and as hemispheres to 2 mm with malachite, parnauite, and copper arsenate minerals at the Majuba Hill Mine in the Antelope district. (Smith and Gianella, 1942; MacKenzie and Bookstrom, 1976; Jensen, 1985)

WASHOE COUNTY With malachite, chalcanthite, azurite, chrysocolla, and copper arsenates in oxidized copper deposits on Peavine Mountain in the Peavine district. (Hudson, 1977; Jenkins, 1981)

As radiating clusters of grass-green crystals in open vugs and on fractures at the Burrus Mine in the Pyramid district; also as powdery pale-green coatings on fractures with chalcanthite as a post-mining efflorescence. (Jensen, 1994)

Reported in a specimen from the Steamboat Springs district. (HMM)

With malachite, chalcanthite, azurite, and chrysocolla in the Wedekind district. (Hudson, 1977)

WHITE PINE COUNTY Probably present in copper deposits of the Robinson district. (Spencer et al., 1917)

With langite, serpierite, and elyite in the Caroline tunnel in the Ward district. (Williams, 1972)

BROMARGYRITE

AgBr

Bromargyrite, originally *bromyrite,* a mineral of oxidized silver-bearing deposits in arid regions, is less common than the chlorine analog chlorargyrite. The variety *embolite* is chlorine-bearing; *iodobromite* contains chlorine and iodine.

CHURCHILL COUNTY Reported in quartz veins with chlorargyrite, gold, and silver at the Fairview Silver Mine in the Fairview district. (Greenan, 1914; Schrader, 1947; Saunders, 1977)

Reported as light- to dark-olive-green, translucent, crystalline coatings and imperfect loose crystals of the variety *iodobromite* with minor wulfenite and iodargyrite in oxidized silver veins in rhyolite and dacite at the Nevada Wonder Mine in the Wonder district. (Burgess, 1917; Schilling, 1979)

CLARK COUNTY Reported as *bromyrite* with chlorargyrite at the Lincoln Mine in the Goodsprings district. (Melhase, 1934e)

ELKO COUNTY Reported at the Protection Mine in the Mountain City district. (Emmons, 1910)

ESMERALDA COUNTY Identified in a sample from an unspecified locality in the Gold Point district. (JHS)

With rosasite and hemimorphite at the Palmetto Mine in the Palmetto district. (M. Jensen, w.c., 1999)

EUREKA COUNTY Reported from an unspecified locality in the Mineral Hill district. (Emmons, 1910)

LANDER COUNTY With antimony minerals and other silver minerals at the Hard Luck-Pradier Mine in the Big Creek district. (Lawrence, 1963)

LINCOLN COUNTY Reported with *embolite*

and chlorargyrite at the Jackrabbit Mine in the Bristol district. (Melhase, 1934d)

Questionably identified at the Apex Mine in the Pioche district. (Tschanz and Pampeyan, 1970)

MINERAL COUNTY With *electrum,* acanthite, and naumannite in the Juniata quartz-*adularia* vein system in the Aurora district. (Osborne, 1991)

With conichalcite at the Headley Mine in the Garfield district. (M. Jensen, w.c., 1999)

NYE COUNTY Reported in a specimen from an unknown locality in the Tonopah district. (CSM)

Reported in a specimen from the Old Belmont Mine in the Belmont district. (CSM)

As waxy blebs at the San Rafael Mine in the Lodi district. (B. and J. Runner, w.c., 1998)

PERSHING COUNTY With chlorargyrite, *electrum,* and naumannite in the Gold Boulder zone 2.8 miles west-northwest of Scossa. (W. Fusch, o.c., 2000)

WHITE PINE COUNTY With antimony minerals and chlorargyrite at the Crown Point Mine in the Bald Mountain district. (Lawrence, 1963)

Reported at the Star Mine in the Cherry Creek district. (Schrader, 1931b)

Doubtfully reported from mines on Treasure Hill in the White Pine district. (Hague, 1870)

BROOKITE

TiO$_2$

Brookite, which is trimorphous with rutile and anatase, is a moderately rare mineral that occurs in trace quantities in igneous rocks, as a constituent of contact-metamorphic rocks, and in hydrothermal veins.

CLARK COUNTY As microcrystals in miarolitic cavities and pegmatites in the Eldorado and Newberry Mountains. (S. Scott, w.c., 1999)

LANDER COUNTY As rare, flattened, black crystals to 0.05 mm with arsenopyrite and quartz crystals at the McCoy Mine (underground) in the McCoy district. (M. Jensen, w.c., 1999)

NYE COUNTY In scheelite-bearing skarn at the Victory Tungsten Mine in the Lodi district. (Humphrey and Wyatt, 1958)

PERSHING COUNTY As rare single crystals on *tourmaline* needles at the Majuba Hill Mine in the Antelope district. (Jensen, 1985)

BRUCITE

Mg(OH)$_2$

Brucite occurs mainly in contact-metamorphosed dolomite, but is also found in low-temperature hydrothermal veins in magnesium-rich rocks. Deposits near Gabbs have been a commercial domestic source for this industrial mineral since 1935.

MINERAL COUNTY Reported with magnesite from an unspecified locality in the Santa Fe district. (WMK)

NYE COUNTY With calcite, dolomite, hydromagnesite, and fluorite in Paleozoic carbonate rocks in the Pocopah district in the Calico Hills on the Nevada Test Site, Sec. 29, T12S, R51E. (Simonds, 1989; Castor et al., 1999b)

At the Premier Chemicals Mine (formerly Basic Refractories Mine) also known as the Gabbs Magnesite-Brucite Mine, Gabbs district, with magnesite and hydromagnesite and local chondrodite, grossular, and *apatite* as alteration products of dolomite adjacent to granitic intrusions. Generally found as white to pale-brown masses with the appearance of freshly broken laundry soap; also as bluish translucent veins and veinlets in massive brucite. Commercial brucite deposits at Gabbs were as much as 1000 feet long and 300 feet wide. (Callaghan, 1933; Callaghan and Vitaliano, 1947; Schilling, 1968a)

Reported as magnesium ore at the Springer property in the Lodi district. (FUMT; AMNH; WMK)

BUDDINGTONITE

(NH$_4$)AlSi$_3$O$_8$

Buddingtonite, a rare member of the *feldspar* group, occurs in some altered areas in felsic volcanic rocks.

ELKO COUNTY As microscopic crystals on larger sulfide and quartz crystals in the Ivanhoe district. (Krohn et al., 1993)

ESMERALDA COUNTY Reported in the Cuprite district. (Felzer et al., 1994)

Identified in the Cedar Mountains. (Baugh and Kruse, 1994)

HUMBOLDT COUNTY Identified in rich mercury ore at the Cordero Mine in the Opalite district. (Erd et al., 1964)

BURANGAITE

$(Na,Ca)_2(Fe^{2+},Mg)_2Al_{10}(PO_4)_8(OH,O)_{12} \cdot 4H_2O$

This is the second world occurrence of this extremely rare mineral found originally in a pegmatite in Rwanda. Absence of detectable iron by SEM/EDX suggests that the Nevada mineral may be a magnesium end member of burangaite.

EUREKA COUNTY As pale-green radiating groups of simple prismatic crystals and hemispheres to 0.3 mm with englishite, carbonate-fluorapatite, and variscite in vugs in partially silicified, pyrite-rich, alunite-bearing barite breccia at the Gold Quarry Mine in the Maggie Creek district. (Jensen et al., 1995)

BURSAITE

$Pb_5Bi_4S_{11}$

A rare sulfosalt mineral reported elsewhere from skarn and greisen tungsten deposits.

LANDER COUNTY Crystalline bursaite was identified with heyrovskýite from an unspecified prospect near Austin in the Reese River district. (F. Cureton, w.c., 1999)

BUTLERITE

$Fe^{3+}(SO_4)(OH) \cdot 2H_2O$

A rare sulfate that is found in fumarolic and hydrothermal deposits; at the type locality in Arizona it formed as the result of a mine fire.

WASHOE COUNTY Reported in a specimen from Steamboat Hot Springs in the Steamboat Springs district. (RMO)

BUTTGENBACHITE

$Cu_{19}Cl_4(NO_3)_2(OH)_{32} \cdot 2H_2O$

Buttgenbachite is a rare mineral formed by the oxidation of copper-containing substances under arid conditions in the presence of nitrate.

ESMERALDA COUNTY As crusts of minute blue crystals on quartz from a prospect 4-1/2 miles north of Coaldale Junction; the locality may be associated with nitrate-bearing marsh deposits in Fish Lake Valley. (Jenkins, 1981)

CACOXENITE

$(Fe^{3+},Al)_{25}(PO_4)_{17}O_6(OH)_{12} \cdot 75H_2O$

Cacoxenite, an uncommon secondary mineral formed where iron minerals have been oxidized in the presence of phosphorus, occurs in a variety of precious- and base-metal hydrothermal deposits in Nevada, but is found elsewhere in iron deposits and pegmatite.

EUREKA COUNTY Reported from the 5000-foot level of the Goldstrike Mine in the Lynn district. (GCF)

As radiating, yellow-orange to orange-brown acicular crystals in druses and isolated spherules, locally with variscite and gypsum at the Gold Quarry Mine in the Maggie Creek district; also abundant in places as a coating on joints and fractures and easily mistaken for jarosite or goethite. (Jensen et al., 1995)

LYON COUNTY Reported on quartz at the Dyke adit in the Talapoosa district. (R. Thomssen, w.c., 1999; UAMM)

In yellowish-brown radial masses on gossan with turquoise at the McArthur deposit in the Yerington district. (UAMM; Jenkins, 1981)

Also as microcrystals with wavellite, turquoise, and jarosite from an unnamed prospect on the south side of Carson Hill in the Yerington district. (M. Jensen, w.c., 1999)

MINERAL COUNTY In a specimen from Hooligan Hill in the Rawhide district. (R. Thomssen, w.c., 1999)

NYE COUNTY As golden-yellow radiating tufts on drusy quartz that is coated with velvety black layers of manganese oxide at the Montana-Tonopah Mine (500-foot level) in the Tonopah district. (Melhase, 1935c)

PERSHING COUNTY Reportedly the state's finest specimens of the species as bright yellow-orange to rust-brown velvety crusts and spherules to 0.1 mm with wavellite, variscite, and crandallite at the Willard Mine in the Willard district. (M. Jensen, w.c., 1999; Jensen and Leising, 2001)

CADMIUM

Cd

Native cadmium is extremely rare, reported elsewhere only from three deposits in the former USSR.

EUREKA COUNTY As microscopic grains in ore from the Goldstrike Mine in the Lynn district; identified by SEM/EDX. (Ferdock et al., 1997; GCF)

CALAVERITE

$AuTe_2$

Calaverite is a relatively uncommon mineral that occurs in hydrothermal precious-metal deposits, generally with other telluride minerals and sulfide.

ESMERALDA COUNTY Reported with free gold in ore that assayed $250,000 per ton (when gold was worth about $25 per troy ounce) at a depth of 200 feet at the Frances Mohawk lease in the Goldfield district; also reported from a small prospect about 0.3 mile west of the Florence Mine. (Rickard, 1940; Searls, 1948; JHS)

HUMBOLDT COUNTY In microscopic grains with coloradoite and gold-bearing arsenian pyrite at the Twin Creeks Mine in the Potosi district. (Simon et al., 1999)

MINERA COUNTY Reported at the Nevada Rand Mine in the Rand district. (Schrader, 1947)

NYE COUNTY Identified by SEM/EDX as tiny striated crystals with scorodite in a specimen from the dump at a shaft in a small mineralized area between the Gold Crater and Jamestown districts on the Nellis Air Force Range. (Tingley et al., 1998a)

CALCITE

CaCO$_3$

A widespread and abundant mineral that occurs in large amounts as limestone and in metamorphosed carbonate rocks. Also found in hydrothermal and supergene veins, as dripstone in caves (speleal limestone), and in carbonatite intrusions. In Nevada, spring-deposited *travertine* and *tufa* precipitated in alkaline lakes are common. Only a small percentage of Nevada occurrences of this common mineral are listed below.

CHURCHILL COUNTY In veins that range in composition from massive magnetite to calcite at the Black Joe prospect in the Copper Kettle district. (Shawe et al., 1962)

In veins as coarsely crystalline bodies nearly a foot in diameter with scalenohedrons up to 1.5 inches long at the Fairview Mine in the Fairview district. (Greenan, 1914; Schrader, 1947; Quade and Tingley, 1987)

As elongated acicular crystals in thin calcite veins in limestone at the Hazel Mine in the Lake district; also noted at the Green prospect. (R. Walstrom, w.c., 1999; R. Thomssen, w.c., 1999)

As *travertine* that formed rapidly in cased drill holes at Brady's Hot Springs. (Garside and Schilling, 1979)

CLARK COUNTY As micritic gray Devonian limestone that has been mined in large amounts to make lime at the Apex Quarry in the Apex district. (Longwell et al., 1965)

In specimens with *limonite* and vanadinite at the Calamine and Prairie Flower Mines in the Goodsprings district. (WMK; S. Pullman, w.c., 1999)

In Paleozoic limestone mined to make lime from 1910 to the 1970s at the Sloan Quarry in the Sloan district. (Longwell et al., 1965)

As crystals to 3 cm in miarolitic cavities and pegmatites in the Eldorado and Newberry Mountains. (S. Scott, w.c., 1999)

DOUGLAS COUNTY Reported in a specimen crusted with *chalcedony* from an occurrence on a volcanic ridge 1 mile south of Wellington in the Wellington district. (UCB)

ELKO COUNTY In limestone, red clay, and breccia as massive veins at the Rain Mine in the Carlin district; locally black or in small transparent to milky crystals. (D. Heitt, o.c., 1991)

In gray to colorless modified rhombs to 6 mm on diopside at the Victoria Mine in the Dolly Varden district. (M. Jensen, w.c., 1999)

As *travertine* in a 500-foot-wide dome with a 200-foot-wide central hole (the Hot Hole) filled with hot water at the Elko Hot Springs in the Elko district. (Garside and Schilling, 1979)

With quartz, *adularia,* pyrite, fluorite, and barite in gold- and silver-bearing veins at the Ken Snyder Mine in the Gold Circle district. (Casteel and Bernard, 1999)

The Meikle Mine in the Lynn district is reported as the source of Nevada's finest calcite specimens in thick crusts of lustrous crystals to 25 cm in pockets or crystal-lined caverns to more than 40 m across. The crystals occur in a variety of habits, from serrated flattened rhombs, to equant truncated scalenohedrons, to elongated sharp scalenohedrons and contact twins on (0001) and rarer "fishtail" twins on (1012). Crystal colors are colorless to white or pale gray and even black (due to clay inclusions) and a peculiar pale green. Also reported in veins that crosscut all rock types at the nearby Rossi Mine. (Jensen, 1999; Snyder, 1989)

Gray Devonian limestone that is mined at the Pilot Peak Quarry in the Proctor district in large amounts to make lime is nearly pure calcite. (Castor, 1991b; SBC)

As large white crystals in faults in fossiliferous limestone and shale at the Polar Star Mine in the Warm Creek district. (Hill, 1916)

As rhombohedrons to 4 inches at Bull Run. (Hoffman, 1878)

ESMERALDA COUNTY In a 600-foot-wide *travertine* mound at an extinct hot spring at the Klondyke Hills in the Klondyke district. (Garside and Schilling, 1979)

As dull gray scalenohedrons to 3/4 inch at the 3 Metals Mine in the Weepah district. (B. and J. Runner, w.c., 1998)

Cadmium-bearing calcite occurs in a vein at the Palmetto Mine in the Palmetto district. (M. Jensen, w.c., 1999)

EUREKA COUNTY As transparent, colorless, equant crystals to 1 cm at the Horse Canyon

Mine open pit in the Cortez district; also in specimens from the Garrison Mine and in gold ore with arsenic minerals, stibnite, quartz, and pyrite at an unspecified locality in Mill Canyon. (M. Jensen, w.c., 1999; WMK)

As *travertine* that is presently being formed and as old *travertine* that contains barite and fluorite at Bruffey's Hot Springs, Sec. 14, T27N, R52E. (White, 1955; Garside and Schilling, 1979)

A blue variety is reported at the Diamond and Lord Byron Mines in the Eureka district. (BMC; WMK; S. Pullman, w.c., 1999)

As a black variety in a specimen from 2 to 3 miles east-northeast of the foot of Lone Mountain in the Lone Mountain district. (WMK)

Common in the Goldstrike Mine in the Lynn district as massive veins to 1 m thick and as scalenohedrons and rhombohedrons to 1.2 cm in vugs and open fractures, commonly with barite; also an alteration mineral in diorite and a major component of contact-metamorphic and sedimentary rock assemblages. (GCF)

Widespread at the Gold Quarry Mine in the Maggie Creek district as coarsely cleavable white crystals in solid veins to 50 cm thick, lining vugs in a wide variety of habits, and rarely as colorless rhombs to 0.2 mm with wavellite and other phosphate minerals. Also reported as white or black (with included hydrocarbon) veins to 5 cm wide, and as "dog tooth" or "nailhead spar" crystals to 2.5 cm at the Maggie Creek Mine. (Jensen et al., 1995; Rota, 1991)

One of the principal gangue minerals in ore at the Modarelli Mine in the Modarelli-Frenchie Creek district. (Shawe et al., 1962)

Along fractures in ore and as an accessory in host rocks at the Barth Iron Mine in the Safford district; also reported as clear to brown scalenohedrons to 3.5 cm that are locally tinted red by hematite inclusions. (Shawe et al., 1962; B. and J. Runner, w.c., 1998; M. Jensen, w.c., 1999)

As calcareous sinter forming in active hot springs at the Hot Springs Point Sulfur Mine; and with sulfur and rare cinnabar, stibnite, and sideronatrite, metasideronatrite, metacinnabar, and gypsum in old sinter. (Garside and Schilling, 1979; GCF)

HUMBOLDT COUNTY Elongated colorless crystals to 1 cm at the Lone Tree Mine in the Buffalo Mountain district. (M. Jensen, w.c., 1999)

Reported at the Silver Coin Mine in the Iron Point district. (Thomssen, 1998)

As tan doubly terminated scalenohedrons to 1 mm on cinnabar at the Cordero Mine in the Opalite district. (GCF)

As white massive material enclosing other minerals at the Getchell Mine in the Potosi district; also as white or pale-yellow to brown rhombs to 5 cm in vugs in the Gertrude vein and South pit underground. Occurs with curetonite at the Redhouse Barite Mine. (JHS; CSM; F. Cureton, w.c., 1999; R. Thomssen, w.c., 1999)

With cinnabar, other mercury minerals, stibnite, quartz, and gypsum at the Cahill Mine in the Poverty Peak district. (Bailey and Phoenix, 1944; F. Cureton, w.c., 1999)

LANDER COUNTY As crystals to 10 cm associated with and partly replaced by blue *chalcedony* at the Mount Airy Mine in the New Pass district in the New Pass Range. (Rose, 1997)

In plates to 40 cm of scalenohedral crystals to 2 cm at the Betty O'Neal Mine in the Lewis district; some crystals are dark gray to black due to included carbonaceous material. (M. Jensen, w.c., 1999; J. McGlasson, w.c., 2001)

Flattened gray crystals to 20 cm at the Kingston Mine in the Kingston district. (M. Jensen, w.c., 1999)

As crude rhombs to 5 cm in vugs at the McCoy Mine in the McCoy district. (M. Jensen, w.c., 1999)

As scalenohedrons in the Reese River district. (Hoffman, 1878)

LINCOLN COUNTY As black calcite in a vein at the May Day Mine, Bristol district; the color is due to minute grains of manganese oxide that are so dispersed that they rarely show any relation to cleavage surface or outward crystal form. Elsewhere in the district, white calcite was reported surrounding argentiferous *wad* in the Jackrabbit Mine; large caverns lined with calcite crystals and stalagtites were found with lead and copper carbonate minerals in the Gypsy Mine; and calcite was noted with aragonite, aurichalcite, and malachite and as root casts in the Bristol Mine. (Melhase, 1934d; Jackson, 1963; UCB; WMK)

As coarse crystalline masses bordering iron-manganese carbonate orebodies at the Caselton and Summit shaft of the Pioche Mines in the Pioche district. (Gemmill, 1968)

LYON COUNTY As *travertine* in a terrace deposit built up by a now-inactive hot spring in Eldorado Canyon 2 miles southeast of Dayton in the Eldorado district; the rock was burned to make lime in stone kilns. (Garside and Schilling, 1979)

As brown to white "dogtooth spar" crystals to 3 cm long in vugs in carbonate rock in a road cut below the Boulder Hill Fluorite Mine in the Wellington district. (GCF)

As crystalline white masses, some filled with pyrite crystals at the Ludwig Mine in the Yerington district. (Melhase, 1935e)

As freshwater limestone 5 miles southeast of Fernley that is quarried to make cement at the Nevada Cement Co. plant. (Castor, 1988)

"Iceland spar" occurs with gypsum at an unspecified locality in the Desert Mountains district. (Klein, 1983)

MINERAL COUNTY As *travertine* in Sodaville Hot Springs in the Sodaville district. (Garside and Schilling, 1979)

NYE COUNTY Reported from the Bare Mountain district; specifically noted in a specimen from the Daisy Fluorspar Mine. (Melhase, 1935b; R. Thomssen, w.c., 1999)

Reported as manganese-bearing in the Bullfrog district. (Ransome et al., 1910; Melhase, 1935b)

Travertine occurs in a mound 2600 feet across and 25 feet high from which hot springs issue at the Chimney Hot Springs in the Butterfield Marsh district. (Garside and Schilling, 1979)

Reported as stained dogtooth crystals from the Papas(?) Gold Mine in the Currant district; also noted as mammillary and coarsely crystalline spelean calcite from the Gold Point Mine and in a large mass of coarsely crystalline breccia about 1 km southwest of the Gold Point Mine. (UCB; Castor and Hulen, 1996)

As coarsely crystalline bodies containing cinnabar, scheelite, and fluorite at the Scheebar Mine in the Fairplay district. (Bailey and Phoenix, 1944)

As small crystals and coatings, and reported with aragonite, at the San Rafael Mine in the Lodi district. (LANH; B. and J. Runner, w.c., 1998)

Manganese-bearing calcite occurs with gypsum at the White Caps Mine in the Manhattan district, where huge masses were formerly found on dumps with realgar and orpiment. Also noted as crystals on stibnite and lustrous flattened rhombs to 2 cm. (Melhase, 1935d; CSM; SM; S. Pullman, w.c., 1999; M. Jensen, w.c., 1999)

Travertine makes up a 600-foot-wide domelike hill that contains a 50-foot-wide circular hole filled with warm water at the Diana's Punch Bowl in the Northumberland district. (Garside and Schilling, 1979)

As fracture fillings and clear crystals to 2 cm at the Northumberland Mine in the Northumberland district. (Kokinos and Prenn, 1985)

Abundant and widespread in lower mine workings, but rare in the upper ore zones at the Round Mountain and Blue Centipede Mines in the Round Mountain district. (Tingley and Berger, 1985; WMK)

In the Mizpah vein at the 200-foot level in the Tonopah district as rhombohedral crystals, some colored green by included malachite; also noted as light-brown *manganocalcite* on the 1000-foot level of the Belmont Mine. (Melhase, 1935c)

Black calcite was an abundant gangue mineral in silver deposits in the Tybo district. (Humphrey, 1960)

Colorless equant crystals to 1 cm at the Indianapolis Mine in the Union district. (M. Jensen, w.c., 1999)

As *tufa* at Lockes Hot Springs near the Butterfield Marsh district. (Garside and Schilling, 1979)

PERSHING COUNTY As rhomohedral crystals to 2 mm with cinnabar at the Red Bird Mine in the Antelope Springs district. (GCF)

As curving aggregates to 4 cm of spelean crystals in the Black Jack Mine in the Imlay district, also as sharp scalenohedrons to 2 cm in the Star Mine. (M. Jensen, w.c., 1999)

Common in tungsten ore in the Mill City district; reported as tubular calcite from the 500 North Drift of the Humboldt Mine. (JHS; WMK)

Part of the gangue in several iron mines in the Mineral Basin district. (Reeves and Kral, 1955)

As contact-twinned, stubby scalenohedrons to 5 cm with barite at the Hecla Rosebud Mine in the Rosebud district. (M. Jensen, w.c., 1999)

With cinnabar in veins at the Cinnabar King, Harris, King George, and Walker Mines in the Spring Valley district. (Bailey and Phoenix, 1944; S. Pullman, w.c., 1999)

Dull white to colorless twinned scalenohedrons to 70 cm from a prospect near the Hycroft

Mine in the Sulphur district. (M. Jensen, w.c., 1999)

As translucent white crystals to 3 mm, some twinned, at the Willard Mine in the Willard district. (GCF)

STOREY COUNTY In the Comstock district as a sparse to abundant mineral in the Comstock Lode and as a precipitate on timbers; more abundant in the Occidental-Brunswick and Flowery lodes. (Gianella, 1936a; WMK; Nichols, 1991; Schilling, 1990)

Spiky to rounded, equant, translucent white crystals to 2 cm on drusy quartz crystals at the Gooseberry Mine in the Ramsey district. (M. Jensen, w.c., 1999)

WASHOE COUNTY As *travertine* in a tower formed over a well drilled in 1916 at the Fly Ranch Hot Springs in the Deephole district. (Garside and Schilling, 1979)

A gangue mineral as white veins to 6 cm wide at the Olinghouse Gold Mine in the Olinghouse district; occurs with gold, goethite, quartz, epidote, *adularia,* and zeolite minerals. (Jones, 1998; GCF)

As "Iceland spar," some marketed for optical use in microscopes, from an unnamed property a few miles north of Pyramid Lake. (Melhase, 1934a; Fulton and Smith, 1932; CSM; AMNH)

As *tufa* in towers, ridges, and in benchlike or rounded masses formed at or near subaqueous hot springs around Pyramid Lake. The Needles at the north end of the lake are *tufa* towers near active hot springs. In some occurrences as "thinolite," crude rectangular pseudomorphs of calcite after an unspecified mineral, possibly gaylussite. Also noted in chara deposits north of Sutcliffe, and as beds of small oncolites near The Needles. (Purkey and Garside, 1995; K. Papke, o.c., 2002; SBC)

WHITE PINE COUNTY Fluorescent calcite is found in road cuts at Conners Pass in the Cooper district. (Klein, 1983)

As *travertine* in a mound over 10 acres in area and as much as 40 feet high at Monte Neva Hot Springs in the Granite district. (Garside and Schilling, 1979)

As speleal formations with aragonite at the Lehman Caves. (WMK)

As "fishtail" twins and colorless razor-sharp rhombs to 2 cm in vugs in gossan at the Ward

Mine open pit and underground in the Ward district. (M. Jensen, w.c., 1999)

Abundant and widespread black calcite occurs in silver deposits in carbonate host rocks with aurorite and other manganese oxide minerals at the Hidden Treasure and Aurora Mines in the White Pine district. (Hewett and Radtke, 1967)

Colorless crystals to 5 mm enclose aurichalcite at the Grand Deposit Mine in the Muncy Creek district. (M. Jensen, w.c., 1999)

CALEDONITE

$Pb_5Cu_2(CO_3)(SO_4)_3(OH)_6$

Caledonite is an uncommon mineral found in the oxidation zones of deposits containing primary lead and copper minerals; however, the mineral is widespread in Nevada, particularly in occurrences with linarite, and excellent crystals have come from Goodsprings.

CLARK COUNTY Very uncommon as a secondary mineral with linarite, which it replaces at the Yellow Pine Mine, 1100-foot level, in the Goodsprings district; also reported at the Root Mine. (Hewett, 1931)

ELKO COUNTY Reported from an unnamed locality on the east slope of Long Canyon in the Lee district. (WMK)

ESMERALDA COUNTY As tiny crystals with a large variety of other minerals at the 3 Metals Mine in the Weepah district. (B. and J. Runner, w.c., 1998)

EUREKA COUNTY With linarite and malachite on barite at an unspecified locality in the Eureka district. (Jenkins, 1981)

HUMBOLDT COUNTY Reported at the Silver Coin Mine in the Iron Point district. (Thomssen, 1998)

LYON COUNTY Reported at the Old (Lost) Soldier Mine in the Churchill district. (S. Pullman, w.c., 1999)

Present with linarite, brochantite, and anglesite at the Jackpot claim in the Wilson district. (Jenkins, 1981)

MINERAL COUNTY Reported in a specimen from the Atherton Mine in Esmeralda County; probably from the Garfield (Atherton) Mine, Garfield district in Mineral County. (WMK)

Reported from an unspecified locality in the Rawhide district. (Schrader, 1947)

Reported with linarite in a specimen from an

unnamed locality in the Silver Star district. (WMK)

NYE COUNTY Rarely with linarite and brochantite in vugs in quartz at the Transylvania Mine in the Belmont district. (Jenkins, 1981)

With linarite and malachite in quartz vugs and in cavities in oxidized galena at the San Rafael Mine in the Lodi district. (Jenkins, 1981)

WASHOE COUNTY Reported from near Wadsworth; may be from one of the mines around Black Warrior Peak. (BMC)

WHITE PINE COUNTY As greenish-blue crystals to 0.1 mm in cavities in earthy galena-stephanite intergrowths with linarite, langite, and brochantite at the Silver King Mine in the Ward district. (Jenkins, 1981)

CALLAGHANITE

$Cu_2Mg_2(CO_3)(OH)_6 \cdot 2H_2O$

This Nevada type mineral has been recognized nowhere else in the world, but many specimens have come from the type locality. It was apparently derived by alteration of brucite-magnesite-*serpentine* rocks along contacts with rock containing primary copper sulfides.

NYE COUNTY Type locality at the Premier Chemicals Mine (formerly Basic Refractories Mine) also known as the Gabbs Magnesite-Brucite Mine, Gabbs district, Secs. 26 and 35, T12N, R36E, as azure-blue crystals to 1 mm as encrustations on and veinlets in brucite and hydromagnesite in a *serpentine*-forsterite-brucite zone along the contact between brucite and granodiorite. Associated minerals include mcguinnessite, nakauriite, and the discredited species *cuproartinite* and *cuprohydromagnesite*. (Beck and Burns, 1954; Schilling, 1968a; Oswald and Crook, 1979)

CALOMEL

Hg_2Cl_2

A secondary mineral that occurs sparsely in some mercury deposits and is generally formed by replacement of cinnabar.

CHURCHILL COUNTY Reported from the Red Bird Mine area in the Bernice district. (R. Thomssen, w.c., 1999)

ELKO COUNTY With cinnabar in a small opalite-type mercury deposit at the Rimrock and Homestake claim groups in the Ivanhoe district. (Bailey and Phoenix, 1944)

EUREKA COUNTY Reported with cinnabar in silicified rhyolite at an unspecified locality in the Beowawe district. (WMK)

HUMBOLDT COUNTY After cinnabar and metacinnabar on Buckskin Mountain in the National district. (Vikre, 1985)

With eglestonite and native mercury at the McDermitt Mine in the Opalite district. (Rytuba and Glanzman, 1979; McCormack, 1986)

In a specimen with eglestonite, cinnabar, and montroydite at the Cahill Mine in the Poverty Peak district. (AMNH; M. Jensen, w.c., 1999)

LANDER COUNTY Identified in mercury ore with cinnabar at the Wild Horse Mine in the Wild Horse district. (Dane and Ross, 1943; Bailey and Phoenix, 1944)

NYE COUNTY As dull black microcrystalline coatings with small colorless crystals of barite at the Paradise Peak Mine in the Paradise Peak district. (Dobak, 1988; UMNH)

PERSHING COUNTY With cinnabar in calcite veins at the Cinnabar King, King George, and Walker Mines in the Spring Valley district. (Bailey and Phoenix, 1944)

STOREY COUNTY A minor species in mercury ore with cinnabar and mercury at the Castle Peak Mine in the Castle Peak district. (Bailey and Phoenix, 1944; Holmes, 1965)

CANFIELDITE

Ag_8SnS_6

Canfieldite is a moderately rare low-temperature hydrothermal mineral that occurs in deposits containing silver and tin.

LANDER COUNTY Reported from the Dean Mine in the Lewis district; identified by EMP. (Ream, 1990; J. McGlasson, w.c., 2001)

Reported in stockwork veins with other sulfides at the Cove Mine in the McCoy district. (Emmons and Coyle, 1988)

NYE COUNTY As tiny bronze-tarnished octahedra with pyrargyrite and diaphorite in quartz-rhodochrosite veins in the Morey district. (Williams, 1968)

CARBONATE-CYANOTRICHITE

$Cu_4^{2+}Al_2(CO_3,SO_4)(OH)_{12} \cdot 2H_2O$

A very rare mineral that is generally associated with other secondary copper minerals.

EUREKA COUNTY As azure-blue, radiating clusters to 2 mm of fan-shaped crystals in druses in the southwest corner of the Gold Quarry Mine in the Maggie Creek district. (Jensen et al., 1995)

CARBONATE-FLUORAPATITE

$Ca_5(PO_4,CO_3)_3F$

A low-temperature *apatite*-group species that includes the variety *francolite,* which is typically found in sedimentary rocks. Also known from high-temperature environments such as in skarn and altered volcanic rock. Many of the Nevada occurrences are in Carlin-type gold deposits.

ELKO COUNTY Reported from the Dee Mine in the Bootstrap district. (R. Thomssen, w.c., 1999)

As white crystalline clusters to 1 mm in specimens from the East pit in the Ivanhoe district. (R. Thomssen, w.c., 1999; GCF)

As colorless microcrystals on barite at the Meikle Mine in the Lynn district. (Jensen, 1999)

ESMERALDA COUNTY Reported as white crystals to 0.5 mm with variscite and metavariscite at unnamed prospects north of Columbus. (M. Jensen, w.c., 1999)

EUREKA COUNTY Clear to white or pale-yellow acicular crystals to 7 mm with whitlockite, variscite, and hyaline opal in jasperoid at the Goldstrike Mine in the Lynn district. (GCF)

As widespread but rare radiating clusters of barrel-shaped crystals in mineralized rock, less commonly as drusy crusts, at the Gold Quarry Mine in the Maggie Creek district. (Jensen et al., 1995)

HUMBOLDT COUNTY As needles to 2 mm and stubby hexagonal crystals to 1 mm with pharmacosiderite at the Lone Tree Mine in the Buffalo Mountain district. (M. Jensen, w.c., 1999)

As colorless crystals to 1.5 mm at the Silver Coin Mine in the Iron Point district; copper-bearing green and lead-bearing varieties have also been noted. (Thomssen, 1998; M. Jensen, w.c., 1999; S. White, w.c., 2000; J. McGlasson, w.c., 2001)

At the Redhouse Barite Mine in the Potosi district as microscopic spherulitic tufts or acicular crystals; also at the Twin Creeks Mine as blue-green spherules to 3 mm, and at the Getchell Mine as white to blue mammillary crusts. (S. Sears, w.c., 1999; M. Jensen, w.c., 1999; R. Thomssen, w.c., 1999)

As pale-blue to white mammillary crusts and crystal clusters from a mine near Golconda Summit in the Golconda district. (R. Thomssen, o.c., 2000)

LANDER COUNTY In a specimen from the Wilson-Independence Mine in the Battle Mountain district. (R. Thomssen, w.c., 1999)

As compact groups of crystals to 4 mm in a quartz vein at the Snowstorm Mine in the North Battle Mountain district. (R. Thomssen, w.c., 1999; S. White, w.c., 2000)

LINCOLN COUNTY Reported in a specimen from the Worthington district. (R. Thomssen, w.c., 1999)

LYON COUNTY As colorless spherules to 1 mm with libethenite crystals at a prospect on the south side of Carson Hill in the Yerington district. (M. Jensen, w.c., 1999)

MINERAL COUNTY Reported in a specimen from the Candelaria district. (R. Thomssen, w.c., 1999)

As colorless hexagonal crystals to 1 mm with orange-brown jarosite crystals at an unnamed mine 2.5 miles south of Kinkaid in the Fitting district. (M. Jensen, w.c., 1999)

NYE COUNTY Reported as *francolite* with vashegyite, variscite, and green opal at the Manhattan Mine in the Manhattan district. (Melhase, 1935d; MM)

As pale-green botryoidal crusts to 1 cm from a prospect near Warm Springs. (M. Jensen, w.c., 1999)

Reported as *francolite* in a specimen from near Beatty in the Bare Mountain (Fluorine) district. (CSM)

PERSHING COUNTY As white to colorless hexagonal crystals to 1 mm with beudantite at a prospect near Imlay Canyon. (M. Jensen, w.c., 1999)

CARBONATE-HYDROXYLAPATITE

$Ca_5(PO_4,CO_3)_3(OH)$

An *apatite*-group mineral that generally occurs in sedimentary rocks; also known as *dahllite* or *podolite.*

ESMERALDA COUNTY In well-developed, doubly terminated, white hexagonal crystals to 0.5 mm with jarosite at an unnamed prospect in the Candelaria district in the Candelaria Hills northeast of Columbus. (R. Walstrom, w.c., 1999; GCF)

HUMBOLDT COUNTY Reported as *dahllite* or *podolite* in sedimentary rocks at the Lone Tree Mine in the Buffalo Mountain district. (Panhorst, 1996)

NYE COUNTY Reported as *dahllite* from an unspecified locality in the Manhattan district. (CSM)

Reported from the Mizpah Mine in the Tonopah district as *dahllite* in minute hexagonal crystals in a white drusy coating with iodargyrite, hyalite, quartz, and manganese oxide; also in the Valley View vein as white crystals with clear hyalite or crystalline iodargyrite coatings. (Rogers, 1912b; Melhase, 1935c)

CARLINITE

Tl$_2$S

Carlinite is an extremely rare Nevada type mineral, thus far found at only two Carlin-type gold deposits in one district. It is of hydrothermal origin and is considered a product of late introduction of fluids rich in thallium, arsenic, mercury, and other metals.

EUREKA COUNTY At the type locality in the east pit of the Carlin Mine in the Lynn district, Sec. 13, T35N, R50E, as subhedral to anhedral, dark-gray, metallic grains less than 0.5 mm long disseminated in black, brecciated fragments of carbonaceous limestone in shear zones with pyrite, avicennite, fine-grained quartz, and hydrocarbon compounds. (Radtke and Dickson, 1975)

Also as 10-micron grains in altered diorite with zircon, silica, and clay minerals in the Deep Post orebody at the Goldstrike Mine in the Lynn district. (GCF)

CARMINITE

PbFe$_2^{3+}$(AsO$_4$)$_2$(OH)$_2$

Carminite is a relatively rare mineral that occurs in the oxidation zones of arsenic-rich deposits, commonly with arseniosiderite, beudantite, and other arsenates.

ELKO COUNTY As aggregates of fine needles, typically in bundles, or in small fibrous spherical groups in cavities in partially leached massive bayldonite at the Killie Mine in the Spruce Mountain district. (Dunning and Cooper, 1987)

ESMERALDA COUNTY Reported in a specimen from the Carrie Mine in the Gilbert district. (R. Thomssen, w.c., 1999)

Mostly as coatings with some very tiny reddish microcrystals at the Weepah Mine in the Weepah district. (S. Pullman, w.c., 1999; B. and J. Runner, w.c., 1998)

EUREKA COUNTY Reported from an unspecified locality in the Eureka district. (Palache et al., 1951)

In the oxidation zone at the Carlin Mine in the Lynn district. (Hausen and Kerr, 1968)

LANDER COUNTY Reported from the Gray Eagle Mine in the Shoshone Range. (R. Thomssen, w.c., 1999)

Reported from the Wilson-Independence Mine in the Battle Mountain district. (J. McGlasson, w.c., 2001)

LINCOLN COUNTY Reported from the Silver King Mine, Silver King district. (R. Thomssen, w.c., 1999)

MINERAL COUNTY As good microcrystals at the Simon Mine in the Bell district. (B. and J. Runner, w.c., 1998; S. White, w.c., 1999)

NYE COUNTY As microcrystals with many associated phases at the San Rafael Mine in the Lodi district. (B. and J. Runner, w.c., 1998)

CARNOTITE

K$_2$(UO$_2$)$_2$V$_2$O$_8$•3H$_2$O

Carnotite is a widespread secondary mineral in uranium deposits that contain vanadium.

CLARK COUNTY Disseminated in Tertiary sandstone on the First Chance claims in the Gold Butte district; also as yellow coatings with opal in joints in the Chinle Formation on the Horse Spring claims. (Garside, 1973)

As yellow to yellowish-green coatings with *limonite*, hydrozincite, galena, and oxidized copper, lead, and zinc minerals at the Singer Mine in the Goodsprings district. (Garside, 1973)

Reported from Boulder Canyon in the McClanahan district. (WMK)

With carbonaceous trash in the Jurassic Aztec Sandstone at the Carnotite No. 1 claim in the Moapa district. (Garside, 1973)

As stains with carbonaceous material, petrified logs, opal, and calcite in the Muddy Mountains district in the Muddy Mountains. (Garside, 1973)

As tiny yellow specks, pebble coatings, and aggregates in gravelly caliche in the Sloan district in cuts along the Union Pacific Railroad from Sloan to Jean. (Garside, 1973)

ELKO COUNTY As small yellow to yellow-green flakes in fractures in fault breccia in black shale at the Deerhead prospect in the Carlin district. (Bentz et al., 1983; LaPointe et al., 1991)

ESMERALDA COUNTY Tentatively identified with meta-autunite as disseminations in lacustrine tuffs at the Cap Strike claims in the Coaldale district. (Garside, 1973)

Questionably identified at the Mustang claims in the Red Mountain district. (Garside, 1973)

HUMBOLDT COUNTY Reported as powdery yellow fracture coatings or thin layers in opal in lacustrine and tuffaceous beds with schröckingerite near the Virgin Valley Ranch in the Virgin Valley district; also identified in spherical cavities with carbonaceous material. (Staatz and Bauer, 1954; Castor et al., 1996; Castor and Henry, 2000)

LINCOLN COUNTY Reported as fracture coatings in uranium ore at the Atlanta Mine in the Atlanta district. (Hill, 1916; Garside, 1973)

As impregnations in carbonaceous material and as coatings on joint surfaces and mud cracks at the Dorothy claim and White Cloud prospect in the Panaca district. (Garside, 1973)

LYON COUNTY Tentatively identified at the Halloween Mine in the Washington district. (Garside, 1973)

With gypsum and possible native sulfur at the White Rose and White Rose No. 1 claims in the Wilson district. (Garside, 1973)

MINERAL COUNTY Golden yellow carnotite is present between grains and as bedding and fracture coatings in Tertiary water-laid tuffs and tuffaceous sandstones at the Amalgamated Uranium/Carol R Mine in the Pamlico district. (Ross, 1961; Garside, 1973)

In abundant opalized wood, logs, and plant material at the Robinson and Bubbles claim groups in the Red Ridge district. (Garside, 1973)

NYE COUNTY Reported in a specimen from the Manhattan district. (WMK)

WASHOE COUNTY Questionably reported with metatorbernite and meta-autunite at the Black Hawk claims in the State Line district. (Garside, 1973)

Reported in a specimen from the Nat Kearns Feldspar Mine in the Stateline Peak district. (WMK)

Reportedly as coatings with some microcrystals at the Tick Canyon Mine in the Dogskin Mountain district. (B. and J. Runner, w.c., 1998)

CASSITERITE

SnO_2

Cassiterite, the principal ore mineral of tin, is a widespread mineral worldwide in vein, greisen, pegmatite, skarn, and placer deposits. However, in Nevada it mainly occurs in hydrothermal deposits of the base and precious metals, and in rhyolite. *Wood tin* is finely crustiform, botryoidal cassiterite that generally occurs in rhyolite.

ELKO COUNTY As fine-grained yellow to tan fracture coatings adjacent to silver-lead ore in vein or replacement deposits in Paleozoic sedimentary rocks in the Delano district. (Olsen, 1960, 1965; Hewett et al., 1968)

Reported as small crystals in gold placers in the Tuscarora district; also reported from rhyolite flows. (Hoffman, 1878; Gianella, 1941)

ESMERALDA COUNTY As a minor species in copper-rich ore in the Goldfield district. (Hewett et al., 1968)

HUMBOLDT COUNTY Reported from an unspecified locality in the Jungo district. (Melhase, 1934b)

Identified in ore at the Getchell Mine in the Potosi district. (Unpublished company report)

LANDER COUNTY A small amount of *wood tin* was reported in base-metal ore at the Independence Mine in the Hilltop district. (Gilluly and Gates, 1965)

As globular, botryoidal, mammillary, and reniform masses or layers of *wood tin* with specular hematite and cristobalite in discontinuous veins of little lateral or vertical extent in Tertiary rhyolite lava in a few prospect pits; also found in nearby placer deposits in the Izenhood district. (Knopf, 1916a, 1916b; Fries, 1942; Castor et al., 1999b)

Identified by EMP with sulfides at the Dean Mine in the Lewis district. (J. McGlasson, w.c., 2001)

As yellow-brown granular masses to 1 mm in sulfide ore at the Cove Mine in the McCoy district. (M. Jensen, w.c., 1999)

MINERAL COUNTY Reported from scheelite veins in the Mina district. (Hess, 1917)

NYE COUNTY As tiny euhedral grains along vein walls in various mines in the Morey district. (Williams, 1968)

With specular hematite in irregular veins in silicified tuff on Yucca Mountain. (Castor et al., 1999b)

PERSHING COUNTY In tin ore at the Majuba Hill Mine in the Antelope district as shiny crystals to 2 mm on quartz and *tourmaline* crystals, and with quartz, *sericite, tourmaline,* and fluorite as irregular aggregates to 60 cm; also associated with chalcocite, chalcopyrite, pyrite, and arsenopyrite. Locally, quartz-cassiterite veinlets cut *tourmaline* crystals. (Smith and Gianella, 1942; Gianella, 1946; Trites and Thurston, 1958; MacKenzie and Bookstrom, 1976; Jensen, 1985)

Reported in a specimen from an unknown locality in the Mill City district. (WMK)

In gold placer deposits with cinnabar, scheelite, and magnetite in the Rabbit Hole district. (Johnson, 1977)

With hübnerite, gold, and cinnabar in placer gravels in the Rosebud district. (Bailey and Phoenix, 1944)

WASHOE COUNTY With pyrite, acanthite, gold, and possible tsumebite at the Wet prospect in the Pyramid district, SE 1/4 NE 1/4 Sec. 21, T23N, R21E. (SBC)

ČECHITE

$Pb(Fe^{2+},Mn)(VO_4)(OH)$

A very rare member of the descloizite group identified on old mine dumps of a polymetallic deposit in the former Czechoslovakia.

HUMBOLDT COUNTY As tiny orange crystals with hinsdalite and pyromorphite at the Silver Coin Mine in the Iron Point district; identified by SEM/EDX. (Thomssen, 1998; R. Thomssen, w.c., 2000)

CELADONITE

$KFe^{3+}(Mg,Fe^{2+})\square Si_4O_{10}(OH)_2$

Celadonite, a relatively common illite-group clay mineral, occurs as amygdule fillings in mafic to intermediate volcanic rocks and as an alteration product after mafic minerals in the same rock types.

NYE COUNTY In the groundmass of intermediate lavas near Urania Peak in the Cactus Springs district. (Ekren et al., 1971)

In the Halligan Mesa tuffs in the Hot Creek Mountains on the Morey Peak and Moores Station Quadrangles. (Jenkins, 1981)

PERSHING COUNTY As amygdule fillings in volcanic rocks in Mazuma Canyon in the Seven Troughs Range. (Jenkins, 1981)

STOREY COUNTY Reported as an alteration mineral in rocks of the Alta Formation in the northern Virginia Range near Clark Station 36 miles east of Reno. (Rose, 1969)

WASHOE COUNTY Identified in a specimen from Steamboat Hot Springs in the Steamboat Springs district. (JHS)

In vugs in rhyolite, mostly as coatings, at Through the Wall. (B. and J. Runner, w.c., 1998)

CELESTINE

$SrSO_4$

Celestine (*celestite*) is the principal commercial source of strontium. It occurs mainly in deposits in sedimentary rocks, generally in association with gypsum, but is also known from hydrothermal veins.

CLARK COUNTY Identified by XRD in limestone and marl in and near borate beds, and also identified microscopically in stromatolitic limestone at the Anniversary Mine in the Muddy Mountains district. (Castor, 1993)

Reported from an unspecified locality in the Sloan district. (M. Jensen, w.c., 1999)

EUREKA COUNTY As pale-blue, transparent to translucent, pyramidal, tabular crystals to 4 mm in a small vug in a vein, and as massive veins in the Betze-Post open pit, 5480-level, at the Goldstrike Mine in the Lynn district. (GCF)

NYE COUNTY Identified by XRD in small amounts with witherite in bedded barite at the

Ann claims in the Northumberland district. (Papke, 1984)

CELSIAN
BaAl$_2$Si$_2$O$_8$

Celsian is a relatively uncommon *feldspar* mineral that forms a continuous series with orthoclase. It mainly occurs in metamorphosed manganese deposits, but has been found in barite deposits and in hydrothermal occurrences.

ELKO COUNTY As tabular crystals to 0.2 mm with *apatite* and sulfide minerals in quartz that filled open spaces in skarn at the Summit View Mine in the Corral Creek district. (Berger and Oscarson, 1998)

As rounded fragments to 0.3 mm in mylonite, locally as much as 3 vol. percent of the rock, at the American Beauty Mine in the Lee district. (Berger and Oscarson, 1998)

HUMBOLDT COUNTY As microscopic, rectangular crystal aggregates after barite from a horizon in the Preble Formation above a bedded barite deposit in the north pit of the Little Britches Barite Mine in the Golconda district; associated minerals include cymrite, quartz, pyrite, goethite, barite, and *sericite*. (Hsu, 1994; M. Jensen, w.c., 1999)

LANDER COUNTY In hydrothermally altered rocks cut in a deep drill hole in the Iron Canyon area of the Battle Mountain district. (Theodore and Roberts, 1971)

NYE COUNTY Tentatively identified with cristobalite and *smectite* in a bismuth- and base-metal-bearing sample from a fumarolic deposit in ash-flow tuff underground at Yucca Mountain. (SBC)

CERUSSITE
PbCO$_3$

Cerussite, a very common mineral in oxidized zones of base-metal deposits, generally occurs with galena and commonly replaces the less stable sulfate anglesite. Found at many localities in Nevada, of which a partial list follows. Good specimen material has come from the Goodsprings, Eureka, Gabbs, and Marietta districts in Nevada.

CHURCHILL COUNTY As gray masses surrounding relict galena in oxidized quartz veins at the Chalk Mountain Mine in the Chalk Mountain district; associated with leadhillite crystals and a variety of other minerals including wulfenite, anglesite, chlorargyrite, plumbojarosite, and vanadinite; locally noted as glassy sixling-twinned crystals to 1 cm. (Vanderburg, 1940; Schilling, 1962a, 1962b; Bryan, 1972; Jensen, 1982, 1990b)

Reported at the Fairview Mine in the Fairview district. (Schrader, 1947; Saunders, 1977)

With chlorargyrite at the Terrell claim group in the Holy Cross district. (Vanderburg, 1940)

Reported in gold-silver-lead ore in the Jessup district. (WMK)

As glassy contact twins on (130) to 1.5 cm at the Westgate Mine in the Westgate district. (R. Walstrom, w.c., 1999)

CLARK COUNTY With galena at the Lead King Mine in the Dike district. (Longwell et al., 1965)

Reported as gold-bearing in a specimen from the Eldorado district. (WMK)

In the Goodsprings district, common in the Shenandoah and Yellow Pine Mines with galena, smithsonite, hydrozincite and wulfenite; good penetration twins were found in the latter to its deepest levels. With galena, hemimorphite, chrysocolla, malachite, aurichalcite, and wulfenite in the Smithsonite Mine. In the Singer Mine as blocky gray crystals to 1 mm on fracture surfaces with hemimorphite and bromargyrite. (Hewett, 1931; Adams, 1998)

Rare in manganese oxide ore at the Three Kids Mine in the Las Vegas district. (Hewett and Webber, 1931)

Abundant as large masses in oxidized ore, often enriched in gold, at the Quartette Mine in the Searchlight district. (Callaghan, 1939; JHS)

DOUGLAS COUNTY With anglesite, galena, jamesonite, and other minerals at the Iron Pot prospect in the Red Canyon district; also in replacement deposits in limestone at the Longfellow Mine. (Lawrence, 1963; Dixon, 1971)

ELKO COUNTY In minor amounts with pyrite at the Aura Queen Mine in the Aura district. (Bentz et al., 1983)

With iron oxides in the oxidized portion of a banded quartz-calcite vein in andesite at the Mint and Mirat Mines in the Burner district. (Emmons, 1910)

In oxidized, lead-rich quartz-calcite veins in

the Contact district; specifically with *jasper,* barite, and quartz in breccia at the Hice prospect. (Schrader, 1935b; LaPointe et al., 1991)

With bindheimite and other minerals at the 86 and Cleveland Mines in the Delano district. (LaPointe et al., 1991)

With anglesite, galena, pyrite, *limonite,* wulfenite, bismuthinite and copper minerals in veins and irregular replacement deposits in limestone at the Dolly Mine in the Dolly Varden district. (Hill, 1916; Melhase, 1934d; Schilling, 1979)

Generally massive with galena, but also as rare crystals in small cavities, at the Killie Mine in the Spruce Mountain district. (Schrader, 1931b; Schilling, 1979; Dunning and Cooper, 1987)

Silver-bearing with canary-yellow anglesite as pods in the footwall of a fault in limey shale at the Durham claims in the Tecoma district. (LaPointe et al., 1991)

ESMERALDA COUNTY Reported with pyromorphite at the Nevada-Goldfield and Courbet Mines in the Divide district. (Ransome et al., 1907)

With molybdenite and other minerals in quartz veins at the Bullfrog-George prospect in the Gold Point district; also with sphalerite at the Hornsilver Mine. (Schilling, 1979; WMK)

As brownish granular masses, and to a lesser extent in crystals at the South Klondyke Mine in the Klondyke district; also with jarosite, malachite, chrysocolla, and other minerals at an unspecified locality. (JHS; WMK; Chipp, 1969)

With hemimorphite, smithsonite and other oxidized minerals at the General Thomas, Alpine, and other mines in the Lone Mountain district. (Ball, 1906; Phariss, 1974)

Reported from 6 miles west of Goldfield in the Montezuma district. (Gianella, 1941)

With quartz, galena, malachite, and wulfenite at the McNamara Mine in the Palmetto district. (Schilling, 1979)

In specimens from the Broken Toe and Rock Hill Mines in the Rock Hill district. (R. Thomssen, w.c., 1999)

With anglesite, chlorargyrite, smithsonite, and hemimorphite at the Sylvania Mine in the Sylvania district. (Schilling, 1979)

As microcrystals at the 3 Metals and Weepah Mines in the Weepah district. (B. and J. Runner, w.c., 1998)

EUREKA COUNTY In quartz from along Mill Creek on the east flank of Tenabo Peak in the Cortez district. (WMK)

With galena, iron oxides, and plumbojarosite at the Diamond Mine in the Diamond district. (WMK)

Abundant with galena, anglesite, and other minerals at the Albion, Black Diamond, Richmond, Richmond-Eureka, and 76 Mines in the Eureka district. (Nolan, 1962; Schilling, 1979; WMK; S. Pullman, w.c., 1999)

With galena in a vein and as small crystals in jasperoid at the Gold Quarry Mine in the Maggie Creek district. (Jensen et al., 1995)

Reportedly silver- and gold-bearing with anglesite, chlorargyrite, hematite, galena, and pyromorphite in replacement and breccia ores in the Mineral Hill district. (Emmons, 1910; Schilling, 1979; WMK)

HUMBOLDT COUNTY With silver-bearing galena at an unspecified locality in the Gold Run district. (WMK)

As colorless plates at the Silver Coin Mine in the Iron Point district. (Thomssen, 1998; J. McGlasson, w.c., 2001)

With anglesite, galena, and other minerals at the Silver Hill Mine in the Dutch Flat district. (Willden, 1964)

LANDER COUNTY With galena and other minerals at the numerous mines in the Battle Mountain district. (Vanderburg, 1939; Roberts and Arnold, 1965)

With malachite, azurite, and chlorargyrite at the Betty O'Neal and Highland Chief Mines in the Lewis district; also noted with owyheeite at the Eagle vein. (Vanderburg, 1939; J. McGlasson, w.c., 2001)

As glassy crystals with calcite, pyrite, and galena at the McCoy-Cove Mine in the McCoy district. (R. Walstrom, w.c., 1999)

In bonanza silver ore in the Reese River district. (Ross, 1953)

LINCOLN COUNTY With secondary copper and lead minerals at the Jackrabbit Mine in the Bristol district. (Melhase, 1934d)

With plumbojarosite in jasperoid at the Lucky Chief Mine in the Chief district. (Callaghan, 1936)

In oxidized ore with anglesite and residual galena at the Groom and other mines in the Groom district. (Humphrey, 1945; Tingley et al., 1998a)

As elongated six-rayed crystals with malachite, rosasite, hemimorphite, galena, and quartz at an unspecified locality in the Pahranagat (Hiko) district. (M. Jensen, w.c., 1999)

With bindheimite at the Eagle Rock claims in the Patterson district. (Schrader, 1931a)

Common in lead-zinc deposits in the Pioche district; specifically reported in the Florence and Virtue Mines. (Westgate and Knopf, 1932; WMK)

LYON COUNTY With linarite, wulfenite, and other minerals at the Old (Lost) Soldier Mine in the Churchill district. (Jenkins, 1981)

With other lead minerals at the Jackpot claim in the Wilson district. (Jenkins, 1981)

MINERAL COUNTY Reported at the Simon Mine in the Bell district. (S. White, w.c., 1999)

Transparent prisms to 1 mm with alunite and other minerals in vugs in silver ore at the Broken Hills Mine in the Broken Hills district. (Jenkins, 1981)

As colorless prismatic crystals at the Potosi Mine in the Candelaria district. (Page, 1959; WMK)

Reported with jamesonite in quartz veins at the Mendora Mine in the Garfield district. (Ponsler, 1977)

With galena, tetrahedrite, and other minerals at the Lucky Boy Mine in the Lucky Boy district. (Hill, 1915)

Reported at the Greek Hills Mine in the Pamlico district. (WMK)

With other lead minerals in the Santa Fe district. (Jenkins, 1981)

NYE COUNTY With silver-bearing galena in the Orizaba Mine in the Cloverdale district; also in a quartz vein with pyrite, galena, anglesite, acanthite, sphalerite, and malachite at the Iron Rail group. (Kral, 1951; Hamilton, 1993)

As yellow crystals to 2 mm with malachite and quartz at the Big Chief Mine in the Ellsworth district. (GCF)

As crystals to 1 cm with wulfenite, mimetite, and other minerals in gossan at the Downieville and Lizzie Mines in the Gabbs district. (Kral, 1951; UCB)

In quartz veins with galena and other minerals at the War Eagle and Star of the West Mines in the Jackson district. (Kral, 1951; Hamilton, 1993)

With galena and pyrite on a dump on the Valley Claim group in the Jett district. (Kral, 1951)

With galena at the Aunt Ethel prospect and Silver King-Resolution claims in the Lodi district; also as micromount specimens and some larger crystals at the San Rafael Mine. (Kral, 1951; B. and J. Runner, w.c., 1998)

Minor as crystals and twins to 1 mm with bindheimite, valentinite, and native sulfur in oxidized ore in the Morey district. (Jenkins, 1981)

In veins in the Oak Spring district on the Nevada Test Site. (Melhase, 1935b; Kral, 1951)

Reported at the Antimonial Mine in the Reveille district. (Lawrence, 1963)

As colorless crystals on quartz fractures near altered aikinite at the Outlaw Mine in the Round Mountain district. (Dunning et al., 1991)

With silver-bearing galena at the Florence Mine in the San Antone district. (Kral, 1951)

In a quartz vein with galena, sphalerite, and pyrite at the Vanderhoef claims in the Troy district. (Hill, 1916)

With anglesite after galena at the Two G Mine in the Tybo district. (Jenkins, 1981; WMK)

In ore from the Reveille, Union, and Washington districts. (Kral, 1951)

PERSHING COUNTY With smithsonite and galena at the Last Chance Mine in the Antelope district. (MacKenzie and Bookstrom, 1976)

With cinnabar and stibnite at the Red Bird Mine in the Antelope Springs district. (Bailey and Phoenix, 1944; Berry et al., 1952)

Reported in earthy form with bindheimite at the Montezuma and other mines in the Arabia district. (Knopf, 1918a; Lawrence, 1963; WMK)

In quartz veins with acanthite and other minerals at the Arizona Mine in the Buena Vista district. (Cameron, 1939)

WASHOE COUNTY With smithsonite, hemimorphite, malachite, and chalcanthite in quartz-calcite veins in the Union (Commonwealth) Mine in the Galena district; with secondary zinc and silver minerals, Galena Hill Mine. (Geehan, 1950; Bonham and Papke, 1969)

Reported with anglesite at the Leadville Mine in the Leadville district. (Gianella, 1941)

In veins in the Pyramid district. (Gianella, 1941; Overton, 1947; Wallace, 1980)

With anglesite, silver halide minerals, and gold in the Wedekind district. (Overton, 1947)

WHITE PINE COUNTY With chlorargyrite and anglesite at a locality in the Cherry Creek district. (Hose et al., 1976; JHS)

In gold-silver-lead ore in the Duck Creek district. (Melhase, 1934d; WMK)

With anglesite, galena, and other minerals at the Cuba Mine in the Granite district. (Melhase, 1934d)

With anglesite after galena at an unspecified locality in the Hunter district. (Melhase, 1934d)

Reported at the Grand Deposit Mine in the Muncy Creek district. (Hose et al., 1976)

With pyromorphite at the Vietti Mine and Betty Jo claim in the Nevada district. (Roberts, 1943; Williams, 1973)

Reported in a specimen of lead ore from the Osceola district. (WMK)

In lead-silver-zinc replacement deposits in the Robinson district, specifically from the Pullcart claim and as jackstraw crystals to 1 cm and clusters with crystals to 2.5 cm at the Elijah Mine. (Spencer et al., 1917; AMNH; M. Jensen, w.c., 1999; GCF)

With a variety of other oxidized minerals in the Taylor district. (Lovering and Heyl, 1974)

Reported from the Caroline tunnel and elsewhere in the Ward district. (Hill, 1916; Williams, 1972)

With anglesite, galena, and secondary copper and zinc minerals in the lead belt in the Mount Hamilton area in the White Pine district. (Humphrey, 1960; Hose et al., 1976)

CERVANTITE

$Sb^{3+}Sb^{5+}O_4$

An oxidation product of stibnite, distinguished from similar antimony oxide minerals such as stibiconite by XRD.

CHURCHILL COUNTY Reported in a specimen from an unknown locality in the Alpine district. (WMK)

ESMERALDA COUNTY Reported in specimens from an unknown localities in the Gilbert and Silver Peak districts. (WMK)

EUREKA COUNTY Reported at the various mines and prospects in the Eureka district. (S. Pullman, w.c., 1999)

Identified as a yellow coating on quartz crystals with stibnite and stibiconite in the Goldstrike Mine in the Lynn district. (GCF)

LANDER COUNTY Reported at an unspecified locality in the Big Creek district. (S. Pullman, w.c., 1999)

MINERAL COUNTY Reported at an unspecified locality in the Mina district. (UAMM)

NYE COUNTY As yellow coatings on stibnite at the White Caps Mine in the Manhattan district. (Melhase, 1935d; Gibbs, 1985)

Reported in a specimen with stibnite at the Page Brothers property in the Tybo district. (WMK)

PERSHING COUNTY As needlelike inclusions to 1 inch long with stibnite in well-formed, transparent to cloudy quartz crystals to 2 inches long at the Bottomley prospect in the Antelope district. (UMNH; AMNH; JHS)

In specimens with stibnite from unspecified localities in the Arabia and Rye Patch districts; also noted as golden-brown inclusions with stibnite in quartz crystals. (WMK; GCF)

With stibnite as inclusions in quartz crystals from an unspecified locality in the Trinity district. (WMK)

Identified by XRD in a specimen from Oreana as a component of a yellow-brown earthy coating with stibnite, senarmontite, stibiconite, cervantite, and gypsum; coating was originally identified as roméite. (GCF)

WHITE PINE COUNTY Tentatively identified as an oxidation product of stibnite at the Bald Mountain Mine in the Bald Mountain district. (Hitchborn et al., 1996)

Reported in the Taylor district. (Lovering and Heyl, 1974)

CHABAZITE

$(Ca_{0.5},K,Na)_4[Al_4Si_8O_{24}] \cdot 12H_2O$ (generalized)

A common zeolite mineral group that has recently been subdivided into three species depending on the most abundant alkali or alkali-earth element. *Chabazite* occurs in amygdules in mafic volcanic rocks, as a diagenetic lacustrine mineral in volcanic tuff, and in hydrothermal veins. Species have not been determined for Nevada localities.

CARSON CITY Pale-green glassy crystals to 0.5 mm with *stilbite* and natrolite at an unnamed copper prospect on Prison Hill in the Carson City district. (R. Walstrom, w.c., 1999; M. Jensen, w.c., 1999)

ESMERALDA COUNTY Reported as *herschelite* in a specimen from Montgomery Pass. (R. Thomssen, w.c., 1999)

HUMBOLDT COUNTY Reported in a specimen from a road cut about 10 miles north of Paradise Valley in the Toiyabe National Forest in the Paradise Valley district. (UCB)

NYE COUNTY With *clinoptilolite*, analcime, and mordenite in the Yucca Flat area, Nevada Test Site. (Hoover, 1968)

STOREY COUNTY Reported in a specimen from the Gould and Curry Mine in the Comstock district. (UCB)

WASHOE COUNTY In amygdules in basalt on Virginia Mountain west of Nixon. (LANH; WMK)

CHAIDAMUITE
$(Zn,Fe^{2+})Fe^{3+}(SO_4)_2(OH) \cdot 4H_2O$

A rare mineral found with other sulfate minerals in an oxidized galena-sphalerite-pyrite deposit in China.

HUMBOLDT COUNTY As brown bladed crystals to 1 mm in specimens from the Getchell Mine in the Potosi district. (HUB; F. Cureton, o.c., 2000; GCF)

CHALCANTHITE
$Cu^{2+}SO_4 \cdot 5H_2O$

Chalcanthite, a common water-soluble mineral that occurs as efflorescences in underground mines in many of the world's copper districts, generally crumbles in dry atmospheres due to water loss. Among Nevada localities, the Yerington district has produced good specimens.

CLARK COUNTY With malachite, azurite, and other secondary copper minerals at the Key West Mine in the Bunkerville district. (Beal, 1965)

Reported in a specimen from an unknown locality near Las Vegas. (WMK)

EUREKA COUNTY With marcasite in veins at the Goldstrike Mine in the Lynn district. (Ferdock et al., 1997)

As thin, blue to blue-green crusts of poorly crystalline efflorescent material on freshly exposed sulfide-bearing rock at the Gold Quarry Mine in the Maggie Creek district. (Jensen et al., 1995)

HUMBOLDT COUNTY Reported in a specimen from an unnamed occurrence at an elevation of 7000 feet on the west slope of Delong Peak in the Jackson Mountains district. (WMK)

LANDER COUNTY As encrustations or stalactites and stalagmites on mine walls in copper deposits in the Battle Mountain district; specifically at the Copper Canyon and Copper Basin Mines. (Roberts and Arnold, 1965; JHS)

Reported at the Lucky Rocks claims in the Lewis district. (J. McGlasson, w.c., 2001)

LYON COUNTY Reported in mineable quantities at the Bluestone and other mines in the Yerington district. (Knopf, 1918b; Melhase, 1935e; AMNH; JHS)

MINERAL COUNTY Locally in oxidized portions of mineralized quartz-*adularia* veins in underground workings in the Aurora district. (D. Lippoth, o.c., 1992)

In veinlets and as coatings on fractures and cavities at the Pine Tree prospect in the Pilot Mountains district. (Ross, 1961)

NYE COUNTY As fine-grained, light-blue to green coatings on quartz samples from a small cave near the Outlaw Mine in the Round Mountain district. (Dunning et al., 1991)

PERSHING COUNTY As abundant large masses of translucent blue fibers, rare crystals to 4 cm, and "ram's horns" on mine workings and timbers at the Majuba Hill Mine in the Antelope district. (Smith and Gianella, 1942; MacKenzie and Bookstrom, 1976; Jensen, 1985)

STOREY COUNTY Reported from mine workings in the Comstock district. (Gianella, 1941; UCB)

Reported in a specimen from an unspecified locality in the Eldorado district in Eldorado Canyon; possibly in Lyon County. (UCB)

WASHOE COUNTY With cerussite, smithsonite, hemimorphite, and malachite in oxidized ore at the Union (Commonwealth) Mine in the Galena district. (Geehan, 1950; Bonham and Papke, 1969)

In quartz and adjacent rocks with bornite, malachite, cornetite, and rare chalcocite at the Red Metals Mine in the Peavine district. (Overton, 1947; Bonham and Papke, 1969; Hudson, 1977)

Reportedly an important copper mineral in oxidized ore with kaolinite, enargite, gold, and other secondary minerals in the Pyramid district; specifically noted at the Burrus Mine as encrustations, lenticular masses, aggregates of delicate feathery crystals, and as prismatic crystals to 3 cm, locally doubly terminated, on mine workings and timbers. (Overton, 1947; Bonham and Papke, 1969; Jensen, 1994; WMK; GCF)

Reported at Steamboat Hot Springs in the Steamboat Springs district. (Gianella, 1941)

With azurite, malachite, brochantite, and chrysocolla in the Wedekind district. (Hudson, 1977)

WHITE PINE COUNTY A common mineral in mines in the Robinson district. (Spencer et al., 1917; Melhase, 1934d; UCB)

CHALCOALUMITE

$Cu^{2+}Al_4(SO_4)(OH)_{12} \cdot 3H_2O$

A very rare secondary mineral in oxidized copper deposits.

CHURCHILL COUNTY Reported as sky-blue botryoidal crusts of tiny (<0.1 mm) platy crystals after brochantite at the Lovelock Mine in the Table Mountain district. (R. Thomssen, w.c., 1999)

CHALCOCITE

Cu_2S

Chalcocite, generally uncommon as a primary species, is an important ore mineral in rich secondary supergene zones in porphyry copper systems, as in the Robinson district. It also occurs in metal deposits in which copper is not the most important product.

CARSON CITY Reported as gold- and silver-bearing in contact-metamorphic rock at the Armenian Mine in the Voltaire district. (Schrader et al., 1917; WMK)

CHURCHILL COUNTY A secondary mineral in veins at the Fairview and other mines in the Fairview district. (Schrader, 1947)

With copper carbonates and oxides at the Lakeview Mine in the Holy Cross district. (Peters et al., 1996)

CLARK COUNTY Reported in a specimen of silver ore in the Eldorado district. (WMK)

At most of the copper mines in the Goodsprings district; specifically the Boss, Blue Jay, Copperside, Copper Chief, and Ninety-Nine Mines. (Hewett, 1931; Melhase, 1934e)

Reported at an unspecified locality in the Searchlight district. (Callaghan, 1939)

DOUGLAS COUNTY Reported in copper ore at the Ruby Hill and other mines in the Gardnerville district. (Hill, 1915)

ELKO COUNTY Reported in veins with other copper minerals in the Contact district; specifically at the Bellevue, Nevada-Bellevue, and Warsaw-June Mines. (Purington, 1903; Bailey, 1903; Schrader, 1935b; Schilling, 1979; LaPointe et al., 1991)

As fine veinlets in pyritic contact-metamorphosed rock at the First Chance claims in the Dolly Varden district. (Hill, 1916; Melhase, 1934d)

Reported with chrysocolla, native copper, and copper carbonates in lens-shaped masses in *garnet*-epidote skarn at the Salt Lake claims in the Ferber district. (Hill, 1916)

In appreciable amounts in copper ore in the Kinsley district. (Hill, 1916)

As sooty, gray-black material after pyrite and chalcopyrite immediately below the water table at the Rio Tinto Mine in the Mountain City district. (Coats and Stephens, 1968)

Generally uncommon in the Railroad district, but reported as an important ore mineral at the Standing Elk Mine. (Ketner and Smith, 1963; Schrader et al., 1917)

With wulfenite and other base-metal minerals in copper-rich ore in the Spruce Mountain district; probably of supergene origin at the Killie Mine. (Hill, 1916; Schrader, 1931b; Trites and Thurston, 1958; Schilling, 1979)

ESMERALDA COUNTY Reported as silver-bearing in some mines in the Cuprite and Montezuma districts. (Schrader et al., 1917)

In quartz veins at the Bullfrog-George prospect in the Gold Point district. (Schilling, 1979)

Questionably reported with digenite, covellite, and bornite at an unspecified locality in the Klondyke district. (Chipp, 1969)

EUREKA COUNTY After chalcopyrite in a base-metal vein with malachite, azurite, pyrite, covellite, quartz, and *limonite* at the Goldstrike Mine in the Lynn district. (GCF)

As thin, sooty black patches to 3 mm across with gypsum on fracture surfaces in brecciated and silicified rocks at the Gold Quarry Mine in the Maggie Creek district. (Jensen et al., 1995)

HUMBOLDT COUNTY Present at the Gracie and Big Pay Mines in the Battle Mountain district. (Willden, 1964)

A minor secondary mineral at the Lone Tree Mine in the Buffalo Mountain district. (R. Braginton, o.c., 1995)

Reported in a specimen from the west slope of Delong Peak at an elevation of 7000 feet in the Jackson Mountains district. (WMK)

With azurite from a prospect in the King Lear Formation on the east side of Navajo Peak in the Red Butte district. (Willden, 1963)

LANDER COUNTY With other copper minerals at various localities, including the Copper Canyon, Little Giant, and Black Jack Mines, and the Sweet Marie lease in the Battle Mountain district. (Roberts and Arnold, 1965; Theodore and Blake, 1978; Schilling, 1979; WMK)

In small amounts with quartz, pyrite, and marcasite in the Hilltop district. (Gilluly and Gates, 1965)

With covellite in near-surface material in the Reese River district. (Ross, 1925, 1953)

With malachite and azurite in some Tertiary deposits in the Washington district. (Hill, 1915)

LINCOLN COUNTY Reported with numerous secondary copper minerals at the Jackrabbit Mine in the Bristol district. (Melhase, 1934d)

In cherty quartz at the Delamar Mine in the Delamar district. (Callaghan, 1937)

In small amounts with tenorite, malachite, and chrysocolla at the Arrowhead (Southeastern) Mine in the Southeastern district. (Tingley et al., 1998a)

LYON COUNTY In quartz veins in the Washington district. (Garside, 1973)

As aggregates to 4 mm in quartz veins in small prospects and mines at the base of the east flank of the Wellington Hills in the Wellington district. (GCF)

A major source of copper with chalcopyrite, chrysocolla, and native copper in the Yerington district; specifically noted at the Hilltop, Lucky Strike, Ludwig, and Yerington Mines. (Ransome, 1909e; WMK)

MINERAL COUNTY Reported in ore in the Mountain View district. (Hill, 1915)

Reported in a specimen from the Turk claim in the Santa Fe district. (WMK)

Reported with magnetite, bornite, and chalcopyrite from Whisky Flat in the Silver Star district. (WMK)

NYE COUNTY With tetrahedrite, galena, and other sulfides at the Transylvania Mine in the Belmont district. (Jenkins, 1981)

In small amounts at the Original Bullfrog and other mines in the Bullfrog district. (Ransome et al., 1910; Castor et al., 1999b)

With covellite as coatings on chalcopyrite at the Fairday Mine in the Cactus Springs district. (Tingley et al., 1998a)

After pyrite in the Gold Crater district. (Tingley et al., 1998a)

Reported in specimens from the Esta Buena and Osborne Copper Mines in the Lodi district. (WMK)

Reported at the Oswald Mine in the Queen City district. (Cornwall, 1972)

With chalcopyrite in sulfide veins at the Wagner Mine in the Wagner district. (Tingley et al., 1998a)

PERSHING COUNTY Abundant with pyrite, chalcopyrite, and rare enargite in pods to 25 cm in rhyolite porphyry in the Copper stope of the Majuba Hill Mine in the Antelope district; also as seams in fissures that cut chalcopyrite, pyrite, and arsenopyrite. (Smith and Gianella, 1942; MacKenzie and Bookstrom, 1976; Jensen, 1985)

In quartz veins at the Arizona Mine in the Buena Vista district. (Cameron, 1939)

Reported in a specimen from the unknown locality in the Imlay district. (WMK)

In *tourmaline*-quartz veins at the Bonanza King Mine in the Spring Valley district. (Campbell, 1939)

STOREY COUNTY Rarely with quartz within a few hundred feet of the surface in the California and other mines along the Comstock Lode in the Comstock district. (Koschmann and Bergendahl, 1968; Bastin, 1922; HUB)

WASHOE COUNTY With bornite, schorl, and gold in quartz veins at the Red Metal, Consolidated Peavine, Dogskin Mountain, Los Angeles, Sylvanite, and other mines in the Peavine district; also noted with covellite, malachite, chrysocolla, pyrite, and cornetite. (Hill, 1915; Overton, 1947; Hudson, 1977; WMK)

As a secondary mineral at the Jumbo Mine in the Pyramid district. (Overton, 1947; Wallace, 1980)

WHITE PINE COUNTY With chalcopyrite and covellite at the Bald Mountain Mine in the Bald Mountain district. (Hitchborn et al., 1996)

The most important copper mineral in rich supergene deposits in the Robinson district; reported as both massive and sooty forms. (Spencer et al., 1917; Melhase, 1934d; WMK)

In sooty form with covellite, chrysocolla, and a variety of other minerals in the Caroline tunnel in the Ward district. (Williams, 1964; WMK)

CHALCOCYANITE
$Cu^{2+}SO_4$

A soluble sublimate found at fumaroles at Mount Vesuvius in Italy and at a volcano in El Salvador.

WHITE PINE COUNTY Reported as a thin olive-green coating on rock containing hematite, galena, and sphalerite from the Ward district. (Weissman and Nikischer, 1999)

CHALCOMENITE
$Cu^{2+}Se^{4+}O_3 \cdot 2H_2O$

A rare mineral formed by the oxidation of primary copper- and selenium-bearing minerals.

PERSHING COUNTY Very rare blue blebs to 5 mm with quartz, *tourmaline,* and *limonite* in the Copper stope at the Majuba Hill Mine in the Antelope district. (Jensen, 1985)

CHALCOPHANITE
$(Zn,Fe^{2+},Mn^{2+})Mn_3^{4+}O_7 \cdot 2H_2O$

Chalcophanite, a relatively rare secondary mineral found in some manganese-bearing lead-silver-zinc deposits, is usually associated with other manganese oxide minerals.

EUREKA COUNTY Reported from an unspecified locality in the Antelope (Baldwin) district. (Heyl and Bozion, 1962)

As rosettes to 0.1 mm of thin, tabular, trigonal crystals with calcite on goethite in vuggy, iron-stained jasperoid breccia at the Gold Quarry Mine in the Maggie Creek district. (Jensen et al., 1995)

LINCOLN COUNTY Reported in the Pioche district. (Heyl and Bozion, 1962)

Reported in a specimen from the Silver King Mine in the Silver King district. (R. Thomssen, w.c., 1999)

MINERAL COUNTY With other manganese and iron oxides and hydroxides in silver ore along shear zones at the NERCO Minerals Candelaria Mine in the Candelaria district. (Moeller, 1988)

WHITE PINE COUNTY Accompanying aurorite with black calcite and other minerals at the Hidden Treasure and North Aurora Mines in the White Pine district. (Hewett and Radtke, 1967)

CHALCOPHYLLITE
$Cu_{18}^{2+}Al_2(AsO_4)_3(SO_4)_3(OH)_{27} \cdot 33H_2O$

Chalcophyllite is a rare mineral that occurs in the oxidation zones of copper deposits that are rich in arsenic.

CHURCHILL COUNTY Questionably reported with brochantite and nickel and cobalt minerals at the Lovelock Mine in the Table Mountain district; also reported from the Nickel Mine. (Ferguson, 1939; R. Thomssen, w.c., 1999)

ESMERALDA COUNTY Rare but good microcrystals at the 3 Metals Mine in the Weepah district. (B. and J. Runner, w.c., 1998)

LANDER COUNTY Reported at the Garrison Mine in the Cortez district. (S. Pullman, w.c., 1999)

MINERAL COUNTY Reported from an unspecified locality near Sodaville. (Palache et al., 1951)

PERSHING COUNTY Reportedly some of the finest specimens in North America are from the Copper stope of the Majuba Hill Mine in the Antelope district as shiny, blue-green hexagonal plates to 3 mm, rarely transparent, with azurite, brochantite, and spangolite in small cavities in rhyolite; also found on the dump of the highest adit. (Smith and Gianella, 1942; Trites and Thurston, 1958; MacKenzie and Bookstrom, 1976; Jensen, 1985)

WASHOE COUNTY Bright green scales occur with tyrolite on phyllite and quartz at an unnamed mine on Peavine Mountain in the Peavine district. (Jenkins, 1981)

As bright blue-green, micaceous, platy crystals to 2 mm from the Burrus and other mines in the Pyramid district. (Jensen, 1994; GCF)

CHALCOPYRITE
$CuFeS_2$

Chalcopyrite, a very common and widespread mineral, occurs as the principal primary sulfide in porphyry, vein, and replacement copper deposits and in many other metallic deposits.

The following is a partial list of many occurrences in Nevada.

CHURCHILL COUNTY With molybdenite, pyrite, and galena in unoxidized portions of lead-silver veins in the Chalk Mountain district. (Schilling, 1979)

In veins in marble at the Coppereid prospect in the Copper Kettle district. (GCF)

In quartz veins in the Fairview district. (Greenan, 1914; Schrader, 1947; Saunders, 1977)

As massive vein material at the White Cloud Mine in the White Cloud district. (GCF; UCB)

In quartz-*adularia* veins with fluorite, pyrite, galena, sphalerite, and molybdenite at the Wonder Mine in the Wonder district. (Schilling, 1979)

Reported from the Holy Cross and Jessup districts. (Schrader, 1947; WMK)

CLARK COUNTY With pyrrhotite, pyrite, pentlandite, and other minerals in and near a large diabase dike at the Great Eastern and Key West Mines in the Bunkerville district. (Lindgren and Davy, 1924; Beal, 1965)

In trace amounts at the Double Standard and Sundog Mines in the Crescent district; also as sparse grains in the Crescent Peak molybdenum porphyry deposit. (Longwell et al., 1965)

A minor mineral in ore at the Ole Mine in the Gold Butte district. (Longwell et al., 1965)

The principal primary copper mineral at the Boss and other mines in the Goodsprings district. (Hewett, 1931; Jedwab, 1998b)

Reported in the Eldorado and Searchlight districts. (Callaghan, 1939; WMK)

DOUGLAS COUNTY With magnetite, particularly near ore deposit margins, in the Minnesota Mine in the Buckskin district; also reported from the Antelope and Center Lode Mines. (Reeves et al., 1958; Overton, 1947; WMK)

Reported at the Ruby Hill Mine in the Gardnerville district. (Overton, 1947)

With bornite in a specimen of copper ore from the Genoa district. (WMK)

With gold and other sulfide minerals at the Longfellow, Winters, Red Canyon, and Lucky Bill Mines in the Red Canyon district. (Overton, 1947; Dixon, 1971)

Reported in copper ore with magnetite in the Wellington district. (WMK)

ELKO COUNTY Disseminated with pyrite in skarn at the Golden Eagle and Tiger Mines in the Aura district. (LaPointe et al., 1991; WMK)

As blebs to 0.1 mm in gold ore with pyrite at the Rossi Mine in the Bootstrap district. (Snyder, 1989)

With sphalerite, pyrite, arsenopyrite, and galena in a banded quartz-calcite vein in andesite at the Mint and Mirat Mines in the Burner district. (Emmons, 1910)

With bornite in quartz veins at many mines, including the productive Nevada-Bellevue and Palo Alto Mines in the Contact district; also in less productive skarn deposits. (Purington, 1903; Bailey, 1903; Schrader, 1912, 1935b; Schilling, 1979; LaPointe et al., 1991)

In the Dolly Varden district at the Victoria Mine in quartz veins with cerussite and in brassy crystals to 6 cm with bismuthinite, pyrite, and sphalerite in calcite; also at other properties with pyrite in quartz veins and in skarn. (Schilling, 1979; M. Jensen, w.c., 1999; Hill, 1916; Melhase, 1934d)

With molybdenite in quartz veins at the Indian Creek prospects in the Edgemont district. (Schilling, 1979)

In contact-metamorphic deposits in the Elk Mountain district. (Schrader, 1912)

With bornite, chrysocolla, and copper carbonates in skarn at the Martha Washington Mine in the Ferber district. (Hill, 1916)

As a replacement or exsolution of sphalerite or as free grains in host rocks at the Big Springs Mine in the Independence Mountains district; also with pyrite, quartz, barite, calcite, and secondary minerals in massive sulfide beds in black shale at the Black Beauty claims. (Youngerman, 1992; LaPointe et al., 1991)

In trace amounts as small grains in pyrite and marcasite aggregates at the Hollister gold deposit in the Ivanhoe district; also as micron-sized inclusions in freibergite. (Deng, 1991)

In quartz veins at the Norman Mine and Bullion prospect in the Jarbidge district. (Schrader, 1923)

In trace amounts at the Meikle Mine in the Lynn district. (Jensen, 1999)

With pyrite, purple fluorite, and other copper minerals in a quartz vein at the Baltimore Mine in the Merrimac district. (LaPointe et al., 1991)

With pyrite and other copper sulfide minerals

in high-grade, lenticular, massive sulfide ore-bodies in Ordovician black shale that produced more than 200 million pounds of copper at the Rio Tinto Mine in the Mountain City district; also a minor mineral in quartz veins elsewhere in the district. (Coats and Stephens, 1968; Emmons, 1910; Schilling, 1979)

With pyrite, bornite, galena, and sphalerite in *garnet*-tremolite skarn at the Delmas, Nevada Bunker Hill, and Storm King Mines in the Railroad district. (Emmons, 1910; Ketner and Smith, 1963; WMK)

At the Killie Mine in the Spruce Mountain district with bornite in veins, but generally oxidized to malachite, azurite, rosasite, chrysocolla, and tenorite; also in vein and skarn deposits elsewhere in the district. (Schrader, 1931b; Schilling, 1979; Dunning and Cooper, 1987; Hill, 1916; LaPointe et al., 1991)

In unoxidized ore with pyrite and bornite in the Tuscarora district. (Emmons, 1910; WMK)

Reported in the Charleston, Gold Circle, Harrison Pass, and White Horse districts. (Lawrence, 1963; Emmons, 1910; LaPointe et al., 1991; Hill, 1916)

ESMERALDA COUNTY With tetrahedrite, molybdenite, galena, and other minerals at the Tonopah-Divide and Gold Zone Mines in the Divide district. (Bonham and Garside, 1979)

In quartz veins with molybdenite and other minerals at the Bullfrog-George prospect in the Gold Point district. (Schilling, 1979)

In unoxidized ore with tetrahedrite and sphalerite at the Atlanta, Combination, and Mohawk Mines in the Goldfield district. (Collins, 1907a, 1907b)

In specimens from the Benton, Gold Crown, Pandora, and Savage Mines in the Montezuma district. (WMK)

Mostly massive at the Weepah Mine in the Weepah district. (B. and J. Runner, w.c., 1998)

Reported at unspecified localities in the Cuprite, Lone Mountain, and Rock Hill districts. (Ball, 1907; Schilling, 1979; WMK)

EUREKA COUNTY Reported at the Mountain View Mine in the Lone Mountain district. (Roberts et al., 1967b)

As brassy to iridescent anhedral grains and masses with other sulfides in quartz veins in diorite and skarn at the Goldstrike Mine in the Lynn

district; also as inclusions in and coating pyrite in gold ore. (GCF)

With sphalerite, galena, and pyrrhotite at the Mount Hope Mine in the Mount Hope district. (Roberts et al., 1967b)

With cuprite, chrysocolla, and galena at the Keystone and Kingston lodes in the Roberts district. (WMK)

With other sulfides in barite-carbonate veins at the Zenoli Mine; also with galena at the Rains Mine in the Safford district. (Vanderburg, 1938b; WMK)

At unspecified localities in the Antelope (Baldwin), Eureka, Cortez, and Modarelli-Frenchie districts. (WMK; Nolan, 1962; Emmons, 1910; Muffler, 1964)

HUMBOLDT COUNTY As tiny inclusions in iron sulfide at the Lone Tree Mine in the Buffalo Mountain district. (Panhorst, 1996)

In a specimen from an unknown locality in the Edna Mountains in the Golconda district. (WMK)

In quartz veinlets with molybdenite and other sulfides at the Moly prospect and Adelaide Mine in the Gold Run district. (Melhase, 1934c; Willden, 1964; Schilling, 1979)

An accessory mineral in magnetite ore at the Delong (Iron King) Mine in the Jackson Mountains district; also at copper prospects in the district. (Shawe et al., 1962; WMK)

With molybdenite and pyrite in quartz veins at the Desert View prospect in the Leonard Creek district. (Schilling, 1979)

In quartz veins at the Cheffoo claim and other properties in the National district. (Lindgren, 1915)

Locally abundant in skarn at the Granite Creek and Riley Mines in the Potosi district; also noted at the Getchell Mine. (Schilling, 1962a; Hotz and Willden, 1964; D. Tretbar, o.c., 1998)

LANDER COUNTY As fracture fillings and disseminated grains with other sulfide minerals at the Copper Canyon and Copper Basin Mines in the Battle Mountain district; also reported in gold-bearing skarn at the Fortitude Mine. (Roberts and Arnold, 1965; Theodore and Blake, 1978; Schilling, 1979; Doebrich, 1993)

A minor member of the sulfide assemblage at the Buffalo Valley and Honeycomb Mines in the Buffalo Valley district. (Kizis et al., 1997)

With tetrahedrite and galena at an unspecified locality in the Ravenswood district. (Hill, 1916)

Widely distributed in silver deposits in the Reese River district; also with uraninite and other uranium minerals at the Apex (Rundberg) Uranium Mine. (Melhase, 1934c; Ross, 1953; Garside, 1973; UAMM)

LINCOLN COUNTY In primary ore from Hiko in the Pahranagat district. (Tschanz and Pampeyan, 1970)

Reported in a specimen from the Lincoln Tungsten Mine in the Tem Piute district. (JHS)

Reported from mines in the Comet, Delamar, Groom, Highland, and Pennsylvania districts. (Tschanz and Pampeyan, 1970; Callaghan, 1937; Humphrey, 1945)

LYON COUNTY Reported at an unspecified locality in the Benway district. (Schrader, 1947)

Locally abundant in quartz veins in the Silver City district. (Gianella, 1936a)

Reported in some uranium occurrences in the Washington district. (Garside, 1973)

With pyrite, pyrargyrite, and native gold in the Wilson district. (Dircksen, 1975)

The most important source of copper in many mines in the Yerington district. In the large porphyry orebody at the Yerington (Anaconda) Mine as minute disseminated grains and narrow seams with pyrite, bornite, and covellite. (Knopf, 1918b; Melhase, 1935e; Moore, 1969)

MINERAL COUNTY In minor amounts in silver- and gold-bearing quartz-*adularia* veins in the Aurora district. (Ross, 1961)

In veins with pyrite, galena, and jamesonite at the Candelaria (NERCO Minerals Co.) Mine in the Candelaria district. (Ross, 1961)

With galena, sphalerite, acanthite, and other minerals at the Mabel Mine in the Garfield district. (Ponsler, 1977)

In quartz-calcite veins at the Lucky Boy Mine in the Lucky Boy district. (Lawrence, 1963)

With pyrite, bornite, and secondary copper minerals in deposits in veins and skarn in the Santa Fe district; specifically reported at the Copper Chief, Emma, and Turk Mines. (Clark, 1922; Ross, 1961; WMK)

Reported at the Badger, Dispozitch, and Endowment Mines in the Silver Star district. (Ross, 1961; WMK)

Reported from the Broken Hills, Eagleville,

Rand, Rawhide, Calico Hills, and Whisky Flat districts. (Schrader, 1947; JHS; WMK)

NYE COUNTY Reported at the Antelope View Mine in the Antelope Springs district. (Tingley et al., 1998a)

Minor in quartz at the Transylvania Mine in the Belmont district. (Jenkins, 1981)

In veins with pyrite, acanthite, galena, and hessite at the Fairday Mine in the Cactus Springs district. (Tingley et al., 1998a)

Along a contact between *serpentine* and brucite-magnesite at the Premier Chemicals Mine (Gabbs Magnesite-Brucite Mine) in the Gabbs district. (Oswald and Crook, 1979)

As inclusions in pyrite in the Gold Crater district. (Tingley et al., 1998a)

In quartz veins with galena, cerussite, and pyrite at the Star of the West Mine in the Jackson district. (Hamilton, 1993)

With galena at the Lost Indian Mine in the Jett district. (WMK)

With galena, pyrite, and gold in veins in the Johnnie district. (Ivosevic, 1978)

In small amounts in ore at the San Rafael Mine in the Lodi district. (Jenkins, 1981; B. and J. Runner, w.c., 1998)

Reportedly gold-bearing at the Manhattan Oxford Mining Co. in the Manhattan district. (WMK)

With bornite, chrysocolla, and copper carbonates at a prospect north of the Northumberland Mine in the Northumberland district; also with pyrite, *tourmaline,* and quartz in contact-metamorphosed rock elsewhere in the district. (Kleinhampl and Ziony, 1984)

In milky quartz veins with pyrite, covellite, benjaminite, aikinite, and other minerals at the Outlaw Mine in the Round Mountain district. (Shannon, 1925; Melhase, 1935d; DeMouthe, 1985; Dunning et al., 1991)

With aguilarite, pyrite, and pyrrhotite at the Blue Horse and Hillside Mines and Catlin claim in the Silverbow district. (Tingley et al., 1998a)

In small grains with other sulfides in veins in the Tonopah district; reportedly common as inclusions in polybasite. (Bastin and Laney, 1918; Melhase, 1935c; CSM)

With tetrahedrite, pyrite, and galena in a quartz vein at the Richmond Mine in the Union district. (Hamilton, 1993; NBMG mining district file)

In quartz veins with bornite and copper oxides

at an unnamed prospect on Limestone Ridge near Belted Peak in the Nellis Air Force Range. (Tingley et al., 1998a)

With chalcocite at the Wagner Mine and prospects in the Wagner district. (Tingley et al., 1998a)

In microscopic grains with luzonite and other sulfides at the Golden Chariot Mine in the Jamestown district in the Nellis Air Force Range. (Tingley et al., 1998a)

Reported in the Bare Mountain, Oak Spring, Paradise Peak, Troy, and Tybo districts. (Melhase, 1935b; WMK; Kral, 1951; Ferguson, 1933)

PERSHING COUNTY At the Majuba Hill Mine in the Antelope district as veinlets with pyrite, arsenopyrite, bornite, and pyrrhotite and disseminated with pyrite, *tourmaline,* and enargite in altered porphyry; also at the Antelope Mine and the Arsenic prospect. (Vanderburg, 1936b; Smith and Gianella, 1942; Lawrence, 1963; MacKenzie and Bookstrom, 1976; Jensen, 1985)

As traces in ore at iron mines in the Mineral Basin district. (Johnson, 1977)

In minor amounts in scheelite-bearing skarn at the Nightingale and Ranson Mines in the Nightingale district. (Smith and Guild, 1944)

With scheelite and pyrite at the Copper King claims in the Ragged Top district. (Johnson, 1977)

In quartz-*tourmaline* veins with other sulfides and gold at the Bonanza King and Pacific Matchless Mines in the Spring Valley district. (Campbell, 1939)

With bornite and digenite in massive sulfide ore at the Big Mike Copper Mine in the Tobin and Sonoma Range district. (Johnson, 1977)

With scheelite and powellite in skarn at the Esther Mine in the Trinity district. (Johnson, 1977)

Reported from the Buena Vista, Kennedy, Willard, and Rochester districts. (Cameron, 1939; Johnson, 1977; Knopf, 1924)

STOREY COUNTY In the Comstock district common in bonanza ore from the Comstock Lode, but less common in low-grade ore; also locally abundant in the Flowery lode. (Bastin, 1922; BMC; Schilling, 1990)

With silver-bearing tetrahedrite and pyrite at the Gooseberry Mine in the Ramsey district. (Rose, 1969)

WASHOE COUNTY Disseminated in *hornblende* gabbro with nickel-bearing pyrrhotite in the vicinity of Wild Horse Canyon, Fox Range, in the Cottonwood district. (Bonham and Papke, 1969)

In quartz-calcite veins with galena, sphalerite, arsenopyrite, and pyrite at the Union (Commonwealth) Mine in the Galena district. (Overton, 1947; Geehan, 1950)

A minor mineral in lead-zinc ore at the Black Panther and Leadville Mines in the Leadville district. (Overton, 1947; Bonham and Papke, 1969; WMK)

In unnamed prospects in skarn with scheelite and in granodiorite in the McClellan district. (Bonham and Papke, 1969)

With pyrite, pyrrhotite, and scheelite in skarn in the Nightingale district. (Bonham and Papke, 1969)

With gold, pyrite, bornite, petzite, and coloradoite in the Olinghouse district. (Overton, 1947; Jones, 1998)

In cavities in sinter with stibnite, cinnabar, and arsenopyrite in the Steamboat Springs district. (Lawrence, 1963)

Reported in the Dogskin Mountain, Peavine, Pyramid, and Wedekind districts. (WMK; Hill, 1915; Hudson, 1977; Wallace, 1980)

WHITE PINE COUNTY With pyrite, sphalerite, galena, gold, and copper carbonates at the Bald Mountain Mine in the Bald Mountain district. (Hitchborn et al., 1996; SBC)

The most abundant copper mineral in porphyry, contact-metasomatic, and vein deposits in the Robinson district. (Spencer et al., 1917; Melhase, 1934d; CSM; WMK)

With pyrite, sphalerite, and other minerals in skarn at the Ward Mine in the Ward district; also reported at the Defiance Mine. (Hasler et al., 1991; Hill, 1916)

Reported from the Cherry Creek, Eagle, Muncy Creek, and White Pine districts. (Hose et al., 1976; WMK; Humphrey, 1960)

CHALCOSIDERITE

$Cu^{2+}Fe_6^{3+}(PO_4)_4(OH)_8 \cdot 4H_2O$

Chalcosiderite, a rare iron analog of turquoise, is only reported from a few localities in the United States, where it occurs in oxidized zones of copper-bearing deposits.

HUMBOLDT COUNTY As aluminum-bearing, pale-green to blue-green crystals to 0.4 mm at the Silver Coin Mine in the Iron Point district; also in yellow-green sheaves with other phosphates. (Thomssen, 1998; GCF; J. McGlasson, w.c., 2001)

Identified by XRD and SEM/EDX as a pale-green phase with libethenite, pseudomalachite, and carbonate-fluorapatite near Golconda Summit in the Golconda district. (R. Thomssen, o.c., 2000)

LANDER COUNTY As bright emerald-green nuggets with faustite and variscite at the Damele Mine just north of Hickison Summit. (Rose and Ferdock, this publication)

As rounded white masses at the Lander Ranch Mine several miles north of Tenabo. (Rose and Ferdock, this publication)

LYON COUNTY With brochantite in skarn at the Douglas Hill Mine in the Yerington district. (Jenkins, 1981; CSM)

CHALCOSTIBITE

$CuSbS_2$

A relatively rare mineral that occurs with other sulfides in hydrothermal deposits.

ESMERALDA COUNTY Reported in the Goldfield district. (Von Bargen, 1999)

HUMBOLDT COUNTY Reported in trace amounts as *wolfsbergite* with arsenopyrite and tetrahedrite in veins at the Lone Tree Mine in the Buffalo Mountain district. (Bloomstein et al., 1993)

WASHOE COUNTY Reported in a specimen from an unknown locality in the Peavine district. (WMK)

CHAMOSITE

$(Fe^{2+},Mg,Fe^{3+})_5Al(Si_3Al)O_{10}(OH,O)_8$

A *chlorite*-group mineral that is generally found in laterite and sedimentary ironstone deposits.

ELKO COUNTY Reported as part of the alteration assemblage at the Jerritt Canyon Mines in the Independence Mountains district. (Hofstra, 1994)

PERSHING COUNTY As white talclike vein fillings with clinozoisite and laumontite at the lower adit of the Majuba Hill Mine in the Antelope district. (Jensen, 1993b)

CHAPMANITE

$Sb^{3+}Fe_2^{3+}(SiO_4)_2(OH)$

A rare hydrothermal mineral that is generally found in veins with sulfide minerals or graphite.

HUMBOLDT COUNTY A primary hydrothermal mineral as dull greenish-yellow microgranular bands and veinlets with cinnabar, stibnite, and pyrite at the McDermitt and Cordero Mines in the Opalite district. (McCormack, 1986; UMNH)

MINERAL COUNTY Reportedly ubiquitous at the Candelaria Silver Mine in the Candelaria district. (Chavez and Purusotam, 1988)

CHATKALITE

$Cu_6Fe^{2+}Sn_2S_8$

A rare tin-bearing sulfide that occurs in hydrothermal deposits.

LANDER COUNTY Identified by EMP in stockwork veins with other sulfides at the Cove Mine in the McCoy district. (Emmons and Coyle, 1988; W. Fuchs, o.c., 2000)

CHENEVIXITE

$Cu_2^{2+}Fe_2^{3+}(AsO_4)_2(OH)_4 \cdot H_2O$

Chenevixite, which occurs in the oxidation zones of arsenic-bearing copper deposits, is a rather nondescript mineral and may be more common than now recognized.

CLARK COUNTY As pale-green masses to 7 mm in a specimen from the Goodsprings district. (GCF)

ESMERALDA COUNTY As yellow microcrystals with olivenite at the 3 Metals Mine in the Weepah district. (B. and J. Runner, o.c., 2000)

MINERAL COUNTY As coatings on fractures at the Simon Mine in the Bell district. (B. and J. Runner, w.c., 1998)

PERSHING COUNTY As olive-green to brown dusty coatings on rhyolite breccia with olivenite, pharmacosiderite, and metazeunerite in the Copper stope, Majuba Hill Mine, Antelope district. (Jensen, 1985)

CHERALITE-(Ce)

(Ce,Ca,Th)(P,Si)O_4

An uncommon radioactive mineral described from a pegmatite in India.

NYE COUNTY Tentative identification of a Th-Ca-Ti-REE silico-phosphate found in gouge with barite, *electrum,* and illite at the Life Preserver Mine in the Tolicha district. (Tingley et al., 1998a)

CHERVETITE

$Pb_2V_2^{5+}O_7$

An uncommon mineral that occurs in the oxidized zone of vanadium-bearing deposits.

CLARK COUNTY Reported with vanadinite from the Ruth Mine in the Goodsprings district. (Kepper, this volume)

CHEVKINITE-(Ce)

(Ce,La,Ca,Na,Th)$_4$(Fe^{2+},Mg)$_2$(Ti,Fe^{3+})$_3$Si$_4O_{22}$

Chevkinite, a relatively uncommon mineral that is dimorphous with perrierite, occurs in some pegmatites with other rare earth–bearing minerals and also occurs as an accessory mineral in granite.

LINCOLN COUNTY Reported with fayalite, hedenbergite, zircon, and other minerals in the Kane Wash Tuff in the Kane Springs Wash volcanic center; also reported in other igneous rocks of the center. (Novak, 1985)

WASHOE COUNTY Reported as *tscheffkinite* with *allanite* in pegmatite at the Deer Lodge claims in the Stateline Peak district, NE 1/4 SE 1/4 Sec. 27, T22N, R18E. (Garside, 1973)

CHLORAPATITE

Ca$_5$(PO$_4$)$_3$Cl

A member of the *apatite* group that occurs in contact-metamorphosed rocks with *scapolite,* in granite pegmatite, and in veins in mafic igneous rock.

EUREKA COUNTY As yellow to yellow-green prismatic hexagonal crystals in magnetite ore at the Barth Iron Mine in the Safford district. (Shawe et al., 1962; GCF)

CHLORARGYRITE

AgCl

Chlorargyrite, also known as *cerargyrite* or *horn silver,* is a generally uncommon mineral formed by the oxidation of primary silver minerals in arid environments. However, it is widespread in Nevada in oxidation zones of many silver deposits and was an important ore of silver, particularly during early mining. *Embolite* is intermediate in composition between chlorargyrite and bromargyrite, *iodyrite* between chlorargyrite and iodargyrite.

CHURCHILL COUNTY With native silver at the Nevada Gold Mine in the Alpine district. (Vanderburg, 1940)

With other silver minerals in ore at the Bell Mountain, Fairview, Nevada Hills, and other mines in the Fairview district. (Greenan, 1914; Schrader, 1947; Quade and Tingley, 1987)

With cerussite at the Terrell Mine in the Holy Cross district. (Vanderburg, 1940)

With gold and scheelite at the Valley King claims in the Jessup district. (Vanderburg, 1940)

With acanthite and wire silver at the Kinney Mine in the Sand Springs district. (Schrader, 1947)

Reported in a specimen from the Westgate district. (R. Thomssen, w.c., 1999)

As grayish-green, waxy, transparent coatings and groupings of deformed crystals, commonly with wulfenite and locally in loose crystals of cubic form at the Nevada Wonder and other mines in the Wonder district. (Burgess, 1917; Vanderburg, 1940; Schilling, 1962a, 1962b; WMK)

CLARK COUNTY Reported in a specimen from the Little Eoluppus Mine in the Eldorado district. (WMK)

As perfect, minute, transparent yellow to tan crystals, and as green, white, and tan sectile masses in silicified dolomite with hemimorphite, pyromorphite, wulfenite, and vauquelinite at an unnamed prospect 2 miles south of Crystal Pass in the Goodsprings district; originally reported as *cerargyrite* with *iodyrite.* (Hewett, 1931; Melhase, 1934e; J. Kepper, w.c., 1998)

ELKO COUNTY Reported in silver-gold ore at the Maggie Mine in the Aura district. (WMK)

As blebs and stringers in late quartz veins at

the Rossi Mine in the Bootstrap district. (Snyder, 1989)

With other silver minerals and pyromorphite in gold-bearing quartz veins at an unnamed prospect on the north flank of Silver Peak in the Cornucopia district. (Emmons, 1910; Lawrence, 1963)

In a number of mines in the Gold Circle district; specifically reported in a specimen from the Link Mine. (Emmons, 1910; WMK)

Reported at the Long Hike and other mines in the Jarbidge district. (Schrader, 1923)

Reported with bromargyrite, stromeyerite, and gold in quartz at the Excelsior, Protection, and Nelson Mines in the Mountain City district. (Emmons, 1910; WMK)

In rich veins with silver and gold in the Tuscarora district, which yielded a single block of chlorargyrite worth $30,000; specifically reported at the Commonwealth and Independence Mines; also reported in float and placer deposits and as superb crystals from the Poorman Mine. (Emmons, 1910; Vanderburg, 1936a; WMK; M. Jensen, w.c., 1999)

ESMERALDA COUNTY As minute, waxy, olive-green crystals on *limonite* or rusty ledge matter with gold, kaolinite, and probable emmonsite at the Gold Coin claim, Diamondfield district. (Ransome, 1909a)

As large masses in brecciated quartz veins with *sericite* in the Tonopah-Divide Mine in the Divide district, reportedly the site of a 5-ton pod of pure chlorargyrite. (Knopf, 1921a; Melhase, 1935c; Schilling, 1979)

With anglesite, cerussite, smithsonite, hemimorphite, acanthite, and pyrargyrite at the Mammoth and Sylvania Mines in the Gilbert district. (Ferguson, 1928; Schilling, 1979)

Reported in a specimen from the Toby Mine in the Gold Point district. (WMK)

Reported from the Broken Toe Mine in the Rock Hill district. (R. Thomssen, w.c., 1999)

Reported at the 3 Metals Mine in the Weepah district. (R. Thomssen, w.c., 1999)

Reported from unspecified localities in the Klondyke, Lida, and Silver Peak districts. (Spurr, 1906; JHS; WMK)

EUREKA COUNTY Reported in a specimen of silver-copper ore at the Garrison Mine in the Cortez district. (WMK)

Plentiful in oxidized ore in the Eureka district. (Nolan, 1962)

With other silver minerals at an unspecified locality in the Mineral Hill district. (Emmons, 1910)

HUMBOLDT COUNTY Reported at the Sleeper Mine in the Awakening district. (Wood, 1988)

As yellow-green to greenish-gray bromine-bearing crystals to 2 mm at the Silver Coin Mine in the Iron Point district. (Vanderburg, 1938a; Thomssen, 1998; M. Jensen, w.c., 1999; J. McGlasson, w.c., 2001)

Reported with "ruby silver" and native silver in quartz-*adularia* veins at the Neversweat claim in the National district. (Lindgren, 1915)

Reported at the Silver Butte claims in the Paradise Valley district. (Vanderburg, 1938a)

In seams to 4 inches wide at the Hornsilver claims in the Sulphur district. (Vanderburg, 1938a)

LANDER COUNTY As greenish-yellow masses with iron oxides and arsenic and bismuth minerals at the Wilson-Independence Mine in the Battle Mountain district; also at other properties in the district. (Roberts and Arnold, 1965)

With gold in oxidized ore at the Campbell, Grey Eagle, and Little Gem Mines in the Bullion district. (Vanderburg, 1939; WMK)

In oxidized ore in the Cortez district. (Vanderburg, 1939)

Reported in a specimen from the Bunker Hill Mine in the Kingston district. (WMK)

With malachite and azurite in rich surface ore in quartz veins at the Betty O'Neal, Eagle, Highland Chief, and Pittsburg Mines in the Lewis district. (Vanderburg, 1939)

Reported at the Nevada Gold Dome Mine in the McCoy district. (Vanderburg, 1939)

Reported from the Snowstorm Mine in the North Battle Mountain district. (R. Thomssen, w.c., 1999)

Very important as a silver ore mineral with pyrargyrite during early mining in the Reese River district. (Ross, 1953)

LINCOLN COUNTY With cerussite and copper carbonates at the Silver Park Mine in the Atlanta district. (Tschanz and Pampeyan, 1970)

Reported as *cerargyrite* with *bromyrite* and cerussite in rich oxidized ore at the Black Metals Mine in the Bristol district. (Melhase, 1934d; Tschanz and Pampeyan, 1970)

Reported at the Little Emma and Oso Mines in the Delamar district. (WMK)

In a vein on the White Horse claim in the Eagle Valley district. (Tschanz and Pampeyan, 1970)

With silver-bearing galena, stetefeldtite, and secondary lead minerals at Hiko in the Pahranagat district. (Tschanz and Pampeyan, 1970)

Common with malachite, azurite, and anglesite at the Raymond, Ely, and other mines in the Pioche district. (Westgate and Knopf, 1932; UCD)

As films with acanthite at the Silver Dale Mines in the Silverhorn district. (Tschanz and Pampeyan, 1970)

LYON/MINERAL COUNTY Reported with acanthite and silver in the Benway district. (Schrader, 1947)

LYON COUNTY Reported in specimens from the Como district. (R. Thomssen, w.c., 1999)

Reported from the Dyke adit in the Talapoosa district. (R. Thomssen, w.c., 1999)

With silver, pyrargyrite, and covellite in a zone of supergene enrichment at an unspecified locality in the Wilson district. (Dircksen, 1975)

MINERAL COUNTY Reported at the Broken Hills Mine in the Broken Hills district. (Schrader, 1947)

In oxidized high-grade silver ore with quartz at the Candelaria (NERCO Minerals Co.) Silver Mine in the Candelaria district; also from the Lucky Hills and Mt. Diablo Mines. (Moeller, 1988; WMK)

With acanthite and gold at an unspecified locality in the Eagleville district. (Schrader, 1947)

Reported at the Mabel, Old Little, and Garfield Mines and Silver State lode in the Garfield district. (Couch and Carpenter, 1943; Ponsler, 1977; NMBM; WMK)

Reported at the Lucky Boy Mine in the Lucky Boy district. (Hill, 1915)

Reported at the Gold Pen and Nevada Rand Mines in the Rand district. (Schrader, 1947)

Reported as nuggets in rich surface ore in the Rawhide district; specifically noted at the Phoenix Mine, Silver Bell claim, and Truett vein; also reported with *embolite* in oxidized ore at the Rawhide Mine. (Vanderburg, 1937a; WMK; Black, 1991)

NYE COUNTY In quartz veins in rhyolite at the South Star Mine in the Bellehelen district; also

reported on gold at the Clifford Brothers Mine. (Kral, 1951; WMK)

As fine cinnamon-brown crystals at the Philadelphia Mine in the Belmont district; also with stetefeldtite and other minerals at the Highbridge and other mines. (Hoffman, 1878; Jenkins, 1981)

Reported with native gold in high-grade ore at the Montgomery-Shoshone Mine in the Bullfrog district. (Ransome et al., 1910; Melhase, 1935b)

Brown to light-green in veinlets with jarosite and silver at the Clifford Mine in the Clifford district. (Ferguson, 1916)

With cerussite in oxidized ore at the Illinois Mine in the Lodi district; also as waxy blebs at the San Rafael Mine. (Kral, 1951; B. and J. Runner, w.c., 1998)

With gold in a gangue of quartz, iron and manganese oxide, halloysite, and fluorite in a shear zone at the Wall Mine in the Manhattan district. (Ferguson, 1924)

In silver ore with "ruby silver," malachite, and azurite at the Northumberland and other mines in the Northumberland district. (Raymond, 1869)

With native silver and stephanite near the surface in veins in the Liberty area at the Juniper and other mines in the San Antone district. (Kleinhampl and Ziony, 1984; WMK)

With *electrum,* stephanite, and acanthite at the Silver Bow Mine in the Silverbow district. (Kral, 1951; WMK)

In the Tonopah district as a secondary mineral at many sites such as the Midway workings and the Mizpah and Silver Top Mines; noted as scales, coatings, and fissure fillings or long, slender, twisted and curved fibers. Also reportedly abundant and widespread in kaolinized *feldspar* as minute, translucent, pale-gray cubes and octahedrons. (Spurr, 1903; Burgess, 1911; Bastin and Laney, 1918; Melhase, 1935c; ASDM; WMK)

In oxidized ore at the Mountain View Mine (400-foot level) in the Tybo district. (Jenkins, 1981; WMK)

Reported from unspecified localities in the Morey, Reveille, and Wilsons districts. (Williams, 1968; Kral, 1951; JHS)

PERSHING COUNTY As very rare waxy brown grains to 1 mm with olivenite and mixite at the

Majuba Hill Mine in the Antelope district. (Jensen, 1985, 1993b)

In near-surface ore at the Arizona and other mines in the Buena Vista district. (Cameron, 1939)

Disseminated in silver ore in altered rhyolite at the Nevada Packard Mine in the Rochester district. (Knopf, 1924; Johnson, 1977)

In near-surface ore at the De Soto, J. C. Stageman, and Queen of Sheba Mines in the Star district. (Cameron, 1939; WMK)

STOREY COUNTY Reported from the Comstock Lode in the Comstock district; specifically reported in a specimen from the Golden Winney Mine. (Smith, 1985; CSM)

WASHOE COUNTY As surface ore with gold, cerussite, and anglesite at the Arkell Mine in the Wedekind district. (Overton, 1947; S. Pullman, w.c., 1999)

WHITE PINE COUNTY Reported at the Star and Imperial Mines in the Cherry Creek district. (Schrader, 1931b)

Reported in the Cooper district on Rattlesnake Knoll. (White, 1871; Hose et al., 1976)

In a specimen from the Red Hills Mine in the Eagle district. (R. Thomssen, w.c., 1999)

Reported at the Siegel and Gold Crown Mines in the Muncy Creek district. (Hose et al., 1976)

Reported at the Monitor and Argus Mines in the Taylor district. (Hill, 1916)

With native silver and secondary copper and lead minerals at the Ward Beecher Mine in the Ward district. (Hill, 1916)

Exceptional occurrences were reported in mines on Treasure Hill in the White Pine district in very high-grade silver ore with silica, calcite, silver, and manganese oxide in brecciated limestone. The Eberhardt Mine reportedly contained one of the most remarkable occurrences on record as a 6-ton mass. The mineral reportedly occurred in some cases as small grayish- to yellowish-green crystals at the Eberhardt and other Treasure Hill mines. (Hague, 1870; Humphrey, 1960; Smith, 1970; WMK)

Reported in specimens from the Bald Mountain and Robinson districts. (JHS; WMK)

CHLORITE

$(Al,Fe^{2+},Fe^{3+},Li,Mg,Mn^{2+},Ni,Zn)_{4-6}(Al,B,Fe^{3+},Si)_4$ $O_{10}(OH,O)_8$ (general)

The mineral group *chlorite* is widespread in Nevada as a hydrothermal alteration product. Species have not been determined for the following localities, which represent a small portion of occurrences in the state.

CHURCHILL COUNTY As an alteration mineral at the Fairview Silver Mine in the Fairview district. (Schrader, 1947)

CLARK COUNTY A common alteration mineral in the Bunkerville district. (Beal, 1965)

ELKO COUNTY An alteration mineral in diorite and in skarn in the Contact district. (Purington, 1903; Schilling, 1979)

ESMERALDA COUNTY The principal mineral at the Lida Talc, Ace in the Hole, and White Horse Mines in the Palmetto district as bodies to 40 feet wide; also present in minor amounts in some talc deposits such as the Shaw and Crystal White Mines. (Papke, 1975)

The primary mineral in several commercial deposits to 600 feet long, such as the White King Extension and Gates Mines in the Sylvania district, as soft, very fine-grained, massive ore with minor talc, quartz, and calcite; also as a minor mineral in some talc deposits, such as at the Oasis and Reed Mines. (Papke, 1975)

LANDER COUNTY A constituent of gold-bearing skarn at the Fortitude Mine in the Battle Mountain district. (Doebrich, 1993)

LINCOLN COUNTY As crystals in cavities in metamorphosed volcanic breccia with *garnet, pyroxene,* and magnetite in the Bristol district. (Westgate and Knopf, 1932; Melhase, 1934d)

LYON COUNTY In greenish-gray, very fine-grained mixtures with *sericite* at the Talco prospect in the Yerington district. (Papke, 1975)

NYE COUNTY As disseminated flakes after antigorite, forsterite, and other silicate minerals at the Premier Chemicals Mine (Gabbs Magnesite-Brucite Mine) in the Gabbs district. (Schilling, 1968a)

PERSHING COUNTY Part of the gangue in iron deposits in the Mineral Basin district. (Reeves and Kral, 1955)

STOREY/LYON COUNTY Common as radiating fibers in the andesite wall rock along the Comstock Lode in the Comstock district. (Gianella, 1936a)

WASHOE COUNTY Reported at Washoe Hot Springs. (White, 1955)

CHLORITOID

$(Fe^{2+},Mg,Mn)_2Al_4Si_2O_{10}(OH)_4$

A widely distributed mineral that mainly occurs in regionally metamorphosed sedimentary rock, but has also been found in hydrothermal veins.

CHURCHILL COUNTY As abundant, black, flaky material with *chiastolite* in hornfels in road cuts 28 miles east of Fallon on the south side of U.S. Highway 50 on the west slope of the Sand Springs Range. (Horton et al., 1962)

CHONDRODITE

$(Mg,Fe^{2+})_5(SiO_4)_2(F,OH)_2$

Chondrodite, a species in the humite group, is a relatively uncommon mineral that occurs in some contact-metamorphic assemblages in magnesian rocks with minerals such as diopside, tremolite, and grossular.

NYE COUNTY With forsterite in selvage zones to 6 inches thick along a contact between brucite and serpentinized granodiorite at the Premier Chemicals Mine (Gabbs Magnesite-Brucite Mine) in the Gabbs district. (Schilling, 1968a)

CHRISTITE

$TlHgAsS_3$

Christite, a Nevada type mineral, is a rare hydrothermal sulfide that occurs in the thallium-, mercury-, and arsenic-rich gold ores of Carlin-type deposits.

EUREKA COUNTY Type locality as grayish-white subhedral grains less than 1 mm long with lorandite, realgar, getchellite, ellisite, barite, and orpiment near the base of the oxidized zone at the Carlin Mine in the Lynn district, Sec. 13, T35N, R50E. (Radtke et al., 1977; Dickson et al., 1979)

HUMBOLDT COUNTY Microscopic grains in ore from the north pit of the Getchell Mine in the Potosi district. (F. Cureton, w.c., 1999)

CHROMITE

$Fe^{2+}Cr_2O_4$

Chromite, a common mineral in ultramafic and layered mafic intrusive rocks, occurs as an accessory mineral or as segregations that may constitute chromium ore. It is somewhat rare in Nevada because such rocks are relatively uncommon.

CARSON CITY As a detrital mineral in placer deposits near Carson City with gold, *garnet,* and zircon. (Garside, 1973)

CHURCHILL COUNTY As 0.5 mm crystals in basalt on Upsal Hogback. (M. Jensen, w.c., 1999)

CLARK COUNTY Reported in hornblendite at the Great Eastern and Key West Mines in the Bunkerville district. (Bancroft, 1910)

Reported at the Boss Mine in the Goodsprings district. (Jedwab, 1998b)

EUREKA COUNTY Reported in gabbroic intrusions at the Goldstrike Mine in the Lynn district. (Berry, 1992)

MINERAL COUNTY As disseminated specks in a *serpentine* mass in the Candelaria district. (Lincoln, 1923; Page, 1959; WMK)

NYE COUNTY Reported as black grains to 2 mm at the Premier Chemicals Mine (Gabbs Magnesite-Brucite Mine) in the Gabbs district. (S. Pullman, w.c., 1999; GCF)

Reported as chrome-rich spinel in brown to reddish-brown grains in ultramafic nodules in basalt at the Lunar Craters volcanic field. (Bergman, 1982)

CHRYSOBERYL

$BeAl_2O_4$

Chrysoberyl is an uncommon mineral that occurs in granite pegmatite, generally in association with beryl.

CLARK COUNTY As yellow to apple-green plates to 3/4 inch with beryl crystals in pegmatite at the Taglo (Teagle) Mine in the Bunkerville district; also locally comprises as much as 10 percent of quartz-rich pegmatite with beryl in dikes in Precambrian schist and gneiss at the Virgin Mountain Chrysoberyl property. (Beal, 1965)

ELKO COUNTY Reported in a specimen with beryl, *tourmaline,* and muscovite in the Ruby Mountains south of Wells. (WMK)

CHRYSOCOLLA

$(Cu^{2+},Al)_2H_2Si_2O_5(OH)_4 \cdot nH_2O$

A common oxidation product in copper deposits, mainly in arid climates; abundant enough to be a significant ore mineral in some deposits.

CARSON CITY Reported in a specimen from the Brunswick Consolidated Mines in the Eldorado district. (WMK)

CHURCHILL COUNTY After well-developed crystals of hemimorphite at a prospect on the east slope of Chalk Mountain in the Chalk Mountain district. (R. Walstrom, w.c., 1999)

In veins at the Fairview Silver Mine in the Fairview district. (Warren, 1973)

In crystal sprays after malachite at the Westgate Mine in the Westgate district. (R. Walstrom, w.c., 1999)

CLARK COUNTY Reported from the Azure Ridge (Bonelli) Mine in the Gold Butte district. (S. Pullman, w.c., 1999)

In mammillary crusts with other secondary minerals at the Boss, Platina, Smithsonite, and other mines in the Goodsprings district; also on fractures with fornacite, mimetite, and pyromorphite at the Singer Mine. (Hewett, 1931; Adams, 1998; WMK; CSM)

As coatings at the Duplex and Quartette Mines in the Searchlight district. (Callaghan, 1939; WMK; S. Pullman, w.c., 1999)

Reported from unspecified localities in the Bunkerville, Eldorado, and Muddy Mountains districts. (Beal, 1965; WMK)

ELKO COUNTY With other secondary copper minerals at the Blue Bird Mine in the Contact district; also with cerussite in skarn at the Ivy Wilson Mine. (Purington, 1903; Bailey, 1903; Schrader, 1935b; Schilling, 1979)

With malachite, azurite, and chalcocite along a contact between quartz monzonite and limestone with quartzite at the Delker Mine in the Delker district. (Hill, 1916; LaPointe et al., 1991)

With malachite, chalcocite, and other copper minerals in skarn at the Eugene patent, Victoria claim, and Lewis claims in the Dolly Varden district. (Hill, 1916; Melhase, 1934d; Schilling, 1979)

In skarn at the Great Eastern and Sidlong claims in the Ferber district. (Hill, 1916; LaPointe et al., 1991)

With other copper minerals in quartz veins at the Kinsley Consolidated and Morning Star Mines in the Kinsley district. (Hill, 1916)

With malachite and azurite at the Silver Star group in the Loray district. (Hill, 1916)

With magnetite, *limonite*, and malachite as irregular pods in green *garnet*-epidote skarn at the Lone Mountain Mine in the Merrimac district. (LaPointe et al., 1991)

Reported as silver-bearing with galena, copper and lead carbonates, and hemimorphite in chimneys at the Tripoli (Aladdin) Mine in the Railroad district; also reported with secondary copper and lead minerals at the Sylvania Mine. (Emmons, 1910; LaPointe et al., 1991; WMK)

After malachite and rosasite with wulfenite and other minerals in oxidized copper ore at the Killic Mine in the Spruce Mountain district. (Schrader, 1931b; Schilling, 1979)

ESMERALDA COUNTY As pseudomorphs after other microminerals at the 3 Metals Mine in the Weepah district. (B. and J. Runner, w.c., 1998)

As pseudomorphs at the Big 3 Mine in the Railroad Springs district. (B. and J. Runner, w.c., 1998)

Reported in specimens from the Cuprite, Klondyke, and Lone Mountain districts. (WMK; JHS)

EUREKA COUNTY Identified in a sample from an unspecified locality in the Lone Mountain district. (JHS)

In light-blue-green masses with quartz and dussertite at the Goldstrike Mine in the Lynn district. (GCF)

In light-blue, waxy, microcrystalline veins and fracture coatings at the Copper King Mine in the Maggie Creek district, where it was also reported as *medmontite*, an invalid species, with faustite; also as porcelaneous, pale-blue-green coatings with other secondary copper minerals at the Gold Quarry Mine. (Rota, 1991; EMP; Jensen et al., 1995)

In a specimen with cuprite, galena, and chalcopyrite from the Roberts district. (WMK)

HUMBOLDT COUNTY As light-blue coatings at the Silver Coin Mine in the Iron Point district. (Thomssen, 1998; J. McGlasson, w.c., 2001)

LANDER COUNTY In oxidized ore at the Copper Canyon Mine in the Battle Mountain district. (Schilling, 1979)

In a specimen with pyrolusite from an unknown locality in the Bullion district. (WMK)

LINCOLN COUNTY With malachite, atacamite, tenorite, and cuprite at the Bristol and May Day Mines in the Bristol district; also noted as emerald-green botryoids with calcite. (Melhase, 1934d; CSM; WMK; S. Pullman, w.c., 1999; GCF)

In a specimen with azurite, malachite, and *psilomelane* in the Pioche district. (WMK)

With tenorite, chalcocite, and malachite along structural zones at the Arrowhead (Southeastern) Mine in the Southeastern district. (Tingley et al., 1998a)

LYON COUNTY With other copper minerals at some radioactive occurrences in the Washington district. (Garside, 1973)

Reported at the Jackpot Mine in the Wellington district. (LANH)

A common mineral at many localities such as the Bluestone, Douglas Hill, and Yerington (Anaconda) Mine in the Yerington district. Commonly associated minerals are cuprite, tenorite, chalcocite, and native copper. At the Yerington Mine, an important ore mineral in clay-altered phenocrysts in porphyry and as narrow seams along fractures. At the Douglas Hill Mine, as pseudomorphs after brochantite. At the MacArthur Mine, as pale-blue-green pseudomorphs to 1 cm after azurite crystals. (Melhase, 1935e; Wilson, 1963; CSM; WMK; R. Walstrom, w.c., 1999; M. Jensen, w.c., 1999)

MINERAL COUNTY In micromount specimens as pseudomorphs after other minerals at the Simon Mine in the Bell district. (B. and J. Runner, w.c., 1998)

After pseudomalachite at an unnamed copper prospect in the Buckley district on Copper Mountain. (R. Walstrom, w.c., 1999)

In a shear zone with quartz, *limonite*, chlorargyrite, and other minerals at the Mabel Mine in the Garfield district. (Couch and Carpenter, 1943)

In skarn at the Julie claim in the Pamlico district. (Smith and Benham, 1985)

In the Santa Fe district with azurite, malachite, and cuprite in veinlike orebodies in limestone at the Wallstreet and Turk Mines; also at other localities. Chrysocolla from this district has been used in jewelry. (Ross, 1961; LANH; WMK)

Reported in a specimen with chalcopyrite and silver- and gold-bearing *limonite* at the Stageman Mine and Moonlight claims in the Silver Star district. (WMK)

NYE COUNTY With *electrum,* gold, acanthite, and uytenbogaardtite in high-grade gold ore at the Bullfrog and Original Bullfrog Mines in the Bullfrog district. (Castor and Weiss, 1992)

In micromount specimens as pseudomorphs after other minerals at the San Rafael Mine in the Lodi district. (B. and J. Runner, w.c., 1998)

At an unspecified locality in the Oak Spring district on the Nevada Test Site as verdigris-green to robin's-egg-blue material that takes an excellent polish and closely resembles turquoise; several hundred pounds have been sold. (JHS)

As green coatings in quartz veins at the Green Top claim in the Round Mountain district. (Lovering, 1954)

In gold- and silver-bearing specimens from the Black Jack prospect in the San Antone district. (Kral, 1951; WMK)

As stains on quartz veins at unnamed prospects on Limestone Ridge near Belted Peak on the Nellis Air Force Range. (Tingley et al., 1998a)

With conichalcite, malachite, tenorite, and azurite at the Wagner Mine and prospects in the Wagner district. (Tingley et al., 1998a)

PERSHING COUNTY Widespread as greenish-blue coatings and stains on rhyolite and as pseudomophs after zeunerite in the Copper stope at the Majuba Hill Mine in the Antelope district. (Smith and Gianella, 1942; MacKenzie and Bookstrom, 1976; Jensen, 1985; R. Walstrom, w.c., 1999)

Reported in a specimen from the Organ Mine in the Mill City district. (WMK)

WASHOE COUNTY A rare mineral in pegmatites near Incline Village. (M. Jensen, w.c., 1999)

With azurite, malachite, brochantite, chalcanthite, and other secondary copper minerals in the Peavine and Wedekind districts. (Hudson, 1977)

WHITE PINE COUNTY Identified in a specimen from an unknown locality in the Bald Mountain district. (JHS)

Reported at the Betty Jo claim in the Nevada district. (S. Pullman, w.c., 1999)

As fracture coatings in the Liberty pit in the Robinson district at Ruth. (Spencer et al., 1917; LANH; GCF)

With chalcocite from the Caroline tunnel in the Ward district. (WMK)

CINNABAR

HgS

Cinnabar, the principal ore mineral of mercury, is very common in Nevada, generally in hydrothermal deposits related to ancient hot spring activity and commonly in finely crystalline silica referred to as "opalite." *Myrickite* is red or orange cinnabar-bearing opal or *chalcedony*. The McDermitt and Cordero Mines have been among the most productive mercury mines in the nation. Many cinnabar localities in Nevada are described by Bailey and Phoenix (1944).

CARSON CITY As small disseminated crystals with jarosite in a fracture zone at the Valley View prospect in the Carson City district. (Bailey and Phoenix, 1944)

Identified in a sample from an unspecified locality in the Delaware district. (JHS)

CHURCHILL COUNTY As thin coats with sulfur and gypsum at the Erway prospect in the Desert district. (Bailey and Phoenix, 1944)

With metacinnabar, sulfur, and pyrite in sinter at the Senator Mine and presently forming at two active fumaroles at the nearby Senator Fumaroles in Dixie Valley, SW 1/4, Sec. 31, T25N, R37E. (Lawrence, 1971; Garside and Schilling, 1979)

Reported at the Wild Horse Mine in the Eastgate district. (Bailey and Phoenix, 1944; WMK)

In gold ore at the Hidden Treasure and Gold Bug properties in the Fairview district. (Bailey and Phoenix, 1944)

With pyrite, barite, gypsum, and jarosite in calcite veinlets at the Cinnabar Hill Mine in the Holy Cross district. (Vanderburg, 1940; Bailey and Phoenix, 1944)

From the Quick-Tung Mine in the Shady Run district. (Lawrence, 1963)

With other mercury minerals at the Red Bird Mine in the Bernice district. (B. and J. Runner, w.c., 1998)

As disseminated crystals and veinlets in iron-stained sandstone at the Rosebud prospect in the Mountain Wells district. (Bailey and Phoenix, 1944)

CLARK COUNTY Mined at the Patsy prospect in the Eldorado district. (Bailey and Phoenix, 1944; WMK).

Reported as coatings on lead minerals at the Kirby, Fredrikson, Yellow Pine, Red Cloud, and Argentine Mines in the Goodsprings district; also with copper minerals, potarite, and gold at the Boss Mine. (Hewett, 1931; UCB; Jedwab, 1998a)

Reported in a specimen from the Harry Howell property in the Muddy Mountains district. (WMK)

ELKO COUNTY As blebs to 0.05 mm in gold ore at the Rossi Mine in the Bootstrap district. (Snyder, 1989)

Earthy red fracture coatings in siltstone and mudstone at the Rain Mine in the Carlin district. (D. Heitt, o.c., 1991)

With marcasite in a banded silica vein in rhyolite flows and tuffs at the Rex and Rand Mines in the Gold Circle district. (Bentz et al., 1983; Klein, 1983)

With lorandite, native arsenic, realgar, orpiment, arsenopyrite, and marcasite at the Jerritt Canyon Mines in the Independence Mountains district. (Hofstra, 1994)

Occurs at more than a dozen properties in the Ivanhoe district, generally in silicified rock including *agate, chalcedony*, and petrified wood, and locally with calomel. At the Hollister gold deposit, with pyrite, marcasite, chalcopyrite, stibnite, realgar, tetrahedrite-tennantite, sphalerite, and micron-sized particles of native gold. (Bailey and Phoenix, 1944; Bartlett et al., 1991; LaPointe et al., 1991; Klein, 1983; WMK)

Reported in opalite at the Bellehelen Mine in the Mud Springs district. (WMK)

With pyrite in quartz veins and as disseminations in andesite at the Horse Mountain Mine in the Rock Creek district; also as botryoidal masses along cracks at the Rock Creek prospect, and in crystalline form with pyrite and stibnite in a shear zone at the Teapot prospect. (Bentz et al., 1983; LaPointe et al., 1991; Lawrence, 1963; Bailey and Phoenix, 1944)

With silica, barite, pyrite, and marcasite in a shear zone in andesite on the Berry Creek claims in the Tuscarora district; and with native mercury and pyrite in a shear zone in altered andesite on the Red Bird group. (Bailey and Phoenix, 1944)

ESMERALDA COUNTY Disseminated in *alum* (kalinite) with sulfur and gypsum at the Alum Mine in the Alum district. (Spurr, 1906)

In opalized and alunized volcanic ash at the

Ralston prospect in the Cuprite district. (Bailey and Phoenix, 1944)

In several deposits in opalite, limestone and phyllite, and rhyolite with pyrite, barite, stibnite, stibiconite, sulfur, and gypsum in the Fish Lake Valley district. (Bailey and Phoenix, 1944)

In silicified andesite at the Castle Rock prospect in the Gilbert district. (Ferguson, 1928; Bailey and Phoenix, 1944)

Reported in a specimen from an unknown locality in the Goldfield district. (WMK)

Abundant in opalite-type deposits at the Montezuma prospect in the Montezuma district. (Bailey and Phoenix, 1944; JHS)

In minor amounts with kaolinite and native sulfur in the Silver Peak district. (Spurr, 1906)

EUREKA COUNTY As thin coatings, veinlets, and disseminated crystals with calomel at the Beowawe prospect in the Beowawe district. (Vanderburg, 1938b; Bailey and Phoenix, 1944; WMK)

Reported on a single property in the Cortez district in the Mill Canyon area. (Bailey and Phoenix, 1944)

As minute disseminated grains and as rims coating other sulfides at the Goldstrike Mine; also in placer gravels in the Lynn district. (GCF; Vanderburg, 1936a)

As red masses to 2 cm at the Gold Quarry Mine in the Maggie Creek district. (M. Jensen, w.c., 1999)

Tentatively identified with sulfur in calcareous sinter at the Hot Springs Point Sulfur Mine in Crescent Valley. (Garside and Schilling, 1979)

HUMBOLDT COUNTY At many properties in the Bottle Creek district, including the productive Blue Can and McAdoo Mines, as veinlets and disseminations in opalite, Tertiary volcanic rocks, and pre-Tertiary sedimentary rocks with pyrite, stibnite, *chalcedony*, and calcite; rarely as nuggets in stream gravels. (Roberts, 1940a; Bailey and Phoenix, 1944)

Coating pyrite, generally as anhedral grains <200 microns, at the Lone Tree Mine in the Buffalo Mountain district. (Panhorst, 1996)

A minor mineral in radioactive breccia at the Moonlight Uranium Mine in the Disaster district; also at the Disaster Peak property. (Dayvault et al., 1985; Bailey and Phoenix, 1944)

As tiny crystals and veinlets in volcanic rocks

with minor pyrite and quartz in the Dutch Flat district; also in gold-bearing placer gravels. (Bailey and Phoenix, 1944)

Reported in a specimen from the Thornton Mine in the Golconda district. (WMK)

Reported in altered rhyolite at the Plymouth property in the Harmony district. (Bailey and Phoenix, 1944)

Reported at the Silver Coin Mine in the Iron Point district. (Thomssen, 1998)

With chalcedonic silica and quartz in a blanketlike deposit at the Buckskin Quicksilver Mine in the National district; also reported in quartz veins elsewhere in the district. (Lindgren, 1915; Melhase, 1934c; Bailey and Phoenix, 1944; Vikre, 1985; WMK)

The Opalite district was the most productive mercury mining area in Nevada at the Cordero and McDermitt Mines. Here, cinnabar was found in small high-grade orebodies under opalite blankets, disseminated in large orebodies in altered volcanic rock, and in large low-grade bodies in opalite. It was described as occurring in clusters of tiny crystals, veinlets, and rare twisted needles. Associated minerals included native mercury, mercury halide and oxychloride minerals, pyrite, hematite and other iron oxides, stibnite, fluorite, and *heulandite*. (Bailey and Phoenix, 1944; Fisk, 1968; Roper, 1976)

As tiny dark-red crystals with quartz, realgar, and stibnite at the Getchell Mine in the Potosi district. (Stolburg and Dunning, 1985)

Reportedly the best specimens of cinnabar in the United States as crystals to 2.5 cm came from the Poverty Peak district, where it was reported as nearly pure veinlets and pods with minor stibnite, calcite, quartz, and gypsum in sandstone and limestone at the Cahill and Snowdrift Mines and other properties; also found with eglestonite, terlinguaite, mercury, and calomel. (Bailey and Phoenix, 1944; Sinkankas, 1967; F. Cureton, w.c., 1999; S. Pullman, w.c., 1999)

Reported with native mercury at the Rattlesnake Canyon prospect in the Red Butte district. (Bailey and Phoenix, 1944)

With native sulfur and kaolinite at the Nevada Sulphur Company Mine in the Sulphur district. (Bailey and Phoenix, 1944)

Reported in specimens at the Silent Kid and Lowe Mines in the Winnemucca district. (WMK)

LANDER COUNTY In placers with native mercury at the Rast prospect in the Callaghan Ranch district. (Bailey and Phoenix, 1944)

Reported in a specimen from the Shoshone Quicksilver Co. Mine in the Reese River district. (WMK)

In a vein with eglestonite, pyrite, barite, *chalcedony*, and jarosite at the Warm Springs prospect in the Warm Springs district. (Bailey and Phoenix, 1944)

As crystals in calcite veins, as crystalline masses in silicified limestone, and as crusts in silicified sandstone, with calomel, pyrite, calcite, barite, and stibnite, at the McCoy and Wild Horse Mines in the Wild Horse district. (Dane and Ross, 1943; Bailey and Phoenix, 1944)

LINCOLN COUNTY Reported from the Cammon claims in the Tem Piute district. (Bailey and Phoenix, 1944)

LYON COUNTY With stibnite at the DeLongchamps prospect in the Ramsey district. (Lawrence, 1963)

Reported from an unspecified locality in the Talapoosa district. (Van Nieuwenhuyse, 1991)

MINERAL COUNTY As high-grade lenses in a fault at the Lou prospect in the Bell district. (Bailey and Phoenix, 1944)

In faults at the Wild Rose Mine and other properties in the Buena Vista district. (Ross, 1961; Bailey and Phoenix, 1944; WMK)

Reported at the Clack Quicksilver Mine in the Fitting district. (AMNH; S. Pullman, w.c., 1999)

At the Cinnabar King and Drew Mines and many other properties in the Pilot Mountains district, mostly in limestone; associated minerals include quartz, calcite, antimony minerals, hemimorphite, azurite, malachite, manganese oxides, olivenite, and metacinnabar. (Foshag, 1927; Ross, 1961; WMK)

In veinlets with gypsum and sulfur at the Poinsettia property in the Poinsettia district. (Bailey and Phoenix, 1944)

Reported with metacinnabar in silicified rock at the Stockton prospect in the Rawhide district; also in a specimen from the Cye Cox property and as red crystalline material in barite at the Regent Mine. (Bailey and Phoenix, 1944; WMK; F. Cureton, w.c., 1999)

Reported in specimens from several properties in the Silver Star district. (WMK)

NY COUNTY In phyllite at the Flower Quicksilver Mine and at the nearby Flower placer property in the Barcelona district (originally reported as the Belmont district); also purplish-red cinnabar was reported from the Van Ness Mine. (Bailey and Phoenix, 1944; Lawrence, 1963)

In the Bare Mountain district in veinlets and disseminated in opal (as *myrickite*) from the Thompson and other mines; also as tiny crystals lining cavities in calcite veins in fluorspar ore at the Daisy Mine. (Melhase, 1935b; Bailey and Phoenix, 1944; Papke, 1979)

As isolated crystals along a shear zone in altered andesite at the Brooklyn claim in the Diamondfield district. (Ransome, 1909a; Bailey and Phoenix, 1944)

In altered rhyolite in a prospect about 1 mile south of Ellsworth in the Ellsworth district. (Kral, 1951)

As small disseminated crystals with jarosite in silicified and iron-stained breccia and tuff at the Finger Rock prospect in the Fairplay district; also at the Nody prospect, which became part of the main ore zone at the Paradise Peak gold mine. (Bailey and Phoenix, 1944; John et al., 1991)

In altered rhyolite at the Antelope prospect in the Gabbs district. (Kral, 1951; Vitaliano and Callaghan, 1963)

With pyrite, enargite, chalcocite, and mckinstryite(?) in a sample from a dump at a shaft in the Gold Crater district on the Nellis Air Force Range. (Tingley et al., 1998a)

With purple fluorite in a fault in ash-flow tuff at a radioactive occurrence at the Dottie Lee claim in the Jackson district; also as sparse crystals in calcite-opal veins with fluorite and autunite at the War Cloud prospect. (Bonham, 1970; Garside, 1973; Hamilton, 1993)

As veinlets and disseminated crystals in shear zones at the Horse Canyon Mine in the Jett district. (Bailey and Phoenix, 1944)

With gold and silver at the Bristol claims in the Kawich district. (Bailey and Phoenix, 1944)

Relatively widespread in gold ore in the White Caps Mine in the Manhattan district, particularly above the 200-foot level; also in a fault at the Pee Wee claims and in placer deposits in Manhattan Gulch. (Ferguson, 1924; Melhase,

1935d; Bailey and Phoenix, 1944; Gibbs, 1985; WMK; Kleinhampl and Ziony, 1984; Vanderburg, 1936a)

As pale-pink coatings on stibiconite and sulfur at the Titus prospect in the Squaw Hills east of Moores Station. (Jenkins, 1981)

In crystals to 1/4 inch in white calcite with local scheelite at the Scheebar Mine in the Paradise Peak district; also reported from the Snowshoe claim. (Bailey and Phoenix, 1944; LANH)

As botryoidal masses at the Black Hawk Mine in the Queen City district. (Bailey and Phoenix, 1944)

Reported with calcite in a specimen from the Fisherman Mine in the Reveille district. (WMK)

As purplish-red veins with black metacinnabar in quartz-barite veins at the Senator Mine in the Round Mountain district (originally reported in the Belmont district); also noted at the Mariposa Canyon prospect. (Bailey and Phoenix, 1944; GCF)

As thin streaks and in botryoidal form on pyrite or marcasite at the West End Mine in the Tonopah district. (Burgess, 1911; Bastin and Laney, 1918; Melhase, 1935c)

With jarosite and clay minerals along faults in altered tuff at the A & B and M & M Mines in the Tybo district. (Bailey and Phoenix, 1944)

In the Union district, which produced more than 10,000 flasks of mercury prior to 1919, mainly from the Mercury Mining Company and Nevada Cinnabar Mines, as veinlets, disseminated crystals, and replacements in sedimentary rock and silicified tuff with local native mercury, manganese oxide, pyrite, and calcite. (Bailey and Phoenix, 1944)

PERSHING COUNTY At many locations in the Antelope Springs district, which produced more than 12,000 flasks of mercury, mostly from the Pershing and Juniper Mines. Notable crystals have come from the Red Bird and Juniper Mines, and cherry-red crystals to 2 cm have been reported on silicified matrix. Associated minerals at the Red Bird Mine include pyrite, sphalerite, stibnite, quartz, calcite, bindheimite, cerussite, hemimorphite, aragonite, and the Nevada type minerals dadsonite and robinsonite. (Bailey and Phoenix, 1944; Berry et al., 1952; Jambor, 1969; AMNH; GCF; WMK; FUMT)

With dolomite and ankerite at the Paymaster prospect in the Black Knob district. (Bailey and Phoenix, 1944)

With calcite in shale at the Mercury Mine in the Buena Vista district. (WMK)

As banded veinlets and disseminated grains in opalite and limestone at the Goldbanks Quicksilver Mine in the Goldbanks district. (Dreyer, 1940; Bailey and Phoenix, 1944)

Reported with gypsum from a sulfur deposit near Humboldt in the Imlay district; also as veinlets, films, and disseminated crystals in black shale and limestone at the Eldorado Mine. (Russell, 1885; Bailey and Phoenix, 1944)

At the Mount Tobin Mine in the Mount Tobin district, which produced about 1500 flasks of mercury; also at the Polkinghorne Mine as grains in barite and limestone and locally as tiny crystals on yellow antimony oxide in cavities with stibnite. (Bailey and Phoenix, 1944; Lawrence, 1963)

In placer gravels in the Placerites district. (Klein, 1983)

In mine dumps and gold placers with cassiterite, scheelite, and magnetite in the Rabbit Hole district. (Melhase, 1934b; Johnson, 1977)

In placer gravel with gold, cassiterite, and hübnerite 3 miles south of Sulphur in the Rosebud district. (Bailey and Phoenix, 1944)

As coarsely crystalline masses in white crystalline calcite veins with small amounts of native mercury, calomel, stibnite, and quartz at the King George Mine in the Spring Valley district; also found in other mines and prospects and as tiny nuggets in placer deposits with gold. (Bailey and Phoenix, 1944; WMK)

In quartz and altered granite with sulfur at the Black Dyke prospect; also in calcite-quartz veins with stibnite at the Fencemaker Mine in the Table Mountain district. (Bailey and Phoenix, 1944; Lawrence, 1963)

STOREY COUNTY As veinlets, crystals in vugs, and disseminated grains in altered andesite with mercury, calomel, and other minerals at the Castle Peak Mine in the Castle Peak district; also as disseminated crystals and small veins at the Washington Hill Mine. (Bailey and Phoenix, 1944; Holmes, 1965; Bonham and Papke, 1969; WMK)

As scattered crystals in silicified andesite tuff at the Taylor-Branch prospect in the Clark district.

(Bailey and Phoenix, 1944; Bonham and Papke, 1969)

Reported in specimens from the Crystal Peak Mine, Virginia City; possibly the Castle Peak Mine in the Castle Peak district. (LANH; WMK)

WASHOE COUNTY Reported from the Hog Ranch Mine in the Cottonwood district. (Thomssen, w.c., 1999)

In altered andesitic agglomerate with pyrite, magnetite, and trace amounts of gold at various prospects in the Lone Pine district in and near the Sheldon National Antelope Refuge. (Bailey and Phoenix, 1944; Bonham and Papke, 1969)

Reported from the Olinghouse district. (Overton, 1947)

Reported at an unknown location on Peavine Mountain in the Peavine district. (Bailey and Phoenix, 1944)

With sulfur, gypsum, quartz, and opal in a 2-mile-long altered zone in Pleistocene alluvial deposits at the San Emidio prospect in the San Emidio district. (Bonham and Papke, 1969)

In a famous occurrence at Steamboat Hot Springs in the Steamboat Springs district as streaks, veinlets, and disseminations in sinter with sulfur, stibnite, metastibnite, pyrite, arsenopyrite, chalcopyrite, and anomalous amounts of gold and silver; also reported from the Silica pit and with pyrite, jarosite, gypsum, and mercury at the Wheeler Ranch prospect. (Becker, 1888; Bailey and Phoenix, 1944; White et al., 1964; White, 1980; Wilson and Thomssen, 1985; Thomssen, w.c., 1999; Overton, 1947)

CLAUDETITE

As_2O_3

A dimorph of arsenolite that forms as an oxidation product of realgar or other primary arsenic minerals; found in Arizona and Utah as a product of mine fires.

NYE COUNTY Reported from the White Caps Mine, 600-foot level, in the Manhattan district; much arsenolite from this mine is thought to be mine-fire related, but the claudetite is believed to be of non-fire origin. (Gibbs, 1985)

CLAUSTHALITE

PbSe

A mineral of precious-metal deposits and mercury deposits that generally occurs with other selenide minerals.

ELKO COUNTY As micron-sized grains with aguilarite, *electrum,* and naumannite in quartz veins at the Ken Snyder Mine in the Gold Circle district. (Casteel and Bernard, 1999)

Inclusions in stibnite from the Meikle Mine in the Lynn district. (D. Tretbar, w.c., 2002)

LANDER COUNTY Identified by EMP with sulfides and other selenium-bearing minerals in a sample from the Dean Mine in the Lewis district. (J. McGlasson, w.c., 2001)

MINERAL COUNTY Reported at the Borealis Mine, Freedom Flat deposit, in the Borealis district. (Eng, 1991)

CLINOBISVANITE

$BiVO_4$

A rare oxidation product that forms in bismuth-bearing hydrothermal deposits.

CHURCHILL COUNTY As small yellow crystals with vanadinite, bismutite, and pottsite at an unnamed locality northeast of Chalk Mountain in the Clan Alpine Mountains. (R. Walstrom, w.c., 1999)

LANDER COUNTY As orange to yellow-orange, bladed microcrystalline aggregates and pulverulent coatings on skarn with duhamelite, scheelite, bismutite, and vanadinite at the Linka Mine in the Spencer Hot Springs district. (Williams, 1988; MV; S. Pullman, w.c., 1999; S. Sears, w.c., 1999)

CLINOCHLORE

$(Mg,Fe^{2+})_5Al(Si_3Al)O_{10}(OH)_8$

Clinochlore, a widespread member of the *chlorite* group, occurs in a variety of metamorphic rocks, in ultramafic igneous rocks, and in veins. Well-crystallized specimens generally come from magnesium-rich contact-metamorphic rocks.

CLARK COUNTY Reported as microcrystals in miarolitic cavities and pegmatites in the Eldorado and Newberry Mountains. (S. Scott, w.c., 1999)

Reported in a specimen as the chromium-bearing variety *kammererite* from an unspecified locality in the Las Vegas district. (WMK)

ELKO COUNTY In dark-green masses on hematite crystals at the Victoria Mine in the Dolly Varden district. (M. Jensen, w.c., 1999)

An alteration mineral at the Jerritt Canyon Mines in the Independence Mountains district. (Hofstra, 1994)

EUREKA COUNTY As light- to grass-green micaceous masses after biotite in diorite at the Goldstrike Mine in the Lynn district; also in skarn as veins and masses with crystals to 1 cm and with phlogopite and calcite in selvages. (GCF)

An alteration product at the Modarelli Mine in the Modarelli-Frenchie Creek district. (Shawe et al., 1962)

HUMBOLDT COUNTY A primary constituent of chloritic schist at the Delong (Iron King) Mine in the Jackson Mountains district. (Shawe et al., 1962)

LYON COUNTY Reported as *penninite* in a specimen from an unknown locality in the Wellington district. (WMK)

Reported in a specimen from the Yerington Consolidated Mines in the Yerington district. (HMM)

NYE COUNTY As a fine-grained alteration product at the Clarkdale and Yellow Gold Mines in the Clarkdale district. (Tingley et al., 1998a)

Reported as *leuchtenbergite* with scheelite in a vein at the Betty O'Neal tungsten prospect in the Gabbs district; also reported from the Linda talc prospect. (Kerr and Callaghan, 1935; Schilling, 1968a; AMNH; JHS)

A widespread alteration mineral in the Tonopah district. (Spurr, 1903)

WASHOE COUNTY Part of the propylitic alteration assemblage in host rocks at the Burrus and other mines in the Pyramid district. (Park-Li, o.c., 1996)

As green-black blocks to 2 cm in pegmatites near Incline Village. (Jensen, 1993a)

CLINOCHRYSOTILE

$Mg_3Si_2O_5(OH)_4$

A mineral of the kaolinite-*serpentine* group, commonly asbestiform, that generally occurs with orthochrysotile in serpentinite.

LYON COUNTY As olive-green tufts of acicular crystals to 3 mm at the Douglas Hills Mine in the Yerington district. (S. Pullman, o.c., 2000; GCF)

NYE COUNTY Reported with orthochrysotile and parachrysotile in brucite deposits at the Premier Chemicals Mine (Gabbs Magnesite-Brucite Mine) in the Gabbs district; also reported as *deweylite*. (AMNH; WMK)

WHITE PINE COUNTY Reported from magnesite deposits in altered tuff in the Currant Creek deposit in the Currant district, also reported as *deweylite*. (Hose et al., 1976; Vitaliano, 1951; Gaines et al., 1997)

CLINOCLASE

$Cu_3^{2+}(AsO_4)(OH)_3$

Clinoclase is a rare mineral that occurs as an oxidation product in some arsenic-bearing copper deposits.

CHURCHILL COUNTY Reported in specimens from the Fireball Mine in the Truckee district. (R. Thomssen, w.c., 1999)

EUREKA COUNTY With malachite on quartz-bearing gossan from an unspecified locality in the Eureka district. (FUMT; Jenkins, 1981)

LYON COUNTY Reported at the Empire-Nevada Mine in the Yerington district. (LANH)

PERSHING COUNTY In world-class specimens as abundant, widespread, shiny, deep-blue crystals to 2 mm from the Tin stope and radiating groups of crystals in rhyolite breccia from the Copper stope at the Majuba Hill Mine in the Antelope district, Sec. 2, T32N, R31E; associated minerals are *limonite*, clay, olivenite, cornubite, and cornwallite. (Smith and Gianella, 1942; Gianella, 1946; Palache and Berry, 1946; MacKenzie and Bookstrom, 1976; Jensen, 1985)

WASHOE COUNTY As tiny, deep-greenish-blue crystals or crusts on fractures in quartz-*tourmaline* rock from an unnamed mine on Peavine Mountain in the Peavine district. (AMNH; Jenkins, 1981)

Identified in a single specimen from the Pyramid district. (Jensen, 1994)

CLINOENSTATITE

$Mg_2Si_2O_6$

A *pyroxene*-group mineral, dimorphous with enstatite, that occurs in intermediate volcanic rocks, ultramafic igneous rocks, high-grade metamorphic rocks, and meteorites.

PERSHING COUNTY Reported in a specimen from the Nightingale district. (MV)

CLINOHEDRITE

$CaZnSiO_4 \cdot H_2O$

An extremely rare mineral that has only been reported from Franklin, New Jersey, and the Christmas Mine, Arizona.

PERSHING COUNTY Reported in a specimen; no other locality data provided. (LANH)

CLINOPTILOLITE

$(Na,K,Ca_{0.5})_6[Al_6Si_{30}O_{72}] \cdot {\sim}20H_2O$

A common zeolite mineral group that has recently been subdivided into three species depending on the most abundant alkali or alkali earth element. *Clinoptilolite* deposits have been mined commercially in Arizona, California, and Oregon. The mineral is quite common in Nevada as an alteration product in volcanic and volcaniclastic rocks; detailed descriptions of some Nevada deposits are in Papke (1972a) and other descriptions are in Holmes (1994).

CHURCHILL COUNTY In an 11-m-thick deposit in zeolitized ash-fall tuff with *erionite* and *phillipsite* in the Copper Valley district. (Holmes, 1994)

With *erionite,* mordenite, and other minerals in zeolitized volcaniclastic rock at the Eastgate deposit in the Eastgate district. (Papke, 1972a)

With *ferrierite* and mordenite at an unnamed deposit in the Sand Springs district about 30 miles southeast of Fallon. (Papke, 1972a)

Reported in two areas north of Fallon near Interstate 80 and the Pershing County line. (Papke, 1972a)

As pale-green crystals to 0.6 mm with cinnabar, cristobalite, and other minerals at the Red Bird Mine in the Bernice district. (B. and J. Runner, w.c., 1998; GCF)

CLARK COUNTY In an unnamed deposit near the Overton Arm of Lake Mead in the Muddy Mountains district; also reported with *heulandite* in tuffs in the Horse Spring Formation. (Papke, 1972a; Bohannon, 1984)

ELKO COUNTY Reported in two unnamed deposits near U.S. Highway 93, 30 and 50 miles north of Wells; also reported in a deposit just south of Elko. (Papke, 1972a)

ESMERALDA COUNTY With mordenite at unnamed localities near Goldfield and in Clayton Valley. (Papke, 1972a)

EUREKA COUNTY Locally in minor amounts in large deposits of *phillipsite* and *erionite* in zeolitized tuffs near Nevada Highway 278 in the Pine Valley district. (Papke, 1972a)

HUMBOLDT COUNTY With *erionite* and mordenite as a product of alteration of volcaniclastic rocks adjacent to mercury and uranium deposits in the McDermitt caldera. (Rytuba and Glanzman, 1979)

In small vugs along calcite veins northeast of Sulphur in the low basalt hills of the Jackson Range in the Sulphur district. (M. Rogers, o.c., 1999)

A common diagenetic replacement mineral in tuffaceous deposits near the Virgin Valley Ranch in the Virgin Valley district. (Castor et al., 1996)

LANDER COUNTY With *erionite, chabazite,* and local *phillipsite* in widespread deposits in tuffaceous beds in the Reese River Valley, Sec. 26, T24N, R43E. (Papke, 1972a)

Reported in a thick deposit in the Fish Creek Mountains 80 km southwest of Battle Mountain. (Papke, 1972a; Holmes, 1994)

LINCOLN COUNTY In a deposit just south of Caliente in the Caliente district. (Papke, 1972a)

LYON COUNTY With mordenite several miles south of Fernley in the Fernley district. (Papke, 1972a)

MINERAL COUNTY Replacing glass shards in thin tuffs in playa lakebeds in Teels Marsh in the Teels Marsh district. (JHS)

NYE COUNTY With mordenite and analcime in Tertiary tuffaceous rocks on the Nevada Test Site; comprises as much as 77 volume percent of drill hole samples from Yucca Mountain. (Hoover, 1968; Bish and Vaniman, 1985)

With mordenite in an unnamed deposit near U.S. Highway 6 about 25 miles east of Tonopah. (Papke, 1972a)

Abundant in tuff, in some cases with mordenite, in localities near Beatty in the Beatty district. (Papke, 1972a; Holmes, 1994)

In a large deposit in green ash-flow and lapilli tuffs at Ash Meadows; the deposit is mined just across the state line in California, where it contains 80 percent or more *clinoptilolite*. (Papke, 1972a; Holmes, 1994)

PERSHING COUNTY With *erionite* and minor *phillipsite* and mordenite in the large Jersey Valley deposits in the Jersey district, from which zeolite was mined in the late 1960s and early 1970s. (Deffeyes, 1959; Papke, 1972a)

As an alteration mineral with barrerite in rhyolite at an unspecified locality in the Seven Troughs district. (D. Hudson, w.c., 1999)

With *ferrierite* and mordenite in a large deposit 17 km northwest of Lovelock in the Lovelock district. (Papke, 1972a; Rice et al., 1992)

STOREY COUNTY With *heulandite* in argillized volcanic rocks in the Comstock district. (Whitebread, 1976)

WASHOE COUNTY Reported from a deposit in Hungry Valley about 38 km north of Reno near State Highway 445. (Papke, 1972a)

As crystals to 2 mm in veins and coatings on boulders in colluvium along West 5th Street in Reno. (R. Monnar, o.c., 2002)

CLINOSAFFLORITE

(Co,Fe,Ni)As$_2$

Clinosafflorite, which is dimorphous with safflorite, is a rare mineral of hydrothermal Co-Ni ores. The Humboldt County locality is unusual because Co-Ni ore is not known to be present in the area.

CHURCHILL COUNTY Identified by SEM/EDX and XRD in a sample with annabergite, rammelsbergite, and millerite from the Lovelock Mine in the Table Mountain district. (M. Coolbaugh, o.c., 2001)

HUMBOLDT COUNTY Identified by XRD in a specimen from the London Mine in the National district. (CSM)

CLINOTYROLITE

Ca$_2$Cu$_9^{2+}$[(As,S)O$_4$]$_4$(O,OH)$_{10}$•10H$_2$O

A rare secondary mineral described from a locality in China.

CHURCHILL COUNTY Reported as yellow to yellowish-green microcrystalline coatings on quartz-rich matrix and as lustrous greenish crystal aggregates to 2 mm at the Lovelock (Bolivia)

Mine in the Table Mountain district. (FUMT; HUB; M. Jensen, w.c., 1999)

CLINOZOISITE

Ca$_2$Al$_3$(SiO$_4$)$_3$(OH)

A widespread epidote-group mineral that is dimorphous with zoisite. It occurs in low- to medium-grade regionally metamorphosed rocks and in contact-metamorphosed calcium-rich rocks.

PERSHING COUNTY As smooth lime-green coatings on fault slickensides in the lower adit of the Majuba Hill Mine in the Antelope district. (Jensen, 1993b)

Intergrown with grossular, epidote, and quartz in skarn at the Nightingale and other mines in the Nightingale district; also found as glassy greenish-gray crystals to 2 cm on and in quartz crystals. (AMNH; LANH; M. Rogers, o.c., 1999; Crowley, 2000; WMK; F. Cureton, o.c., 2000)

WASHOE COUNTY Reported as acicular crystals in calcite in vesicles in andesite in the Olinghouse district. (M. Jensen, w.c., 1999)

A product of propylitic alteration in the Peavine district. (Hudson, 1977)

Reported from an unnamed occurrence at Pyramid Lake, possibly the Nightingale occurrence listed above under Pershing County. (AMNH)

CLINTONITE

CaMg$_2$AlAl$_3$SiO$_{10}$(OH)$_2$

A mica-group mineral that occurs in *chlorite* schist, metasomatically altered limestone, and skarn.

DOUGLAS COUNTY Reported in a specimen from an unknown locality. (RMO)

LYON COUNTY Reported from Ludwig. (Anthony et al., 1995)

COBALTITE

CoAsS

A mineral of high-temperature hydrothermal deposits and skarn deposits; generally occurs with Ni minerals and other cobalt minerals.

CHURCHILL COUNTY As grains to 1 mm with siegenite, hemimorphite, köttigite, and austinite from an unnamed prospect near the southwest end of the Clan Alpine Mountains on the east side of Stingaree Valley. (M. Jensen, w.c., 1999)

COERULEOLACTITE

$(Ca,Cu^{2+})Al_6(PO_4)_4(OH)_8 \cdot 4-5H_2O$

A relatively rare member of the turquoise group that occurs in some turquoise deposits.

NYE COUNTY From the Royal Blue Mine, Royston district, Sec. 3, T5N, R40E, with turquoise and other phosphates; specimens in museums are incorrectly credited to Lyon County. (Jenkins, 1981; MV; AM; EMP; M. Jensen, w.c., 1999)

COFFINITE

$U(SiO_4)_{1-x}(OH)_{4x}$

Coffinite is a moderately common mineral in sandstone-type and hydrothermal vein uranium deposits.

HUMBOLDT COUNTY Reported with uraninite at the Moonlight Uranium Mine in the Disaster district. (Sharp, 1955; Williams, 1980)

LANDER COUNTY Reported in veins with uraninite, sulfide minerals, and secondary uranium minerals at the Apex (Rundberg) Uranium Mine in the Reese River district. (Garside, 1973)

COHENITE

$(Fe,Ni,Co)_3C$

A mineral found in shocked iron meteorites.

CLARK COUNTY As rounded, branched crystals with kamacite, taenite, troilite, graphite, and schreibersite in the Las Vegas meteorite, which was found about 30 miles northeast of Las Vegas and may be a piece of the Canyon Diablo, Arizona meteorite. (Buchwald, 1975)

COLEMANITE

$Ca_2B_6O_{11} \cdot 5H_2O$

The borate colemanite occurs at many localities in California, where it was first described. It is found with borax, ulexite, and other evaporite minerals; in some deposits it is an important commercial mineral.

CLARK COUNTY Reported in specimens from Las Vegas; probably from the Anniversary Mine (see below). (WMK)

Occurs at the Anniversary Mine in the Muddy

Mountains district in a large deposit from which about 200,000 tons averaging 20 percent B_2O_3 were mined in the 1920s. It is also in smaller deposits in the White Basin. It occurs at both sites as irregular layers and nodules interbedded with marl, dolomitic limestone, algal limestone, and gypsum of the Horse Spring Formation. (Noble, 1923; Castor, 1993)

ESMERALDA COUNTY Tentatively identified with ulexite in the Columbus Marsh district. (UCB)

COLLINSITE

$Ca_2(Mg,Fe^{2+})(PO_4)_2 \cdot 2H_2O$

A rare mineral found in altered phosphate nodules in a pegmatite in South Dakota.

HUMBOLDT COUNTY As elongate, bipyramidal, colorless microcrystals with hisingerite, englishite, curetonite, and other minerals at the Redhouse Barite Mine in the Potosi district. (F. Cureton, w.c., 1999; S. Sears, w.c., 1999)

COLORADOITE

HgTe

Coloradoite, a relatively uncommon mineral, occurs in hydrothermal tellurium-bearing precious-metal deposits.

HUMBOLDT COUNTY As minute grains with realgar at the Getchell and Twin Creeks Mines in the Potosi district. (Dunning, 1988; Simon et al., 1999)

NYE COUNTY As irregular rounded inclusions in an unnamed sulfosalt mineral at the Outlaw Mine in the Round Mountain district. (Foord et al., 1988; Dunning et al., 1991)

WASHOE COUNTY Reported with petzite in gold-rich quartz veins at the Gus Schave property in the Olinghouse district. (Overton, 1947)

COLUMBITE

$(Fe,Mn,Mg)(Nb,Ta)_2O_6$ (general)

A mineral group name, the former species *columbite* has been subdivided into four minerals, all of which are most common in granite pegmatite. The species has not been identified for the following occurrences.

CLARK COUNTY Reported in a specimen from the Superior-Nevada Mine in the Crescent district. (LANH)

Reported with *allanite* in pegmatites near schist-granite contacts in the Gold Butte district; specifically noted to occur about 3000 m south of Jumbo Peak. (Volborth, 1962b; Garside, 1973)

ELKO COUNTY Reported as *columbite-tantalite* with uraninite, beryl, and *garnet* in pegmatite in the Gilbert Canyon district. (Olson and Hinrichs, 1960; Garside, 1973)

Reported in a specimen from the M. W. Young property in Ruby Valley. (WMK)

Reported as *columbite-tantalite* with quartz, beryl, and other minerals in pegmatite dikes in the Dawley Canyon area of the Valley View district. (Olson and Hinrichs, 1960; Garside, 1973)

ESMERALDA COUNTY With xenotime, euxenite, *monazite,* and other minerals in gold-bearing Tule Canyon placer gravels in the Tokop district. (Garside, 1973)

COLUSITE

$Cu_{26}V_2(As,Sn,Sb)_6S_{32}$

Colusite is an uncommon mineral in copper-bearing hydrothermal deposits with other sulfosalt and sulfide minerals.

ESMERALDA COUNTY Reported in vanadium-bearing ore from an unspecified locality in the Goldfield district. (Levy, 1968)

LANDER COUNTY Reported with rhodostannite at the Lucky Rocks claims in the Lewis district. (J. McGlasson, w.c., 2001)

CONICHALCITE

$CaCu^{2+}(AsO_4)(OH)$

Conichalcite, a relatively common copper arsenate mineral, occurs in the oxidized zone of hydrothermal deposits.

CHURCHILL COUNTY Reported in specimens from Fireball Ridge in the Truckee district. (R. Thomssen, w.c., 1999)

CLARK COUNTY As light-green botryoidal balls to 1.0 mm composed of acicular crystals to 0.3 mm wide on fracture surfaces at the Singer Mine in the Goodsprings district. (Palache et al., 1951; Adams, 1998)

ESMERALDA COUNTY As apple-green stains and crusts on quartz at unnamed prospects north of Columbus. (M. Jensen, w.c., 1999)

As apple-green crusts and spherules with *agardite* at the Big 3 Mine in the Railroad Springs district. (M. Jensen, w.c., 1999)

Reported in a specimen from the Broken Toe Mine in the Rock Hill district. (R. Thomssen, w.c., 1999)

As microcrystalline, apple-green hemispheres in the Weepah Mine open pit in the Weepah district. (S. Pullman, w.c., 1999; M. Jensen, w.c., 1999)

EUREKA COUNTY As green crystalline spherules to 1 mm in the Genesis pit in the Lynn district. (M. Jensen, w.c., 1999)

As pale-green radiating needles to 0.1 mm with malachite, barite, and carbonate-fluorapatite in iron-stained jasperoid at the Gold Quarry Mine in the Maggie Creek district; also at the Copper King Mine. (Jensen et al., 1995; MV)

HUMBOLDT COUNTY As yellow-green spheroids at the Silver Coin Mine in the Iron Point district. (Thomssen, 1998; J. McGlasson, w.c., 2001)

In green coatings at the Twin Creeks Mine in the Potosi district. (M. Jensen, w.c., 1999)

LINCOLN COUNTY With chrysocolla and tenorite at the Bristol Silver Mine in the Bristol district; also with pyromorphite and as pale-green clusters to 1 mm on jarosite. (Palache et al., 1951; T. Potucek, w.c., 1999; S. White, o.c., 2000; GCF)

Reported in a specimen from the Silver King Mine in the Silver King district. (R. Thomssen, w.c., 1999)

LYON COUNTY Reported at the Empire-Nevada Mine in the Yerington district. (Palache et al., 1951)

MINERAL COUNTY As apple-green spherules to 1 mm with adamite in the Simon Mine in the Bell district. (M. Jensen, w.c., 1999; S. White, w.c., 1999)

As yellowish-green blocky crystals on quartz at the Potosi Mine in the Candelaria district; also with annabergite in crusts in the Candelaria Mine open pit. (Jenkins, 1981; M. Jensen, w.c., 1999)

Reported from the Copper Queen and Calavada Mines in the Santa Fe district; also as bright apple-green coatings at an unnamed mine 3 miles due west of Calavada Summit. (R. Thomssen, w.c., 1999; S. Pullman, w.c., 1999; M. Jensen, w.c., 1999)

Brilliant green crystal aggregates are found on chrysocolla at the Bataan Mine in the Garfield district. (Jenkins, 1981; S. White, w.c., 1999)

NYE COUNTY As individual olive-green spherules to 2 mm in quartz vugs at the Illinois Mine in the Lodi district; also as olive to grass-green mammillary crusts in the San Rafael and Hasbrouck Mines. (Jenkins, 1981)

In breccia with chrysocolla and malachite at the Wagner Mine and prospects in the Wagner district. (Tingley et al., 1998a)

PERSHING COUNTY As light-green microcrystalline coatings on fractures in quartz at the Majuba Hill Mine in the Antelope district. (Jensen, 1985)

In apple-green spherules to 1 mm with bindheimite at an unnamed mine near the Green Antimony Mine in the Wild Horse district. (M. Jensen, w.c., 1999)

WASHOE COUNTY Reported from the Burrus Mine in the Pyramid district. (R. Thomssen, w.c., 1999)

WHITE PINE COUNTY In oxidized veins with a large number of other minerals in the Taylor district. (Lovering and Heyl, 1974)

CONNELLITE
$Cu_{19}^{2+}Cl_4(SO_4)(OH)_{32} \bullet 3H_2O$

A rare oxidation product that forms in copper-bearing deposits.

CHURCHILL COUNTY Reported from the Lovelock Mine in the Table Mountain district. (R. Thomssen, w.c., 1999)

ESMERALDA COUNTY Reported from an unspecified locality in the Candelaria district. (S. Pullman, w.c., 1999)

Reported from the Broken Toe Mine in the Rock Hill district. (R. Thomssen, w.c., 1999)

Reported in specimens as 0.7 mm blue sprays of radiating acicular crystals from the 3 Metals Mine in the Weepah district. (R. Thomssen, w.c., 1999; B. Tunner, o.c., 2000; GCF)

Identified by S. Williams in a specimen from a locality in the Monte Cristo Range 4-1/2 miles north of Coaldale Junction. (R. Thomssen, w.c., 1999)

PERSHING COUNTY As 0.1 mm brilliant blue acicular microsprays in altered sulfide pods at the Majuba Hill Mine, middle adit in the Antelope district. (Jensen, 1985)

COPIAPITE
$Fe^{2+}Fe_4^{3+}(SO_4)_6(OH)_2 \bullet 20H_2O$

Copiapite is a relatively common secondary mineral that occurs in small amounts with melanterite and other sulfates in many ore deposits where it has resulted from the oxidation of pyrite.

CLARK COUNTY Identified in a specimen from an unspecified locality near Las Vegas. (WMK)

ESMERALDA COUNTY Reported from an unspecified locality in the Coaldale district. (WMK)

Reported from near the Sweeney lease in the Goldfield district. (Ashley and Albers, 1975)

HUMBOLDT COUNTY As yellowish crusts formed by rapid oxidation of melanterite on the 700-foot level of the Cordero Mine in the Opalite district. (Fisk, 1968; WMK)

NYE COUNTY As crystals to 0.2 mm on vug surfaces in oxidized silver ore at the upper Magnolia workings in the Morey Mine in the Morey district. (Jenkins, 1981)

STOREY COUNTY As an efflorescent mineral in mines on the Comstock Lode in the Comstock district. (Palache et al., 1951)

WASHOE COUNTY Reported from oxidized portions of lead-silver orebodies at the Union (Commonwealth) Mine in the Galena district. (Overton, 1947)

Reported at Steamboat Hot Springs. (Palache et al., 1951)

COPPER
Cu

Native copper is a relatively common mineral in oxidized copper-bearing deposits. Good specimens have come from the Robinson, Battle Mountain, Yerington, and Mountain City districts in Nevada.

CLARK COUNTY Sparse in some ore at the Boss and Columbia Mines in the Goodsprings district. (Hewett, 1931)

DOUGLAS COUNTY Reported at the Ruby Hill Mine in the Gardnerville district. (S. Pullman, w.c., 1999)

ELKO COUNTY As rare seams in the oxidation zone with bornite and copper carbonates in copper-rich quartz-calcite veins in skarn at the Brooklyn Mine in the Contact district. (Schrader, 1935b; Schilling, 1979)

As crystalline dendrites to 12 mm with malachite from the Victoria Mine in the Dolly Varden district. (GCF)

Small beads and nuggets in a copper- and gold-bearing quartz vein in granodiorite at the Austean prospect in the Elk Mountain district. (Schrader, 1912)

As crystallized fronds and arborescent masses with cuprite, malachite, and azurite from the oxidized zone of the massive sulfide orebody at the Rio Tinto Mine in the Mountain City district; also found in other mines in the district. (Coats and Stephens, 1968; Emmons, 1910; Jenkins, 1981; WMK)

ESMERALDA COUNTY Reported from surface pannings in the Goldfield district. (Schrader et al., 1917)

In sheets and small nodules with malachite and azurite at an unspecified locality in the Montezuma district. (Lincoln, 1923; JHS)

EUREKA COUNTY Reported at the Lord Byron Mine in the Eureka district. (BMC)

As brown grains and wires to 2 mm on fractures in the upper plate of the Roberts Mountains thrust at the Goldstrike Mine in the Lynn district. (GCF)

In bright grains to 60 microns with spongy cuprite and other minerals lining fractures at the Gold Quarry Mine in the Maggie Creek district; also reported from the Mike deposit. (Jensen et al., 1995; GCF)

HUMBOLDT COUNTY As small dendrites to 2 mm with carbonate-fluorapatite crystals and siderite at the Lone Tree Mine in the Buffalo Mountain district. (M. Jensen, w.c., 1999)

Reported at the Getchell Mine in the Potosi district. (D. Tretbar, o.c., 1999)

Reported from an unspecified location in the Red Butte district. (Ransome, 1909b)

As blebs to 1 cm in ore at the Roberts Copper Mine in the Varyville district. (R. Walstrom, w.c., 1999)

LANDER COUNTY With other copper minerals at Copper Canyon and Copper Basin Mines in the Battle Mountain district. A piece reportedly originally weighing more than 50 pounds was recovered from one site. (Roberts and Arnold, 1965; JHS)

Reported at the Morning Star and Pittsburg Mines in the Lewis district. (Emmons, 1910)

LYON COUNTY In arborescent masses to 3 cm at the Empire-Nevada Mine in the Yerington district; also at the Ludwig Mine on iron rails. (Schrader et al., 1917; Melhase, 1935e; LANH; M. Jensen, w.c., 1999; WMK)

MINERAL COUNTY Reported with malachite, azurite, antlerite, and cuprite at the Northern Light and Republic Mines in the Mountain View district. (Ross, 1961; WMK)

NYE COUNTY Rare as small specks with hübnerite and *monazite* in placer gravels at the Red Top claim in the Round Mountain district. (Ferguson, 1922; Garside, 1973)

Reported in a specimen from the Tonopah district. (WMK)

Reported in a specimen from the Wagner claims in the Wagner district. (LANH)

PERSHING COUNTY Rare with cuprite in the Copper stope at the Majuba Hill Mine in the Antelope district. (Jensen, 1985)

Reported from the Iron Hat district. (Von Bargen, 1999)

STOREY COUNTY In a specimen from the Comstock district as branching wires to 11 mm coated with malachite. (J. McGlasson, o.c., 2000; GCF)

WASHOE COUNTY With malachite, cuprite, and tenorite in laumontite veins on the north side of the Truckee River Canyon near the Painted Rock interchange on I-80 in the Olinghouse district. (Rose, 1969)

In oxidized veins with malachite and cuprite at mines in the Pyramid district; also as clumps and wires to 1 cm, possibly formed after mining, on the dump for the lower adit of the Jones-Kincaid Mine. (Overton, 1947; Bonham and Papke, 1969; GCF)

WHITE PINE COUNTY Reported in a specimen from the Bald Mountain district. (WMK)

Reported from many localities in the Robinson district with other copper minerals, particularly from the Alpha and Richard Mines in the Kimberly area, which produced arborescent specimens coated with crystalline cuprite. (Spencer et al., 1917; Melhase, 1934d; WMK)

COQUIMBITE
$Fe_2^{3+}(SO_4)_3 \cdot 9H_2O$

Coquimbite, a relatively rare mineral that is soluble in water, occurs in ore deposits where pyrite has undergone oxidation; it also occurs in hot spring deposits.

ESMERALDA COUNTY Reported in a specimen from the Coaldale district. (WMK)

Reported at the Sweeney lease in the Goldfield district. (Ashley and Albers, 1975)

Reported in a specimen with melanterite and pyrite from the Lida district. (WMK)

WASHOE COUNTY As white to lavender crystalline masses with melanterite and bonattite from Steamboat Hot Springs in the Steamboat Springs district. (Jenkins, 1981)

CORDEROITE

$Hg_3S_2Cl_2$

A Nevada type mineral, corderoite is a secondary mineral that occurs with cinnabar in hydrothermal deposits; at the type locality it was an ore mineral found in large quantities.

EUREKA COUNTY As globular coatings in brecciated jasperoid with phosphate minerals such as variscite at the Gold Quarry Mine in the Maggie Creek district; also intergrown with tiemannite as black spongy masses to 2 mm. (Jensen et al., 1995)

HUMBOLDT COUNTY Type locality at the Cordero Mine in the Opalite district, Sec. 33, R37E, T47N, in mercury ore in altered lacustrine beds and volcanic rocks with cinnabar, montmorillonite, quartz, cristobalite, and other minerals; reportedly after cinnabar as pale-pink to orange coatings to 2 mm that turn black on exposure to sunlight; also common at the McDermitt Mine. (Foord et al., 1974; Roper, 1976)

NYE COUNTY In ore as dodecahedrons to 5 microns with an unnamed Sb-Bi-Fe oxide mineral at the Paradise Peak Mine in the Fairplay district. (Rytuba and Heropoulos, 1992)

CORDIERITE

$Mg_2Al_4Si_5O_{18}$

A widely distributed mineral that occurs in thermally metamorphosed aluminum-rich sedimentary rocks; also found in high-grade gneiss, mafic igneous rocks, granite, and granite pegmatite.

CLARK COUNTY With almandine, sillimanite, and other minerals in gneiss in the Gold Butte district. (Volborth, 1962b)

HUMBOLDT COUNTY With biotite and andalusite in contact-metamorphosed pelitic rocks in the Osgood Mountains. (Hotz and Willden, 1964)

In hornfels with andalusite and *feldspar* in the Bloody Run Hills. (WMK)

LINCOLN COUNTY As poikilitic patches to 0.5 mm in metamorphosed shale on the south slope of McCullough Hill, Highland Range. (Westgate and Knopf, 1932)

MINERA COUNTY With biotite in altered diorite at the Green Talc Mine in the Fitting district. (Hudson, 1983)

NYE COUNTY In hornfels at the Victory tungsten deposit in the Lodi district. (Humphrey and Wyatt, 1958)

STOREY COUNTY Reported as porphyroblasts in hornfels at the South End shaft in the Comstock district. (HMM; WMK)

CORKITE

$PbFe_3^{3+}(PO_4)(SO_4)(OH)_6$

An uncommon secondary mineral in oxidized base-metal deposits.

ELKO COUNTY Rare coating cavities in ore rich in iron oxide minerals at the Killie Mine in the Spruce Mountain district. (Dunning and Cooper, 1987)

HUMBOLDT COUNTY As brown microcrystals at the Silver Coin Mine in the Iron Point district. (Thomssen, 1998; M. Jensen, w.c., 1999; J. McGlasson, w.c., 2001)

LYON COUNTY As microcrystals in shrinkage cracks in nodules and as rough crystals on the outside of some nodules in jasperoid on Boulder Hill in the Wellington district. (M. Rogers, o.c., 1999)

NY COUNTY Identified by SEM/EDX and habit as minute pseudocubic crystals lining cavities in quartz at the Outlaw Mine in the Round Mountain district. (Dunning et al., 1991)

CORNETITE

$Cu_3^{2+}(PO_4)(OH)_3$

Cornetite is a rare mineral that occurs in the oxidation zones of copper deposits. Specimens from the Yerington district are found in many mineral collections.

LYON COUNTY As single, dark-blue, prismatic

and tabular crystals and as crusts in oxidized copper ore with pseudomalachite at the Blue Jay and Empire-Nevada Mines in the Yerington district; also reported at the Douglas Hill Mine. (Palache et al., 1951; AMNH; S. White, w.c., 1999: R. Thomssen, w.c., 1999)

PERSHING COUNTY Reported at the Majuba Hill Mine in the Antelope district. (MacKenzie and Bookstrom, 1976)

WASHOE COUNTY In quartz and wall rock with bornite, malachite, and rare chalcocite at the Red Metal Mine in the Peavine district. (Overton, 1947)

CORNUBITE
$Cu_5^{2+}(AsO_4)_2(OH)_4$

Cornubite is a rare secondary copper mineral that has been reported from only a few localities. Specimens from Majuba Hill are considered to be the world's best.

LYON COUNTY Reported with malachite at the Empire-Nevada Mine in the Yerington district. (TUB)

PERSHING COUNTY As light-green botryoidal crusts on rhyolite or as balls and crusts on cornwallite at the Majuba Hill Mine in the Antelope district; also as coatings on clinoclase and pseudomorphs after parnauite. (Jensen, 1985)

CORNWALLITE
$Cu_5^{2+}(AsO_4)_2(OH)_4$

A rare mineral in oxidized copper deposits.

CHURCHILL COUNTY In specimens from the Fireball Mine in the Truckee district. (R. Thomssen, w.c., 1999)

ESMERALDA COUNTY As dark-green microcrystalline patches with olivenite at the 3 Metals Mine in the Weepah district. (B. Runner, o.c., 2000)

LANDER COUNTY As deep-green botryoidal crusts from the Little Gem Mine, Cortez district. (Excalibur Mineral Co., 2001c)

MINERAL COUNTY As bright green crystalline hemispheres to 0.5 mm in vugs in quartz at an unnamed prospect near Rhyolite Pass in the Fitting district. (LANH; M. Jensen, w.c., 1999)

PERSHING COUNTY Dark-olive to forest-green

botryoidal crusts on rhyolite with cornubite at the Majuba Hill Mine in the Antelope district. (MacKenzie and Bookstrom, 1976; Jensen, 1985)

WASHOE COUNTY With enargite, malachite, chrysocolla, and olivenite as grass-green spherules and globular aggregates to 1 mm in vuggy, iron oxide–stained, ash-flow tuff at the Burrus Mine in the Pyramid district. (Jensen, 1994)

CORONADITE
$Pb(Mn^{4+},Mn^{2+})_8O_{16}$

Coronadite is a widespread mineral that occurs in bedded manganese deposits, veins, and hot spring deposits as a hydrothermal phase; it also occurs as an oxidation product of primary manganese minerals.

CLARK COUNTY In layered manganese oxide deposits in tuffs and sedimentary rocks with pyrolusite, cryptomelane, hollandite, *phillipsite,* and celadonite at the Three Kids Mine in the Las Vegas district. (Hewett and Fleischer, 1960; Van Gilder, 1963; Hewett, 1971)

LYON COUNTY As thin black coatings with hemimorphite and fluorite at the Boulder Hill Mine in the Wellington district, Sec. 30, T10N, R24E. (R. Walstrom, w.c., 1999)

MINERAL COUNTY Locally abundant in oxidized rock in the Candelaria Mine open pit in the Candelaria district. (Chavez and Purusotam, 1988)

NYE COUNTY In massive black coatings with duftite and hemimorphite at the Hasbrouck shaft in the Lodi district. (M. Jensen, w.c., 1999)

CORUNDUM
Al_2O_3

Corundum is common mineral in aluminum-rich and silica-poor rocks such as syenite, quartz-free pegmatite, high-grade aluminous metamorphic rocks, and in some hydrothermal alteration zones. Good crystals have come from localities in Douglas and Mineral Counties.

CLARK COUNTY Reported as a minor mineral in Precambrian gneiss in the Bunkerville district in the Virgin Mountains. (Beal, 1965)

DOUGLAS COUNTY As blue crystals and crystal aggregates with andalusite, pyrophyllite, diaspore, and rutile at the Blue Metal (Blue Danube)

Mine in the Buckskin district. (Gianella, 1941; Binyon, 1946a; Hudson, 1983; WMK)

HUMBOLDT COUNTY Reported in a specimen from the Donnelly property in the Donnelly district. (WMK)

LINCOLN COUNTY As anhedral metacrysts to 0.3 mm with *tourmaline* in hornfels in the Highland district on the west spur of Blind Mountain in the Highland Range. (Westgate and Knopf, 1932)

MINERAL COUNTY Dark blue, generally granular, and commonly in veins with andalusite, pyrophyllite, and other minerals at the Dover and Green Talc Mines in the Fitting district; also reported in a sample from the Bismark prospect. (Vanderburg, 1937a; Klinger, 1952; Hudson, 1983; JHS)

As crystals or crystalline aggregates in specimens at the Deep and Kenjay properties and Hawthorne Ferretti claim in the Pamlico district. (WMK; JHS)

COSALITE

$Pb_2Bi_2S_5$

Cosalite is an uncommon mineral that has been found in hydrothermal vein deposits and in tungsten- and bismuth-rich skarn.

LINCOLN COUNTY In small amounts with galenobismutite, native bismuth, and bismuthinite at the Free tunnel in the Tem Piute district. (Buseck, 1967; JHS)

WASHOE COUNTY Reportedly silver-bearing in quartz veins that cut scheelite-bearing skarn at the Crosby Mine in the Nightingale district; also reported with sphalerite and marcasite from the White Caps Tungsten Mine. (Bonham and Papke, 1969; M. Jensen, w.c., 1999)

COULSONITE

$Fe^{2+}V_2^{3+}O_4$

Coulsonite is a very rare spinel-group species that occurs with magnetite at only a few localities worldwide, including occurrences in iron ore from India and in a mantle xenolith from basalt in China. Titanium-bearing coulsonite has been identified in Australia.

CHURCHILL COUNTY As bluish-gray exsolution lamellae in magnetite and disseminated subhedral grains with *sphene, apatite,* and chlorine-rich

scapolite (meionite) in metamorphosed igneous rocks with *chlorite, hornblende, pyroxene,* and muscovite at the Buena Vista Mine in the Mineral Basin district about 20 miles southeast of Lovelock; also noted as crude black octahedra. (Radtke, 1962; Perry, 1963; EMP; Minerals Unlimited specimens in AMNH)

COVELLITE

CuS

Covellite is a common copper sulfide species that occurs both as a supergene secondary phase and as a low-temperature primary phase.

CHURCHILL COUNTY Reported at the Dromedary Hump Mine and Consadao shaft in the Fairview district. (Schrader, 1947)

CLARK COUNTY Reported in bitumen from the Boss Mine in the Goodsprings district. (Kepper, this volume)

ELKO COUNTY As minute blebs after chalcopyrite at the Rossi Mine in the Bootstrap district. (Snyder, 1989)

Rare in quartz-calcite veins at the Helen B. Smith Mine in the Contact district. (Schrader, 1935b; Schilling, 1979)

Common as small irregular grains to 0.03 mm with iron sulfide and sphalerite at the Hollister deposit in the Ivanhoe district. (Deng, 1991)

Common after pyrite and chalcopyrite immediately below the groundwater table at the Rio Tinto Mine in the Mountain City district. (Coats and Stephens, 1968)

Common as tiny grains with chalcopyrite and galena in the Railroad district in the Pinon Range. (Ketner and Smith, 1963)

With chalcopyrite and galena, Killie Mine in the Spruce Mountain district. (Dunning and Cooper, 1987)

ESMERALDA COUNTY With chalcopyrite and other minerals at the Carrie Mine in the Gilbert district. (Ferguson, 1928)

With digenite after bornite in the Klondyke district. (Chipp, 1969)

HUMBOLDT COUNTY With other sulfides at the Lone Tree Mine in the Buffalo Mountain district. (Bloomstein et al., 1993)

A purple secondary mineral at the Silver Coin Mine in the Iron Point district. (Thomssen, 1998; J. McGlasson, w.c., 2001)

Reported in quartz veins on Auto Hill in the National district. (Lindgren, 1915)

Reported in ore from the Red Butte district. (Ransome, 1909b; Melhase, 1934c)

With chalcopyrite at the Silver Hill Mine in the Dutch Flat district. (Willden, 1964)

LANDER COUNTY A minor mineral in ore from the Copper Canyon, Little Giant, and Sweet Marie Mines in the Battle Mountain district. (Roberts and Arnold, 1965; Theodore and Blake, 1978; Schilling, 1979)

In surficial material at various mines in the Reese River district. (Ross, 1925, 1953)

Reported from the Gray Eagle Mine in the Shoshone Range. (R. Thomssen, w.c., 1999)

LINCOLN COUNTY With bornite and chalcopyrite at the Pennsylvania Mine in the Pennsylvania district. (Tschanz and Pampeyan, 1970)

LYON COUNTY In supergene-enriched ore after chalcopyrite in the Wilson district. (Dircksen, 1975)

Widespread in small amounts in primary ore from mines at Ludwig in the Yerington district. (Knopf, 1918b; Melhase, 1935e)

MINERAL COUNTY As crusts in quartz vugs at the Broken Hills Mine in the Broken Hills district. (Jenkins, 1981)

In relatively unoxidized rock in the Candelaria Mine open pit in the Candelaria district. (Chavez and Purusotam, 1988)

Reported from the Rawhide district. (Schrader, 1947)

NYE COUNTY Partially replacing stromeyerite and chalcopyrite in gold ore at the Bullfrog and Original Bullfrog Mines in the Bullfrog district. (Castor and Sjoberg, 1993)

With chalcocite coating chalcopyrite at the Fairday Mine in the Cactus Springs district on the Nellis Air Force Range. (Tingley et al., 1998a)

In quartz veins with pyrite, chalcopyrite, and molybdenite at the Outlaw Mine in the Round Mountain district. (Melhase, 1935d; JHS)

With bornite and acanthite after chalcopyrite at the Tonopah-Belmont Mine in the Tonopah district. (Bastin and Laney, 1918)

As microscopic grains with luzonite, tetrahedrite, and other sulfide minerals at the Golden Chariot Mine in the Jamestown district on the Nellis Air Force Range. (Tingley et al., 1998a)

Reported from the Lodi and Union districts. (Jenkins, 1981; JHS)

PERSHING COUNTY As indigo-blue plates in tiny balls with chalcopyrite at the Majuba Hill Mine in the Antelope district. (Smith and Gianella, 1942; Jensen, 1985)

In quartz veins with pyrite and other minerals at the Arizona Mine in the Buena Vista district. (Cameron, 1939)

Reported in the Iron Hat district (Von Bargen, 1999)

In quartz-*tourmaline* veins at the Bonanza King Mine in the Spring Valley district. (Campbell, 1939)

STOREY COUNTY After chalcopyrite in shallow bonanza ore in the Andes and Hale and Norcross Mines in the Comstock district; also reported as rare in drill cuttings from the Flowery lode. (Bastin, 1922; Koschmann and Bergendahl, 1968; Schilling, 1990)

WASHOE COUNTY A minor product of supergene enrichment and as bluish iridescent tarnishes and very thin coatings on enargite at the Burrus Mine in the Pyramid district. (Wallace, 1980; Jensen, 1994)

WHITE PINE COUNTY With chalcopyrite and chalcocite at the Bald Mountain Mine in the Bald Mountain district. (Hitchborn et al., 1996)

Coating pyrite and chalcopyrite in some disseminated ore in the Robinson district. (Spencer et al., 1917)

In the supergene assemblage in the Caroline tunnel in the Ward district. (Williams, 1964)

CRANDALLITE
$CaAl_3(PO_4)_2(OH)_5 \cdot H_2O$

Crandallite is a relatively common mineral in phosphate deposits, turquoise and variscite deposits, and in some pegmatites.

ELKO COUNTY Identified in thin section at the Rain Mine in the Carlin district. (D. Heitt, o.c., 1991)

Reported from the Jerritt Canyon Mines in the Independence Mountains district. (Hofstra, 1994)

ESMERALDA COUNTY As white microcrystals with turquoise and variscite in the Candelaria district. (Palache et al., 1951; M. Jensen, w.c., 1999)

EUREKA COUNTY In white balls to 4 mm at the

center of radiating crystal clusters of variscite at the Goldstrike Mine in the Lynn district. (GCF)

As white to yellow-orange concentric bands in fracture linings to 1 cm in silicified siltstone at the Gold Quarry Mine in the Maggie Creek district. (Jensen et al., 1995)

HUMBOLDT COUNTY As spherulitic aggregates locally coated with ferrous sulfate at the Getchell Mine in the Potosi district. (S. Sears, w.c., 1999)

NYE COUNTY As white spherules to 1 mm in surface exposures near Warm Springs. (M. Jensen, w.c., 1999)

PERSHING COUNTY In white spherules and as flattened hexagonal crystals to 0.1 mm at the Willard Mine in the Willard district. (M. Jensen, w.c., 1999; Jensen and Leising, 2001)

CREASEYITE

$Pb_2Cu_2^{2+}Fe_2^{3+}Si_5O_{17} \cdot 6H_2O$

Creaseyite is a rare mineral that occurs in association with other secondary copper minerals in few base-metal deposits.

CLARK COUNTY Tentatively identified as pale-yellow-green sprays to 0.2 mm in vuggy quartz from an unnamed prospect about 9 miles northwest of Nelson. (R. Thomssen, w.c., 1999)

ESMERALDA COUNTY As crudely radial masses of minute, greenish-yellow crystals on iron-stained quartz at an unnamed prospect in the Gold Point district. (Jenkins, 1981; EMP; UAMM)

CREEDITE

$Ca_3Al_2(SO_4)(F,OH)_{10} \cdot 2H_2O$

Creedite is a relatively uncommon mineral found with other aluminofluoride minerals in the oxidation zone of base-metal or other deposits containing fluorine.

CHURCHILL COUNTY As stout crystals, sprays of colorless crystals in nodules, and ball-like crystal groups to 4 cm at an unnamed prospect in the Mopung Hills, West Humboldt Range, in the Lake district. (R. Walstrom, w.c., 1999; M. Jensen, w.c., 1999)

Colorless microcrystals in spherical clusters to 3 cm with gearksutite and fluorite at the Michigan prospect in the Mountain Wells district. (R. Walstrom, w.c., 1999; M. Jensen, w.c., 1999)

ESMERALDA COUNTY In nodules and small individual sprays of colorless crystals on drusy quartz with purple fluorite at the Broken Toe Mine in the Rock Hill district; at the Rock Hills mine as 0.1 mm glassy crystals in powdery white rock. (R. Walstrom, w.c., 1999; M. Jensen, w.c., 1999)

HUMBOLDT COUNTY As colorless microcrystals with brochantite at the Silver Coin Mine in the Iron Point district. (Thomssen, 1998; F. Cureton, w.c., 1999)

LYON COUNTY As small pale-violet crystals in a clayey white vein in a road cut below fluorite-bearing jasperoid at the Boulder Hill Mine in the Wellington district; also reported as microcrystals in nodules, colorless to pale-lavender crystals on fluorite, and as spherical clusters to 3 cm. (HMM; M. Rogers, o.c., 1999; R. Walstrom, w.c., 1999; M. Jensen, w.c., 1999)

NYE COUNTY In the southern Lodi (Granite Camp) district at an unnamed gold prospect as colorless, pink, and pale-buff, reticulated, glassy needles to 2 mm in fluorite-quartz veins with free gold; also reported as single crystals in clay and multiple crystals in crusts, drusy cavity linings, wartlike masses, and radiating aggregates. Some specimens are erroneously reported from Tonopah. (Foshag, 1932; Melhase, 1935d; Palache et al., 1951; Kral, 1951)

Nevada's and one of the world's best occurrence is lavender crystal groups and radiating clusters to 2 cm on rhyolite matrix at the Hall Molybdenum (Anaconda, Liberty) Mine in the San Antone district. (GCF; M. Jensen, w.c., 1999)

CRISTOBALITE

SiO_2

Cristobalite, a polymorph of quartz and tridymite, is a common mineral that forms in many environments; in Nevada it generally occurs in lithophysae or amygdules in volcanic rocks, as a constituent of diagenetic or hydrothermal alteration of volcanic rocks, and at hot springs.

EUREKA COUNTY White crystals to 1 mm lining amygdules in a large boulder of silicified rhyolite 2 miles east of Palisade in Palisade Canyon. (Dean Heitt o.c., 1999; GCF)

HUMBOLDT COUNTY A product of alteration of volcanic rocks adjacent to mercury and uranium deposits in the McDermitt caldera. (Rytuba and Glanzman, 1979)

A diagenetic mineral in bedded tuff and sedimentary rock in the Virgin Valley district. (Castor et al., 1996)

LANDER COUNTY Reported with cassiterite, hematite, and tridymite in veins in rhyolite in the Izenhood district. (Knopf, 1916b; Anonymous, 1938; Fries, 1942)

MINERAL COUNTY Reported in a specimen from Powell Canyon near Hawthorne in the Wassuk Range. (AMNH; HMM)

NYE COUNTY Ubiquitous in tuff above the water table at Yucca Mountain; also common in veins. (Bish and Vaniman, 1985; Carlos et al., 1995)

STOREY COUNTY In argillized volcanic rocks in the Virginia City area in the Comstock district. (Whitebread, 1976)

WASHOE COUNTY As pervasive replacement and crustiform, open-space filling in tuff at a uranium prospect pit southwest of the DeLongchamps Mine in the Pyramid district. (Castor et al., 1996)

Common in older sinter terraces and as an alteration product in granitic rock in the Steamboat Springs district. (D. Hudson, w.c., 1999)

WHITE PINE COUNTY In lithophysal cavities in rhyolite as small crystals with *garnet* at Garnet Hill 5 miles east of Ely and 2 miles north of Ruth in the Robinson district. (S. Kleine, o.c., 1999)

CRONSTEDTITE
$Fe_2^{2+}Fe^{3+}(SiFe^{3+})O_5(OH)_4$

A mineral in the kaolinite-*serpentine* group that is found in low-temperature hydrothermal veins.

NYE COUNTY Jet black brilliant vitreous crystals to 0.5 mm with siderite, sphalerite, and quartz in veins at the Cornucopia Mine; two polytypes determined (1H and 9R). (Frondel, 1962)

CRYOLITE
Na_3AlF_6

Cryolite, a rare mineral that was once mined from deposits in Greenland and Russia for use in aluminum processing, is most commonly found in granite pegmatite.

MINERAL COUNTY In a small late-stage breccia pipe with other Na-Li-Al-F minerals in an *amazonite*-topaz pegmatite at the Zapot claims in the Fitting district about 25 km northeast of Hawthorne. (Foord et al., 1998, 1999b)

CRYOLITHIONITE
$Na_3Li_3Al_2F_{12}$

A rare mineral found with cryolite in granite pegmatite.

MINERAL COUNTY With cryolite and other minerals in pegmatite at the Zapot claims in the Fitting district; described as a colorless mineral that fluoresces white to pale-cream in ultraviolet light and is phosphorescent. (Foord et al., 1998, 1999b)

CRYPTOMELANE
$K(Mn^{4+},Mn^{2+})_8O_{16}$

Cryptomelane is a widespread mineral in oxidized manganese deposits, but it is not always recognized; there are undoubtedly more localities in Nevada than listed below.

CARSON CITY Reported at the Dixon Mine in the Delaware district. (Hewett and Fleischer, 1960)

CLARK COUNTY With hollandite and other manganese oxide minerals at the Three Kids manganese deposit in the Las Vegas district. (Hewett and Fleischer, 1960)

HUMBOLDT COUNTY Black, rounded, stalactitic growths and masses to 4 cm at Hall Creek. (Weissman and Nikischer, 1999)

MINERAL COUNTY In silver-bearing shear zones with other iron and manganese oxides and hydroxides at the Candelaria Mine in the Candelaria district. (Moeller, 1988)

WHITE PINE COUNTY With aurorite, todorokite, native silver, and other minerals at the North Aurora, Hidden Treasure, and Ward Beecher Mines in the White Pine district. (Radtke et al., 1967)

CUBANITE
$CuFe_2S_3$

Cubanite, generally a high-temperature hydrothermal mineral, occurs in some porphyry copper deposits and more commonly in massive sulfide deposits, where it may be an important ore mineral.

LANDER COUNTY Tentatively identified in

quartz-pyrrhotite-chalcopyrite veins at the Copper Canyon porphyry deposit in the Battle Mountain district. (Theodore and Blake, 1978)

PERSHING COUNTY Identified in polished sections with chalcocite at the Majuba Hill Mine in the Antelope district. (Jensen, 1993b)

CUPRITE
$Cu_2^{1+}O$

Cuprite is a common mineral in the oxidation zones of many base-metal deposits that contain copper. Exceptional specimen material has come from the Robinson, Mountain City, and Yerington districts in Nevada.

CHURCHILL COUNTY Identified by XRD in a specimen from the Nickel Mine area, Table Mountain district. (S. Lutz, w.c., 2000)

CLARK COUNTY With other secondary copper minerals at the Key West Mine in the Bunkerville district. (Beal, 1965)

Reported at the Sundog Mine in the Crescent district. (Longwell et al., 1965)

Noted in surface ore in the Gold Butte district; specifically reported at the Lakeview and Tramp Mines. (Hill, 1916; Longwell et al., 1965)

Reported at the Rose and other mines in the Goodsprings district. (Hewett, 1931)

As earthy aggregates and coatings at an unspecified locality in the Searchlight district. (Callaghan, 1939; Gianella, 1941)

DOUGLAS COUNTY Abundant as irregular masses and stringers in ore at the Ruby Hill Mine in the Gardnerville district. (Hill, 1915)

ELKO COUNTY With malachite and other minerals in oxidized ore in the Contact district; specifically reported from the Rattler and Silver Circle Mines. (Purington, 1903; Bailey, 1903; Schrader, 1912, 1935; Schilling, 1979)

Common with chalcocite at the Rio Tinto and other mines in the Mountain City district; exceptional crystals were found in some of the mines. (Coats and Stephens, 1968; Jenkins, 1981)

In small quantities in the Railroad (Bullion) district. (Ketner and Smith, 1963; Palache et al., 1944)

ESMERALDA COUNTY Reported in a specimen from an unknown locality in the Coaldale district. (WMK)

With copper carbonates and tenorite in the Cuprite district. (Ball, 1907)

Reported as silver-bearing in a specimen with galena at the Geraldine Mine in the Lida district. (WMK)

EUREKA COUNTY Reported in a specimen from the west slope of the Sulphur Springs Range about 5 miles east of Alpha in the Alpha district. (WMK)

With malachite in a specimen from an unknown locality in the Antelope (Baldwin) district. (WMK)

Reported in a specimen with barite at the Carlin Mine in the Lynn district. (WMK)

As crystal druses in fractures and as small veinlets in silicified siltstone at the Copper King Mine in the Maggie Creek district; also common at the Gold Quarry Mine as deep purplish-red masses to 2 mm with barite, alunite, carbonate-fluorapatite, and other minerals in fractures in silicified rock. (Roberts et al., 1967b; Rota, 1991; Jensen et al., 1995)

With malachite, azurite, and chalcopyrite as breccia matrix at an unspecified locality in Sheep Creek Canyon in the Modarelli-Frenchie Creek district. (Muffler, 1964)

With galena, chrysocolla, and chalcopyrite at an unspecified locality in the Roberts district. (WMK)

In a specimen with *limonite* at the Onondaga Mine in the Safford district. (WMK)

HUMBOLDT COUNTY Rare at the Lone Tree Mine in the Buffalo Mountain district. (R. Braginton, o.c., 1995)

In several prospects in the Red Butte district with covellite, native copper, and other minerals. (Ransome, 1909b)

With malachite, azurite, and chrysocolla in the Cove Meadow deposit in the Varyville district. (Trengove, 1959)

LANDER COUNTY A constituent of oxidized enriched ore with malachite, copper, and azurite at the Copper Canyon, Copper Basin, and Copper Queen Mines in the Battle Mountain district. (Roberts and Arnold, 1965; Schilling, 1979; WMK)

LINCOLN COUNTY With malachite, atacamite, tenorite, and chrysocolla at the Ida May and Mayflower Mines in the Bristol district. (CSM; WMK)

LYON COUNTY As superb red cubes with tenorite and copper carbonates in the Empire-Nevada

Mine in the Yerington district; also reported from the Ludwig, Mason, and other mines. (Ransome, 1909e; Melhase, 1935e; UCB; JHS; WMK)

MINERAL COUNTY Reported at the Northern Light and Republic Mines in the Mountain View district. (Ross, 1961; WMK)

With other secondary copper minerals from the Champion, St. Patrick, Wall Street, and other mines in the Santa Fe district. (Hill, 1915; Ross, 1961; WMK)

In a specimen with chrysocolla and malachite from the Dunlap Mine in the Silver Star district. (WMK)

In specimens from unknown localities in the Buckley and Garfield districts. (WMK)

NYE COUNTY In small masses in the oxidized zone in the Tonopah district. (Melhase, 1935c)

PERSHING COUNTY Abundant as scattered grains, veinlets, and masses, and more rarely as crystals in rhyolite with chalcocite, chrysocolla, and malachite at the Majuba Hill Mine in the Antelope district. (Trites and Thurston, 1958; MacKenzie and Bookstrom, 1976; Jensen, 1985)

Reported in the Iron Hat district. (Von Bargen, 1999)

With tenorite in veinlets to 1 cm in the oxidized zone of the massive sulfide deposit at the Big Mike Copper Mine in the Tobin and Sonoma Range district. (Johnson, 1977)

WASHOE COUNTY With tenorite, malachite, and native copper in laumontite veins in the Truckee River Canyon north of Interstate 80 in the Olinghouse district. (Rose, 1969)

On mine dumps with native copper and malachite in the Pyramid district. (GCF)

WHITE PINE COUNTY Reported from an unspecified locality in the Bald Mountain district. (Hill, 1916)

Reported at the Grand Deposit Mine in the Muncy Creek district. (Gianella, 1941)

As magnificent red crystals with native copper in oxidized ore in the Kimberly area in the Robinson district, particularly at the Alpha and Richard Mines; also reported as bright red filiform *chalcotrichite* and as massive replacements of chalcocite. (Spencer et al., 1917; Melhase, 1934d; CSM; UCB; GCF; WMK)

Reported from the lead belt in the White Pine district. (Hose et al., 1976)

CUPROPAVONITE
$AgPbCu_2Bi_5S_{10}$

A rare mineral that occurs with other bismuth minerals in hydrothermal vein deposits.

LINCOLN COUNTY As silvery metallic grains with other sulfides in quartz at an unspecified locality in the Chief district. (AMNH; Jenkins, 1981)

Reported at the April Fool Mine in the Delamar district. (Anthony et al., 1990)

Reported at an unspecified locality in the Pioche district. (S. Pullman, w.c., 1999)

CUPROSKLODOWSKITE
$(H_3O)_2Cu^{2+}(UO_2)_2(SiO_4)_2 \cdot 2H_2O$

A secondary mineral that occurs in copper-bearing uranium deposits.

MINERAL COUNTY Reported with zeunerite and metatorbernite at the Sunday Mining Company claims in the Marietta district. (Garside, 1973)

CUPROTUNGSTITE
$Cu_3^{2+}(WO_4)(OH)_2$

Cuprotungstite is a rare secondary mineral formed by the alteration of scheelite in tungsten deposits that contain copper.

ESMERALDA COUNTY As greenish masses with *wolframite* at an unspecified mine in the Lida district. (Palache et al., 1951)

MINERAL COUNTY Reported in a specimen as *cuproscheelite* at the Frank R. Channing property in the Leonard district. (WMK)

CURETONITE
$Ba(Al,Ti)(PO_4)(OH,O)F$

Curetonite, a Nevada type mineral, occurs as a secondary species in a barite prospect. It may be found in other barite occurrences in phosphatic rocks.

HUMBOLDT COUNTY Type locality as minute, pale- to yellowish-green crystals with jarosite, along vein walls and as inclusions in *adularia* and barite veinlets that cut bedded barite near a contact with phosphatic shale at the Redhouse Barite Mine in the Potosi district, Sec. 12, T37N, R41E; associated minerals include hisingerite, opal, and phosphates such as montgomeryite and wavellite. (Williams, 1979; S. Sears, w.c., 1999)

CYANOTRICHITE

$Cu_4^{2+}Al_2(SO_4)(OH)_{12} \cdot 2H_2O$

Cyanotrichite, a rare secondary mineral that occurs in the oxidation zones of copper deposits, is frequently associated with spangolite, brochantite, and other sulfate minerals.

CHURCHILL COUNTY Blue acicular crystals to 1 mm with brochantite at the Lovelock and Nickel Mines in the Table Mountain district. (M. Jensen, w.c., 1999; S. Pullman, w.c., 1999)

ESMERALDA COUNTY Reported at the 3 Metals Mine in the Weepah district. (R. Thomssen, w.c., 1999)

EUREKA COUNTY As pale-blue, finely fibrous, radiating sprays up to 3 mm with malachite, chrysocolla, azurite, jarosite, barite, goethite, and quartz at the Gold Quarry Mine in the Maggie Creek district. (Jensen et al., 1995)

HUMBOLDT COUNTY As fibrous blue microcrystals at the Silver Coin Mine in the Iron Point district. (Thomssen, 1998; J. McGlasson, w.c., 2001)

LYON COUNTY At some properties with brochantite and spangolite in the Yerington district; specifically reported at the Shaw, Harris, Milne and Baker Mines. (Palache et al., 1951; HMM; M. Jensen, w.c., 1999)

NYE COUNTY As blue acicular crystals to 1 mm with gypsum at the Stokes Iron Mine in the Ellsworth district. (M. Jensen, w.c., 1999)

PERSHING COUNTY Rare as sky-blue, needlelike crystals to 1 mm in velvety coatings with other copper minerals and pyrite at the Majuba Hill Mine in the Antelope district. (Palache et al., 1951; Jensen, 1985)

WASHOE COUNTY Tentatively identified as blue needles in radial groups at the Red Metal Mine in the Peavine district. (Hill, 1915)

As deep sky-blue radiating needles that form drusy crusts with malachite and chrysocolla at the Burrus Mine in the Pyramid district. (Jensen, 1994)

CYMRITE

$BaAl_2Si_2(O,OH)_8 \cdot H_2O$

A relatively uncommon mineral that has been found elsewhere in barite deposits, metamorphic rocks, and hydrothermal metal deposits. The Nevada occurrence is associated with bedded barite.

HUMBOLDT COUNTY As microscopic rectangular crystal aggregates replacing barite crystals with quartz, pyrite, goethite, and *sericite* in siliceous rock in a horizon above bedded barite at the north pit of the Little Britches Barite Mine in the Golconda district, NE 1/4 Sec. 31, T36N, R41E. (Hsu, 1994)

CYRILOVITE

$NaFe_3^{3+}(PO_4)_2(OH)_4 \cdot 2H_2O$

An uncommon mineral found with other phosphate minerals in pegmatite.

HUMBOLDT COUNTY Glassy yellow crystals less than 0.5 mm with pharmacosiderite at the Lone Tree Mine in the Buffalo Mountain district. (M. Jensen, w.c., 1999)

Reported from a prospect to the east of the Silver Coin Mine in the Iron Point district. (Thomssen, w.c., 1999)

d

DADSONITE

$Pb_{10+x}Sb_{14-x}S_{31-x}Cl_x$

The Nevada type mineral, dadsonite, has been identified at only six localities in the world, four of which were simultaneously recognized in the type description. It occurs in hydrothermal veins with other lead sulfosalt minerals.

PERSHING COUNTY Type locality as minute gray fibers with robinsonite (another Nevada type mineral), pyrite, sphalerite, stibnite, cinnabar, boulangerite, and various secondary oxides and sulfates at the Red Bird Mine in the Antelope Springs district, Sec. 33, T27N, R34E. (Jambor, 1967, 1969; DeMouthe, 1985)

DAUBRÉEITE

BiO(OH,Cl)

Daubréeite is an extremely rare mineral that forms by oxidation of primary bismuth sulfides and sulfosalts.

NYE COUNTY Reported as crusts of minute yellowish crystals at an unspecified locality in the Manhattan district. (Jenkins, 1981; HMM)

DAUBRÉELITE

$Fe^{2+}Cr_2S_4$

A mineral found in meteorites.

NYE COUNTY Comprises about 10 percent by volume of the Quartz Mountain meteorite in association with taenite, kamacite, schreibersite, and troilite. (Buchwald, 1975)

DELAFOSSITE

$Cu^{1+}Fe^{3+}O_2$

Delafossite generally occurs as a relatively uncommon secondary mineral near the base of oxidation zones of some copper deposits, commonly with cuprite, tenorite, copper, and copper carbonates.

EUREKA COUNTY Reported at Eureka as botryoidal crusts with *limonite* and malachite; identified by C. Frondel using XRD. (Palache et al., 1944)

LYON COUNTY Reported at the Empire-Nevada (Yerington, Weed Heights) Mine in the Yerington district. (S. Pullman, w.c., 1999)

PERSHING COUNTY With malachite, pseudomalachite, zeunerite, and tyrolite at the Majuba Hill Mine in the Antelope district. (Jenkins, 1981; UCB)

WHITE PINE COUNTY In the Robinson district as metallic black spherical aggregates to 3 mm with surficial, poorly developed, triangular crystals, in white to pale-gray clay in oxidized jasperoid at the Alpha Mine; also noted with native copper, cuprite, azurite, and malachite at other locations, such as the Kimberly pit, Richard, and Giroux Mines. (Rogers, 1922; Pabst, 1946; Bauer et al., 1966; UCB)

DELRIOITE

$CaSrV_2^{5+}O_6(OH)_2 \cdot 3H_2O$

A very rare mineral that is reported elsewhere only as an efflorescence at a uranium-vanadium deposit in Colorado.

LINCOLN COUNTY In specimens reported only as from Pahranagat. (FUMT; EMP)

DESCLOIZITE

$PbZn(VO_4)(OH)$

Descloizite is a widespread secondary mineral that occurs in small amounts in base-metal deposits; the Chalk Mountain and Goodsprings localities in Nevada are well-known occurrences.

CHURCHILL COUNTY As bright orange, yellow, brown, and black microcrystals and botryoidal coatings in vugs in gossan and altered dolomite that provide a contrasting background for overgrowths of vanadinite crystals at the Chalk Mountain Mine in the Chalk Mountain district. (Bryan, 1972; Jensen, 1990b)

CLARK COUNTY The copper-bearing variety *cuprodescloizite* occurs as crusts of dark-green to nearly black crystals at many sites in the Goodsprings district, including the Canner, Highline, Mobile, Mountain Top, Ninety-Nine, Prairie Flower, Whale, Williams, and Valley Forge Mines; also reported as *psittacinite*. (Hewett, 1931; Melhase, 1934e; Longwell et al., 1965; AMNH; S. Pullman, w.c., 1999)

Reported in a specimen from the Quartette Mine in the Searchlight district. (HMM)

ESMERALDA COUNTY With phoenicochroite, fornacite, wulfenite, and other minerals in quartz veins in the Candelaria district a mile north of the site of Columbus. (R. Walstrom, w.c., 2000)

Reported at the Broken Toe Mine in the Rock Hill district. (R. Thomssen, w.c., 1999)

EUREKA COUNTY As druses of brilliant black crystals to 1 mm with heyite pseudomorphs and partially corroded vanadinite at the Gold Quarry Mine in the Maggie Creek district; also intergrown with selenium and vanadinite on marcasite pods. (Jensen et al., 1995)

HUMBOLDT COUNTY As dark brown microcrystals at the Silver Coin Mine in the Iron Point district. (Thomssen, 1998; J. McGlasson, w.c., 2001)

LINCOLN COUNTY As light-yellowish-gray powder in veins and as orange microcrystalline crusts on gossan at the eastern shaft of the Republic Mine in the Chief district. (Callaghan, 1936; M. Jensen, w.c., 1999)

LYON COUNTY Reported in a specimen from the Lost Soldier Mine in the Churchill district. (LANH)

NYE COUNTY As a copper-bearing variety in yellow-green coatings of small rosettes on quartz at the Indianapolis Mine in the Union district. (M. Jensen, w.c., 1999)

In yellow-orange sprays of microcrystals with wulfenite crystals at the Downieville Lead Mine in the Gabbs district. (M. Jensen, w.c., 1999)

WHITE PINE COUNTY As a copper-bearing variety in tungsten mines at Minerva in the Shoshone district. (Hose et al., 1976)

DEVILLINE

$CaCu_4^{2+}(SO_4)_2(OH)_6 \cdot 3H_2O$

Devilline is a rare mineral found in the oxidation zones of copper and other base-metal deposits.

ELKO COUNTY In blue crusts of microcrystals with langite on quartz at the Victoria Mine in the Dolly Varden district. (M. Jensen, w.c., 1999)

LANDER COUNTY Reported from Whisky Gulch in the Shoshone Range. (R. Thomssen, w.c., 1999)

PERSHING COUNTY As microcrystals with luetheite at the Majuba Hill Mine in the Antelope district. (Jensen, 1993b)

WHITE PINE COUNTY Reported in the Robinson district. (Von Bargen, 1999)

As crusts of pale-greenish-blue crystals to 0.1 mm with langite and other minerals at the Silver King Mine in the Ward district. (Williams, 1964)

DIADOCHITE

$Fe_2^{3+}(PO_4)(SO_4)(OH) \cdot 5H_2O$

A secondary mineral that generally occurs in gossan or as an efflorescence in mines.

EUREKA COUNTY In soft, pale-brown nodules to 2 cm at the Van-Nav-San claim in the Gibellini district. (M. Jensen, w.c., 1999)

DIAPHORITE

$Pb_2Ag_3Sb_3S_8$

Diaphorite is a relatively rare mineral that occurs with other sulfides in moderate-temperature silver deposits.

MINERA COUNTY With owyheeite, jamesonite, and sphalerite in ore at the Broken Hills Mine in the Broken Hills district. (M. Jensen, w.c., 1999)

NYE COUNTY Moderately abundant in quartz-rhodochrosite veins with sphalerite, pyrargyrite, and canfieldite at the Keyser and other mines in the Morey district; reported as black or tarnished metallic blue to bronze euhedral crystals to 2 mm after andorite in comb quartz vugs, and as crystalline masses to 3 cm. (Williams, 1968; Williams and Millett, 1970; HUB; Jenkins, 1981)

DIASPORE

AlO(OH)

Diaspore occurs abundantly in bauxite deposits and in smaller amounts in other settings, in Nevada it is generally associated with high-sulfidation precious-metal deposits as at Goldfield.

ESMERALDA COUNTY Widely distributed with quartz, alunite, and pyrite as rounded, colorless, anhedral grains that are generally <0.5 mm in diameter in the Goldfield district; specifically reported from Vindicator and Preble Mountains and from the Sandstorm and Florence Mines. (Ransome, 1909a; Ashley and Albers, 1975)

Reported from an unspecified locality near Hawthorne. (AMNH)

LYON COUNTY Locally abundant in the southeastern part of the Como district as clear, shiny, prismatic crystals to 1 mm and as aggregates in quartz veins. (Vikre and McKee, 1994)

MINERAL COUNTY With pyrophyllite, andalusite, corundum, and other minerals in metamorphosed Mesozoic volcanic rocks at the Dover, Bismark, and Green Talc claims in the Fitting district. (Ross, 1961; Tatlock, 1964)

With pyrophyllite and quartz in altered ash-flow tuff in the Gabbs Valley Range. (Hudson, 1983)

NYE COUNTY As massive replacement of wall rock with pyrophyllite, kaolinite, and dickite at the Cactus Nevada Mine in the Cactus Springs district on the Nellis Air Force Range. (Tingley et al., 1998a)

STOREY COUNTY With alunite, anatase, ilmenite, and other minerals in altered volcanic rocks in the Virginia City area in the Comstock district. (Whitebread, 1976)

With quartz and alunite in advanced argillic alteration at the Ramsey-Comstock Mine in the Ramsey district. (Rose, 1969)

WASHOE COUNTY In trace amounts as a fine-grained product of advanced argillic alteration in hydrothermal breccia with pyrophyllite, quartz,

pyrite, and dickite in the Peavine and Wedekind districts. (Hudson, 1977)

With pyrophyllite, rutile, and other minerals in advanced argillic alteration associated with enargite-luzonite-pyrite mineralization at the Burrus Mine in the Pyramid district; noted as pale-yellow to clear crystals to 2 mm in vugs in silicified ledges near the Good Hope Mine. (Wallace, 1980; SBC)

DICKITE

$Al_2Si_2O_5(OH)_4$

A clay mineral of the kaolinite group, dickite is polymorphic with kaolinite, halloysite, and nacrite, and is most commonly found in Nevada as a mineral in hydrothermally altered zones.

CHURCHILL COUNTY With annabergite and other nickel minerals at the Nickel Mine in the Table Mountain district. (Ransome, 1909b)

ELKO COUNTY Reported as an alteration mineral in the Tuscarora district, but not identified by XRD analysis. (Nolan, 1936; SBC)

ESMERALDA COUNTY A widespread alteration product in the Cuprite district. (Swaze et al., 1993)

EUREKA COUNTY A widespread and common massive alteration mineral at the Goldstrike Mine in the Lynn district; also noted as satiny stacked platelets in a vug with calcite and blue barite in contact-metamorphic rock. (GCF)

NYE COUNTY As massive replacement of wall rock with pyrophyllite, kaolinite, and diaspore at the Cactus Nevada Mine in the Cactus Springs district on the Nellis Air Force Range. (Tingley et al., 1998a)

PERSHING COUNTY Reported with cinnabar in a specimen from American Canyon in the Humboldt Range; probably from the Kaolinite Mine in the Spring Valley district. (AMNH)

STOREY COUNTY A common alteration phase on the basis of spectroscopic surveys in the Comstock district. (R. Bedell, o.c., 1999)

WASHOE COUNTY In altered rock peripheral to gold ore at the Hog Ranch Mine in the Cottonwood district. (Bussey, 1996)

A very fine-grained phase with quartz, pyrophyllite, diaspore, and pyrite in hydrothermal breccia in Tertiary stocks and older rocks in the Peavine and Wedekind districts. (Hudson, 1977)

DIGENITE

Cu_9S_5

Digenite is a seldom recognized but probably common supergene copper sulfide that is commonly associated with other supergene phases such as chalcocite and covellite.

CLARK COUNTY With chrysocolla, cinnabar, bornite, tenorite, potarite, and gold at the Boss Mine in the Goodsprings district. (Jedwab, 1998a)

ESMERALDA COUNTY With covellite after bornite in ore in the Klondyke district. (Chipp, 1969)

HUMBOLDT COUNTY A minor mineral in the sulfide assemblage at the Lone Tree Mine in the Buffalo Mountain district. (R. Braginton, o.c., 1995)

LANDER COUNTY In stockwork veins with other sulfides at the Cove Mine in the McCoy district. (Emmons and Coyle, 1988)

PERSHING COUNTY With chalcocite, chalcopyrite, enargite, and pyrite at the Majuba Hill Mine in the Antelope district. (MacKenzie and Bookstrom, 1976; Jensen, 1985)

Reported in the Iron Hat district. (Von Bargen, 1999)

With chalcopyrite and bornite in massive sulfide ore at the Big Mike Copper Mine in the Tobin and Sonoma Range district. (Johnson, 1977)

STOREY COUNTY Rare with quartz in drill-hole cuttings from the Flowery lode in the Comstock district. (JHS)

DIOPSIDE

$CaMgSi_2O_6$

A *pyroxene*-group mineral that is common in regionally and contact-metamorphosed rocks and less common in ultramafic to mafic igneous rocks. In Nevada, most often recognized in skarn associated with base-metal or tungsten deposits.

CLARK COUNTY With biotite and microcline in foliated granite near Ruby Spring in the Gold Butte district. (Volborth, 1962b)

ELKO COUNTY With other skarn minerals in the Contact district. (Schilling, 1979)

EUREKA COUNTY A common mineral in calc-silicate metamorphic rock as light-green to black crystals and compact masses at the Goldstrike Mine in the Lynn district; as crystals to 3 cm in diopside veins with calcite. (GCF)

LANDER COUNTY In the skarn assemblage at the West orebody in the Battle Mountain district. (Theodore and Blake, 1978)

Reported from the Linka Mine in the Spencer Hot Springs district. (R. Thomssen, w.c., 1999)

LINCOLN COUNTY Identified in skarn at the Lincoln Tungsten Mine in the Tem Piute district. (JHS)

LYON COUNTY Gray-green crystals to 1.5 cm in calcite at the Douglas Hill Mine in the Yerington district. (Melhase, 1935e; UCB; M. Jensen, w.c., 1999; AMNH)

MINERAL COUNTY Common in skarn at the Julie claim in the Pamlico district. (Smith and Benham, 1985)

Reported at the Champion Mine in the Santa Fe district. (UCB)

NYE COUNTY As gray-green crystals to 1.5 cm in calcite at the Great Basin (Victory) Mine in the Lodi district. (M. Jensen, w.c., 1999)

Chromium-bearing diopside occurs in glassy green crystal fragments to 15 cm at Lunar Craters in the Marcath Cinder Cone, at Easy Chair Crater, and in basalt flows. (Bergman, 1982)

PERSHING COUNTY Reported in hornfels with wollastonite and grossular at the MGL and Nightingale Tungsten Mines in the Nightingale district. (WMK)

Identified in a sample from an unspecified locality in the Rye Patch district. (JHS)

WHITE PINE COUNTY A common skarn mineral with andradite garnet in the Robinson district. (James, 1976)

DIOPTASE

$Cu^{2+}SiO_2(OH)_2$

A relatively uncommon mineral that occurs sparsely in the oxidation zones of some copper-bearing deposits.

CLARK COUNTY In the Goodsprings district as small, perfect, emerald-green crystals on chrysocolla at the Mountain Top Mine, and as crusts at the Blue Jay and Rose Mines; also noted at the Columbia Mine, and as crystals to 0.1 mm in the Boss mine. (Hewett, 1931; Melhase, 1934e; Jenkins, 1981)

LINCOLN COUNTY Questionably reported at the Bristol Silver Mine in the Bristol district. (UAMM)

DJURLEITE

$Cu_{31}S_{16}$

Djurleite is a secondary sulfide mineral that is found with other sulfides in supergene enriched ore, particularly in porphyry copper deposits.

PERSHING COUNTY Reported in the Iron Hat district. (Von Bargen, 1999)

WASHOE COUNTY Identified by XRD in silver-gray masses with schorl and malachite in quartz veins in the Peavine district. (GCF)

DOLOMITE

$CaMg(CO_3)_2$

A very common mineral in sedimentary carbonate rocks, but also present in some hydrothermal veins. Recrystallized rock that contains alternating white and dark bands of dolomite is referred to as "zebra rock." Only a few of the occurrences of this widespread mineral in Nevada are reported below.

CHURCHILL COUNTY In veins and wall rock at the Fairview Silver Mine in the Fairview district. (Schrader, 1947)

In a specimen from near the Silver Coin Mine in the Iron Point district. (R. Thomssen, w.c., 1999)

CLARK COUNTY In "zebra rock" that occurs locally in Devonian limestone at the Chemical Lime Apex Quarry in the Apex district. (SBC)

As small white to gray rhombs to 3 mm at the Green Monster, Yellow Pine, and Prairie Flower Mines in the Goodsprings district; also reported as cobalt-bearing. (B. Hurley, w.c., 1999; J. Kepper, w.c., 2000)

Mined since 1929 from nearly pure deposits in the Dawn and Bullion Members of the Monte Cristo Limestone at Sloan in the Sloan district. (Deiss, 1952; Longwell et al., 1965)

ELKO COUNTY As a primary and replacement mineral and in cross-cutting veinlets at the Rossi Barite Mine in the Bootstrap district. (Snyder, 1989)

In host rocks at the Jerritt Canyon Mines in the Independence Mountains district. (Hofstra, 1994)

Reported with "zebra" texture in ore and as white pearly rhombs to 2 mm at the Meikle Mine in the Lynn district. (Jensen, 1999)

EUREKA COUNTY A common component of carbonate rocks at the Goldstrike Mine in the Lynn district; also in veins with quartz. (GCF)

As pale, orange to pink crystals lining calcite vugs in carbonate rocks at the Gold Quarry Mine in the Maggie Creek district. (Jensen et al., 1995)

HUMBOLDT COUNTY Reportedly as an iron-bearing alteration mineral and in late stage cross-cutting veinlets at the Twin Creeks Mine in the Potosi district. (Simon et al., 1999)

MINERAL COUNTY With quartz as a replacement mineral in *serpentine* at the Candelaria Mine in the Candelaria district. (Moeller, 1988)

Reported as *rhombospar* with scheelite, quartz, *garnet,* and epidote at the W. H. Leonard property in the Leonard district. (WMK)

NYE COUNTY Reported in a deposit on the east slope of Bare Mountain southeast of Beatty; possibly a reference to marble quarries near Cararra in the Bare Mountain district. (Melhase, 1935b)

In massive dolostone of the Triassic Luning Formation, the host rock for magnesite and brucite deposits in the Gabbs district; also reported in white to gray veinlets and disseminated crystals at the Premier Chemicals (Gabbs Magnesite-Brucite) Mine. (Callaghan, 1933; Vitaliano and Callaghan, 1956; Schilling, 1968a)

Associated with ore at the White Caps Mine in the Manhattan district. (Melhase, 1935d)

Reported with magnetite in a specimen from the Gossan Iron Mine in the Wellington district. (WMK)

As rhombs to 1 cm at the Hasbrouck shaft in the Lodi district. (M. Jensen, w.c., 1999)

PERSHING COUNTY As pearly white crystals to 2 mm in vugs with cinnabar crystals at the Pershing Quicksilver Mine in the Antelope Springs district. (M. Jensen, w.c., 1999)

With cinnabar and ankerite at the Paymaster prospect in the Black Knob district. (Bailey and Phoenix, 1944)

STOREY COUNTY In quartz veins with pyrite at the Gooseberry Mine in the Ramsey district. (JHS)

WASHOE COUNTY Reported in a specimen from Leadville in the Leadville district. (LANH)

WHITE PINE COUNTY In host rock at the Vantage deposit, Alligator Ridge Mine, in the Bald Mountain district. (Ilchick, 1990)

DOYLEITE

$Al(OH)_3$

A rare mineral reported elsewhere only in veins with calcite and other minerals in nepheline syenite in Quebec.

EUREKA COUNTY Idenified by X RD in botryoids on siderite crystals at the Gold Quarry Mine in the Maggie Creek district. (GCF)

DRAVITE

$NaMg_3Al_6(BO_3)_3Si_6O_{18}(OH)_4$

Dravite, a magnesium-bearing member of the *tourmaline* group, generally occurs in metamorphosed limestone or mafic igneous rocks; it is rare in pegmatite.

WASHOE COUNTY Reported as euhedral needles of *ferroan-dravite* to schorl with quartz in breccia matrix in the Peavine district; thought to have formed during Tertiary hydrothermal activity. (Hudson, 1977)

DUFRÉNITE

$Fe^{2+}Fe_4^{+3}(PO_4)_3(OH)_5 \cdot 2H_2O$

An uncommon mineral found with other phosphate minerals in pegmatite and in phosphate deposits.

HUMBOLDT COUNTY Possibly the best crystals of this species in the world occur with pharmacosiderite, jarosite, barite, and arseniosiderite as fracture coatings in oxidized rock in the Sequoia pit at the Lone Tree Mine in the Buffalo Mountain district. (M. Jensen, w.c., 1999)

In a dark-green microcrystalline crust at the Silver Coin Mine in the Iron Point district. (M. Jensen, w.c., 1999)

PERSHING COUNTY As dark-green to black botryoidal coatings and vug linings with strengite, meurigite, and pharmacosiderite at the Willard Mine in the Willard district. (Jensen and Leising, 2001)

DUFRÉNOYSITE

$Pb_2As_2S_5$

An extremely rare mineral that occurs with other sulfides in moderate- to low-temperature hydrothermal deposits.

LINCOLN COUNTY Tentatively identified in siliceous lead ore at the Prince Mine in the Pioche district. (Westgate and Knopf, 1932; Melhase, 1934d)

DUFTITE

$PbCu(AsO_4)(OH)$

Duftite is a relatively rare mineral that occurs in oxidized base-metal deposits; associated minerals include other copper arsenates such as conichalcite and olivenite.

ELKO COUNTY As bright-yellowish-green to brown crusts on the surfaces of veins and cavities with cerussite, chlorargyrite, secondary copper and other minerals, in quartz, calcite, barite, and clay gangue at the Standing Elk (Aladdin) Mine in the Railroad district. (Ketner and Smith, 1963; Smith, 1976)

ESMERALDA COUNT Reported in the Candelaria district. (S. Pullman, w.c., 1999)

In specimens from the Broken Toe and Rock Hill Mines in the Rock Hill district. (R. Thomssen, w.c., 1999)

Reported at the Weepah Mine in the Weepah district; also reported at the 3 Metals Mine. (S. Pullman, w.c., 1999; R. Thomssen, w.c., 1999)

HUMBOLDT COUNTY As apple-green microcrystalline spheroids at the Silver Coin Mine in the Iron Point district; also reported as *duftite-ß*. (Thomssen, 1998; J. McGlasson, w.c., 2001)

LANDER COUNTY Reported in the Reese River district. (Von Bargen, 1999)

MINERAL COUNTY As apple-green microcrystals at the Headley Mine in the Garfield district. (M. Jensen, w.c., 1999)

NYE COUNTY In dark-green botryoidal masses to 1 mm on mammillary crusts of conichalcite at the Hasbrouck Mine in the Lodi district; also as apple-green microcrystals. (Jenkins, 1981; M. Jensen, w.c., 1999)

DUHAMELITE

$Pb_2Cu_4^{2+}Bi(VO_4)_4(OH)_3 \cdot 8H_2O$

A very rare mineral that occurs in quartz veins in metamorphic rocks.

LANDER COUNTY Identified by electron microprobe as small green crystals with pottsite, clinobisvanite, and bismutite at the Linka Mine in the Spencer Hot Springs district; also noted in massive form. (R. Walstrom, w.c., 1999; S. Sears, w.c., 1999)

DUMONTITE

$Pb_2(UO_2)_3O_2(PO_4)_2 \cdot 5H_2O$

Dumontite is a rare secondary uranium mineral that generally occurs with kasolite.

CLARK COUNTY With kasolite, iron oxides, hydrozincite, and other base-metal minerals at the Green Monster and Desert Valley Mines in the Goodsprings district. (Garside, 1973)

DUMORTIERITE

$Al_7(BO_3)(SiO_4)_3O_3$

Dumortierite is a relatively common mineral that occurs in aluminum-rich regionally metamorphosed rocks and in pegmatite.

CLARK COUNTY As lavender crystals in pegmatite from the Virgin Mountains in the Bunkerville district. (S. Pullman, w.c., 1999)

Reported from Las Vegas; possibly from the occurrence listed above. (BMC)

MINERAL COUNTY In minor amounts with andalusite and other aluminum-rich minerals at the Dover and Green Talc Mines in the Fitting district. (Vanderburg, 1937a; WMK)

NYE COUNTY Reported without further notation in the Manhattan district. (Gianella, 1941)

Reported as a "sodalite blue" mineral after quartz in granite in the Round Mountain district. (Grawe, 1928b)

PERSHING COUNTY In metarhyolite on Lincoln Hill in the Rochester district. (Knopf, 1924; Grawe, 1928b; Johnson, 1977)

Mined as an industrial mineral between 1925 and 1949 from a deposit at the Champion Mine in the Sacramento district near Oreana, Sec. 36, T29N, R33E; it occurs as cobalt-blue to deep pink crystalline masses, euhedral blue or pink crystals, and lavender to pink fibrous veins in metamorphosed tuff with andalusite, *sericite,* quartz, and coarse muscovite. (Grawe, 1928b; Jones and Grawe, 1928; Carpenter, 1928; Kerr and Jenney, 1935; Johnson, 1977; WMK)

WASHOE COUNTY As pink masses with quartz and muscovite in pegmatite dikes at the south end of the Granite Range 8 miles northwest of Gerlach. (Grawe, 1928b; Gianella, 1941; WMK)

DUNDASITE

$PbAl_2(CO_3)_2(OH)_4 \cdot H_2O$

An uncommon mineral that occurs in oxidized base-metal deposits.

WHITE PINE COUNTY As feathery white crystals in clusters to 0.2 mm in a specimen from the Success Mine in the Schell Creek Range; identified by S. Williams. (R. Thomssen, w.c., 1999)

DUSSERTITE

$BaFe_3^{3+}(AsO_4)_2(OH)_5$

A rare secondary mineral that occurs with other arsenate species.

ELKO COUNTY In minor amounts with barite, hematite, jarosite, cinnabar, and quartz in jasperoid breccia at the Rain Mine in the Carlin district. (Bentz et al., 1983; LaPointe et al., 1991)

EUREKA COUNTY As massive brown vein material with quartz, chrysocolla, and iron hydroxides at the Goldstrike Mine in the Lynn district; also with jarosite crystals, goethite, and hematite in silicified breccia. (GCF)

MINERAL COUNTY Spherules to 1 mm on barite crystals near Rhyolite Pass in the Fitting district. (M. Jensen, w.c., 1999)

DYSCRASITE

Ag_3Sb

Dyscrasite, a relatively uncommon intermetallic compound, occurs in hydrothermal silver vein deposits and in massive sulfide deposits; it may be of primary or secondary origin.

LANDER COUNTY Reported with quartz at an unspecified locality in the Reese River district. (Palache et al., 1944)

NYE COUNTY Intergrown with tetrahedrite and polybasite at the Highbridge Mine in the Belmont district. (Jenkins, 1981)

STOREY COUNTY Reported in a specimen from the Comstock Lode in the Comstock district. (HMM)

EDGARBAILEYITE

$Hg_6^{1+}Si_2O_7$

A newly named secondary mineral that occurs in mercury deposits.

PERSHING COUNTY Reported as a brownish orange crust of tiny crystals on hyalite opal with cinnabar, calomel, and *clinoptilolite* in a spherulite in a specimen from the Red Bird Mine area in the Antelope Springs district. (R. Thomssen, w.c., 1999)

EGLESTONITE

$Hg_6^{1+}Cl_3O(OH)$ or $Hg_4^{1+}Cl_2O$

Eglestonite is a rare mineral that occurs as an oxidation product of cinnabar or other mercury minerals; it is photosensitive and darkens with exposure to light.

HUMBOLDT COUNTY At the Cordero Mine in the Opalite district with terlinguaite and calomel; also at the McDermitt Mine with corderoite, cinnabar, calomel, and native mercury. (Yates, 1942; Jenkins, 1981; McCormack, 1986)

A dull yellow coating on cinnabar and matrix with terlinguaite, mercury, and calomel at the Cahill Mine in the Poverty Peak district. (UMNH; MM; EMP; AMNH; F. Cureton, w.c., 1999)

LANDER COUNTY Tentatively identified as a yellow mineral that darkens rapidly on exposure to sunlight in a narrow cinnabar vein with pyrite, barite, *chalcedony,* and jarosite at the Warm Springs mercury prospect in the Warm Springs district. (Bailey and Phoenix, 1944)

ELBAITE

$Na(Li,Al)_3Al_6(BO_3)_3Si_6O_{18}(OH)_4$

A mineral in the *tourmaline* group, elbaite occurs in granite, granite pegmatite, metamorphic rocks, and in hydrothermal veins. *Rubellite,* a pink variety, is in some cases a valuable gem.

MINERAL COUNTY Reported at the Zapot claims in the Fitting district. (S. Pullman, w.c., 1999)

ELECTRUM

(Au,Ag)

Electrum is a nonspecies name for native metal that is intermediate between silver and gold. A silver-gold alloy that contains more than 50 at. percent (65 wt. percent) gold should be classified as gold, and an alloy with less as silver. In many cases, identifications in this mineral series are made in the field or laboratory on the basis of color; the alloy at the following occurrences has not been identified as either gold or silver, or it has compositions that straddle or are near the intermediate composition.

CHURCHILL COUNTY As blebs and stringers in quartz veins at the Fairview Silver Mine in the Fairview district. (Greenan, 1914; Schrader, 1947)

As tiny discrete grains in quartz breccia and veins and intergrown with chalcopyrite in gabbro at the Dixie Comstock Mine in the Mountain Wells district; gold contents are 53–63 wt. percent. (Vikre, 1994)

ELKO COUNTY In dendritic form with naumannite, aguilarite, fischesserite, and other precious metal minerals at the Ken Snyder Mine in the Gold Circle district. (Saunders et al., 1996; Casteel and Bernard, 1999)

As grains to 88 microns of 61–63 wt. percent gold with pyrite, marcasite, sphalerite, chalcopyrite, and other sulfides at the Hollister gold deposit in the Ivanhoe district. (Bailey and Phoenix, 1944; Bartlett et al., 1991; Deng, 1991)

Reported in a specimen with gold at the Grand Prize Mine in the Tuscarora district. (WMK)

HUMBOLDT COUNTY Dendritic with opal, *adularia,* naumannite, and acanthite at the Jumbo Mine in the Awakening district. (Saunders et al., 1996)

In oxidized ore at the Lone Tree Mine in the Buffalo Mountain district. (R. Braginton, o.c., 1995)

As tiny grains in powellite at the Moonlight Uranium Mine in the Disaster district. (Castor et al., 1996)

In dendritic form in a rich ore shoot with naumannite, tetrahedrite, and chalcopyrite at the National Mine in the National district; contains 49–52 at. percent gold. (Lindgren, 1915, 1933; Melhase, 1934c; Vikre, 1985; Saunders et al., 1996)

Dendritic in quartz-*adularia* veins in the Ten Mile district. (Saunders et al., 1996)

LANDER COUNTY With gold, silver, and sulfides at the Dean Mine in the Lewis district. (J. McGlasson, w.c., 2001)

LYON COUNTY With pyrargyrite, tetrahedrite, and other sulfides in quartz-calcite-*adularia* veins in the Como district. (Vikre and McKee, 1994)

Reported with silver, gold, and other minerals in veins in the Silver City district, modern probe analyses show contents of 48–64 at. percent gold; also in a placer deposit with a calculated gold content of 66 wt. percent. (Gianella, 1936a; Vikre, 1989; Vanderburg, 1936a)

Reported in specimens at the Talapoosa Mine and Prosky property in the Talapoosa district. (WMK)

MINERAL COUNTY With acanthite, naumannite, and bromargyrite in banded quartz-*adularia* veins in andesitic rocks in the Aurora district; also reported with silver-bearing tetrahedrite. (Hill, 1915; Osborne, 1985, 1991)

Finely divided with pyrite in quartz veins in Tertiary volcanic rocks at the Olympic Mine in the Bell district. (Knopf, 1921b)

Associated with chlorargyrite and acanthite in iron- and manganese-stained quartz veins at the Randall property in the Rand district. (Vanderburg, 1937a; Ross, 1961)

With acanthite and chlorargyrite in quartz veins in argillized rhyolite at the Rawhide Mine in the Rawhide district; contains significant silver in sulfide-bearing ore, but is almost exclusively gold in oxidized ore. (Lincoln, 1923; Black, 1991)

NYE COUNTY Reported with gold and silver in a cellular quartz specimen collected some 60 feet downslope from the Antelope View Mine in the Antelope Springs district. (Schrader, 1913)

Electrum with 42–57 at. percent gold occurs with acanthite, uytenbogaardtite, and gold at the Bullfrog and Original Bullfrog Mines in the Bullfrog district. (Castor and Weiss, 1992; Castor and Sjoberg, 1993)

With pyrite in fine-grained quartz at the Clarkdale and Yellow Gold Mines in the Clarkdale district on the Nellis Air Force Range. (Tingley et al., 1998a)

Visible upon panning of ore in the Golden Arrow district; occurs in closely spaced quartz veinlets and in association with pyrite. (Ferguson, 1916; WMK)

Reported to contain about 60 wt. percent gold at the productive Round Mountain Mine in the Round Mountain district; although most production is from low-grade disseminated ore in ash-flow tuff, many fine specimens of leaf, crystalline, and dendritic form have been produced, particularly in recent years, from nearly solid veins to 1 inch thick that have yielded masses of spongy *electrum* in excess of 200 ounces. Also in coarse, angular grains with 36 wt. percent Ag in placer gravels that were mined profitably for many years. (Mills, 1984; L. McMaster, o.c., 1996; Vanderburg, 1936a)

One of the principal ore minerals at the Blue Horse and Hillside Mines and Catlin claim in the Silverbow district on the Nellis Air Force Range. (Tingley et al., 1998a)

Commonly with acanthite in quartz veins in rich ore from the Tonopah district; may be selenium-bearing at the Mizpah and other mines. (Bastin and Laney, 1918; Nolan, 1935; S. Pullman, w.c., 1999)

PERSHING COUNTY As gray intergrowths with quartz in veins with alunite at an unspecified locality in the Scossa district. (Vanderburg, 1936b)

Reported as an alloy with 45.7 percent gold, 37.6 percent silver, and 13.9 percent mercury (in wt. percent) in silicified boulders with naumannite and silver halides at the Gold Boulder zone

2.8 miles west-northwest of Scossa. (W. Fusch, o.c., 2000)

Reported in a specimen from the Fairview claim in the Seven Troughs district. (WMK)

STOREY COUNTY Reported to contain 49–67 at. percent gold in specimens of bonanza ore with acanthite and other sulfide minerals from the Comstock Lode and other vein systems in the Comstock district; specifically reported from the Hale and Norcross, Savage, Con Virginia, and Ophir Mines. (Bastin, 1922; Vikre, 1989; H. Bonham, o.c., 1993; HUB)

WHITE PINE COUNTY *Electrum* with approximately 65 wt. percent gold occurs with bitumen in quartz veins at the Gold Point Mine in the Currant district. (Castor and Hulen, 1996)

ELLISITE

Tl_3AsS_3

A Nevada type mineral, ellisite has only been identified in the thallium-arsenic sulfide assemblage at the Carlin Mine.

EUREKA COUNTY Type locality as minute, dark-gray grains with lorandite, getchellite, christite, native arsenic, and realgar in small discontinuous patches along bedding planes in mineralized, silty, carbonaceous dolomite beds in the east pit at the Carlin Mine in the Lynn district, Sec. 13, T35N, R50E. (Radtke et al., 1977; Dickson et al., 1979)

ELPASOLITE

K_2NaAlF_6

A very rare mineral that has been identified in granite pegmatite and in a hydrothermal antimony-bearing quartz vein.

MINERAL COUNTY As anhedral grains, generally less than 50 microns long, in a pegmatite with cryolite, cryolithionite, and other Na-Al-Li-F minerals at the Zapot claims in the Fitting district. (Foord et al., 1998, 1999b)

ELYITE

$Pb_4Cu^{2+}(SO_4)(OH)_8$

A Nevada type mineral, elyite is a rare oxidation zone species from a lead-silver deposit containing copper. In the original description, it was stated that there is probably less than 50 mg of the material in existence.

WHITE PINE COUNTY Type locality in the Caroline tunnel in the Ward district as sprays of violet crystals to 0.15 mm in and on oxidized cavity fillings with galena, langite, and serpierite; the underground locality was destroyed by subsequent open pit mining; Sec. 15, T14N, R63E. (Williams, 1972)

EMBOLITE

Ag(Cl,Br)

Embolite is a non-species name for silver halide compositions intermediate between chlorargyrite and bromargyrite.

CHURCHILL COUNTY Reported in oxidized lead-silver veins with wulfenite and other minerals at the Chalk Mountain Mine in the Chalk Mountain district. (Vanderburg, 1940; Schilling, 1962a, 1962b)

Reported in quartz veins at the Fairview Silver Mine in the Fairview district. (Schrader, 1947; Saunders, 1977)

CLARK COUNTY Rare as gray equant crystals to 1 mm on fractures with hemimorphite and cerussite at the Singer Mine in the Goodsprings district. (Adams, 1998)

HUMBOLDT COUNTY As yellow-green clots to 1 mm at the Silver Coin Mine in the Iron Point district (T. Loomis, o.c., 2000; GCF)

MINERAL COUNTY Reported as an important silver phase in oxidized ore with chlorargyrite at the Rawhide Mine in the Rawhide district. (Black, 1991)

NYE COUNTY As olive-green coatings and small groups of imperfect cubes and octahedrons on massive *psilomelane* or other gangue minerals in the Tonopah district. (Burgess, 1911; Melhase, 1935c)

STOREY COUNTY On quartz in a specimen from the Chollar Mine in the Comstock district. (UCB)

EMMONSITE

$Fe_2^{3+}Te_3^{4+}O_9 \cdot 2H_2O$

A rare secondary mineral that occurs in oxidized tellurium-rich gold ore; reported as *durdenite* at some localities.

ESMERALDA COUNTY As greenish-yellow crusts and stains or as fibrous rosettes in oxidized ore with gold, alunite, *limonite,* and possible kaolinite at the Black Butte Mine in the Diamondfield district, also with gold, chlorargyrite, and *limonite* at the Gold Coin claim. (Ransome, 1909a; EMP; WMK)

Relatively abundant as bright green to yellowish-green fibrous botryoidal and finely drusy crusts, dense microcrystalline masses, and rare radial groups of rough, serrated, acicular crystals in oxidized material with tellurite, mackayite, and *blakeite* at the Jumbo, Clermont, and Mohawk Mines in the Goldfield district. (Frondel and Pough, 1944; Gaines, 1972)

NYE COUNTY Reported as the hydrous ferric tellurite *durdenite* with malachite, cerussite, pyrolusite, and other minerals in a vein in limestone in the Oak Spring district. (Melhase, 1935b)

EMPLECTITE

$CuBiS_2$

A mineral that occurs with other sulfides in hydrothermal veins.

ESMERALDA COUNTY Reported in the Goldfield district. (Von Bargen, 1999)

ENARGITE

Cu_3AsS_4

Enargite, a relatively common mineral, occurs as a primary sulfosalt mineral in many deposits of copper and other base metals. It is also a typical phase in high-sulfidation, epithermal preciousmetal deposits.

ELKO COUNTY Reported at several mines in the Good Hope district; a Dana locality. (M. Jensen, w.c., 1999)

Reported in unoxidized ore in the Tuscarora district. (Emmons, 1910; Melhase, 1934c)

ESMERALDA COUNTY As dark-gray, brittle crusts at the Victor and Gold Bar Mines in the Goldfield district; also noted at the Merger shaft and as crystalline masses with famatinite and as crystals with longitudinal striations from unspecified localities. (Ransome, 1909a; Albers and Stewart, 1972; Hsu, o.c., 1997; R. Thomssen, w.c., 1999)

With galena, iron-rich sphalerite, and proustite in primary ore at the Alpine Mine in the Lone Mountain district. (Phariss, 1974)

HUMBOLDT COUNTY Reported from various mines in the northern part of the Pine Forest Range in the Vicksburg district. (Gianella, 1941)

LANDER COUNTY One of several sulfosalt minerals in minor quantities in quartz-rhodochrosite veins in the Reese River district. (Ross, 1953)

MINERAL COUNTY In trace amounts in unoxidized samples of quartz-pyrite alteration at the Freedom Flats orebody in the Borealis district. (Eng, 1991)

Reported in a specimen from the Leonard Mine in the Leonard district. (WMK)

NYE COUNTY As disseminated grains and stringers with galena, hessite, sphalerite, freibergite, and other sulfides at the Fairday Mine in the Cactus Springs district. (Tingley et al., 1998a)

After pyrite in the Gold Crater district. (Tingley et al., 1998a)

PERSHING COUNTY A rare mineral in pods of chalcocite and chalcopyrite in the Copper stope at the Majuba Hill Mine in the Antelope district. (Jensen, 1985)

WASHOE COUNTY With other sulfides in quartz, quartz-calcite, and quartz-*tourmaline* veins in the Peavine district; specifically reported from the Black Panther, Fravel, and Golden Fleece Mines. (Hill, 1915; Gianella, 1941; Bonham and Papke, 1969; Hudson, 1977)

Intergrown with luzonite in silver-bearing veins with quartz, pyrite, and barite at several mines in the Pyramid district; specifically reported at the Burrus Mine, Sec. 15, T23N, R21E, as steel-gray masses to 15 cm thick, disseminated crystals, or free-standing crystals to 1.5 cm in vugs. (Overton, 1947; Bonham and Papke, 1969; Wallace, 1980; Jensen, 1994; GCF)

Reported with pyrite in the Wedekind district. (Hudson, 1977)

WHITE PINE COUNTY Reported with sphalerite at the Glencoe Mine in the Eagle district. (Hill, 1916; Melhase, 1934d)

Fine crystals were reported with andorite, bournonite, and tennantite in quartz vugs at the

Monitor Mine in the Taylor district. (Williams, 1965)

Tentatively reported in the White Pine district. (Raymond, 1869)

ENGLISHITE

$K_3Na_2Ca_{10}Al_{15}(PO_4)_{21}(OH)_7 \cdot 26H_2O$

A rare mineral reported in pegmatite and from variscite nodules.

ELKO COUNTY As pale-yellow microcrystalline spherules on quartz with barite and metastibnite in the Meikle Mine in the Lynn district. (Jensen, 1999)

EUREKA COUNTY In groups to 2 mm of delicate, micaceous, colorless to white crystals in vugs with other phosphate minerals at the Gold Quarry Mine in the Maggie Creek district. (Jensen et al., 1995)

HUMBOLDT COUNTY With curetonite, montgomeryite, and collinsite at the Redhouse Barite Mine in the Potosi district. (F. Cureton, w.c., 1999)

ENSTATITE

$Mg_2Si_2O_6$

A common *pyroxene*-group mineral that mainly occurs in ultramafic and mafic igneous rocks and in high-grade metamorphic rocks. The variety *bronzite* has bronzelike, submetallic luster on cleavage surfaces.

CLARK COUNTY Reported as enstatite-*hypersthene* or *bronzite* with diopside, *hornblende,* biotite, *plagioclase,* and *olivine* in ultramafic intrusive rock at the vermiculite mine northeast of Gold Butte in the Gold Butte district. (Leighton, 1967; Volborth, 1962b)

LANDER COUNTY Reported in a specimen from the Bunker Hill Mine in the Kingston district. (WMK)

PERSHING AND CHURCHILL COUNTIES Reported as *bronzite* in widespread gabbroic rocks of the Humboldt lopolith. (Speed, 1962)

EPIDOTE

$Ca_2(Fe^{3+},Al)_3(SiO_4)_3(OH)$

A common mineral that occurs in regionally and contact-metamorphosed rocks and in hydrothermally altered rocks.

CARSON CITY In decomposed granite in Vicee Canyon on Winnie Lane northwest of Carson City. (Klein, 1983)

With andradite and scheelite at the Alex Eske and Valley View Mines in the Delaware district. (D. Hudson, w.c., 1999)

CHURCHILL COUNTY In skarn with powellite, scheelite, and pyrite in the Chalk Mountain district. (Schilling, 1979)

An alteration mineral at the Fairview Mine in the Fairview district; also noted as euhedral green prisms to 1 cm in massive white calcite veins at the Slate Mine. (Schrader, 1947; Saunders, 1977; B. Hurley, w.c., 1999)

In massive to euhedral grains with grossular in scheelite-bearing skarn at prospects in the Sand Springs district. (B. Hurley, w.c., 1999)

CLARK COUNTY A common metamorphic and alteration mineral in the Bunkerville district. (Beal, 1965)

As crystals to 1 cm in miarolitic cavities and pegmatites in the Newberry Mountains; reportedly most common in the Christmas Tree Pass area. (S. Scott, w.c., 1999; W. Lombardo, o.c., 2001)

DOUGLAS COUNTY Reported in a specimen from the Red Canyon Copper Mine in the Red Canyon district. (WMK)

ELKO COUNTY With other skarn minerals along a limestone-granodiorite contact in the Contact district. (Schilling, 1979)

Plentiful in contact-metamorphic zones in the Dolly Varden district. (Melhase, 1934d)

With *garnet* in skarn associated with copper-iron ore at the Lone Mountain Mine in the Merrimac district. (LaPointe et al., 1991)

ESMERALDA COUNTY A major constituent of skarn near Cucomungo Spring in the Sylvania district. (Schilling, 1979)

EUREKA COUNTY With diopside, calcite, and grossular in calc-silicate rock at the Goldstrike Mine in the Lynn district. (GCF)

HUMBOLDT COUNTY Reported from an unspecified locality in the Black Rock Range in the Black Rock district. (BMC)

In a specimen with scheelite, quartz, *garnet,* calcite, and pyrite from the Hood Tungsten property in the Golconda district. (WMK)

In contact-metamorphic zones with *pyroxene* and *garnet* in the Gold Run district. (Melhase, 1934c)

With arsenopyrite and other minerals at the Juanita Mine in the Varyville district. (M. Jensen, w.c., 1999)

LANDER COUNTY In skarn in the West orebody at the Copper Canyon Mine and at the Fortitude Gold Mine in the Battle Mountain district. (Roberts and Arnold, 1965; Doebrich, 1993)

LYON COUNTY Common in metavolcanic rock at the Owen prospect in the Wilson district. (Reeves et al., 1958)

With wollastonite, *apatite,* quartz, and chalcanthite at the Bluestone Mine in the Yerington district; also at the Ludwig Gypsum Mine. (Melhase, 1935e; WMK; S. Pullman, w.c., 1999)

MINERAL COUNTY With hematite, magnetite, and other metasomatic minerals at the Iron Crown prospect in the Fitting district. (Reeves et al., 1958)

With magnesio-axinite, ferro-axinite, prehnite, vesuvianite, and actinolite in skarn near Luning in the Garfield district. (Dunn et al., 1980)

Euhedral crystals to 5 cm with quartz crystals in skarn at the Hooper No. 2 shaft in the Leonard district; also in specimens of scheelite-bearing skarn from the Sunnyside and Sutton No. 1 Mines. (M. Jensen, w.c., 1999; WMK)

As a skarn mineral at the Lucky Boy Mine in the Lucky Boy district. (Lawrence, 1963)

With magnetite in metavolcanic rocks at the Walker Lake prospect, Mountain View district. (Reeves et al., 1958)

In world-class specimens of deep-green crystals to 15 cm long in pockets with quartz crystals in skarn at the Julie claim in the Pamlico district near Hawthorne, NW 1/4 Sec. 9, T7N, R32E. (Smith and Benham, 1985)

Widespread in the Excelsior Formation in the county. (Ross, 1961)

NYE COUNTY A fine-grained alteration product at the Clarkdale and Yellow Gold Mines in the Clarkdale district, Nellis Air Force Range. (Tingley et al., 1998a)

With magnetite-hematite replacement bodies at the Engle-Stouder prospect in the Ellsworth district. (Reeves et al., 1958)

PERSHING COUNTY Common in skarn with magnetite at the Basalt prospect in the Copper Valley district. (Shawe et al., 1962)

Reported in a specimen with smoky quartz in the Gerlach district. (WMK)

Abundant in scheelite-bearing skarn in the Juniper Range district. (Johnson, 1977)

Common in skarn at the Nevada-Massachusetts Co. Mines in the Mill City district. (JHS)

As crystals in scheelite-bearing skarn at the Nightingale and Ranson Mines in the Nightingale district. (Smith and Guild, 1944; WMK)

As crystals in quartz and with quartz crystals in specimens reported to be from Winnemucca Lake. (WMK)

STOREY COUNTY An alteration mineral in andesite along the Comstock Lode in the Comstock district. (Gianella, 1936a)

WASHOE COUNTY At the Olinghouse Mine in the Olinghouse district as an alteration mineral after mafic phenocrysts, with quartz and calcite in cavities, and in veins with quartz, scheelite, gold, zeolite minerals, and calcite. (Jones, 1998; SBC)

A product of propylitic alteration in Tertiary rocks with calcite, clinozoisite, *chlorite,* and other minerals in the Peavine and Wedekind districts. (Hudson, 1977)

As feathery masses to 10 cm in pegmatite at Incline Village. (Jensen, 1993a)

WHITE PINE COUNTY In the Robinson district as the dominant mineral of propylitic alteration in porphyry; also in minor amounts in skarn. (Spencer et al., 1917; James, 1976; Maher, 1996)

EPISTILBITE

$(Ca,Na)_2[Al_2Si_4O_{12}] \cdot 4H_2O$

A relatively rare zeolite mineral that generally occurs in cavities in mafic volcanic rocks.

WASHOE COUNTY Identified by XRD in veins with quartz, calcite, gold, and other minerals at the Olinghouse Mine in the Olinghouse district. (GCF)

EPSOMITE

$MgSO_4 \cdot 7H_2O$

Epsomite, which is very soluble in water, is relatively common as an efflorescent mineral on mine walls or as encrustations on playas.

CHURCHILL COUNTY As fibrous crusts coating mine walls in the Webber shaft of the Fairview Mine in the Fairview district. (Schrader, 1947)

CLARK COUNTY Reported from the Goodsprings district. (Kepper, this volume)

ESMERALDA COUNTY Reported in a specimen from Coaldale in the Coaldale district. (LANH)

EUREKA COUNTY An efflorescent mineral in areas of sulfide-rich altered diorite at the Goldstrike Mine in the Lynn district. (GCF)

HUMBOLDT COUNTY As fibrous masses in abandoned workings at the Getchell Mine in the Potosi district. (S. Sears, w.c., 1999)

NYE COUNTY In carpets of white needles to 5 cm on floors of old workings at the White Caps Mine in the Manhattan district. (Foshag and Clinton, 1927)

Deposited on cracks from water used to wet down stopes at the West End Mine in the Tonopah district. (Bastin and Laney, 1918)

PERSHING COUNTY Common as efflorescent white fibers to 1 cm long on timbers and walls of mine workings with chalcanthite and gypsum at the Majuba Hill Mine in the Antelope district. (Jensen, 1985)

STOREY COUNTY On mine walls along the Comstock Lode in the Comstock district. (Bastin, 1922; Palache et al., 1951)

WHITE PINE COUNTY In leached caps of porphyry deposits in the Robinson district. (Spencer et al., 1917)

ERIOCHALCITE
$Cu^{2+}Cl_2 \cdot 2H_2O$

A rare soluble mineral that occurs as encrustations around fumaroles and as a weathering product in copper deposits in arid climates.

MINERAL COUNTY Reported as a light-blue mineral from an unspecified locality near Hawthorne. (Anthony et al., 1997; AMNH)

ERIONITE
$(Na,K,Ca_{0.5})_{10}[Al_{10}Si_{26}O_{72}] \cdot \sim 30H_2O$ (general)

A zeolite group mineral that has recently been subdivided into three species depending on the most abundant alkali or alkali earth element. *Erionite* is relatively common, particularly in altered areas in rhyolite tuff; it occurs more rarely in amygdules in mafic volcanic rocks. *Erionite* is reportedly carcinogenic.

CHURCHILL COUNTY Abundant in widespread deposits in altered tuffaceous rocks with *clinoptilolite,* mordenite, *phillipsite,* calcite, and other minerals at the Eastgate deposit. (Papke, 1972a; JHS)

ELKO COUNTY With *phillipsite* and *clinoptilolite* in tuff beds on the east side of Pine Valley. (Papke, 1972a)

EUREKA COUNTY In large amounts in tuffaceous lacustrine beds with *clinoptilolite, phillipsite,* calcite, jarosite, and other minerals in Pine Valley 34 miles south of Carlin in the Pine Valley district. (Deffeyes, 1959; Papke, 1972a)

HUMBOLDT COUNTY With *clinoptilolite* and volcanic glass in narrow alteration zones between fresh glassy rock and mercury deposits in the McDermitt caldera in the Opalite district. (Rytuba and Glanzman, 1979)

LANDER COUNTY As relatively pure beds and as mixtures with *clinoptilolite* in early Pliocene sediments in the Mountain Springs district. (Deffeyes, 1959)

Abundant with *clinoptilolite, phillipsite,* mordenite, analcime, and other minerals in large deposits in the Steiner Canyon district in the Reese River Valley, Sec. 26, T24N, R43E. (Deffeyes, 1959; Papke, 1972a)

PERSHING COUNTY Abundant in the large Jersey Valley zeolite deposits in the Jersey district in tuffaceous sedimentary rocks; also with *clinoptilolite, phillipsite,* and opal in fractures in rhyolite tuff. (Deffeyes, 1959; JHS)

ERYTHRITE
$Co_3(AsO_4)_2 \cdot 8H_2O$

Erythrite, or *cobalt bloom,* is a relatively uncommon mineral that is typically formed by the oxidation of cobalt and nickel arsenides.

CARSON CITY In cobalt ore from an unspecified locality south of Carson City that reportedly consists of black cobalt ochre—a mixture of erythrite and manganese oxide. (WMK)

CHURCHILL COUNTY Reported as *cobalt bloom* in crystal clusters to 1 mm with tetrahedrite, azurite, annabergite, brochantite, chalcophyllite, and morenosite at the Lovelock and Nickel Mines in the Table Mountain district. (Ransome, 1909b; S. Pullman, w.c., 1999)

CLARK COUNTY With azurite, chalcanthite, brochantite, chrysocolla, cuprite, malachite, and *garnierite* in oxidized ore in the open pit at the Key West Mine in the Bunkerville district. (Beal, 1965)

With copper and other cobalt minerals at the Highline Mine in the Goodsprings district. (Melhase, 1934e)

ESKEBORNITE

$CuFeSe_2$

A mineral of low-temperature hydrothermal vein deposits, eskebornite forms a series with chalcopyrite.

ELKO COUNTY Reported with other selenide minerals and *electrum* in veins in the Gold Circle (Midas) district. (P. Goldstrand, o.c., 2000)

ESKIMOITE

$Ag_7Pb_{10}Bi_{15}S_{36}$

Eskimoite is a very rare mineral that is found with other bismuth-bearing sulfosalts in hydrothermal veins and in a cryolite-bearing pegmatite.

NYE COUNTY In quartz at an unspecified prospect in the Manhattan district. (Anthony et al., 1990; FUMT; AMNH; Jenkins, 1981)

EUCAIRITE

CuAgSe

A widely distributed mineral in selenium-bearing hydrothermal deposits.

CLARK COUNTY Identified by J. Jedwab in a specimen from the Azurite Mine in the Goodsprings district. (J. Kepper, this volume)

ELKO COUNTY In ore from the Ken Snyder Mine in the Gold Circle (Midas) district as micron- to sub-millimeter-sized inclusions in quartz veins with aguilarite, *electrum,* gold, and naumannite. (Casteel and Bernard, 1999)

HUMBOLDT COUNTY With *electrum,* selenium, and base-metal sulfide minerals at the Crofoot-Lewis Mine in the Sulphur district. (Ebert et al., 1996)

EULYTITE

$Bi_4(SiO_4)_3$

An uncommon mineral reported from a few localities in eastern Europe, also known as *eulytine.*

LANDER COUNTY As pale-green masses with bismutite at the Linka Mine in the Spencer Hot Springs district; confirmed by XRD. (Excalibur Mineral Co., 2002b)

EUXENITE-(Y)

$(Y,Ca,Ce,U,Th)(Nb,Ta,Ti)_2O_6$

A relatively common complex oxide mineral that occurs in granite pegmatite and in detrital deposits.

ESMERALDA COUNTY With *uranothorite,* xenotime, and *columbite* in gold-bearing placer gravels in Tule Canyon in the Tokop district. (Garside, 1973)

MINERAL COUNTY Tentatively identified with *samarskite* in a pegmatite body in granitic rock at the Lucky Susan No. 1 claim in the Buena Vista district. (Ross, 1961; Garside, 1973)

Acanthite, 3.5 cm wide, Belcher Mine, Comstock district, Storey County. B. and F. Cureton collection. Photograph by J. Scovil.

Agardite, 2.5-mm spray, Big 3 Mine, Railroad Springs district, Esmeralda County. S. White collection. Photograph by S. White.

Almandine-Spessartine, 1.3-cm crystal, Garnet Hill, Robinson district, White Pine County. W. Lombardo collection. Photograph by J. Scovil.

Annabergite, 1-mm crystals,
Nickel Mine, Table Mountain
district, Churchill County.
S. White collection.
Photograph by S. White.

Arsenolite, crystals to 3 mm,
Eureka district, Eureka County.
W. Lombardo collection.
Photograph by J. Scovil.

Aurichalcite with Hemimorphite, 3.7-cm-wide field, Dauntless Mine, Robinson district, White Pine
County. S. Kleine collection. Photograph by J. Scovil.

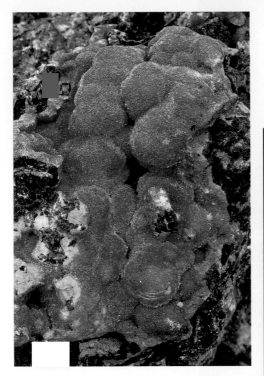

Azurite, 4.6-cm-high field, Robinson
district, White Pine County. W. M. Keck
Museum. Photograph by J. Scovil.

Barite, 8 cm high, Meikle Mine,
Lynn district, Elko County. S. Kleine
collection. Photograph by J. Scovil.

Barite, 7.3 cm wide,
Northumberland Mine,
Northumberland district, Nye
County. S. Kleine collection.
Photograph by J. Scovil.

Barite, crystals to 2.2 cm, Rosebud Mine, Rosebud district, Pershing County. K. Allen collection. Photograph by S. Castor.

Barite with Orpiment inclusions, 5-mm crystal, Regent Mine, Rawhide district, Mineral County. S. White collection. Photograph by S. White.

Boltwoodite, 8-mm-wide field, Green Monster Mine, Goodsprings district, Clark County. J. Kepper collection. Photograph by S. White.

Calcite, 22.5 cm wide, Meikle Mine, Lynn district, Elko County. S. Kleine collection.
Photograph by S. Castor.

Calcite, crystals to 1.5 cm, Getchell Mine,
Potosi district, Humboldt County. W. M. Keck
Museum. Photograph by J. Scovil.

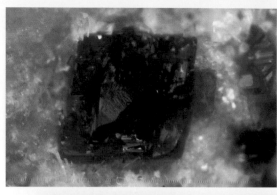

Callaghanite,
0.7-mm crystal, Gabbs
Magnesite-Brucite Mine,
Gabbs district, Nye County.
S. White collection.
Photgraph by S. White.

Cassiterite and Hematite,
3.4 cm high, Izenhood
district, Lander County.
W. M. Keck Museum.
Photograph by J. Scovil.

Chalcophyllite and Adularia, 2.7 cm wide, Majuba Hill Mine, Antelope district,
Pershing County. S. Kleine collection. Photograph by J. Scovil.

Cinnabar, crystals to 2 cm, Poverty Peak district, Humboldt County. W. M. Keck Museum. Photograph by J. Scovil.

Clinoclase (blue) and Cornubite (green), 1-cm-wide field, Majuba Hill Mine, Antelope district, Pershing County. S. Kleine collection. Photograph by J. Scovil.

Clinozoisite on Quartz, 8.3 cm high, Nightingale district, Washoe County. W. M. Keck Museum. Photograph by J. Scovil.

Conichalcite on Olivenite, 2-mm field, Simon Mine, Bell district, Mineral County. S. White collection. Photograph by S. White.

Conichalcite (green) on Jarosite (brown), jarosite crystals to 1.7 mm, Bristol Silver Mine, Bristol district, Lincoln County. S. White collection. Photograph by S. White.

Copper, 10.5 cm wide, Rio Tinto Mine, Mountain City district, Elko County. S. Kleine collection. Photograph by J. Scovil.

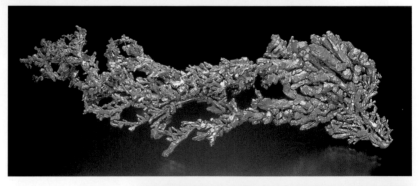

Cornetite, crystals to 0.1 mm, Empire-Nevada Mine, Yerington district, Lyon County. S. White collection. Photograph by S. White.

Creedite, 2-cm field, Hall Mine, San Antone district, Nye County. S. Kleine collection. Photograph by J. Scovil.

Cuprite on Copper, cuprite crystals to 0.9 cm, Kimberly, Robinson district, White Pine County. W. M. Keck Museum. Photograph by J. Scovil.

Epidote, 3.6 cm high, Julie claim, Pamlico district, Mineral County. W. Lombardo collection. Photograph by J. Scovil.

Faustite, crystals to 0.1 mm, Copper King Mine, Eureka County. S. White collection. Photograph by S. White.

Fluorite on Barite, fluorite crystals to 2 cm, Boulder Hill Mine, Wellington district, Lyon County. S. Kleine collection. Photograph by S. Castor.

Galkhaite, crystals to 3 mm, Getchell Mine, Potosi district, Humboldt County. D. Tretbar collection. Photograph by S. Sears.

Getchellite and Orpiment on Stibnite, getchellite crystals to 2 cm, Twin Creeks Mine, Potosi district, Humboldt County. G. Ferdock collection. Photograph by S. Castor.

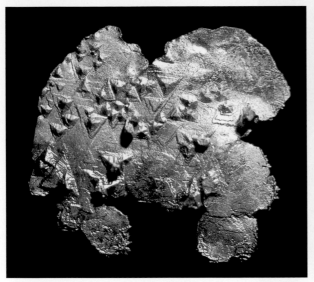

Gold, 1 cm wide, Mad Mutha Mine, Ten Mile district, Humboldt County. E. Muceus collection. Photograph by S. Castor.

Gold, 0.4 cm high, Humboldt Mountain, Mill City district, Humboldt County. E. Coogan collection. Photograph by J. Scovil.

Gold, 3.1 cm high, Mexican Mine, Comstock district, Storey County. G. Witters collection. Photograph by J. Scovil.

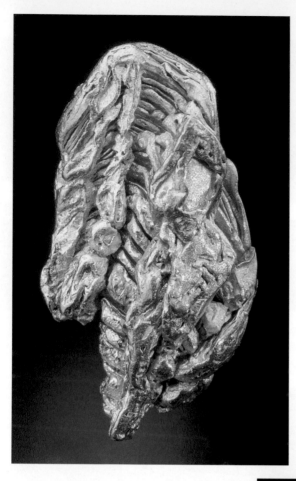

Gold, 3.2 cm high, Majuba placer,
Pershing County. S. Kleine collection.
Photograph by J. Scovil.

Gold, 3.3 cm high, Round Mountain
Mine, Round Mountain district, Nye
County. L. McMaster collection.
Photograph by J. Scovil.

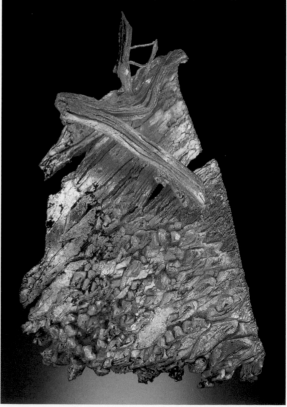

Grossular, crystals to
2.4 cm, Nightingale
district, Pershing County.
W. M. Keck Museum.
Photograph by J. Scovil.

Grossular, crystals to 1.7 cm,
Buena Vista district, Mineral
County. Clark County Museum.
Photograph by J. Scovil.

Gypsum, crystals to 12 cm, Robinson district,
White Pine County. W. M. Keck Museum.
Photograph by J. Scovil.

Heterogenite, 2 cm high, Blue Jay Mine, Goodsprings district, Clark County. W. Lombardo collection. Photograph by J. Scovil.

Hewettite on Opal, 5-mm spray, Gold Quarry Mine, Maggie Creek district, Eureka County. J. Weissman collection. Photograph by J. Weissman.

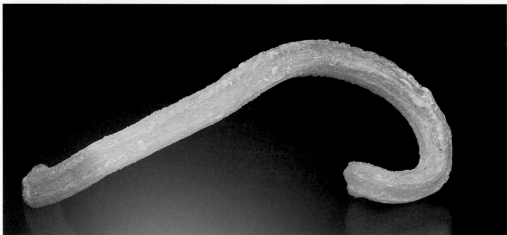

Kalinite "ram's horn," 12.1 cm wide, Alum Mine, Alum district, Esmeralda County. B. and F. Cureton collection. Photograph by J. Scovil.

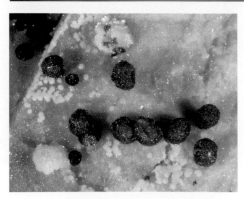

Kazakhstanite (spheroids) and Strengite (yellow), 0.2-cm-wide field, Gold Quarry Mine, Maggie Creek district, Eureka County. J. Weissman collection. Photograph by J. Weissman.

Kleinite on Opal, crystals to 2 mm, McDermitt Mine, Opalite district, Humboldt County. S. White collection. Photograph by S. White.

Lavendulan, 0.7-mm spray, San Rafael Mine, Lodi district, Nye County. S. White collection. Photograph by S. White.

Libethenite, 2.5-mm spray, Snowstorm Mine, North Battle Mountain district, Lander County. R. Wahlstrom collection. Photograph by S. White.

Linarite, 3.5-cm-wide field, Goodsprings district, Clark County. W. M. Keck Museum. Photograph by J. Scovil.

Mackayite, cystals to 1 mm, Mohawk Mine, Goldfield district, Esmeralda County. R. Thomssen collection. Photograph by S. White.

Magnesio-axinite, 1.6 cm high, unnamed prospect near Luning, Mineral County. S. Pullman collection. Photograph by J. Scovil.

Malachite, 10 cm high, Bristol Silver Mine, Bristol district, Lincoln County. T. Potucek collection. Photograph by J Scovil.

Malachite and Azurite, 4.6-cm-wide field, Robinson district, White Pine County. W. M. Keck Museum. Photograph by J. Scovil.

Malachite after Azurite,
2 cm high, Mount Potosi,
Goodsprings district, Clark
County. S. Scott collection.
Photograph by J. Scovil.

Mcguinnessite, 2.5-mm-wide field, West Gate district, Churchill
County. S. White collection.
Photograph by S. White.

Marcasite on Barite, 3 cm high,
Rosebud Mine, Rosebud district,
Pershing County. N. Prenn collection.
Photograph by J. Scovil.

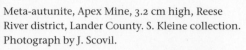

Meta-autunite, Apex Mine, 3.2 cm high, Reese River district, Lander County. S. Kleine collection. Photograph by J. Scovil.

Meurigite, 1.4-mm spray, Silver Coin Mine, Humboldt County. S. White specimen. Photograph by S. White.

Miargyrite, 5-cm-wide cluster, Rosebud Mine, Rosebud district, Pershing County. K. Allen specimen. Photograph by S. Castor.

Microcline and Quartz, 6 cm high, Newberry Mountains, Clark County. S. Scott collection. Photograph by J. Scovil.

Millerite, 1.8 cm wide, Meikle Mine, Lynn district, Elko County, Geoprime Minerals specimen. Photograph by J. Scovil.

Mimetite, crystals to 2.2 mm, Simon Mine, Bell district, Mineral County. S. White collection. Photograph by S. White.

Naumannite, crystals to 1 cm, Ken Snyder Mine, Gold Circle (Midas) district, Elko County. Ken Snyder Mine specimen. Photograph by S. Castor.

Nevadaite, 9 mm wide, Gold Quarry Mine, Maggie Creek district, Eureka County. S. Kleine Collection. Photograph by J. Scovil.

Opal after wood, 2.5-cm diameter, Virgin Valley district, Humboldt County. B. and F. Cureton collection. Photograph by J. Scovil.

Opal, faceted, 20 x 23 mm, Virgin Valley district, Humboldt County. B. and F. Cureton collection. Photograph by J. Scovil.

Orpiment, 1.1 cm high, Getchell Mine, Potosi district, Humboldt County. G. Ferdock collection. Photograph by J. Scovil.

Orpiment, 10.2 cm high, Twin Creeks Mine, Potosi district, Humboldt County. Collector's Edge specimen. Photograph by J. Scovil.

Powellite, 2.5-cm vertical field, Tonopah-Divide Mine, Divide district, Nye County. B. and F. Cureton collection. Photograph by J. Scovil.

Pyrargyrite, crystals to 6 mm, Tonopah district, Nye County. W. M. Keck Museum. Photograph by J. Scovil.

Pyrite on Barite, 5.3 cm wide, Meikle Mine, Lynn district, Elko County. S. Kleine collection.
Photograph by J. Scovil.

Pyromorphite, 2.4-cm field, Bull Run Canyon,
Elko County. W. M. Keck Museum.
Photograph by J. Scovil.

Pyromorphite after Galena with Conichalcite,
3.6-cm vertical field, Bristol Silver Mine, Bristol
district, Lincoln County. T. Potucek collection.
Photograph by J. Scovil.

Quartz, 8.5 cm high, Lovelock area, Pershing County. B. and F. Cureton collection. Photograph by J. Scovil.

Quartz epimorph after Calcite, 11.4 cm high, Aurora, Esmeralda County. N. Prenn collection. Photograph by J. Scovil.

Quartz (amethyst) and Calcite, crystals to 2 cm, Comstock district, Storey County. W. M. Keck Museum. Photograph by J. Scovil.

Quartz (amethyst), faceted, 38 x 57 mm, Petersen Mountain, Washoe County. J. Johnson collection. Photograph by J. Scovil.

Quartz with Spessartine inclusions, 2-cm field, Newberry Mountains, Clark County. W. Lombardo collection. Photograph by J. Scovil.

Quartz, 13.2-cm high, Petersen Mountain, Washoe County. Hallelujah Mining specimen. Photograph by J. Scovil.

Quartz with Stibnite inclusions, 3 cm wide, Trinity Mountains, Pershing County. J. Gray collection. Photograph by S. Castor.

Realgar and Fluorite, 6.5 cm wide, Getchell Mine, Potosi district, Humboldt County. S. Kleine collection. Photograph by J. Scovil.

Rhodochrosite spheroids (orange) and Barite (white), spheroids to 5 mm, Lone Tree Mine, Buffalo Mountain district, Humboldt County. S. White collection. Photograph by S. White.

Richelsdorfite, 1.6-mm sphere, Burrus Mine, Pyramid district, Washoe County. S. White collection. Photograph by S. White.

Scheelite and Gold on Quartz, white scheelite crystals to 0.6 cm, Olinghouse Mine, Olinghouse district, Washoe County. Olinghouse Mine specimen. Photograph by J. Scovil.

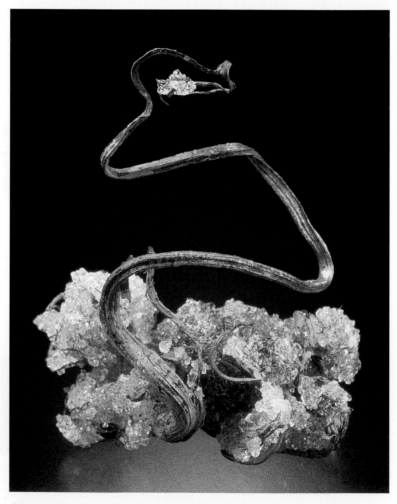

Silver and Quartz, 2 cm high, Hale and Norcross Mine, Comstock district, Storey County. G. Witters collection. Photograph by J. Scovil.

Silver, 5 cm high, Reese River district,
Lander County. B. and F. Cureton collection.
Photograph by S. Castor.

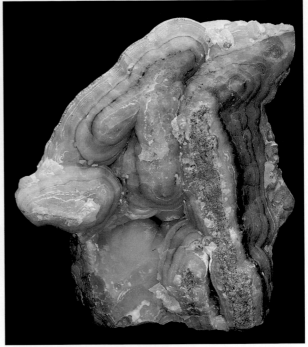

Smithsonite, 12 cm high, Goodsprings district, Clark
County. W. M. Keck Museum. Photograph by J. Scovil.

Sonoraite, 1-mm sprays, McGinnity shaft,
Goldfield district, Esmeralda County.
S. White collection. Photograph
by S. White.

Sphalerite and Galena, crystals to
2 cm, Sixteen-to-One Mine, Silver
Peak district, Esmeralda County.
W. Lombardo collection.
Photograph by J. Scovil.

Stephanite, crystals to 7 mm, Comstock district, Storey County. W. M. Keck Museum. Photograph by J. Scovil.

Stibnite, 6.7 cm high, Betze-Post Mine, Lynn district, Eureka County. G. Ferdock collection. Photograph by J. Scovil.

Stibnite and Calcite, 3.3-cm spray, White Caps
Mine, Manhattan district, Nye County. B. and F.
Cureton collection. Photograph by J. Scovil.

Stibnite, 15.3 cm high, Murray Mine,
Independence Mountains district, Elko County.
S. Kleine collection. Photograph by J. Scovil.

Sulfur, 6.6 cm high,
Alum Mine, Alum
district, Esmeralda
County. W. M.
Keck Museum.
Photograph by
J. Scovil.

Topaz and Quartz, 13.3 cm wide, Zapot claim, Fitting district, Mineral County. H. Gordon Minerals specimen. Photograph by J. Scovil.

Torbernite, 6-mm crystal, Majuba Hill Mine, Antelope district, Pershing County. S. Kleine collection. Photograph by J. Scovil.

Turquoise after Orthoclase, 1.5 cm high, Mina area, Pilot Mountains district, Mineral County. B. and F. Cureton collection. Photograph by J. Scovil.

Turquoise, 1-mm spheroids, Silver Coin Mine, Iron Point district, Humboldt County. S. White collection. Photograph by S. White.

Vanadinite on Descloizite, crystals to 3 mm, Chalk Mountain Mine, Chalk Mountain district, Churchill County. S. Pullman collection. Photograph by J. Scovil.

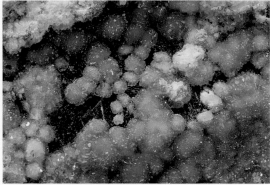

Vanadinite, crystals to 2 mm, Ruth Mine, Goodsprings district, Clark County. S. Scott collection. Photograph by S. White.

Variscite, 2.4-cm-wide field, Betze-Post Mine, Lynn district, Eureka County. S. Kleine collection. Photograph by J. Scovil.

Vivianite, 12-cm-wide field, Robinson district, White Pine County. W. M. Keck Museum. Photograph by J. Scovil.

Wakabayashilite, fibers to 2 cm, White Caps Mine, Manhattan district, Nye County. B. and F. Cureton collection. Photograph by J. Scovil.

Wulfenite, crystals to 7 mm, Ruth pit, Robinson district, White Pine County. W. Lombardo collection. Photograph by J. Scovil.

FAMATINITE

Cu_3SbS_4

Famatinite is a relatively uncommon mineral that is found in small amounts in hydrothermal copper deposits and in some high-sulfidation epithermal precious-metal deposits such as at Goldfield.

ESMERALDA COUNTY Reported in high-grade ore at the Clermont, Florence, Jumbo, Jumbo Extension, January, Merger (St. Ives), Mohawk, and Combination Mines in the Goldfield district. Described as metallic masses and crude crystals of pinkish- to dark-gray color with a tinge of copper-red; associated with quartz, pyrite, bismuthinite, goldfieldite, and gold. Also noted with barite at the Sandstorm Mine. (Ransome, 1909a, 1910a, 1910b; FUMT; AMNH; HMM; SBC)

FAUSTITE

$(Zn,Cu^{2+})Al_6(PO_4)_4(OH)_8 \bullet 4H_2O$

The Nevada type mineral, faustite, apparently forms in lieu of its relative turquoise in some settings where zinc is available.

ELKO COUNTY In pale-blue crusts with variscite crystals at the Rain Mine in the Carlin district. (M. Jensen, w.c., 1999)

EUREKA COUNTY Type locality as apple-green veins and tiny crystals to 0.25 inch or as nodules in argillized, fine-grained Paleozoic sedimentary rock at the Copper King Mine in the Maggie Creek district, SE 1/4 SE 1/4 Sec. 28, T34N, R51E; associated minerals include montmorillonite, quartz, alunite, chrysocolla, malachite, azurite, and cuprite. (Erd et al., 1953)

HUMBOLDT COUNTY As blue-green radiating crystals on limonitic quartz at an unspecified locality near Valmy. (Jenkins, 1981)

LANDER COUNTY In bright yellow-green nuggets with turquoise at the Carico Lake Mine. (Rose and Ferdock, this publication)

FAYALITE

$Fe_2^{2+}SiO_4$

Fayalite, the iron end member of the *olivine* group, is less common than the magnesium end member forsterite. It occurs in a wide variety of volcanic rocks as a phenocryst mineral and a product of vapor-phase alteration; it is also found in plutonic rocks and in some metamorphic rocks.

ELKO COUNTY As phenocrysts and microlites in topaz rhyolite in the Toano Range. (Price et al., 1992)

HUMBOLDT COUNTY As phenocrysts in peralkaline ash-flow tuff and intrusive rocks of the McDermitt caldera. (Conrad, 1984; Rytuba and McKee, 1984; Castor et al., 1996)

LINCOLN COUNTY As phenocrysts in peralkaline rocks of the Kane Springs Wash caldera. (Noble, 1968)

NYE COUNTY As a phenocryst mineral in intermediate to acidic lavas in the Mellan Hills in the Mellan Mountain district on the Nellis Air Force Range. (Ekren et al., 1971)

As phenocrysts in a number of peralkaline tuffs, such as the tuff of the Belted Range, and in flow rocks such as the Kawich Valley Rhyolite on the Nellis Air Force Range. (Noble et al., 1968; Ekren et al., 1971)

FEDOTOVITE

$K_2Cu_3^{2+}O(SO_4)_3$

A mineral that has been reported elsewhere as occurring with tenorite and other minerals as encrustations around fumaroles at a volcano in Kamchatka, Russia.

CHURCHILL COUNTY Identified by XRD in a specimen from the Nickel Mine area, Table Mountain district. (S. Lutz, w.c., 2000)

FERBERITE

$Fe^{2+}WO_4$

Ferberite, which occurs in both skarn and vein tungsten deposits, is the iron end member of the *wolframite* group and is less common than the manganese end member hübnerite. Some Nevada occurrences listed under *wolframite* may be of ferberite.

LINCOLN COUNTY Reported from an unspecified locality near Panaca; possibly the Comet Mine in the Comet district. (Gianella, 1941)

NYE COUNTY Reported with scheelite in a specimen from an unknown locality near Beatty; possibly the Grand Junction Mine in the Bare Mountain district. (Gianella, 1941; WMK)

WHITE PINE COUNTY Reported with ferrocolumbite and ilmenite in cavities in *garnet*-bearing rhyolite near Ruth in the Robinson district. (M. Rogers, w.c., 1999; M. Jensen, w.c., 1999)

FERGUSONITE-(Y)

$YNbO_4$

Fergusonite-(Y) is widespread in small amounts in granite pegmatite and placer deposits. The Nevada occurrence is unique.

PERSHING COUNTY Reported as yellow-orange microcrystals with pharmacosiderite and devilline at the Majuba Hill Mine in the Antelope district. (M. Jensen, w.c., 1999)

FERRIERITE

$(Na,K,Mg_{0.5},Ca_{0.5})_6[Al_6Si_{30}O_{72}] \bullet 18H_2O$
(general)

A zeolite group mineral that has recently been subdivided into three species depending on the most abundant alkali or alkali earth element. *Ferrierite* is relatively rare; it is found in amygdules in mafic volcanic rocks and as a product of alteration in felsic tuffaceous rocks. Species have not been determined for Nevada occurrences.

CHURCHILL COUNTY Minor with *clinoptilolite* and mordenite in altered pyroclastic rocks about 35 miles southeast of Fallon in the Sand Springs district. (Papke, 1972a)

PERSHING COUNTY In a large zeolite deposit in bedded tuff northwest of Lovelock, Sec. 29, T28N, R30E, as aggregates of crystals to 12 microns in a large deposit in bedded tuffs with mordenite, *clinoptilolite*, tridymite, and cristobalite; some of the tuff is nearly pure *ferrierite*. (Sand and Regis, 1968; Rice et al., 1992)

FERRIOMOLYBDITE

$Fe_2^{3+}(Mo^{6+}O_4)_3 \bullet 8H_2O$ (?)

Ferrimolybdite is a common mineral in molybdenum deposits, where it is an oxidation product of molydenite.

ESMERALDA COUNTY As yellow blooms directly above rocks that contain powellite at the Tonopah-Divide Mine in the Divide district; also reported as bright-yellow aggregates of minute needles on the 165-foot level. (Knopf, 1921a; Schilling, 1962b)

As yellow aggregates in quartz veins with powellite at the Slate Ridge prospects, 3 to 4 miles south of Goldpoint in the Gold Point district. (Schilling, 1979)

In *limonite* and coating fractures at the Red Hill (Redlich) Mine in the Rock Hill district. (Schilling, 1979)

As a widespread bright yellow oxidation product of molybdenite at the Cucomungo deposit in the Tule Canyon district. (Schilling, 1979)

HUMBOLDT COUNTY As yellow coatings on quartz veins with molybdenite, chalcopyrite, and other minerals at the Desert View prospect in the Leonard Creek district. (Kirkemo et al., 1965; Schilling, 1979)

In pure yellow masses to 2 cm in the Moly pit at the Getchell Mine in the Potosi district. (Dunning, 1987; M. Jensen, w.c., 1999; AMNH)

NYE COUNTY As rare coatings near oxidized molybdenite flakes in quartz at the Outlaw Mine in the Round Mountain district. (Dunning et al., 1991)

Abundant in the oxidized zone at the Hall property in the San Antone district. (Michell, 1945)

WHITE PINE COUNTY As yellow masses and crystals to 1.5 mm on oxidized skarn with molybdenite at the Mt. Hamilton Mine in the White Pine district. (GCF)

FERRITUNGSTITE

$(W^{6+},Fe^{3+}_2)(O,OH)_6 \cdot pH_2O$; p up to 1.75

Ferritungstite is a rare mineral that occurs in the oxidation zones of some tungsten deposits.

HUMBOLD COUNTY Clove-brown bladed microcrystals with scheelite, fluorapatite, and laumontite at the Mountain King Mine in the Potosi district. (S. Sears, w.c., 1999)

MINERAL COUNTY As bright-yellow powder, microscopic dipyramidal crystals, and tiny fibers in cavities in limonitic gossan with bismutite and jarosite in oxidized scheelite-bearing skarn at the Nevada Scheelite Mine in the Leonard district. (Richter et al., 1957; Ross, 1961)

Reported with magnesio-axinite, prehnite, vesuvianite, actinolite, and epidote at an unspecified locality near Luning. May be the same locality as *axinite* with zoisite, *tourmaline,* prehnite, and other minerals reported by Melhase (1935e) about 7 miles north-northwest of Luning. (Dunn et al., 1980; S. Pullman, w.c., 1999)

FERRO-AXINITE

$Ca_2Fe^{2+}Al_2BSi_4O_{15}(OH)$

The common iron-rich end member of the *axinite* group; found in regionally and contact-metamorphosed rocks and in granite pegmatite.

HUMBOLDT COUNTY Reported in a specimen from the Mountain King Mine in the Potosi district. (R. Thomssen, w.c., 1999)

FERROCOLUMBITE

$Fe^{2+}Nb_2O_6$

A *columbite*-group mineral that generally occurs in granite pegmatite and is less common in carbonatite.

CLARK COUNTY As black crystals to 2 cm with microcline in the Newberry prospect near Ireteba Peak; also reported as *columbite-tantalite* microcrystals in miarolitic cavities and pegmatites in the Eldorado and Newberry Mountains. (M. Jensen, w.c., 1999; S. Scott, w.c., 1999)

WHITE PINE COUNTY As microcrystals with ilmenite and ferberite in lithophysal cavities in *garnet*-bearing rhyolite near Ruth in the Robinson district. (M. Rogers, w.c., 1999; M. Jensen, w.c., 1999)

FERRO-EDENITE

$NaCa_2Fe^{2+}_5Si_7AlO_{22}(OH)_2$

An amphibole-group mineral that occurs in alkaline igneous rocks.

LINCOLN COUNTY As phenocrysts in late rhyolite flows of the Kane Springs Wash volcanic center. (Novak, 1985)

FERROHORNBLENDE

$\square Ca_2[Fe^{2+}_4(Al,Fe^{3+})]Si_7AlO_{22}(OH)_2$

Ferrohornblende is a very widespread member of the *hornblende* group in igneous and metamorphic rocks, but many localities lack chemical confirmation.

WASHOE COUNTY Black crystals to 2 cm in wall rock of pegmatites at Incline Village. (M. Jensen, w.c., 1999)

As black phenocrysts to 4 cm in dacite on the southeast flank of Peavine Mountain northwest of Reno. (GCF)

FERVANITE

$Fe^{3+}_4(VO_4)_4 \cdot 5H_2O$

A secondary vanadium mineral that is best known from uranium-vanadium mines of the Colorado Plateau.

EUREKA COUNTY Possibly the world's best specimens of the species occur as radiating clusters to 0.5 mm of brown to yellow-brown or pale-green, lathlike crystals and thin botryoidal coatings on black kazakhstanite at the Gold Quarry Mine in the Maggie Creek district. (Jensen et al., 1995)

FIBROFERRITE

$Fe^{3+}(SO_4)(OH) \cdot 5H_2O$

Fibroferrite is a relatively uncommon mineral that occurs in hot-spring deposits and as an oxidation product after pyrite in some ore deposits with other iron sulfates such as jarosite.

ESMERALDA COUNTY Reported in a specimen from an unnamed occurrence near Coaldale. (WMK)

HUMBOLDT COUNTY As light-brown crystalline masses with copiapite, melanterite, and other supergene iron minerals at the Cordero Mine in the Opalite district. (Fisk, 1968)

NYE COUNTY As yellow acicular prisms to 0.1 mm in quartz vugs in the silver-tin ore with botryogen, bindheimite, copiapite, rozenite, valentinite, and native sulfur in the Upper Magnolia workings at the Morey Mine in the Morey district. (Jenkins, 1981)

WASHOE COUNTY As a partial replacement product of melanterite stalactites to 40 cm long and 15 cm in diameter in lower workings at the Commonwealth Mine in the Galena district. (Scull, 1951)

FISCHESSERITE

Ag_3AuSe_2

A rare mineral that occurs in hydrothermal precious-metal veins, generally in combination with gold and naumannite.

ELKO COUNTY With gold, aguilarite, naumannite, and other minerals in precious-metal veins at the Ken Snyder and Bonberger Mines in the Gold Circle (Midas) district; also identified by SEM/EDX as grains to 100 microns (22.4 percent Se, 47.55 percent Ag, and 30.09 percent Au by weight) with naumannite as rims around gold in a specimen from the Eastern Star Mine. (Casteel and Bernard, 1999; L. Larson, o.c., 1998; SBC)

FIZÉLYITE

$Pb_{14}Ag_5Sb_{21}S_{48}$ (?)

Fizélyite is a rare mineral that occurs in hydrothermal veins. The Nevada localities listed are the only ones in the United States.

CLARK COUNTY As micron-sized grains with potarite, gold, and other minerals at the Boss Mine in the Goodsprings district. (Jedwab et al., 1999)

NYE COUNTY As deeply striated or fluted prisms to 12 mm in quartz-rhodochrosite veins in the Morey district; fizélyite is an intermediate step in the replacement of andorite by owyheeite. (Williams, 1968)

FLORENCITE

$(La,Nd,Ce)Al_3(PO_4)_2(OH)_6$ (general)

A mineral group that generally occurs in alkaline igneous rock and carbonatite. *Florencite* is subdivided into three species depending on the dominant rare earth element present. Definitive analyses are not available for the occurrence listed below.

CLARK COUNTY Reported as LaCe-aluminosulfophosphate with Ba and Sr in association with bitumen at the Boss Mine in the Goodsprings district. (Jedwab et al., 1999)

FLUELLITE

$Al_2(PO_4)F_2(OH) \cdot 7H_2O$

A relatively uncommon mineral that occurs in pegmatite and vein deposits.

EUREKA COUNTY As brilliant, transparent, colorless, yellow, or dark-purple to black crystals to 2 mm in radiating, flattened clusters on fracture surfaces with other phosphate minerals, hewettite, torbernite, and anatase at the Gold Quarry Mine in the Maggie Creek district. (Jensen et al., 1995)

PERSHING COUNTY Nevada's finest specimens and likely the largest crystals of the species in the world are euhedral, colorless to very-pale-purple, glassy crystals to 1 cm and partial crystals to 1.5 cm with wavellite and other phosphates in quartz vugs at the Willard Mine in the Willard district. (M. Jensen, w.c., 1999; Jensen and Leising, 2001)

FLUOBORITE

$Mg_3(BO_3)(F,OH)_3$

Fluoborite is a relatively rare mineral that occurs in trace amounts in some contact-metamorphic deposits.

LINCOLN COUNTY Reported with ludwigite and szaibélyite in contact-metamorphosed rock in a copper prospect on the west base of Blind Mountain in the Bristol district. (Westgate and Knopf, 1932)

FLUORAPATITE

$Ca_5(PO_4)_3F$

The most common member of the *apatite* group, fluorapatite is a widespread accessory mineral in igneous and metamorphic rocks, and also occurs in hydrothermal veins.

CHURCHILL COUNTY Relatively abundant as flesh-colored opaque crystals to 15 cm in vugs with calcite and actinolite at the Minerals Materials (Buena Vista) Iron Mine in the Mineral Basin district. (M. Jensen, w.c., 1999)

CLARK COUNTY In granite in the Eldorado and Newberry Mountains. (S. Scott, w.c., 1999)

ESMERALDA COUNTY Reported in a specimen from the Tonopah-Divide Mine in the Divide district. (R. Thomssen, w.c., 1999)

EUREKA COUNTY As pearly white crystals to 5 mm in shallow oxidized zones at the Goldstrike Mine in the Lynn district. (GCF)

As white to pale-green spheroids to 2.5 mm at the Gold Quarry Mine in the Maggie Creek district. (GCF)

HUMBOLDT COUNTY A minor component of gangue at the Moonlight Uranium Mine in the Disaster district. (Castor et al., 1996)

Reported as sulfur-bearing at the Silver Coin Mine in the Iron Point district. (Thomssen, 1998)

As prismatic, light-blue microcrystals with scheelite, *axinite,* and laumontite at the Mountain King Mine in the Potosi district; also reported in a specimen from the south pit of the Getchell Mine. (S. Sears, w.c., 1999; R. Thomssen, w.c., 1999)

LANDER COUNTY In the skarn assemblage at the West orebody in the Battle Mountain district. (Theodore and Blake, 1978)

LYON COUNTY Reported in a specimen from the Flyboy Mine in the Cambridge Hills. (R. Thomssen, w.c., 1999)

As gray euhedral crystals to 1.5 cm in clusters in pegmatites at an unnamed prospect near Sweetwater. (M. Jensen, w.c., 1999)

NYE COUNTY With variscite, strengite, and wardite at the Vashegyite Gem Mine in the Manhattan district. (WMK)

As minute white balls on *limonite*-stained

jasperoid at the Northumberland Mine in the Northumberland district. (Kokinos and Prenn, 1985)

PERSHING COUNTY As pale-pink to flesh-colored, opaque, euhedral crystals to 5 cm with magnetite crystals and actinolite in vugs at the Segerstrom-Heizer Mine in the Mineral Basin district. (M. Jensen, w.c., 1999)

WASHOE COUNTY Rare as euhedral green crystals to 8 mm with microcline in pegmatite at Incline Village. (Jensen, 1993a)

FLUORAPOPHYLLITE

$KCa_4Si_8O_{20}(F,OH) \cdot 8H_2O$

A species in the apophyllite group; occurs as a secondary mineral in cavities in basalt, in contact-metamorphosed rocks, and as a late-stage hydrothermal phase.

HUMBOLDT COUNTY Noted as clear euhedral microcrystals on calcite at the Getchell Mine in the Potosi district. (S. Sears, w.c., 1999)

FLUORITE

CaF_2

Fluorite is the most common mineral that contains fluorine. It is found in many ore deposits, both as a gangue species and as a principal economic mineral, and it is also found as an accessory mineral in some igneous rocks. The mineral has been mined from many Nevada deposits; detailed descriptions of these may be found in Papke (1979).

CHURCHILL COUNTY As purple cubes to 2 mm with quartz at an unnamed prospect on Chalk Mountain in the Chalk Mountain district. (M. Jensen, w.c., 1999)

As lavender masses with hübnerite at an unnamed occurrence in the Eastgate district. (Hess, 1917)

Reported in veins at the Fairview Mine in the Fairview district. (Schrader, 1947)

About 1900 tons of fluorspar were produced from a vein along a fault in Mesozoic slate and limestone at the Revenue Mine in the I.X.L. district; purple and black fluorite has been reported, as well as white to gray cubic crystals to 4 cm and cubo-octahedral green crystals to 1 cm. (Vanderburg, 1940; Papke, 1979; GCF)

As white crystals to 7 mm at the Green prospect in the Lake district. (R. Thomssen, w.c., 1999)

Mined from veins and vugs along an a contact between granitic rock and Mesozoic sedimentary rock at the Dixie Mine in the Mountain Wells district, where it occurs locally as pale-purple cubic crystals to 1 cm; also found as orange-brown cubic crystals to 1.5 cm in veinlets in limestone at the La Plata (Michigan) prospect. (Papke, 1979; Vanderburg, 1940; M. Jensen, w.c., 1999)

As pods in the footwall of a fault at the Merkt prospect in the Westgate district; also reported as colorless cubic crystals to 5 mm near Westgate. (Papke, 1979; M. Jensen, w.c., 1999)

Reported in skarn at the Nevada United Mine in the White Cloud district. (Vanderburg, 1940)

In the Wonder district in veins and masses in volcanic rock at the Little Jim and Purple Spar prospects, in quartz-*adularia* veins with sulfides at the Wonder Mine, and as purple octahedra to 5 mm in clusters on quartz at the Spider and Wasp claim. (Papke, 1979; Schilling, 1979; M. Jensen, w.c., 1999)

CLARK COUNTY In veins in gneiss at the Fluoric Mine in the Crescent district. (Vanderburg, 1937b; WMK)

As crystals to 3 cm in miarolitic cavities in the Eldorado and Newberry Mountains; also in the host granite. (S. Scott, w.c., 1999)

As an accessory mineral in Precambrian rapakivi granite in the Gold Butte district; purple fluorite occurs with other minerals in pegmatite at the Hilltop Mine and greenish fluorite occurs with clear quartz, sulfides, and beryllium minerals at the Webster (Radio Crystal) Mine. (Volborth, 1962b; Longwell et al., 1965; M. Jensen, w.c., 1999)

Reported in a specimen from an unnamed locality in the Goodsprings district. (LANH)

ELKO COUNTY In precious-metal veins with calcite, *adularia,* quartz, pyrite, and barite at the Ken Snyder Mine in the Gold Circle district. (Casteel and Bernard, 1999)

In a specimen from the Aljo prospect in the Merrimac district. (R. Thomssen, w.c., 1999)

As microlites in topaz rhyolite vitrophyre in the Toano Range. (Price et al., 1992)

ESMERALDA COUNTY Lining vugs as purple masses and cubes to 8 mm in a quartz vein at the Bullfrog-George prospect in the Gold Point district. (Ball, 1906; Schilling, 1979)

In a quartz vein with sulfide at the Flora prospect in the Rock Hill district; also as purple aggregates in quartz veins with *adularia* at the Red Hill (Redlich) Mine and in specimens from the Broken Toe and Rock Hill Mines. (Horton, 1916; Papke, 1979; Schilling, 1979; R. Thomssen, w.c., 1999)

In skarn at the Amry prospect in the Sylvania district. (Horton, 1916; Papke, 1979)

In gold-bearing placer gravels in the Tule Canyon district. (Garside, 1973)

EUREKA COUNTY With beryl and hematite in large bodies delineated by drilling in Ordovician sedimentary rock at the Bisoni prospect in the Fish Creek district; also with beryl at the Reese and Berry prospect. (Griffiths, 1964; Papke, 1979; Holmes, 1963)

As purple, pale-green, and clear masses and cubes to 3 cm with realgar and orpiment in calcite veins at the Goldstrike Mine in the Lynn district. (GCF)

HUMBOLDT COUNTY Pale-purple to gray in a quartz vein at the Sunset prospect in the Black Rock district. (Papke, 1979)

White to dark-purple with quartz, pyrite, zircon, and secondary uranium minerals at the Moonlight Uranium Mine in the Disaster district. (Rytuba, 1976; Castor et al., 1996; Castor and Henry, 2000)

As colorless and purple cubic crystals to 1 mm with brochantite on fractures at the Silver Coin Mine in the Iron Point district. (Thomssen, 1998; M. Jensen, w.c., 1999; J. McGlasson, w.c., 2001)

Rare at the Cordero Mine in the Opalite district. (Bailey and Phoenix, 1944)

Common as zoned clear to violet cubes to 1 cm with orpiment, realgar, and galkhaite at the Getchell Mine in the Potosi district. (Joralemon, 1951; Stolburg and Dunning, 1985; F. Cureton, w.c., 1999)

LANDER COUNTY About 900 tons were mined from the Iowa Canyon Mine in the Iowa Canyon district, where it occurred in a brilliant green to pale-gray vein to 4.5 feet wide; smaller amounts of pale-purple to grayish-green or white fluorite were mined from the Nevada Fluorspar Mine; also as disseminated grains with quartz, calcite, dolomite, *sericite,* montmorillonite, and ralstonite, and as green octahedral crystals to 4 cm in vugs at the Blazer prospect. (Papke, 1979; M. Jensen, w.c., 1999)

Reported in a specimen from the Cirac property in the Reese River district. (LANH)

LINCOLN COUNTY At six prospects in a 5-square-mile area in veins, pods, and masses, commonly with quartz, in silicified and altered volcanic and volcaniclastic rocks; noted in vuggy masses of crystals to 1 cm at the Blue Bell prospect in the Quinn Canyon district. (Sainsbury and Kleinhampl, 1969; Papke, 1979)

In tungsten-bearing skarn at the Lincoln Tungsten Mine in the Tem Piute district. (Buseck, 1967; JHS)

In disseminated grains with nodules and veinlets of gearksutite in altered quartz latite at the Cougar prospect in the Vigo district. (Papke, 1979)

As fine-grained, mostly light-gray to white material in a replacement body in Paleozoic dolomitic rock from which about 45,000 tons was mined at the Carp (Wells Cargo) Mine in the Viola district. (Tschanz and Pampeyan, 1970; Papke, 1979)

As a replacement body with chalcedonic quartz in tuff at the Fluorite Basin prospect in the Viola district. (Papke, 1979)

LYON COUNTY In light-blue, pale-gray, and pale-pinkish-brown crystals to 6 cm in jasperoid replacing dolomitic limestone at the Boulder Hill Mine in the Wellington district. (Moore, 1969; Papke, 1979; GCF)

MINERAL COUNTY With quartz at the Prospectus vein in the Aurora district. (UCB; S. Pullman, w.c., 1999)

About 182,000 tons of fluorspar were produced from the Baxter (Kaiser) Mine in the Broken Hills district, mainly from a single vein in Tertiary andesite. The fluorite has been described as very fine-grained; layered white, pale-green, and lavender aggregates; and as crystals to 2.5 cm across. Smaller deposits are at the Spardome Mine and the Little Fluorspar and West Slope prospects. (Ross, 1961; Papke, 1979; JHS)

In veins and masses in Cambrian schist at the Fluorite and Linda prospects in the Buena Vista district, with cubes to 3 cm reported at the latter; also in veins of pale-purple to clear cubes with quartz and calcite in rhyolite at the Fluorspar King and Blue Bell Mine. (Papke, 1979; Melhase, 1935e; Ross, 1961)

As colorless, green, and purple coatings, masses, and crystals in pegmatitite at the Zapot claims in the Fitting district; also noted as purple microcrystalline crusts on blue microcline. (Foord et al., 1998; Foord et al., 1999b; M. Jensen, w.c., 1999)

NYE COUNTY In the Bare Mountain district at the Daisy Mine, NW 1/4 Sec. 23, T12S, R47E, the largest fluorite producer in Nevada at more than 250,000 tons, as purple earthy to crystalline masses and white to yellow crystals in complex, brecciated replacement bodies with local calcite and cinnabar in Cambrian dolomite; also in breccia bodies in dolomite at the Diamond Queen (Goldspar) and Mary Mines. (Melhase, 1935b; Thurston, 1949; Lovering, 1954; Cornwall and Kleinhampl, 1961, 1964; Papke, 1979)

In a vein in quartz latite at the Colton prospect in the Cloverdale district. (Papke, 1979)

In a replacement body and minor associated veinlets at the White Pine prospect in the Currant district. (Papke, 1979)

With cinnabar and scheelite at the Scheebar Mine in the Fairplay district. (Bailey and Phoenix, 1944)

In calcite-opal veins with sparse cinnabar and autunite at the War Cloud prospect in the Jackson district. (Hamilton, 1993)

A minor mineral with creedite and halloysite in scheelite-bearing skarn at the Victory tungsten deposit in the Lodi district. (Melhase, 1935d; Humphrey and Wyatt, 1958)

In large masses and in cavities as small but well-formed yellow or brown crystals locally with leaves or wires of gold at the White Caps Mine in the Manhattan district; also noted in cavities at the Manhattan Consolidated and Union Amalgamated Mines, and in a breccia pipe at the Keystone Mine. (Melhase, 1935d; Gibbs, 1985; Ferguson, 1924; Papke, 1979)

In the large Quinn Canyon district, which extends south into Lincoln County, as complex orebodies that replaced Ordovician limestone at the Horseshoe Mine, which produced 28,000 tons of fluorspar; also in Tertiary volcanic rock at the less productive Nyco and Rainbow Mines and in a number of prospects in either Paleozoic carbonate or Tertiary volcanic rock. (Papke, 1979; Sainsbury and Kleinhampl, 1969)

As isolated, deep-purple to colorless crystals to

4 cm in granular masses of muscovite at the Outlaw Mine in the Round Mountain district; also rarely in gold-bearing veins and disseminations at the Round Mountain Mine, and as pink aggregates and tiny crystals in quartz-hübnerite veins from unspecified localities. (Shannon, 1925; Melhase, 1935d; Dunning et al., 1991; Tingley and Berger, 1985; Ferguson, 1922)

Reported in a specimen from an unknown locality in the Tonopah district. (WMK)

In veins and pods in silicified Triassic sedimentary rocks (limestone, claystone, and conglomerate) at several properties in the Union district; minor production has come from the Union Canyon and Chicago Lode Mines. (Kral, 1951; Papke, 1979)

With pyrite and quartz in the Limestone Ridge area on the Nellis Air Force Range. (Tingley et al., 1998a)

Reported in a specimen from the Wahmonie district on the Nevada Test Site. (WMK)

As purple material with hübnerite at unspecified locality in the Ellsworth district. (Hess, 1917)

PERSHING COUNTY Locally abundant as purple cubes to 2 mm and intergrown in masses with black *tourmaline* and quartz at the Majuba Hill Mine in the Antelope district. (Jensen, 1985)

In a brecciated zone in Triassic sedimentary rocks, Emerald Spar prospect in the Antelope Springs district; also replacing limestone at the Bohannon prospect, and as violet and green crystalline veins to 2 cm at the Relief Canyon Mine. (Papke, 1979; Johnson, 1977; M. Jensen, w.c., 1999)

As small lenses and seams in shale south of Black Knob in the Black Knob district. (Horton, 1961)

With aragonite, calcite, and *chalcedony* in replacement bodies, breccia, and minor veins, locally as purple and green octahedral crystals to 2 cm, in Triassic siltstone at the Valery (Valerie) Mine in the Imlay district; also in a quartz vein at the Piedmont Mine, and in small pegmatite pods with beryl at Lakeview Mine. (Kral, 1947; Johnson, 1977; Papke, 1979; M. Jensen, w.c., 1999)

In a replacement deposit at the Needle Peak Mine in the Mount Tobin district. (Horton, 1961; Papke, 1979)

In pegmatite with scheelite and other minerals at the Oreana Tungsten Mine in the Rye Patch district. (Olson and Hinrichs, 1960)

In pods and veinlets in Triassic shale and sandstone at the Nevada Fluorspar Mine and Suzie prospect in the Table Mountain district. (Papke, 1979)

With dolomite and calcite in veins at the Mammoth prospect in the Washiki district. (Horton, 1961; Papke, 1979)

WHITE PINE COUNTY With quartz in veins at the Hilltop prospect in the Cherry Creek district. (Papke, 1979; Horton, 1961)

In veins in volcanic breccia on Rattlesnake Knoll at the Rattlesnake Heaven prospect in the Cooper district. (Papke, 1979; Hose et al., 1976)

In purple cubes to 1 cm with hübnerite, bismuthinite, and triplite from an unspecified locality in the Kern Mountains in the Eagle district. (Hill, 1916; Melhase, 1934d)

In a vein in rhyolite at the Sawmill Canyon prospect in the Ellison district. (Papke, 1979)

As small purple grains with phenakite, beryl, pyrite, *sericite,* and scheelite in stratabound masses in limestone, and with scheelite and other minerals in quartz veins at the Mount Wheeler Mine and Jeppson claims in the Lincoln district, Sec. 15, T12N, R68E. (Hose et al., 1976; Griffiths, 1964)

Locally with manganese oxides, calcite, and quartz in replacement manganese orebodies in the Nevada district. (Horton, 1961; Hose et al., 1976)

As a gangue mineral and as purple aggregates in seams in the Robinson district. (Horton, 1961; GCF)

Reported as a gangue mineral in the Sacramento and Tungsten districts. (Horton, 1961)

As pale-purple modified cubes to 1 cm with pyrite in the underground Ward Mine in the Ward district. (M. Jensen, w.c., 1999)

In a specimen with barite from the White Pine district. (UCB)

FORNACITE

$(Pb_2,Cu^{2+})_3[(Cr,As)O_4]_2(OH)$

A relatively rare mineral in the oxidized zones of some hydrothermal base-metal deposits.

CLARK COUNTY In small vugs and on fracture surfaces as 0.15–0.5 mm olive-green to brown, thin, wedge-shaped plates to chisel-shaped crystals that commonly form v-shaped twins at the

Singer Mine in the Goodsprings district; also reported at the Alice Mine and a prospect near Crystal Pass. (Adams, 1998; J. Kepper, this volume)

In specimens from a prospect 9 miles northwest of Nelson. (R. Thomssen, w.c., 1999)

ELKO COUNTY In masses of pale-brown to dark-yellow crystals in cavities with bayldonite and hemimorphite crystals in cerussite veins at the Killie Mine in the Spruce Mountain district. (Dunning and Cooper, 1987)

ESMERALDA COUNTY As groups of green crystals with wulfenite, mimetite, phoenicochroite, and other minerals in quartz veins in the Candelaria district 1 mile north of the site of Columbus. (R. Walstrom, w.c., 2000)

HUMBOLDT COUNTY As green microcrystals with duftite and pyromorphite at the Silver Coin Mine in the Iron Point district. (Thomssen, 1998; M. Jensen, w.c., 1999)

NYE COUNTY Reported with skinnerite, stetefeldtite, molybdofornacite, and vauquelinite in a specimen from the Belmont Mine in the Belmont district. (EMP)

FORSTERITE

Mg_2SiO_4

Forsterite, the magnesium end member of the *olivine* group, is very common in mafic igneous rocks and in contact-metamorphosed rocks. Many occurrences are listed under *olivine* because chemistry has not been determined.

HUMBOLDT COUNTY As equant grains to 5 mm that are mostly altered to serpentine in ultramafic rock from the Twin Creeks Mine, Potosi district. (SBC)

LINCOLN COUNTY Largely altered to *serpentine* as minute specks in coarse marble near McCullough Hill in the Pioche district. (Westgate and Knopf, 1932)

NYE COUNTY In 1- to 6-inch-thick selvage zones with chondrodite, and in veinlets of colorless crystals along a contact between brucite and serpentinized granodiorite at the Premier Chemicals (Gabbs Magnesite-Brucite) Mine in the Gabbs district. (Schilling, 1968a)

In the Lunar Craters volcanic field as glassy green phenocrysts to 1 cm and nodules to 15 cm in basalt flows and loose crystals to 8 cm at Easy Chair Crater. (Bergman, 1982; M. Jensen, w.c., 1999)

FRAIPONTITE

$(Zn,Al)_3(Si,Al)_2O_5(OH)_4$

An uncommon clay mineral of the kaolinite-*serpentine* group that is generally found in zinc-bearing mineral deposits.

ELKO COUNTY A common alteration mineral at the Meikle Mine in the Lynn district. (Jensen, 1999)

EUREKA COUNTY As light-green blebs in sheared and argillized limestone at the Maggie Creek Mine in the Maggie Creek district. (Rota, 1991)

NYE COUNTY In white coatings with duftite and coronadite at the Hasbrouck shaft in the Lodi district. (M. Jensen, w.c., 1999)

FRANKDICKSONITE

BaF_2

Frankdicksonite is a single locality Nevada type mineral that formed in an environment rich in barium. Its existence as a naturally occurring substance was predicted before the mineral was actually discovered.

EUREKA COUNTY Type locality in quartz veinlets collected from two sites in the east pit of the Carlin Mine in the Lynn district: as small shattered grains to 2 mm in a 5-cm-wide quartz vein in silicified, carbonaceous, arsenic-rich, gold-bearing limestone; and as cubic crystals to 4 mm in a small quartz veinlet along a contact between an altered dike and silicified limestone. (Radtke and Brown, 1974)

FRANKLINITE

$(Zn,Mn^{2+},Fe^{2+})(Fe^{3+},Mn^{3+})_2O_4$

Franklinite is a very rare spinel-group mineral that is best known from metamorphosed zinc and manganese deposits; the Nevada locality is questionable.

CHURCHILL COUNTY Questionably identified in the Webber shaft in the Fairview district. (Schrader, 1947)

FREIBERGITE

$(Ag,Cu,Fe)_{12}(Sb,As)_4S_{13}$

A silver-bearing sulfosalt mineral that is increasingly recognized in hydrothermal deposits;

for many years considered as a variety of tetrahedrite.

ELKO COUNTY With pyrargyrite, stibnite, arsenopyrite, and pyrite in a quartz vein in ash-flow tuff at the Buckeye and Ohio Mine in the Good Hope district. (Emmons, 1910; Bentz et al., 1983)

As small (<0.2 mm) inclusions in pyrite and marcasite aggregates or filling gaps between grains with chalcopyrite, *electrum,* sphalerite, and pyrrhotite at the Hollister gold deposit in the Ivanhoe district. (Deng, 1991)

EUREKA COUNTY As brownish anhedral grains either within galena or at the contact between galena and bournonite in base-metal sulfide and quartz veins in diorite at the Goldstrike Mine in the Lynn district. (GCF)

Reported at the Garrison Mine in the Cortez district. (S. Pullman, w.c., 1999)

With owyheeite in the Eagle vein in the Lewis district. (*Mineral News,* February 1990)

A minor but widely distributed mineral in ore at the Cove Mine in the McCoy district. (Rick Streiff, unpublished company report, 1991)

HUMBOLDT COUNTY Reported from the Sheba vein in the Star district. (HUB; USGS Mineral Resource Data System)

LANDER COUNTY In a specimen from an unnamed prospect in the Imlay district. (M. Jensen, o.c., 2000)

LYON COUNTY As brown masses with owyheeite in quartz from the Eagle vein in the Lewis district. (GCF)

NYE COUNTY Disseminated grains and stringers with galena, hessite, sphalerite, enargite, and other sulfides at the Fairday Mine in the Cactus Springs district on the Nellis Air Force Range. (Tingley et al., 1998a)

With pyrite, galena, sphalerite, and chalcopyrite on dumps at the Jefferson and Prussian Mines in the Jefferson Canyon (Jefferson) district. (Ferguson and Cathcart, 1954)

Reported at the Northumberland Mine in the Northumberland district. (Kokinos and Prenn, 1985)

PERSHING COUNTY Reported in a specimen from a locality near Lovelock. (NMBM)

In a specimen from an unnamed prospect in the Imlay district. (M. Jensen, o.c., 2000)

WASHOE COUNTY As 5-micron inclusions in pyrite with roquesite and other sulfides at the Sure Fire Mine in the Pyramid district. (SBC)

FREIESLEBENITE

AgPbSbS$_3$

Freieslebenite is a relatively rare mineral that has been found in small quantities with other sulfides in silver-bearing hydrothermal deposits.

LANDER COUNTY Tentatively identified by EMP in a sample from the Dean Mine in the Lewis district. (J. McGlasson, w.c., 2001)

Reported from an unknown locality near Austin in the Reese River district. (AMNH)

NYE COUNTY As tiny hairlike tufts of black or silver color in quartz and quartz-rhodochrosite veins in the Morey district. (Williams, 1968)

WHITE PINE COUNTY Reported at the Paymaster Mine in the White Pine district. (Von Bargen, 1999)

FRIEDRICHITE

Pb$_5$Cu$_5$Bi$_7$S$_{18}$

A rare mineral that occurs in hydrothermal veins with other sulfide phases.

NYE COUNTY Identified by XRD in a specimen from the Silver Bismuth claim in the Round Mountain district near the Outlaw Mine. (R. Walstrom, w.c., 2000)

GAHNITE

$ZnAl_2O_4$

A spinel-group mineral that occurs as an accessory in granite and granite pegmatite, in metamorphic rocks and metamorphosed zinc and aluminum deposits, and as a heavy mineral in placers.

LYON COUNTY Reported in a specimen from the Ludwig Mine in the Yerington district. (UCB)

NYE COUNTY Identified by SEM/EDX in a sample of silicified tuff from Yucca Mountain. (Castor et al., 1999b)

GAITITE

$Ca_2Zn(AsO_4)_2 \bullet 2H_2O$

A very rare mineral found with other arsenate species at the Tsumeb Copper Mine in Africa.

HUMBOLDT COUNTY Reportedly the world's best specimens as white spherules to 1 mm with calcite crystals at the Twin Creeks Mine in the Potosi district. (M. Jensen, w.c., 1999)

GALENA

PbS

Galena is one of the most common of the sulfide minerals and the principal ore mineral of lead; there are many occurrences in hydrothermal ore deposits in Nevada.

CHURCHILL COUNTY In massive veins to 2 inches wide, commonly surrounded by cerussite and wulfenite, at the Chalk Mountain Mine in the Chalk Mountain district; also noted with molybdenite, pyrite, and chalcopyrite in lead-silver veins in the district. (Bryan, 1972; Schilling, 1979; Jensen, 1982)

Reported with chalcopyrite in quartz veins at the Fairview and Nevada Hills Mines in the Fairview district. (Greenan, 1914; Schrader, 1947; WMK)

Sparse in quartz-*adularia* veins with other sulfide minerals and fluorite at the Wonder and Hercules Mines in the Wonder district. (Schilling, 1979)

Reported from unspecified localities in the Holy Cross, I.X.L., and Jessup districts. (Schrader, 1947; Willden and Speed, 1974; WMK)

CLARK COUNTY Reported as auriferous at the Morton Mine in the Eldorado district. (Longwell et al., 1965; WMK)

With cerussite, wulfenite, and other minerals in the Goodsprings district; specifically reported at the Anchor, Boss, and Yellow Pine Mines. (Hewett, 1931; LANH; B. Hurley, w.c., 1999)

Locally common in quartz-calcite veins with *adularia* and other minerals at the Duplex and Quartette Mines in the Searchlight district. (Callaghan, 1939; HMM; WMK)

Reported from unspecified localities in the Crescent, Dike, and Gold Butte districts. (Longwell et al., 1965)

DOUGLAS COUNTY With other sulfides from several properties in the Red Canyon district, including the Longfellow, Winters, and Lucky Bill Mines. (Lawrence, 1963; Moore, 1969; Dixon, 1971; WMK)

Reported from unspecified localities in the Genoa and Wellington districts. (WMK)

ELKO COUNTY With other sulfide minerals at mines in the Aura district. (Emmons, 1910; LaPointe et al., 1991)

Reported at the Dig claims and Pass claims in the Black Mountain district. (LaPointe et al., 1991)

Reported at the Mint and Mirat Mines in the Burner district. (Emmons, 1910; WMK)

With pyrite and arsenopyrite in a quartz vein at the Slattery Mine in the Charleston district. (LaPointe et al., 1991)

With partially oxidized pyrite in gossan at the Coal Canyon Mine in the Coal Mine district; also with pyrite and stibnite at the Hunter prospect. (Bentz et al., 1983; LaPointe et al., 1991)

Widespread with other base-metal and silver minerals in the Contact district, as at the Silver King and Silkworm Mines. (Schrader, 1935b; Schilling, 1979)

Reported in a specimen from the Leonard Mine in the Cornucopia district. (WMK)

With other sulfides and secondary minerals such as hemimorphite in skarn at the Ruby Mountain claims and the Summit View Mine in the Corral Creek district. (Bentz et al., 1983; LaPointe et al., 1991)

With other sulfides, cerussite, anglesite, and bindheimite in a replacement orebody in limestone at the Delano Mine in the Delano district. (LaPointe et al., 1991)

Reported from the Burns Mine and the Indian Creek prospect in the Edgemont district. (LaPointe et al., 1991; Schilling, 1979)

As tiny grains with freibergite, sphalerite, and pyrite at the Hollister deposit in the Ivanhoe district. (Deng, 1991)

In quartz veins with molybdenite, copper minerals, and scheelite at the Southam Mine in the Kinsley district. (Schilling, 1979)

With other sulfides in a quartz vein at the American Beauty Mine in the Lee district; also reported at the Knob Hill Mine. (Bentz et al., 1983; Berger and Oscarson, 1998)

With cerussite at the Betty Lou Mine in the Loray district. (Hill, 1916; LaPointe et al., 1991)

As cubes to 0.5 mm on yellow sphalerite at the Meikle Mine in the Lynn district. (Jensen, 1999)

With pyrite and sphalerite in finely banded horizons in the Vinini Formation as possible sedimentary-exhalative mineralization at the LAJ claims in the Merrimac district. (NBMG mining district file)

At the Blue Bird No. 1 and Lead prospects in the Moor district. (Erickson et al., 1966; LaPointe et al., 1991)

In minor amounts in massive sulfide orebodies at the Rio Tinto Mine in the Mountain City district; also reported at other localities in the district. (Coats and Stephens, 1968; Emmons, 1910)

With anglesite and cerussite at the Deadhorse property in the Mud Springs district. (Hill, 1916)

With chalcopyrite, pyrite, and quartz at the Western Star Mine in the Pilot Peak district. (LaPointe et al., 1991)

Reported at many properties in the Railroad district; in silver ore at the Aladdin Mine and with pyromorphite at the Lady of the Lake Mine. (Ketner and Smith, 1963; LaPointe et al., 1991; Emmons, 1910; WMK)

With copper minerals and sphalerite in quartz veins at the Battle Creek Mine in the Ruby Valley district. (Hill, 1916; LaPointe et al., 1991)

As the major primary ore mineral with other sulfides in replacement orebodies and veins in the Spruce Mountain district; specifically reported at the Bullshead, Killie, Monarch, and Spruce Standard Mines. (Schrader, 1931b; Schilling, 1979; Hill, 1916; Dunning and Cooper, 1987; LaPointe et al., 1991)

With anglesite and cerussite in replacement bodies at the Jackson Mines in the Tecoma district. (Hill, 1916)

Reported at unspecified locations in the Dolly Varden, Ferber, Gold Circle, and Tuscarora districts. (Hill, 1916; Schilling, 1979; WMK; Emmons, 1910)

ESMERALDA COUNTY Reported with other sulfides in specimens from dumps at the Tonopah-Divide and Gold Zone Mines in the Divide district; also reported from the Revert and Brougher Mines. (Bonham and Garside, 1979; WMK)

Reported at the Carrie Mine in the Gilbert district. (Ferguson, 1928)

In quartz veins with molybdenite and other minerals at the Bullfrog-George Mine in the Gold Point district. (Albers and Stewart, 1972; Schilling, 1979)

Reported in specimens at the Antelope and Florence Mines in the Goldfield district. (WMK)

Identified in a specimen from the General Thomas Mine in the Lone Mountain district. (JHS)

Reported from several mines in the Montezuma district. (Albers and Stewart, 1972; JHS; WMK)

Locally abundant with *limonite*, cerussite, malachite, and wulfenite at the McNamara Mine in the Palmetto district. (Schilling, 1979)

Reportedly silver-bearing at the Texas No. 1, Helen, and Richmond lodes, Silver Peak district. (LANH; WMK)

Reported as silver-bearing at the Sylvania and Bilk Mines in the Sylvania district. (Schilling, 1979; WMK)

Reported as modified cubes to 2 cm and as cubes to 10 cm with pyrite crystals at the Gold Eagle Mine in the Weepah district. (M. Jensen, w.c., 1999)

Reported from unspecified localities in the Cuprite, Dyer, Good Hope, Klondyke, Lida, and Red Mountain districts. (JHS; WMK; Albers and Stewart, 1972; Keith, 1977)

EUREKA COUNTY Reported as silver-bearing with other sulfides from several properties in the Cortez district, including the Garrison (Cortez) Mine. (Emmons, 1910; WMK)

With cerussite, iron oxides, and plumbo-jarosite in the Diamond district; specifically at the Black Diamond Mine and the Will McBride claims. (Roberts et al., 1967b; F. Cureton, w.c., 1999; WMK)

Reportedly abundant and silver-bearing from many properties in the Eureka district, including the Richmond Mine and deep workings in the Ruby Hill and Adams Hill areas. (Nolan, 1962; Schilling, 1979; WMK; S. Pullman, w.c., 1999)

With other base-metal sulfides in quartz veins in diorite and in trace amounts elsewhere in the Goldstrike Mine in the Lynn district. (GCF)

Widespread in minor amounts as masses and coarse crystals in discontinuous veins and stock-works, Gold Quarry Mine in the Maggie Creek district; also reported from the Nevada Star (Good Hope) and Richmond Mines. (Roberts et al., 1967b; Jensen et al., 1995; Rota, 1991; GCF; WMK)

Reported at the Mineral Hill Consolidated, Live Yankee, and Mary Ann Mines in the Mineral Hill district. (Emmons, 1910; Schilling, 1979; JHS)

With other sulfides in the Keystone Mine in the Roberts district; also with cuprite, chryso-colla, and chalcopyrite in a specimen from the Kingston lode. (Roberts et al., 1967b; WMK)

At the Zenoli Mine in barite-carbonate veins in the Safford district; also reported in specimens from other properties. (Roberts et al., 1967b; WMK)

Reported from unspecified localities in the Antelope (Baldwin), Fish Creek, Lone Mountain, and Mount Hope districts. (WMK; Roberts et al., 1967b; JHS)

HUMBOLDT COUNTY A minor mineral at the Lone Tree Mine in the Buffalo Mountain district. (Bloomstein et al., 1993)

Minor in uranium ore at the Moonlight Uranium Mine in the Disaster district. (Dayvault et al., 1985)

Reported as silver-bearing with other sulfides at the Hot Springs and Wilson Mines in the Golconda district. (B. Hurley, w.c., 1999; WMK)

Reported as silver-bearing in quartz-*adularia* veins at the National and Stahl National Mines in the National district. (Lindgren, 1915)

With molybdenite, pyrite, and chalcopyrite at the Riley Mine in the Potosi district; also at the Twin Creeks Mine in quartz veins with tetra-hedrite and stibnite, and reported with jarosite and yellow sphalerite from the Getchell Mine. (Hotz and Willden, 1964; Simon et al., 1999; GCF)

With malachite, tetrahedrite, and azurite at the Prodigal Mine in the Red Butte district. (WMK)

With arsenopyrite, sphalerite, and tetra-hedrite in quartz-calcite veins at the Pansy Lee Mine in the Ten Mile (Winnemucca) district. (Lawrence, 1963; WMK)

As tiny grains in uranium-rich opal in the Virgin Valley district. (Castor et al., 1996)

Reported from unspecified localities in the Awakening, Gold Run, Jungo, Rebel Creek, and Sherman districts. (Calkins, 1938; Melhase, 1934c; WMK)

LANDER COUNTY Reported in large cubes from Galena in the Battle Mountain district; also common in small amounts with other sulfides at the Copper Canyon, Eastern, and other mines. (Hoffman, 1878; Roberts and Arnold, 1965; Schilling, 1979; WMK)

Reported from various mines in the Bullion district. (WMK)

Reported from the Hill Top, Independent, and Kimberly Mines, Hilltop district. (Stewart et al., 1977a; WMK)

Reported from various mines in the Kingston district. (Stewart et al., 1977a; WMK)

With other base-metal minerals, silver minerals, and gold in quartz veins at the Betty O'Neal Mine in the Lewis district, where it has been described as having a "melted" appearance; also reported from the Highland Chief Mine. (Vanderburg, 1939; Stewart et al., 1977a; J. McGlasson, w.c., 2001)

Reported at the Patterson Pass Mine in the Ravenswood district. (Hill, 1916)

Reported as crystals to 2 inches with other sulfides in veins at mines in the Reese River district. (Hoffman, 1878; Melhase, 1934c; Ross, 1953; WMK)

Reported from unspecified localities in the Cortez, Jackson, Jersey, McCoy, and New Pass districts. (Stewart et al., 1977a; Johnson, 1977)

LINCOLN COUNTY Reported as silver-bearing from many properties in the Bristol district, including the Bristol and Jackrabbit Mines. (Melhase, 1934d; Tschanz and Pampeyan, 1970; Klein, 1983; WMK)

Reported at the Cave Valley Mine in the Patterson district. (Schrader, 1931a)

In quartz veins at an unspecified locality in the Chief district. (Callaghan, 1936)

Reported at the Comet, Lyndon, Silver Star, and West Mines, Comet district. (Tschanz and Pampeyan, 1970; WMK)

In ore at the Mt. View and Shoute Mines in the Freiburg district. (WMK)

Common in silver-bearing masses in quartz-calcite veins and replacing limestone at the Groom, Black Metal, and other mines in the Groom district. (Humphrey, 1945; Tingley et al., 1998a)

Reported at the Manhattan and other mines in the Highland district. (Tschanz and Pampeyan, 1970; WMK)

Reported as silver-bearing at the Illinois and other mines on the Hyko lode in the Pahranagat district. (Tschanz and Pampeyan, 1970; WMK)

In ore from many mines of the Pioche district, the most prolific source of lead and zinc in Nevada; found with sphalerite and pyrite in large replacement orebodies in limestone as at the Caselton and Ely Valley Mines, and with sphalerite in veins in quartzite on Prospect Mountain as at the Raymond and Ely Mine. (Westgate and Knopf, 1932; Tschanz and Pampeyan, 1970; Gemmill, 1968; WMK)

With bismuth minerals at the Lincoln Tungsten Mine in the Tem Piute district. (Buseck, 1967; JHS)

Reported at the Shamrock Mine in the Viola district. (Tschanz and Pampeyan, 1970; WMK)

LYON COUNTY Reported at the Old (Lost) Soldier Mine in the Churchill district. (WMK)

Reportedly common in the Washington district. (Garside, 1973)

MINERAL COUNTY With other sulfides in replacement deposits and in sheeted quartz veins at the Simon and other mines in the Bell district. (Knopf, 1921b; Ross, 1961; WMK)

Reported as traces in veins at the Gravity and Curley Barite Mines in the Buckley district. (JHS)

In several mines, including the New Potosi Mine, with pyrite, bindheimite, and jamesonite in the Candelaria district; also reported from the Candelaria Silver Mine with other sulfides in carbonate veins in serpentinite. (Knopf, 1923; Ross, 1961; Chavez and Purusotam, 1988)

In gold- and silver-bearing quartz veins at many properties such as the Mabel Mine in the Garfield district. (Vanderburg, 1937a; Ross, 1961; Ponsler, 1977; WMK)

Common with other sulfides in quartz-calcite veins as at the Lucky Boy Mine in the Lucky Boy district. (Hill, 1915; Ross, 1961; Lawrence, 1963; WMK)

Reported at the Endowment and other mines in the Marietta district. (Ross, 1961; WMK)

Reportedly silver-bearing at the Sunset Mine in the Pamlico district. (Ross, 1961; WMK)

Reported from several properties, including the Redemption and Todd Mines, in the Santa Fe district. (Ross, 1961; WMK)

In a specimen from the Garfield lease in the Silver Star district. (WMK)

Reported in the Broken Hills, Eagleville, Fitting, Rand, and Rawhide districts. (Schrader, 1947)

NYE COUNTY In quartz veins with sphalerite and pyrite at the Antelope View Mine in the Antelope Springs district on the Nellis Air Force Range. (Tingley et al., 1998a)

In a specimen with tetrahedrite, rhodochrosite, and other minerals at the Spanish Belt (Barcelona) Mine in the Barcelona district. (UCB)

As silver-bearing cubic crystals at the LeMoine Mine in the Bullfrog district. (WMK)

With other sulfides in veins at the Fairday Mine in the Cactus Springs district on the Nellis Air Force Range. (Tingley et al., 1998a)

In a quartz vein with cerussite, anglesite, and other sulfides at the Iron Rail group in the Cloverdale district. (Kral, 1951; Hamilton, 1993)

In nodules in calcareous shale at the Nut Pine prospect in the Ellsworth district. (Kral, 1951)

Locally in quartz veins at the Jim claims in the Fairplay district. (Kral, 1951)

With cerussite in pipelike bodies at the Pius Kaelin Mine in the Gold Crater district on the Nellis Air Force Range; also with other sulfides at an unnamed mine shaft. (Kral, 1951; Tingley et al., 1998a)

In quartz veins with cerussite, pyrite, and chalcopyrite at the Star of the West Mine in the Jackson district. (Hamilton, 1993)

In microscopic grains with luzonite and other sulfides at the Golden Chariot Mine in the Jamestown district. (Tingley et al., 1998a)

With chalcopyrite and pyrite at the Lost Indian Mine in the Jett district. (Kral, 1951; WMK)

In large replacement deposits and in a quartz vein at the San Rafael Mine in the Lodi district; also reported in copper ore from quartz veins in the Hasbrouck Mine. (Schrader, 1947; Kral, 1951; WMK)

With pyrite in irregular pipes and chimneys at the Last Chance claims in the Longstreet district. (Kral, 1951; Kleinhampl and Ziony, 1984)

With other sulfides in veins and lenticular masses of quartz at the Nemo prospect and other properties in the Manhattan district. (Ferguson, 1924; WMK)

Reportedly silver-bearing at the Michigan Boy group in the Oak Spring district. (Melhase, 1935b; Kral, 1951)

Reported at the Antimonial, Crescent, and North Star Mines, Reveille district. (Kral, 1951; Lawrence, 1963; WMK)

Silver-bearing in veins at the Royston Mines and other properties in the Royston district. (Kral, 1951)

In base-metal veins with other sulfides at the Florence and Hall Mines in the San Antone district. (Kral, 1951; Shaver, 1991)

In veins with pyrite, barite, acanthite, and *electrum* at the Life Preserver and Landmark Mines in the Tolicha district on the Nellis Air Force Range. (Tingley et al., 1998a)

In rich sulfide ore with sphalerite, chalcopyrite, pyrite, and silver sulfides in many mines in the Tonopah district; specifically reported at the Montana-Tonopah Mine and the Desert Queen shaft. (Spurr, 1905; Bastin and Laney, 1918; Melhase, 1935c)

With pyrite as replacements in limestone and shale at the Leadhill claim in the Troy district. (Kral, 1951)

Reportedly silver-bearing with other sulfides at the Buckeye and Murphy (Twin River) Mines in the Twin River district; also at the Buckeye, Gruss, Murphy and Teichert Mines, and Dallimore-Douglas claims. (Hague, 1870; Ferguson and Cathcart, 1954; Stager and Tingley, 1988; Kral, 1951; Kleinhampl and Ziony, 1984)

Reportedly silver-bearing with sphalerite and other sulfides in veins at the Tybo Mine in the Tybo district; also reported at the Bunker Hill mine and other properties. (Ferguson, 1933; WMK)

In quartz veins with other sulfides at the Richmond Mine in the Union district; also with sphalerite in replacement orebodies at the Grantsville Mine and reported at other properties. (Hamilton, 1993; Kral, 1951; Kleinhampl and Ziony, 1984; WMK)

In silver-lead deposits at the Warner (Werner) Mine and other properties in the Washington district. (Hill, 1915; Kral, 1951; Kleinhampl and Ziony, 1984)

Reported from unspecified localities in the Belmont, Danville, Gabbs, Jefferson Canyon, Johnnie, Millett, Morey, Willow Creek, and Wilsons districts. (Kral, 1951; Kleinhampl and Ziony, 1984; Ivosevic, 1978; WMK; Williams, 1968; JHS)

PERSHING COUNTY In a quartz vein at the Antelope Mine in the Antelope district; also reported at the Majuba Hill and Superior Mines. (Lawrence, 1963; MacKenzie and Bookstrom, 1976; WMK)

Reported at the Pflenger Mine in the Buena Vista district. (Cameron, 1939; JHS; WMK)

Reported in lead-silver-gold ore at the Etchegoyhen and Guthries Mines in the Mill City district. (WMK)

In silver-bearing quartz veins and bodies at the Packard and Wabash Mines in the Rochester district. (Knopf, 1924; WMK)

In quartz-calcite veins with scheelite and other sulfide minerals at the Rye Patch Mine in the Rye Patch district. (Ransome, 1909b; Johnson, 1977; WMK)

In quartz-*tourmaline* veins at the Bonanza King Mine in the Spring Valley district. (Campbell, 1939)

Reported in silver-gold-iron ore at the Keystone Mine in the Willard district. (WMK)

Reported in the Antelope Springs, Iron Hat, Kennedy, Mount Tobin, Muttlebury, Nightingale, Sacramento, and Sierra districts. (Lawrence, 1963; WMK; Johnson, 1977; JHS; Ferguson et al., 1951)

STOREY COUNTY With other sulfides in specimens of bonanza and low-grade ore from the Comstock Lode in the Comstock district as at the California, Consolidated Virginia, and Ophir Mines; also reported in the Flowery lode. (Bastin, 1922; Gianella, 1936a; WMK; Schilling, 1990)

WASHOE COUNTY With other sulfides in quartz-calcite veins at the Mountain View Mine in the Deephole district. (Bonham and Papke, 1969)

Reportedly silver-bearing with sphalerite and other sulfides in quartz-calcite veins in the Galena district; specifically noted from the Imperial and Union Mines. (Geehan, 1950; Bonham and Papke, 1969; WMK)

Reportedly silver-bearing in quartz-calcite veins with other sulfides at the Leadville Mine in the Leadville district. (Burgess, 1926; Bonham and Papke, 1969; WMK)

In trace amounts with ore at the Olinghouse Gold Mine in the Olinghouse district; intergrown with gold wires and crystals in some specimens. (Jones, 1998; GCF)

With enargite and other sulfides in veinlets and stringers in the Fravel Mines in the Peavine district and in prospects along Peavine Creek. (Bonham and Papke, 1969)

In veins with pyrite, sphalerite, and enargite at localities in the Pyramid district, including the Franco American (Nevada Dominion), Burrus, and Ruth Mines. (Bonham and Papke, 1969; Wallace, 1975; SBC)

With sphalerite and pyrite at the Wedekind Mine in the Wedekind district. (Overton, 1947; Bonham and Papke, 1969; WMK; J. Tingley, o.c., 1999)

Reported at unspecified localities in the McClellan and Nightingale districts. (Bonham and Papke, 1969)

WHITE PINE COUNTY With other sulfides at the Bald Mountain Mine in the Bald Mountain district. (Hitchborn et al., 1996)

Reportedly silver-bearing in a specimen from the Racine Mine in the Chase district. (WMK)

With pyrite and malachite at the Star Mine in the Cherry Creek district. (Schrader, 1931b; WMK; JHS)

Reported as argentiferous and auriferous from the Eureka claim in the Duck Creek district. (WMK)

With anglesite, cerussite, and other secondary lead minerals at the Cuba Mine in the Granite district. (Hill, 1916; Melhase, 1934d)

Reported in lead-silver ore at the Shorty Mine in the Lincoln district. (WMK)

Reportedly silver-bearing at the Poor Man, Success, and Telephone Mines in the Robinson district. (Melhase, 1934d; WMK)

Reportedly silver-bearing in silver-copper ore at the Look-Out Mine in the Shoshone district. (WMK)

Reported with anglesite at the Monitor Mine in the Taylor district. (Lovering and Heyl, 1974; WMK)

With other sulfides in lead-silver-copper deposits in the Ward district; specifically as an ore mineral with sphalerite and chalcopyrite in skarn at the Silver King (Ward) Mine. (Hill, 1916; Hose et al., 1976; WMK; Hasler et al., 1991)

With secondary lead, zinc, and copper minerals and commonly in nodules surrounded by cerussite and anglesite from mines in the lead belt in the White Pine district; also described with other sulfides in quartz veins that cut skarn in the Centennial zone at Mount Hamilton, and reported in placer deposits. (Humphrey, 1960; Myers et al., 1991; WMK)

Reported at unspecified localities in the Eagle, Hunter, Muncy Creek, Nevada, and Osceola districts. (Hill, 1916; Melhase, 1934d; WMK)

GALENOBISMUTITE

$PbBi_2S_4$

Galenobismutite is a relatively rare mineral that occurs in hydrothermal metal deposits and in tungsten- and bismuth-bearing skarn.

EUREKA COUNTY In veins with boulangerite, jamesonite, tetrahedrite, and galena at the Grant Mine and the Eureka tunnel in the Eureka district. (Vikre, 1998)

LINCOLN COUNTY Identified on the basis of optical properties as the most common bismuth-bearing phase with cosalite, bismuthinite, native bismuth, and galena in tungsten skarn at the Free tunnel in the Tem Piute district. (Buseck, 1967)

GALKHAITE

$(Cs,Tl)(Hg,Cu,Zn)_6(As,Sb)_4S_{12}$

Galkhaite is a rare mineral that occurs in a few arsenic- and thallium-bearing gold deposits. To date, it has been reported from eight localities in the world, five of which are in Nevada.

ELKO COUNTY Identified by SEM/EDX in a specimen from the Jerritt Canyon Mines in the Independence Mountains district. (G. Arehart, o.c., 2000)

EUREKA COUNTY With weissbergite, getchellite, and other mercury and thallium minerals such as carlinite and christite at the Carlin Gold Mine in the Lynn district; also reported with realgar and other minerals at the Goldstrike and Rodeo Mines. (Dickson and Radtke, 1978; Ferdock et al., 1997; G. Arehart, o.c., 2000)

HUMBOLDT COUNTY The world's best locality as steel-gray to deep-red, striated cubes in vugs and fractures and as irregular grains with associated getchellite, realgar, orpiment, arsenic-bearing pyrite, stibnite, fluorite, and quartz in carbonaceous shale at the Getchell Mine in the Potosi district; found locally underground as red cubes to 5 mm and gray crystals to 1.2 cm, some displaying interpenetration twinning. (Botinelly et al., 1973; Jungles, 1974; Chen and Szymanski, 1981; Thomssen, 1996; GCF)

GARNET

$(Ca,Fe^{2+},Mg,Mn^{2+})_3(Al,Cr^{3+},Fe^{3+},Mn^{3+},$
$Si,Ti,V^{3+},Zr)_2(SiO_4)_3$ (general formula)

A mineral group name. *Garnet* group minerals are very common in metamorphic rocks in Nevada and occur in many tungsten skarn deposits. Species have not been identified at the following occurrences.

ELKO COUNTY Very abundant with other skarn minerals along a limestone-granodiorite contact in the Contact district. (Schrader, 1912, 1935b; Schilling, 1979)

Green *garnet* was reported from an unspecified locality with other metamorphic minerals in the Dolly Varden district. (Melhase, 1934d)

Common in a pegmatite body with quartz and beryl at the Harrison Pass prospect in the Harrison Pass district. (Garside, 1973)

Green *garnet*-epidote skarn hosts copper-iron mineralization at the Lone Mountain Mine in the Merrimac district. (LaPointe et al., 1991)

With quartz, beryl, and other pegmatite minerals in Dawley Canyon in the Valley View district. (Olson and Hinrichs, 1960)

ESMERALDA COUNTY Abundant with epidote, quartz, and *pyroxene* in skarn at the Peg Leg (Greenstock) Mine in the Tonopah district. (Bonham and Garside, 1979)

A major constituent of skarn at the Cucomungo deposit in the Tule Canyon district. (Schilling, 1979)

HUMBOLDT COUNTY With epidote, quartz, scheelite, calcite, and pyrite in a specimen from the Hood Tungsten property in the Golconda district. (WMK)

LINCOLN COUNTY As green crystals lining cavities and joint seams in metamorphosed volcanic breccia near the Bristol Mine in the Bristol district. (Westgate and Knopf, 1932; Melhase, 1934d; AMNH)

As small, zoned, honey-yellow crystals with quartz crystals, amphibole, and other minerals in a small prospect near McCullough Hill. (Westgate and Knopf, 1932)

LYON COUNTY Large quantities of *garnet*, in places almost pure, occur in skarn in the Yerington district. Garnetized fossil halobia (a Triassic bivalve) were found at the McConnell Mine. (Knopf, 1918b; Melhase, 1935e; JHS)

MINERAL COUNTY A major constituent of Mo- and W-bearing skarn at the Queen (Indiana Queen) Mine in the Buena Vista district. (Hess and Larsen, 1920)

In a copper-tungsten deposit at the Bataan Mine in the Garfield district. (Ponsler, 1977)

With epidote, scheelite, quartz, dolomite, and pyrite at the Sutton No. 1 Mine in the Leonard district. (WMK)

A skarn mineral at the Lucky Boy Mine in the Lucky Boy district. (Lawrence, 1963)

With hessite, petzite, pyrite, and other minerals in white vein quartz at the Silver Dyke Tungsten Mine in the Silver Star district; also with epidote and tremolite near Pepper Spring in the Excelsior Mountains. (Sigurdson, 1974; Garside, 1979)

NYE COUNTY A metasomatic mineral associated with magnetite-hematite replacement bodies at the Engle-Stouder prospect in the Ellsworth district. (Reeves et al., 1958)

A constituent of skarn mined for tungsten at the Scheebar Mine in the Fairplay district. (Bailey and Phoenix, 1944)

Reported as an unusual mixture of equal amounts of grossular and spessartine at the Victory Tungsten Mine in the Lodi district. (Humphrey and Wyatt, 1958)

PERSHING COUNTY In skarn with epidote and magnetite at the Basalt prospect in the Copper Valley district. (Shawe et al., 1962)

The most abundant skarn mineral, locally as excellent crystals to 3 cm, at the Nevada-Massachusetts Co. Mines in the Mill City district. (JHS; NMNH collections)

A major component of scheelite-bearing skarn at the M.G.L. Tungsten Mine in the Nightingale district. (Smith and Guild, 1944)

Reported at the Ragged Top Mine in the Ragged Top district. (WMK; AMNH)

GARNIERITE

$Ni_3Si_2O_5(OH)_4$

Not a mineral name, but a general term for hydrous nickel silicate; species have not been identified in specimens from the following localities.

CHURCHILL COUNTY Reported in a specimen from an unknown locality, possibly from the Nickel or Lovelock Mines in the Table Mountain district. (AMNH)

CLARK COUNTY In oxidized ore with erythrite and secondary copper minerals at the Key West Mine in the Bunkerville district. (Beal, 1965)

HUMBOLDT COUNTY Reported in nickel ore from an unspecified locality in the Winnemucca district. (WMK)

NYE COUNTY In serpentinite from southwest of Manhattan and east of State Route 376; probably from a large area of nickel-rich rock near Willow Spring. (D. Hudson, w.c., 1999; Kleinhampl and Ziony, 1984)

Reported in nickel ore from an unspecified locality in the Toiyabe Range south of Austin; probably from one of the serpentinite exposures in the county. (WMK; Kleinhampl and Ziony, 1984)

GAYLUSSITE

$Na_2Ca(CO_3)_2 \cdot 5H_2O$

Gaylussite is an uncommon mineral that occurs in evaporite deposits.

CHURCHILL COUNTY As encrustations of yellowish-white, tubular crystals on rocks, pebbles, and organic matter with trona and other minerals on the shores of Soda and Little Soda Lakes near Fallon in the Soda Lake district. (Blake, 1866; Silliman, 1866; Vanderburg, 1940; BMC)

Reported as crystals from a drill hole in Dixie Marsh (Humboldt Salt Marsh). (Papke, 1976; M. Jensen, w.c., 1999)

ESMERALDA COUNTY Reported in Fish Spring Valley. (Hoffman, 1878)

NYE COUNTY Abundant in soda-bearing beds at depths of 660 to 2880 feet beneath Butterfield Marsh in the Butterfield Marsh district in Railroad Valley; also in lesser quantities at shallower depths. (Lincoln, 1923; Kral, 1951)

GEARKSUTITE

$CaAl(OH)F_4 \cdot H_2O$

Gearksutite is a relatively uncommon mineral that occurs in small quantities in a number of fluorspar deposits; it also occurs in granite pegmatite and in hot-spring deposits.

CHURCHILL COUNTY As white spherical masses to 3 cm at the Michigan prospect in the Mountain Wells district. (R. Walstrom, w.c., 1999)

With creedite as large white nodules in a fault zone at an unnamed prospect in the northern part of the Mopung Hills in the Lake district. (R. Walstrom, w.c., 1999)

LINCOLN COUNTY In fine-grained, dull white nodules and veinlets to 15 mm wide at the Cougar prospect in the Vigo district. (Papke, 1979)

LYON COUNTY As white nodules at the Boulder Hill Mine in the Wellington district. (R. Walstrom, w.c., 1999)

NYE COUNTY As white porcelaneous masses to 15 cm and nodules to 8 cm in gouge zones in the Hall (Anaconda) Molybdenum Mine in the San Antone district; also reported from the Liberty Mine at Tonopah, but this locality is in the San Antone district less than a mile south of the Hall Mine. (R. Walstrom, w.c., 1999; M. Jensen, w.c., 1999; Anthony et al., 1997)

GEDRITE

$\square Mg_5Al_2Si_6Al_2O_{22}(OH)_2$

Gedrite, an amphibole-group mineral related to anthophyllite, is most common in high-grade regionally metamorphosed rocks.

CLARK COUNTY Reported as *magnesiogedrite* from an unspecified locality in the Moapa area. (HUB)

GERSDORFFITE

NiAsS

A mineral that generally occurs in hydrothermal deposits with other nickel minerals.

CHURCHILL COUNTY Reportedly the principal primary constituent of nickel ore at the Nickel Mine in the Table Mountain district. (Ransome, 1909b)

EUREKA COUNTY As 30-micron inclusions in pyrite in the Betze deposit at the Goldstrike Mine in the Lynn district. (GCF)

Rare in sulfide-rich siliceous barite breccias with pyrite, sphalerite, tetrahedrite, and arsenopyrite at the Gold Quarry Mine in the Maggie Creek district. (Jensen et al., 1995)

HUMBOLDT COUNTY A minor sulfide at the Lone Tree Mine in the Buffalo Mountain district. (R. Braginton, o.c., 1995)

LANDER COUNTY Identified by EMP at the Cove Mine in the McCoy district. (W. Fusch, o.c., 2000)

MINERAL COUNTY Rare with other sulfides and chromite in *serpentine* at the Candelaria Mine in the Candelaria district; also reported as silvery grains to 1 cm with annabergite. (Moeller, 1988; UCB; M. Jensen, w.c., 1999)

GETCHELLITE

$AsSbS_3$

A Nevada type mineral that has been found in Carlin-type gold deposits.

EUREKA COUNTY With christite, lorandite, and realgar in ore in carbonaceous, silty dolomite at the Carlin Mine in the Lynn district. (Radtke et al., 1977)

HUMBOLDT COUNTY Identified in a specimen from an unknown locality in the Opalite district. (JHS)

Type locality as transparent, dark-red-orange, sectile, anhedral to rare euhedral, micaceous grains and masses in orpiment and realgar with stibnite, cinnabar, quartz, galkhaite, and laffittite along a shear zone in Paleozoic sedimentary rocks at the Getchell Mine in the Potosi district, Sec. 33, T39N, R42E. (Weissberg, 1965)

NYE COUNTY Reported in a specimen from an unknown locality in the Manhattan district. (HUB)

GIBBSITE

$Al(OH)_3$

Gibbsite is a common mineral in laterite and bauxite deposits; it also occurs in low-temperature hydrothermal and metamorphic environments.

EUREKA COUNTY A common constituent of fault gouge at the Goldstrike Mine in the Lynn district. (Lavkulich, 1993)

HUMBOLDT COUNTY Mixed with manganese oxide minerals at the Golconda manganese-tungsten deposits in the Golconda district. (Hewett et al., 1968)

NYE COUNTY Reportedly a hypogene mineral with local kaolinite, alunite, and barite in and near veins that cut contact-metamorphosed limestone at the Queen City prospect in the Queen City district; comprises about 80 percent of the main vein. (Hewett et al., 1968)

GILALITE

$Cu_5^{2+}Si_6O_{17} \bullet 7H_2O$

A rare metamorphic or mesothermal mineral at the type locality in Arizona.

CLARK COUNTY Reported as green globules from the Boss Mine in the Goodsprings district; identified by P. Hlava using EMP. (J. Kepper, this volume)

GIRAUDITE

$(Cu,Zn,Ag)_{12}(As,Sb)_4(Se,S)_{13}$

A very rare mineral previously found in veins in granite at a locality in France; the Nevada locality is the second occurrence in the world.

LYON COUNTY Rare in quartz stockwork veins and breccias with other sulfide minerals at the Talapoosa Mine in the Talapoosa district. (Van Nieuwenhuyse, 1991)

GIRDITE

$Pb_3H_2(Te^{4+}O_3)(Te^{6+}O_6)$

A very rare species formed by oxidation of tellurium-bearing minerals.

WASHOE COUNTY Identified by XRD as yellowish coatings on altered volcanic rock from an unspecified locality in the Olinghouse district. (GCF)

GISMONDINE

$Ca[Al_2Si_2O_8] \cdot 4.5H_2O$

Gismondine is a zeolite mineral that is typically found in cavities in alkaline volcanic rocks.

HUMBOLDT COUNTY Reported from the Getchell Mine in the Potosi district. (D. Tretbar, w.c., 2000)

GLAUBERITE

$Na_2Ca(SO_4)_2$

Glauberite is a soluble mineral that occurs in evaporite deposits and rarely at hot springs.

CLARK COUNTY Reported from about 6 miles east of Muddy Peak in the Muddy Mountains district. (Longwell et al., 1965)

As lenticular masses in a large deposit in the Muddy Creek Formation at the Stewart property in the Saint Thomas district, now beneath Lake Mead; also noted as crystals in salt deposits along the Virgin River. (Vanderburg, 1937; Papke, 1976)

MINERAL COUNTY With mirabilite, thenardite, trona, and other minerals in the Rhodes Marsh district; the occurrence is unusual for the variety of saline minerals in a relatively small playa. (Melhase, 1935e; Ross, 1961)

GLAUCONITE

$K_{0.8}R^{3+}_{1.33}R^{2+}_{0.67}\square Al_{0.13}Si_{3.87}O_{10}(OH)_2$
where R = Na,Fe,Al,Mg

Glauconite is a widespread mineral in marine sedimentary rocks; it forms by diagenetic alteration of biotite and other detrital minerals in sandstone, siltstone, and impure limestone.

CLARK COUNTY Reported in a sandy bed in the lower member of the Cambrian Bright Angel (Pioche) Shale on Frenchman Mountain; also identified by XRD as pellets in dolomite in the lower part of the Nopah Formation on Frenchman Mountain. (Palmer, 1981; Matti et al., 1993)

WHITE PINE COUNTY In an 8-foot-thick section in the Devonian Pilot Shale at an unspecified occurrence in the Pancake district. (Kleinhampl and Ziony, 1984)

GLAUKOSPHAERITE

$(Cu,Ni)_2(CO_3)(OH)_2$

A member of the rosasite group; occurs in oxidized copper-nickel deposits.

CLARK COUNTY As green spherules to 1 mm with heterogenite at the Key West Mine in the Bunkerville district. (Jensen, 1990a; S. Pullman, w.c., 1999)

GODLEVSKITE

$(Ni,Fe)_7S_6$

A rare mineral that occurs in hydrothermal veins or magmatic deposits with other nickel-bearing sulfides.

CLARK COUNTY Reported from an unspecified locality near Moapa, probably from the Bunkerville district. (Anthony et al., 1990; FUMT; HUB; S. Kleine, o.c., 1999)

GOETHITE

$\alpha\text{-}Fe^{3+}O(OH)$

Goethite is a very common oxidation product derived from iron-bearing minerals such as pyrite or magnetite; it is also formed as a primary precipitate in some bogs and springs. *Limonite* is a general term for hydrous iron oxides that are generally partly or wholly composed of goethite. Occurs at many localities in Nevada, only a few of which are listed below.

CHURCHILL COUNTY As pseudomorphs after cubic pyrite crystals to 1 cm from an unnamed prospect in the Chalk Mountain district. (M. Jensen, w.c., 1999)

As a replacement of pyrite at the Fairview Mine in the Fairview district. (Schrader, 1947; Saunders, 1977)

CLARK COUNTY As pseudomorphs after pyrite crystals to 5 cm in miarolitic cavities and pegmatites in the Eldorado and Newberry Mountains. (S. Scott, w.c., 1999)

Pseudomorphs after pyrite cubes to 1 cm at the Boss Mine in the Goodsprings district. (Jedwab, 1998b; CIT)

ELKO COUNTY In oxidized rock at the Rossi Mine in the Bootstrap district. (Snyder, 1989)

Along fractures and bedding and disseminated in oxidized rock at the Rain Mine in the Carlin district. (D. Heitt, o.c., 1991)

An oxidation product of pyrite, marcasite, and arsenopyrite at the Big Springs Mine in the Independence Mountains district. (Youngerman, 1992; Hofstra, 1994)

Common as a gangue mineral in the Gold Circle (Midas) district. (P. Goldstrand, w.c., 1999)

Widespread in the oxidized zone at the Hollister gold deposit in the Ivanhoe district. (Deng, 1991)

In oxidized zones at the Meikle Mine in the Lynn district. (Jensen, 1999)

Abundant with silica in hard brown gossan that contains plentiful small cavities within which many secondary minerals occur at the Killie Mine in the Spruce Mountain district. (Dunning and Cooper, 1987)

Identified by XRD in pseudomorphs after pyrite in nodules from bedded rock near Willow Creek about 6 km east of Willow Creek Reservoir. (P. Goldstrand, w.c., 2000; SBC)

ESMERALDA COUNTY Reported as *blakeite,* a discredited species considered to be a mixture of goethite and tellurite, in brown microcrystalline crusts with mackayite and emmonsite in a tellurium-rich vein at the McGinnity shaft of the Mohawk Mine in the Goldfield district; also reported in a specimen from the Lone Star Mine. (Frondel and Pough, 1944; DeMouthe, 1985; WMK)

From an unknown type locality for the discredited species *esmeraldite,* named for Esmeralda County but later determined to be goethite with absorbed water. (Eakle, 1901; Strunz, 1970)

EUREKA COUNTY As pseudomorphs after pyrite at the Garrison Mine in the Cortez district. (WMK)

Reported in the Eureka district. (S. Pullman, w.c., 1999)

As pale- to dark-brown and orange coatings and masses, some after pyrite crystals to 1.5 cm, at the Goldstrike Mine in the Lynn district. (GCF)

In a layer to 5 feet thick of "steel rock" in the footwall of the orebody at the Barth Iron Mine in the Safford district; also reported with cuprite at the Onondaga Mine. (Shawe et al., 1962; WMK)

HUMBOLDT COUNTY Dominant iron oxide mineral at the Lone Tree Mine in the Buffalo Mountain district. (R. Braginton, o.c., 1995)

In oxidized radioactive breccia at the Moonlight Uranium Mine in the Disaster district. (Castor et al., 1996)

Reportedly tungsten-bearing in a specimen from an unknown locality in the Edna Mountains in the Golconda district. (WMK)

As reddish-brown coatings on specimens from the Silver Coin Mine in the Iron Point district. (Thomssen, 1998)

Common as an alteration product of marcasite at the Cordero Mine in the Opalite district. (Fisk, 1968)

LANDER COUNTY Attached to placer gold in the Battle Mountain district. (Vanderburg, 1936a)

With galena, quartz, and silver at the Lady of the Lake Mine in the Bullion district. (WMK)

LINCOLN COUNTY Reported after *selenite* in a specimen from the Bristol Silver Mine in the Bristol district. (CSM)

MINERAL COUNTY As brown cubes at the Lowman Mine and the Julie claim in the Pamlico district. (Lawrence, 1963; S. Pullman, w.c., 1999)

NYE COUNTY Common as stains and coatings along cracks at the Round Mountain Mine in the Round Mountain district. (Tingley and Berger, 1985)

PERSHING COUNTY As dark–brownish-orange crusts and boxworks formed by surface weathering of sulfide minerals at the Majuba Hill Mine in the Antelope district. (Jensen, 1985)

After pyrite with gold inclusions in quartz-*tourmaline* veins at the Bonanza King Mine in the Spring Valley district. (JHS)

Brownish stains, coatings, and veinlets at the Willard Mine, Willard district. (Jensen and Leising, 2001)

WASHOE COUNTY In quartz veins with gold, generally coating quartz crystals, at the Olinghouse Mine in the Olinghouse district. (Jones, 1998; GCF)

As pseudomorphs after pyritohedral pyrite crystals in pegmatite near Incline Village. (Jensen, 1993a)

WHITE PINE COUNTY A replacement mineral

in the oxidized zone at the Vantage deposit on Alligator Ridge in the Bald Mountain district. (Ilchick, 1990)

GOLD

Au

A widespread native metal in small amounts, gold occurs in many of the epithermal vein deposits in Nevada and in placer gravels. Commonly referred to as *electrum* when alloyed with silver.

CHURCHILL COUNTY As 0.4 mm cubes, dodecahedrons, and octahedrons in partly carbonized logs in tuffaceous agglomerate at an unspecified locality in the Aspen district; possibly the same locality for gold in petrified cypress near Chalk Wells. (M. Jensen, w.c., 1999; WMK)

As blebs and stringers in banded quartz veins at the Fairview Mine in the Fairview district; also in placer deposits at the Eagleville and Nevada Hills Mines. (Greenan, 1914; Schrader, 1947; Quade and Tingley, 1987; Klein, 1983)

Reported from an 8-inch-thick vein at the Gold King Mine in the Jessup district. (WMK)

In veins and placers 2 miles north of Sand Springs in the Sand Springs district. (Klein, 1983; WMK)

In quartz-*adularia* veins at the Wonder, Horgan, and other mines in the Wonder district. (Schilling, 1979; Klein, 1983; WMK)

CLARK COUNTY Limited amounts have been panned from Nickel Creek at the Prima Donna Mine in the Bunkerville district. (Beal, 1965)

In wire and granular forms in fracture fillings at the Keystone, Red Cloud, Yellow Jacket, and Potosi Mines in the Goodsprings district; also reported as palladium-bearing with cinnabar, potarite, and copper minerals in a specimen from the Boss Mine. (Hewett, 1931; Klein, 1983; LANH; WMK; Jedwab, 1998a)

Reported with bismutite and hematite from an unspecified locality in the Newberry district. (WMK)

With *limonite* and specular hematite in a quartz-cemented breccia vein at the Blossom Mine in the Searchlight district; also in veins in gneiss at the Duplex, Quartette, and Searchlight Mines, and in placers. (Longwell et al., 1965; Klein, 1983)

Reported in a specimen from the Lucy Gray Mine in the Sunset district. (LANH)

In placer gravels in the Eldorado, Gold Butte, and Moapa districts. (Vanderburg, 1936a; Klein, 1983)

DOUGLAS COUNTY As dust and nuggets to 1 ounce that contain 18–22 wt. percent silver at the Spring Canyon and Guild-Bovard placers in the Buckskin district. (Vanderburg, 1936a; WMK)

In placer and lode deposits in the Genoa district. (Klein, 1983)

As fine particles and nuggets to 1 ounce with 12 wt. percent silver in placer gravels along Buckeye Creek in the Mount Siegel district. (Vanderburg, 1936a; Klein, 1983)

Reported in placer concentrates from Dogtown. (WMK)

ELKO COUNTY In the Lost Gulch placer in the Alder district. (Vanderburg, 1936a; LaPointe et al., 1991)

As inclusions in goethite pseudomorphs after pyrite and disseminated in silicified rock at the Doby deposit in the Aura district. (Klein, 1983; LaPointe et al., 1991)

As wires to 0.2 mm and disseminated grains in oxidized ore at the Rossi Mine in the Bootstrap district. (Snyder, 1989)

As particles to 10 microns at the Rain Mine in the Carlin district. (D. Heitt, o.c., 1991)

In placer gravels along 76 Creek, the Bruneau River, and other streams in the Charleston district. (Vanderburg, 1936a; Klein, 1983; LaPointe et al., 1991)

In quartz veins at the Bull Run and Lucky Girl Mines in the Edgemont district. (Emmons, 1910; Klein, 1983; WMK)

In minor amounts from placers in the Gold Basin district. (Klein, 1983)

With acanthite, naumannite, and other minerals in quartz veins at the Eastern Star, Ken Snyder, and other mines in the Gold Circle (Midas) district; also reported as *electrum*. (Emmons, 1910; Klein, 1983; LaPointe et al., 1991; Casteel and Bernard, 1999; WMK)

At the Hawthorne Placer Mine in the Halleck district. (Whitehill, 1875)

As grains to 4 microns, generally included in or bordering arsenic-bearing pyrite or goethite after pyrite, at the Big Springs Mine in the

Independence Mountains district. (Youngerman, 1992; Hofstra, 1994)

Coarse in gravel at the Penrod Placer Mine in the Island Mountain district. (Vanderburg, 1936a; Klein, 1983; LaPointe et al., 1991)

In quartz-*adularia* veins in micron sizes to plates larger than a man's hand and thick as a knife blade in the Jarbidge district at many properties, including the productive Long Hike and Bourne Mines; associated with silver, naumannite, chlorargyrite, pyrite, and silver-bearing sulfides. (Schrader, 1912, 1923; Klein, 1983; LaPointe et al., 1991)

With quartz, iron oxide, silver, and chlorargyrite near the surface and with sulfides at depth in the Nelson Mine in the Mountain City district; also in placer gravels as dust to nuggets with 18 wt. percent Ag. (Emmons, 1910; Vanderburg, 1936a; Klein, 1983; LaPointe et al., 1991)

Reported in the Tuscarora district as gold or *electrum* in quartz-*adularia* veins at the Dexter Mine, in replacement veins with pyrargyrite and other sulfides at the Navajo Mine, and from other lode mines; also in placer deposits as dust to nuggets as large as 9 pounds. Analyses of *electrum* from some lode deposits show that it contains 70 at. percent silver, and should technically be classified as silver. (Emmons, 1910; Nolan, 1936; Vanderburg, 1936a; Klein, 1983; LaPointe et al., 1991; WMK; SBC)

ESMERALDA COUNTY In small amounts from placers and lode mines in the Gilbert district. (Albers and Stewart, 1972; Klein, 1983; WMK)

In quartz veins with pyrite, molybdenite, and other minerals at the Bullfrog-George prospect in the Gold Point district. (Schilling, 1979; Klein, 1983)

In the Goldfield district, which produced more than 4 million ounces, as tiny grains intergrown with quartz or as scattered large particles with famatinite, bismuthinite, and goldfieldite in primary ore and as fine grains with quartz, alunite, bismite, emmonsite, tellurite, and other minerals in oxidized ore. Specimens from the Mohawk Mine were particularly rich, containing zones or shells of very finely divided gold, and such material was called "jewelry ore." (Ransome, 1909a; Melhase, 1935b; Klein, 1983; CSM; LANH)

In minor amounts from lodes and placers in the Klondyke district; one nugget worth $1,200 was reported. (Vanderburg, 1936a; Klein, 1983)

Reported as delicate, arborescent forms in quartz from an unspecified locality in the Silver Peak district; also noted in ore from mines in the Mineral Ridge area. (Hoffman, 1878; Spurr, 1906; WMK)

Reported from both lode and placer deposits in the Sylvania district, which includes the historic Pigeon Springs district. (Vanderburg, 1936a; Klein, 1983)

Reported as filiform in malachite in the Tokop (Gold Mountain) district; also noted in float and placer deposits. (Hoffman, 1878; Vanderburg, 1936a; Klein, 1983; WMK)

Coarse, well-rounded to angular in placers with 20–35 wt. percent silver at the Eagle and other mines in the Tule Canyon district. (Vanderburg, 1936a; Klein, 1983)

Reported from the Weepah Mine in the Weepah district. (Albers and Stewart, 1972; Klein, 1983; WMK)

EUREKA COUNTY In a placer specimen at an unspecified locality in the Beowawe district. (WMK)

Said to occur in pyrite at the Buckhorn Mine in the Buckhorn district; reported as free particles to 18 microns in the North Buckhorn deposit. (Klein, 1983; Jennings, 1991; NMBM; WMK)

Reported in hard-rock ore in the Eureka district; small amounts were also reported in placers. (Nolan, 1962; Klein, 1983; NMBM; WMK)

The Lynn district has produced more gold than any other district in Nevada from the modern Carlin, Genesis, Goldstrike, and other mines; but most has come from arsenic-rich pyrite rather than discrete particles of native metal. However, tiny particles of gold (<0.1 micron) occur with pyrite in some ore, and free gold has been reported in oxidized ore. In addition, about 9000 ounces were mined from placers along Lynn, Simon, and Rodeo Creeks, where it ranged from 92 to 96 wt. percent Au, possibly the purest in Nevada. (Arehart et al., 1993; Klein, 1983; GCF; Vanderburg, 1936a)

Large amounts of gold have come from the Gold Quarry Mine in the Maggie Creek district, but it is mostly in sulfide as in the Lynn district; however, very pure native gold has been noted in spongy nests on iron oxide–coated surfaces. Less than 700 ounces of placer gold were mined from placers near the turn of the century. (Jensen et al., 1995; Klein, 1983)

HUMBOLDT COUNTY At the Sleeper Mine in the Awakening district with 69 wt. percent Au and 31 wt. percent Ag in bands to 5 mm thick in quartz-*adularia* veins, with 86 wt. percent Au with chlorargyrite at shallow levels, and replacing stibnite crystals to 1 cm in carbonate-banded veins; also reported at the Jumbo and other mines. (Saunders, 1994; Nash and Trudel, 1996; Saunders et al., 1996; M. Jensen, w.c., 1999; Klein, 1983; WMK)

As minute particles (<5 microns) with pyrite in unoxidized ore and in grains to 60 microns in oxidized ore at the Lone Tree Mine in the Buffalo Mountain district. (Kamali and Norman, 1996; Panhorst, 1996)

From placers along Chinese and Horse Creeks in the Disaster (Kings River) district. (Vanderburg, 1936a)

Coarse and angular with 6 wt. percent Ag in placer and slope wash deposits in the Dutch Flat district, in some cases attached to quartz and in places with cinnabar; also in minor amounts from veins. (Vanderburg, 1936a; Klein, 1983)

Minor from placers in the Gold Run district. (Vanderburg, 1936a; Klein, 1983; WMK)

Reported from placer workings in the Jungo district. (Tingley, 1992)

Reported as *electrum* with 51–62 at. percent Au from the Bell vein at the Buckskin-National Mine in the National district. (Vikre, 1985)

In the Getchell Mine in the Potosi district as tiny grains; in the Twin Creeks Mine as microscopic grains in arsenic-rich pyrite. (Joralemon, 1951; Stolburg and Dunning, 1985; Simon et al., 1999)

Reported in quartz veins in the Pueblo district. (Klein, 1983)

More than $30,000 worth was found in quartz float at Pole Creek in the Rebel Creek district; elsewhere, minor in placers. (Vanderburg, 1936a)

In reticulated branching dendrites, as crystals in arborescent growths, and as single leaves in the Ten Mile district with 73 wt. percent Au and 27 wt. percent Ag; specifically noted in quartz veins at the Mad Mutha Mine as leaves and spongy mats, some sharply crystalline, and as

specimens of modified octahedrons to 1 cm from Blue Mountain. (DeMouthe, 1987; Saunders et al., 1996; J. Kepper, w.c., 1998; E. Muceus, o.c., 2000)

In small placer deposits in the Varyville district. (Vanderburg, 1936a; Klein, 1983; WMK)

In quartz veins at the Ashdown Mine in the Vicksburg district. (Vanderburg, 1936a; Klein, 1983)

In veins with silver-bearing minerals and in small amounts in placers in the Winnemucca district. (Klein, 1983; WMK)

LANDER COUNTY In the Battle Mountain district as inclusions in pyrrhotite and *garnet* in skarn at the Fortitude Mine, and as sponge gold at other lode localities; also in placers as dust to 3.5-ounce nuggets with 11–17 wt. percent Ag. (Doebrich, 1993; DeMouthe, 1987; Vanderburg, 1936a)

At the Gold Acres and Pipeline Mines as submicroscopic particles with iron oxide; also in placer deposits as angular dust and small nuggets with 8 wt. percent Ag in the Bullion district. (Stewart et al., 1977a; Foo et al., 1996; Vanderburg, 1936a)

As grains to 10 microns in the oxidized zones of a Carlin-type deposit at the Cortez Mine in the Cortez district. (Stewart et al., 1977a; NMBM)

In quartz veins with pyrite and other sulfides at the Pittsburg, Morning Star, Independence, and other mines in the Hilltop district. (Stewart at al., 1977a; Klein, 1983; WMK)

Reported in placers in the Iowa Canyon district. (Vanderburg, 1936a)

Reported in a specimen from the Dean Mine in the Lewis district. (R. Thomssen, w.c., 1999)

Well crystallized with quartz and pyrite in the open pit at the McCoy-Cove Mine in the McCoy district; also reported in placers. (Klein, 1983; R. Walstrom, w.c., 1999)

In quartz veins at the New Pass and other mines in the New Pass district. (Stewart at al., 1977a; Klein, 1983)

Reported with quartz at the Pittsburg Mine in the Reese River district. (Klein, 1983; NYS)

Reportedly minor in placers in the Steiner Canyon district. (Klein, 1983)

LINCOLN COUNTY Reported in shale in a specimen from the Irmin Mine in the Chief district; possibly the Irmine Mine between the Ely Springs and Highland districts. (WMK)

In a vein with chlorargyrite in the Jumbo Mine in the Delamar district; also with comb quartz, cerussite, mimetite, and other minerals in the Magnolia Mine. (Tschanz and Pampeyan, 1970; WMK)

Reportedly minor in placers in the Eagle Valley district. (Klein, 1983)

In nuggets to the size of wheat grains in placer gravel in gulches below the Hanus shaft, 5 miles northwest of the Groom Mine in the Groom district, now on the Nellis Air Force Range. (Humphrey, 1945)

Reported with silver-bearing galena at the Tasko Mine in the Highland district. (WMK)

Reported in specimens from the Alps Mine, Castleton shaft, and Raymond Ely vein in the Pioche district. (HMM; WMK)

LYON COUNTY Reportedly minor in placers in northern Smith Valley on the border with Douglas County, possibly in the Buckskin district. (Klein, 1983)

In specimens from the Gooseberry and Ramsey Mines in the Ramsey district. (WMK; P. Vikre, o.c., 2000)

Reported with *electrum,* silver, and sulfide minerals in veins in the Silver City district; specifically in specimens from the Buckeye, Dayton, and Haywood Mines. The first recorded discovery of gold in Nevada, in 1848, was in the sands of the Carson River at the mouth of Gold Canyon; Gold Canyon placers in the Silver City district later yielded over $300,000. (Gianella, 1936a; WMK; Vanderburg, 1936a)

Identified in a specimen from an unknown locality in the Wellington district. (JHS)

In quartz stringers, lenses, and veins in granodiorite at the New York, Pine Grove, Rockland, and Wilson Mines in the Wilson district. (Klein, 1983)

From placers in the Yerington district in northwestern Smith Valley and the northern Singatse Range as angular to well-rounded particles with 6–10 wt. percent silver, mostly dust and flakes but also as nuggets to 1/2 ounce; also reported in siliceous sinter from the Mountain View Mine. (Vanderburg, 1936a; WMK)

MINERAL COUNTY In quartz veins with pyrite and silver in the Ashby Gold Mines in the Ashby district; also in siliceous gossan at the La Panta Mine. (Ross, 1961)

Reported from rich ore shoots in the Aurora district, particularly on Last Chance Hill, with *electrum,* tetrahedrite, pyrite, chalcopyrite, and silver selenides in quartz-*adularia* veins. (Hill, 1915; WMK; S. Pullman, w.c., 1999)

Reported with barite and unspecified bismuth- and tellurium-bearing minerals at the Douglas property in the Bell district; also reported in placers. (WMK; Klein, 1983)

With silver-bearing minerals in quartz veins in Tertiary volcanic rocks in the Buena Vista district. (Klein, 1983)

As minor placer production in the Eagleville district. (Vanderburg, 1936a)

As leaves and flakes to 3 cm from an undisclosed locality in the Fitting district. (WMK; GCF)

Visible with *limonite* and chrysocolla in quartz veins at the Star prospect in the Mount Grant district; also in an old upland placer at the Grant Mountain Gold Mine. (Ross, 1961; Klein, 1983; WMK)

As wires and nuggets with silver-bearing galena in quartz veins at the Pamlico Mine in the Pamlico district. (Hill, 1915; Ross, 1961; WMK)

In quartz-calcite veins with manganese oxide at the Belleville Mine in the Pilot Mountains district; also in limited amounts of fine grains in placers in Telephone Canyon downstream from the Belleville Mine. (Vanderburg, 1936a, 1937a; Klein, 1983)

Reported in a specimen from the Gold Pen Mine in the Rand district. (WMK)

Reported at many mines in the Rawhide district in vein networks, specifically noted as *electrum* composed almost exclusively of gold in oxidized ore at the Rawhide Mine; also in placers from which nuggets to 3.5 ounces were recovered. (Klein, 1983; Black, 1991; AMNH; WMK; Vanderburg, 1936a)

In brecciated and altered zones in andesitic tuff at the Moho and Roosevelt Mines in the Silver Star district; also in quartz-calcite-*adularia* veins at Douglas Camp. (Hill, 1915; Vanderburg, 1937a; WMK)

NYE COUNTY Reported at an unspecified locality about 25 miles south of Westgate. (Whitehill, 1873)

Reported as tiny crystals, generally octahedrons, in calcite veins in marble from Carrara in the Bare Mountain district; also noted as micron-sized particles with arsenic-rich iron sulfide at the Mother Lode and Sterling Mines, and with *limonite* and calcite at the Diamond Queen mine. (Melhase, 1935a; Klein, 1983; Castor et al., 1999b; GCF; WMK)

Coarse with pyrite in quartz veins in volcanic rock at the Duluth and Penelas Mines in the Bruner district. (Klein, 1983; Kleinhampl and Ziony, 1984)

Reported as coarse wires and leaves to 10 mm with chrysocolla in quartz at the Old Bullfrog Mine in the Bullfrog district; also described with 66-81 at. percent gold with acanthite, uytenbo-gaardtite, *electrum* with 42-57 at. percent gold, and other minerals at the Bullfrog and Original Bullfrog Mines. (Melhase, 1935b; WMK; Castor and Weiss, 1992; Castor and Sjoberg, 1993)

Found by panning rich ore in the Clifford district; also reported with chlorargyrite at the Clifford Mine. (Ferguson, 1916; WMK)

In the El Primo placer on Cloverdale Creek in the Cloverdale district. (Kral, 1951; Klein, 1983)

As leaves on fractures at the Cooper and Nagle lease in the Ellendale district; also reported in iron-stained quartz veins in rhyolite at the Wilkerson property. (Kleinhampl and Ziony, 1984; Klein, 1983; WMK)

Reported with chlorargyrite and acanthite in breccia at the Paradise Peak Mine in the Fairplay district, mostly as <20-micron particles with no detectable silver commonly in vugs on quartz and overgrown by barite; also noted as visible gold with jarosite, cinnabar, and sulfur. (Thomason, 1986; John et al., 1991)

As tiny grains with scorodite, barite, iodar-gyrite, and iron oxide in a sample from an unnamed shaft in the Gold Crater district on the Nellis Air Force Range. (Tingley et al., 1998a)

Identified in a sample from an unknown local-ity in the Goldfield district, possibly the Free Gold Mine. (JHS)

As microscopic inclusions in sulfosalt miner-als at the Golden Chariot Mine in the Jamestown district. (Tingley et al., 1998a)

Reported with 12 wt. percent silver in angular grains commonly encrusted with quartz from placers at the Johnnie, Congress, and other mines in the Johnnie district. (Vanderburg, 1936a)

Identified in a sample from an unspecified locality in the Kawich district. (JHS)

As nuggets in the Longstreet placer workings in the Longstreet district. (Kral, 1951)

From many mines in the Manhattan district, in some cases as fine crystalline specimens; the best are of platy flakes in narrow quartz-*adularia*-fluorite-calcite veins, fine arborescent gold with small crystals in illite-kaolinite with quartz in pipelike bodies in limestone, and tiny crystals of with *limonite* and manganese oxide. In addition, as fine to coarse, arborescent masses and nuggets to 5 ounces with 23-30 wt. percent Ag from placer gravels up to 100 feet thick. (Ferguson, 1924; WMK; AMNH; S. Pullman, w.c., 1999; Vanderburg, 1936a)

As microscopic grains at the Northumberland Mine in the Northumberland district. (Kokinos and Prenn, 1985)

Reportedly in many mines in the Round Mountain district in feathery to crystalline form with quartz, pyrite, *adularia,* and other minerals; also in placers (see *electrum*). (Ransome, 1909d; Ferguson, 1917, 1922; Vanderburg, 1936a; Klein, 1983; Tingley and Berger, 1985)

As irregular grains to 0.1 mm with 65-98 wt. percent Au in quartz veins at the Life Preserver Mine in the Tolicha district. (Tingley et al., 1998a)

In various mines in the Tonopah district as grains, flakes, and arborescent masses with acan-thite, chlorargyrite, and pyrargyrite in quartz veins; generally more yellow and coarser in oxi-dized ore. (Spurr, 1903; Bastin and Laney, 1918; Melhase, 1935c; Klein, 1983; AM)

Reported in ore in the No. 1 vein at the Irwin group in the Troy district. (Hill, 1916)

With pyrite in a quartz vein in unnamed work-ings near an old stamp mill near the forks of the South Twin River in the Twin River district; also reported in gravels from Clear Creek and Ophir Canyon. (Ferguson and Cathcart, 1954; Klein, 1983)

With marcasite, pyrite, chalcopyrite, and other sulfides in a banded quartz vein at the Berlin Mine in the Union district. (NBMG mining district file)

Reported in quartz from San Juan Canyon in the Washington district. (WMK)

Reported as beautiful specimens with green

talc in fractures in a drusy quartz-calcite vein at the Melbourn Mine in the Gold Spring group in the Willow Creek district, Sec. 20, T4N, R56E; also with arsenopyrite in a quartz vein in the Colorado group. (Hill, 1916; Kral, 1951)

PERSHING COUNTY From the Majuba placer in the Antelope district; also in a specimen from the No. 1 adit of the Majuba Hill Mine. (Klein, 1983; R. Thomssen, w.c., 1999)

In placers in Indian, Buena Vista, and Congress Canyons near Unionville in the Buena Vista district. (Whitehill, 1877; Klein, 1983)

Reported in quartz veins in the Imlay district, specifically in a specimen from the Betty La Verne Mine. (Johnson, 1977; WMK)

In quartz-pyrite veins with sulfide minerals at the Sunnyside, Gold Note, and Imperial Mines in the Kennedy district. (Johnson, 1977; WMK)

Coarse with 10–27 wt. percent silver in placers in the Placerites district. (Vanderburg, 1936a)

From float and placer deposits in the Rabbit Hole district, mainly in Coarse Gold Canyon in gravels to a depth of 60 feet; described as nuggety in coarse, flat particles with 10 percent silver. (Melhase, 1934b; Vanderburg, 1936a; Klein, 1983)

As rough, crystalline, placer gold with 12–20 wt. percent Ag in gravel to 100 feet thick in several canyons in the Rochester district; also coarse with *limonite* at the Hagan lode and reported in *tourmaline*-bearing veins. (Vanderburg, 1936a; Johnson, 1977)

In placers in Rosebud Canyon and its tributaries in the Rosebud district; noted with cinnabar, cassiterite, and hübnerite in gravel near Barrel Springs. (Vanderburg, 1936a; Bailey and Phoenix, 1944; WMK)

As coarse, rough, placer gold with 12 percent silver in the Sawtooth district; partly located in Humboldt County. (Vanderburg, 1936a)

Reported as yellow gold on surface vein quartz and as gray *electrum* with alunite at depth in the Scossa district, Sec. 10, T33N, R30E. (Vanderburg, 1936b)

As visible clusters of tiny particles and coarse, crystalline aggregates in clear sugary quartz veins and veinlets in the Seven Troughs district. (Ransome, 1909b; Johnson, 1977)

Coarse, with 12 wt. percent silver, from placers in the Sierra district that yielded over $4 million, mainly in Barber, Wright, Auburn, and Rock Hill Canyons. Also in milky-white quartz veins with sulfides at the Auburn, Auld Lang Syne, and White Bear Mines. (Vanderburg, 1936a; Ransome, 1909b; Johnson, 1977)

As dust and small nuggets with 30 wt. percent silver in placer deposits, such as the Bonanza King Mine, in the Spring Valley district that were reportedly the most productive in Nevada with a total yield of over 500,000 ounces; also as visible flakes in *limonite* in quartz-*tourmaline* veins with sulfide minerals. (Ransome, 1909b; Vanderburg, 1936a; Campbell, 1939; Klein, 1983)

In small amounts in placers in Trinity Canyon in the Trinity district. (Klein, 1983)

As nuggets to 11 ounces in placers along Willow Creek and in Spaulding Canyon in the Willow Creek district. (Johnson, 1977; Klein, 1983; WMK)

Identified in specimens from unknown localities in the Antelope Springs and Nightingale districts. (JHS; R. Thomssen, w.c., 1999)

STOREY COUNTY Reported in specimens of bonanza ore from the Comstock Lode in the Comstock district (see *electrum*). (Bastin, 1922; Vikre, 1989; H. Bonham, o.c., 1993; HUB)

WASHOE COUNTY Very minor placer gold was reported from Galena Creek in the Galena district. (Klein, 1983)

In oxidized ore from mines in the Jumbo district; reported with 61 at. percent Au from the Dunlop Mine. (Bonham and Papke, 1969; Vikre, 1989)

In gravels of an ancient river channel in the Little Valley district; total production was estimated at $100,000, but a single pocket 5 miles southwest of Franktown is said to have yielded $60,000 worth. (Vanderburg, 1936a; Klein, 1983)

As wires, crystals, and flakes in veins in intermediate volcanic rock at the Babe Mine and other properties in the Olinghouse district. At the modern Olinghouse Mine open pit as fine specimens of wire and crystalline gold in veins with quartz, calcite, epidote, scheelite, *heulandite,* epistilbite, and scolecite; fineness is approximately 700. In addition, produced from placer deposits in drainages below the lode mines as dust and nuggets to 1.5 ounces with 32 wt. percent Ag. (Overton, 1947; Bonham and Papke, 1969; Jones, 1998; SBC; S. Kleine, o.c., 1999; Vanderburg, 1936a)

In the Peavine district in small amounts from placer deposits, as bright masses to several mm with copper minerals in quartz veins near Cold Springs, and in ore at the Mortensen and Wall Mine. (Vanderburg, 1936a; S. Kleine, o.c., 1999; WMK)

In oxidized veins in the Pyramid district; specifically noted in the Ruth vein. (Overton, 1947; SBC)

In a natural gold-silver-mercury alloy in a specimen from an unknown locality in the hills northwest of Pyramid Lake. (WMK)

Very minor from placers along Steamboat Creek in the Steamboat Springs district. (Klein, 1983)

With anglesite, cerussite, and silver halide minerals in the Wedekind district. (Overton, 1947)

WHITE PINE COUNTY As needles and octahedra to 2 mm in oxidized rock in the Vantage deposit on Alligator Ridge in the Bald Mountain district; also in small visible grains with *limonite,* azurite, and malachite in the Top pit, Bald Mountain Mine and in nuggets to 0.5 ounce in placer deposits in Water Canyon and its tributaries. (Ilchick, 1984; SBC; Vanderburg, 1936a; JHS)

In the Osceola district as dust to coarse nuggets with 15 percent Ag in alluvial fan and gulch placers that produced as much as $3.5 million. The largest nugget discovered in Nevada (24 pounds), was found here in 1878 and mining was ongoing in 1988, when ounce-sized nuggets were still being extracted. (Vanderburg, 1936a; Klein, 1983; Schilling, 1988)

GOLDFIELDITE

$Cu_{12}(Te,Sb,As)_4S_{13}$

A Nevada type mineral, goldfieldite is a rare mineral that occurs in epithermal precious-metal veins and in veins associated with porphyry copper deposits. Since the original work by Ransome, it has been discredited and reinstated as a valid mineral species.

ESMERALDA COUNTY Type locality as crude gray tetrahedrons and anhedral grains in dark crusts with gold, marcasite, polybasite, famatinite, alunite, and other minerals in rich ore from several properties in the Goldfield district, notably the Claremont and Mohawk Mines. (Ransome, 1909a, 1910a, 1910b; Levy, 1968; DeMouthe, 1985; MV; J. Kepper, w.c., 1998; CSM; FUMT)

GOLDQUARRYITE

$CuCd_2Al_3(PO_4)_4F_2(H_2O)_{10}\bullet(H_2O)_2$

A new Nevada type species and the first identified cadmium-rich phosphate mineral. To date, it is known only from a single Carlin-type gold deposit, originally reported as "unknown #2" by Jensen et al. (1995).

EUREKA COUNTY Radiating, acicular, glassy blue crystals in sprays to 3 mm associated with hewettite and opal in a specimen from a freshly basted muckpile at the Gold Quarry Mine in the Maggie Creek district. (Jensen et al., 1995; Excalibur Mineral Co., 2002a; A. Roberts, o.c., 2002; Roberts et al., 2003)

GORTDRUMITE

$(Cu,Fe)_6Hg_2S_5$

A very rare mineral described with other sulfides from a hydrothermal vein in limestone in Ireland.

EUREKA COUNTY Tentatively identified by SEM/EDX as microscopic grains in ore at the Goldstrike Mine in the Lynn district. (GCF)

GOSLARITE

$ZnSO_4\bullet7H_2O$

Goslarite is a relatively common mineral that occurs as efflorescences on mine walls in base-metal deposits.

LANDER COUNTY As white efflorescences on mine walls at various properties in the Battle Mountain district. (Roberts and Arnold, 1965)

LYON COUNTY Reported at the Jackpot Mine in the Wellington district. (S. Pullman, w.c., 1999)

STOREY COUNTY Reported at an unspecified locality in the Comstock district. (S. Pullman, w.c., 1999)

GOUDEYITE

$(Al,Y)Cu_6^{2+}(AsO_4)_3(OH)_6\bullet3H_2O$

A Nevada type mineral, goudeyite is a member of the *agardite* group that occurs as a secondary mineral in a copper- and arsenic-bearing hydrothermal deposit.

PERSHING COUNTY Type locality as greasy, pale- to dark-yellow-green patches irregularly coating fracture surfaces of iron-stained and altered rhyolite in the Copper stope at the Majuba Hill Mine in the Antelope district, Sec. 2, T32N, R31E; also as tufts and masses of very fine-grained, yellow-green to bright green, acicular crystals in a mineralized fault zone. (MacKenzie and Bookstrom, 1976; Wise, 1978; Jensen, 1985; J. Kepper, w.c., 1998)

GRAPHITE
C

Graphite is a relatively common mineral that mostly occurs in schist as a product of metamorphism of primary carbonaceous substances; it is less common in igneous rocks and meteorites. In Nevada, graphite has been identified at several Carlin-type gold deposits, where it may have been formed by hydrothermal alteration of carbonaceous material.

CARSON CITY As extensive deposits in schist at the Chedic Mine and Carson Black Lead Company in the Voltaire district, where it was mined for use in paint pigment and foundry facings. (Lincoln, 1923; Overton, 1947)

CLARK COUNTY With kamacite, taenite, troilite, and schreibersite in the Las Vegas meteorite. (Buchwald, 1975)

ELKO COUNTY Widespread in trace amounts in the Independence Mountains district, and locally abundant as at the Murray Mine. (Hofstra, 1994; GCF)

Reported in a specimen from the Ruby Mountains. (WMK)

EUREKA COUNTY As clots and crystalline masses bordering, and associated with, gold ore in the Betze orebody at the Goldstrike Mine in the Lynn district. (GCF)

HUMBOLDT COUNTY As microscopic grains and rarely as coatings on ore minerals at the Getchell Mine in the Potosi district. (Stolburg and Dunning, 1985)

LANDER COUNTY Reported as a gangue mineral in ore from the Reese River district. (Stewart at al., 1977a)

WASHOE COUNTY Reported in a specimen from the Black Eagle Mine near Verdi. (WMK)

Reported in a specimen from an unknown locality in the Steamboat Springs district. (WMK)

A minor constituent in metamorphosed carbonate and argillaceous rocks of the Nightingale

sequence in the central part of the county. (Bonham and Papke, 1969)

GRATONITE
$Pb_9As_4S_{15}$

A sulfosalt mineral that occurs in hydrothermal base-metal and copper ores; the occurrence in a Carlin-type deposit in Nevada is unusual.

EUREKA COUNTY Reported with other sulfosalts from the Carlin Mine in the Lynn district. (Kuehn, 1989)

GRAYITE
$(Th,Pb,Ca)PO_4 \cdot H_2O$

A radioactive secondary mineral that mainly occurs in lithium-bearing pegmatite.

MINERAL COUNTY As fine-grained yellow coatings on thorite in pegmatite at the Zapot claims in the Fitting district. (Foord et al., 1999b)

GREENOCKITE
CdS

Greenockite is widespread in many base-metal deposits where it occurs in minor amounts as a secondary mineral, generally coating sphalerite.

CLARK COUNTY Reported with sphalerite at the Potosi Mine in the Goodsprings district. (Takahashi, 1960)

EUREKA COUNTY As yellowish, acicular, zinc-bearing crystals on fractures in sulfide-bearing and argillically altered limestone at the Goldstrike Mine in the Lynn district. (GCF)

On fracture surfaces as small yellow-orange grains that resemble orpiment at the Gold Quarry Mine in the Maggie Creek district; also as brilliant yellow acicular crystals to 0.1 mm in vugs in silicified rock with minor *allanite* and carbonate-fluorapatite. (Jensen et al., 1995)

LANDER COUNTY As greenish-yellow cavity coatings in oxidized lead-zinc ore at the Lucky Strike Mine in the Battle Mountain district. (Roberts and Arnold, 1965)

WHITE PINE COUNTY With paratellurite and devilline in oxidized rock at the Silver King Mine in the Ward district. (Williams, 1964)

GREIGITE

$Fe^{2+}Fe_2^{3+}S_4$

A relatively rare sulfide that is generally found in lacustrine sediments, but also in hydrothermal vein deposits.

CHURCHILL COUNTY Identified by XRD in a specimen from the Nickel Mine area, Table Mountain district. (S. Lutz, w.c., 2000)

GROSSULAR

$Ca_3Al_2(SiO_4)_3$

A *garnet*-group mineral that generally occurs in metamorphosed calcareous rocks; common in tungsten skarn deposits in Nevada.

CARSON CITY Reported in a specimen from Signal Peak (possibly Mount Siegel) in the Pine Nut Range. (WMK)

ELKO COUNTY Reported in a specimen from Gilbert Canyon in the Ruby Range. (AMNH)

EUREKA COUNTY As subhedral orange crystals with vesuvianite, calcite, and quartz in skarn at the Goldstrike Mine in the Lynn district. (Walck, 1989)

HUMBOLDT COUNTY As honey-colored to dark-brown crystalline masses in skarn in limestone along the contact with granitic rock in the Gold Run district; some crystals show brilliant iridescence on crystal faces and in thin section. (Ingerson and Barksdale, 1943; JHS)

Common with diopsidic *pyroxene* and *plagioclase* in calc-silicate rocks adjacent to granodiorite in the Potosi district. (Hotz and Willden, 1964)

LANDER COUNTY With vesuvianite, calcite, and quartz in tungsten-bearing skarn at the Linka Mine in the Spencer Hot Springs district. (McKee, 1976)

LINCOLN COUNTY With vesuvianite and other minerals in float near Blind Mountain Spring and with diopside and carbonate in marble near McCullough Hill. (Westgate and Knopf, 1932)

LYON COUNTY Reported at the Ludwig Gypsum and Douglas Hill Mines in the Yerington district. (Harris, 1980; S. Pullman, w.c., 1999)

MINERAL COUNTY White grossular was reported with diopside, tremolite, and other minerals at granodiorite-limestone contacts in the Hawthorne district. (Hill, 1915)

Reported in a specimen from the Gunmetal Mine in the Pilot Mountains district. (HMM)

NYE COUNTY White, translucent, and greenish grossular was reported with a variety of other minerals in wall rocks at the Queen City Mine in the Queen City district. (Hewett et al., 1968)

As dark-peach-tan grossular-spessartine in a vein in granodiorite at the Victory Mine in the Gabbs district; also with diopside in masses in silicated rock adjacent to granodiorite stocks in the vicinity of magnesite deposits. (Lee, 1962; Schilling, 1968a)

PERSHING COUNTY As lustrous orange to orange-brown crystals to 8 cm at the Nightingale Mine in the Nightingale district; also identified by XRD in massive garnetite in the north part of the district and found elsewhere in the district with clinozoisite, epidote, wollastonite, diopside, and quartz in skarn. (Crowley, 2000; GCF; WMK)

Tentatively identified as tiny veinlets in andradite in the Panther Canyon area in the Rye Patch district. (Vitaliano, 1944)

STOREY COUNTY Reported with wollastonite and diopside in calc-silicate rocks near Ramsey. (Rose, 1969)

WASHOE COUNTY With epidote, diopside, scheelite, and other minerals in skarn at the Garnet tungsten prospect in the Cottonwood district. (Bonham and Papke, 1969)

GROUTITE

$Mn^{3+}O(OH)$

A secondary mineral that occurs in banded iron formations, hydrothermal metal deposits, and other environments.

MINERAL COUNTY Antimony-bearing groutite was reported in oxidized rock at the Candelaria Mine in the Candelaria district. (Chavez and Purusotam, 1988)

GRUZDEVITE

$Cu_6Hg_3Sb_4S_{12}$

A rare sulfosalt mineral found outside Nevada in veins in Kyrgyzstan and Mexico.

EUREKA COUNTY Tentatively identified by SEM/EDX as microscopic grains in ore at the Goldstrike Mine in the Lynn district. (GCF)

GUANAJUATITE

Bi_2Se_3

A rare mineral that occurs in hydrothermal deposits with other bismuth minerals.

LANDER COUNTY Reported as a constituent of gold-silver ore at the Red Top Mine in the Hilltop district. (WMK)

GUÉRINITE

$Ca_5H_2(AsO_4)_4 \cdot 9H_2O$

A very rare mineral described elsewhere only in museum specimens from Germany.

HUMBOLDT COUNTY Found on ore samples as clear, pearly scales to 0.5 mm and as subhedral crystals coating calcite and realgar with pharmacolite at the north pit of the Getchell Mine in the Potosi district. (Dunning, 1988)

GUMMITE

Gummite is a generic term for colorful uranium oxide minerals, mostly hydrated, whose identity is unknown. It is quite common in uranium deposits throughout the world.

HUMBOLDT COUNTY Reported with meta-autunite and metatorbernite at the Moonlight Mine in the Disaster district. (Garside, 1973)

PERSHING COUNTY In yellow, orange, reddish-brown to black massive crusts with *pitchblende,* smoky quartz, calcite, and pyrite at the Stalin's Present prospect in the Buena Vista district. (Lovering, 1954; Garside, 1973)

WASHOE COUNTY Reported as an oxidation product of uraninite with meta-autunite and uranophane at the Thunder Bird claims and other prospects in the Pyramid district. (Brooks, 1956; Garside, 1973)

Reported in veinlets with autunite and uranophane at the Buckhorn Mine in the Stateline Peak district. (Hetland, 1955; Garside, 1973; Castor et al., 1996)

GUNNINGITE

$(Zn,Mn^{2+})SO_4 \cdot H_2O$

A rare species reported elsewhere from mines in Canada.

EUREKA COUNTY Identified by XRD and SEM/ EDX analyses in light- to dark-blue efflorescent coatings on silicified and sulfide-bearing limestones at the Goldstrike Mine in the Lynn district; intimately mixed with szomolnokite and lesser amounts of starkeyite, anhydrite, and gypsum. Most of the blue coating is an intermediate phase between gunningite and szomolnokite. (GCF)

GUSTAVITE

$PbAgBi_3S_6$ (?)

A rare mineral that occurs in hydrothermal deposits and in pegmatite.

NYE COUNTY Identified with benjaminite and aikinite in a quartz vein at the Outlaw Mine in the Round Mountain district. (Nuffield, 1975; Karup-Møeller and Makovicky, 1979; Dunning et al., 1991; CSM)

GYPSUM

$CaSO_4 \cdot 2H_2O$

Gypsum is a very common mineral that occurs in evaporite deposits, in caves, and as a secondary mineral in some hydrothermal deposits. Large deposits are mined in Nevada for use in wallboard, plaster of Paris, Portland cement, and other uses. Only a portion of the large number of occurrences in the state are reported below. *Selenite* is a varietal term for clear crystalline gypsum, *satin spar* denotes fibrous gypsum, and *alabaster* is fine-grained white gypsum.

CARSON CITY Reported in a sample from a limestone quarry at the Carson City Prison in the Carson City district. (UCB)

CHURCHILL COUNTY As *selenite* from the Humboldt Salt Marsh in the Dixie Marsh district. (WMK)

Reported at the Fairview Silver Mine in the Fairview district. (Schrader, 1947)

In calcite veins with cinnabar at the Cinnabar Hill Mine in the Holy Cross district. (Vanderburg, 1940)

CLARK COUNTY Mined in large quantities from nearly pure beds lying on Permian Kaibab Limestone at the Blue Diamond Mine in the Arden district; also once mined from bedded deposits in the Toroweap Formation, and reported as *selenite* from the Great Western Quarry. (Papke, 1987; Longwell et al., 1965; WMK)

Relatively pure, white, massive, fine-grained, and friable with interbedded dolomite in the Permian Toroweap Formation at the Bunkerville Ridge deposit in the Bunkerville district; also in a 65-foot-thick white, massive, fine-grained bed in the Pennsylvanian-Permian Pakoon Formation in the Virgin Mountains deposit. (Papke, 1987)

In a light-gray, 15-foot-thick bed in dolomite of the Pennsylvanian-Permian Bird Spring Formation at the White Beauty (Londo) deposit in the Charleston district. (Papke, 1987)

As aggregates of white tabular crystals in a hematitic matrix at the Boss and Milford Mines in the Goodsprings district. (Jedwab, 1998b)

Once mined from impure and iron-stained gypsum beds in the Miocene Horse Spring and the Triassic Moenkopi Formations at the North Rainbow Gardens, East Rainbow Gardens, and White Eagle deposits in the Las Vegas district. (Papke, 1987; LANH)

Reported as the variety *alabaster* from relatively pure, white, fine-grained gypsum beds at the Anderson (January) gypsum deposit in the Triassic Chinle Formation; also once mined from the White Star and White Star No. 2 deposits in the Permian Kaibab Formation in the Moapa district. (WMK; Papke, 1987)

Mined in large quantities from fine-grained, friable, porous material and masses of *selenite* crystals with some local layers of fine-grained, sugary, compact gypsum in a flat-lying deposit at least 120 feet thick in the Miocene Muddy Creek Formation at the PABCO Mine in the Muddy Mountains district; found as *selenite* masses and encrustations in nearby Gypsum Cave; and found as pieces of *satin spar* at the Anniversary Colemanite Mine. (Papke, 1987; CIT; SBC)

In beds to 100 feet thick with quartz, calcite, dolomite, and clay minerals in the Miocene Horse Spring Formation at the Wechech Basin deposit. (Papke, 1987)

ELKO COUNTY As fracture coatings above ore at the Rain Mine in the Carlin district. (D. Heitt, o.c., 1991)

As acicular crystals to 2 cm with barite crystals at the Meikle Mine in the Lynn district. (Jensen, 1999)

ESMERALDA COUNTY Reported at the Alum Mine in the Alum district. (S. Pullman, w.c., 1999)

Widely distributed in ore and altered rocks in the Goldfield district. (Ransome, 1909a)

Reported as *satin spar* from an unspecified locality in the Rock Hill district. (WMK)

Identified as recently formed crystals on Clayton Playa in the Silver Peak Marsh district. (JHS)

EUREKA COUNTY As small clear to white blades on freshly exposed rock at the Goldstrike Mine in the Lynn district; also in veins to 30 cm thick and with variscite and other phosphate minerals. (GCF)

Widespread coating other minerals as colorless, flat-lying crystals and intergrowths at the Gold Quarry Mine in the Maggie Creek district. (Jensen et al., 1995)

As crystals and crystalline masses with native sulfur at the Hot Springs Point Sulfur Mine. (GCF)

HUMBOLDT COUNTY As colorless prisms at the Silver Coin Mine in the Iron Point district. (Thomssen, 1998; J. McGlasson, w.c., 2001)

As rare, small, white, tabular or needlelike crystals or fibrous bundles at the Cordero Mine in the Opalite district. (Fisk, 1968)

As clear radiating needles and crystal groups in cavities and cracks in quartz-realgar ore in the north pit at the Getchell Mine in the Potosi district. (Stolburg and Dunning, 1985)

With stibnite, calcite, quartz, cinnabar, and other mercury minerals at the Cahill Mine in the Poverty Peak district. (Bailey and Phoenix, 1944; F. Cureton, w.c., 1999)

LANDER COUNTY In a specimen from the Betty O'Neal Mine in the Lewis district. (R. Thomssen, w.c., 1999)

LINCOLN COUNTY Fine-grained, sugary, and porous with quartz, calcite, clay minerals, and iron oxides at the White Queen and Snowhite deposits. (Papke, 1987)

LYON COUNTY Mined from a fine-grained, white deposit with anhydrite, calcite, and pyrite in Mesozoic rocks at the Adams Mine in the Mound House district; also as an impure, poorly consolidated Quaternary deposit at Mound House. (Louderback, 1903; Papke, 1987; SBC)

As clear, transparent *selenite* crystals scattered over the surface in mud deposited at the bottom of ancient Lake Lahontan along U.S. Highway Alternate 50, 8 miles east of Fernley. (JHS)

Reported in a specimen from the Wabuska

Marsh district; also in a specimen from the Wabuska Smelter. (WMK; R. Thomssen, w.c., 1999)

Associated with Jurassic limestone in a large pure body with anhydrite at depth at the Ludwig Mine in the Yerington district; also in a specimen from Carson Hill. (Rogers, 1912a; Stoddard and Carpenter, 1950; Papke, 1987; R. Thomssen, w.c., 1999)

MINERAL COUNTY In pure, white, fine-grained beds in Jurassic sedimentary rocks at the Regan Mine in the Mountain View district. (Ross, 1961; Papke, 1987)

Reported in a specimen from the Santa Fe Mine in the Santa Fe district; also in a specimen from 4 miles east of Luning. (R. Thomssen, w.c., 1999; WMK)

As clear crystals of *selenite* at Sodaville Hot Springs in the Sodaville district. (Garside and Schilling, 1979)

NYE COUNTY As crystals to 0.5 inch in the Belmont district. (Hoffman, 1878)

Well crystalized and plentiful along fractures in ore at the Phelps Stokes Mine in the Ellsworth district. (Reeves et al., 1958)

Reported in a specimen from the San Rafael Mine in the Lodi district. (LANH)

Reported in a specimen from the White Caps Mine in the Manhattan district. (LANH)

As fibrous aggregates and crystals to 5 mm at the Northumberland Mine in the Northumberland district. (Kokinos and Prenn, 1985)

Common as veinlets and fracture encrustations in the Tonopah district. (Spurr, 1903; Melhase, 1935c)

PERSHING COUNTY As white crusts formed by oxidation subsequent to mining at the Majuba Hill Mine in the Antelope district. (Jensen, 1985)

Mined in large amounts from a massive, white, fine-grained deposit in the Jurassic-Triassic Nightingale Sequence at the Empire Gypsum Mine (Selenite Quarry) in the Gerlach (Hooker) district; includes a 100-foot-long *selenite* vein

that yields clear, cleaved masses to 3 feet across. (Papke, 1967; F. Cureton, w.c., 1999; M. Jensen, w.c., 1999; Price et al., 1999)

With cinnabar and native sulfur in siliceous sinter a half mile south of the old townsite of Humboldt in the Imlay district. (Russell, 1885)

In faulted and folded sedimentary beds with Jurassic limestone at the North Lovelock, Lovelock, Crystal Placer, and Muttlebury Canyon deposits in the Muttlebury district. (Louderback, 1903; JHS; Papke, 1987)

In a bedded, snow-white, sugary mass of probable Jurassic age at the Corn Beef deposit in the Table Mountain district. (Louderback, 1903; JHS; Papke, 1987)

As colorless euhedral crystals to 5 cm from the Willard Mine in the Willard district. (Jensen and Leising, 2001)

STOREY COUNTY As clear doubly terminated and in some cases twinned crystals to 15 cm northwest of Washington Hill in the Castle Peak district. (B. Park-Li, o.c., 1999; GCF)

Reported in the Comstock district. (Bastin, 1922)

WASHOE COUNTY As *selenite* in corroded crystals to 8 cm, some with "fishtail" terminations, in the Peavine district. (GCF; WMK)

As *selenite* with chalcanthite formed by post-mine oxidation in the Pyramid district. (Overton, 1947)

Reported as *satin spar* in the Steamboat Springs district. (BMC)

In fractures with pyrite and as single corroded crystals of *selenite* to 5 cm on mine dumps in the Wedekind district. (Morris, 1903; GCF; WMK; S. Pullman, w.c., 1999)

WHITE PINE COUNTY As rare veins in oxidized ore at the Vantage deposit on Alligator Ridge in the Bald Mountain district. (Ilchick, 1990)

As fine *selenite* crystals to 5 inches, some with "fishtail" terminations, from the Richards Mine in the Robinson district. (WMK; S. Pullman, w.c., 1999)

HÄGGITE

$V_2O_2(OH)_3$

A relatively uncommon mineral that is generally found in the oxidized zone in sandstone-type uranium deposits.

EUREKA COUNTY Widespread as an efflorescence on freshly exposed pit walls with hummerite at the Gold Quarry Mine in the Maggie Creek district; also noted as botryoidal crusts and disseminated radiating crystal groups. (Jensen et al., 1995)

HAIDINGERITE

$CaHAsO_4 \cdot H_2O$

Haidingerite, a relatively rare mineral, occurs in the oxidation zones of some mineral deposits that contain primary arsenic minerals.

HUMBOLDT COUNTY On specimens as thin white coatings and masses with pharmacolite, from which it was derived by partial dehydration, in the north pit of the Getchell Mine in the Potosi district. (Dunning, 1988)

NYE COUNTY With pharmacolite and pitticite on broken rock at the White Caps Mine in the Manhattan district; formed by post-mining oxidation. (Foshag and Clinton, 1927; Palache et al., 1951)

HALITE

NaCl

Halite is a very common mineral that occurs in playa deposits, in salt domes, in bedded evaporites, and in hydrothermal deposits as a daughter phase in fluid inclusions. It has been mined from several Nevada playas as described in Papke (1976).

CHURCHILL COUNTY Produced from brines and reported in a specimen from the Carson Sink district. (Papke, 1976; WMK)

Mined in the 1860s from a 1-foot-thick surficial layer at the Humboldt Salt Marsh in the Dixie Marsh district; also in subsurface layers to depths of 15 feet and noted as crystals in a layer with gaylussite and ulexite. (Papke, 1976; M. Jensen, w.c., 1999)

Produced from brine at the Eagle (Leete) Marsh until 1913; also in Lyon County. (Papke, 1976)

Harvested from 1863 to the present from a solid crust several inches thick that is deposited following surface water accumulation on Fourmile Flat in the Sand Springs Marsh district; noted locally as pink cubes to 2.5 cm and clusters of crystals. (Papke, 1976; Schilling, 1963; WMK)

Produced until 1915 from brine and a thin surficial crust at Alkali Flat in the White Plains Flat district. (Papke, 1976)

CLARK COUNTY Reported in a specimen from an unknown locality in the Virgin Mountains; possibly from along the Virgin River (see below). (WMK)

As thick, coarsely crystalline, pure beds at the Salvation and Big Salt Cliff Mines and other localities once exposed along both sides of a 12-mile-long section of the Virgin River Valley in the Saint Thomas district, now mostly under the waters of Lake Mead; mined by Native Americans prior to A.D. 1150 and in modern times until 1937. (Papke, 1976; BMC; AMNH)

ESMERALDA COUNTY Mined in small amounts in the Alkali Spring Valley district. (Papke, 1976; Tingley, 1992)

Identified chemically in surface samples in the Columbus Marsh district. (Papke, 1976)

Reported from drill core in Clayton Valley and produced as a by-product of evaporation of brines for lithium in the Silver Peak Marsh district. (Papke, 1976; UCB; JHS)

EUREKA COUNTY Mined from surficial deposits and produced from brine in Williams Marsh in the Diamond Marsh district. (Papke, 1976)

As microscopic cubes in silica-flooded ore at the Goldstrike Mine in the Lynn district. (GCF)

LANDER COUNTY As daughter crystals in fluid inclusions at the West orebody in Copper Canyon in the Battle Mountain district. (Theodore and Blake, 1978)

LINCOLN COUNTY Reported in large masses at Hyko (Hiko). (Hoffman, 1878)

LYON COUNTY Produced from 1871 to 1913 at the Eagle Salt Works in the Leete district. (Papke, 1976; WMK)

NYE COUNTY Some production was reported in Railroad Valley in the Butterfield Marsh district. (Papke, 1976)

Minor production was reported in Big Smoky Valley in the Spalding Marsh district. (Papke, 1976)

MINERAL COUNTY Reported with other salts in surface crusts in the Double Springs Marsh district. (Horton, 1964b)

Mined from shallow pits in a nearly pure surficial crust and a subsurface bed in the Rhodes Marsh district; other evaporite minerals are mirabilite, thenardite, ulexite, and borax. (Vanderburg, 1937a; Papke, 1976)

Mined in small amounts in the Teels Marsh district. (Papke, 1976)

WASHOE COUNTY Mined from a surface crust and produced in larger quantities from brine pumped from wells at the Buffalo Springs Salt Works on the west side of the Smoke Creek Desert. (Russell, 1885; Papke, 1976)

WHITE PINE COUNTY Mined from a surface crust as much as 8 inches thick in the Spring Valley district. (Papke, 1976)

A daughter mineral in fluid inclusions in jasperoid in the Robinson district. (Lovering, 1972)

HALLOYSITE

$Al_2Si_2O_5(OH)_4$

A clay mineral that is related to kaolinite and endellite, halloysite is a product of alteration or protracted weathering of aluminosilicate minerals such as *feldspar.*

CLARK COUNTY Reportedly rare as an alteration mineral in the Bunkerville district. (Beal, 1965)

ELKO COUNTY Identified along structures in siliceous sediments at the Rain Mine in the Carlin district; also with kaolinite in red clay. (D. Heitt, o.c., 1991)

Reported in a specimen as *severite* or halloysite at the Central Pacific Mine in the Lucin district. (UCB)

ESMERALDA COUNTY A widespread alteration product in the Cuprite district. (Swaze et al., 1993)

EUREKA COUNTY Reported in basalt at the Buckhorn Mine in the Buckhorn district. (WMK)

Common as white microgranular to tubelike crystals in fault gouge and altered rock at the Goldstrike Mine in the Lynn district; also in overlying Tertiary ash beds. (Lavkulich, 1993; Ferdock et al., 1997; GCF)

With malachite and azurite as a whitish clayey fracture coating on oxidized siltstone at the Gold Quarry Mine in the Maggie Creek district. (Jensen et al., 1995)

LANDER COUNTY Identified in a sample from an unspecified locality in the Bullion district. (JHS)

NYE COUNTY In quartz-fluorite veins with creedite at an unnamed gold mine in the Lodi district. (Melhase, 1935d)

Reported in minor amounts in open spaces in quartz veins on the top of Round Mountain in the Round Mountain district. (Ferguson, 1922)

STOREY COUNTY Identified by XRD with quartz and pyrite at the Chollar Mine in the Comstock district. (GCF)

WASHOE COUNTY Reported as *metahalloysite* in an open pit on Geiger Grade in the Comstock district; used for making bricks. (D. Hudson, w.c., 1999)

In deposits formed by hydrothermal alteration of Tertiary lapilli tuff in the Sand Pass district in the Terraced Hills, Sec. 13, T27N, R19E; mined for use in Portland cement. (Bonham and Papke, 1969)

WHITE PINE COUNTY In copper-bearing blue clay at the Richard Mine in the Robinson district. (UCB)

HALOTRICHITE

$Fe^{2+}Al_2(SO_4)_4 \cdot 22H_2O$

A relatively uncommon water-soluble mineral found as efflorescences on mine walls, generally with gypsum.

ELKO COUNTY On mine walls with szomolnokite at the Meikle Mine in the Lynn district. (Jensen, 1999)

ESMERALDA COUNTY As white or slightly yellowish fibrous encrustations in disused workings above the water table with coquimbite, rozenite, alunogen, melanterite, and kaolinite at the January, Combination, and Red Top Mines and other locations in the Goldfield district. (Ransome, 1909a; Ashley and Albers, 1973; WMK; S. Pullman, w.c., 1999)

EUREKA COUNTY As vanadium-bearing, pale-olive-green coatings and fibrous spherules on joint surfaces in bleached and altered siltstone at the Gold Quarry Mine in the Maggie Creek district. (Jensen et al., 1995)

STOREY COUNTY Reported in a specimen from Virginia City in the Comstock district. (AMNH)

WASHOE COUNTY Reported in a specimen from an unknown locality in the Peavine district. (WMK)

HAMBERGITE

$Be_2BO_3(OH)$

An uncommon mineral that has been found in granite and syenite pegmatite.

MINERAL COUNTY As creamy-tan to white, tabular crystals with earthy to silky luster as much as 30 cm long with smoky quartz, *amazonite,* topaz, and albite at the Zapot claims in the Fitting district. (Foord et al., 1999b)

HAMMARITE

$Pb_2Cu_2Bi_4S_9$

A rare sulfosalt mineral found in hydrothermal deposits.

NYE COUNTY As minute grains formerly identified as krupkaite at the Outlaw Mine in the Round Mountain district. (Dunning et al., 1991)

HANKSITE

$KNa_{22}(SO_4)_9(CO_3)_2Cl$

A relatively rare evaporite mineral first found in playa deposits in California.

CHURCHILL COUNTY Reported in a specimen from Ragtown in the Soda Lake district. (CSM)

HAUSMANNITE

$Mn^{2+}Mn_2^{3+}O_4$

A relatively common manganese oxide mineral that occurs as a primary phase in veins and as an economic mineral in metamorphosed manganese deposits.

WHITE PINE COUNTY Reported as thick black coatings in the Nevada district. (S. Kleine, o.c., 1999)

HAWLEYITE

CdS

Hawleyite is a relatively uncommon mineral that is dimorphous with greenockite, with which it is easily confused. It is considered to be a secondary sulfide mineral.

EUREKA COUNTY Reportedly common as orange to yellow microcrystalline coatings that were formerly identified as greenockite on sphalerite and metamorphosed limestone in the Eureka district; specifically identified in specimens from the Silver Queen Mine. (Jenkins, 1981; S. Pullman, w.c., 1999; F. Cureton, w.c., 1999)

Reported from the Mount Hope Mine in the Mount Hope district. (R. Thomssen, w.c., 1999)

WHITE PINE COUNTY Reported at the Ward Mine in the Ward district. (S. Pullman, w.c., 1999)

HECTORITE

$Na_{0.3}(Mg,Li)_3Si_4O_{10}(F,OH)_2$

A relatively rare clay mineral in the *smectite* group, hectorite is generally found in altered tuffaceous lacustrine deposits.

CLARK COUNTY Identified in clay-rich beds in the Lovell Wash Member of the Horse Spring Formation on the east side of Frenchman Mountain in the Las Vegas district. (SBC)

ESMERALDA COUNTY Identified in drill core from

beneath Clayton Valley in the Silver Peak Marsh district. (Kunasz, 1970; JHS)

HUMBOLDT COUNTY Reported in volcanic sediments of the McDermitt caldera; specifically at the Disaster Peak and Montana Mountains deposits in the Disaster district. (Rytuba and Glanzman, 1979; Odom, 1992)

HEDENBERGITE
$CaFe^{2+}Si_2O_6$

Hedenbergite, a *pyroxene*-group mineral, is found in skarn, metamorphosed iron formations, alkaline igneous rocks, and kimberlite.

ELKO COUNTY Reported in a replacement body mined for lead and silver at the Spruce Standard Mine in the Spruce Mountain district. (Papke, 1979)

ESMERALDA COUNTY A manganiferous variety occurs as seams and coarsely bladed crystals with andradite, calcite, magnetite, galena, and sphalerite in a pyrometasomatic deposit in Paymaster Canyon in the Weepah district, Sec. 2, T1N, R40E. (Gulbrandsen and Gielow, 1960)

EUREKA COUNTY Reported in calc-silicate rock at the Goldstrike Mine in the Lynn district. (Berry, 1992)

HUMBOLDT COUNTY Reported as *ferrohedenbergite* phenocrysts in the tuff of Long Ridge of the McDermitt caldera at an unspecified locality in the Opalite district. (Conrad, 1984; Rytuba and McKee, 1984)

Reported in a specimen from an unknown locality in the Virgin Valley district. (SM)

LINCOLN COUNTY As phenocrysts with fayalite and sanidine in the Kane Wash tuff in the Kane Springs Wash volcanic center. (Novak, 1985)

LYON COUNTY In skarn at the Mason Valley Mine in the Yerington district. (Melhase, 1935e; JHS)

PERSHING COUNTY In skarn with *garnet,* epidote, and quartz at the Esther Mine in the Trinity district. (Johnson, 1977)

HEDLEYITE
Bi_7Te_3

An uncommon mineral that generally occurs with bismuth, gold, and sulfide minerals in skarn, but has also been identified in gold placers.

LANDER COUNTY With marcasite, pyrrhotite, arsenopyrite, and other sulfide minerals in gold-bearing skarn at the McCoy Mine in the McCoy district. (Brooks et al., 1991)

HELIOPHYLLITE
$Pb_6As^{3+}_2O_7Cl_4$ (?)

A rare secondary mineral in base-metal deposits.

CLARK COUNTY Tentatively identified by J. McCormack with pyromorphite from the Milford No. 1 Mine in the Goodsprings district. (J. Kepper, this volume)

HELVITE
$Mn^{2+}_4Be_3(SiO_4)_3S$

An uncommon mineral that mostly occurs in granite pegmatite, but has also been reported in granite, syenite, skarn, and gneiss.

CLARK COUNTY Reported as microcrystals in miarolitic cavities in the Eldorado and Newberry Mountains. (S. Scott, w.c., 1999)

HEMATITE
α-Fe_2O_3

Hematite is a common mineral that occurs in a variety of settings. It is most widespread as a weathering product of primary iron minerals, but also occurs as a primary hydrothermal mineral, as a vapor-phase mineral in volcanic rocks, and in commercially important sedimentary and metamorphic iron deposits. *Specularite* is a metallic crystalline variety, and *martite* inherits its cubic crystal form from replaced magnetite. Only a portion of the Nevada occurrences of this common mineral are reported below.

CARSON CITY The principal ore mineral with quartz and magnetite in veinlike bodies along fault zones at the Capitol and Bessemer (Brunswick Canyon, Iron King) prospects in the Delaware district. (Reeves et al., 1958)

CHURCHILL COUNTY With magnetite in pods 10 to 20 feet across at the White Rock prospect in the Corral Canyon district. (Shawe et al., 1962)

As *specularite* in replacement pods, locally more than 100 feet wide, and as veins in contact-metamorphosed limestone at the Coppereid Mine in the White Cloud district; also reported as silvery plates to 1 cm in copper-rich zones. (Shawe et al., 1962; GCF; WMK; M. Jensen, w.c., 1999)

CLARK COUNTY As microcrystals in miarolitic cavities in the Eldorado and Newberry Mountains. (S. Scott, w.c., 1999)

Widespread as a secondary mineral and as nodules in the Goodsprings district; specifically reported at the Boss, Milford, and Mohawk Mines. (Jedwab, 1998b; J. Kepper, w.c., 1998)

DOUGLAS COUNTY Reported as *martite* in minor quantities on magnetite crystals and as selvages parallel to crystallographic axes in magnetite at the Minnesota Mine in the Buckskin district. (Reeves et al., 1958)

ELKO COUNTY Reported as a replacement of pyrite in oxidized zones at the Rossi Mine in the Bootstrap district. (Snyder, 1989)

With manganese oxide in jasperoid at the Willow claims in the Burner district. (Bentz et al., 1983)

Widespread as red to maroon earthy masses on fractures, along bedding, and as Liesegang bands at the Rain Mine in the Carlin district. (D. Heitt, o.c., 1991)

As *specularite* in quartz-calcite veins in the Contact district. (Bailey, 1903; Schilling, 1979)

As "iron roses" to 1.5 cm in calcite with bismuthinite and other minerals at the Victoria Mine in the Dolly Varden district. (M. Jensen, w.c., 1999)

Widespread after pyrite at the Hollister deposit in the Ivanhoe district. (Deng, 1991)

With *limonite* in silicified gossan in a fault zone at the north end of the Toano Range in the Loray district. (LaPointe et al., 1991)

As replacement deposits in limestone on the Silver King and Sun claims in the Railroad district. (Shawe et al., 1962)

As very small spheroidal masses with corkite in the oxidized zone at the Killie Mine in the Spruce Mountain district. (Dunning and Cooper, 1987)

Specularite occurs as black flakes with quartz, beryl, and other pegmatite minerals in Dawley Canyon in the Valley View district. (Olson and Hinrichs, 1960)

As pseudomorphs after pyrite in a specimen from Camp Creek. (AMNH)

ESMERALDA COUNTY With magnetite at the

Boak prospect (Iron King Mine) in the Candelaria district. (Reeves et al., 1958)

As earthy red crusts with *limonite,* and as glistening black films of microscopic spherules and crystals of "specular" hematite, at the Hat lease in the Goldfield district. (Ransome, 1909a)

Reported as *specularite* in a sample from an unspecified locality in the Klondyke district. (JHS)

EUREKA COUNTY Red, with beryl and fluorite, at the Bisoni prospect in the Fish Creek district. (Griffiths, 1964)

In bright red to brown fracture coatings and in calcite veins with galena, augite, and stibnite at the Goldstrike Mine in the Lynn district. (GCF)

As black, pseudocubic, vanadium-bearing, 20-micron crystals in vugs in barite breccia with fluorapatite, alunite, and kaolinite at the Gold Quarry Mine in the Maggie Creek district. (Jensen et al., 1995)

The principal ore mineral at the Modarelli Mine in the Modarelli-Frenchie Creek district; also after magnetite in breccia cement, veins, and replacement deposits in volcanic rock. (Shawe et al., 1962)

An important ore mineral as black to red massive material at the Barth Iron Mine in the Safford district. (Emmons, 1877; Jones, 1913; Shawe et al., 1962; Jensen, 1988a)

HUMBOLDT COUNTY Widespread on fractures at the Lone Tree Mine in the Buffalo Mountain district. (Panhorst, 1996)

In minor amounts along faults and fractures in metavolcanic rocks at the Iron King Mine in the Jackson Mountains district; also the principal ore mineral, in part as *specularite* in veins, at the Low prospect. (Shawe et al., 1962)

Common as an oxidation product of marcasite at the Cordero Mine in the Opalite district. (Fisk, 1968)

LANDER COUNTY In skarn at the West orebody in Copper Canyon in the Battle Mountain district. (Theodore and Blake, 1978)

As *specularite* with cassiterite and tridymite in irregular veins in rhyolite at several prospects in the Izenhood district; considered to be of vapor-phase origin. (Knopf, 1916a; Fries, 1942)

As lustrous crystals to 1 cm associated with *adularia* crystals at the Surprise pit of Battle

Mountain Gold Company in the Battle Mountain district. (M. Jensen, w.c., 1999)

LINCOLN COUNTY In a specimen reported from the Zero Mine in the Pioche district. (WMK)

LYON COUNTY In jasperoid in metavolcanic rock at the Owen prospect in the Wilson district. (Reeves et al., 1958)

LYON/STOREY COUNTY With magnetite, epidote, *garnet*, and other minerals in skarn at the Dayton iron deposit in the Red Mountain district. (Reeves et al., 1958)

MINERAL COUNTY With magnetite at the Black Butte and Gillis prospects in the Fitting district. (Reeves et al., 1958)

With magnetite and unspecified copper minerals in a replacement deposit in limestone at the King David prospect in the Pamlico district. (Reeves et al., 1958)

Reported as gold-bearing in a specimen from near Sand Springs in the Rawhide district. (WMK)

With magnetite in the Foster copper prospect in the Red Ridge district. (Reeves et al., 1958)

With magnetite in skarn at the Iron Butte Mine in the Santa Fe district; also as veins and replacement bodies in dolomite at the Iron Gate Mine. (Reeves et al., 1958)

Reported as *specularite* in a specimen from near Mina. (WMK)

NYE COUNTY In oxidized magnetite ore from the Phelps Stokes Mine in the Ellsworth district; also along shear zones with *garnet,* epidote, and quartz at the Engle-Stouder prospect. (Reeves et al., 1958)

An oxidation product of sulfides in the Oak Spring district. (Melhase, 1935b)

With other oxidation minerals at the Round Mountain Mine in the Round Mountain district; reportedly gold-bearing in a specimen from the Saddle Rock Mine. (Tingley and Berger, 1985; WMK)

As *specularite* in veins with rare *electrum* at the Life Preserver Mine in the Tolicha district. (Tingley et al., 1998a)

With *limonite,* jarosite, iodargyrite, and cacoxenite in oxidized rock in the Tonopah district; also reported as *turgite,* hematite with absorbed water, in a specimen from the district. (Melhase, 1935c; WMK)

Reported in a specimen from the Gossan Iron Mine in the Wellington district. (WMK)

As *specularite* with cassiterite, tridymite, and other minerals in a vein in bedded tuff on Yucca Mountain; also with other minerals in veins in drill core and as a minor mineral in lithophysal cavities in ash-flow tuff. (Castor et al., 1999b; Carlos et al., 1995)

PERSHING COUNTY Abundant to common as *martite* with magnetite in veins and orebodies in gabbro at several localities such as the Thomas and Segerstrom-Heizer Mines in the Mineral Basin district; reported as rosettes to 2 cm of thin-bladed crystals in calcite vugs at the latter. (Johnson, 1977; Reeves and Kral, 1955; M. Jensen, w.c., 1999)

In gossan formed by oxidation of magnetite-pyrite bodies in the Wild Horse district. (Johnson, 1977)

STOREY COUNTY Reported in ore specimens from the Sutro and Overman Mines in the Comstock district. (NYS; WMK)

WASHOE COUNTY Reported in a specimen from the Ritters Iron Mine in the Peavine district. (WMK)

With secondary uranium minerals in uranium ore from the DeLongchamps, Lowary, and Red Bluff Mines in the Pyramid district. (Overton, 1947; Brooks, 1956)

Reported in a specimen from the Wedekind Mine in the Wedekind district. (WMK)

As lustrous crystals to 1 mm on quartz in pegmatite at Incline Village. (Jensen, 1993a)

WHITE PINE COUNTY After iron sulfides at the Vantage deposit on Alligator Ridge in the Bald Mountain district. (Ilchick, 1990)

Reported from the Robinson district. (Melhase, 1934d; JHS)

As tabular crystals to 1 mm with tridymite in lithophysal cavities in rhyolite in a road cut along U.S. Highway 50 near Little Antelope Summit. (M. Jensen, w.c., 1999)

HEMIHEDRITE
$Pb_{10}Zn(CrO_4)_6(SiO_4)_2F_2$

Hemihedrite is a rare mineral that occurs with other chromate minerals in oxidized base-metal deposits.

CLARK COUNTY Orange to orange-brown crusts and microcrystals after cerussite with galena, minium, pyromorphite, and mimetite at the Milford Mine in the Goodsprings district. (J. Kepper, w.c., 1998)

Reported as orange crystals to 2 mm with phoenicochroite and wulfenite at a prospect 9 miles northwest of Nelson. (R. Thomssen, o.c., 1999)

As yellow-orange to red-orange crystalline coatings on cavernous iron-stained quartz near Boulder City, probably the same as the above locality. (Jenkins, 1981)

HEMIMORPHITE

$Zn_4Si_2O_7(OH)_2 \cdot H_2O$

Hemimorphite is a very common mineral in many western United States base-metal deposits. Excellent specimens have come from the Eureka, Goodsprings, and Downieville districts in Nevada.

CHURCHILL COUNTY Very rare as crystals on the 510-foot level of the Chalk Mountain Mine in the Chalk Mountain district, and as crystals to 1 cm at an unnamed prospect. (M. Jensen, w.c., 1999)

As a massive white replacement of highly altered skarn with Co-Cu-Ni mineralization at an unnamed prospect at the southwest end of the Clan Alpine Mountains near Westgate. (M. Jensen, w.c., 1999; S. White, w.c., 1999)

CLARK COUNTY Reported from an unspecified locality in the Gass Peak district. (Gianella, 1941)

With smithsonite and aurichalcite at the Azure Ridge (Bonelli) Mine in the Gold Butte district. (Longwell et al., 1965; LANH)

At many localities in the Goodsprings district as colorless crystals to 1 cm, generally in cavities in *limonite;* associated minerals include galena, cerussite, hydrozincite, wulfenite, and secondary copper minerals. Noted as bladed, sheathlike aggregates and acicular to stubby crystals at the Milford Mine; in chert layers in limestone at the Monte Cristo Mine; as small, perfect crystals on wulfenite at the Mobile Mine; and as colorless blades to 2 mm on fractures at the Singer Mine. Also from the Prairie Flower and Addison Mines. (Hewett, 1931; Adams, 1998; J. Kepper, w.c., 1998; CSM; CIT; LANH; F. Cureton, w.c., 1999; S. Pullman, w.c., 1999; R. Thomssen, w.c., 1999)

Abundant as radiating crystal aggregates in oxidized ore in the Searchlight district. (JHS)

ELKO COUNTY As crystals on fractures in rhyolite adjacent to quartz veins at the Jackson Mine in the Gold Circle district. (LaPointe et al., 1991)

With galena and pyrite in vugs and as open space fillings in manganese-rich gossan at the Little J. W. claims and the Monarch Mine in the Merrimac district. (LaPointe et al., 1991; Bentz et al., 1983)

With secondary copper minerals in gossan and with brown calcite in veins and pods in marble and *garnet*-diopside skarn at the Sun Mine in the Railroad district. (LaPointe et al., 1991)

As sharp crystals with rosasite in cavities in veins that cut gossan at the Killie Mine in the Spruce Mountain district; also reported as small crystals on limestone, and in white and blue banded veins to 4 cm thick. (Schilling, 1979; Dunning and Cooper, 1987; GCF)

ESMERALDA COUNTY Reported with smithsonite, cerussite, chrysocolla, and tenorite at an unspecified locality in the Lone Mountain district; also from the General Thomas Mine. (Phariss, 1974; M. Jensen, w.c., 1999)

As crystals with bromargyrite, Palmetto Mine in the Palmetto district. (M. Jensen, w.c., 1999)

Reported in specimens from the Broken Toe and Rock Hill Mines in the Rock Hill district. (R. Thomssen, w.c., 1999)

In a specimen from the Big 3 Mine in the Railroad Springs district. (R. Thomssen, w.c., 1999)

With anglesite, cerussite, chlorargyrite, and smithsonite at the Sylvania Mine in the Sylvania district. (Schilling, 1979)

As crystals with chrysocolla in an open cut at the Gold Eagle Mine in the Weepah district; also reported from the 3 Metals Mine. (M. Jensen, w.c., 1999; R. Thomssen, w.c., 1999)

EUREKA COUNTY Reported as minute crystals in oxidized lead-silver-zinc ore at several localities in the Eureka district; noted as crystals to 1 cm in the Lawton Mine. (Nolan, 1962; LANH; CIT; F. Cureton, w.c., 1999; M. Jensen, w.c., 1999; S. Pullman, w.c., 1999)

In a specimen from the Tusc Mine in the Maggie Creek district. (R. Thomssen, w.c., 1999)

HUMBOLDT COUNTY As small colorless crystal clusters with chrysocolla at the Silver Coin Mine in the Iron Point district. (Thomssen, 1998; J. McGlasson, w.c., 2001)

LANDER COUNTY Reported in a number of mines in the Battle Mountain district. (Roberts and Arnold, 1965)

Reported from the Garrison Mine and elsewhere in the Cortez district. (S. Pullman, w.c., 1999; Gianella, 1941)

Reported from the Jersey Valley Mine in the Jersey Valley district. (R. Thomssen, w.c., 1999)

LINCOLN COUNTY As branching aggregates of bladed crystals with barite in an unspecified mine in the Chief (Caliente) district. (Callaghan, 1936)

With hydrozincite and smithsonite in oxidized zinc ore in the Pioche district. (Jenkins, 1981)

Reported from the Silver King Mine in the Silver King district. (R. Thomssen, w.c., 1999)

LYON COUNTY In the Wilson district as clear acicular crystals in radiating groups with rosasite at the Jackpot claim and as platy crystals with coronadite at the Boulder Hill Mine. (LANH; Jenkins, 1981; B. Runner, o.c., 2000)

MINERAL COUNTY Reportedly abundant as crystals with adamite in the Bell district; also noted with smithsonite at the Simon Mine. (Knopf, 1921b; M. Jensen, w.c., 1999; S. White, w.c., 1999)

Reported in minor amounts in the Candelaria district. (Page, 1959)

As crystals with duftite at the Headley Mine in the Garfield district. (M. Jensen, w.c., 1999)

With cinnabar and stibnite at the Drew Mine in the Pilot Mountains district. (Foshag and Clinton, 1927; Foshag, 1927)

Reported as colorless crystals at the Anderson and Champion Mines in the Santa Fe district. (Jenkins, 1981; LANH; S. Pullman, w.c., 1999)

NYE COUNTY As white to clear prisms to 1.5 cm on gossan in the Downieville area in the Gabbs district. (Jenkins, 1981)

As white to clear, radiating prisms to 8 mm in cavities at the Hasbrouck and San Rafael Mines in the Lodi district; also reported in a specimen with wulfenite, adamite, and mimetite. (Jenkins, 1981; ASDM)

Sparse in bunches of acicular white prisms to 0.3 mm in vugs in quartz with sphalerite at the Lower Magnolia workings of the Morey Mine in the Morey district. (Jenkins, 1981)

Found between the 500 and 600 levels on iron oxide at the West End Mine in the Tonopah district. (Bastin and Laney, 1918)

PERSHING COUNTY In mercury ore at the Red Bird Mine in the Antelope Springs district. (Berry et al., 1952)

As crystals in vugs in barite-quartz-calcite veins at the Land Rover claims in the Iron Hat district. (Papke, 1984)

WASHOE COUNTY With cerussite, smithsonite, malachite, and chalcanthite in oxidized ore at the Union (Commonwealth) Mine in the Galena district. (Geehan, 1950; Bonham and Papke, 1969)

Reported at an unspecified locality in the Pyramid district. (Wallace, 1980)

WHITE PINE COUNTY As abundant crystals to 7 mm at the Grand Deposit Mine in the Muncy Creek district. (Gianella, 1941; M. Jensen, w.c., 1999)

As white bladed crystals to 2 cm in the Robinson district; noted at the Dauntless, Willard, Monroe, and Elijah Mines. (Hose et al., 1976; S. Kleine, o.c., 1999; R. Thomssen, w.c., 1999)

In oxidized ore in the Taylor district. (Lovering and Heyl, 1974)

Found in ore from the lead belt in the White Pine district. (Hose et al., 1976)

HENTSCHELITE
$Cu^{2+}Fe_2^{3+}(PO_4)_2(OH)_2$

A very rare secondary mineral in the lazulite group; occurs in a barite vein in Germany, in oxidized copper deposits in England, Australia, and Argentina, and in pegmatite in Portugal.

HUMBOLDT COUNTY As 0.1 mm green crystals with kidwellite and jarosite in a specimen from the Silver Coin Mine in the Iron Point district; identified by E. Foord with SEM/EDX and XRD. (Thomssen, 1998)

HESSITE
Ag_2Te

Hessite occurs in hydrothermal vein deposits with other telluride minerals, silver, and gold.

ESMERALDA COUNTY Reported with petzite and sylvanite in tellurium-rich ore from the Goldfield district. (Tolman and Ambrose, 1934)

HUMBOLDT COUNTY A silver ore mineral at the Sleeper Mine in the Awakening district. (Wood, 1988)

MINERAL COUNTY With petzite, chalcopyrite, and other minerals in vein quartz at the Silver Dyke Tungsten Mine in the Silver Star district. (Sigurdson, 1974)

NYE COUNTY Reported with altaite at the Bullfrog Mine in the Bullfrog district; also identified by SEM/EDX in a sample with silver-gold telluride, *electrum,* and possible tetrahedrite. (Eng et al., 1996; SBC)

Identified by SEM/EDX as small disseminated grains and aggregates with pyrite, acanthite, chalcopyrite, and galena in a vein specimen from the Fairday Mine in the Cactus Springs district. (Tingley et al., 1998a)

In high-grade ore from an unnamed prospect in upper Jefferson Canyon in the Jefferson Canyon (Jefferson) district. (Kral, 1951; WMK)

WASHOE COUNTY Reported as microscopic anhedral metallic grains in propylitized rock from the Green Hill area in the Olinghouse district. (S. Kleine, O.C., 1999)

With chalcocite in veins in a Cu-Ag-Au deposit in diorite porphyry in the Lake Range district on the Pyramid Lake Indian Reservation. (Satkoski and Berg, 1982)

Identified by SEM/EDX as inclusions to 15 microns in tetrahedrite at the Blue Bird Mine in the Pyramid district. (SBC)

WHITE PINE COUNTY Intergrown with altaite and galena in primary lead-silver ore at the Silver King Mine and Caroline tunnel in the Ward district; also reported at the Ward Mine. (Everett, 1964; HMM; R. Thomssen, w.c., 1999)

HETAEROLITE
$ZnMn_2^{3+}O_4$

A rare secondary mineral that is generally associated with other manganese oxide minerals.

MINERAL COUNTY Locally abundant in oxidized rock at the Candelaria Mine in the Candelaria district. (Chavez and Purusotam, 1988)

WHITE PINE COUNTY As black microcrystals on aurichalcite crystals at the Grand Deposit Mine in the Muncy Creek district. (Jensen, 1993c)

HETEROGENITE
$Co^{3+}O(OH)$

Heterogenite, or *stainierite,* is a relatively rare mineral that occurs in the oxidation zones of cobalt-bearing base-metal deposits.

CHURCHILL COUNTY In black coatings on calcite with cobalt-bearing köttigite at an unnamed prospect near the southwest end of the Clan Alpine Mountains. (M. Jensen, w.c., 1999)

CLARK COUNTY Reported as *stainierite* at many localities in the Goodsprings district and said to be widespread as black spots or dendritic growths in dolomite and rare as black mammillary crusts or stalactites; specifically noted at the Blue Jay, Boss, Columbia, Contact, Copper Chief, Copper Glance, Copperside, Highline, Mountain King, Swansea, and Yellow Pine Mines. (Hewett, 1931; Cooke and Doan, 1935; F. Cureton, w.c., 1999)

HUMBOLDT COUNTY Reported at the Silver Coin Mine in the Iron Point district. (Thomssen, 1998)

HEULANDITE
$(Na,K,Ca_{0.5},Sr_{0.5},Mg_{0.5})_9[Al_9Si_{27}O_{72}] \bullet {\sim}24H_2O$ (general)

A common zeolite mineral group that has recently been subdivided into three species depending on the most abundant alkali or alkali earth element. *Heulandite* occurs in amygdules in intermediate to mafic volcanic rocks, as an alteration product in felsic volcanic rocks and volcaniclastic sediments, and rarely in hydrothermal veins. It is difficult to distinguish from related zeolites such as *clinoptilolite,* and is undoubtedly present at many Nevada localities that are not listed below.

CHURCHILL COUNTY Identified as an alteration product in tuff from an unspecified locality in the Desatoya Mountains near Eastgate. (JHS)

HUMBOLDT COUNTY Reported at the Cordero Mine in the Opalite district. (Bailey and Phoenix, 1944)

In a specimen from the Getchell Mine in the Potosi district. (R. Thomssen, w.c., 1999)

LANDER COUNTY Reported from Mule Canyon in the Shoshone Range. (R. Thomssen, w.c., 1999)

As translucent reddish-orange tablets to 2 mm at the Mount Airy Blue Chalcedony Mine in the New Pass district. (SBC)

LYON COUNTY Reported with *stilbite* at Mason Pass in the Yerington district. (LANH; S. Pullman, w.c., 1999)

Glassy euhedral crystals to 1 mm lining fractures in petrified wood in tuff at Wilson Canyon. (M. Jensen, w.c., 1999)

NYE COUNTY As tiny tabular and prismatic crystals on fractures in ash-flow tuff in drill core from

Yucca Mountain, locally with stellerite. (Carlos et al., 1995)

STOREY COUNTY A secondary mineral in rocks of the Old Gregory Formation south of the Truckee River near Clark Station. (Rose, 1969)

Reported with *clinoptilolite* in altered volcanic rocks in the Virginia City area in the Comstock district. (Whitebread, 1976)

WASHOE COUNTY As massive to transparent crystals to 1 cm in amygdules in intermediate volcanic rock in the Olinghouse district; also as white to pink cleavable masses with gold, calcite, quartz, and other zeolite minerals in veins at the Olinghouse Gold Mine. (L. Garside, o.c., 1998; Jones, 1998; S. Kleine, o.c., 1999)

Pearly crystals to 1 cm in pegmatite at Incline Village. (Jensen, 1993a)

HEWETTITE

$CaV_6^{5+}O_{16} \cdot 9H_2O$

A rare secondary mineral that is found in oxidized vanadium deposits and in sandstone uranium deposits. The Gold Quarry Mine locality is the first reported occurrence in a hydrothermal gold deposit.

EUREKA COUNTY As sparse well-developed brown crystals with metahewettite and minyulite at the Van-Nav-San claim in the Gibellini district; also reported at the Bisoni prospect. (R. Walstrom, w.c., 1999; UAMM)

Possibly the best crystals in the world as maroon-red, hairlike needles in sprays in breccia at the Gold Quarry Mine in the Maggie Creek district, associated with opal, tyuyamunite, kazakhstanite, and fervanite; also noted as velvety red layers on fractures in jasperoid. (Jensen et al., 1995; MV)

HEXAHYDRITE

$MgSO_4 \cdot 6H_2O$

Hexahydrite is a moderately rare mineral that occurs in some saline deposits and as efflorescences in old mine workings.

LANDER COUNTY In fault gouge with jarosite, clays, and a variety of other minerals at the Gold Acres Mine open pit in the Bullion district. (Armbrustmacher and Wrucke, 1978)

STOREY COUNTY Reported as a white crystalline

efflorescence on mine walls at the Crown Point Mine in the Comstock district. (Hsu, unpublished data; M. Jensen, w.c., 1999)

WASHOE COUNTY As white to silvery-white silky fibers to 10 cm long in the Nevada Dominion adit in the Pyramid district; also with alunogen in a specimen from the Burrus Mine. (B. Park-Li, o.c., 1996; LANH)

With apjohnite and pickeringite at the Wedekind Mine in the Wedekind district. (GCF)

HEYITE

$Pb_5Fe_2^{2+}(VO_4)_2O_4$

A Nevada type mineral, heyite is very rare, having been recognized at only a few localities in oxidized hydrothermal deposits.

EUREKA COUNTY As orange-brown crystals with descloizite and vanadinite in gossan in a fault zone at the Gold Quarry Mine in the Maggie Creek district. (Jensen et al., 1995)

HUMBOLDT COUNTY Bright yellow to yellow-orange microcrystalline crusts and small flakes with stolzite and hinsdalite at the Silver Coin Mine in the Iron Point district. (Thomssen, 1998; M. Jensen, w.c., 1999; J. McGlasson, w.c., 2001)

WHITE PINE COUNTY Type locality at the Betty Jo claim, 8 miles southeast of Ely in the Nevada district, Sec. 10, T15N, R64E, as yellow-orange prisms to 0.4 mm on and replacing tungsten-bearing wulfenite in quartz in silicified limestone; also with pyromorphite, cerussite, mimetite, shattuckite, and chrysocolla. (Williams, 1973)

HEYROVSKÝITE

$(Pb,Ag)_5Bi_3S_8$

Heyrovskýite is a rare mineral that occurs with other bismuth minerals in tungsten skarn and in high-temperature quartz veins.

LANDER COUNTY As tiny disseminated gray grains with bismuthinite, bismuth, bursaite, galena, and some unspecified metallic phases in skarn near Austin; the locality has recently been identified as the Reward prospects in the Ravenswood district. (Jenkins, 1981; EMP; HUB; HMM; F. Cureton, w.c., 1999; R. Thomssen, w.c., 1999)

HIDALGOITE

$PbAl_3(AsO_4)(SO_4)(OH)_6$

A rare secondary mineral that occurs with other secondary lead minerals.

NYE COUNTY As 1.2-mm, clear, prismatic crystals with powdery yellow mimetite and beudantite and powdery green bayldonite at the San Rafael Mine in the Lodi district. (B. Runner, o.c., 2000)

HINSDALITE

$(Pb,Sr)Al_3(PO_4)(SO_4)(OH)_6$

An uncommon secondary mineral that occurs with other secondary lead minerals and has also been reported as a supergene phase in sulfide-bearing veins.

HUMBOLDT COUNTY As yellow-green rhombs with pyromorphite, stolzite, and mimetite at the Silver Coin Mine in the Iron Point district. (Thomssen, 1998; M. Jensen, w.c., 1999; J. McGlasson, w.c., 2001)

As microcrystals with whitlockite, beudantite, mitridatite, and other minerals at the Twin Creeks Mine in the Potosi district. (M. Jensen, w.c., 1999)

NYE COUNTY As copper-bearing, olive-green masses at the San Rafael Mine in the Lodi district. (B. and J. Runner, o.c., 2000)

HISINGERITE

$Fe_2^{3+}Si_2O_5(OH)_4 \cdot 2H_2O$

A secondary mineral formed by weathering or late-stage alteration of iron-bearing silicate or sulfide minerals; may be a primary mineral in some sulfide deposits.

CHURCHILL COUNTY Reported with *sericite* and kaolinite in breccia at the Fairview Mine in the Fairview district. (Saunders, 1977)

CLARK COUNTY As thin, shiny, medium- to dark-brown crusts lining openings in cellular goethite at the Milford No. 1 Mine in the Goodsprings district. (J. Kepper, w.c., 1998)

ELKO COUNTY An alteration mineral at the Jerritt Canyon Mines in the Independence Mountains district. (Hofstra, 1994)

HUMBOLDT COUNTY With opal, montgomeryite, evansite(?), and curetonite at the Redhouse barite prospect in the Potosi district. (Williams, 1979)

MINERAL COUNTY Reported as a gangue mineral in the Rawhide district. (Schrader, 1947)

WHITE PINE COUNTY A minor dark-brown, brittle, resinous phase near porphyry-skarn contacts in the Robinson district. (James, 1976)

HOCARTITE

Ag_2FeSnS_4

An uncommon mineral that occurs with other sulfides in tin-bearing veins.

LANDER COUNTY Reported at the Dean Mine in the Lewis district. (Anthony et al., 1990)

HODRUŠHITE

$Cu_8Bi_{12}S_{22}$

Hodrušhite is an extremely rare mineral that occurs in bismuth-bearing hydrothermal vein and replacement deposits. The locality listed is only the fourth reported occurrence in the world, but its exact location is speculative.

LINCOLN COUNTY Reported as intergrowths with pavonite in a specimen from an unspecified locality near Pioche, possibly the Black Metals (Jackrabbit) Mine in the Bristol district. (Jenkins, 1981)

HOLLANDITE

$Ba(Mn^{4+},Mn^{2+})_8O_{16}$

Hollandite, which is structurally and chemically related to cryptomelane, is a relatively common mineral in manganese deposits where it occurs both as a primary mineral and as a weathering product.

CLARK COUNTY In manganese ore at the Three Kids Mine in the Las Vegas district. (Hewett and Fleischer, 1960)

ESMERALDA COUNTY Reported with calcite, barite, and fluorite at the Mohawk Mine in the Silver Peak district. (Hewett and Radtke, 1967)

HUMBOLDT COUNTY Reported with *psilomelane* at the Golconda manganese-tungsten deposit in the Golconda district. (Warner et al., 1959)

Reported as black crystalline material with cryptomelane at an unspecified locality in the Virgin Valley district. (UCB; SM; AMNH; F. Cureton, w.c., 1999)

NYE COUNTY In drill core from Yucca Mountain

as tiny rods on lithiophorite on fractures; also reported as arsenic-bearing in veins with quartz and possible lithiophorite. (Carlos et al., 1995; Castor et al., 1999b)

WASHOE COUNTY Tentatively identified as tiny lenses with crustiform cristobalite and *chalcedony* in altered tuff from a radioactive mineral prospect near the DeLongchamps adit in the Pyramid district. (Castor et al., 1996)

HORNBLENDE

$Ca_2(Mg,Fe,Al)_5(Si,Al)_8O_{22}(OH)_2$ (general)

Hornblende is a group name for several amphibole species that occur in a wide variety of igneous and metamorphic rocks. Species have not been identified for the following localities.

CLARK COUNTY In Precambrian rock types in the Virgin Mountains: with *plagioclase* and titanite in amphibolite; with biotite and other minerals in gneiss; and with augite, pyrite, pyrrhotite, and other minerals in hornblendite dikes and sills as at the Great Eastern Mine in the Bunkerville district. (Beal, 1965)

EUREKA COUNTY In quartz monzonite at the Barth Iron Mine in the Safford district. (Shawe et al., 1962)

LYON COUNTY Reported as actinolitic *hornblende* in granodiorite at the Wheeler Mine in the Wilson district. (Princehouse and Dilles, 1996)

MINERAL COUNTY In skarn along a limestone-granodiorite contact at the Lucky Boy Mine in the Lucky Boy district. (Lawrence, 1963)

NYE COUNTY Reported in a specimen from Easy Chair Crater in the Lunar Craters volcanic field, possibly the amphibole species kaersutite, which has been identified from here. (HMM)

HÖRNESITE

$Mg_3(AsO_4)_2 \cdot 8H_2O$

A rare mineral that occurs outside Nevada in xenoliths of metamorphosed limestone in tuff and as encrustations with other secondary metallic minerals in a cave.

CHURCHILL COUNTY In a specimen from the Nickel Mine in the Table Mountain district. (R. Thomssen, w.c., 1999)

NYE COUNTY Reported in a specimen from the White Caps Mine in the Manhattan district; may have formed during a mine fire. (RMO)

HOWIEITE

$Na(Fe^{2+},Mn)_{10}(Fe^{3+},Al)_2Si_{12}O_{31}(OH)_{13}$

A rare metamorphic mineral that is reported from several localities in California and Oregon.

Reported in a specimen from a locality listed as Nevada without further notation; possibly a mislocated specimen from California. (TUB)

HÜBNERITE

$Mn^{2+}WO_4$

A Nevada type mineral, hübnerite is a relatively common *wolframite*-group species that occurs in both tungsten-bearing skarn deposits and in hydrothermal vein deposits.

CHURCHILL COUNTY Reported with fluorite at an unspecified locality in the Eastgate district on the west side of the Desatoya Mountains. (Hess, 1917)

ELKO COUNTY Reported in a specimen with scheelite at an unnamed beryl deposit in the Ruby Mountains. (WMK)

ESMERALDA COUNTY Reported in a specimen from the unknown locality in the Coaldale district near Coaldale. (WMK)

In quartz veins in the Sylvania district. (Schilling, 1964b)

HUMBOLDT COUNTY Reported in minor amounts with scheelite in mill concentrates at the Getchell Mine in the Potosi district. (Hotz and Willden, 1964)

LANDER COUNTY Reported as brown-black crystals to 2 cm in quartz from near Austin. (TUB; M. Jensen, w.c., 1999)

LINCOLN COUNTY Reported in a specimen from Panaca, possibly from the Comet Mine 12 miles northwest of Panaca, where *wolframite* is reportedly abundant. (HMM)

MINERAL COUNTY Reportedly mined near Columbus. (Hess, 1917)

Reported at an unspecified locality in the Mina district near Mina. (Hess, 1917)

NYE COUNTY Reported with scheelite in a specimen from an unknown locality in the Belmont district; *wolframite* and/or hübnerite occurs in quartz veins at the Falcon and Wolframite prospects. (WMK; Stager and Tingley, 1988)

Reported in a specimen from an unknown locality in the Ellendale district; may be a mislabeled Ellsworth district specimen. (WMK)

Type locality as coarse red-brown blades with scheelite and purple fluorite in the Erie and Enterprise quartz veins at the Ellsworth Mine in the Ellsworth district, Sec. 19, T13N, R38E; also reported at the Commodore Mine. (Hess, 1917; Palache et al., 1951; Stager and Tingley, 1988)

With tetrahedrite in quartz veins in granitic rock at the Spanish Spring Mine in the Manhattan district; also in placer deposits. (Hess, 1917; Ferguson, 1924; Kleinhampl and Ziony, 1984)

With fluorite, muscovite, and tetrahedrite in quartz veins in granite at the Stevenson and Schuppy Mine in the Round Mountain district; also with native copper and *monazite* from the placer gravels at the Red Top prospect, and with *wolframite* and scheelite in quartz veins at the Lucky Seven Mine. (Ferguson, 1922; Garside, 1973; Stager and Tingley, 1988)

As black platy masses in quartz, thin plates in cavities, and deep red transparent crystals from unspecified sites in the Tonopah district; reported from the Tonopah-Belmont Mine, and noted as brown-black crystals to 1.5 cm from the Mizpah Silver group. (Melhase, 1935c; CSM; WMK; M. Jensen, w.c., 1999)

PERSHING COUNTY Reported in a specimen from an unknown locality in the Mill City district; *wolframite* or hübnerite was noted at the Florence Hill prospects. (WMK; Stager and Tingley, 1988)

Reported in tungsten ore with tungstite at the Indian Peak (Royal Ann) group in the Toy district. (WMK)

In small amounts in gold placers near Scossa in the Rosebud district. (Bailey and Phoenix, 1944; JHS)

WHITE PINE COUNTY As bladed crystals with cubes of purple fluorite, bismuthinite, and triplite in white quartz veins in the Eagle district; also reported as the only tungsten mineral at the Tungstonia Mine in reddish-brown crystals to several inches with sulfide minerals. (Hill, 1916; Melhase, 1934d; WMK; Stager and Tingley, 1988)

The principal tungsten mineral as crystals in quartz at the Deer Trail Mine in the Geyser district. (Hose et al., 1976; M. Jensen, w.c., 1999)

Reported in specimens from unknown localities in the Osceola district. (NMBM; WMK)

As crystals to 3 inches with minor fluorite and scheelite in quartz veins at the Hub Mine in the Tungsten district; also reported in quartz veins and veinlets at the Eagle and Side Issue claims, including a specimen of solid hübnerite weighing 114 pounds from the latter property; also reported in a specimen from the McGregor group. (Hose et al., 1976; Stager and Tingley, 1988; M. Jensen, w.c., 1999; CSM)

HUEMULITE
$Na_4MgV_{10}^{5+}O_{28} \bullet 24H_2O$

A mineral that has been found with other vanadate species in uranium deposits associated with organic material in conglomerate and sandstone in Argentina.

EUREKA COUNTY As bright yellow earthy masses and crude microscopic crystals on organic-rich shale with bokite, barnesite, metahewettite, and other minerals at the Gibellini vanadium deposit and the Van-Nav-San claim in the Gibellini district; also reported as yellow powder at the Bisoni prospect. (Jenkins, 1981; MV; AM)

HUMMERITE
$KMgV_5^{5+}O_{14} \bullet 8H_2O$

A rare water-soluble secondary efflorescent mineral.

EUREKA COUNTY As bright yellow to orange efflorescences on joint surfaces with kaolinite, gypsum, and haggite at the Gold Quarry Mine in the Maggie Creek district. (Jensen et al., 1995)

HUNTITE
$CaMg_3(CO_3)_4$

A Nevada type mineral, huntite is a relatively rare species that occurs in cavities in magnesian carbonates as a low-temperature mineral, possibly a weathering product.

LANDER COUNTY A white chalky mineral in a specimen from a locality near Austin. (Weissman and Nikischer, 1999)

NYE COUNTY In small quantities with hydromagnesite in brucite-magnesite orebodies at the Premier Chemicals (Gabbs Magnesite-Brucite) Mine in the Gabbs district. (Vitaliano and Beck, 1963)

WHITE PINE COUNTY Type locality at the Ala-Mar (Currant Creek) deposit, Sec. 34, T12N, R59E, in the Currant district as white microcrystalline, porcelaneous masses in vugs and along fractures with magnesite and other minerals in nodules in tuff; also reported from the Snowball Mine as a late-stage mineral in vugs to 10 cm with calcite and iron oxides and as white powder at the base of altered tuff. (Faust, 1953; Hose et al., 1976)

HURÉAULITE
$Mn_5^{2+}(PO_4)_2[PO_3(OH)]_2 \cdot 4H_2O$

A mineral that generally occurs in as a secondary mineral after primary phosphate species in pegmatite.

EUREKA COUNTY Identified by XRD as a microgranular alteration mineral in diorite at the Goldstrike Mine in the Lynn district. (GCF)

HUTTONITE
$ThSiO_4$

Huttonite, a dimorph of thorite, is a rare mineral that occurs in pegmatites with other radioactive minerals and in beach sands.

MINERAL COUNTY As tiny glassy grains in brecciated and hydrothermally altered granite with thorite and other minerals at the Holiday (Holliday) Mine in the Fitting district; distinguished in thin section or by characteristic white fluorescence. (Ross, 1961; UCB)

HYDROBIOTITE
$[K(Mg,Fe^{2+})_3(Al,Fe^{3+})Si_3O_{10}(OH,F)_2] \cdot [(Mg,Fe^{2+}, Al)_3(Si,Al)_4O_{10}(OH)_2 \cdot 4H_2O]$

Hydrobiotite, a 1:1 interstratified mixture of biotite and vermiculite, is an alteration product of other mica minerals.

CLARK COUNTY Reported as brownish-black micaceous flakes in quartz stockworks and adjacent rock with molybdenite, pyrite, chalcopyrite, and bornite at the Crescent Peak porphyry molybdenum deposit in the Crescent district. (Schilling, 1979)

HUMBOLDT COUNTY With *adularia* in quartz veins at the Ashdown Mine in the Vicksburg district. (Schilling, 1979)

HYDROMAGNESITE
$Mg_5(CO_3)_4(OH)_2 \cdot 4H_2O$

Hydromagnesite is a relatively scarce secondary mineral that is formed by the alteration of magnesium-rich rocks such as dolomite.

CHURCHILL COUNTY White crystalline crusts to 1 cm thick with mcguinnessite at the Skarn No. 1 claim in the Chalk Mountain district. (M. Jensen, w.c., 1999; R. Thomssen, w.c., 1999)

LINCOLN COUNTY In minor amounts with ludwigite, szaibélyite, and fluoborite in an unnamed copper prospect on the west base of Blind Mountain in the Bristol district. (Gillson and Shannon, 1925; Westgate and Knopf, 1932)

NYE COUNTY As white, fibrous to earthy aggregates in weathered brucite and as crystal groups to 1.5 cm with mcguinnessite at the Premier Chemicals Mine (also Basic Refractories or Gabbs Magnesite-Brucite Mine) in the Gabbs district. (Vitaliano and Beck, 1963; Schilling, 1968a; M. Jensen, w.c., 1999)

Reported as white tufted groups of crystals in the Lodi district. (CIT; UAMM)

Reported with brucite, calcite, and dolomite in the Pocopah district in the Calico Hills on the Nevada Test Site. (Castor et al., 1999b)

HYDRONIUM JAROSITE
$(H_3O^{1+})_2Fe_6^{3+}(SO_4)_4(OH)_{12}$

A secondary mineral formed by rapid breakdown of sulfide minerals; *carphosiderite* is an invalid name for the species.

CLARK COUNTY With plumbojarosite and jarosite at the Boss Mine in the Goodsprings district. (Jedwab, 1998b)

EUREKA COUNTY Yellow to orange botryoidal coatings with dussertite at the Goldstrike Mine in the Lynn district. (GCF)

As brilliant-orange, transparent, euhedral crystals with greenish spherules of phosphatic scorodite at the Gold Quarry Mine in the Maggie Creek district. (Jensen et al., 1995)

HUMBOLDT COUNTY Reported in a specimen from the Sulphur district. (UAMM)

LINCOLN COUNTY Reported as *carphosiderite* in pyrite-uraninite pods at the Atlanta Mine in the Atlanta district. (Garside, 1973)

As brown crystals to 5 mm in specimens from

the Bristol Silver Mine in the Bristol district. (Weissman and Nikischer, 1999)

HYDROTUNGSTITE

$H_2WO_4 \cdot H_2O$

A rare mineral found in oxidized tungsten deposits with scheelite and ferberite.

CHURCHILL COUNTY Reported at the St. Anthony Mine in the Toy district. (AMNH)

HYDROXYAPOPHYLLITE

$KCa_4Si_8O_{20}(OH,F) \cdot 8H_2O$

A relatively rare mineral that is generally found with zeolite minerals in amygdules in basalt, but also in contact-metamorphosed limestone.

EUREKA COUNTY As clear pseudocubic crystals to 2 mm in vugs in veins to 4 mm thick in calc-silicate rock and marble at the Goldstrike Mine in the Lynn district. (GCF)

HYDROXYLAPATITE

$Ca_5(PO_4)_3(OH)$

A member of the *apatite* group that occurs in hydrous metamorphosed rocks such as talc schist and in granite pegmatite.

EUREKA COUNTY At the Goldstrike Mine in the Lynn district as clear acicular crystals with whitlockite, other phosphates, and iron oxide minerals in silicified and brecciated zones in calcareous sedimentary rock; as tiny crystals in altered diorite; and with quartz and pyrite in gold ore. Identified by XRD and SEM/EDX. (GCF)

LINCOLN COUNTY Reported in a specimen from an unknown locality in the Pioche district. (RMO)

HYDROZINCITE

$Zn_5(CO_3)_2(OH)_6$

Hydrozincite, a relatively common mineral, occurs as a replacement of sphalerite in oxidized base-metal deposits with hemimorphite, smithsonite, cerussite, and other secondary minerals.

CLARK COUNTY Very common at the Sampson claim in the Gass Peak district. (Longwell et al., 1965)

The most abundant zinc mineral at many localities in the Goodsprings district, generally as earthy, white to dark-brown masses, but locally as delicate encrustations of very small crystals in radiating masses and lining vugs with hemimorphite crystals; associated with calcite, cerussite, galena, smithsonite, and wulfenite, and specifically reported at the Anchor, Argentena, Bill-Nye, Boss, Milford, Monte Cristo, Pilgrim, Potosi, Prairie Flower, Rover, Shenandoah, Yellow Pine, and other mines. (Ford and Bradley, 1916; Hewett, 1931; LANH; CSM; NMBM; S. Pullman, w.c., 1999)

ELKO COUNTY Reported from unspecified localities in the Ruby Valley and Spring Mountain districts. (Heyl and Bozion, 1962)

ESMERALDA COUNTY Reported from an unspecified locality in the Gold Point district. (Heyl and Bozion, 1962)

Reported from the Alpine Mine in the Lone Mountain district; also in small but fine crystals from the General Thomas Mine. (Heyl and Bozion, 1962; R. Thomssen, w.c., 1999)

EUREKA COUNTY At the Mountain View (Lone Mountain) Mine with smithsonite, zincite, cerussite, malachite, and azurite. (Roberts et al., 1967b)

LANDER COUNTY As crystalline masses with smithsonite in the oxidized portions of zinc orebodies in the Battle Mountain district. (Roberts and Arnold, 1965)

LINCOLN COUNTY Abundant in the ore from the 300-foot level of the Gypsy workings at the Bristol Silver Mine in the Bristol district; also reported at the Fortuna prospect just north of Jackrabbit. (Westgate and Knopf, 1932)

Reported with dufrénoysite and other minerals at the Prince Mine in the Pioche district. (Melhase, 1934d)

With cerussite and zincite(?) at the Lucky prospect in the Viola district. (Tschanz and Pampeyan, 1970)

LYON COUNTY Reported in a specimen from the Jackpot Mine in the Wellington district. (LANH)

MINERAL COUNTY Reported at unspecified localities in the Pilot Mountains and Silver Star districts. (Heyl and Bozion, 1962)

NYE COUNTY Reported from unspecified localities in the Paradise Range and near Pahrump. (Heyl and Bozion, 1962)

A minor mineral in oxidized ore at the San Rafael Mine in the Lodi district. (Jenkins, 1981)

WHITE PINE COUNTY With other zinc minerals at the Grand Deposit Mine in the Muncy Creek district. (Hose et al., 1976; WMK)

Reported from an unspecified locality in the Robinson district. (Heyl and Bozion, 1962)

HYPERCINNABAR

HgS

A high-temperature polymorph of cinnabar and metacinnabar that was first identified in a mercury deposit in California.

NYE COUNTY Identified by S. Williams in black pulverulent masses at the White Caps Mine in the Manhattan district. (HUB; F. Cureton, w.c., 1999)

HYPERSTHENE

$(Mg,Fe^{2+})_2Si_2O_6$

An abandoned name for intermediate compositions between enstatite and ferrosilite, members of the *pyroxene* group. *Hypersthene* occurs in a variety of igneous rocks and in high-grade regionally metamorphosed rocks. Species have not been identified at the following occurrences.

CLARK COUNTY With augite, *hornblende,* and *andesine* in Precambrian amphibolite in the Bunkerville district. (Beal, 1965)

In Precambrian rocks in the Gold Butte district; reported with *andesine* and quartz in charnockite gneiss, and with *bronzite, olivine,* biotite, and other minerals in ultramafic rocks. (Volborth, 1962b)

In Precambrian gneiss on the west flank of Frenchman Mountain near Las Vegas. (Matti et al., 1993)

With biotite and augite in crystal-rich lava 2 miles northwest of Squaw Peaks. (Anderson, 1978)

EUREKA COUNTY As phenocrysts in andesite that hosts ore at the Barth Iron Mine in the Safford district. (Shawe et al., 1962)

NYE COUNTY As phenocrysts in the tuff of Wilson's Camp on the Nellis Air Force Range; also in lavas in the Belted and Kawich ranges. (Ekren et al., 1971)

i

ICE

H_2O

Though not often thought of as a mineral, as a naturally occurring crystalline form of water, permanent occurrences of ice clearly qualify.

WHITE PINE COUNTY As a permanent ice field on the upper slopes of Wheeler Peak. (JHS)

ILLITE

$K_{0.65}Al_{2.0}\square Al_{0.65}Si_{3.35}O_{10}(OH)_2$

A group of mica-clay minerals that occur abundantly as alteration products in many Nevada ore deposits.

ELKO COUNTY As coatings on fractures and faults throughout the pit at the Rain Mine in the Carlin district, but most common in basal red clay. (D. Heitt, o.c., 1991)

A major constituent of the alteration assemblage at the Jerritt Canyon Mines in the Independence Mountains district. (Hofstra, 1994)

A common alteration mineral at the Meikle Mine in the Lynn district. (Jensen, 1999)

HUMBOLDT COUNTY The major clay alteration mineral in the gold-bearing assemblage at the Twin Creeks Mine in the Potosi district. (Simon et al., 1999)

NYE COUNTY An alteration mineral at the Clarkdale and Yellow Gold Mines in the Clarkdale district on the Nellis Air Force Range. (Tingley et al., 1998a)

In gouge zones along with barite, *electrum,* and cheralite at the Life Preserver Mine in the Tolicha district on the Nellis Air Force Range. (Tingley et al., 1998a)

An alteration mineral with alunite, kaolinite, and quartz at the Franz Hammel, Golden Chariot, and other mines in the Jamestown district on the Nellis Air Force Range. (Tingley et al., 1998a)

STOREY COUNTY A widespread alteration and vein mineral in the Comstock district. (D. Hudson, w.c., 1999)

WASHOE COUNTY Reported at Washoe Hot Springs. (White, 1955)

WHITE PINE COUNTY A primary constituent of the sediments at the Vantage deposit, on Alligator Ridge in the Bald Mountain district. (Ilchick, 1990)

ILMENITE

$Fe^{2+}TiO_3$

Ilmenite is a common mineral that occurs as an accessory phase in many types of igneous rock, as a heavy mineral in placers, and in primary ore deposits.

CLARK COUNTY Reported in the Goodsprings district. (Kepper, this volume)

ELKO COUNTY Very rare in gravity concentrates at the Hollister deposit in the Ivanhoe district. (Deng, 1991)

With magnetite in stream gravels along Dixie Creek in the Railroad district. (Beal, 1963)

HUMBOLDT COUNTY As grains to 10 microns in uranium-rich opal in the Virgin Valley district. (Castor et al., 1996)

MINERAL COUNTY As crystals to 10 cm long in the Zapot pegmatite in the Fitting district; contains as much as 50 percent of the manganese-rich end member pyrophanite. (Foord et al., 1999b)

STOREY COUNTY A constituent of altered volcanic rocks in the Comstock district. (Whitebread, 1976)

WHITE PINE COUNTY As black microcrystals with quartz in lithophysae on Garnet Hill in the Robinson district. (M. Jensen, w.c., 1999)

ILSEMANNITE

$Mo_3O_8 \cdot nH_2O$ (?)

Ilsemannite occurs in small quantities as a secondary mineral in deposits that contain molybdenum.

ELKO COUNTY Reportedly common as blue stains and coatings on mine walls at the Meikle Mine in the Lynn district. (Jensen, 1999)

ESMERALDA COUNTY In the Cucomungo molybdenum deposit in the Tule Canyon district; presently forming on the Roper adit dump and reported with ferrimolybdite, molybdenite, pyrite, melanterite, and other sulfates at the McBoyle prospect. (Schilling, 1962a, 1962b, 1979)

EUREKA COUNTY Blue to blue-black, massive, earthy crusts along shears and in altered areas in silicified limestone at the Goldstrike Mine in the Lynn district; occurs near, and possibly with, morenosite. (GCF)

HUMBOLDT COUNTY As blue stains with marcasite at the Lone Tree Mine in the Buffalo Mountain district. (M. Jensen, w.c., 1999)

Reported as dark-blue coatings along fractures in carbonaceous shale at the Getchell Mine in the Potosi district; said to occur with stibnite and pascoite in the south pit, but not found during subsequent research. (Joralemon, 1951; S. Sears, w.c., 1999)

MINERAL COUNTY Reported at an unspecified locality in the Rand district. (Schilling, 1980)

WASHOE COUNTY Blue molybdenum oxide stains, presumably ilsemannite, occur on quartz stringers at the Hill-Johnson prospect in the Freds Mountain district. (Schilling, 1962a, 1962b)

Reported as blue stains coating the walls of a pit in hot spring–altered basaltic rock in the Steamboat Springs district. (Schilling, 1964a)

ILVAITE

$CaFe_2^{2+}Fe^{3+}SiO_7O(OH)$

Ilvaite is a relatively rare mineral found in contact-metamorphic aureoles with *garnet,* vesuvianite, and other minerals.

ELKO COUNTY Reported from an unspecified mine in the Contact district. (Schrader, 1912)

LINCOLN COUNTY With pyrrhotite and molybdenite in the contact aureole at the Lincoln Tungsten Mine in the Tem Piute district; also reported from the Free tunnel. (Buseck, 1967; JHS)

INDIALITE

$Mg_2Al_4Si_5O_{18}$

Indialite, a dimorph of cordierite, has been found in sedimentary rock altered by a coal fire, in veins in metamorphosed aluminum-rich rock, and in volcanic rock.

WASHOE COUNTY As massive, cream-colored material that makes up 30 vol. percent of a volcanic flow at Pyramid Lake. (S. Pullman, w.c., 1999)

IODARGYRITE

AgI

Iodargyrite is the rarest of the silver halides, although there are probably more localities for the mineral in Nevada than in any other state. Like chlorargyrite and bromargyrite, it occurs as an oxidation product of primary silver sulfides.

CHURCHILL COUNTY In the oxidized zone of silver sulfide veins in rhyolite and dacite as pockets of loose, sulfur-yellow crystals in cracks and cavities, and in occasional crystals associated with *iodobromite,* at the Nevada-Wonder Mine in the Wonder district, Sec. 4, T18N, R35E. (Burgess, 1917; Schilling, 1979)

CLARK COUNTY Reported as *iodyrite* with *cerargyrite* at the Grizzly prospect in the Goodsprings district 2 miles south of Crystal Pass. (Hewett, 1931; Melhase, 1934e)

ESMERALDA COUNTY Reported with bismoclite at the Goldfield Consolidated Mine in the Goldfield district; also reported at the Grizzly Bear Mine. (S. Pullman, w.c., 1999; R. Thomssen, w.c., 1999)

HUMBOLDT COUNTY In minor quantities at the Sleeper Mine in the Awakening district. (Wood, 1988)

LANDER COUNTY Reported as *iodyrite* coating quartz and *argentite* in the Reese River district. (Hoffman, 1878)

LINCOLN COUNTY Reported at the Charles Ross Mine in the Eagle Valley district. (Schrader et al., 1917)

MINERAL COUNTY Reported at the Penglaze Mine in the Rawhide district. (Schrader et al., 1917)

Occurs rarely with chlorargyrite at the Regent Mine in the Rawhide district. (M. Jensen, w.c., 1999)

NYE COUNTY With *electrum* and other silver minerals at the Montgomery-Shoshone Mine in the Bullfrog district. (Eng et al., 1996)

Associated with disseminated gold and silver mineralization at the Paradise Peak Mine in the Fairplay district. (Dobak, 1988)

With scorodite and barite in the Gold Crater district on the Nellis Air Force Range. (Tingley et al., 1998a)

Common as brilliant, loose, sulfur-yellow crystals to 6 mm in fissures and vugs in veins, as drusy crystalline crusts, and as powder in the Tonopah district; associated with jarosite and gypsum. (Burgess, 1911; Bastin and Laney, 1918; Melhase, 1935c)

PERSHING COUNTY Yellow microcrystals on clinoclase crystals from the Tin stope at the Majuba Hill Mine in the Antelope district. (Jensen, 1993b)

WHITE PINE COUNTY Doubtfully reported with bromargyrite and chlorargyrite in the mines on Treasure Hill in the White Pine district. (Hague, 1870)

IRANITE

$Pb_{10}Cu(CrO_4)_6(SiO_4)_2(F,OH)_2$

A close relative to hemihedrite found in small amounts with dioptase, fornacite, and other minerals at an ancient mine in Iran.

CLARK COUNTY As dark tips on tiny orange hemihedrite crystals in specimens from a prospect 9 miles northwest of Nelson; identified by S. Williams. (R. Thomssen, w.c., 1999)

IRIDIUM

(Ir,Os,Ru)

A very rare native metal that is found with other platinum-group metals.

CLARK COUNTY With platinum in mixtures with jarosite-group minerals at the Boss Mine in the Goodsprings district. (Jedwab, 1998b; Jedwab et al., 1999)

JAHNSITE-(CaMnMg)

$$CaMn^{2+}(Mg,Fe^{2+})_2Fe_2^{3+}(PO_4)_4(OH)_2 \cdot 8H_2O$$

Jahnsite, which was named after pegmatite scholar R. H. Jahns, is a secondary phosphate mineral that has been found with a host of other phosphate species in pegmatite at localities in South Dakota and Brazil. It has apparently been subdivided chemically into species, but not on the basis of the most abundant cation.

EUREKA COUNTY Light-green, botryoidal and spherical clusters on hematite and goethite fracture coatings with vésigniéite, jarosite, whitlockite, and other phosphate minerals at the Goldstrike Mine in the Lynn district. (GCF)

JALPAITE

Ag_3CuS_2

A low-temperature silver-bearing sulfide that is found in hydrothermal precious-metal deposits.

MINERAL COUNTY Reported with other silver minerals at the Rawhide Mine, Rawhide district. (Black, 1991)

STOREY COUNTY Identified in ore from the Savage and Yellow Jacket Mines in the Comstock district. (Vikre, 1989)

JAMESONITE

$Pb_4FeSb_6S_{14}$

Jamesonite occurs in small quantities in many base-metal deposits, generally as mats and masses of tiny, needlelike crystals.

CHURCHILL COUNTY In minor amounts with stibnite, tetrahedrite, and other minerals in quartz and calcite veins at the Lofthouse Mine in the Bernice district. (Lawrence, 1963)

Reported in calcite-quartz vein material from

a waste dump with stibnite and multicolored antimony oxides at the Hazel (Green) Mine in the Mopung Hills, Lake district, SW 1/4 Sec. 9, T23N, R29E. (Lawrence, 1963)

DOUGLAS COUNTY With galena, stibnite, arsenopyrite, and pyrite at the Iron Pot prospect in the Red Canyon district. (Lawrence, 1963)

HUMBOLDT COUNTY Reported at an unspecified locality in the Santa Rosa Range. (Gianella, 1941)

LANDER COUNTY Reported with quartz in a specimen from the Reese River district. (AMNH)

MINERAL COUNTY As gray to black bunches of acicular crystals to 5 mm in stockwork quartz veins at the Broken Hills Mine in the Broken Hills district, Sec. 21, T14N, R35E. (Jenkins, 1981)

With pyrite, chalcopyrite, galena, and arsenopyrite in quartz-carbonate gangue at the Candelaria Mine in the Candelaria district; also as fine-grained, fibrous cores up to 30 cm across surrounded by banded bindheimite in a quartz vein with pyrite, tetrahedrite, galena, and calcite at the Potosi Mine. (Moeller, 1988; Knopf, 1923; Page, 1959; Lawrence, 1963)

In quartz veins with bindheimite, pyrite, and *limonite* at the Lowman Mine in the Pamlico district. (Lawrence, 1963)

NYE COUNTY Reported in a specimen from Wall Canyon in the Jett district. (WMK)

Reported at several mines in the Morey district as broken crystals with sphalerite on early rhodochrosite and intergrown with boulangerite and zinkenite in coarsely crystalline aggregates. (Williams, 1968; Jenkins, 1981)

PERSHING COUNTY In the mercury-antimony ores at the Red Bird Mine in the Antelope Springs district. (Jambor, 1969)

With bindheimite in quartz veins in the Arabia district, Trinity Range. (Knopf, 1918a; Lawrence, 1963)

Rare in quartz stringers at the Nevada-Massachusetts Co. Mine in the Mill City district. (Lawrence, 1963)

With minor galena and resinous sphalerite in a narrow quartz vein hosted in aplite on the west side of the Black Range in the Rochester district. (Knopf, 1924)

Reported in a specimen from the De Soto Mine in the Star district. (R. Thomssen, w.c., 1999)

As small pods, streaks, and single crystals, locally oxidized to bindheimite and related antimony oxides, with quartz, pyrite, gypsum, and *limonite* at the Green Antimony Mine in the Wild Horse district. (Lawrence, 1963)

WASHOE COUNTY With sphalerite and tetrahedrite in quartz veins at the Modoc Mine and Silver Fox prospect in the Cottonwood district. (Bonham and Papke, 1969)

JAROSITE

$K_2Fe_6^{3+}(SO_4)_4(OH)_{12}$

Jarosite, the most common member of the alunite-jarosite group, frequently goes unrecognized and is usually lumped with other species under the term *limonite*. It is an extremely common mineral in Nevada ore deposits as a product of surface oxidation of primary iron sulfide minerals.

CARSON CITY With cinnabar at the Valley View prospect in the Carson City district. (Bailey and Phoenix, 1944)

Identified in a sample from 9 miles east of Carson City in the Delaware district. (JHS)

CHURCHILL COUNTY Reported from an unspecified locality in the Chalk Mountain district. (Jenkins, 1981)

Along fractures in the wall rock and in veins at the Fairview Mine in the Fairview district. (Warren, 1973)

Identified in a specimen from 25 miles southsouthwest of Fallon in the Holy Cross district. (JHS)

Reported from the Green prospect in the Lake district. (R. Thomssen, w.c., 1999)

Reported in a specimen from Fireball Ridge in the Truckee district. (R. Thomssen, w.c., 1999)

CLARK COUNTY Millimeter-sized, dark-reddish-brown, tan, yellow, or gray crystals with plumbojarosite and other minerals at the Boss, John, Kirby, Yellow Pine, and other mines and prospects in the Goodsprings district. (Hewett, 1931; Jedwab, 1998b; J. Kepper, w.c., 1998; UAMM; S. Pullman, w.c., 1999)

Reported 5 miles north of Overton in the Moapa district. (HMM)

ELKO COUNTY Locally common as yellowish microcrystals at the Meikle Mine in the Lynn district. (Jensen, 1999)

As crystals to 5 mm with clay and barite in fracture fillings, druses, and breccia at the Rain Mine in the Carlin district. (D. Heitt, o.c., 1991)

In a specimen from the Bell Mine in the Independence district. (R. Thomssen, w.c., 1999)

Reported at the Killie Mine in the Spruce Mountain district. (S. Pullman, w.c., 1999)

ESMERALDA COUNTY As dark-brown crystals to 1 cm in comb-textured veins to 5 cm thick in the Gilbert district. (Ferguson, 1928; SBC; GCF)

Brown drusy coating on dacite at the Mohawk Mine in the Goldfield district; also noted in some specimens as a lustrous brown coating on alunite. (Keith et al., 1980; CSM; S. Pullman, w.c., 1999; T. Loomis, o.c., 2000)

In a specimen from the Big 3 Mine in the Railroad Springs district. (R. Thomssen, w.c., 1999)

Reported at the Weepah Mine in the Weepah district. (S. Pullman, w.c., 1999)

EUREKA COUNTY Widespread as lustrous, brown to orange-brown, barrel-shaped crystals to 4 mm on fractures at the Goldstrike Mine in the Lynn district. (GCF)

Common as yellow fine-grained coatings on fractures at the Gold Quarry Mine in the Maggie Creek district; also as lustrous crystals in druses with barite, scorodite, and kazakhstanite. (Jensen et al., 1995)

Locally abundant at the Modarelli Mine in the Modarelli-Frenchie Creek district. (Shawe et al., 1962)

HUMBOLDT COUNTY Reported from the Valmy area in the Battle Mountain district. (Jenkins, 1981)

With pharmacosiderite, dufrénite, and arseniosiderite on barite at the Lone Tree Mine in the Buffalo Mountain district. (GCF)

In oxidized radioactive breccia at the Moonlight Uranium Mine in the Disaster district. (Castor et al., 1996)

Reported from the Edna Mountains, 10 miles east of Golconda in the Golconda district. (WMK)

Reported as yellow-brown rhombs and plates with phosphate minerals at the Silver Coin Mine in the Iron Point district. (Thomssen, 1998; J. McGlasson, w.c., 2001)

Widespread with iron oxides and cinnabar at the Cordero Mine in the Opalite district. (JHS)

LANDER COUNTY Reported from an unspecified locality in the Battle Mountain district. (Roberts and Arnold, 1965)

Reported at the Warm Springs prospect in the Warm Springs district. (Bailey and Phoenix, 1944)

LINCOLN COUNTY Reported at the Bristol and May Day Mines in the Bristol district; at the Bristol Mine as lustrous brown crystals to 2 mm with conichalcite. (S. Pullman, w.c., 1999; S. White, w.c., 1999)

As yellow-brown powdery coatings and lumps in scorodite-rich quartz veins at the Advance Mine in the Chief district. (Callaghan, 1936; WMK)

As yellow druses from 8 miles southeast of Carp Station in the Mormon Mountains district. (WMK)

Fine crystals were found on the dumps of the Meadow Valley 1 and Meadow Valley 2 Mines in the Pioche district; also noted in other mines in the district. (Melhase, 1934d; Gemmill, 1968; CIT)

In specimens from the Silver King Mine in the Silver King district. (R. Thomssen, w.c., 1999)

LYON COUNTY In specimens from the Dyke adit in the Talapoosa district. (R. Thomssen, w.c., 1999)

Reported from Carson Hill in the Yerington district. (R. Thomssen, w.c., 1999)

MINERAL COUNTY As brown crystals to 1 mm at the Simon Mine in the Bell district. (S. White, w.c., 2000)

Reported from the western base of the southern Monte Cristo Mountains. (Schrader, 1947)

In oxidized portions of a shear zone with quartz, *limonite*, chlorargyrite, silver, anglesite, chrysocolla, and malachite at the Mabel Mine in the Garfield district. (Couch and Carpenter, 1943)

As brown crystals at the Homestake–Santa Fe Mine in the Santa Fe district. (R. Thomssen, w.c., 1999; GCF)

As lustrous orange-brown crystals to 1 mm

with carbonate-fluorapatite crystals from a mine 4 km south of Kinkaid in the Pamlico district. (M. Jensen, w.c., 1999)

Reported from unspecified localities in the Buckley, Candelaria, Rand, and Rawhide districts. (JHS; R. Thomssen, w.c., 1999; Schrader, 1947)

NYE COUNTY Reported from the Sterling Mine in the Bare Mountain district. (R. Thomssen, w.c., 1999)

Reported from unspecified localities in the Belmont and Ellendale districts. (Jenkins, 1981; Ferguson, 1917)

Crystalline jarosite occurs with cinnabar, native sulfur, and possible supergene gold as fracture coatings at the Paradise Peak Mine in the Fairplay district. (John et al., 1991)

Reported from an unspecified locality in the Lodi district. (Jenkins, 1981)

As tiny crystals in cracks at the Northumberland Mine in the Northumberland district. (Kokinos and Prenn, 1985)

With other oxidation minerals at the Round Mountain Mine in the Round Mountain district; also as dark-brown, massive and crystallized material along fractures in the oxidized zone at the Outlaw Mine. (Tingley and Berger, 1985; Dunning et al., 1991)

Abundant as light-ochre-yellow to reddish-brown flaky masses in the lower part of the oxidized zones with iodargyrite at the Montana-Tonopah Mine in the Tonopah district. (Melhase, 1935c; UAMM; WMK)

PERSHING COUNTY In small amounts in oxidized rock at the Majuba Hill Mine in the Antelope district. (Jensen, 1985)

As pseudomorphs after pyrite, arsenopyrite, and sillimanite(?), and as well-developed single crystals at the Arabia and Montezuma Mines in the Arabia district. (LANH; R. Walstrom, w.c., 1999)

Reported from unspecified localities in the Rabbit Hole and Rosebud districts. (WMK; JHS)

As rust-orange crystals to 1 mm at the Willard Mine in the Willard district. (Jensen and Leising, 2001)

Reported in a specimen from the Arsenic Mine. (FUMT)

WASHOE COUNTY Reported in specimens from the Crista pit at the Hog Ranch Mine in the Cottonwood district. (R. Thomssen, w.c., 1999)

Reported from an unspecified locality in the Olinghouse district. (Bailey and Phoenix, 1944)

As drusy crusts and small clusters to 1 mm of yellow-orange crystals; associated with goethite and olivenite at the Burrus Mine in the Pyramid district. (Jensen, 1994)

WHITE PINE COUNTY An oxidation product along quartz veins with alunite and barite at the Vantage deposit on Alligator Ridge in the Bald Mountain district. (Ilchick, 1984)

Reported from an unspecified locality in the Taylor district. (Lovering and Heyl, 1974)

JORDANITE

$Pb_{14}(As,Sb)_6S_{23}$

Jordanite is a rare mineral that occurs in a few hydrothermal deposits that contain lead-arsenic sulfides. The Nevada occurrences are somewhat unusual because of the general lack of base metals in Carlin-type deposits.

EUREKA COUNTY As microcrystalline replacement of galena in pods to several mm in barite at the Carlin Mine in the Lynn district; also associated with stibnite and possible tennantite. (Hausen and Kerr, 1968)

HUMBOLDT COUNTY As inclusions in galena and yellow sphalerite at the Getchell Mine in the Potosi district. (D. Tretbar, o.c., 1999)

JORDISITE

MoS_2

Jordisite is dimorphous with molybdenite and occurs in low- to moderate-temperature epithermal deposits.

EUREKA COUNTY As 10-micron inclusions in pyrite at the Betze deposit, Goldstrike Mine in the Lynn district. (GCF)

JOSÉITE-B

Bi_4Te_2S

An uncommon mineral that occurs with gold, native bismuth, and other sulfides in quartz veins and skarn; also from placer deposits.

WHITE PINE COUNTY Reported as tiny silver-colored crystals in quartz from the Caroline tunnel in the Ward district. (RMO; HMM; T. Bee, o.c., 2002)

JUNOITE

$Pb_3Cu_2Bi_8(S,Se)_{16}$

A rare mineral that generally occurs with sulfide and sulfosalt minerals in hydrothermal deposits.

LANDER COUNTY As adamantine, light-green crystals in quartz with bismutite, pottsite, and other minerals in tungsten skarn at the Linka Mine in the Spencer Hot Springs district. (Williams, 1988; S. Pullman, w.c., 1999; GCF)

KAERSUTITE

$NaCa_2(Mg_4Ti)Si_6Al_2O_{23}(OH)$

An amphibole-group mineral that occurs in alkalic igneous rocks.

NYE COUNTY As phenocrysts to 8 cm in the Marcath flow of the Lunar Craters volcanic field. (Bergman, 1982)

KALINITE

$KAl(SO_4)_2 \cdot 11H_2O$

Kalinite, or potash *alum,* is a relatively common mineral in the western United States. In many places it is found with native sulfur.

EUREKA COUNTY In clear to white transparent botryoidal masses from an undisclosed locality near Eureka in the Eureka district. (Nevada Gem and Mineral Society Mineral Show, 1998)

ESMERALDA COUNTY As white, banded fibrous masses with cinnabar, polyhalite, and native sulfur at the Alum Mine in the Silver Peak district, Sec. 28, T1N, R39E; in some specimens as curved white "ram's horns." (Spurr, 1906; UAMM; F. Cureton, o.c., 1999)

HUMBOLDT COUNTY As white, banded masses of long fibers with cinnabar and native sulfur at the Nevada Sulphur Company Mine in the Sulphur district, Sec. 36, T36N, R29E. (Adams, 1904; Bailey and Phoenix, 1944)

Reported with other *alum* minerals from an unspecified locality in the Virgin Valley district. (WMK)

WASHOE COUNTY Reported from an unspecified locality in the Steamboat Springs district. (WMK)

KAMACITE

(Fe,Ni)

A common nickel-poor mineral in iron meteorites; generally forms a crosshatched lamellar texture with taenite known as Widmanstätten structure.

CHURCHILL COUNTY With taenite and possible troilite in the Hot Springs meteorite. (Buchwald, 1975; Kring, 1997)

CLARK COUNTY With taenite, troilite, schreibersite, graphite, and cohenite in the Las Vegas meteorite. (Buchwald, 1975)

NYE COUNTY With taenite, schreibersite, troilite, and daubréelite in the Quartz Mountain meteorite. (Buchwald, 1975)

With taenite, schreibersite, and probable troilite in the Quinn Canyon meteorite, an iron-nickel meteorite that weighed about 1.5 tons when found. (Buchwald, 1975)

WHITE PINE COUNTY Meteoric iron is reported in a specimen from an unnamed locality in the county. (WMK)

KAMCHATKITE

$KCu_3^{2+}(SO_4)_2OCl$

A mineral that has been found in yellow-brown crystals with hematite and other minerals as a product of fumarolic activity at a volcano in Kamchatka, Russia.

CHURCHILL COUNTY Identified by XRD in a specimen from the Nickel Mine area, Table Mountain district. (S. Lutz, w.c., 2000)

KAOLINITE

$Al_2Si_2O_5(OH)_4$

A clay species that occurs as an alteration mineral in many of Nevada's mining districts. Kaolinite is an important commercial clay mineral outside of Nevada, but has not been mined extensively in the state.

CHURCHILL COUNTY In veins and wall rock at the Fairview Mine in the Fairview district. (Schrader, 1947)

CLARK COUNTY After orthoclase at Crystal Pass in the Goodsprings district; also found as pseudomorphs after orthoclase crystals to 1.5 cm at the Yellow Pine Mine. (GCF; G. Reynolds, w.c., 2000)

ELKO COUNTY As white earthy veins and in pods to 12 feet across with alunite at the Rain Mine in the Carlin district. (D. Heitt, o.c., 1991)

A principal alteration product in granodiorite in the Contact district. (Bailey, 1903)

A common alteration mineral at the Meikle Mine in the Lynn district. (Jensen, 1999)

ESMERALDA COUNTY Reported from an unspecified locality in the Buena Vista district. (LANH)

As a common, widespread alteration product in the Cuprite district. (Swaze et al., 1993)

Abundant in the Goldfield district, particularly in oxidized ore, as minutely crystalline aggregates with quartz, alunite, and pyrite, and locally with barite and gold. Also noted in sulfide zones and as an alteration mineral in unoxidized country rock near orebodies. (Ransome, 1909a; S. Pullman, w.c., 1999)

As beds generally less than 2 feet thick in tuffaceous Tertiary lakebeds west of Tonopah. (Olson, 1964a)

EUREKA COUNTY As white powdery masses commonly associated with stibnite and barite in silicified limestone and altered diorite at the Goldstrike Mine in the Lynn district; also in veins to 5 cm wide, particularly in sulfide-rich areas. (Ferdock et al., 1997; GCF)

Widespread as white coatings, vug fillings, and thin veinlets at the Gold Quarry Mine in the Maggie Creek district. (Jensen et al., 1995)

An alteration mineral at the Modarelli Mine in the Modarelli-Frenchie Creek district. (Shawe et al., 1962)

After *feldspar* phenocrysts at the Barth Iron Mine in the Safford district. (Shawe et al., 1962)

HUMBOLDT COUNTY An alteration mineral at a uranium prospect on Buff Peak in the Bottle Creek district. (Castor et al., 1996)

An alteration mineral found principally in fault zones at the Lone Tree Mine in the Buffalo Mountain district. (Panhorst, 1996)

White, earthy masses at the Silver Coin Mine in the Iron Point district. (Thomssen, 1998; J. McGlasson, w.c., 2001)

A local alteration phase at the National Mine in the National district. (Lindgren, 1915, 1933)

The most common alteration mineral at the Cordero Mine in the Opalite district; locally in relatively pure veinlets. (Fisk, 1968)

LANDER COUNTY In fairly large bodies associated with turquoise deposits along the White Horse fault near Mount Tenabo in the Cortez district. (Olson, 1964a)

LINCOLN COUNTY Reported as the discredited species *leverrierite* with dufrénoysite and other minerals at the Prince Mine in the Pioche district. (Melhase, 1934d)

Creamy white, hard kaolinite clay with alunite in altered rhyolite was mined in the 1920s at the Boyd Mine, NE 1/4 Sec. 2, T6S, R66E. (Tschanz and Pampeyan, 1970)

MINERAL COUNTY Identified in samples from the Sodaville montmorillonite deposit. (Papke, 1970)

NYE COUNTY An alteration product of *feldspar* at the Antelope View Mine in the Antelope Springs district on the Nellis Air Force Range. (Tingley et al., 1998a)

Kaolinite colored by iron and manganese oxides occurs in an irregular pod in Paleozoic carbonate at the Bond and Marks clay deposit in the Bare Mountain district 6 miles east of Beatty. (Olson, 1964a)

Reported from an unknown locality in the Bullfrog district. (WMK)

Widespread as an alteration mineral in the Cactus Springs district on the Nellis Air Force Range; associated with pyrophyllite, diaspore, and dickite at the Cactus Nevada Mine. (Tingley et al., 1998a)

With alunite at the A and O Clay Mine in the Currant district. (Kral, 1951)

A primary alteration mineral in the Gold

Crater district on the Nellis Air Force Range; locally as crystals to 0.5 mm. (Tingley et al., 1998a)

Reported as *leverrierite,* a discredited name for kaolinite or clay mineral mixtures, at the White Caps Mine in the Manhattan district. (Melhase, 1935d; Bailey and Phoenix, 1944)

As soft white clay common with chlorargyrite in the Tonopah district; also found with cinnabar. (Melhase, 1935c; WMK)

Part of the alteration assemblage with alunite, illite, and quartz at the Franz Hammel, Golden Chariot, and other mines in the Jamestown district on the Nellis Air Force Range. (Tingley et al., 1998a)

PERSHING COUNTY As an alteration mineral in rhyolite at the Majuba Hill Mine in the Antelope district; also as white powdery coatings on pharmacosiderite and scorodite crystals. (Jensen, 1985)

Reported in a sample from the Antelope Springs district 15 miles east of Lovelock. (JHS)

At the Stoker (Adamson-Dickson, New York Canyon) deposit in the Table Mountain district, Sec. 11, T25N, R35E, the largest known kaolin clay deposit in Nevada; the clay is stained red and is thought to have formed by hydrothermal alteration of the host phyllite. (Olson, 1964a)

With phosphate minerals in vugs and fractures at the Willard Mine, Willard district. (Jensen and Leising, 2001)

STOREY COUNTY Reportedly common in the Comstock district. (Bastin, 1922)

WASHOE COUNTY Widespread with illite in altered rock in the Geiger Grade area; common kaolinitic clay used in brick manufacture was mined from the Geiger clay pit in the Castle Peak district. (Bonham and Papke, 1969)

An alteration phase adjacent to gold ore at the Hog Ranch Mine in the Cottonwood district. (Bussey, 1996)

In hydrothermally altered rock in the Peavine and Wedekind districts in the Reno-Sparks area. (Bonham and Papke, 1969)

Widespread and locally abundant at the Burrus and other mines in the Pyramid district as vug fillings and fracture coatings of white powdery to flaky crystals. (Overton, 1947; Jensen, 1994)

As an alteration product of basaltic andesite with quartz, cristobalite, opal, and alunite at the Faith clay pit in the Steamboat Springs district, NW 1/4 Sec. 32, T18N, R20E. (Bonham and Papke, 1969)

WHITE PINE COUNTY Commonly found in quartz veins with stibnite at the Vantage deposit on Alligator Ridge in the Bald Mountain district. (Ilchick, 1984, 1990)

Reported as *miloschite* (a chromium-bearing variety) in jasperoid at an unspecified locality in the Robinson district 8 miles northwest of Ely. (Lovering, 1972; WMK)

KASOLITE

$Pb(UO_2)SiO_4 \cdot H_2O$

Kasolite, a relatively rare secondary uranium mineral, occurs in the oxidized zones of uranium deposits.

CLARK COUNTY As yellow–brown, crystalline masses with *limonite,* chrysocolla, hydrozincite, dumontite, and other lead and zinc minerals at the Green Monster Mine in the Goodsprings district. (Garside, 1973)

LYON COUNTY As ochre-yellow, monoclinic crystals in quartz veins with base-metal and silver sulfides, pyrite, chrysocolla, uraninite, torbernite, and other uranium minerals at the West Willys claim group in the Washington district. (Lovering, 1954; Garside, 1973)

MINERAL COUNTY Reported at the Lucky Horseshoe, 4D, and Silver State claims in the Marietta district. (Garside, 1973)

KAWAZULITE

$Bi_2(Te,Se,S)_3$

A rare hydrothermal selenium-bearing mineral that is found in veins and breccia.

MINERAL COUNTY As silvery micaceous flecks to 1 mm with gold and chalcopyrite in quartz in an unnamed prospect in Montreal Canyon in the Fitting district. (Jensen, 1993d; S. Pullman, w.c., 1999)

WHITE PINE COUNTY Reported at the Ward Mine in the Ward district. (Anthony et al., 1990)

KAZAKHSTANITE

$Fe_5^{3+}V_3^{4+}V_{12}^{5+}O_{39}(OH)_9 \cdot 9H_2O$

A very rare mineral found in Kazakhstan in black shale; to date the Nevada occurrence is the mineral's only locality in the world outside Kazakhstan.

EUREKA COUNTY In radiating groups of brown-black, bladed crystals, smaller than 0.5 mm, with variscite, fluellite, cacoxenite, hewettite, and other minerals in a fractured zone in altered carbonate rocks at the Gold Quarry Mine in the Maggie Creek district, Sec. 35, T34N, R51E; also as sooty black patches to 10 cm with fervanite, hewettite, tyuyamunite, and schubnelite on fractures in jasperoid. (Jensen et al., 1995)

KENHSUITE

γ-$Hg_3S_2Cl_2$

A recently described Nevada type mineral that was named for the well-known earth scientist Dr. K. J. Hsu.

HUMBOLDT COUNTY As canary-yellow fibers, tablets, and blades with cinnabar and corderoite in montmorillonite at the McDermitt Mine in the Opalite district. (McCormack and Dickson, 1998)

KENYAITE

$Na_2Si_{22}O_{41}(OH)_8 \cdot 6H_2O$

A rare mineral found in concretions from a saline lake in Africa and in altered volcanic rock in California.

NYE COUNTY Tentatively identified with *erionite* in ash-flow tuff drill core from Yucca Mountain. (Carlos et al., 1995)

KERMESITE

Sb_2S_2O

Kermesite is the first step in the oxidation of stibnite. It is relatively rare elsewhere in the United States, but is common in Nevada at antimony deposits that have been oxidized under arid conditions.

CHURCHILL COUNTY As an oxidation product of stibnite at the Lofthouse Mine in the Bernice district. (Lawrence, 1963)

ELKO COUNTY As red stains at the Burns Basin Mine in the Independence Mountains district. (Lawrence, 1963)

Reported at the Gribble Antimony Mine in the Island Mountain district. (Lawrence, 1963)

ESMERALDA COUNTY As red stains at the B & B Mine in the Fish Lake Valley district. (Lawrence, 1963)

EUREKA COUNTY As red stains with stibnite and yellow and white antimony oxides at the Blue Eagle Mine in the Antelope (Baldwin) district. (Lawrence, 1963)

Red coatings on euhedral barite crystals at the Goldstrike Mine in the Lynn district. (GCF)

HUMBOLDT COUNTY With stibnite and other antimony oxides in quartz veins at the Apex Antimony and Cottonwood Canyon Mines in the Battle Mountain district. (Lawrence, 1963)

With stibnite in quartz veins at the Antimony King and other mines in the Big Creek district. (Lawrence, 1963)

As haloes around stibnite at the Kattenhorn Mine in the Hilltop district. (Lawrence, 1963)

Reported from the Getchell Mine in the Potosi district. (R. Thomssen, w.c., 1999)

As red stains at the W. P. Mine in the Ten Mile district. (Lawrence, 1963)

LANDER COUNTY With stibnite at the Dean Mine in the Lewis district; identified by EMP. (J. McGlasson, w.c., 2001)

LYON COUNTY Common at the De Longchamps Mine in the Ramsey district. (Lawrence, 1963)

MINERAL COUNTY Reported at the Lithia Mine in the Poinsettia district. (Lawrence, 1963)

NYE COUNTY Reported at the White Caps Mine in the Manhattan district. (Lawrence, 1963; Gibbs, 1985)

Coating fractures at the Antimonial Mine in the Reveille district. (Lawrence, 1963)

Along cleavage planes in stibnite at the Blackbird Mine in the Silverton district. (Lawrence, 1963)

Reported at the Lucky Tramp prospect and Page (Hot Creek) Mine in the Tybo district. (Lawrence, 1963)

PERSHING COUNTY Reported in small amounts at the Sutherland Mine in the Black Knob district. (Lawrence, 1963)

As red stains in a quartz vein at the Black Warrior Mine in the Buena Vista district. (Lawrence, 1963)

As small pods in stibnite at the Polkinghorne prospect in the Mount Tobin district. (Lawrence, 1963)

On cleavage planes in stibnite at the Muttlebury Canyon Mine in the Muttlebury district. (Lawrence, 1963)

As red stains at the Fencemaker Mine in the Table Mountain district. (Lawrence, 1963)

Present at the Ore Drag Mine in the Tobin and Sonoma Range district. (Lawrence, 1963)

Doubtfully reported at the Rosal Mine in the Willard district. (Lawrence, 1963)

With stibiconite along cleavage in stibnite at the Hollywood Mine in the Antelope Springs district; also as coatings on stibnite crystals at the Red Bird Mine. (Lawrence, 1963; M. Jensen, w.c., 1999)

WASHOE COUNTY With stibnite at the Choates and Donatelli Mines in the McClellan district. (Lawrence, 1963)

WHITE PINE COUNTY Along cleavage planes in stibnite at the Dees Mine in the Bald Mountain district. (Lawrence, 1963)

With stibnite at the Lage Mine in the Butte Valley district. (Lawrence, 1963)

Reported at the Enterprise Mine in the Taylor district. (Lawrence, 1963)

KËSTERITE

$Cu_2(Zn,Fe)SnS_4$

A relatively rare sulfide mineral that generally occurs in tin deposits.

LANDER COUNTY Identified by EMP as a minor but widely distributed mineral in silver-gold ore at the Cove Mine in the McCoy district. (W. Fusch, o.c., 2000)

KIDWELLITE

$NaFe_9^{3+}(PO_4)_6(OH)_{10} \cdot 5H_2O$

Kidwellite is formed as a late-stage replacement of other phosphate minerals; all other domestic occurrences are in the southeastern United States.

ESMERALDA COUNTY Reported from the Tonopah-Divide Mine in the Divide district. (R. Thomssen, w.c., 1999)

Reported in a specimen from Coaldale. (R. Thomssen, w.c., 1999)

HUMBOLDT COUNTY Aluminum- and arsenic-bearing varieties, along with normal kidwellite, occur as gray to apple-green spherules and crude crystals to 1 mm on jarosite and with lipscombite at the Silver Coin Mine in the Iron Point district; also found at a prospect 1.2 miles north of I-80 near Valmy. (Thomssen, 1998; F. Cureton, w.c., 1999; Jenkins, 1981)

KIESERITE

$MgSO_4 \cdot H_2O$

Kieserite is common in salt deposits in Europe; the single Nevada occurrence is apparently a mine efflorescence and may not be of natural origin.

EUREKA COUNTY White coatings on fractures exposed to air by mining at the Goldstrike Mine in the Lynn district; an alteration/dehydration product of starkeyite that occurs in intrusive breccia and skarn. (GCF)

KINGITE

$Al_3(PO_4)_2(OH,F)_3 \cdot 9H_2O$

A rare secondary mineral deposited by meteoric water; presently known from only six occurrences in the world.

EUREKA COUNTY As white, flat-lying, radiating, silky crystalline spherules or as spongy masses of small sugary crystals in heavily brecciated siliceous siltstone associated with hewettite, kazakhstanite, and variscite-strengite at the Gold Quarry Mine in the Maggie Creek district. (Jensen et al., 1995)

KINGSMOUNTITE

$(Ca,Mn^{2+})_4(Fe^{2+},Mn^{2+})Al_4(PO_4)_6(OH)_4 \cdot 12H_2O$

Kingsmountite is rare mineral found elsewhere only as an oxidation product of primary phosphate minerals in granite pegmatites.

HUMBOLDT COUNTY As white to pink radiating microcrystals with mitridatite at the Twin Creeks Mine in the Potosi district. (M. Jensen, w.c., 1999)

KIPUSHITE

$(Cu^{2+},Zn)_5Zn(PO_4)_2(OH)_6 \cdot H_2O$

A very rare mineral that occurs with other secondary species in base-metal deposits at four sites worldwide.

LANDER COUNTY Reported in a specimen from the Snowstorm Mine in the North Battle Mountain district. (LANH)

KLEBELSBERGITE

$Sb_4^{3+}O_4(OH)_2(SO_4)$

Klebelsbergite is a very rare mineral that forms as an oxidation product of stibnite or other antimony-bearing minerals. The locality listed below is the only one known in the United States.

WHITE PINE COUNTY Tentatively identified in jasperoid at the North Star claim in the Taylor district. (Lovering, 1972)

KLEINITE

$Hg_2N(Cl,SO_4) \cdot nH_2O$

A rare mineral that is thought to have formed by reaction of cinnabar with oxidized meteoric water from which it derives its nitrogen. Kleinite is photosensitive, darkening on exposure to light.

HUMBOLDT COUNTY As fine, bright-yellow, hexagonal crystals in fractures in opalite breccia, commonly associated with gypsum at the McDermitt and Cordero Mines in the Opalite district. (McCormack, 1986; UMNH)

KOECHLINITE

Bi_2MoO_6

A rare alteration product in the oxidation zone of bismuth-molybdenum deposits.

NYE COUNTY Rare as white to light-gray spheroids composed of flattened, subhedral crystals in fractures in quartz at the Outlaw Mine in the Round Mountain district. (Dunning et al., 1991)

KÖTTIGITE

$Zn_3(AsO_4)_2 \cdot 8H_2O$

A rare secondary mineral in base-metal deposits.

CHURCHILL COUNTY As purple cobalt-bearing crystals to 1 mm with austinite at an unnamed prospect near the southwest end of the Clan Alpine Mountains on the east side of Stingaree Valley. (M. Jensen, w.c., 1999)

KOTULSKITE

Pd(Te,Bi)

A platinum-group metal mineral found in magmatic and hydrothermal deposits.

CLARK COUNTY As microscopic, anhedral, reddish-brown grains in silicified matrix at the Key West Mine in the Bunkerville district, Sec. 22, T15S, R70E. (S. Kleine, o.c., 1999; S. Pullman, o.c., 2000)

KRENNERITE

$(Au,Ag)Te_2$

Krennerite, a rare mineral that occurs in precious-metal telluride deposits, is associated with such minerals as sylvanite, calaverite, petzite, and native gold.

ESMERALDA COUNTY A gray telluride mineral containing no silver was tentatively identified as krennerite at the Jumbo Extension and Goldfield-Belmont Mines in the Goldfield district. (Ransome et al., 1907)

KRUPKAITE

$PbCuBi_3S_6$

A rare sulfosalt that occurs with other bismuth minerals in vein and pegmatite deposits.

ESMERALDA COUNTY As prismatic crystals intergrown with aikinite at a site near Cucomungo Spring in the Tule Canyon district. (F. Cureton, w.c., 1999)

KRYZHANOVSKITE

$Mn^{2+}Fe_2^{3+}(PO_4)_2(OH)_2 \cdot H_2O$

A rare, secondary phosphate mineral found in pegmatite; the Nevada occurrence may be unique because it is in a hydrothermal deposit.

LYON COUNTY Reported as blocky reddish crystals to 0.2 mm in a specimen from Carson Hill near Mason Pass in the Yerington district. (R. Thomssen, w.c., 1999)

KURAMITE

Cu_3SnS_4

A rare tin-bearing mineral that occurs in high-sulfidation epithermal deposits.

ESMERALDA COUNTY Reported from the Gold-field district. (Von Bargen, 1999)

NYE COUNTY Tentatively identified by SEM/EDX as sparse grains with hessite, enargite, freibergite, and other sulfides at the Fairday Mine in the Cactus Springs district on the Nellis Air Force Range. (Tingley et al., 1998a)

Tentatively identified as tiny grains in pyrite with enargite and an unidentified copper sulfo-salt in a sample of silicified and alunitized tuff from a shaft dump in NE 1/4 NE 1/4 Sec. 1, T5S, R45E in the Gold Crater district on the Nellis Air Force Range. Chemistry, in wt. percent by SEM/EDX is Sn = 24.8, Cu = 31.7, Fe = 3.7, Zn = 9.3, and S = 30.4. (Tingley et al., 1998a; SBC)

KURUMSAKITE

$(Zn,Ni,Cu^{2+})_8Al_8V_2^{5+}Si_5O_{35} \cdot 27H_2O$ (?)

Very rare mineral described from a single occurrence in cavities and fissures in bituminous schist in Kazakhstan.

CHURCHILL COUNTY Identified by XRD in a specimen from the Nickel Mine area, Table Mountain district. (S. Lutz, w.c., 2000)

KUTNOHORITE

$Ca(Mn^{2+},Mg,Fe^{2+})(CO_3)_2$

Kutnohorite (kutnahorite) is a relatively uncommon carbonate mineral that forms series with both dolomite and ankerite; it has been reported from base-metal deposits and other environments.

LANDER COUNTY Reported in a specimen from the Henry Logan Mine in the Lewis district. (R. Thomssen, w.c., 1999)

NYE COUNTY Identified by X-ray diffraction in drill core from Yucca Mountain. (Castor et al., 1992)

KYANITE

Al_2SiO_5

A common aluminum silicate mineral in high-grade metamorphic rocks.

ELKO COUNTY Scarce in gneiss in the northern part of the East Humboldt Range; mostly replaced by sillimanite. (Lush, 1982)

As conspicuous porphyroblasts in gneiss with a little sillimanite in the Clover Hill area. (Snoke, 1992)

As crystals with *garnet* in Gilbert Canyon in the Ruby Range. (NYS)

MINERAL COUNTY Reported from an unspecified locality in the Silver Star district. (WMK)

NYE COUNTY Reported in a specimen from 18 miles southeast of Beatty in Ryan Canyon at the American Progress Mine in the Bare Mountain district. (WMK)

LAFFITTITE

$AgHgAsS_3$

A rare mineral in hydrothermal deposits with other arsenic sulfides.

HUMBOLDT COUNTY As brilliant, dark-red, anhedral grains and crystals to 1 mm with getchellite, orpiment, realgar, stibnite, and sphalerite at the Getchell Mine in the Potosi district, Sec. 33, T39N, R42E. (Stolburg and Dunning, 1985)

LANGITE

$Cu_4^{2+}(SO_4)(OH)_6 \cdot 2H_2O$

Langite is a rare mineral found as an oxidation species in copper deposits. It is usually associated with such species as devilline, brochantite, cyanotrichite, and spangolite.

LANDER COUNTY Reported in specimens with antlerite and chalcanthite from Whisky Gulch in the Shoshone Range. (R. Thomssen, w.c., 1999; J. McGlasson, w.c., 2001)

LYON COUNTY As crusts of blue-green crystals to 0.75 mm at the Empire-Nevada Mine in the Yerington district. (Jenkins, 1981)

Reported in specimens from the Wabuska Smelter. (R. Thomssen, w.c., 1999)

PERSHING COUNTY Deep-blue crusts and single crystals to 0.5 mm at the Majuba Hill Mine in the Antelope district. (Jensen, 1993b)

WHITE PINE COUNTY Reported in the Robinson district. (Von Bargen, 1999)

Reported as flaky aggregates of pale-greenish-blue crystals to 0.25 mm with serpierite, elyite, devilline, linarite, brochantite, and cerussite in the Caroline tunnel in the Ward district. (Williams, 1964, 1972; Jenkins, 1981)

LAUMONTITE

$Ca_4[Al_8Si_{16}O_{48}] \cdot 18H_2O$

Laumontite is a relatively scarce member of the zeolite group. It is found as amygdule fillings in mafic volcanic rocks, in some pegmatites, and rarely as a product of hydrothermal alteration.

EUREKA COUNTY As fibrous, vitreous to pearly pink coatings on fractures in diorite at the Goldstrike Mine in the Lynn district. (GCF)

HUMBOLDT COUNTY As acicular white microcrystals up to several mm long at the Mountain King Mine in the Potosi district. (S. Sears, w.c., 1999)

PERSHING COUNTY As white crystals to 1 cm with chamosite and clinozoisite in the lower adit of the Majuba Hill Mine in the Antelope district. (Jensen, 1993b)

WASHOE COUNTY As fibrous, opaque white crystals to 1 cm in vugs at the Olinghouse Mine in the Olinghouse district. (M. Jensen, w.c., 1999)

In an unusual occurrence with native copper, cuprite, and malachite in veinlets that cut Alta Formation andesite on the north side of the Truckee River Canyon near the Painted Rock interchange on I-80. (Rose, 1969; M. Jensen, w.c., 1999)

LAVENDULAN

$NaCaCu_5^{2+}(AsO_4)_4Cl \cdot 5H_2O$

A relatively uncommon secondary copper mineral that occurs in base-metal deposits with other secondary species such as malachite and olivenite.

CHURCHILL COUNTY Blue crystalline crusts to 0.5 mm thick with zeunerite crystals at the Nickel Mine in the Table Mountain district in Cottonwood Canyon of the Stillwater Range. (M. Jensen, w.c., 1999)

Reported from the Fireball Mine in the Truckee district. (R. Thomssen, w.c., 1999)

ESMERALDA COUNTY As small' blue crystal groups at the 3 Metals Mine in the Weepah district. (M. Jensen, w.c., 1999)

LINCOLN COUNTY Reported from the Silver King Mine in the Silver King district. (R. Thomssen, w.c., 1999)

NYE COUNTY Reported as brilliant-blue sprays to 0.7 mm at the San Rafael Mine in the Lodi district. (S. White, w.c., 2000)

PERSHING COUNTY As microscopic blue masses and rare blue crystal clusters to 2 mm on silicified matrix at the Majuba Hill Mine in the Antelope district. (Jensen, 1985, 1993b)

LAZULITE

$MgAl_2(PO_4)_2(OH)_2$

Lazulite is a deep-blue mineral that is usually found in quartz or pegmatite veins.

DOUGLAS COUNTY Reported to have been found near Gardnerville. (WMK)

NYE COUNTY Reported in specimens from the Bare Mountain district; said to be associated with kyanite at the American Progress Mine. (LANH; AMNH; HMM; WMK)

LAZURITE

$(Na,Ca)_{7-8}(Al,Si)_{12}(O,S)_{24}[(SO_4),Cl_2,(OH)_2]$

Lazurite, known as "lapis lazuli" when richly colored and suitable for gemstones, is a contact-metamorphic mineral that occurs in limestone near contacts with granitic rock.

CLARK COUNTY Reported as "lapis lazuli" in a specimen from an unspecified locality in the Searchlight district; this is an unlikely locality for this mineral. (WMK)

LEADHILLITE

$Pb_4(SO_4)(CO_3)_2(OH)_2$

Leadhillite, a relatively rare mineral, is found in small quantities in the oxidation zones of base-metal deposits.

CHURCHILL COUNTY Reportedly Nevada's best specimens as vitreous, colorless, euhedral flattened hexagonal crystals to 2 cm and microcrystals in vugs at the Chalk Mountain Mine (510-foot level) in the Chalk Mountain district. (Jensen, 1990b; UMNH)

CLARK COUNTY In oxidized lead-copper vein deposits with quartz, calcite, cerussite, wulfenite, and other secondary minerals at the Quartette, Duplex, and Good Hope Mines in the Searchlight district. (Callaghan, 1939)

EUREKA COUNTY Tentatively reported in oxidized ore at the Richmond-Eureka (Ruby Hill) Mine in the Eureka district. (Curtis, 1884; S. Pullman, w.c., 1999)

LINCOLN COUNTY As pseudomorphs after calcite rhombs to 1 cm at an unspecified locality. (Jenkins, 1981)

NYE COUNTY Olive- to gray-green prisms to 1 mm with phosgenite in vugs in altered Luning Formation carbonate rocks at the Illinois Mine in the Lodi district, Sec. 17, T13N, R37E. (Jenkins, 1981)

LECONTITE

$(NH_4,K)Na(SO_4) \cdot 2H_2O$

A mineral found in cave deposits of bat guano.

WASHOE COUNTY Reported from an unspecified locality near Reno. (UCB)

LEPIDOCROCITE

$\gamma\text{-Fe}^{3+}O(OH)$

Lepidocrocite is a common but seldom recognized hydrated iron oxide mineral that occurs along with goethite in *limonite*. It is undoubtedly present in many places in Nevada as an oxidation product of pyrite.

EUREKA COUNTY As fracture coatings in the oxidized zone at the Goldstrike Mine in the Lynn district. (Sampson, 1993)

HUMBOLDT COUNTY Reported at the Silver Coin Mine in the Iron Point district. (Thomssen, 1998)

With goethite and amorphous *limonite* minerals in iron-rich layers of tungsten-manganese deposits near Golconda in the Golconda district. (Berger, 1982)

LEPIDOLITE

$K(Li,Al)_3(Si,Al)_4O_{10}(F,OH)_2$

Lepidolite is a relatively scarce member of the mica group. It occurs in lithium-bearing pegmatites and is associated with such minerals as beryl, spodumene, and elbaite.

ELKO COUNTY Small amounts with beryl, green microcline, and other minerals in pegmatite at the head of Corral Creek in the Corral Creek district in the Ruby Mountains. (Olson and Hinrichs, 1960)

ESMERALDA COUNTY As large crystals and massive aggregates in a quartz-rich pegmatite dike with orthoclase and biotite at the Susan prospect in the Tokop district. (Olson and Hinrichs, 1960; Garside, 1973)

MINERAL COUNTY Iron-bearing lepidolite is a late-stage mineral in pegmatite at the Zapot claims in the Fitting district. (Foord et al., 1999b)

Reported near Marietta at the Mineral Jackpot prospect in the Marietta district, Sec. 20, T4N, R32E. (JHS)

PERSHING COUNTY Thin laminae of lepidolite occur with *tourmaline* and *garnet* in pegmatite in Crusoe Canyon in the Seven Troughs Range. (Olson and Hinrichs, 1960)

LEUCOPHOSPHITE

$KFe_2^{3+}(PO_4)_3(OH) \cdot 2H_2O$

A mineral that occurs elsewhere in the world in pegmatites and sedimentary phosphate deposits. The Nevada occurrences in Carlin-type gold deposits are unusual.

ELKO COUNTY As lustrous, pale-tan, glassy crystals to 1 mm associated with pharmacosiderite and other minerals at the Rain Mine in the Carlin district. (M. Jensen, w.c., 1999)

EUREKA COUNTY Reported in a wide variety of colors with metahewettite and minyulite from the Cockalorum Wash Quadrangle in the Gibellini district. (Zientek et al., 1979)

As isolated golden-yellow to brown-spherical and flattened-crystal aggregates on fracture surfaces with variscite, fluellite, tinticite, and phosphofibrite at the Gold Quarry Mine in the Maggie Creek district; contains some aluminum and vanadium. (Jensen et al., 1995)

HUMBOLDT COUNTY As microcrystals associated with cacoxenite and variscite at the Lone Tree Mine in the Buffalo Mountain district. (M. Jensen, w.c., 1999)

In tan- to cream-colored, platy to massive crystalline aggregates to 2 mm at the Silver Coin Mine in the Iron Point district. (Thomssen, 1998; M. Jensen, w.c., 1999; J. McGlasson, w.c., 2001)

LIBETHENITE

$Cu_2^{2+}(PO_4)(OH)$

Libethenite is a moderately rare mineral found in the oxidation zones of some copper deposits. Some good specimens have come from the Yerington district.

CLARK COUNTY As olive-green crystals and crusts on silica and as loosely coherent masses on the dump at the Boss Mine in the Goodsprings district. (Hewett, 1931; Melhase, 1934e)

EUREKA COUNTY As transparent pale-green crystals associated with cuprite, vésigniéite, and barite in vugs along fractures in siliceous siltstone at the Gold Quarry Mine in the Maggie Creek district. (Jensen et al., 1995)

Reported from the Bisoni claims in the Gibellini district. (R. Thomssen, w.c., 1999)

HUMBOLDT COUNTY Reported from the Silver Coin Mine in the Iron Point district. (Thomssen, 1998)

As pale-olive-green, prismatic microcrystals with chalcosiderite, pseudomalachite, and carbonate-fluorapatite near Golconda Summit. (R. Thomssen, o.c., 2000)

LANDER COUNTY Reported in specimens from the Snowstorm Mine in the North Battle Mountain district. (R. Thomssen, w.c., 1999)

LYON COUNTY In oxidized ore from the Douglas Hill, Empire-Nevada, Ludwig, and Blue Jay Mines in the Yerington district; described as olive- to dark-green crusts and prismatic crystals to 4 mm. Also noted in crystals to 1 mm on white breccia fragments with carbonate-fluorapatite, native copper, cuprite, and chrysocolla at an unnamed prospect about 0.5 km north of the MacArthur Mine on the south side of Carson Hill. (Smith, 1904; Ransome, 1909e; Knopf, 1918b; Melhase, 1935e; UAMM; S. Pullman, w.c., 1999; M. Jensen, w.c., 1999)

MINERAL COUNTY In pale-green, acicular sprays

with brochantite and pseudomalachite at an unnamed prospect on Copper Mountain in the Buckley district near the north end of the Gillis Range. (R. Walstrom, w.c., 1999)

PERSHING COUNTY Reported at the Majuba Hill Mine in the Antelope district. (MacKenzie and Bookstrom, 1976)

WASHOE COUNTY Reported as an emerald-green mineral with malachite in altered quartz monzonite near the Guanomi Mine in the Pyramid district. (Satkoski and Berg, 1982)

LILLIANITE

$Pb_3Bi_2S_6$

Lillianite is a rare mineral that occurs with other lead-bismuth sulfosalt minerals in some hydrothermal deposits.

LANDER COUNTY Reported in a drill hole specimen from the Wilson-Independence Mine in the Battle Mountain district. (R. Thomssen, w.c., 1999)

NYE COUNTY As silvery metallic stringers with galena, sphalerite, and pyrite in quartz veinlets that cut altered volcanic rock in the Manhattan district. (CIT; Jenkins, 1981)

LINARITE

$PbCu^{2+}(SO_4)(OH)_2$

Linarite, a deep-blue mineral that may be misidentified as azurite, is widespread in small amounts in the oxidation zones of many western base-metal deposits. Some Nevada localities have produced good specimens, particularly the Goodsprings district.

CHURCHILL COUNTY As an alteration product of galena in the veins at the Webber shaft and Nevada Hills Mine in the Fairview district. (Schrader, 1947; WMK)

As blue, microcrystalline masses at the Westgate Mine in the Westgate district. (R. Walstrom, w.c., 1999; GCF)

CLARK COUNTY Uncommon as a secondary mineral with caledonite in the Goodsprings district; exceptional crystals largely altered to caledonite, including one described as tabular-prismatic and 10 cm long, were reportedly found on the 900 level of the Yellow Pine

Mine; also reported from the Root Mine. (Hewett, 1931)

Linarite is reported at the Quartette Mine in the Searchlight district. (Callaghan, 1939)

ELKO COUNTY Reported from a site on the western flank of the Ruby Mountains on the east side of Long Canyon in the Lee district. (WMK)

ESMERALDA COUNTY Massive to crystalline linarite was reported to occur with unidentified associated minerals near Silver Peak. (CIT)

As radiating, flat, blue crystals to 1 mm in specimens from the 3 Metals Mine in the Weepah district; also reported from the Weepah Mine. (R. Thomssen, w.c., 1999; B. Runner, o.c., 2000)

EUREKA COUNTY As crystalline coatings on quartz and barite, with malachite and caledonite at an unspecified site in the Eureka district. (Jenkins, 1981)

HUMBOLDT COUNTY Bright- to deep-blue blades to 5 mm with brochantite, malachite, and chlorargyrite at the Silver Coin Mine in the Iron Point district. (Thomssen, 1998; M. Jensen, w.c., 1999; J. McGlasson, w.c., 2001)

LINCOLN COUNTY Reported from an unspecified locality in the Pioche district. (HMM)

LYON COUNTY As royal-blue crystalline coatings and prismatic crystals to 4 mm with cerussite, wulfenite, and other minerals at the Old (Lost) Soldier Mine in the Churchill district, SE 1/4 SW 1/4 Sec. 22, T17N, R24E. (CIT; Jenkins, 1981)

In minor amounts with malachite, caledonite, and brochantite at the Jackpot Mine in the Wilson district. (Jenkins, 1981)

MINERAL COUNTY Reported in copper-gold-silver ore from an unspecified locality in the Candelaria district. (WMK)

In fine specimens from the Atherton (Garfield) Mine in the Garfield district, Sec. 21, T7N, R33E; erroneously reported in Esmeralda County for some specimens. (Melhase, 1935e; AMNH)

With brochantite as a major oxide mineral at the Lucky Boy Mine in the Lucky Boy district. (Ross, 1961)

With base-metal sulfides and other secondary copper minerals at the Dispozitch, Silver Gulch, and Spence Mines in the Silver Star district. (Ross, 1961; NMBM; WMK)

As blue crystals to 5 mm on fractures with

brochantite at the Dunlop Mine in the Pilot Mountains district. (M. Jensen, w.c., 1999)

NYE COUNTY As prismatic crystals to 2 mm with caledonite, anglesite, and cerussite crystals in vugs in quartz at the Transylvania Mine in the Belmont district. (Jenkins, 1981)

Sparingly with caledonite in oxidized lead ore at the San Rafael Mine in the Lodi district. (Jenkins, 1981)

WHITE PINE COUNTY As prisms to 0.5 mm with brochantite, langite, and caledonite in cavities in galena and stephanite at the Silver King Mine in the Ward district. (Williams, 1968)

LINDACKERITE

$H_2Cu_5^{2+}(AsO_4)_4 \cdot 8\text{-}9H_2O$

Lindackerite is an extremely rare mineral known from only a few localities in the world. The Majuba Hill locality is the only known occurrence of the mineral in the United States.

PERSHING COUNTY As apple–green, radial masses to 1 cm with tyrolite on altered rhyolite in the copper and tin deposit at the Majuba Hill Mine in the Antelope district. (CIT; Jenkins, 1981)

LINDSTRÖMITE

$Pb_3Cu_3Bi_7S_{15}$

Lindströmite occurs with other sulfosalts in very few hydrothermal and contact-metasomatic deposits.

NYE COUNTY As gray metallic grains to 2 mm in specimens from an unspecified locality in the Manhattan district. (Jenkins, 1981; Anthony et al., 1990; GCF)

Reported at the Outlaw Mine in the Round Mountain district. (Dunning, 1987)

LIPSCOMBITE

$(Fe^{2+},Mn^{2+})Fe_2^{3+}(PO_4)_2(OH)_2$

Lipscombite, which is formed by alteration of primary iron oxides or sulfides, occurs with other iron phosphates or secondary minerals.

ELKO COUNTY In dark-green, microcrystalline coatings on fractures in the oxidized zone; associated with barite crystals and wavellite at the Rain Mine in the Carlin district. (Dean Heitt, o.c., 1991)

HUMBOLDT COUNTY As gray-green, microcrys-

talline spherules to 3 mm on iron-stained quartz with turquoise, jarosite, and kidwellite at the Silver Coin Mine in the Iron Point district. (Thomssen, 1998; Jenkins, 1981; J. McGlasson, w.c., 2001)

LITHARGE

PbO

Litharge, an uncommon or seldom-recognized mineral, occurs with other oxidized lead compounds in base-metal deposits.

CLARK COUNTY As yellow aggregates in vugs in silicified dolomite at the Boss Mine in the Goodsprings district. (J. Kepper, w.c., 1998)

NYE COUNTY In massive white crusts on cerussite, anglesite, and residual galena at the San Rafael Mine in the Lodi district. (Jenkins, 1981)

LITHIOPHORITE

$(Al,Li)Mn^{4+}O_2(OH)_2$

Lithiophorite, a moderately rare mineral, is found in manganese-oxide deposits but more often in suites containing other lithium species.

NYE COUNTY With pyrolusite and gibbsite in thermally metamorphosed limestone adjacent to quartz latite at the Queen City prospect in the Queen City district, Sec. 26, T2S, R53E. (Hewett et al., 1968)

Identified in veins in drill core from Yucca Mountain. (Carlos et al., 1991)

LITIDIONITE

$KNaCuSi_4O_{10}$

Litidionite (*lithidionite* of E. S. Dana) is a rare mineral found elsewhere only as tiny blue plates in lapilli with tridymite and wollastonite in the crater of Mount Vesuvius, Italy.

LANDER COUNTY As pale-blue crystals to 0.5 mm with chrysocolla in altered shale at the Pyro pit, Snowstorm Mine, Sec. 25, T33N, R45E, North Battle Mountain district; identified by S. Williams. (S. White, w.c., 2000; J. McGlasson, w.c., 2001)

LIVINGSTONITE

$HgSb_4S_8$

A rare mineral found in low-temperature hydrothermal veins with cinnabar and other minerals.

PERSHING COUNTY In a specimen from the Adamson property 15 miles east of Lovelock in the Antelope Springs district. (WMK)

LIZARDITE

$Mg_3Si_2O_5(OH)_4$

Probably the most common *serpentine*-group mineral; typically found as a replacement of *olivine* or other minerals in ultramafic rock.

CHURCHILL COUNTY Identified by XRD in a specimen from the Nickel Mine area, Table Mountain district; reportedly contains nickel. (S. Lutz, w.c., 2000)

EUREKA COUNTY White, vitreous and resinous to plasterlike with aliettite in fracture coatings at the 5040-foot level of the Goldstrike Mine in the Lynn district. (GCF)

LÖLLINGITE

$FeAs_2$

Löllingite typically occurs in mesothermal deposits with other arsenide minerals such as skutterudite and nickeline.

HUMBOLDT COUNTY As a minor mineral at the Lone Tree Mine in the Buffalo Mountain district. (Bloomstein et al., 1993)

LORANDITE

$TlAsS_2$

Lorandite, a rare hydrothermal mineral that is probably formed at low temperatures, has been found in small quantities in a number of arsenic-thallium-bearing gold deposits worldwide; all Nevada occurrences are in Carlin-type gold deposits.

ELKO COUNTY With cinnabar, native arsenic, realgar, orpiment, arsenopyrite, and marcasite at the Jerritt Canyon Mines in the Independence Mountains district. (Hofstra, 1994)

EUREKA COUNTY As small, subhedral grains to 4 mm that are commonly intergrown with real-

gar, orpiment, and other minerals with barite in veinlets in gold-bearing limestone at the Carlin Mine in the Lynn district. (Radtke et al., 1974c; Radtke et al., 1977)

HUMBOLDT COUNTY As microscopic prisms and needles on realgar at the Getchell Mine in the Potosi district. (CIT)

LUDWIGITE

$Mg_2Fe^{3+}BO_5$

Ludwigite occurs in small amounts in a number of hydrothermal ore deposits and in contact zones in limestones.

LINCOLN COUNTY The variety *magnesioludwigite* occurs with szaibélyite and fluoborite on Blind Mountain in the Bristol district. (Gillson and Shannon, 1925; Westgate and Knopf, 1932; S. Pullman, w.c., 1999)

LYON COUNTY As black needles in skarn 0.5 mile east of the Ludwig gypsum pit in the Yerington district. (M. Jensen, w.c., 1999; S. Pullman, w.c., 1999)

MINERAL COUNTY As brown-black, radiating fibers to 1.5 cm in skarn at unnamed tungsten prospects near Cedar Summit in the Bell district. (LCH)

LUETHEITE

$Cu_2^{2+}Al_2(AsO_4)_2(OH)_4 \cdot H_2O$

A very rare mineral known elsewhere only from occurrences in Arizona and Australia.

PERSHING COUNTY With devilline as gray coatings under pharmacosiderite cubes in specimens from the Copper stope of the Majuba Hill Mine in the Antelope district. (Jensen, 1985)

LUZONITE

Cu_3AsS_4

A sulfosalt that is dimorphous with enargite and occurs in similar settings but is probably less common. The three Nevada localities represent a large portion of a handful of reported occurrences in the United States.

ESMERALDA COUNTY Reported as a massive silvery sulfide in the Goldfield district. (M. Jensen, w.c., 1999)

NYE COUNTY With tetrahedrite, galena, chalcopyrite, covellite, sphalerite, and pyrite at the Golden Chariot Mine in the Jamestown district; in some samples as crudely formed crystals to 5 mm that yield XRD patterns for antimony-bearing luzonite. (Tingley et al., 1998a; SBC)

WASHOE COUNTY In intergrowths with more abundant enargite in the Burrus vein in the Pyramid district; pyrite, chalcopyrite, and arsenopyrite are associated minerals. (Wallace, 1980)

MACKAYITE

$Fe^{3+}Te_2O_5(OH)$

Mackayite, a Nevada type mineral, was named for John W. Mackay (1831–1902), mine owner on the Comstock Lode who endowed the Mackay School of Mines at Reno. It is a very rare oxidation product of telluride-rich, precious-metal ores that has been found at only eight other localities in the world.

ESMERALDA COUNTY Light-green to brownish- or olive-green equant crystals to 0.5 mm that are found in crusts or lining vugs and seams in silicified volcanic rocks at the Jumbo, Clermont, and Mohawk Mines and the McGinnity shaft in the Goldfield district. Mostly as isolated crystals with *limonite*, emmonsite, tellurite, and *blakeite* crusts, and with crystals of alunite, quartz, and barite. (Frondel and Pough, 1944; DeMouthe, 1985)

MAGHEMITE

$\gamma\text{-}Fe_2O_3$

A dimorph of hematite, maghemite generally occurs as a weathering or low-temperature oxidation product of magnetite. It generally occurs with other iron oxide and iron hydroxide species.

EUREKA COUNTY Along grain contacts between hematite and magnetite at the Barth Mine in the Safford district. (GCF)

MAGNESIO-AXINITE

$Ca_2MgAl_2BSi_4O_{15}(OH)$

A rare member of the *axinite* group, typically formed by contact-metamorphism and boron metasomatism.

MINERAL COUNTY Reported as pink to brown crystals to 1.0 cm with prehnite, vesuvianite, actinolite, and epidote from an unspecified locality near Luning. May be the same locality as *axinite* with zoisite, epidote, *tourmaline,* hematite, *garnet,* prehnite, and copper minerals described by Melhase about 7 miles north-northwest of Luning. A specimen in the University of California, Berkeley, museum reportedly came from 4 miles northwest of Mina. (Dunn et al., 1980; Melhase 1935e; S. Pullman, w.c., 1999; UCB)

MAGNESIOHASTINGSITE

$NaCa_2(M_4Fe^{3+})Si_6Al_2O_{22}(OH)_2$

A member of the *hornblende* group that occurs in intermediate volcanic rocks and alkalic basalts.

CLARK COUNTY Reported in a specimen from an unknown locality in the Moapa area. (HUB)

MAGNESIOHORNBLENDE

$\square Ca_2[Mg_4(Al,Fe^{3+})]Si_7AlO_{22}(OH)_2$

Very widespread member of the *hornblende* group that probably occurs, but has not been specifically identified, in many granitic rocks in Nevada.

EUREKA COUNTY As black prisms to 5 mm that are an uncommon component of diorite at the Goldstrike Mine in the Lynn district; commonly altered to iron oxides and other phases. (GCF)

MAGNESITE

$MgCO_3$

Magnesite is a common mineral found in replacement deposits in carbonate or volcanic rocks and as a minor constituent of hydrothermal alteration assemblages.

CLARK COUNTY Reported in a specimen from an unknown locality in the Crescent district. (WMK)

Very fine-grained white magnesite occurs in beds of the Rainbow Gardens Member of the Horse Spring Formation about 4 miles southwest of Overton in the Moapa district; minor amounts were reportedly mined for use in pigments and fillers. (Hewett et al., 1936; Olson, 1964b; Bohannon, 1984)

In Horse Spring Formation beds at the Bauer Magnesite Mine, Sec. 24, T18S, R70E. (Hewett et al., 1936)

DOUGLAS COUNTY As a gangue mineral with dolomite and serpentine minerals at the Minnesota Mine in the Buckskin district. (Reeves et al., 1958)

ESMERALDA COUNTY In a specimen from the Snow Ball claim 2.5 miles southeast of the General Thomas Mine in the Lone Mountain district. (WMK)

NYE COUNTY As nodules and lenses in calcareous tuff in the Currant district; larger occurrences are in White Pine County. (Hewett et al., 1936; Vitaliano, 1951)

As large masses of gray to white crystal aggregates with brucite and other magnesium-bearing minerals in replacement deposits in Triassic dolomite that is intruded by granodiorite at the Premier Chemicals (Gabbs Magnesite-Brucite) Mine in the Gabbs district, Sec. 35, T12N, R36E; magnesite and brucite have been mined in this area for many years. (Callaghan, 1933; Martin, 1956; Schilling, 1968a)

Reported in a specimen from an unknown locality in the Lodi district. (WMK)

"Bone" magnesite was reported in limestone about 14 miles north of U.S. Highway 95 and 2 miles northeast of the Oak Spring road; possibly the same location as that for brucite and hydromagnesite in the Pocopah district in the Calico Hills on the Nevada Test Site. (Kral, 1951; Castor et al., 1999b)

STOREY COUNTY Questionably identified as an alteration mineral in andesite in the Truckee River canyon. (Rose, 1969)

WHITE PINE COUNTY As nodules and grains replacing altered tuff and as veins and lenses along fractures in tuff with huntite, serpentine, stevensite, dolomite, calcite, *deweylite, chalcedony,* opal, and quartz in the Currant district. (Hewett et al., 1936; Vitaliano, 1951; Faust, 1953; Hose et al., 1976)

MAGNETITE
$Fe^{2+}Fe_2^{3+}O_4$

A common accessory mineral in igneous and metamorphic rocks, and a common heavy mineral in stream sediments. In economic concentrations in magmatic, contact-metamorphic, and banded iron formation deposits.

CARSON CITY In placer sand in the Carson City district. (WMK)

In veinlike orebodies along fault zones with quartz and hematite at the Capitol and Bessemer (Brunswick Canyon, Iron King) prospects in the Delaware district. (Reeves et al., 1958)

CHURCHILL COUNTY In veins that range in composition from massive magnetite to calcite at the Black Joe prospect in the Copper Kettle district. (Shawe et al., 1962)

As pods 10 to 20 feet across with hematite at the White Rock prospect in the Corral Canyon district. (Shawe et al., 1962)

As an accessory mineral in wall rock in the Fairview district. (Schrader, 1947; Warren, 1973)

In a specimen from IXL Canyon in the IXL district. (WMK)

CLARK COUNTY Reported at the Key West Mine in the Bunkerville district. (Beal, 1965)

As crystals to 5 mm in miarolitic cavities and pegmatites in the Eldorado and Newberry Mountains. (S. Scott, w.c., 1999)

As octahedrons at the Mohawk Mine and in massive form in *garnet*/diopside skarn at the Iron Gold Mine in the Goodsprings district; also noted in specimens from the Boss Mine. (Jedwab, 1998b; J. Kepper, w.c., 1998)

DOUGLAS COUNTY As the principal, and, in many places, the only, ore mineral in a productive, irregular, replacement orebody in dolomite with magnesite, *serpentine, chlorite,* and *sericite,* as well as minor hematite, pyrite, and chalcopyrite at the Minnesota Mine in the Buckskin district, Sec. 19, T14N, R24E. (Reeves et al., 1958)

In copper ore with chalcopyrite from an unspecified locality in the Wellington district. (WMK)

ELKO COUNTY In skarn along a porphyry dike at the Maverick Springs prospect in the Cave Creek district. (Smith, 1976)

In quartz-calcite veins and skarn in the Contact district. (Schilling, 1979)

With malachite, *limonite,* and chrysocolla as pods in green *garnet*-epidote skarn at the Lone Mountain Mine in the Merrimac district. (LaPointe et al., 1991)

Reported from an unspecified locality in the Tuscarora district. (UCB)

ESMERALDA COUNTY The principal iron mineral with minor hematite at the Boak prospect and Iron King Mine in the Candelaria district. (Reeves et al., 1958)

EUREKA COUNTY At depth below the hematitic orebody at the Modarelli Mine in the Modarelli–Frenchie Creek district; also the principal ore mineral in brecciated volcanic rock and in veins to 20 feet wide at the Frenchie Creek, Imperial Jackson, and Big Pole Creek prospects. (Shawe et al., 1962)

Principal primary ore mineral at the Barth Mine in the Safford district, Sec. 7, T31N, R51E; locally as crystals to 8 mm in vugs in otherwise massive ore. (Shawe et al., 1962; GCF)

HUMBOLDT COUNTY In host rock at the Lone Tree Mine in the Buffalo Mountain district. (R. Braginton, o.c., 1995)

With hematite in replacement bodies in metavolcanic rock at the Iron King, Black Jack, and Redbird Mines and adjacent prospects in the Jackson Mountains district. (Shawe et al., 1962)

LANDER COUNTY In skarn at the West orebody in the Battle Mountain district. (Theodore and Blake, 1978)

The ore mineral in dolomite-hosted deposits at the Hancock (Mickey) and Uhalde–New World (Nevada Iron) Mines in the McCoy district. (Shawe et al., 1962)

LINCOLN COUNTY As splendid crystals in cavities in volcanic rocks on Blind Mountain with green *garnet,* clinochlore, and *pyroxene* in the Bristol district. (Melhase, 1934d)

Reported from unspecified localities in the Chief and Tem Piute districts. (LANH; JHS)

LYON COUNTY In veins in metavolcanic rock at the Owen prospect in the Wilson district. (Reeves et al., 1958)

In a small veinlike body in granodiorite at the Easter prospect at the Ludwig Mines in the Yerington district; also in skarn with chalcopyrite, pyrite, and other minerals at the Bear–Lagomarsino Airport deposit. (Melhase, 1935e; Reeves et al., 1958; Doebrich et al., 1996)

LYON/STOREY COUNTIES The primary iron ore mineral in irregular lenses and podlike bodies of contact-metamorphic origin with hematite, *limonite,* epidote, *garnet,* quartz, pyrite, mica, gypsum, and calcite at the Dayton Iron Mine in the Red Mountain district. (Reeves et al., 1958)

MINERAL COUNTY In contact-metamorphic deposits at the Black Butte, Gillis, Iron Crown, Last Chance, and Black Hawk prospects in the Fitting district. (Reeves et al., 1958)

Disseminated in epidotized metavolcanic rocks and concentrated along cross faults on both sides of U.S. Highway 95 at the Walker Lake prospect in the Mountain View district. (Reeves et al., 1958)

With hematite and unspecified copper minerals in a replacement deposit in limestone at the King David prospect in the Pamlico district. (Reeves et al., 1958)

With hematite along a contact between diorite and limestone in a copper prospect at the Foster prospect in the Red Ridge district. (Reeves et al., 1958)

In the Santa Fe district with hematite in skarn and limestone at the Iron Gate and Iron Butte Mines; also with smoky quartz and *feldspar* crystals in cavities in granite 7 miles east of Luning. (Reeves et al., 1958; UCB)

With copper sulfides at unspecified localities in the Silver Star and Whisky Flat districts. (WMK)

NYE COUNTY With hematite, *garnet,* epidote, and quartz along shear zones at the Engle-Stouder prospect in the Ellsworth district. (Reeves et al., 1958)

Principal ore mineral with minor hematite in a pit from which 400,000 tons of ore were mined at the Phelps-Stokes Mine in the Gabbs district, Sec. 21, T12N, R37E. (Reeves et al., 1958)

In veins with quartz and sulfide at the Hall Molybdenum Mine in the San Antone district. (Shaver, 1991)

Reported with dolomite 20 miles east of the Cuprite district at the Gossan Iron Mine. (WMK)

PERSHING COUNTY In massive pods in skarn at the Basalt prospect in the Copper Valley district. (Shawe et al., 1962)

The major iron ore mineral with local hematite in several irregular to tabular replacement bodies in scapolitized diorite and metavolcanic rocks with *chlorite,* calcite, *apatite,* titanite, and actinolite-tremolite in the Mineral Basin district; reported as lustrous, black, octahedral crystals to 1 cm at the American Ore Company Mines, and as euhedral crystals to 15 cm with fluorapatite crystals and actinolite in vugs at the Segerstrom-Heizer Mine. (Reeves and Kral, 1955; Shawe et al., 1962; Johnson, 1977; M. Jensen, w.c., 1999)

In gold placers with cassiterite, cinnabar, and scheelite in the Rabbit Hole district. (Johnson, 1977)

In large pods at the Basalt prospect in the Ragged Top district. (Johnson, 1977)

In placers with titanium-bearing minerals in the Scossa district. (Beal, 1963)

As massive bodies, veins, and disseminated grains with pyrite at unspecified localities in the Wild Horse district. (Johnson, 1977)

STOREY COUNTY Replacing shaly limestone at the Iron Blossom prospect in the Red Mountain district. (Reeves et al., 1958)

WASHOE COUNTY As irregular veinlets to 10 inches wide in diorite with local hematite, pyrite, and chalcopyrite at the Red Metal Mine in the Peavine district; also in skarn with copper-oxide minerals. (Hudson, 1977)

In massive veins and pods to 50 feet long in an area nearly half a mile square at the Black Hawk prospect in the State Line district. (Shawe et al., 1962; Bonham and Papke, 1969)

Common accessory as crude crystals to 3 mm in pegmatite in the Incline Village area. (Jensen, 1993a)

WHITE PINE COUNTY Reported from unspecified localities in the Robinson district. (Melhase, 1934d; JHS)

MALACHITE

$Cu_2^{2+}(CO_3)(OH)_2$

Malachite is an extremely common mineral, occurring as an oxidation product of primary copper minerals with azurite, chrysocolla, and other secondary copper minerals. It is found in variable quantities as stains and crusts in many mining districts in Nevada, and fine specimens have come from the Robinson, Santa Fe, Bristol, and Goodsprings districts.

CARSON CITY With other copper minerals at the Brunswick Consolidated Mining Co. in the Delaware district. (Schrader et al., 1917; WMK)

CHURCHILL COUNTY An alteration product of chalcopyrite at the Coppereid prospect in the Copper Kettle district. (GCF)

At the Fairview Mine in the Fairview district. (Schrader, 1947)

Reported from the Lovelock Mine in the Table Mountain district. (R. Thomssen, w.c., 1999)

With cerussite at the Westgate Mine in the Westgate district. (R. Walstrom, w.c., 1999)

Reported from the White Cloud and Holy Cross districts. (Ransome, 1909b; Schrader, 1947)

CLARK COUNTY With azurite, brochantite, and other secondary copper minerals at the Key West Mine in the Bunkerville district. (Beal, 1965)

The most abundant copper mineral in the Goodsprings district with galena, cerussite, hemimorphite, chrysocolla, aurichalcite, and wulfenite; it occurs as crusts and nodular masses of radiating crystals, as loosely coherent masses of small crystals, and as thin films or veinlets. At the Blue Jay Mine it has been found as perfect, centimeter-sized pseudomorphs after azurite crystals, and the Singer Mine as botryoidal balls to 2 mm on brecciated dolomite. (Hewett, 1931; CIT; Adams, 1998)

Reported from unspecified localities in the Eldorado, Searchlight, and Gold Butte districts. (WMK; Callaghan, 1939; Hill, 1916)

DOUGLAS COUNTY At the Ruby Hill Mine in the Gardnerville district. (Hill, 1915; S. Pullman, w.c., 1999)

Reported from unspecified localities in the Red Canyon and Wellington districts. (Hill, 1915)

ELKO COUNTY With other copper minerals, molybdenite, and *specularite* in silicified limestone at the Alice Mine in the Contact district; also in gossan at the Green Monster claim and in a vein at the Haskney-Shoun property. (Purington, 1903; Bailey, 1903; Schrader, 1912; Schilling, 1979; LaPointe et al., 1991)

As deep-green, delicate, fuzzy balls to 2 cm in diameter in quartz crystal–lined vugs in limonitic breccia; also in quartz veins with azurite, cerussite, and other minerals at the Victoria Mine in the Dolly Varden district. (Hill, 1916; Melhase, 1934d; Schilling, 1979; GCF)

With other copper minerals at the Red Elephant incline in the Elk Mountain district. (Schrader, 1912)

As stains in gossan at the Copper Valley prospect in the Ferguson Spring district. (Hill, 1916)

With azurite in a quartz-*adularia* vein at the Long Hike Mine in the Jarbidge district. (Schrader, 1923)

With azurite and *limonite* in a quartz vein at the Silver prospect in the Lafayette district. (LaPointe et al., 1991)

In skarn at the Lone Mountain Mine and Aljo claims in the Merrimac district; also in altered porphyry at the IM claims. (LaPointe et al., 1991)

Common as green stains, coatings, and masses with azurite at the Rio Tinto Mine in the Mountain City district. (Coats and Stephens, 1968)

With other copper minerals in skarn at the Blue Belle, Sylvania, and Hoffman Mines in the Railroad district. (Ketner and Smith, 1963; Granger et al., 1957; Hill, 1916)

Abundant in shallow oxidized ore at the Killie and Spence Mines in the Spruce Mountain district, where it coats limestone fractures and forms veins with cerussite. (Hill, 1916; Schilling, 1979)

With hemimorphite and minor scheelite as vug fillings, small lenses, and stringers in *garnet*-epidote skarn at the High Top claims in the Swales Mountain district. (Stager and Tingley, 1988)

Reported from unspecified localities in the Delker, Ferber, Kinsley, Loray, Ruby Valley, and White Horse districts. (Hill, 1916)

ESMERALDA COUNTY In a specimen from the Simon Mine in the Bell district. (R. Thomssen, w.c., 1999)

Reported at the Johnson Mine in the Crow Springs district. (WMK)

With azurite, cuprite, chalcocite, and chrysocolla in the Cuprite district. (Ball, 1907)

Reported at the Silver King group in the Dyer district. (Albers and Stewart, 1972; WMK)

As green coatings and stains in oxidized portions of quartz veins at the Bullfrog-George prospect in the Gold Point district. (Schilling, 1979)

Identified in a sample from the General Thomas Mine in the Lone Mountain district. (JHS)

Reported with silver-bearing galena at the Benton, Gold Crown, and Pandora Mines in the Montezuma district. (Schrader et al., 1917; WMK)

As minor green stains at the McNamara Mine in the Palmetto district. (Schilling, 1979)

In a specimen from the Big 3 Mine in the Railroad Springs district. (R. Thomssen, w.c., 1999)

Reported from the Broken Toe Mine in the Rock Hill district. (R. Thomssen, w.c., 1999)

In a specimen from the Weepah Mine in the Weepah district. (S. Pullman, w.c., 1999)

Reported from unspecified localities in the Coaldale, Goldfield, Gilbert, Good Hope, and Klondyke districts. (WMK; Schrader et al., 1917; Ferguson, 1928; Albers and Stewart, 1972; Chipp, 1969)

EUREKA COUNTY Reported as silver-bearing with cuprite in a specimen from an unspecified locality in the Antelope (Baldwin) district. (WMK)

As stout sprays to 1.5 cm with azurite at the Lord Byron and Eureka Consolidated Mines in the Eureka district. (Nolan, 1962; FUMT; BMC; WMK)

Light-green stains in diorite at the Goldstrike Mine in the Lynn district. (GCF)

Widespread as stains, thin coatings, and small acicular crystals at the Gold Quarry Mine in the Maggie Creek district; also as bright-green veins at the Copper King Mine. (Jensen et al., 1995; Rota, 1991)

With galena and sphalerite from an unspecified locality in the Mineral Hill district. (WMK)

Reported from Sheep Creek Canyon. (Muffler, 1964)

HUMBOLDT COUNTY As coatings, spheroids, and fibrous crystals at the Silver Coin Mine in the Iron Point district. (Thomssen, 1998; J. McGlasson, w.c., 2001)

In quartz veins at the Desert View prospect in the Leonard Creek district. (Schilling, 1979)

With bornite at the Getchell and Felex Mines in the Potosi district. (WMK; D. Tretbar, o.c. 1999)

Reported from unspecified localities in the Jackson Mountains and Red Butte districts. (WMK; Willden, 1963; Ransome, 1909b)

LANDER COUNTY Reported in massive and mammillary forms from Copper Canyon in the Battle Mountain district. (Hoffman, 1878; Schilling, 1979)

Reported as a replacement of tetrahedrite at an unspecified locality in the Big Creek district. (Lawrence, 1963)

Reported from the Tenabo area in the Cortez district. (Gianella, 1941)

At the Betty O'Neal Mine and in the Eagle vein, Lewis district. (Vanderburg, 1939; J. McGlasson, w.c., 2001)

Reported from unspecified localities in the Hilltop, Ravenswood, and Washington districts. (WMK; Hill, 1916)

LINCOLN COUNTY Reported from the Bristol Silver, Cave, Ida May, Jackrabbit, and May Day Mines in the Bristol district with cuprite, atacamite, tenorite, and chrysocolla; the district was the source of some of the state's best specimens of the species as emerald-green, botryoidal, crystalline crusts on *limonite* and stout fibrous crystals to 2 cm. (Hill, 1916; Melhase, 1934d; UMNH; CSM; M. Jensen, w.c., 1999; WMK)

Reported with azurite, chrysocolla, and *psilomelane* from a locality in the Pioche district. (WMK)

As stains, coatings, small pods, and lenses with tenorite, chalcocite, and chrysocolla at the Arrowhead (Southeastern) Mine in the Southeastern district. (Tingley et al., 1998a)

Reported from unspecified localities in the Atlanta and Patterson districts. (Hill, 1916)

LYON COUNTY In quartz-calcite veins with other minerals 10 miles north of Schurz on the south end of Painted Mesa in the Benway district. (Schrader, 1947)

Reported from many properties in the Yerington district, including the Dyer, Ludwig, Empire-Nevada (Yerington Pit), Mason Valley, and Douglas Hill Mines. (Ransome, 1909e; JHS; WMK; S. Pullman, w.c., 1999; R. Thomssen, w.c., 1999)

MINERAL COUNTY In oxidized portions of a shear zone with quartz, *limonite,* chlorargyrite, jarosite, anglesite, chrysocolla, and silver at the Mabel Mine in the Garfield district. (Couch and Carpenter, 1943)

As stains in quartz-calcite veins with azurite at the Lucky Boy Mine in the Lucky Boy district. (Lawrence, 1963)

Reported at the Northern Lights Mine in the Mountain View district. (Ross, 1961; S. Pullman, w.c., 1999)

Reported at the Copper Mountain group in the Rand district. (Schrader, 1947; WMK)

Fibrous green crystals to 2 mm along fractures in skarn at the Moonlight Mine near the Nevada Scheelite Mine in the Leonard district; also noted at the Sunny Side Mine. (M. Jensen, w.c., 1999; Schrader, 1947; WMK)

With azurite, chrysocolla, and cuprite in veins up to 30 feet thick; also in contact-metamorphosed zones; specifically reported at the Del Monte, Good Enough, Wallstreet, and Turk Mines in the Santa Fe district. (Clark, 1922; Ross, 1961; WMK)

With chrysocolla and possible chalcopyrite in a marble roof pendant at the western base of the southern Monte Cristo Mountains. (Kleinhampl and Ziony, 1984)

Reported from the Fermina Mine in the Santa Fe district. (R. Thomssen, w.c., 1999)

Reported from unspecified localities in the Candelaria, Eagleville, Leonard, Marietta, Pilot Mountains, and Silver Star districts. (Ross, 1961; Schrader, 1947; Schrader et al., 1917; Foshag, 1927)

NYE COUNTY At the Antelope View Mine in the Antelope Springs district on the Nellis Air Force Range. (Tingley et al., 1998a)

In a quartz vein at the Iron Rail group near the western base of Mt. Ardivey in the Cloverdale district. (Hamilton, 1993)

As radiating fibers to 3 mm with cerussite on quartz crystals in vugs at the Big Chief Mine in the Ellsworth district. (GCF)

With azurite and manganese oxides in a quartz vein at the Sailor Boy Mine in the Jefferson Canyon (Jefferson) district. (Kleinhampl and Ziony, 1984)

Reported at the Dollar and Last Chance Mines in the Jett district. (Lawrence, 1963)

On fractures with azurite at the Northumberland Mine in the Northumberland district. (Kokinos and Prenn, 1985)

On dumps at the Antimonial and Gila Mines in the Reveille district. (Ball, 1907; Lawrence, 1963)

With azurite, pyrite, and galena at the Florence Mine in the San Antone district. (Kleinhampl and Ziony, 1984)

As delicate acicular masses in the Mizpah vein in the Tonopah district; also with chlorargyrite and coated with transparent calcite on the Tonopah Mining Co. property. (Bastin and Laney, 1918; Melhase, 1935c)

With azurite and other minerals in pods in silicified dolomite just south of the summit of Rawhide Mountain at the Lloyd Trickey claim in the Tybo district. (Kleinhampl and Ziony, 1984)

With conichalcite, azurite, *melaconite,* and

chrysocolla in the Wagner district. (Tingley et al., 1998a)

Reported from unspecified localities in the Bare Mountain, Belmont, Bullfrog, Gabbs, Lodi, Manhattan, Morey, Oak Spring, Silverbow, Troy, Tybo, Union, and Willow Creek districts. (Melhase, 1935b; Jenkins, 1981; Ball, 1907; Hill, 1916; JHS)

PERSHING COUNTY With azurite as green stains and botryoidal crusts of radiating, light-green needles to 3 mm at the Majuba Hill Mine in the Antelope district. (Jensen, 1985)

Reported 15 miles east of Lovelock in the Antelope Springs district. (MacKenzie and Bookstrom, 1976)

Reported from unspecified localities in the Buena Vista, Indian, Iron Hat, Juniper Range, Ragged Top, and Table Mountain districts. (Cameron, 1939; Johnson, 1977; Ferguson, 1939)

STOREY COUNTY Reported from an unspecified locality near Clark Station in the Clark district. (Rose, 1969)

Wood molded in malachite and other copper minerals was found in an open cut at the Sierra Nevada Mine in the Comstock district. (UCB)

WASHOE COUNTY With cerussite, smithsonite, hemimorphite, and chalcanthite in oxidized ore at the Union (Commonwealth) Mine and other localities in the Galena district. (Geehan, 1950; Bonham and Papke, 1969)

With bornite, chalcanthite, cornetite, and rare chalcocite at the Red Metals Mine in the Peavine district; also with azurite, brochantite, enargite, and other copper minerals in quartz-*tourmaline* veins at the Redelius prospects. (Hill, 1915; Overton, 1947; Hudson, 1977; WMK)

On enargite and as green coatings, botryoids, and fibers in the oxidized part of veins and adjacent rocks at the Burrus and other mines in the Pyramid district. (Overton, 1947; Bonham and Papke, 1969; Jensen, 1994)

With azurite and tenorite at the Antelope Mine in the Stateline Peak district. (Bonham and Papke, 1969)

With azurite, brochantite, and other copper minerals in low hills just north of Sparks in the Wedekind district. (Hudson, 1977)

WHITE PINE COUNTY As dark-green botryoidal crusts with azurite, chalcopyrite, and gold in pods of silicified rock in the Top pit at the Bald Mountain Mine in the Bald Mountain district. (SBC)

Reported with galena and malachite at the Star Mine in the Cherry Creek district. (WMK; JHS)

With secondary lead minerals 10 miles northwest of the Steptoe post office in the Hunter district. (Melhase, 1934d)

Reported in oxidized ore at the Alpha Mine in the Robinson district; also found locally as fibrous crusts in open pits such as the Liberty pit. (Spencer et al., 1917; WMK; SBC)

With galena, oxidized lead and zinc minerals, chrysocolla, and cuprite in the Mount Hamilton area in the White Pine district. (Humphrey, 1960)

Reported from unspecified localities in the Duck Creek, Muncy Creek, and Ward districts. (Hill, 1916)

MANDARINOITE

$Fe_2^{3+}Se_3O_9 \cdot 6H_2O$

A rare secondary species that has been found with other selenium-bearing minerals.

EUREKA COUNTY Rare as elongated individuals and radiating clusters of transparent yellow-green crystals in brecciated jasperoid with strengite and barite at the Gold Quarry Mine in the Maggie Creek district. (Jensen et al., 1995)

WASHOE COUNTY As lustrous green crystals to 3 mm, possibly the world's best, in leached vuggy rock with kaolinite, alunite, and native sulfur along the Wind Mountain fault at the Wind Mountain Mine in the San Emidio district. (Wood, 1991; M. Jensen, w.c., 1999)

MANGANAXINITE

$Ca_2Mn^{2+}Al_2BSi_4O_{15}(OH)$

A member of the *axinite* group; typically in contact-metamorphic rocks or pegmatite.

MINERAL COUNTY Reported from a locality near Luning with prehnite, vesuvianite, actinolite, and epidote; ferro-axinite and magnesio-axinite occur in the same deposit. (Dunn et al., 1980)

MANGANITE

$Mn^{3+}O(OH)$

Manganite, a relatively common manganese mineral, is formed in low-temperature hydrothermal deposits, in sedimentary manganese deposits, and as a product of surface weathering.

CLARK COUNTY Rarely as crystals with other manganese minerals in tabular bodies at the Three Kids Mine in the Las Vegas district, Secs. 35 and 36, T21S, R63E. (Hewett and Webber, 1931; Longwell et al., 1965)

MINERAL COUNTY Reported in oxidized ore as lustrous black crystals to 2 mm on fractures at the Candelaria Mine in the Candelaria district. (Chavez and Purusotam, 1988; M. Jensen, w.c., 1999)

NYE COUNTY Abundant in the zone of oxidation as vertically striated rods or needles in pockets in the Tonopah district. (Melhase, 1935c)

After sooty manganese oxide with high silver contents with quartz and calcite in a shear zone in limestone at the Larsh prospect about one-half mile west of Tybo in the Tybo district. (Ferguson, 1933)

MANGANOCOLUMBITE

$(Mn^{2+},Fe^{2+})(Nb,Ta)_2O_6$

A member of the *columbite* group that mostly occurs as a primary mineral in granite pegmatite and is rarely found in placer deposits.

MINERAL COUNTY A rare mineral in the Zapot pegmatite in the Fitting district. (Foord et al., 1999b)

MANGANOTANTALITE

$Mn^{2+}Ta_2O_6$

An accessory and primary mineral in granite pegmatites.

CLARK COUNTY As microcrystals in miarolitic cavities and pegmatites in the Eldorado and Newberry Mountains. (S. Scott, w.c., 1999)

MANSFIELDITE

$AlAsO_4 \cdot 2H_2O$

A rare mineral that has been found with other arsenic-bearing species such as realgar and scorodite.

ELKO COUNTY As massive white to light-green material mixed with variscite at the Rain Mine in the Carlin district. (D. Heitt, o.c., 1991)

EUREKA COUNTY Tentatively identified in phosphate veins in the upper portions of the deposit at the Goldstrike Mine in the Lynn district. (GCF)

MARCASITE

FeS_2

A low-temperature mineral in hydrothermal veins and in sedimentary rocks. Dimorphous with pyrite.

CHURCHILL COUNTY With nickel minerals at the Nickel Mine in the Table Mountain district. (Ransome, 1909b)

ELKO COUNTY Disseminated throughout the barite deposit at the Rossi Mine in the Bootstrap district as a partial replacement of pyrite. (Snyder, 1989)

As large tabular crystals in veins and as disseminated grains with pyrite at the Jerritt Canyon and Big Springs Mines in the Independence Mountains district. (Youngerman, 1992; Hofstra, 1994)

Common as an intergrowth with pyrite, in places with *electrum,* at the Hollister deposit in the Ivanhoe district. (Deng, 1991)

With pyrite, realgar, and stibnite in gold ore at the Pony Creek deposit on the crest of the southern Pinon Range about halfway between Robinson Mountain and Pony Creek in the Larrabee district. (Westmont Mining Inc., unpublished data, 1990)

Common in ore at the Meikle Mine in the Lynn district (Jensen, 1999)

Reported at the Red Cow prospect in the Rock Creek district. (Lawrence, 1963)

ESMERALDA COUNTY As late-stage mammillary crusts and concentric, radially fibrous shells in ore with gold and quartz in the Goldfield district; specifically noted at the Combination, Gold Bar, Reilly, Red Top, Florence, and Mohawk Mines. (Ransome, 1909a; LANH; HMM)

EUREKA COUNTY Reported from a locality about 5 miles east of Cortez in the Buckhorn district. (JHS)

As metallic gray to brassy stalactitic deposits, disseminated masses with pyrite, and crystals in silicified breccia zones at the Goldstrike Mine in the Lynn district; also noted as radiating sprays to 3 mm of spearlike crystals after millerite on sphalerite. (GCF)

Rarely as spherules in sulfide-rich silicified barite breccias, locally with vanadinite and descloizite, at the Gold Quarry Mine in the Maggie Creek district. (Jensen et al., 1995)

HUMBOLDT COUNTY With pyrite in gold-silver quartz veins at the Sleeper Mine in the Awakening district. (Saunders, 1994)

As fine-grained particles, lath-shaped crystals to 3 mm that are partially replaced by pyrite, and veins at the Lone Tree Mine in the Buffalo Mountain district. (Kamali and Norman, 1996; Panhorst, 1996)

A minor component in radioactive breccia from the Moonlight Uranium Mine in the Disaster district. (Dayvault et al., 1985)

An accessory in a magnetite orebody at the Delong (Iron King) Mine in the Jackson Mountains district. (Shawe et al., 1962)

As a secondary sulfide along with stibnite and pyrargyrite in fractures and joints in quartz veins at the National Mine in the National district. (Lindgren, 1915; Melhase, 1934c)

Common as disseminated grains, crystals in vugs, and veins to 20 cm thick in rhyolite at the Cordero Mine in the Opalite district. (Fisk, 1968)

In botryoidal form with stibnite and cinnabar on microcrystalline quartz in the south pit of the Getchell Mine in the Potosi district; also at the Twin Creeks Mine as inclusions in pyrite. (Joralemon, 1951; Simon et al., 1999; F. Cureton, w.c., 1999)

LANDER COUNTY In fractures and as small disseminated grains at the Copper Canyon Mine in the Battle Mountain district; also in skarn at the Fortitude Gold Mine. (Theodore and Blake, 1978; Schilling, 1979; Doebrich, 1993)

In stockwork veins with pyrite, sphalerite, galena, and other sulfides at the Cove Mine in the McCoy district. (Emmons and Coyle, 1988)

Reported with pyrite, quartz, and chalcocite from an unspecified locality in the Hilltop district. (WMK)

LINCOLN COUNTY Identified in a specimen from the Lincoln Tungsten Mine in the Tem Piute district. (JHS)

NYE COUNTY In unoxidized gold ore at the Sterling Mine in the Bare Mountain district. (J. Marr, o.c., 1996)

As botryoidal coatings at the West End Mine in the Tonopah district. (JHS)

After pyrrhotite at the Tybo Mine in the Tybo district. (Ferguson, 1933)

PERSHING COUNTY Locally in iron ore in the Mineral Basin district. (Reeves and Kral, 1955)

As six-sided plates, generally with barite, at the Hecla Rosebud Mine in the Rosebud district; considered by some to be pseudomorphs after hexagonal pyrrhotite crystals. (N. Prenn, o.c., 1999; M. Jensen, o.c., 2000)

WASHOE COUNTY Reported as crystals in a specimen from the Perry Mine in the Pyramid district. (WMK)

Reported in a specimen from the Steamboat Springs district. (LANH)

WHITE PINE COUNTY With other sulfides in unoxidized gold ore in the RBM deposit at the Bald Mountain Mine in the Bald Mountain district. (Hitchborn et al., 1996)

MARIALITE

$3NaAlSi_3O_8 \cdot NaCl$

Sodium-rich member of the *scapolite* group that typically occurs in regionally metamorphosed rocks and skarns.

EUREKA COUNTY As cream-white fibers to 1 cm with calcite at the Gold Quarry Mine in the Maggie Creek district. (M. Jensen, w.c., 1999)

LYON COUNTY In a specimen from an unspecified locality in the Yerington district. (HMM)

MINERAL COUNTY As opaque, white, elongated crystals to 2 cm in skarn in prospects near the top of Indian Head Peak in the Fitting district. (M. Jensen, w.c., 1999)

WASHOE COUNTY With *chlorite,* quartz, actinolite, epidote, and albite in wall rocks of magnetite veins at the Red Metal Mine in the Peavine district. (Hudson, 1977)

MAROKITE

$CaMn_2^{3+}O_4$

A mineral found in hydrothermal deposits with other manganese-oxide species.

MINERAL COUNTY Reported as one of a suite of manganese-oxide minerals at the Candelaria Mine in the Candelaria district. (Chavez and Purusotam, 1988)

MASSICOT

PbO

Massicot, a relatively rare mineral, occurs with other lead minerals in the oxidation zones of lead deposits.

CLARK COUNTY As yellow coatings on galena at the Argentena and Monte Cristo Mines in the Goodsprings district. (TUB; CIT)

ESMERALDA COUNTY Tentatively identified as a yellow oxide at the Commonwealth Mine in the Goldfield district (Ransome, 1909a)

EUREKA COUNTY Reported at the Richmond and Windfall Mines in the Eureka district. (LANH; UCB)

HUMBOLDT COUNTY Yellow-orange crystals of massicot have been found growing on the surface of discarded assay sheet lead in the waste dump at the Getchell Mine in the Potosi district. (Scott, 1974)

MINERAL COUNTY As yellow scaly plate, possibly artificial, on specimens from an unspecified site. (CIT)

NYE COUNTY Reported in a specimen from Belmont in the Belmont district. (AMNH)

PERSHING COUNTY As yellow-orange, locally banded crusts on quartz from the mine dumps at the Queen of Sheba and Silver Creek Mines in the Star district. (CIT)

MATILDITE

$AgBiS_2$

A high-temperature sulfosalt mineral.

MINERAL COUNTY Reported in the Freedom Flats deposit at the Borealis Mine in the Borealis district. (Eng, 1991)

NYE COUNTY With benjaminite and aikinite at the Outlaw Mine in the Round Mountain dis-

trict. (Nuffield, 1953, 1975; Karup-Møeller, 1972, 1977; Karup-Møeller and Makovicky, 1979)

MATLOCKITE

PbFCl

Matlockite is a rare mineral that occurs in the oxidation zones of lead deposits.

ELKO COUNTY Doubtfully reported at the Killie (Latham) Mine in the Spruce Mountain district. (Jenkins, 1981)

MATULAITE

$CaAl_{18}(PO_4)_{12}(OH)_{20} \cdot 28H_2O$

A rare mineral that has been found elsewhere with other phosphate species in iron deposits and in a pegmatite.

MINERAL COUNTY In pearly, white, translucent microspherules at the Candelaria Mine in the Candelaria district. (UMNH)

MCGUINNESSITE

$(Mg,Cu^{2+})_2(CO_3)(OH)_2$

Mcguinnessite is a rare carbonate mineral that was recently described from a locality in California. The Gabbs occurrence was the second reported world locality.

CHURCHILL COUNTY As turquoise-blue spherules to 0.5 mm with hydromagnesite and magnetite at the Chalk Mountain Mine and Dixie No. 5 and Corey No. 2 claims in the Chalk Mountain district; with hemimorphite at the Skarn No. 1 claim. (F. Cureton, w.c., 1999; M. Jensen, w.c., 1999; R. Walstrom, w.c., 1999; R. Thomssen, w.c., 1999; GCF)

As 0.5-mm, light-blue, botryoidal masses with wulfenite at the Westgate Mine in the Westgate district. (R. Thomssen, w.c., 1999; GCF)

NYE COUNTY As silky, robin's-egg-blue, translucent microspherules with hydromagnesite and aragonite, and as powdery-blue coatings of minute acicular crystals on dolomite or *serpentine* in one of the abandoned brucite pits at the Premier Chemicals (Gabbs Magnesite-Brucite) Mine in the Gabbs district. (Oswald and Crook, 1979; UMNH)

MCKINSTRYITE

(Ag,Cu)$_2$S

A low-temperature hydrothermal species that occurs in veins with other sulfide minerals; difficult to distinguish from stromeyerite in the absence of accurate chemical analyses.

NYE COUNTY Tentatively identified as small grains with enargite, cinnabar, and chalcocite in a sample from a silicified ledge in the Gold Crater district on the Nellis Air Force Range. (Tingley et al., 1998a)

MEIONITE

3CaAl$_2$Si$_2$O$_8$•CaCO$_3$

Calcium end member of the *scapolite* group; forms a series with marialite.

PERSHING COUNTY Identified with coulsonite in a specimen from the Buena Vista Hills in the Mineral Basin district. (EMP)

MELANTERITE

Fe^{2+}SO$_4$•7H$_2$O

Melanterite is a common efflorescent mineral in mines in which pyrite has been oxidized. It may be enriched in copper, zinc, or magnesium; *pisanite* is a copper-bearing variety.

ESMERALDA COUNTY In small amounts as bluish-green veinlets and silky white efflorescences in the lower part of the oxidation zone, in places with halotrichite, at the Combination, Red Top, Reilly, Florence, and Mohawk Mines in the Goldfield district. (Ransome, 1909a; Ashley and Albers, 1975; WMK; S. Pullman, w.c., 1999)

In a specimen with pyrite and coquimbite from an unspecified locality in the Lida district. (WMK)

EUREKA COUNTY As silvery-white tufts on sulfide-rich rock at the Goldstrike Mine in the Lynn district. (GCF)

HUMBOLDT COUNTY As greenish crusts formed by the oxidation of marcasite as soon as mine workings are opened at the Cordero Mine in the Opalite district. (Fisk, 1968; WMK)

In sea-green botryoids at the Getchell Mine in the Potosi district. (S. Sears, w.c., 1999)

LANDER COUNTY As efflorescences and spectacular stalactites and stalagmites in various mines in the Battle Mountain district; reported as *pisan-*

ite efflorescences in some mines. (Roberts and Arnold, 1965)

With calcite, quartz, azurite, realgar, and other minerals in gold ore at the Gold Acres Mine in the Bullion district. (Hays and Foo, 1991)

Reported as *pisanite* on mine walls at the Watt Mine in the Reese River district. (Ross, 1953)

LINCOLN COUNTY Reported from a locality that may be in the Pioche district. (Palache et al., 1951)

NYE COUNTY With apjohnite and alunogen on mine walls at the Urania Mine in the Cactus Peak district on the Nellis Air Force Range. (J. Price, o.c., 1999)

Found as long, slender, green stalactites formed by post-mining oxidation on the 210- and 310-foot levels of the White Caps Mine in the Manhattan district. (Ferguson, 1924; Foshag and Clinton, 1927)

Copper-bearing with copiapite, chlorargyrite, bindheimite, and other minerals in oxidized silver ore in the Morey district. (Jenkins, 1981)

PERSHING COUNTY Massive *pisanite* was identified in the matrix of a specimen from the Majuba Hill Mine in the Antelope district. (NYS)

As blue and green coatings on vein material at the Pflum Mine in the Star district. (Lawrence, 1963)

STOREY COUNTY In a specimen from an unknown locality in the Castle Peak district. (WMK)

Reported at the Sutro Tunnel in the Comstock district. (Palache et al., 1951)

WASHOE COUNTY Reported in underground workings at the Red Metal Mine in the Peavine district. (Hudson, 1977)

One of the sulfate minerals in sinter deposits at Steamboat Hot Springs in the Steamboat Springs district. (Jenkins, 1981)

In many underground workings in the Wedekind district as an oxidation product of pyrite. (Hudson, 1977)

WHITE PINE COUNTY Reported in a specimen from an unknown locality in the Duck Creek district. (WMK)

Alone or with chalcanthite as efflorescences in a number of older mine workings in the Robinson district. (Spencer et al., 1917)

MENDOZITE

$NaAl(SO_4)_2 \cdot 11H_2O$

A relatively uncommon water-soluble mineral that occurs in evaporite deposits and as an efflorescent phase in mine workings.

ESMERALDA COUNTY Reported in a specimen from the Fish Lake Marsh district. (RMO)

MENEGHINITE

$Pb_{13}CuSb_7S_{24}$

Meneghinite is a rare mineral that occurs in some base-metal deposits with other lead-antimony sulfosalts.

EUREKA COUNTY As minute (<0.1 mm) silvery-gray exsolution lamellae in galena and bournonite in base-metal veins in diorite at the Goldstrike Mine in the Lynn district. (GCF)

HUMBOLDT COUNTY In superb specimens of wooly masses at the National Mine in the National district. (Melhase, 1934c)

MINERAL COUNTY Doubtfully reported in silver-lead ore in the Broken Hills district. (Schrader, 1947)

PERSHING COUNTY Rare as gray, prismatic-crystalline masses and microcrystals with cinnabar at the Red Bird Mine in the Antelope Springs district. (F. Cureton, w.c., 1999)

Reported with miargyrite at the Queen of Sheba Mine in the Star district. (J. McGlasson, w.c., 2001)

MERCURY

Hg

Native mercury, a relatively uncommon native metal, generally occurs in small drops as a supergene alteration product of cinnabar. It is also associated with mercury chlorides and oxychlorides.

CLARK COUNTY With cinnabar at the Patsy Mine in the Eldorado district. (Longwell et al., 1965)

ELKO COUNTY With cinnabar, marcasite, and other minerals at an unspecified locality in fracture zones in altered andesite in the Tuscarora district. (Bailey and Phoenix, 1944)

HUMBOLDT COUNTY Reportedly panned from the soil at the Stall prospect in the National district. (Bailey and Phoenix, 1944)

At the McDermitt Mine in the Opalite district as isolated blebs and globules in extremely rich corderoite-cinnabar ore with calomel and eglestonite; at the nearby Cordero Mine as tiny disseminated droplets with cinnabar and mercury oxychlorides. (Bailey and Phoenix, 1944; Fisk, 1968; Roper, 1976; McCormack, 1986)

With cinnabar, other mercury minerals, stibnite, calcite, quartz, and gypsum in sandstone and limestone at the Cahill Mine in the Poverty Peak district. (F. Cureton, w.c., 1999)

LANDER COUNTY Abundant with cinnabar as disseminations in limestone and quartzite at the Rast prospect in the Callaghan Ranch district. (Bailey and Phoenix, 1944)

LINCOLN COUNTY Reported at the Homestake and other mines in the Eagle Valley district. (Tschanz and Pampeyan, 1970)

MINERAL COUNTY Reported from an unspecified locality in the Fitting district. (S. Pullman, w.c., 1999)

NYE COUNTY With cinnabar from an unspecified locality in the Union district. (Bailey and Phoenix, 1944)

PERSHING COUNTY In mercury-antimony ore at the Red Bird Mine in the Antelope Springs district. (Bailey and Phoenix, 1944)

Reported by miners in small amounts at the Goldbanks Quicksilver Mine in the Goldbanks district. (Dreyer, 1940)

With cinnabar in calcite veins at the Cinnabar King, King George, and Walker Mines in the Spring Valley district. (Bailey and Phoenix, 1944)

STOREY COUNTY Abundant with calomel and cinnabar at the Castle Peak Mine in the Castle Peak district. (Bailey and Phoenix, 1944; Holmes, 1965)

WASHOE COUNTY With jarosite and cinnabar at the Wheeler Ranch prospect in the Olinghouse district. (Overton, 1947)

Detected as vapor from geothermal wells in the Steamboat Springs district; also as thin coatings on sulfur crystal masses at the current location of the geothermal plant and reported at the Steamboat Sulfur Mine and Theeler Ranch prospect. (Bailey and Phoenix, 1944; Brannock et al., 1948; Wilson and Thomssen, 1985; B. Hurley, w.c., 1999)

MERENSKYITE

$(Pd,Pt)(Te,Bi)_2$

A mineral reported mainly from magmatic deposits with a variety of minerals that contain platinum group metals.

CLARK COUNTY Reported at the Key West Mine in the Bunkerville district, Sec. 22, T15S, R70E. (S. Pullman, w.c., 1999)

Reported with pyrite in a specimen from Moapa; however, this specimen is probably from the Key West Mine locality noted above. (HUB)

META-ANKOLEITE

$K_2(UO_2)_2(PO_4)_2 \cdot 6H_2O$

A secondary uranium mineral that was originally identified in a pegmatite occurrence.

HUMBOLDT COUNTY Identified by SEM/EDX as crystals to 20 microns with autunite along fracture coatings or thin layers in radioactive opal beds near the Virgin Valley Ranch in the Virgin Valley district; also in spherical cavities in tuffaceous lake beds with carbonaceous material in the same area. (Castor et al., 1996; Castor and Henry, 2000)

META-AUTUNITE

$Ca(UO_2)_2(PO_4)_2 \cdot 2-6H_2O$

Meta-autunite, one of the most common secondary minerals of uranium, is formed by dehydration of autunite at ordinary conditions of temperature and humidity. It is likely that most specimens reported as autunite in collections in Nevada are actually of meta-autunite, although possibly they were the higher hydrate when collected.

CARSON CITY On skarn at the 8-Spot and Lucky Bird claims in the Carson City district. (Garside, 1973)

Reported in Tertiary ash-flow tuff at the Sally claim group in the Delaware district. (Garside, 1973)

CHURCHILL COUNTY In Tertiary rhyolite at the Patriot claim group. (Garside, 1973)

CLARK COUNTY Reported at the Yellow Jacket claims. (Garside, 1973)

In conglomerate at the Lucky Bart prospect in the Gold Butte district. (Garside, 1973)

DOUGLAS COUNTY In a breccia zone in granitic rocks at the Julietta prospect in the Delaware district. (Garside, 1973)

ELKO COUNTY Reported in a fault zone in limestone west of Montello in the Montello district. (Garside, 1973)

Reported at several localities in the Mountain City district. (Garside, 1973)

As sparse, flaky crystals on pegmatite in the southern Ruby Mountains. (Olson and Hinrichs, 1960)

ESMERALDA COUNTY With phosphuranylite in crusts and cavity fillings in rhyolitic welded tuff in the Coaldale prospect in the Coaldale district; also reported at the Gap Strike claim. (Garside, 1973)

Along a fault zone with secondary copper minerals in the Precambrian rock at the Red Rock claims in the Tokop district; also in a kaolinized shear zone at the Randolph Mine. (Garside, 1973)

Reported at the Esmeralda Uranium claims in the Windypah district. (Garside, 1973)

HUMBOLDT COUNTY With ilsemannite and pascoite on phyllite from the east wall of the south pit at the Getchell Mine in the Potosi district. (Garside, 1973)

Reported on a number of claims in the Moonlight Mine area in the Disaster district. (Garside, 1973)

LANDER COUNTY As tabular crystals to 3 mm on iron-stained quartz with uraninite, metatorbernite, other uranium minerals, pyrite, bornite, and chalcopyrite at the Apex (Rundberg) Mine, Sec. 1, T18N, R43E, in the Reese River district; also reported from the Bulldog Mine. (Garside, 1973; UCB; AMNH)

LINCOLN COUNTY With metatorbernite at the Peak claims in the Eagle Valley district. (Tschanz and Pampeyan, 1970)

Reported at the Tem Piute Tungsten Mine in the Tem Piute district. (Garside, 1973)

LYON COUNTY Reported at the Lava Cap, Glacier King, and Flyboy claims in the Yerington district. (Garside, 1973)

In decomposed granite at the Quartz Mine, and on the Halloween and Silver Pick claim groups in the Washington district. (Garside, 1973)

MINERAL COUNTY Reported at the Dixie claims in the Fitting district. (Garside, 1973)

Reported with uranophane at the Blue Boy and Marietta claims in the Marietta district; also reported with kasolite and metatorbernite at the Lucky Horseshoe claim. (Garside, 1973)

Reported at the William Johnson claims in the Red Ridge district. (Garside, 1973)

NYE COUNTY With uranophane at an unspecified locality in the Bullfrog district. (Garside, 1973)

With metatorbernite in Paleozoic shale in the Northumberland district. (Garside, 1973)

In fracture zones around the perimeter of a granodiorite intrusion in the Round Mountain and Belmont areas in the Toquima Range. (Garside, 1973)

In tuffaceous rocks at several localities in the Tonopah district. (Garside, 1973)

PERSHING COUNTY Reported on several claim blocks in the Nightingale and Selenite Ranges. (Garside, 1973)

WASHOE COUNTY Reported from prospects in the Dogskin Mountain, Seven Lakes Mountain, Peterson Mountain, Spanish Springs, and Hungry Valley areas north of Reno. (Garside, 1973)

As scaly aggregates in fractures and cavities with sabugalite, phosphuranylite, uraninite, uranospinite, and uranophane at the DeLongchamps Mine in the Pyramid district; also reported at the Red Bluff Mine and other localities. (Brooks, 1956; Bonham and Papke, 1969; Garside, 1973; Castor et al., 1996)

As tabular subhedral crystals to 0.5 mm on fractures in cuts in felsic ash-flow tuff at the Buckhorn and Bastain prospects in the Stateline Peak district. (Garside, 1973; B. Hurley, w.c., 1999)

WHITE PINE COUNTY Reported at the Ruggles Leader claims in the Telegraph district. (Garside, 1973)

METACINNABAR

HgS

Metacinnabar, a dark mineral that is dimorphous with and generally associated with cinnabar, occurs in some mercury deposits in Nevada.

ELKO COUNTY Reported at the Mountain King Mine on Wyoming Hill in the Ivanhoe district. (UCB)

HUMBOLDT COUNTY With cinnabar in thin veins and fractures in a blanketlike deposit of chalcedonic silica at the top of Buckskin Mountain in the National district. (Vikre, 1985)

Reported in a specimen from the Getchell Mine in the Potosi district. (LANH)

MINERAL COUNTY Reported at the Keg prospect in the Pilot Mountains district at the head of Cinnabar Canyon. (Foshag, 1927; Bailey and Phoenix, 1944)

NYE COUNTY Reported in a specimen from an unknown locality in the Bare Mountain district. (WMK)

As coarse crystals with cinnabar and pyrite in quartz-barite veins in the Belmont district. (Bailey and Phoenix, 1944)

As scattered crystals in quartz veins with cinnabar, jarosite, and other minerals at the Mariposa Canyon and Senator Mines in the Round Mountain district. (Bailey and Phoenix, 1944)

PERSHING COUNTY Reported in a specimen from the Red Bird Mine in the Antelope Springs district. (LANH)

WASHOE COUNTY Reported at Steamboat Springs in the Steamboat Springs district. (Wilson and Thomssen, 1985)

METADELRIOITE

$CaSrV_2^{5+}O_6(OH)_2$

A very rare mineral that occurs elsewhere only as an efflorescent phase on sandstone on a dump at a uranium mine in Colorado.

LINCOLN COUNTY Reported in a specimen from Pahranagat. (EMP)

METAHEWETTITE

$CaV_6^{5+}O_{16} \bullet 3H_2O$

A rare mineral occurring as a product of the surface oxidation of primary vanadium-bearing substances such as sulfides or organics.

EUREKA COUNTY As yellow, pink, orange, and reddish-brown crusts and masses with minyulite on black, metalliferous, organic shale at the Gibellini and Bisoni (Bisoni-McKay) claims in the Gibellini district west and southwest of the Gibellini Manganese Mine in the Cockalorum Wash Quadrangle. (Desborough et al., 1979; Zientek et al., 1979; R. Walstrom, w.c., 1999; JHS)

METASCHODERITE

$Al_2(PO_4)(VO_4) \cdot 6H_2O$

A Nevada type mineral, metaschoderite occurs as a dehydration product of schoderite in oxidized vanadium-rich oil shale. Only three occurrences are known worldwide.

EUREKA COUNTY Type locality as yellow-orange crusts and coatings of minute crystals with schoderite, wavellite, and vashegyite in phosphatic chert at the Van-Nav-San claim group near the Gibellini Manganese Mine in the Gibellini district, Sec. 34, T16N, R52E. (Hausen, 1960, 1962; Zientek et al., 1979; DeMouthe, 1985)

METASIDERONATRITE

$Na_2Fe^{3+}(SO_4)_2(OH) \cdot H_2O$

A rare secondary mineral that generally occurs with other sulfate species.

EUREKA COUNTY As orange to yellow-orange, adamantine, radiating, acicular crystal clusters to 2.5 cm in sulfur with an unidentified green micaceous mineral at the Hot Springs Point Sulfur Mine 5 miles east-southeast of the town of Crescent Valley at the southern tip of the Dry Hills; sideronatrite alters to metasideronatrite upon exposure. (GCF)

METASTIBNITE

Sb_2S_3

A Nevada type mineral; appears to be a primary mineral in fumarolic deposits, but also can occur as the result of oxidation of stibnite.

ELKO COUNTY In bright-red coatings with stibnite on altered monzonite at the Meikle Mine in the Lynn district. (Jensen, 1999)

EUREKA COUNTY As red to red-brown stains and thin mammillary coatings on barite and quartz crystals at the Goldstrike Mine in the Lynn district. (GCF)

HUMBOLDT COUNTY As red stains in the Getchell Mine in the Potosi district; also reported as bright-red coatings on orpiment at the Twin Creeks Mine. (Joralemon, 1951; M. Jensen, w.c., 1999)

NYE COUNTY Reported in a specimen from an unknown locality in the Manhattan district. (FUMT)

WASHOE COUNTY Type locality as red stains on sinter terraces at Steamboat Hot Springs in the Steamboat Springs district, Sec. 33, T18N, R20E; associated with cinnabar and arsenic sulfide. (Becker, 1888)

METATORBERNITE

$Cu^{2+}(UO_2)_2(PO_4)_2 \cdot 8H_2O$

A relatively common secondary uranium mineral formed by the dehydration of torbernite; torbernite and metatorbernite are similar in appearance to autunite and meta-autunite, but are not fluorescent and occur in copper-bearing environments.

ELKO COUNTY With autunite and torbernite at the Autunite and October claims in the Mountain City district. (Garside, 1973)

EUREKA COUNTY Deep-green torbernite from the Gold Quarry Mine in the Maggie Creek district in time becomes pale-green metatorbernite by partial dehydration. (Jensen et al., 1995)

LANDER COUNTY With pyrite, bornite, chalcopyrite, uraninite, and other uranium minerals at the Apex (Rundberg) Uranium Mine in the Reese River district. (Garside, 1973)

MINERAL COUNTY As euhedral crystals in andesite breccia at the Sunday Mining Company claims in the Marietta district; also reported as thin coatings and seams in Mesozoic granite and Tertiary volcanic breccia at the Red Stone claims west of Teels Marsh. (Garside, 1973; Ross, 1961)

NYE COUNTY As pale-green, tabular crystals to 1 mm with *sericite* at the Hall Mine in the San Antone district. (M. Jensen, w.c., 1999)

PERSHING COUNTY As pale-grass-green, translucent crystals with torbernite at the Majuba Hill Mine in the Antelope district. (Jensen, 1985)

METATYUYAMUNITE

$Ca(UO_2)_2V_2^{5+}O_8 \cdot 3H_2O$

Dehydration product of tyuyamunite.

EUREKA COUNTY As a dull, yellow-brown dehydration product after bright-yellow tyuyamunite at the Gold Quarry Mine in the Maggie Creek district. (Jensen et al., 1995)

METAVARISCITE

$AlPO_4•2H_2O$

Metavariscite is less common than variscite with which it is dimorphous; the two minerals are usually found together as a product of surface processes in phosphate-rich rocks.

ELKO COUNTY Identified in thin section at the Rain Mine in the Carlin district. (D. Heitt, o.c., 1991)

ESMERALDA COUNTY As radiating colorless to white crystals to 1 mm with variscite and other minerals at an unnamed prospect in the Candelaria district, Sec. 17, T3N, R36E. (Palache et al., 1951; R. Walstrom, w.c., 1999)

HUMBOLDT COUNTY Reported at the Silver Coin Mine in the Iron Point district. (Thomssen, 1998)

NYE COUNTY As white laths to 0.5 mm from an unnamed prospect near Warm Springs. (M. Jensen, w.c., 1999)

METAVOLTINE

$K_2Na_6Fe^{2+}Fe_6^{3+}(SO_4)_{12}O_2•18H_2O$

Metavoltine is a moderately rare mineral found with other sulfates in fumarolic deposits. It has been found at only a few localities in the United States.

WASHOE COUNTY As dull-yellow crystals to 1 mm with coquimbite, bonattite, and other sulfates on siliceous sinter at Steamboat Hot Springs. (Jenkins, 1981)

METAZEUNERITE

$Cu^{2+}(UO_2)_2(AsO_4)_2•8H_2O$

Metazeunerite, a dehydration product of zeunerite, is less common than meta-autunite and metatorbernite, principally occurring as a secondary mineral in base-metal deposits that contain small quantities of uranium.

CHURCHILL COUNTY Reported at the Nickel Mine in the Table Mountain district. (S. Pullman, w.c., 1999)

HUMBOLDT COUNTY As green tetragonal dipyramids to 1 mm in cavities in partially to wholly oxidized radioactive breccia with uranium-rich zircon at the Moonlight Uranium Mine in the Disaster district. (Castor et al., 1996)

PERSHING COUNTY As opaque green crystals

formed by partial dehydration of zeunerite in the Copper stope at the Majuba Hill Mine in the Antelope district. (Trites and Thurston, 1958; Jensen, 1985)

MEURIGITE

$KFe_7^{3+}(PO_4)_5(OH)_7•8H_2O$

A Nevada type mineral, meurigite has been found at only a few localities outside Nevada.

EUREKA COUNTY As pale-yellow, fibrous spherules to 0.5 mm on goethite at the Goldstrike Mine in the Lynn district. (M. Jensen, w.c., 1999)

At the Gold Quarry Mine in the Maggie Creek district, a co-type locality, as pale-yellow aggregates of radial fibers to 0.2 mm in brecciated and crushed jasperoid as fracture fillings with fluellite and other phosphates, hewettite, tiemannite, and corderoite; originally identified as phosphofibrite, but later research showed it to be a new mineral. (Jensen et al., 1995; Birch et al., 1996; Mandarino, 1997; MV)

HUMBOLDT COUNTY As cream-white fibrous hemispheres to 2 mm with leucophosphite at the Silver Coin Mine in the Iron Point district. (Thomssen, 1998; HUB; FUMT; F. Cureton, w.c., 1999; M. Jensen, w.c., 1999)

PERSHING COUNTY In pale-yellow, radiating crystal clusters to 0.2 mm with strengite and dufrénite at the Willard Mine in the Willard district. (Jensen and Leising, 2001)

MIARGYRITE

$AgSbS_2$

Miargyrite, probably the least common of the antimony-bearing "ruby silver" minerals, occurs as both a hypogene and a supergene phase in a few silver deposits; it is usually associated with a variety of other sulfide minerals.

ELKO COUNTY Selenium-bearing as minute black crystal aggregates to 0.5 mm with quartz and barite at the Meikle Mine in the Lynn district. (Jensen, 1999)

Reported from the Tuscarora district. (Von Bargen, 1999)

HUMBOLDT COUNTY In minor amounts in gold-silver veins at the Sleeper Mine in the Awakening district. (Saunders, 1994)

With *electrum,* pyrargyrite, naumannite, and

tetrahedrite at the National Mine in the National district. (Vikre, 1980, 1985)

In unoxidized gold ore at the Crofoot-Lewis Mine in the Sulphur district. (Ebert et al., 1996)

LANDER COUNTY Identified by EMP with other sulfides at the Dean Mine in the Lewis district. (J. McGlasson, w.c., 2001)

A mineral reported as *kenngottite,* an undescribed species from the Isabella Mine in the Reese River district, is probably miargyrite. (FUMT)

PERSHING COUNTY As lustrous crystals to 5 mm with barite at the Rosebud Mine in the Rosebud district. (K. Allen, o.c., 2000)

With meneghinite at the Queen of Sheba Mine in the Star district. (J. McGlasson, w.c., 2001)

STOREY COUNTY Reported from the Comstock Lode in the Comstock district; verified by EMP as brilliant black crystals to 2.5 mm and clusters to 4 mm on quartz in a specimen from the Ophir Mine. (Von Bargen, 1999; F. Cureton, o.c., 2000)

WHITE PINE COUNTY With andorite, tetrahedrite, pyrostilpnite, enargite, bournonite, and other minerals at the Patriot Mine in the Taylor district. (Williams, 1965)

MICHENERITE

PdBiTe

An important palladium mineral in copper-nickel-sulfide deposits.

CLARK COUNTY Reported to occur in a specimen from Moapa; probably from the Bunkerville district east of Moapa. (HUB)

MICROCLINE

KAlSi$_3$O$_8$

The triclinic dimorph of orthoclase; very common in granitic and metamorphic rocks, and the typical *potassium feldspar* in granite pegmatite. *Amazonite* is a blue variety that is found in pegmatite. Microcline occurs as a major phase in many plutonic and metamorphic rock exposures in Nevada, but only the most interesting occurrences are listed below.

CLARK COUNTY A major constituent of beryl- and chrysoberyl-bearing pegmatite at the Taglo Mine in the Bunkerville district, Sec. 9, T15S, R71E. (JHS)

The most common *feldspar* in pegmatitic miarolitic cavities with smoky and clear quartz, albite, and other minerals in the Newberry and Eldorado Mountains; crystals to 25 cm are commonly twinned. Specific localities are as white crystals to 5 cm in the Christmas Tree Pass area and along State Route 163 about 17 miles west of Laughlin; also found about 4 miles east of Searchlight as white to bluish-gray crystals to 4 cm. (S. Scott, w.c., 1999; B. Hurley, w.c., 1999)

ELKO COUNTY With quartz, beryl, and other pegmatite minerals in the Ruby Mountains; a specimen containing *amazonite* is reported to have come from the Valley View district in this area. (Olson and Hinrichs 1960; WMK)

EUREKA COUNTY In green surface coatings on a fault in diorite at the Goldstrike Mine in the Lynn district. (GCF)

MINERAL COUNTY *Amazonite* is found as light–blue, euhedral crystals to 8 cm with white streaks at the Zapot claims in the Fitting district in the Gillis Range approximately 25 km northeast of Hawthorne. (Foord et al., 1999b)

NYE COUNTY Reported in the Belmont district. (WMK)

PERSHING COUNTY In scheelite- and beryl-bearing pegmatite at the Oreana Tungsten Mine in the Rye Patch district. (Johnson, 1977)

WASHOE COUNTY Abundant as pink perthite with quartz and minor muscovite in irregular pegmatite bodies at the Feldspar Mine in the Stateline Peak district, Sec. 13, T22N, R18E. (JHS)

With quartz, albite, and *allanite* in small pegmatite masses at the Red Rock–rare earth prospect in the Stateline Peak district, Sec. 27, T22N, R18E. (JHS)

As flesh-colored crystals to 20 cm in pegmatite at Incline Village. (Jensen, 1993a)

MILLERITE

NiS

Millerite is one of the most widespread nickel minerals. It occurs as acicular crystals in veins with other sulfide minerals and in cavities in limestone; it is also found in *serpentine* and in magmatic ore deposits.

CHURCHILL COUNTY As tiny grains in nickeline with marcasite, annabergite, and other nickel minerals at·the Nickel Mine in the Table Mountain district; also reported at the Lovelock Mine. (Ransome, 1909b; Ferguson, 1939; R. Thomssen, w.c., 1999)

CLARK COUNTY Reported from the 200-foot level, west drift, of the Key West Mine in the Bunkerville district. (LANH)

ELKO COUNTRY Radiating, acicular, brass-colored crystals to 2 cm with sphalerite on pyrite at the Meikle Mine in the Lynn district reportedly comprise the finest specimens of the species from the western United States. (N. Prenn, o. c., 2002)

EUREKA COUNTY With sphalerite, barite, and pyrite as prismatic crystals to 5 mm long on fractures at the Goldstrike Mine in the Lynn district; also in trace amounts in gold ore. (GCF; Ferdock et al., 1997)

LINCOLN COUNTY Reported in a specimen with quartz from an unknown locality. (UAMM)

MIMETITE

$Pb_5(AsO_4)_3Cl$

Mimetite, the arsenic analog of pyromorphite, is relatively widespread but seldom abundant in base-metal deposits. A number of Nevada localities, notably Goodsprings, Chalk Mountain, Downeyville, and Quartz Mountain, have yielded specimens with small but excellent crystals.

CHURCHILL COUNTY Superbly crystallized in white to pale-yellow prisms to 5 mm with wulfenite and vanadinite at the Chalk Mountain Mine in the Chalk Mountain district. (Jenkins, 1981)

CLARK COUNTY On fracture surfaces as 0.5 to 1.0 mm, lemon-yellow to orange, equant to prismatic crystals with fornacite, chrysocolla, malachite, and pyromorphite at the Singer Mine in the Goodsprings district; also reported from the Whale, Argentena, Yellow Pine, and Prairie Flower Mines. (Adams, 1998; LANH; UCB; Hewett, 1931; Melhase, 1934e; F. Cureton, w.c., 1999)

Reported from a prospect 9 miles northwest of Nelson. (R. Thomssen, w.c., 1999)

ELKO COUNTY As rare, small, pale-orange, subhedral crystals in oxidized ore at the Killie Mine in the Spruce Mountain district. (Dunning and Cooper, 1987)

ESMERALDA COUNTY With fornacite, wulfenite, and other minerals near Columbus in the Candelaria district. (R. Walstrom, w.c., 2000)

Reported in specimens from the Broken Toe Mine in the Rock Hill district; also from the Rock Hill Mines. (R. Thomssen, w.c., 1999)

As microcrystalline, yellow needles with carminite at the Weepah Mine open pit in the Weepah district, Sec. 23, T1N, R39E; also as spheroids to 4 mm on chrysocolla at the 3 Metals Mine. (S. Pullman, w.c., 1999; R. Thomssen, w.c., 1999; B. Runner, o.c., 2000; GCF)

EUREKA COUNTY Formerly abundant and a major source of lead in bonanza ore with galena, cerussite, anglesite, and wulfenite in the Eureka district; specifically reported from the Lawton and Ruby Hill Mines. (Hoffman, 1878; Nolan, 1962; Schilling, 1979; LANH; R. Thomssen, w.c., 1999)

As transparent, pale-yellow, hexagonal crystals to 50 microns with cerussite crystals and bindheimite on galena at the Gold Quarry Mine in the Maggie Creek district. (Jensen et al., 1995)

HUMBOLDT COUNTY As yellow-orange to colorless microcrystals on barite with chlorargyrite and fornacite at the Silver Coin Mine in the Iron Point district. (S. Pullman, w.c., 1999; R. Thomssen, w.c., 1999; J. McGlasson, w.c., 2001)

LANDER COUNTY With plumbojarosite in oxidized ore at the Nevada and Lucky Strike Mines in the Battle Mountain district. (Roberts and Arnold, 1965)

In a specimen with arsenopyrite, hemimorphite, and jarosite from the Hoyt Mine in the Reese River district. (UAMM)

Reported from the Gray Eagle Mine in the Shoshone Range. (R. Thomssen, w.c., 1999)

LINCOLN COUNTY As fibrous, yellow-orange material with scorodite in veins in the Lucky Hobo Mine in the Chief (Caliente) district; also as granular green aggregates with hemimorphite and barite in the Gold Chief Mine. (Callaghan, 1936; WMK)

Reported from the Silver King Mine in the Silver King district. (R. Thomssen, w.c., 1999)

LYON COUNTY As tiny crystals to 0.5 mm on manganese-stained siliceous gossan at the Old (Lost) Soldier Mine in the Churchill district. (Jenkins, 1981)

MINERAL COUNTY Locally abundant as greenish-yellow hexagonal crystals to 2 mm at the Simon Mine in the Bell district, Sec. 9, T8N, R37E. (M. Jensen, w.c., 1999; S. White, w.c., 1999)

NYE COUNTY As bright-yellow prisms to 2 mm with wulfenite, hemimorphite, and adamite in oxidized ore at the San Rafael Mine in the Lodi district, Sec. 30, T14N, R36E. (ASDM; Jenkins, 1981)

With wulfenite as orange-brown, barrel-shaped prisms to 5 mm on gossan at Downieville in the Gabbs district. (Jenkins, 1981)

WHITE PINE COUNTY With tungsten-bearing wulfenite, heyite, and a variety of other minerals at the Betty Jo claim in the Nevada district 8 miles southeast of Ely. (Williams, 1973)

MINIUM
$Pb_2^{2+}Pb^{4+}O_4$

Minium is a rare mineral in highly oxidized lead ore. However, it is possible that the mineral is not always recognized because of its superficial similarity to *limonite*. May form in mine fires.

CLARK COUNTY Reported in a specimen from the Monte Cristo Mine in the Goodsprings district. (LANH)

EUREKA COUNTY Reported in a specimen from an unknown locality in the Eureka district. (HMM)

LINCOLN COUNTY Reported from the Pioche district. (Hoffman, 1878)

MINERAL COUNTY Questionably identified in ore in the Broken Hills district. (Schrader, 1947)

With galena and a variety of other minerals at the Dispozitch Mine in the Silver Star district. (Ross, 1961)

NYE COUNTY Questionably identified as red-orange masses with realgar and orpiment in calcite at the White Caps Mine in the Manhattan district. (Jenkins, 1981)

WHITE PINE COUNTY Reported from an unspecified locality near Hamilton in the White Pine district. (UMNH)

MINYULITE
$KAl_2(PO_4)_2(OH,F) \cdot 4H_2O$

A rare secondary mineral that occurs in phosphatic sedimentary deposits; the occurrence at the Willard Mine may be unique.

EUREKA COUNTY As white to pink radiating crystals, fibrous spherulites, and veins at the Gibellini property and Van-Nav-San claim group in the Gibellini district in the Cockalorum Wash Quadrangle; commonly associated with meta-hewettite. (Zientek et al., 1979; R. Walstrom, w.c., 1999)

PERSHING COUNTY As silky, white to yellow spherules to 1.3 cm with goethite and quartz at the Willard Mine in the Willard district, Sec. 26, T28N, R32E; reportedly the state's finest occurrence of the species. (Jensen and Leising, 2001)

MIRABILITE
$Na_2SO_4 \cdot 10H_2O$

Mirabilite is a relatively common mineral in playa and bedded evaporite deposits.

CHURCHILL COUNTY Reported at Ragtown in the Soda Lake district. (S. Pullman, w.c., 1999)

CLARK COUNTY Reported from the Goodsprings district. (S. Pullman, w.c., 1999)

With thenardite and gypsum in an evaporite sequence at least 10 m thick in the Thumb Member of the Horse Spring Formation in the Muddy Mountains district about 5 km west-northwest of Bitter Spring; glauberite was also reported from this area, but not identified from this locality. (SBC)

LYON COUNTY Clear crystals occur at depths of 6–8 feet with thenardite in the Wabuska Salt Marsh. (M. Jensen, w.c., 1999)

MINERAL COUNTY One of many evaporite constituents in a playa deposit at the Rhodes salt marsh in the Rhodes Marsh district; reportedly in a 4.5-m-thick bed beneath a thenardite bed. (Melhase, 1935e; Vanderburg, 1937a)

MITRIDATITE
$Ca_2Fe_3^{3+}(PO_4)_3O_2 \cdot 3H_2O$

A secondary phosphate mineral first identified as nodules and veinlets in iron ore from the USSR.

HUMBOLDT COUNTY As olive-green clusters to 0.5 mm with hinsdalite, montgomeryite, and whitlockite at the Twin Creeks Mine in the Potosi district. (M. Jensen, w.c., 1999)

LANDER COUNTY In thin, olive-green coatings with carbonate-fluorapatite crystals at the Buffalo Valley Mine in the Buffalo Valley district. (M. Jensen, w.c., 1999)

MIXITE

$BiCu_6^{2+}(AsO_4)_3(OH)_6 \cdot 3H_2O$

Mixite is a rare mineral found in the oxidation zones of copper deposits containing abundant arsenic. Occurs at only a few localities in the western United States.

CHURCHILL COUNTY Reported from the Nickel Mine in the Table Mountain district. (R. Thomssen, w.c., 1999)

ESMERALDA COUNTY Reported from the Broken Toe Mine in the Rock Hill district. (R. Thomssen, w.c., 1999)

HUMBOLDT COUNTY As blue-green, acicular microcrystals at the Silver Coin Mine in the Iron Point district; reportedly contains calcium and yttrium and may be *agardite*. (Thomssen, 1998; M. Jensen, w.c., 1999)

MINERAL COUNTY Reported in a specimen from the Simon Mine in the Bell district. (R. Thomssen, w.c., 1999)

NYE COUNTY Reported at the San Rafael Mine in the Lodi district. (S. Pullman, w.c., 1999)

PERSHING COUNTY As rare, extremely minute, blue fibers with olivenite on drusy quartz in the Tin stope at the Majuba Hill Mine in the Antelope district. (MacKenzie and Bookstrom, 1976; Jensen, 1985)

MOHITE

Cu_2SnS_3

A rare sulfide mineral of hydrothermal origin.

LINCOLN COUNTY Reported at the April Fool Mine in the Delamar district. (Anthony et al., 1990; HUB; S. Pullman, w.c., 1999)

MOLYBDENITE

MoS_2

Molybdenite is a common mineral in Nevada, occurring in molybdenum and copper-molybdenum-porphyry deposits, base-metal vein or replacement deposits, tungsten-bearing skarns, and in pegmatites; it has been mined from a few deposits in the state. Descriptions of many Nevada localities are in Schilling (1979).

CHURCHILL COUNTY In quartz-calcite veins and masses with galena, pyrite, and chalcopyrite in the Chalk Mountain district. (Schilling, 1979)

Reported in the IXL district 25 miles northwest of the town of Stillwater. (WMK)

Rarely as flakes in quartz-*adularia* veins at the Wonder Mine in the Wonder district. (Schilling, 1979)

CLARK COUNTY Reported in copper-nickel ore with sphalerite at the Key West Mine in the Bunkerville district. (Lindgren and Davy, 1924; Beal, 1965)

Reported in a minor occurrence in the Charleston district. (Schilling, 1980)

As locally abundant flakes along the borders of stockwork quartz veinlets and disseminated in adjacent rock with hydrobiotite, pyrite, chalcopyrite and bornite at the Crescent Peak porphyry-molybdenum deposit in the Crescent district. (Schilling, 1979)

As crystals to 1 mm in pegmatitic-miarolitic cavities in the Eldorado and Newberry Mountains. (S. Scott, w.c., 1999)

Reported at the Boss Mine in the Goodsprings district. (Jedwab, 1998b)

DOUGLAS COUNTY In stockwork veins at the Pine Nut porphyry-molybdenum deposit in the Gardnerville district. (Schilling, 1980; Doebrich et al., 1996)

In skarn at the Arrowhead and Sweetwater Mines in the Risue Canyon district. (Schilling, 1962a, 1962b; AMNH)

ELKO COUNTY In minor amounts with powellite and scheelite in skarn at the Batholith Mine in the Charleston district. (Schilling, 1964b; Bentz et al., 1983)

With copper oxides and sulfides in metamorphosed limestone and alaskite at the Copper Shield, Bonanza, and Ivy Wilson Mines in the Contact district; also found with other sulfide minerals in quartz-calcite veins in many mines. (Purington, 1903; Schrader, 1935b; Schilling, 1962a, 1979)

As flaky crystalline masses to 3 inches in quartz veins with pyrite, chalcopyrite, galena, sphalerite, and rare nickeline at the Indian Creek prospects in the Edgemont district. (Schilling, 1962a, 1979)

With scheelite in a quartz vein at the Valdez prospect in the Elk Mountain district; also in skarn with *garnet,* pyrite, and copper sulfide minerals at the Robinette prospect. (Lawrence, 1963; Kirkemo et al., 1965)

With powellite and scheelite in a contact-metasomatic deposit at the Batholith and Coon Creek Mines in the Jarbidge district. (Schilling, 1980)

In a deposit that produced minor molybdenum at the Southham Mine in the Kinsley district as abundant (locally 35 percent) flakes and aggregates in quartz veins with scheelite, galena, and copper minerals. (Schilling, 1979)

With pyrite and rare chalcopyrite in quartz veins in quartz monzonite and slate at the Huber Hill and Owyhee prospects in the Mountain City district. (Schilling, 1962a, 1979)

Reported in biotite schist, Northern Ruby Mountains. (Snoke et al., 1979)

ESMERALDA COUNTY Locally in veins with sulfides in the Divide district; specifically reported at the Tonopah-Divide Mine. (Albers and Stewart, 1972; Bonham and Garside, 1979)

With sulfides and scheelite at the Carrie Mine in the Gilbert district. (Ferguson, 1928; Schilling, 1962a)

With other sulfides and fluorite in a 4- to 9-foot-wide quartz vein at the Bullfrog-George claim in the Gold Point district. (Papke, 1979)

As isolated flakes and rosettes in alaskite, underground and on the dump at the Silver Peak shaft in the Goldfield district. (Schilling, 1962a, 1979)

Reported with oxidized-copper minerals in quartz monzonite at the Moly prospect in the Lone Mountain district. (Phariss, 1974; Bonham and Garside, 1979)

As flakes along the margins of quartz-pyrite stockwork veinlets and disseminated in country rock at the Rock Hill (Red Hill) Mine in the Rock Hill district. (Sanford and Stone, 1914; Schilling, 1979)

Locally abundant with other sulfides in quartz veins and lead-zinc orebodies and near the surface with chlorargyrite and secondary base-metal minerals at the Sylvania Mine in the Sylvania district. (Schilling, 1979)

Common and widespread as disseminated flakes and rosettes and along the margins of quartz stockwork veinlets in granite in the Cucomungo (Siskon) deposit, Secs. 2 and 3, T7S, R39E, in the Tule Canyon district; also reported at the McBoyle prospect. (Kirkemo et al., 1965; Schilling, 1962a, 1979)

With powellite in skarn at the Black Horse Mine in the Black Horse district. (Schilling, 1962a)

EUREKA COUNTY In small amounts at unspecified sites in the Eureka district. (Gianella, 1941)

With fluorite, sphalerite, and beryl at the Bisoni claims in the Fish Creek district; also reported with pyrrhotite and sphalerite in contact-metasomatized limestone at the Antelope Mine. (Papke, 1979; Schilling, 1979)

Reported in organic-rich shale near the Gibellini Mine in the Gibellini district. (Desborough et al., 1979)

As anhedral to subhedral crystals in diorite in, or adjacent to, siliceous stibnite-bearing rocks at the Goldstrike Mine in the Lynn district. (GCF)

With quartz and other minerals as replacement bodies in limestone in the Mineral Hill district. (Emmons, 1910; Schilling, 1979)

In stockwork quartz veins in a large, unexploited Climax-type, porphyry-molybdenum deposit with more than 450 million short tons at 0.13–0.32 percent molybdenite about 21 miles northwest of Eureka in the Mount Hope district. (Schilling, 1979; Jones and Schilling, 1982; Westra and Riedell, 1996)

HUMBOLDT COUNTY As small flakes and fracture coatings along and adjacent to quartz veinlets with pyrite, chalcopyrite, and other minerals at the Copper Canyon Mine in the Battle Mountain district. (Theodore and Blake, 1978; Schilling, 1979)

As traces in uranium-rich silicified tuff in the Bottle Creek district. (Castor et al., 1982, 1996)

A minor mineral at the Lone Tree Mine in the Buffalo Mountain district. (R. Braginton, o.c., 1995)

As silvery crystals to 2 cm in quartz from an unnamed prospect on Edna Mountain in the Golconda district. (M. Jensen, w.c., 1999; WMK)

In quartz veinlets with chalcopyrite, pyrite, and scheelite at the Molly prospect in the Gold Run district. (Schilling, 1979)

As hexagonal plates to 4 cm and as bunches of crystals with chalcopyrite and pyrite in quartz veins at the Desert View prospect in the Leonard Creek district. (Kirkemo et al., 1965; Schilling, 1979)

As shiny black flakes and paint with ilsemannite along cracks in the north pit of the Getchell

Mine in the Potosi district; also as very rich crystalline material in masses to 80 pounds in the Moly shaft and in tungsten ore at the Granite Creek and Riley Mines and other properties. (F. Cureton, w.c., 1999; Schilling, 1962a, 1962b, 1979; Hotz and Willden, 1964)

With scheelite in a quartz vein at the Bloody Run Mine in the Sherman district. (Kerr, 1946)

With pyrite and chalcopyrite in quartz veins at the Ashdown Mine in the Vicksburg district, where a large body of molybdenite-rich rock was intersected by a 3000-foot decline shaft in the early 1980s; also reported with scheelite in skarn at the Defense Tungsten Mine. (M. Jensen, w.c., 1999; Schilling, 1962a, 1962b, 1979)

Reported at an unspecified locality in the Winnemucca district. (Schilling, 1979)

LANDER COUNTY As films on fractures and in quartz stockwork veinlets with abundant pyrite at the Buckingham porphyry deposit in the Battle Mountain district, Sec. 30, T32N, R44E; also along fractures in the Copper Canyon and Miss Nevada Mines. (Schilling, 1979; Roberts and Arnold, 1965)

In a small deposit in Trenton Canyon on the west side of Battle Mountain in the Buffalo Valley district. (Schilling, 1979; WMK)

Reported in the Tenabo area and the Violet shaft in the Bullion district. (Emmons, 1910; Schilling, 1979)

A minor mineral in veins in the Lewis district. (Gilluly and Gates, 1965)

In small quantities in veins in the New York Canyon area in the Reese River district. (Ross, 1953)

With scheelite and bismuth minerals in skarn in the Spencer Hot Springs district. (McKee, 1976)

LINCOLN COUNTY In a minor occurrence at an unspecified locality in the Patterson district. (Schilling, 1979)

With pyrite at Stampede Gap southwest of Pioche in the Pioche district. (Gemmill, 1968)

As much as 10 vol. percent of contact-metasomatized rock, in places with ilvaite, at the North Tem Piute Mine in the Tem Piute district; also reported with scheelite, powellite, fluorite, and sulfide minerals in skarn at the Lincoln Tungsten Mine. (Buseck, 1967; Tschanz and Pampeyan, 1970)

Reported in a small deposit at Manhattan Gap, the pass that separates the Highland and Bristol Ranges. (Schilling, 1979)

LYON COUNTY In minor amounts on the south end of Painted Mesa 10 miles north of Schurz in the Benway district. (Schilling, 1979)

Reported in a specimen from the Old (Lost) Soldier Mine in the Churchill district. (WMK)

Noted in a small vein near Devils Gate in the Silver City district; also reported with quartz- and base-metal sulfide minerals in the South End mines. (Gianella, 1936a; Schilling, 1979; Vikre, 1989)

Reported at the W & P Mine in the Wilson district. (Schilling, 1979)

Rare in copper ore at the Yerington Mine in the Yerington district. (Moore, 1969; JHS)

MINERAL COUNTY Reported at the Douglas prospect in the Bell district. (Schilling, 1979)

A major constituent of a tungsten-molybdenum skarn deposit at the Queen (Indiana Queen) Mine in the Buena Vista district. (Papke, 1979)

Reported in the Calico Hills district north of Schurz. (Schilling, 1979)

At various locations in the Garfield Hills in the Garfield district. (Schilling, 1962a)

Reported at the Mineral Jackpot prospect in the Marietta district. (Schilling, 1979)

Disseminated with copper minerals in sedimentary rocks at the Pine Tree prospect in the Pilot Mountains district. (Schilling, 1962a)

With chalcopyrite and other sulfides in altered porphyry and garnetized limestone at the Copper Mountain Mine in the Rand district; also noted at an unspecified site in veinlets in andesite. (Schilling, 1962a; Schrader, 1947)

Reported at unspecified sites in the Rawhide district. (Schrader, 1947)

With pyrite, chalcopyrite, and bornite in narrow quartz veins in the lower adit of the Silver Dyke Tungsten Mine in the Silver Star district. (Garside, 1979)

Reported in Cory Canyon in the Mount Grant district in the Wassuk Range. (Schilling, 1979)

NYE COUNTY In veins and skarn with base-metal sulfide minerals at the Barcelona Mine and Perkins and Warren prospects in the Barcelona (Spanish Belt) district. (Kral, 1951; Ervine, 1972)

As much as 2 vol. percent with copper and silver in a quartz vein in granite at the Belmont Big

Four Mine in the Belmont district. (Lincoln, 1923; Schilling, 1979)

In quartz vein samples from dumps and in veinlets and fractures in drill core of quartz-*sericite-feldspar* rock at the Calico, Hasbrouck, and Quartz Mountain Metals Mines in the Lodi district. (Schilling, 1979; Kleinhampl and Ziony, 1984)

In a slide block of skarn in the Longstreet district, Sec. 25, T7N, R45E. (Kleinhampl and Ziony, 1984)

Reported from an unspecified locality in the Manhattan district. (Schilling, 1979)

With sulfides in a vein at the Superior prospect in Wall Canyon in the Jett district. (Schilling, 1962a; WMK)

With pyrite, pyrrhotite, chalcopyrite, covellite, and *tourmaline* in quartz veins in rhyolitic tuff at the Keyser Mine in the Morey district. (Kleinhampl and Ziony, 1984)

With scheelite and various sulfides in and adjacent to the Climax stock in the Oak Spring district on the Nevada Test Site. (Melhase, 1935b; Kral, 1951; Houser and Poole, 1960)

As scales and rosettes in quartz veins with muscovite and fluorite at the Outlaw Mine in the Round Mountain district; some of the rosettes reach 5 cm in diameter with sharp hexagonal outlines. (Shannon, 1925; Melhase, 1935d; Dunning et al., 1991)

With pyrite and chalcopyrite in quartz veins in a large porphyry-type orebody that was mined between 1981 and 1990 from an open pit at the Hall (Nevada Molybdenum, Cyprus Tonopah) Mine in the San Antone district, Sec. 5, T5N, R42E. (Schilling, 1964a; Kleinhampl and Ziony, 1984; Jones, 1991)

With scheelite in skarn at the Peg Leg (Greenstock) Mine in the Tonopah district. (Bonham and Garside, 1979)

PERSHING COUNTY Common and widespread in strongly silicified rocks with quartz and *tourmaline* at the Majuba Hill Mine in the Antelope district; such mineralization was found to extend to depths of several thousand feet in drill holes. (MacKenzie and Bookstrom, 1976; Jensen, 1985, 1993b)

As scattered flakes to 1 cm in skarn and quartz at the Stormy Day Mine in the Hooker district. (Johnson, 1977; M. Jensen, w.c., 1999)

In drill samples from a large Cu-Mo porphyry

on Granite Mountain in the Kennedy district. (M. Jensen, w.c., 1999)

As flakes in quartz and with pyrrhotite in tungsten ore at the Nevada–Massachusetts Company mines, including the Humboldt and Stank Mines, in the Mill City district. (Johnson, 1977; JHS; WMK)

In minor amounts in skarn with quartz and scheelite in the Nightingale district. (Smith and Guild, 1944)

In tungsten ore in the Wild Horse district. (Johnson, 1977)

In small quartz veins and in skarn at the Rose Creek Mine in the Rose Creek district. (Roberts, 1944)

Reported at the Empire Mine in the Antelope Springs district. (Schilling, 1979)

Reported at the Vernon and Snowstorm claims in the Seven Troughs district. (Schilling, 1979)

STOREY COUNTY In a specimen from the Consolidated Virginia Mine in the Comstock district; also reported in the Flowery lode and South End Group mines. (WMK; Schilling, 1990; Vikre, 1989)

WASHOE COUNTY In *pyroxene*-rich skarn with scheelite on one of the dumps at the Jay Bird Mine in the Nightingale district. (Bonham and Papke, 1969)

As thin films on fractures in granitic rock at the Verdi prospect in the Truckee River Canyon near the Fleish hydroelectric plant. (Schilling, 1962a)

Rare with abundant pyrite in quartz veinlets and along fractures in altered intrusive rock at the Guanomi Mine in the Pyramid district on the Pyramid Lake Indian Reservation. (Bonham and Papke, 1969)

WHITE PINE COUNTY With powellite in milky quartz veins that cut the Bald Mountain stock in the Bald Mountain district. (Schilling, 1980; Hitchborn et al., 1996; WMK)

Reported at the McMurry prospect in the Cherry Creek district. (Hill, 1916)

Widespread in small amounts in porphyry-copper ore in the Robinson district, and recovered since 1941 in concentrate that is exceptionally high in rhenium. (Melhase, 1934d; Hose et al., 1976; WMK)

A minor mineral in tungsten deposits in the Shoshone district. (Schilling, 1980)

Reported from an unspecified locality in the Ward district. (Schilling, 1980)

In quartz veins in and near the Monte Cristo stock in the White Pine district. (Hose et al., 1976)

MOLYBDITE

MoO_3

A rare secondary mineral; most material thought to be molybdite proves to be ferrimolybdite on careful examination. An authentic occurrence in Czechoslovakia is as coatings on molybdenite in a quartz vein.

ESMERALDA COUNTY Reported as abundant bright-yellow aggregates of minute needles at the Tonopah-Divide Mine in the Divide district; mainly on the 165-foot level. (Schilling, 1979)

Reported in small tablets and irregular aggregates in quartz veins with pyrite, gold, base-metal minerals, fluorite, and molybdenite at the Bullfrog-George prospect in the Gold Point district. (Schilling, 1979)

HUMBOLDT COUNTY Reported as yellow fracture coatings in quartz veins with molybdenite at the Molly prospect in the Gold Run district. (B. Hurley, w.c., 1999)

LYON COUNTY Reported in a specimen with molybdenite in the Yerington district. (WMK)

MOLYBDOFORNACITE

$Pb_2Cu^{2+}[(As,P)O_4][(Mo^{6+},Cr^{6+})O_4](OH)$

A secondary mineral related to fornacite and vauquelinite that occurs in oxidized sulfide deposits.

HUMBOLDT COUNTY As minute, yellow-green plates with adamite, duftite, and pyromorphite at the Silver Coin Mine in the Iron Point district. (Thomssen, 1998; M. Jensen, w.c., 1999; J. McGlasson, w.c., 2001)

NYE COUNTY Reported with fornacite, stetefeldtite, skinnerite, and vauquelinite in a specimen from the Belmont Mine, Belmont district. (EMP)

MONAZITE

$(Ce,La,Nd,Th)PO_4$ (general)

Monazite is a common trace constituent of many granitic rocks. It also occurs as a detrital mineral in many placer deposits. *Monazite* species are defined by the dominant rare earth element; where such data are not available the material is assigned to the group.

CARSON CITY With chromite, zircon, *garnet,* and gold in stream placers near Carson City in the Carson City district. (Garside, 1973)

CLARK COUNTY Reported in aplite-pegmatite dikes with *allanite,* xenotime, zircon, *apatite,* and bastnäsite at the Thor and Prospectors uranium claims in the Crescent district. (Garside, 1973)

With magnetite in quartz-*feldspar*-biotite bodies at the Hilltop Mine and Uranium No.1 and Old Dad prospects in the Gold Butte district. (Garside, 1973)

ELKO COUNTY In placer gravels in the Alder, Gold Basin, and Mountain City districts. (Garside, 1973)

ESMERALDA COUNTY In gold placers with fluorite, xenotime, *columbite, uranothorite,* euxenite, zircon, and rutile in Tule Canyon in the Tokop district. (Garside, 1973)

NYE COUNTY With hübnerite and rare native copper in the placer deposits of the Round Mountain district; also reported in placer gravels in the Manhattan district. (Garside, 1973)

WHITE PINE COUNTY With *allanite* and other accessory minerals in quartz monzonite. (Garside, 1973; JHS)

MONAZITE-(CE)

$(Ce,La,Nd,Th)PO_4$

A species of the *monazite* group with Ce as the dominant rare earth element.

CLARK COUNTY With *apatite* in a vein that cuts granitic gneiss in a small prospect pit near Crippled Jack Well in the Crescent district, SE 1/4 SE 1/4 Sec. 35, T29S, R61E. (Castor, 1991a; SBC)

EUREKA COUNTY As minute (<0.1 mm) grains in argillized and silicified limestone at the Goldstrike Mine in the Lynn district. (Berry, 1992)

As euhedral grains to 10 microns with anatase, *apatite,* calcite, illite, *sericite,* pyrite, quartz, and zircon in a possible dike at the Gold Quarry Mine in the Maggie Creek district. (Jensen et al., 1995)

MONAZITE-(LA)

$(La,Ce,Nd)PO_4$

A species of the *monazite* group with La as the dominant rare earth element.

EUREKA COUNTY As 0.1 mm grains in quartz and base-metal sulfide veins in diorite at the Goldstrike Mine in the Lynn district; identified by SEM/EDX. (GCF)

MONCHEITE

$(Pt,Pd)(Te,Bi)_2$

A mineral that is generally found with sulfides and other platinum-group element minerals in magmatic deposits; less commonly in placer deposits.

CLARK COUNTY Reported with other platinum-group minerals such as merenskyite in hornblendite at the Key West Mine in the Bunkerville district. (S. Pullman, w.c., 1999)

Reported in altered peridotite as microinclusions in chalcopyrite at an unspecified locality near Moapa; probably the same as the above occurrence. (EMP; M. Jensen, w.c., 1999)

MONTANITE

$Bi_2Te^{6+}O_6 \cdot 2H_2O$

An uncommon and inadequately described species that occurs as an alteration product of tetradymite.

LANDER COUNTY Reported in a specimen from an unknown locality in the Hilltop district. (WMK)

Reported with chrysocolla as coatings on sheeted quartz veins at the Lucky Rocks claims in Whisky Canyon, Lewis district. (J. McGlasson, w.c., 2001)

MONTBRAYITE

$(Au,Sb)_2Te_3$

A rare mineral that is found with other telluride species and gold in hydrothermal deposits.

LINCOLN COUNTY Reported as silvery grains in quartz at the April Fool Mine in the Delamar dis-

trict. (Anthony et al., 1990; LANH; EMP; HUB; M. Jensen, w.c., 1999)

MONTGOMERYITE

$Ca_4MgAl_4(PO_4)_6(OH)_4 \cdot 12H_2O$

Montgomeryite is a rare mineral that is usually associated with variscite and other phosphate minerals in secondary deposits.

ELKO COUNTY Reported from an unspecified locality in the Wendover area. (TUB; R. Thomssen, w.c., 1999)

HUMBOLDT COUNTY As white crystals to 3 mm with carbonate-fluorapatite, variscite, hisingerite, englishite, collinsite, curetonite, and other minerals at the Redhouse Barite Mine in the Golconda district. (Williams, 1979; F. Cureton, w.c., 1999)

Reported as pale-pink to white crystalline clusters to 0.5 mm at the Twin Creeks Mine in the Potosi district. (M. Jensen, w.c., 1999)

NYE COUNTY In microcrystalline white clusters to 0.5 mm on goethite and barite crystals at the Northumberland Mine in the Northumberland district. (M. Jensen, w.c., 1999)

PERSHING COUNTY Identified by SEM/EDX as cream-colored microcrystalline laths at the Willard Mine in the Willard district. (M. Jensen, o.c., 2000)

MONTICELLITE

$CaMgSiO_4$

A mineral that is typically found in metamorphic or metasomatized rocks; also noted in carbonatites and ultramafic rocks.

PERSHING COUNTY A fine-grained mineral in marble in the Mill City district. (Weissman and Nikischer, 1999)

MONTMORILLONITE

$(Na,Ca)_{0.3}(Al,Mg)_2Si_4O_{10}(OH)_2 \cdot nH_2O$

Aluminum-rich species of the montmorillonite clay mineral group, including most types of *bentonite* clay. Montmorillonite is an important commercial clay mineral in Nevada, where it is most commonly found in altered volcanic or sedimentary rocks. It is very common in alteration zones associated with metal deposits in volcanic rocks; only a partial list of such occurrences is shown below.

CHURCHILL COUNTY In a 5- to 10-foot-thick bed with tuff and mudstone at the Trinity deposit in the Desert district. (Papke, 1970)

In a bed at least 7 feet thick in an adit at the Lahontan deposit. (Papke, 1970)

CLARK COUNTY In a relatively pure bed at the Willow Tank deposit in the Moapa district. (Papke, 1970)

Nearly pure in a 2- to 3-foot-thick bed in an open pit at the Vanderbilt Mine in the Muddy Mountains district. (Papke, 1970)

Reported in a specimen from an unknown locality in the Searchlight district. (LANH)

In pumiceous rock in the Castle Mountains deposit, where a small amount of very pure white to pale-pink waxy clay has been mined and sold as "Nevalite." (Papke, 1970)

ELKO COUNTY In a 2-foot-thick bed in Miocene rocks at the Huntington Creek deposit in the Huntington Creek district. (Papke, 1970)

An alteration mineral at the Jerritt Canyon Mines in the Independence Mountains district. (Hofstra, 1994)

ESMERALDA COUNTY Mined commercially from the Blanco Clay Mine in the Coaldale district, a deposit of dull to slightly waxy, white to very light-gray clay in altered nonwelded ash-flow tuff. (Papke, 1970)

EUREKA COUNTY As pale-green scales disseminated and on fractures at the Goldstrike Mine in the Lynn district. (GCF)

As massive tan material in the Carlin Formation at the Gold Quarry Mine in the Carlin district. (GCF)

An alteration product at the Modarelli Mine in the Modarelli–Frenchie Creek district. (Shawe et al., 1962)

HUMBOLDT COUNTY An alteration mineral at a uranium prospect on the east flank of Buff Peak in the Bottle Creek district. (Castor et al., 1982, 1996)

In lakebeds and breccia at the McDermitt Mine in the Opalite district (Roper, 1976)

Exposed in excavations at the Barret Springs deposit in the Ten Mile district. (Papke, 1970)

Common in tuffaceous lacustrine deposits in the Virgin Valley district. (Castor et al., 1996)

LINCOLN COUNTY With kaolinite in altered igneous rock in a shaft at the Bristol Well deposit in the Bristol district; also in underground workings as fault gouge as much 6 feet thick at the Bristol Silver Mine. (Papke, 1970)

LYON COUNTY As grayish-yellow clay with a minimum thickness of 40 feet in altered volcanic rocks at the Jupiter Mine in the Desert Mountains district; mined intermittently since the 1930s. (Papke, 1970)

MINERAL COUNTY As red-colored clay in trenches at the Broken Hills deposit in the Broken Hills district. (Papke, 1970)

Very pale-orange clay mined from the Miocene-Pliocene Esmeralda Formation at the Walker Lake deposit in the Fitting district. (Papke, 1970)

Yellowish-gray to pale-red clay mined from pits in altered tuff-breccia at the Mina deposit in the Pilot Mountains district. (Papke, 1970)

As pale-olive clay in altered tuffaceous mudstone of the Esmeralda Formation at the Rhodes Marsh deposit in the Rhodes Marsh district. (Papke, 1970)

As very pale-orange clay in rhyolite and darker clay in andesite at the Sodaville deposit in the Silver Star district; mined in the 1920s and 1930s for use in drilling mud. (Papke, 1970)

NYE COUNTY The Ash Meadows district has been the source of most of the montmorillonite group clay that has been mined in Nevada. *Bentonite* has been mined from several pits in Pliocene lake sediments in the western part of Ash Meadows, mostly in Sec. 36, T17S, R49E, and Sec. 1, T17S, R49E; however, the predominant clay mineral in most samples from these deposits is saponite. Montmorillonite also occurs in thin beds in a similar setting in the eastern part of Ash Meadows, Secs. 22, 23, and 26, T17S, R50E; and has been mined from the K-B deposit in the north part of Ash Meadows, Sec. 29, T16S, R50E. (Papke, 1970; SBC)

White clay with local iron oxide has been mined underground from veins and irregular masses in ash-flow tuff since the early 1950s at the New Discovery Mine, SW1/4 Sec. 18, T12S, R47E, in the Bullfrog district; also in similar rocks at prospects on Beatty Peak and at the Roswell deposit. (Papke, 1970)

In altered ash-flow tuff at the Sarcobatus Flat deposit in the Clarkdale district. (Papke, 1970)

In altered tuffs at the Pahute Mesa deposit. (Papke, 1970)

Common in tuffaceous rocks at Yucca Mountain. (Broxton et al., 1987)

PERSHING COUNTY In Tertiary lacustrine sediments at the Rosebud Canyon deposit in the Rabbit Hole district. (Papke, 1970)

Mined from altered tuff north of Coal Canyon at the Secs. 6 and 8 deposits, T27N, R33E. It also occurs south of Coal Canyon in papery Triassic-Jurassic shales at the Sec. 11 deposit, and in altered Tertiary perlite at the Sec. 2 deposit, both in T27N, R32E. (Papke, 1970)

STOREY COUNTY As a common alteration mineral replacing rock breccia fragments and andesite rock adjacent to lodes and faults in the Comstock district. (Vikre, 1989)

WASHOE COUNTY At the Nixon deposit in the Lake Range district on the Pyramid Indian Reservation. (Papke, 1970)

As grayish-white swelling clay in altered volcanic ash at the San Emidio deposit in the San Emidio district. (Papke, 1970)

In alteration associated with uranium prospects in the Dogskin Mountain, Pyramid, and Stateline Peak districts. (Castor et al., 1996)

In a large deposit in Hungry Valley north of Reno. (SBC)

WHITE PINE COUNTY After *oligoclase* in altered monzonite at the Copper Flat pit in the Robinson district. (UCB)

MONTROYDITE

HgO

Montroydite is a rare mineral found in the oxidation zones of mercury deposits.

HUMBOLDT COUNTY Reported at the McDermitt Mercury Mine in the Opalite district. (Rytuba and Glanzman, 1979)

With cinnabar, calomel, and eglestonite at the Cahill Mine in the Poverty Peak district. (M. Jensen, w.c., 1999)

MINERAL COUNTY Reported in the Freedom Flat deposit at the Borealis Mine in the Borealis district. (Eng, 1991)

NYE COUNTY Reported at the Daisy Fluorspar Mine in the Bare Mountain district. (S. Pullman, w.c., 1999; R. Thomssen, w.c., 1999)

Reported at the Paradise Peak Mine in the Fairplay district. (Dobak, 1988)

WHITE PINE COUNTY Reported in a specimen from an unknown locality near Ely. (UAMM)

MOPUNGITE

$NaSb^{5+}(OH)_6$

Mopungite is a Nevada type mineral, named for the Mopung Hills in which the type locality lies. It is an extremely rare oxidation product of stibnite, found to date at only one other occurrence in the world.

CHURCHILL COUNTY Type locality as colorless to white pseudocubic crystals up to 0.3 mm across in spongy crusts on fractures that cut the antimony-oxide minerals stibiconite, senarmontite, roméite, and tripuhyite at the Green prospect in the Lake district, SW1/4 Sec. 9, T23N, R29E. (Williams, 1985)

MORDENITE

$(Na_2,Ca,K_2)_4[Al_8Si_{40}O_{96}] \cdot 28H_2O$

Mordenite is a fibrous zeolite mineral that generally occurs in amygdules in intermediate to basic volcanic rocks and as a product of diagenetic or hydrothermal alteration in pyroclastic rocks.

CHURCHILL COUNTY As white fibers to 1 cm in vugs in an outcrop near Burnt Cabin Summit in the Aspen district, SE1/4 Sec. 10, T14N, R37E. (M. Jensen, w.c., 1999)

With other zeolite minerals in tuff and lacustrine sedimentary rocks at the Eastgate deposit in the Eastgate district. (Papke, 1972a)

With *ferrierite* and *clinoptilolite* in the Sand Springs district. (Papke, 1972a)

ELKO COUNTY Reported in a specimen from 20 miles out of Jarbidge, north of Elko. (UCB)

ESMERALDA COUNTY With *clinoptilolite* in altered volcanic rocks in deposits near Goldfield. (Papke, 1972a)

EUREKA COUNTY As white, hairlike crystals to 6 mm lining vugs with opal and quartz in dike rock at the Goldstrike Mine in the Lynn district. (GCF)

HUMBOLDT COUNTY In altered rock with clay minerals at the Crofoot-Lewis Mine in the Sulphur district. (Ebert et al., 1996)

A diagenetic replacement mineral in lacustrine and fluvial strata in the Virgin Valley district. (Castor et al., 1996)

With *clinoptilolite* as a product of diagenetic or hydrothermal alteration in broad zones in the McDermitt caldera. (Rytuba and Glanzman, 1979)

LANDER COUNTY A minor mineral with *clinoptilolite* and *erionite* at the Jersey Valley zeolite deposit in the Jersey Valley district. (JHS)

With *phillipsite,* analcime, and other minerals in altered tuffs in the Steiner Canyon district in the Reese River Valley, 35 miles north of Austin. (Papke, 1972a)

LYON COUNTY With *clinoptilolite, heulandite,* and *stilbite* in altered pyroclastic rocks in Wilson Canyon, 10 miles south of Fernley, in the Talapoosa district. (Papke, 1972a; S. Pullman, w.c., 1999)

NYE COUNTY In altered tuff 5 miles southeast of Tonopah. (Papke, 1972a)

With *clinoptilolite* about 25 miles east of Tonopah, south of U.S. Highway 6. (Papke, 1972a)

With *clinoptilolite* in zeolitized tuffs southeast of Beatty. (Papke, 1972a)

With analcime, *clinoptilolite,* and *chabazite* in the Yucca Flat area on the Nevada Test Site; also in drill core from Yucca Mountain. (Hoover, 1968; Broxton et al., 1987)

PERSHING COUNTY Tentatively identified as an abundant zeolite mineral in the Long Tom vein in the Seven Troughs district; also a widespread alteration product of rhyolite. (D. Hudson, w.c., 1999)

With *ferrierite* and *clinoptilolite* in a large deposit northwest of Lovelock. (Rice et al., 1992)

MORENOSITE

$NiSO_4 \cdot 7H_2O$

Morenosite is an uncommon mineral occurring as an oxidation product of primary nickel sulfides and arsenides.

CHURCHILL COUNTY In oxidized ore with annabergite coating nickeline, millerite, and other nickel minerals at the Nickel Mine in the Table Mountain district. (Ransome, 1909b; Ferguson, 1939; Gianella, 1941; CSM)

EUREKA COUNTY As pale-green, friable, earthy, efflorescent crusts associated with millerite, sphalerite, violarite, and siegenite at the Goldstrike Mine in the Lynn district. (GCF)

MORINITE

$NaCa_2Al_2(PO_4)_2(F,OH)_5 \cdot 2H_2O$

A relatively rare mineral that generally occurs with other phosphate species in pegmatite.

HUMBOLDT COUNTY Copper-bearing morinite occurs as turquoise-blue to dull-blue botryoids and spherules to 1 mm with wavellite, perhamite, turquoise, and crandallite at the Silver Coin Mine in the Iron Point district. (Thomssen, 1998; M. Jensen, w. c., 1999; J. McGlasson, w.c., 2001)

MOSCHELLANDSBERGITE

Ag_2Hg_3

A mineral that occurs with cinnabar and other sulfides in low-temperature hydrothermal deposits.

CLARK COUNTY Tentatively identified in bitumen from the Azurite Mine in the Goodsprings district; may be the compositionally similar mineral schachnerite. (Kepper, this volume)

MINERAL COUNTY Reported in a specimen from an unknown locality in the Rawhide district. (RMO)

MOSESITE

$Hg_2N(Cl,SO_4,MoO_4,CO_3) \cdot H_2O$

Mosesite is an extremely rare mineral, occurring as an oxidation product of cinnabar or other primary mercury minerals in a few deposits.

HUMBOLDT COUNTY Tentatively identified with other secondary mercury minerals such as calomel, eglestonite, and kleinite at the McDermitt Mine in the Opalite district. (Rytuba and Glanzman, 1979; McCormack, 1986)

MINERAL COUNTY As groups of small yellow crystals intergrown in a very irregular manner with native mercury and cinnabar in veins and small irregular masses in limestone at the Clack Quicksilver Mine in the Fitting district. (Bird, 1932; Palache et al., 1951)

MOTTRAMITE

PbCu^{2+}(VO$_4$)(OH)

Mottramite, a relatively uncommon mineral, is generally found in oxidized parts of base-metal deposits that contain vanadium.

CLARK COUNTY In fine specimens from the Bill Nye, Boss, Whale, and Highline Mines in the Goodsprings district; associated with pyrite, pyromorphite, galena, and gold. (Knopf, 1915b; Hewett, 1931; Mark Rogers, o.c., 1999; J. Kepper, this volume)

As rare greenish-yellow films on other minerals at the Quartette Mine in the Searchlight district. (Callaghan, 1939)

ESMERALDA COUNTY With wulfenite, fornacite, phoenicochroite, and other minerals in quartz veins a mile north of the site of Columbus in the Candelaria district. (R. Walstrom, w.c., 2000)

HUMBOLDT COUNTY As brown-black clusters of microcrystals at the Silver Coin Mine in the Iron Point district. (Thomssen, 1998; M. Jensen, w.c., 1999; J. McGlasson, w.c., 2001)

LANDER COUNTY In a specimen from the Snowstorm Mine in the North Battle Mountain district. (R. Thomssen, w.c., 1999)

LINCOLN COUNTY Reported at the Cave Valley Mine in the Cave Valley district, Sec. 16, T9N, R64E. (Schrader, 1931a)

MINERAL COUNTY As an olive-green crust on bull quartz at an unnamed shaft about 1 km east of the Nevada Scheelite Mine in the Leonard district. (GCF)

As yellow-green to olive-green crusts at the Lithia Mine in the Poinsettia district. (M. Jensen, w.c., 1999)

NYE COUNTY Common as yellow-brown coatings on massive white quartz in unspecified mines near the town of Belmont in the Belmont district. (M. Jensen, w.c., 1999)

As yellow-green microcrystals associated with vanadinite on the 350-foot level at the Downieville Lead Mine in the Gabbs district. (M. Jensen, w.c., 1999)

PERSHING COUNTY As yellow-green coatings at a prospect near the head of Nickel Canyon. (J. McGlasson, w.c., 2001)

WASHOE COUNTY In mustard-yellow coatings on quartz as a secondary mineral after galena at the Red Rock prospect in the Stateline Peak district, SE1/4 Sec. 27, T22N, R18E. (M. Jensen, w.c., 1999)

As rare mustard-yellow coatings on sandy quartz in a pegmatite near Incline Village. (M. Jensen, w.c., 1999)

MUSCOVITE

KAl$_2$□AlSi$_3$O$_{10}$(OH)$_2$

A mica mineral that is found in granite, granite pegmatite, metamorphic rocks, and as the variety *sericite* in hydrothermally altered rock associated with ore deposits. *Mariposite* and *fuchsite* are chrome-bearing varieties.

CHURCHILL COUNTY In veins and wall rock at the Fairview Mine in the Fairview district. (Schrader, 1947; Saunders, 1977)

CLARK COUNTY As a major constituent of beryllium-bearing pegmatites at the Taglo Mine in the Bunkerville district, Sec. 9, T15S, R71E; also at the Santa Cruz prospect in subhedral books to 2 cm with beryl and *garnet*. (Beal, 1965; B. Hurley, w.c., 1999)

As crystals to 8 cm in miarolitic cavities and pegmatites in the Eldorado and Newberry Mountains. (S. Scott, w.c., 1999)

In a specimen from an unidentified locality in the Gold Butte district. (WMK)

ELKO COUNTY Coating fractures and faults throughout the pit at the Rain Mine in the Carlin district, but most common in red clay. (D. Heitt, o.c., 1991)

As *mariposite* in a shaft at the Emigrant Springs prospects in the Delano district. (LaPointe et al., 1991)

Reported from a pegmatite in the Dolly Varden district. (WMK)

An alteration mineral associated with gold- and silver-bearing veins at the Ken Snyder Mine in the Gold Circle (Midas) district. (P. Goldstrand, w.c., 1999)

As the variety *mariposite* in prospects in the Lafayette district. (LaPointe et al., 1991)

Excellent muscovite mica was reported in the Ruby Valley district. (Melhase, 1934c; WMK)

Mined from pegmatitic dikes in the Dawley Canyon area in the Valley View district on the east flank of the Ruby Mountains in books to a foot in diameter, mostly in Secs. 16 and 17, T29N, R58E. The Errington–Thiel Mica Mine was the largest producer. (Olson and Hinrichs, 1960)

EUREKA COUNTY *Sericite* is widespread at the Goldstrike Mine in the Lynn district, most notably in diorite as the major alteration mineral around quartz base-metal veins. (GCF)

An alteration mineral at the Modarelli Mine in the Modarelli–Frenchie Creek district. (Shawe et al., 1962)

LANDER COUNTY In skarn at the West orebody in the Battle Mountain district. (Theodore and Blake, 1978)

MINERAL COUNTY Euhedral hexagonal crystals to 2 cm in pegmatite vugs at the Zapot claims in the Fitting district, where it was also noted as green *sericite;* also at the Dover Andalusite Mine. (Foord et al., 1999b; M. Jensen, w.c., 1999; S. Pullman, w.c., 1999)

As greenish books to 2 inches across and 0.5 inch thick at the Mabel prospect in the Garfield district. (JHS)

As selvages along a diorite dike in limestone in the Pamlico district. (UCB)

Reported as *mariposite* in a specimen from the Rawhide Mine in the Rawhide district. (WMK)

NYE COUNTY As fine-grained *sericite* at the Clarkdale and Yellow Gold Mines in the Clarkdale district. (Tingley et al., 1998a)

As scales and aggregates of scales with a pearly luster in quartz veins at the Outlaw Mine in the Round Mountain district; individual crystals range up to 1 cm. (Melhase, 1935d; Dunning et al., 1991)

As soft pearly flakes of *sericite* in ore in the Tonopah district. (Melhase, 1935c)

PERSHING COUNTY As the variety *fuchsite* with dumortierite in the Champion Mine in the Sacramento district; with quartz, *feldspar,* beryl, and native antimony in the Oreana (Little) Tungsten Mine. (UCB; JHS)

WASHOE COUNTY As *sericite* in altered country rock at the Nevada Dominion and other mines in the Pyramid district; comprises as much as 50 percent of altered rock adjacent to veins. (Wallace, 1975, 1980)

Sparse with quartz and *feldspar* in a pegmatite in the Stateline Peak district. (Bonham and Papke, 1969)

A major constituent of pocket fillings in pegmatite, and as pseudomorphs to 3 cm after schorl crystals, near Incline Village. (Jensen, 1993a)

WHITE PINE COUNTY A defining mineral in widespread *sericite*-quartz alteration in monzonite porphyry in the Robinson district. (Bauer et al., 1966)

NACRITE

$Al_2Si_2O_5(OH)_4$

Nacrite, a clay mineral that is polymorphic with other species in the kaolinite group, occurs in hydrothermal deposits.

ELKO COUNTY In veins in altered ash-flow tuff in the Cornucopia district, Secs. 6 and 7, T41N, R52E. (D. Hudson, w.c., 1999)

WASHOE COUNTY In peripheral alteration zones at the Hog Ranch Mine in the Cottonwood district. (Bussey, 1996)

NAGYÁGITE

$Pb_5Au(Sb,Bi)Te_2S_6$

A gold-bearing sulfide that occurs in epithermal veins.

MINERAL COUNTY Reported with gold and tiemannite in the Freedom Flats deposit at the Borealis Mine in the Borealis district. (Eng, 1991)

NAKAURIITE

$Cu_8^{2+}(SO_4)_4(CO_3)(OH)_6 \bullet 48H_2O$

Nakauriite is a very rare mineral that occurs as an oxidation product of primary copper minerals; the Nevada locality was the second reported occurrence of the mineral in the world.

NYE COUNTY As mats of sky-blue to pale-blue fibers to 3 mm with callaghanite and other minerals on *serpentine* near the contact of a copper sulfide vein with magnesite-brucite rock at the Premier Chemicals (Gabbs Magnesite-Brucite) Mine in the Gabbs district. (Oswald and Crook, 1979; Peacor et al., 1982; DeMouthe, 1985)

NATRITE

Na_2CO_3

A rare carbonate mineral that occurs in alkaline igneous complexes. The Nevada occurrence, which is from a poorly defined location, is suspect.

Reported in a specimen from an unknown locality in southern Nevada. (WMK)

NATROALUNITE

$Na_2Al_6(SO_4)_4(OH)_{12}$

Natroalunite, which is less common than its potassium-analog alunite, occurs as an alteration product of *feldspars* and other minerals in extremely acid environments.

ESMERALDA COUNTY Most material identified as alunite in the Goldfield district reportedly contains a high proportion of natroalunite; it is widespread and generally associated with intense alteration of rocks in the vicinity of orebodies, occurring in quartz veins, as breccia matrix, and as an alteration phase in silicified rock; specifically noted at the January, Combination, Blue Bull, Dixie, and Mohawk Mines. (Ransome, 1907, 1909a; Keith et al., 1979)

STOREY COUNTY In alunite-quartz-pyrite alteration in volcanic rocks in the Comstock district. (Whitebread, 1976)

WHITE PINE COUNTY Locally common in oxidized gold ore at the Bald Mountain Mine in the Bald Mountain district. (Hitchborn et al., 1996)

Reported as white microcrystalline material in the Robinson district. (Jenkins, 1981)

NATROAPOPHYLLITE

$NaCa_4Si_8O_{20}F \cdot 8H_2O$

Natroapophyllite is a rare member of the apophyllite group that occurs in skarn.

EUREKA COUNTY Clear pseudocubic crystals to 2 mm in vugs in veins that cut calc-silicate rock and marble along the southern margin of the gold deposit at the Goldstrike Mine in the Lynn district. (GCF)

HUMBOLDT COUNTY In a drill hole specimen from the Getchell Mine in the Potosi district. (R. Thomssen, w.c., 1999)

NATROJAROSITE

$Na_2Fe_6^{3+}(SO_4)_4(OH)_{12}$

The Nevada type mineral natrojarosite, which is less common than its potassium equivalent jarosite, occurs in similar environments to natroalunite.

CLARK COUNTY Common at many localities as yellow earthy lenses and masses of irregular shape and also found as minute crystals in calcite in the Goodsprings district; may form isomorphous mixtures with jarosite and other members of the group; specifically reported at the Boss and Copperside Mines. (Hewett, 1931)

ESMERALDA COUNTY As orange-brown crystals to 5 mm and as druses in the Gilbert district. (M. Jensen, w.c., 1999)

HUMBOLDT COUNTY Reported in a specimen from an unknown locality in the Sulphur district. (WMK)

LANDER COUNTY In minor amounts with jarosite in weakly mineralized rocks adjacent to the Gold Acres Mine in the Bullion district. (Armbrustmacher and Wrucke, 1978)

LINCOLN COUNTY As lustrous, yellow-orange, platy crystals to 1.5 mm in quartz vugs at the Atlanta Mine open pit in the Atlanta district, Sec. 15, T7N, R68E. (M. Jensen, w.c., 1999)

MINERAL COUNTY The type locality for the mineral, described as a yellowish-brown, glistening powder, was given as the eastern side of the Soda Springs Valley, Nevada, on the road to the Vulcan Copper Mine, Esmeralda County (prior to the formation of Mineral County). The location of the Vulcan Copper Mine is not known, but it was probably in the Santa Fe or Pilot Mountains dis-

tricts. The mineral is also variously reported from Sodaville, Luning, and the Rhodes Marsh district. (Hillebrand and Penfield, 1902; AMNH; WMK; S. Pullman, w.c., 1999)

NYE COUNTY As drusy microcrystalline coatings on barite crystals at the underground Northumberland Mine in the Northumberland district. (M. Jensen, w.c., 1999)

WASHOE COUNTY As lustrous, brown, wedge-shaped crystals to 0.5 mm in ball-like clusters southeast of Winnemucca Lake in the Nightingale district; identified by XRD. (GCF)

NATROLITE

$Na_2[Al_2Si_3O_{10}] \cdot 2H_2O$

Natrolite is a low-temperature, zeolite-group mineral that is most often found in cavities in mafic volcanic rocks; it also occurs in hydrothermally altered alkaline igneous rocks.

EUREKA COUNTY Identified by XRD as a white, microgranular alteration product that commonly retains the morphology of original glass shards in Tertiary tuff at the Goldstrike Mine in the Lynn district. (GCF)

NATRON

$Na_2CO_3 \cdot 10H_2O$

Natron is seldom found in nature, usually occurring as a precipitate from brines taken from playas in desert regions.

CHURCHILL COUNTY Reported as a product of cold-weather preparation of sodium carbonate from the brines of Soda Lake and Little Soda Lake near Fallon; also noted as occurring at Ragtown. (Papke, 1976; Palache et al., 1951)

HUMBOLDT COUNTY Reported in a specimen from an unknown location in the Sulphur district. (WMK)

NAUMANNITE

Ag_2Se

Naumannite is found in epithermal precious-metal deposits, and is common in some low-sulfidation vein deposits in northern Nevada.

ELKO COUNTY An important silver mineral in banded quartz-*adularia* veins with *electrum* at the Ken Snyder Mine in the Gold Circle (Midas) district; rarely found in crystals to more than 1 cm. (Casteel and Bernard, 1999; SBC)

With finely divided gold in altered rhyolite at the Long Hike Mine in the Jarbidge district. (Schrader, 1923; Davidson, 1964; LaPointe et al., 1991)

HUMBOLDT COUNTY A minor mineral in gold-silver veins with acanthite, miargyrite, marcasite, *electrum,* and tetrahedrite at the Sleeper Mine in the Awakening district. (Wood, 1988; Saunders, 1994)

With *electrum,* pyrargyrite, miargyrite, and tetrahedrite at the National Mine in the National district. (Vikre, 1985)

With other selenium minerals and base-metal sulfides in gold ore at the Crowfoot-Lewis Mine in the Sulphur district. (Ebert et al., 1996)

MINERAL COUNTY Widespread with *electrum* and acanthite in quartz-*adularia* veins in the Aurora district. (Osborne, 1985)

Reported from an unspecified location in the Rawhide district. (Schrader, 1947)

NYE COUNTY With *electrum,* acanthite, uyten-bogaardtite, and other sulfide minerals at the Bullfrog Mine in the Bullfrog district. (Castor and Sjoberg, 1993)

PERSHING COUNTY As crystals to 2 mm at the Hecla Rosebud Mine in the Rosebud district; intergrown with native gold in some specimens. (M. Jensen, o.c., 2000; J. Clark, o.c., 2000)

Rich masses were reported with pyrite, quartz, native gold, pyrargyrite, acanthite, and *electrum* in veins at the Fairview Mine in the Seven Troughs district. (Ransome, 1909b; F. Cureton, w.c., 1999)

With mercury-bearing *electrum* in boulders at the Gold Boulder zone 2.8 miles west-northwest of Scossa. (W. Fusch, o.c., 2000)

NEOTOCITE

$(Mn^{2+},Fe^{2+})SiO_3 \cdot H_2O$ (?)

Neotocite is a widespread mineral in bedded manganese deposits. It occurs with *psilomelane,* cryptomelane, pyrolusite, and other minerals.

CLARK COUNTY A black, lustrous, massive phase with other manganese minerals in tabular ore-bodies at the Three Kids Mine in the Las Vegas district. (Hewett and Webber, 1931; Hewett and Fleischer, 1960)

HUMBOLDT COUNTY Reported as a variety with cobalt and nickel at the Silver Coin Mine in the Iron Point district. (Thomssen, 1998)

LYON COUNTY As a copper-bearing variety with tenorite in ore at the MacArthur Mine in the Yerington district. (D. Hudson, w.c., 1999)

NEVADAITE*

$(Cu_2VAl)Al_8(PO_4)_8F_8(OH)_2(H_2O)_{20}(H_2O)_{-1.5}$

A new Nevada type mineral species that has been found in a single Carlin-type gold deposit.

EUREKA COUNTY Type locality as spherules to 1 mm and druses of pale-green to turquoise-blue acicular radiating crystals associated with fluellite, hewettite, strengite, variscite, and other minerals at the Gold Quarry Mine in the Maggie Creek district. (Jensen et al., 1995; M. Jensen, w.c., 1999; F. Hawthorne, w.c., 2003)

NEYITE

$Pb_7(Cu,Ag)_2Bi_6S_{17}$

A rare sulfosalt mineral known outside Nevada from mesothermal veins.

LANDER COUNTY As massive material in pyrite and arsenopyrite at the Whisky Canyon prospect in the Lewis district. (Von Bargen, 1999)

WASHOE COUNTY Tentatively identified by SEM/EDX in a specimen of pyrite and enargite in quartz from the dump at the Blue Bird Mine in the Pyramid district, SE 1/4 NE 1/4 Sec. 15, T23N, R21E, as grains to 10 microns in tetrahedrite with hessite(?), stannite, and an unidentified Cu-Bi sulfide. (SBC)

NICKELINE

NiAs

Nickeline (*niccolite*) is a minor phase in high-temperature hydrothermal veins and in orthomagmatic ore such as in the copper-nickel deposits of Sudbury, Canada.

CHURCHILL COUNTY The most abundant unoxidized nickel mineral in the deposit at the Nickel Mine in the Table Mountain district; associated with annabergite and other nickel minerals. (Ransome, 1909b; Ferguson, 1939)

ELKO COUNTY Rare, gray, metallic mineral with annabergite and molybdenite in quartz veins at the Indian Creek prospects in the Edgemont district, W1/2 Sec. 16, T44N, R51E. (Schilling, 1979; F. Cureton, w.c., 1999)

MINERAL COUNTY Reported in the Freedom Flats deposit at the Borealis Mine in the Borealis district. (Eng, 1991)

NYE COUNTY Tentatively identified in serpentinite at the Willow Springs prospect in the Manhattan district. (Kleinhampl and Ziony, 1984)

NICKEL-SKUTTERUDITE

$(Ni,Co)As_{2-3}$

A mineral that forms a series with skutterudite; occurs in moderate-temperature hydrothermal veins.

CHURCHILL COUNTY Identified with gersdorffite, morenosite, and annabergite at the Lovelock and Nickel Mines in the Table Mountain district. (Ransome, 1909b; M. Jensen, w.c., 1999)

NISSONITE

$Cu_2Mg_2(PO_4)_2(OH)_2 \cdot 5H_2O$

A very rare species; found elsewhere only at one occurrence with other secondary copper minerals at a prospect in California and at two sites in Australia.

CLARK COUNTY As greenish-blue, diamond-shaped, platy crystals coating goethite and altered dolomite in the 400-foot-level dump at the Boss Mine in the Goodsprings district. (Kepper, 1998)

NITER

KNO_3

A soluble mineral found on some playas during hot weather after rain, and in a few cases as a concentrate from bat guano in cave deposits.

CHURCHILL COUNTY Reported with soda niter and halite at Niter Buttes, 25 miles southeast of Lovelock. (Gianella, 1941)

ELKO COUNTY Reported with soda niter at Charleston in the Charleston district. (Gianella, 1941)

ESMERALDA COUNTY Reported from a guano deposit near Silver Peak. (Gianella, 1941)

LANDER COUNTY Identified in a specimen from the north end of Smoky Valley. (JHS)

WASHOE COUNTY Reported at sites along Grass

Valley Creek near Leadville, Secs. 13 and 15, T37N, R22E. (Gale, 1912; Gianella, 1941)

NITRATINE

$NaNO_3$

Nitratine or soda niter is a water-soluble mineral found in guano deposits in caves and in playa accumulations in very arid regions.

CHURCHILL COUNTY Reported as natural sodium nitrate with other salts as encrustations in cavities in volcanic rock at Niter Buttes, Secs. 13 and 14, T25N, R31E. (Gale, 1912)

ELKO COUNTY Reported with niter from Charleston in the Charleston district. (Gianella, 1941)

PERSHING COUNTY Reported to occur in deposits in Humboldt County 25 miles east of Lovelock's Station (apparently a holdover from a reference prior to the creation of Pershing County); may be the occurrence noted above in Churchill County. (Ford, 1932)

WASHOE COUNTY Reported with niter near Leadville. (Gianella, 1941)

NONTRONITE

$Na_{0.3}Fe^{3+}_2(Si,Al)_4O_{10}(OH)_2 \cdot nH_2O$

Nontronite, a member of the montmorillonite group of clay minerals, is found as a product of hydrothermal alteration in a variety of ore deposits.

ELKO COUNTY Reported as a plentiful gangue mineral in contact-metamorphic copper deposits at the Mammoth and other mines in the Contact district. (Schrader, 1912)

With *garnet* in gangue at the Red Elephant Mine in the Elk Mountain district. (Schrader, 1912)

ESMERALDA COUNTY A rare zinc-bearing variety occurs with hedenbergite and andradite in a pyrometasomatic deposit in Paymaster Canyon in the Weepah district, Sec. 2, T1N, R40E. (Gulbrandsen and Gielow, 1960)

EUREKA COUNTY A common granular to massive, waxy green alteration and vein mineral at the Goldstrike Mine in the Lynn district; commonly in veins to 3 cm thick or as fracture coatings in oxidized rock. (GCF)

HUMBOLDT COUNTY Reported as *chloropal* in

masses with stibnite in quartz veins at the Snow-drift Mine in the Red Butte district. (Lawrence, 1963)

LANDER COUNTY Reported in a specimen from an unknown location in the Reese River district. (WMK)

LYON COUNTY Found as the variety *chloropal* at the Ludwig Mine in the Yerington district. (AMNH)

MINERAL COUNTY On fracture surfaces in the Walker Lake montmorillonite deposit in the Fitting district. (Papke, 1970)

Reported from copper mines in the Eagleville district, and from unspecified locations in the Rand and Rawhide districts. (Schrader, 1947; WMK)

WHITE PINE COUNTY Identified as the main clay mineral with sulfides in late alteration of skarn near porphyry contacts in the Robinson district. (James, 1976)

As irregular masses in jasperoid in the Lake portion of the Ward district. (Lovering, 1972)

NOVÁČEKITE

$Mg(UO_2)_2(AsO_4)_2 \cdot 12H_2O$

Nováčekite, a rare member of the autunite group, is found in a few uranium deposits in the western United States.

MINERAL COUNTY With saléeite and metator-bernite in sandstone near the contact with a granitic intrusive; the exact location of this occurrence is not known but is probably in the Fitting district at the Broken Bow and Broken Bow King uranium prospect claims. (Garside, 1973)

OFFRÉTITE

$CaKMg[Al_5Si_{13}O_{36}] \cdot 16H_2O$

A zeolite group mineral generally found lining cavities in basalt.

PERSHING COUNTY Reported with *clinoptilolite* in a specimen from an unknown location. (MM)

OLIVENITE

$Cu_2^{2+}(AsO_4)(OH)$

Olivenite is a widespread mineral in the oxidation zones of copper-bearing hydrothermal deposits that contain arsenic. Excellent specimens have come from the Majuba Hill and Burrus localities in Nevada.

CHURCHILL COUNTY As olive-green crystals to 1 mm with annabergite and malachite crystals at the Lovelock and Nickel Mines in the Table Mountain district. (M. Jensen, w.c., 1999)

In specimens from the Fireball Mine in the Truckee district. (R. Thomssen, w.c., 1999)

CLARK COUNTY As yellow-green crystals on the lowest level of the Azure Ridge (Bonelli) Mine in the Gold Butte district. (Hewett, 1931)

Reported as yellowish-green, feathery coatings at the Azurite Mine in the Goodsprings district; also noted at the Lavina Mine. (Melhase, 1934e)

ESMERALDA COUNTY In specimens from the Broken Toe Mine in the Rock Hill district. (R. Thomssen, w.c., 1999)

As green microcrystals with chenevixite and cornwallite at the 3 Metals Mine in the Weepah district. (B. and J. Runner, o.c., 2000)

EUREKA COUNTY As olive-green, acicular crystals to 1 mm at the Genesis Mine in the Lynn district. (M. Jensen, w.c., 1999)

HUMBOLDT COUNTY A zinc-bearing variety occurs as greenish microcrystals at the Silver Coin Mine in the Iron Point district. (Thomssen, 1998; M. Jensen, w.c., 1999)

LINCOLN COUNTY In specimens from the Silver King Mine in the Silver King district. (R. Thomssen, w.c., 1999)

LYON COUNTY Reported in a specimen from the Empire-Nevada Mine in the Yerington district. (LANH)

MINERAL COUNTY As green sprays to 4 mm with conichalcite and bayldonite at the Simon Mine in the Bell district. (S. White, w.c., 2000; GCF)

Rare as crystals to 0.5 mm with annabergite in the Mount Diablo pit at the Candelaria Mine in the Candelaria district. (M. Jensen, w.c., 1999)

Reported as *erinite* in a sample from the eastern portion of the Garfield Hills in the Garfield district. (WMK)

At the Drew Mercury Mine in the Pilot Mountains district. (Foshag, 1927; Bailey and Phoenix, 1944)

As olive-green crystals to 0.5 mm with barite crystals in a prospect near Rhyolite Pass in the Fitting district. (M. Jensen, w.c., 1999)

NYE COUNTY Pale-olive-green needles to 0.5 mm on conichalcite at the Hasbrouck and San Rafael Mines in the Lodi district. (Jenkins, 1981; S. Pullman, w.c., 1999)

Rare crystals to 1 mm with quartz and pyrargyrite at the Northumberland Mine in the Northumberland district. (M. Jensen, w.c., 1999)

Reported from Grantsville in the Union district. (R. Thomssen, w.c., 1999)

PERSHING COUNTY Among North America's finest specimens as abundant, deep-olive-green to nearly black, transparent crystals to 3 cm with clinoclase, cornubite, and cornwallite from the upper two levels at the Majuba Hill Mine in the Antelope district; also reported as rare, white, fibrous masses of the variety *leucochalcite* with crusts of chenevixite and in a specimen from the middle tunnel dump. (Smith and Gianella, 1942; Trites and Thurston, 1958; Jensen, 1985; R. Thomssen, w.c., 1999)

WASHOE COUNTY As transparent, dark-green to olive-green, single crystals to 4 mm on fracture surfaces and as upright clusters of blades to 2 mm at the Burrus Mine in the Pyramid district, Sec. 15, T23N, R21E, with barium-pharmacosiderite, scorodite, malachite, and other secondary minerals; white, fibrous *leucochalcite* may also be locally common. (Bonham and Papke, 1969; Jensen, 1994)

OLIVINE

$(Mg,Fe,Mn,Ni)_2SiO_4$ (general)

A mineral group that contains fayalite and forsterite. *Olivine* is common in mafic igneous rocks in Nevada; occurrences for which the species has not been identified are listed below.

CLARK COUNTY As phenocrysts to 5 mm in glassy, basalt-flow rock in volcanic rocks of Powerline Road 4 miles northeast of Henderson. (Bell and Smith, 1980)

ELKO COUNTY As phenocrysts in igneous dikes at the Jerritt Canyon Mines in the Independence Mountains district. (Hofstra, 1994)

EUREKA COUNTY As phenocrysts in andesite at the Barth Iron Mine in the Safford district. (Shawe et al., 1962)

NYE COUNTY As phenocrysts in Quaternary basalt in Crater Flat east of the Bare Mountain district. (Cornwall, 1972)

WASHOE COUNTY Widespread as phenocrysts and in the groundmass in basaltic flow rocks of the Peavine Sequence. (Bonham and Papke, 1969)

OPAL

$SiO_2 \bullet nH_2O$

The state gemstone of Nevada, opal is generally considered to be amorphous silica; however, partially crystalline varieties such as *opal-CT* and *opal-T* can be differentiated by X-ray diffraction. The mineral is commonly deposited near or at the surface by thermal water.

CARSON CITY Opalized wood occurs on the floor of Brunswick Canyon, 4 miles east of Carson City in the Delaware district. (Klein, 1983)

CHURCHILL COUNTY Common as hot spring sinter at Brady's Hot Springs in the Leete district. (Garside and Schilling, 1979)

Reported from the Red Bird Mine area in the Bernice district. (R. Thomssen, w.c., 1999)

In a specimen from northeast of the mouth of Cottonwood Canyon in Dixie Valley. (R. Thomssen, w.c., 1999)

CLARK COUNTY A common alteration mineral in the Bunkerville district. (Beal, 1965)

ELKO COUNTY In a specimen from a locality east of Currie in the Dolly Varden district. (CSM)

At a locality a few miles northeast of Elko, in all shades of red, brown, purple, white, and black, mottled or beautifully banded in brilliantly contrasting colors in pieces weighing as much as 50 pounds. (Melhase, 1934c)

With cinnabar inclusions on dumps at the Silver Cloud Mine in the Ivanhoe district. (Klein, 1983)

Reported with varicolored fluorescence from Oasis in the Pequop district. (MV; LANH)

As white to light-gray, circular, siliceous sinter mounds formed by active and extinct hot springs at Sulphur Hot Spring in Ruby Valley, Sec. 11, T31N, R59E. (Garside and Schilling, 1979)

ESMERALDA COUNTY Specimens are reported to have come from both Goldfield and Luning. (LANH)

EUREKA COUNTY In a spring-sinter terrace 250 feet high, 100 feet wide, and 2800 feet long at Beowawe Geysers; also forming at present around certain hot-spring pools. (Garside and Schilling, 1979)

On fractures in the upper portions of the Goldstrike Mine in the Lynn district as clear to milky, botryoidal to mammillary crusts; also as fault breccia cement and lining vesicles in a dike. (GCF)

Presently depositing as sinter at Walti Hot Springs in Grass Valley. (Garside and Schilling, 1979)

HUMBOLDT COUNTY Identified in a specimen from the Cordero Mine in the Opalite district. (JHS)

Amorphous opal (*opal-A*), *opal-CT* and *chalcedony* have been identified with a variety of other minerals, including gold and acanthite, in sinter near the Lewis pit in the Sulphur district. (Ebert et al., 1996)

Some of the world's largest precious fire opal

masses have come from the Virgin Valley district as petrified wood and, rarely, as petrified pine cones in tuffaceous lake beds at the Rainbow Ridge, Royal Peacock, and Bonanza Opal Mines; the largest mass reported weighed 17 ounces and had a reported value (in 1919) of $125,000. Common opal also occurs in varicolored beds, including a bed of white, highly fluorescent material. (Randolph and Dake, 1935; Klein, 1983; Castor et al., 1996; CSM; NMBM; Sinkankas, 1997; F. Cureton, w.c., 1999)

Fire opal occurs in basalt at the Firestone Opal Mine 22 miles north of the town of Paradise Valley in the Santa Rosa Range. (Klein, 1983; Sinkankas, 1997)

As clear to milky-white fire opal in dark-gray andesite and basalt at the Black Rock (Little Jo) Opal Mine, Sec. 5, T37N, R25E. (Mitchell, 1991; Sinkankas, 1997)

Fire opal occurs in basalt at the Royal Rainbow Mine, Sec. 1, T38N, R24E. (Sinkankas, 1997)

LANDER COUNTY Reported in the Izenhood district. (Anonymous, 1938)

LINCOLN COUNTY As white powder in the Chief district. (Callaghan, 1936)

Fire opal occurs at the Webber Mine. (Rose and Ferdock, this publication)

LYON COUNTY With *agate, jasper,* and petrified wood 7 to 12 miles southwest of Fernley; a specimen of green opal reportedly came from 6 miles southwest of Fernley. (Klein, 1983; WMK)

Reported at the Ludwig Mines in the Yerington district; also in a specimen from Carson Hill. (Melhase, 1935e; R. Thomssen, w.c., 1999)

MINERAL COUNTY Found in a ferruginous area approximately 300 feet in diameter with hematite, *chalcedony,* and magnetite at the Gillis prospect in the Fitting district. (Reeves et al., 1958)

Reported in a specimen from Rawhide in the Rawhide district. (WMK)

As opalized wood fragments and logs at the Robinson and Bubbles uranium prospects in the Red Ridge district on the Walker River Indian Reservation. (Garside, 1973)

NYE COUNTY In milky-white to bluish botryoidal aggregates in volcanic rock about 5 miles northeast of Beatty along the west flank of Hogback Ridge south of Fleur De Lis Road; also found as the cinnabar-bearing variety *myrickite* in Beatty Wash. (CSM; B. Hurley, w.c., 1999; WMK)

In thin-bedded Tertiary lacustrine sediments at the Basic Silica Quarry in the Gabbs district approximately 7 miles southwest of Gabbs. (Kleinhampl and Ziony, 1984)

Green opal is associated with vashegyite and variscite at the Tom Molly Mine and the Train prospect in the Manhattan district. (Melhase, 1935d)

Reported at the Northumberland Mine in the Northumberland district. (Kokinos and Prenn, 1985)

Found in the Valley View vein in the Tonopah district as colorless hyalite opal sprinkled with a frosting of white *apatite* crystals which enclose iodargyrite crystals in places; also found as opalized wood in the Tonopah area. (Melhase, 1935c; NMBM)

Identified as both *opal-A* and *opal-CT* in samples from Yucca Mountain. (Castor et al., 1990)

Reported as fire opal with bluish body color at the Starfire Mine. (Nichols, 1979; Sinkankas, 1997)

PERSHING COUNTY As a minor constituent of the silica blanket at the hot springs in the Goldbanks district. (JHS)

In a low mound, 500 feet in diameter, as sinter with sulfur that was, and still is, deposited at Kyle Hot Springs in Buena Vista Valley, Sec. 1, T29N, R36E. (Garside and Schilling, 1979)

STOREY COUNTY With cinnabar along the canyons around Five Mile Flat in the Castle Peak district. (Klein, 1983)

Reported in a specimen from the Comstock district. (UAMM)

WASHOE COUNTY An alteration mineral in tuff adjacent to uranium deposits at the DeLongchamps prospect and Red Bluff Mine in the Pyramid district. (Brooks, 1956; Castor et al., 1996)

Geyserite is found as a thin film of siliceous sinter on the casing of a well at Needle Rocks at the northwestern end of Pyramid Lake, Sec. 12, T26N, R20E. (Garside and Schilling, 1979)

As opalite and sinter in terraces at active and extinct hot springs in the Steamboat Springs district. (JHS)

As opalized wood in a 2-square-mile area near Lost Creek, 30 miles north of Gerlach. (Klein, 1983)

Opalized wood is found with *chalcedony, agate,* and *jasper* in Wall Canyon. (Klein, 1983)

With cinnabar, sulfur, and other minerals in a thermally active zone in the San Emidio desert. (Garside and Schilling, 1979)

In a specimen of petrified wood from Mahogany Peak near Fox Mountain north of Gerlach. (NMBM)

ORPHEITE
$H_6Pb_{10}Al_{20}(PO_4)_{12}(SO_4)_5(OH)_{40} \cdot 11H_2O$ (?)

A very rare species; found elsewhere at only one occurrence with anglesite, pyromorphite, and other secondary minerals at a site in Bulgaria.

HUMBOLDT COUNTY Reported in a specimen from Valmy. (HUB)

ORPIMENT
As_2S_3

Orpiment, a relatively common mineral in Nevada, is particularly abundant in Carlin-type gold deposits, in which it may be associated with relatively high-grade ore. Some of the world's finest specimens come from the Getchell and Twin Creek Mines.

ELKO COUNTY As small yellow veins and blebs with realgar and sulfur in carbonaceous sediments at the Rain Mine in the Carlin district. (D. Heitt, o.c., 1991)

In veins or partially replacing porous rock units at the Jerritt Canyon, West Generator Hill, and Murray Mines in the Independence Mountains district. (Hofstra, 1994)

EUREKA COUNTY In a specimen of gold ore with calcite, arsenopyrite, realgar, stibnite, quartz, and pyrite from Mill Canyon in the Cortez district. (WMK; Gianella, 1941)

Reported from an unspecified locality in the Eureka district. (Gianella, 1941)

As small, brownish-yellow, thallium-bearing, euhedral to subhedral crystals with barite, calcite, quartz, and realgar in veinlets and as radiating crystal groups and massive veins in the Carlin Mine in the Lynn district; also as yellow crystals to 1 cm, radiating aggregates, and felty to micaceous masses with realgar, calcite, fluorite, and quartz in the Goldstrike Mine. (Radtke et al., 1973, 1974a; Rota, 1989; GCF)

As masses to 1 cm in calcite at the Gold Quarry Mine in the Maggie Creek district. (Jensen, 1985)

As yellow-orange crystals to 5 cm in calcite with minor barite at the East Gold Pick pit of the Gold Bar Mine in the Antelope Springs district. (S. Kleine, o.c., 1997)

HUMBOLDT COUNTY Originally reported with cinnabar and realgar in crusts and veinlets at the Lone Tree Mine in the Buffalo Mountain district; later found at the same locality as world-class specimens of gemmy, lemon-yellow to orange crystals to 8 cm, some with barite crystals. (Bloomstein et al., 1993; Lees, 2000)

Reported as an alteration product of arsenopyrite at the National Mine in the National district. (Lindgren, 1915; Melhase, 1934c)

In bright-yellow aggregates in quartz-calcite rock with realgar and as brilliant orange crystals at the Getchell Mine in the Potosi district; also as lustrous, yellow-orange crystals to 3 cm in veins, locally overgrown with stibnite, at the Twin Creeks Mine, and as crystals at the Pinson Mine. (Joralemon, 1951; Weissberg, 1965; Stolburg and Dunning, 1985; M. Doyle-Kunkle, o.c., 1998; M. Jensen, w.c., 1999; Cook, 2000)

LANDER COUNTY With realgar in small pods, and thought to be secondary after arsenopyrite, at the Copper Canyon Mine in the Battle Mountain district. (Roberts and Arnold, 1965)

LYON COUNTY Reported at the Mason Valley Mine in the Yerington district. (Gianella, 1941; WMK)

NYE COUNTY In classic specimens from the White Caps Mine in the Manhattan district as crystals with realgar, stibnite, arsenopyrite, and pyrite; in some specimens identified as "hair orpiment," which was later found to be wakabayashilite. Also noted at the Lucky Boy and Manhattan Consolidated Mines in the same district. (Ferguson, 1924; Palache and Modell, 1930; Melhase, 1935d; Gibbs, 1985)

With realgar, cinnabar, and other minerals in silver-gold ore at the Paradise Peak Mine in the Paradise Peak district. (Thomason, 1986)

PERSHING COUNTY Massive with realgar at the Florida Canyon Mine in the Imlay district. (NYS)

WASHOE COUNTY Found among the sinter deposits at Steamboat Hot Springs in the Steamboat Springs district. (Palache et al., 1951)

WHITE PINE COUNTY Reported in the Vantage deposit on Alligator Ridge in the Bald Mountain district. (Ilchick, 1990)

ORTHOCHRYSOTILE

$Mg_3Si_2O_5(OH)_4$

Orthochrysotile is a *serpentine*-group mineral that is generally found in veins in serpentinite. It is probably not uncommon, but is difficult to distinguish due to intermixture with the more common species clinochrysotile.

EUREKA COUNTY Identified by XRD as green microcrystals with quartz in breccia matrix at the Maggie Creek district. (GCF)

NYE COUNTY In a *nemalite* specimen with clino-chrysotile and parachrysotile at the Premier Chemicals (Gabbs Magnesite-Brucite) Mine in the Gabbs district. (AMNH)

ORTHOCLASE

$KAlSi_3O_8$

Orthoclase is a very common *feldspar* group mineral in felsic igneous rocks; it includes the hydro-thermal variety *adularia,* which is common in epithermal veins in Nevada.

CHURCHILL COUNTY As *adularia* in veins and wall rock at the Fairview and other mines in the Fairview district. (Schrader, 1947; Saunders, 1977)

Adularia is found as euhedral white crystals to 1.5 cm with quartz at the Summit King and Summit Queen Mines in the Sand Springs district. (M. Jensen, w.c., 1999)

Adularia was noted in the Westgate district. (R. Thomssen, w.c., 1999)

As quartz-*adularia* veins with fluorite at the Wonder Mine in the Wonder district. (Schilling, 1979)

CLARK COUNTY In granite in the Eldorado and Newberry Mountains. (S. Scott, w.c., 1999)

In a specimen from Gold Butte. (WMK)

Euhedral crystals, commonly Baveno twinned, occur at several sites in the Goodsprings district. Large quantities of pink crystals to 4 cm have weathered out of granitic rock at Crystal Pass, approximately 2 miles south of Goodsprings. In other occurrences as white to tan crystals to 2.5 cm in a roadbed at the Alice Mine, greenish-tan crystals to 1.5 cm in a felsic dike on the east side of Shenandoah Peak, and white kaolinized crystals to 2 cm at the Prairie Flower Mine in a roadbed and on the mine dump. (Drugman,

1938; UCB; FUMT; UAMM; F. Cureton, w.c., 1999; B. Hurley, w.c., 1999; S. Pullman, w.c., 1999)

As *adularia* in quartz-calcite veins with base-metal minerals in the Searchlight district. (Callaghan, 1939)

ELKO COUNTY *Adularia* is found in gold-bearing quartz veinlets at the Rossi Mine in the Bootstrap district. (Snyder, 1989)

Abundant in a granodiorite stock in the Contact district. (Purington, 1903)

Adularia occurs with calcite and quartz and lesser amounts of pyrite, fluorite, and barite in precious-metal veins at the Ken Snyder Mine in the Gold Circle (Midas) district. (Casteel and Bernard, 1999)

Reported in a specimen from the Harrison Pass district. (LANH)

Adularia, some barium-bearing, occurs in igneous dikes at the Jerritt Canyon Mines in the Independence Mountains district. (Hofstra, 1994)

Adularia to 2 mm is abundant in silicified zones at the Hollister deposit in the Ivanhoe district. (Deng, 1991)

As *adularia* crystals to 1/4 inch in drusy cavities in quartz veins with locally high gold values at the Dexter and Modoc Mines in the Tuscarora district. (Emmons, 1910; McKee et al., 1976)

Reported as *adularia* in pegmatite dikes in the Valley View district. (Olson and Hinrichs, 1960)

ESMERALDA COUNTY As *adularia* crystals to 1 cm on fractures in prospects north of Hasbrouck Mountain in the Divide district. (GCF)

Reported as fine blue and flesh-colored crystals from Fish Lake Valley. (Hoffman, 1878)

As *adularia* in quartz veins with fluorite at the Red Hill (Redlich) Mine in the Rock Hill district. (Schilling, 1979)

Adularia occurs as fine-grained intergrowths with quartz and calcite at the Sixteen-to-One Mine in the Silver Peak district. (Keith et al., 1976)

EUREKA COUNTY As *adularia* in veins at the Buckhorn Mine in the Buckhorn district. (Wells and Silberman, 1973)

Common as white to gray crystals to 5 mm in diorite at the Goldstrike Mine in the Lynn district. (GCF)

As phenocrysts in quartz monzonite at the Barth Iron Mine in the Safford district. (Shawe et al., 1962)

HUMBOLDT COUNTY As small *adularia* crystals along faults at the Lone Tree Mine in the Buffalo Mountain district. (Panhorst, 1996)

As *adularia* in uranium ore and replacing *feldspar* in country rock at the Moonlight Uranium Mine in the Disaster district. (Castor et al., 1996; Castor and Henry, 2000)

Reported as *adularia* in quartz veins at the Crow Mine in the Adelaide district; reportedly in Pershing County, but probably in the equivalent Gold Run district. (Silberman et al., 1973)

As *adularia* microcrystals in the Ivanhoe pit, Ivanhoe district. (J. McGlasson, w.c., 2001)

A gangue mineral in iron ore from the Jackson Mountains district; also found in altered metavolcanic host rock at the Delong (Iron King) Mine. (Shawe et al., 1962)

Adularia occurs with barite in veins in massive barite at the Redhouse prospect in the Potosi district; at the Twin Creeks Mine with native gold, quartz, and pyrite, and also with stibnite, quartz, native tellurium, and coloradoite in later mineralization. (DeMouthe, 1985; Groff, 1996; Simon et al., 1999)

Adularia is found in quartz veins at the Ashdown Mine in the Vicksburg district. (Schilling, 1979)

LANDER COUNTY In skarn at the West orebody in the Battle Mountain district. (Theodore and Blake, 1978)

As sharp *adularia* crystals to 4 mm in the McCoy Mine, McCoy district. (J. McGlasson, w.c., 2001)

LYON COUNTY *Adularia* to 1 mm is abundant in the Comstock Lode in the Silver City district. (D. Hudson, w.c., 1999)

Reported from the Christenson shaft in the Talapoosa district. (R. Thomssen, w.c., 1999)

MINERAL COUNTY *Adularia* is found in silver- and gold-bearing quartz veins in the Aurora district. (Hill, 1915)

In a specimen from Camp Douglas in the Silver Star district. (R. Thomssen, w.c., 1999)

NYE COUNTY Reported about 30 miles southeast of Goldfield in the Antelope Springs district on the Nellis Air Force Range. (Cornwall, 1972)

Adularia crystals to 2.5 cm occur in the Bullfrog district. (C. Henry, o.c., 2002)

As euhedral *adularia* crystals in veins at the Clarkdale and Yellow Gold Mines in the Clarkdale

district on the Nellis Air Force Range. (Tingley et al., 1998a)

As *adularia* in gold-bearing, quartz-*chalcedony* veins in the southwestern Shoshone Mountains. (Hamilton, 1993)

As *adularia* crystals to 5 mm in the White Ore vein in the Manhattan district. (C. Henry, o.c., 2002)

Adularia is common as white to clear crystals in gold-bearing veins with quartz and *electrum* at the Round Mountain Mine and in prospects in the Round Mountain district; it also occurs as an alteration mineral in the rhyolitic host rock. (Lovering, 1954; Tingley and Berger, 1985; Henry et al., 1997)

In veins with calcite and quartz in the Silverbow district. (Tingley et al., 1998a)

Microcrystalline *adularia* is found in banded veins at the Life Preserver Mine in the Tolicha district on the Nellis Air Force Range. (Tingley et al., 1998a)

In quartz-*adularia-sericite* alteration in the Tonopah district. (Spurr, 1903; Melhase, 1935c; Nolan, 1935)

As *adularia* with barite in hematized tuff at Yucca Mountain. (Castor et al., 1999b)

PERSHING COUNTY Sparingly as light-gray, rhombohedral crystals to 4 mm in small pockets in rhyolite with smoky quartz crystals and rare crystals of pharmacosiderite, arthurite, and scorodite at the Majuba Hill Mine in the Antelope district. (Jensen, 1985)

As cinnabar-coated *adularia* with barite and quartz at the Red Bird Mine in the Antelope Springs district. (GCF)

As *adularia* in quartz veins with pyrite and other minerals at the Arizona Mine in the Buena Vista district. (Cameron, 1939)

As *adularia* in quartz veins at the Black Canyon Mine in the Imlay district. (Silberman et al., 1973)

In a vein at the Kindergarten Mine in the Seven Troughs district. (Silberman et al., 1973)

As *adularia* in veins at the Reo Mine in the Ten Mile district. (Silberman et al., 1973)

STOREY COUNTY White, fine-grained *adularia* is one of the main constituents, with quartz and calcite, in veins in the Comstock district; it is also common in the Flowery lode as crystals to 2 mm with quartz and calcite and as white porcelaneous intergrowths with quartz. (Schilling, 1990; M. Jensen, w.c., 1999; D. Hudson, w.c., 1999)

WASHOE COUNTY *Adularia* occurs in quartz veinlets with pyrite in prospects in the Olinghouse district. (Bonham and Papke, 1969)

WHITE PINE COUNTY An important alteration mineral with biotite in altered porphyry in the Robinson district; in addition, montmorillonite pseudomorphs of Carlsbad twins occur near Lane City. (James, 1976; UCB)

OSARIZAWAITE

$Pb_2Cu_2^{2+}Al_4(SO_4)_4(OH)_{12}$

A lead- and copper-bearing member of the alunite group that occurs in oxidized base-metal deposits.

WHITE PINE COUNTY Reported as a thin yellow crystalline crust with conichalcite in a specimen from the Red Hills Mine in the Eagle district. (R. Thomssen, w.c., 1999)

OSMIRIDIUM

(Ir,Os)

Osmiridium is an intermediate alloy of osmium and iridium that is reported from magmatic deposits and placers at several sites in the world. The Nevada site listed below is a unique hydrothermal occurrence for which the dominant metal has not been determined.

CLARK COUNTY With platinum in mixtures with jarosite-group minerals at the Boss Mine in the Goodsprings district. (Jedwab, 1998b)

OWYHEEITE

$Ag_2Pb_7(Sb,Bi)_8S_{20}$

Owyheeite is widespread in Nevada silver deposits, particularly those with manganese-rich gangue. In some deposits it is the chief silver-bearing species and is associated with "ruby silver" minerals, jamesonite, boulangerite, and other sulfides.

EUREKA COUNTY Reported in a specimen from Mill Canyon in the Cortez district. (JHS)

HUMBOLDT COUNTY As black grains in massive quartz at the Silver Coin Mine in the Iron Point district. (Thomssen, 1998; M. Jensen, w.c., 1999)

LANDER COUNTY As 20-mm crystals with freibergite in the Eagle vein in the Lewis district; also reported from the Dean Mine. (Ream, 1990; Von Bargen, 1999; R. Thomssen, w.c., 1999)

MINERAL COUNTY As light-gray, lustrous, acicular prisms to 7 mm in a quartz vein with sphalerite, galena, andorite, pyrargyrite, and other minerals at the Broken Hills Mine in the Broken Hills district, Sec. 21, T14N, R35E. (Jenkins, 1981)

NYE COUNTY The most abundant silver mineral, generally as brownish, gray, or black acicular prisms to 2 mm with andorite in quartz veins at the Morey and Keyser Mines in the Morey district. (Williams, 1968; S. Pullman, w.c., 1999)

As black acicular prisms to 5 mm with sphalerite on quartz at an unspecified mine in the Gabbs area. (Jenkins, 1981)

PERSHING COUNTY A common sulfide at the Arizona, De Soto, and Sheba Mines in the Star district. (Cameron, 1939)

As coarse, silver-gray blades in white quartz in a specimen from the Rochester district. (Jenkins, 1981)

PACHNOLITE

NaCaAlF$_6$•H$_2$O

Pachnolite, an alteration product of cryolite and other alkali aluminum fluorides, is most commonly found in pegmatite.

MINERAL COUNTY With cryolite, cryolithionite, and other related minerals in an *amazonite*-topaz pegmatite at the Zapot claims in the Fitting district. (Foord et al., 1999)

PALLADIUM

Pd

Naturally occurring metallic palladium is rare; it is generally found as a placer mineral or as segregations in mafic and ultramafic igneous rocks.

CLARK COUNTY Tentatively identified with EMP by J. Jedwab in specimens from the Key West Mine in the Bunkerville district. (J. Kepper, o.c., 2001)

PALYGORSKITE

(Mg,Al)$_2$Si$_4$O$_{10}$(OH)•4H$_2$O

An alteration product of magnesium silicate minerals that occurs in a variety of geologic settings, including fault gouge.

CHURCHILL/MINERAL COUNTIES Reported in a specimen from an unknown locality in the Broken Hills district. (WMK)

MINERAL COUNTY As mountain leather masses to 4 cm in vugs at the Silver Dyke Tungsten Mine in the Silver Star district. (M. Jensen, w.c., 1999)

PARACHRYSOTILE

Mg$_3$Si$_2$O$_5$(OH)$_4$

A *serpentine* group mineral that is probably more widespread than generally recognized.

NYE COUNTY In a *nemalite* specimen from a brucite deposit with orthochrysotile and clinochrysotile at the Premier Chemicals (Gabbs Magnesite-Brucite) Mine in the Gabbs district. (AMNH)

PARAGONITE

NaAl$_2$□ AlSi$_3$O$_{10}$(OH)$_2$

Paragonite, the sodium analog of muscovite, is found in the same settings as that mineral but is much less common. Some *sericite* in alteration zones associated with ore deposits in Nevada may be paragonite.

CHURCHILL COUNTY Reported in wall rock and in veins at the Fairview Mine in the Fairview district. (Schrader, 1947)

PARAREALGAR

AsS

Dimorphous with, and occurring as an alteration product of, realgar, generally as fine powder.

EUREKA COUNTY Yellow-orange powder that coats realgar, generally within a few days of mining at the Goldstrike Mine in the Lynn district; does not appear to form on well-crystallized realgar. (GCF)

HUMBOLDT COUNTY As pale-yellow-orange coatings on, and thin halos around, realgar that has been exposed to weathering at the Getchell Mine in the Potosi district; a thin layer of weilite, haidingerite, and picropharmacolite is found coating the pararealgar. (Roberts et al., 1980; Dunning, 1988)

LYON COUNTY As yellow-orange inclusions in barite crystals at the Boulder Hill Mine in the Wellington district, W1/4 Sec. 30, T10N, R24E. (R. Walstrom, w.c., 1999)

PARATACAMITE
$Cu_2^{2+}Cl(OH)_3$

Paratacamite is a rare mineral that has only been recognized at a few localities in the United States in oxidation zones of copper deposits in the arid Basin and Range province.

ESMERALDA COUNTY Reported in a specimen from the 3 Metals Mine in the Weepah district. (R. Thomssen, w.c., 1999)

HUMBOLDT COUNTY In a specimen from the Silver Coin Mine in the Iron Point district. (R. Thomssen, w.c., 1999)

LYON COUNTY Deep-green rhombohedrons showing polysynthetic twinning with atacamite, cerussite, and wulfenite at the Old (Lost) Soldier Mine in the Churchill district, SE1/4 SW1/4 Sec. 22, T17N, R24E. (Jenkins, 1981)

Reported as dark-green crystals and crusts at the Douglas Hill Mine and unspecified prospects near Mason Pass in the Yerington district. (EMP; F. Cureton, w.c., 1999; M. Jensen, w.c., 1999)

PARATELLURITE
TeO_2

Paratellurite is a rare mineral that occurs as an oxidation product of primary tellurium minerals.

WHITE PINE COUNTY As an oxidation product of galena, hessite, and altaite mixtures in the ore at the Silver King Mine in the Ward district. (Williams, 1964)

PARGASITE
$NaCa_2(Mg_4Al)Si_6Al_2O_{22}(OH)_2$

Pargasite is a common amphibole group mineral found in metamorphic rocks and intermediate volcanic rocks; it is generally identified as *hornblende*.

EUREKA COUNTY Potassium-bearing pargasite occurs as a rare microgranular component of

diorite at the Goldstrike Mine in the Lynn district. (GCF)

PARNAUITE
$Cu_9^{2+}(AsO_4)_2(SO_4)(OH)_{10} \cdot 7H_2O$

Parnauite, a Nevada type mineral, was originally identified as one of the suite of copper arsenate minerals derived from the oxidation of primary chalcopyrite-arsenopyrite ore at Majuba Hill.

ESMERALDA COUNTY Reported in a specimen from the 3 Metals Mine in the Weepah district. (R. Thomssen, w.c., 1999)

PERSHING COUNTY As finely acicular, light-blue to blue-green or silvery-white rosettes and fans that occur as crusts on fracture surfaces, generally with brochantite, chalcophyllite, chenevixite, goudeyite, malachite, and spangolite in the Copper stope of the middle adit at the Majuba Hill Mine in the Antelope district, Sec. 2, T32N, R31E; also noted as deep-green crystal clusters to 4 mm with langite. (Wise, 1978; Jensen, 1985)

WASHOE COUNTY Uncommon as thin velvety druses of green to yellow-green crystals in cavities in oxidized enargite ore with brochantite, barium-pharmacosiderite, and azurite at the Burrus Mine in the Pyramid district. (Jensen, 1994)

PARTZITE
$Cu_2^{2+}Sb_2^{2+}(O,OH)_7$ (?)

Partzite, a mineral of the stibiconite group, occurs in oxidized base-metal ore as an oxidation product of antimony-bearing sulfides.

ESMERALDA COUNTY As greenish, brownish, or ochre-colored crusts and masses in milky quartz with oxidized copper minerals and stetefeldtite on an unidentified mine dump southeast of the site of Gilbert in the Gilbert district. (Jenkins, 1981; EMP; NMBM; F. Cureton, w.c., 1999)

LANDER COUNTY As an oxidation mineral in sulfide-bearing quartz veins in the Goat Ridge window. (Prihar et al., 1996)

MINERAL COUNTY Reported in rich silver ore with stetefeldtite from an unspecified locality in the Candelaria district. (UCB)

PASCOITE

$Ca_3V_{10}^{5+}O_{28} \cdot 17H_2O$

Pascoite is a rare mineral except in uranium-vanadium deposits in the Colorado Plateau region of the United States, where it generally occurs as efflorescences in mine workings.

HUMBOLDT COUNTY Intergrown with ilsemannite and associated with stibnite at the Getchell Mine in the Potosi district. (Joralemon, 1951; Jenkins, 1981)

PAVONITE

$(Ag,Cu)(Bi,Pb)_3S_5$

Pavonite is probably the most common of several lead-silver-bismuth sulfosalts. It is found in hydrothermal ore with other bismuth species and has been recognized at several localities in the United States.

LINCOLN COUNTY Reported as silvery grains intergrown with hodrušite in quartz in a specimen from a locality near Pioche, possibly the Black Metal (Jackrabbit) Mine in the Bristol district. (Tschanz and Pampeyan, 1970; Jenkins, 1981; Anthony et al., 1990)

NYE COUNTY Reported from an unspecified locality in the Manhattan district. (Anthony et al., 1990)

In laths to 2 mm at the Outlaw Mine in the Round Mountain district. (Dunning et al., 1991)

PEARCEITE

$(Ag,Cu)_{16}As_2S_{11}$

Pearceite, the arsenic analog of polybasite, may frequently go unrecognized because of its physical resemblance to that mineral. It occurs in small quantities in hydrothermal silver deposits that are relatively rich in arsenic.

LINCOLN COUNTY Reported as a constituent of "copper pitch ore" where it occurs as microscopic silvery grains at the Cave Valley Mine in the Cave Valley district; also reported as inclusions to 1 cm or more in diameter in argentiferous galena on the Streator claims. (Schrader, 1931a)

Reported to occur after galena in fissure ore in the Pioche district. (Westgate and Knopf, 1932)

MINERAL COUNTY Identified with other silver-bearing minerals at the Rawhide Mine in the Rawhide district. (Black, 1991)

STOREY COUNTY Well-crystallized pearceite with chalcopyrite was noted in an ore specimen from the 2100-foot level in the Hardy vein at the Ophir Mine in the Comstock district. (Bastin, 1922; Smith, 1985)

PECTOLITE

$NaCa_2Si_3O_8(OH)$

Pectolite occurs with zeolites in cavities in basic igneous rocks. It is also known as a primary mineral in alkaline igneous rocks, peridotites, and metamorphic rocks.

HUMBOLDT COUNTY Reported in a specimen from Horse Mountain. (AM)

PENTAHYDRITE

$MgSO_4 \cdot 5H_2O$

A water-soluble mineral of the chalcanthite group that occurs as an efflorescence in mine workings.

STOREY COUNTY As an efflorescence in an unspecified mine at Virginia City in the Comstock district. (M. Jensen, w.c., 1999)

PENTLANDITE

$(Fe,Ni)_9S_8$

Pentlandite is a common mineral in nickel-bearing massive sulfide deposits. It usually occurs with pyrrhotite, from which it has commonly exsolved.

CLARK COUNTY With pyrite and chalcopyrite in copper-nickel ore at the Key West and Great Eastern Mines in the Bunkerville district; also with nickel-bearing pyrrhotite in *hornblende* gneiss. (Lindgren and Davy, 1924)

In fractures with galena and an unidentified Pd-Bi-Te mineral in a specimen of friable ore from the Boss Mine in the Goodsprings district. (Jedwab, 1998b)

DOUGLAS COUNTY Questionably identified in intergrowths with pyrrhotite associated with pyrite and chalcopyrite disseminated in, and adjacent to, mafic dikes in the Red Canyon district. (Dixon, 1971)

ELKO COUNTY In small amounts in association with pyrrhotite as disseminations in sedimentary rocks in the Railroad district. (Ketner and Smith, 1963)

EUREKA COUNTY As anhedral inclusions to 30 microns in pyrite with arsenic-rich rims in the silicified core of the Betze deposit at the Goldstrike Mine in the Lynn district. (GCF)

PERHAMITE

$Ca_3Al_7(SiO_4)_3(PO_4)_4(OH)_3 \cdot 16.5H_2O$

A rare mineral that occurs in pegmatite elsewhere in the world. The Nevada occurrence is in a uniquely different setting.

HUMBOLDT COUNTY Abundant as tiny, white, hexagonal crystals in clusters at the Silver Coin Mine in the Iron Point district; also noted as colorless balls and plates. (Thomssen, 1998; S. Kleine, o.c., 1999; J. McGlasson, w.c., 2001)

PERICLASE

MgO

Periclase is a relatively rare mineral found in some high-temperature contact-metamorphic deposits in magnesian limestone and dolomite.

NYE COUNTY As rare disseminated grains in brucite, magnesite, and dolomite at the Premier Chemicals (Gabbs Magnesite-Brucite) Mine in the Gabbs district. (Schilling, 1968a)

PERITE

$PbBiO_2Cl$

A rare mineral that occurs in manganese-rich skarn in Sweden.

ESMERALDA COUNTY As tiny, pale-greenish-yellow crystals after aikinite with beyerite, wulfenite, mimetite, and bismutite in a specimen from a prospect on the west side of Cucomungo Wash in the Tule Canyon district; identified by S. Williams. (R. Thomssen, w.c., 1999)

PERRIERITE

$(Ca,Ce,Th)_4(Mg,Fe^{2+})_2(Ti,Fe^{3+})_3Si_4O_{22}$

A rare earth silicate mineral that is found in tuff, in granitic rocks, and in syenite pegmatite.

NYE COUNTY Identified as a trace accessory in

ash-flow tuff of the Paintbrush Group in the Yucca Mountain area. (Broxton et al., 1989)

PETZITE

Ag_3AuTe_2

Petzite is an uncommon mineral found in precious-metal telluride deposits. It is usually found in association with hessite, calaverite, krennerite, and other telluride minerals.

ESMERALDA COUNTY In tellurium-rich ore with hessite and sylvanite in the Goldfield district. (Tolman and Ambrose, 1934)

LANDER COUNTY Reported with gold and a mineral of the joséite-tetradymite group in the Telluride area at the head of Rocky Canyon in the Battle Mountain district. (Roberts and Arnold, 1965)

MINERAL COUNTY Found with hessite, pyrite, and chalcopyrite in white quartz veins at the Silver Dyke Tungsten Mine in the Silver Star district. (Sigurdson, 1974)

NYE COUNTY Reported in a specimen of gold-silver ore at the Wingfield shaft in the Wahmonie district. (WMK)

WASHOE COUNTY Identified with coloradoite in quartz and calcite veins at the Buster Mine (Gus Schave property) in the Olinghouse district. (Hill, 1911; Overton, 1947)

PHARMACOLITE

$CaHAsO_4 \cdot 2H_2O$

Pharmacolite is a rare mineral found in the oxidation zones of a few deposits that contain primary arsenic minerals.

EUREKA COUNTY As white, crystalline clusters to 1 mm on orpiment at the Gold Bar Mine, east Gold Pick pit, in the Antelope district. (S. Kleine, o.c., 1997; M. Jensen, w.c., 1999)

HUMBOLDT COUNTY As dull, yellow-brown coatings on matrix and white crystalline sprays to 1 mm on realgar crystals in the north pit dumps at the Getchell Mine in the Potosi district; occurs with pararealgar. (Dunning, 1987; UMNH; M. Jensen, w.c., 1999)

NYE COUNTY With haidingerite and pitticite at the White Caps Mine in the Manhattan district. (Palache et al., 1951; Gibbs, 1985)

PERSHING COUNTY Reported with arthurite at

the Majuba Hill Mine in the Antelope district. (Jenkins, 1981)

PHARMACOSIDERITE

$KFe_4^{3+}(AsO_4)_3(OH)_4 \cdot 6\text{-}7H_2O$

Pharmacosiderite occurs as an oxidation product of primary arsenic-bearing ore; specimens from some Nevada localities, such as Majuba Hill, are quite attractive.

CHURCHILL COUNTY In specimens from the Fireball Mine in the Truckee district. (R. Thomssen, w.c., 1999)

CLARK COUNTY Reported in a specimen from near Hoover Dam. (AMNH)

ESMERALDA COUNTY Reported from the Carrie Mine in the Gilbert district. (R. Thomssen, w.c., 1999)

EUREKA COUNTY Associated with alunite veins at the Goldstrike Mine in the Lynn district. (Arehart et al., 1992)

HUMBOLDT COUNTY As brilliant, dark-green cubes to 3 mm on and with jarosite, dufrénite, and arseniosiderite on barite in fractures in oxidized rock in the Sequoia pit at the Lone Tree Mine in the Buffalo Mountain district. (GCF)

As crystals to 20 microns in vugs in oxidized breccia at the Moonlight Uranium Mine in the Disaster district. (Castor et al., 1996)

Reported at the Silver Coin Mine in the Iron Point district. (Thomssen, 1998)

LINCOLN COUNTY Reported from the Silver King Mine in the Silver King district. (R. Thomssen, w.c., 1999)

NYE COUNTY As olive-green and honey-brown cubes, tetrahedrons, and tetrahedral twins to 0.3 mm on jasperoid at the Northumberland Mine in the Northumberland district. (Kokinos and Prenn, 1985)

As small, pale-sea-green crystals lining vugs and fractures in brecciated tuff at the Sunnyside Mine in the Round Mountain district. (Tingley and Berger, 1985)

As light-yellowish-green cubes, diagonally striated, coating quartz on the 370-foot level at the Montana-Tonopah Mine in the Tonopah district. (Melhase, 1935c)

PERSHING COUNTY Locally abundant as transparent, deep-green cubes, associated with scorodite and arthurite, at the Majuba Hill Mine in the Antelope district; many superb specimens have been collected and crystals to 2 cm have been reported. (MacKenzie and Bookstrom, 1976; Jensen, 1985)

Pale-yellow-green cubes to 0.1 mm on dufrénite at the Willard Mine, Willard district. (Jensen and Leising, 2001)

WASHOE COUNTY Reported from the Burrus Mine in the Pyramid district. (R. Thomssen, w.c., 1999)

PHENAKITE

Be_2SiO_4

Phenakite is a relatively rare mineral found in beryllium-bearing pegmatites and hydrothermal deposits.

CLARK COUNTY Tentatively reported with beryl in a pegmatite just west of the Arizona state line in the Virgin Mountains in the Bunkerville district. (Olson and Hinrichs, 1960)

ELKO COUNTY Rare in pegmatite dikes with quartz, beryl, *tourmaline,* and other minerals in the Dawley and Gilbert Canyon areas in the Valley View district of the southern Ruby Mountains. (Olson and Hinrichs, 1960)

MINERAL COUNTY A rare mineral in the Zapot pegmatite in the Fitting district. (Foord et al., 1999b)

WHITE PINE COUNTY As euhedral to subhedral crystals to 1 cm in irregularly shaped veins to 3 cm thick with beryl, bertrandite, *allanite, monazite,* fluorite, quartz, carbonate, muscovite, scheelite, and pyrite in a replacement deposit in limestone at the Wheeler Peak (Mt. Wheeler) Mine in the Lincoln district. (Whitebread and Lee, 1961; Lee and Erd, 1963; Griffiths, 1964)

PHILLIPSITE

$(K,Na,Ca_{0.5}Ba_{0.5})_{4\text{-}7}[Al_{4\text{-}7}Si_{12\text{-}9}O_{32}] \cdot 12H_2O$ (general)

A common zeolite mineral group that has recently been subdivided into three species, depending on the most abundant alkali or alkali earth element. *Phillipsite* is most commonly found in cavities in basalt; however, all Nevada occurrences are in diagenetically or hydrothermally altered tuffaceous rocks.

CHURCHILL COUNTY With *erionite* and other

zeolite minerals at the Eastgate deposit in the Eastgate district. (Papke, 1972a)

CLARK COUNTY Abundant as dull, brownish-black aggregates of tiny crystals in sedimentary and volcaniclastic rocks with gypsum and manganese minerals at the Three Kids Manganese Mine in the Las Vegas district. (Hewett and Weber, 1931)

EUREKA COUNTY With *clinoptilolite,* jarosite, and other minerals in the Pine Valley district. (Papke, 1972a)

LANDER COUNTY With mordenite, analcime, and *clinoptilolite* in tuff 35 miles north of Austin in the Steiner Canyon district in the Reese River Valley. (Papke, 1972a)

MINERAL COUNTY Abundant in rhyolite tuffs and tuffaceous clays of Teels Marsh, a salt-crusted playa in the Teels Marsh district; associated with gaylussite, searlesite, *clinoptilolite,* and analcime in some samples. (Hay, 1964)

In altered volcanic rocks a few miles west of Candelaria. (Papke, 1972a)

PERSHING COUNTY Found with *clinoptilolite* and *erionite* in altered tuff in the Jersey Valley zeolite deposits 42 miles south of Battle Mountain; *phillipsite* identified in a specimen from the Mount Tobin area is probably from the same locality. (Deffeyes, 1959; Papke, 1972a; JHS)

PHLOGOPITE

KMg$_3$AlSi$_3$O$_{10}$(OH)$_2$

A magnesium-rich mica that forms a series with biotite, phlogopite generally occurs in ultramafic igneous rocks or in metamorphosed dolomite or magnesian limestone.

DOUGLAS COUNTY Reported from an unspecified locality along the state line between Nevada and California. (WMK)

ELKO COUNTY Reported with beryl and other minerals in pegmatite in Dawley Canyon in the Valley View district. (Olson and Hinrichs, 1960)

EUREKA COUNTY In intrusive rocks and skarn at the Goldstrike Mine in the Lynn district; also with clinochlore and calcite in selvages. (GCF)

In breccia around fragments of andesite in the footwall of the iron orebody at the Barth Iron Mine in the Safford district. (Shawe et al., 1962)

HUMBOLDT COUNTY In ultramafic rock at the Twin Creeks Mine in the Potosi district; identified by SEM/EDX. (SBC)

LANDER COUNTY In skarn at the West orebody in Copper Canyon in the Battle Mountain district. (Theodore and Blake, 1978)

NYE COUNTY In gangue with *chlorite, sericite,* actinolite, calcite, and dolomite at the Phelps Stokes Mine in the Ellsworth district. (Reeves et al., 1958)

PERSHING COUNTY Reported as a major constituent of pegmatite with scheelite and beryl at the Oreana Tungsten Mine in the Rye Patch district. (Olson and Hinrichs, 1960)

WHITE PINE COUNTY With nontronite and actinolite in skarn adjacent to porphyry at the Tripp open pit in the Robinson district. (Melhase, 1934d; James, 1976)

PHOENICOCHROITE

Pb$_2$(CrO$_4$)O

A rare secondary mineral related to crocoite that occurs in lead-bearing veins.

CLARK COUNTY In the Eldorado district near Nelson as small, crude, bright-red crystals with hemihedrite. (R. W. Thomssen, w.c., 1999)

ESMERALDA COUNTY As red microcrystals with fornacite, wulfenite, and other minerals in quartz veins a mile north of the site of Columbus in the Candelaria district. (R. Walstrom, w.c., 2000)

LANDER COUNTY Reported in a specimen from an unknown locality in the Kingston district. (WMK)

PHOSGENITE

Pb$_2$(CO$_3$)Cl$_2$

Phosgenite is a rare mineral that is generally found with cerussite and other secondary lead minerals in oxidized portions of base-metal deposits.

LINCOLN COUNTY Coating pyrite with tiny crystals of pyromorphite in some of the fissure ore at the Prince Mine in the Pioche district. (Westgate and Knopf, 1932; Melhase, 1934d)

NYE COUNTY As yellow to greenish-yellow prisms to 0.5 mm with leadhillite in vugs in altered limestone at the Illinois Mine in the Lodi district, Sec. 17, T13N, R37E. (Jenkins, 1981)

PHOSPHOFIBRITE

$KCu^{2+}Fe^{3+}_{15}(PO_4)_{12}(OH)_{12} \cdot 12H_2O$

A rare secondary mineral that typically occurs with other iron phosphates; first described from an occurrence in the Black Forest of Germany.

HUMBOLDT COUNTY As blue fibers with lipscombite at the Silver Coin Mine, Iron Point district; identified by XRD. (J. McGlasson, w.c., 2001)

PHOSPHOSIDERITE

$Fe^{3+}PO_4 \cdot 2H_2O$

Phosphosiderite is a rare mineral, generally found with other phosphate species in pegmatites or iron ore deposits. At the Nevada locality it is associated with strengite and other secondary phosphate minerals.

NYE COUNTY In intimate mixtures with strengite as fracture fillings along a limestone-rhyolite contact near Manhattan in the Manhattan district; variscite and vashegyite are also present. (Shannon, 1923; McConnell, 1940)

PHOSPHURANYLITE

$KCa(H_3O)_3(UO_2)_7(PO_4)_4O_4 \cdot 8H_2O$

Phosphuranylite, a relatively uncommon secondary uranium mineral, is found as an alteration product of primary uranium minerals or in occurrences in which no primary species are apparent.

ESMERALDA COUNTY With autunite as cavity fillings in a rhyolitic welded tuff at a prospect south of Coaldale Junction in the Coaldale district. (Garside, 1973)

LYON COUNTY Reported with autunite and torbernite at the Flyboy Mine, Sec. 16, T11N, R26E. (Garside, 1973)

PERSHING COUNTY With autunite at the Four Jacks, Pennies, and Maybeso claims in the Nightingale district. (Garside, 1973)

WASHOE COUNTY Reported with autunite in uranium prospects in or adjacent to Tertiary ashflow tuff at the Sunnyside occurrence in the Dogskin Mountain district. (Bonham and Papke, 1969)

With autunite, sabugalite, uraninite, and other uranium minerals at the DeLongchamps and Red Bluff Mines in the Pyramid district. (Brooks, 1956; Garside, 1973)

PICKERINGITE

$MgAl_2(SO_4)_4 \cdot 22H_2O$

Pickeringite, a relatively rare water-soluble sulfate mineral, is generally found as efflorescence with other *alum* group minerals in arid regions.

CARSON CITY Reported in a specimen from Carson City. (AMNH)

CHURCHILL COUNTY As hard gray masses about half the size of a pea on partings and as fibrous veins in tuff in the low hills south of Lahontan Reservoir in the Camp Gregory district. (Hewett, 1924; Melhase, 1934a)

NYE COUNTY Identified in a specimen reported to be from a location called Alumine, probably the Alum Mine near Silver Peak in Esmeralda County (see kalinite). (LANH)

STOREY COUNTY As an efflorescence on mine walls at an unspecified locality in the Comstock district. (M. Jensen, w.c., 1999)

WASHOE COUNTY Reported in a specimen from Steamboat Springs in the Steamboat Springs district. (MV)

With apjohnite and hexahydrite at the Wedekind Mine in the Wedekind district. (GCF)

PICROPHARMACOLITE

$H_2Ca_4Mg(AsO_4)_4 \cdot 11H_2O$

A rare mineral that is generally found in carbonate rock–hosted ore deposits.

HUMBOLDT COUNTY As white coatings less than 5 mm thick and botryoidal cauliflowers composed of microscopic bladed crystals in and associated with realgar and orpiment at the Getchell Mine, north pit, in the Potosi district. (Stolburg and Dunning, 1985; Dunning, 1988)

PIEMONTITE

$Ca_2(Al,Mn^{3+},Fe^{3+})_3(SiO_4)_3(OH)$

Piemontite, a relatively uncommon manganese-rich member of the epidote group, is generally found in manganese-rich metamorphic rocks.

HUMBOLDT COUNTY Reported from the Getchell Mine in the Potosi district. (R. W. Thomssen, w.c., 1999)

LYON COUNTY Reported as pink masses at the Ludwig gypsum pit in the Yerington district. (S. Pullman, w.c., 1999)

MINERAL COUNTY Reported at the Grutt Brothers Mine in the Rawhide district. (CSM)

PERSHING COUNTY A purplish-red mineral in small amounts in massive manganese-*chalcedony* bodies at the Black Diablo Mine in the Black Diablo district, Sec. 2, T32N, R39E. (Ferguson et al., 1951; Johnson, 1977)

WASHOE COUNTY As bright-red to black crystals to an inch or more in vugs, veins, encrustations, and disseminations in metavolcanic rocks of the Peavine sequence on Peavine Mountain about 8 miles northwest of Reno; also reported on Petersen Mountain to the north. (Gianella, 1937; Bonham and Papke, 1969; Garside, 1998)

PIGEONITE

$(Mg,Fe^{2+},Ca)(Mg,Fe^{2+})Si_2O_6$

Pigeonite, a clinopyroxene group mineral, occurs in quickly chilled igneous rocks, particularly mafic to intermediate volcanic rocks. It is probably much more common in Nevada than recognized.

HUMBOLDT COUNTY Pigeonite and the variety *ferropigeonite* are found as phenocrysts in mafic volcanic rocks along with Fe-rich augite and Fe-rich *hypersthene* in the Opalite district. (Conrad, 1984)

LINCOLN COUNTY With anorthoclase, augite, and other minerals in trachyte at the mouth of Boulder Canyon in Kane Springs Wash. (Novak, 1985)

PINTADOITE

$Ca_2V_2^{5+}O_7 \cdot 9H_2O$

A water-soluble efflorescent mineral that is best known from sandstone-hosted uranium-vanadium deposits in southeastern Utah.

EUREKA COUNTY Identified by EMP and XRD in olive-green coatings on specimens from the Van-Nav-San claims, Gibellini district. (GCF)

PITTICITE

$Fe_x^{3+}(AsO_4)_y(SO_4)_z \cdot nH_2O$

Pitticite, a rare and poorly understood amorphous mineral, occurs in the oxidation zones of arsenic-rich hydrothermal deposits.

EUREKA COUNTY Identified in blue to gray-black crusts and efflorescent coatings on silicified and sulfide-rich limestone with quartz, stibnite, barite, gypsum, kaolinite, and pyrite at the Goldstrike Mine in the Lynn district. (GCF)

NYE COUNTY Formed by post-mining oxidation as thin seams in melanterite, small globules on melanterite stalactites, botryoidal crusts, and ooze from cracks on mine walls at the White Caps Mine in the Manhattan district; also identified as greenish-white scales at the Manhattan Consolidated Mine. (Foshag and Clinton, 1927; Palache et al., 1951; WMK; Jenkins, 1981)

As black, waxy masses at the Northumberland Mine in the Northumberland district. (Kokinos and Prenn, 1985)

PERSHING COUNTY Reported with scorodite at the Majuba Hill Mine in the Antelope district. (EMP; Jenkins, 1981)

PLAGIONITE

$Pb_5Sb_8S_{17}$

A sulfosalt mineral found with other sulfides in hydrothermal vein deposits.

NYE COUNTY As iridescent, dark-gray to black, flattened needles with rhodochrosite and sphalerite in a specimen from the Morey district; identified by S. Williams. (R. Thomssen, w.c., 1999)

PLANCHÉITE

$Cu_8^{2+}Si_8O_{22}(OH)_4 \cdot H_2O$

A relatively uncommon secondary mineral in oxidized copper deposits.

CHURCHILL COUNTY As blue crystals to 1.5 mm on green brochantite at the Lovelock Mine in the Table Mountain district. (B. and J. Runner, w.c., 2000)

PLANERITE

$Al_6(PO_4)_2(PO_3OH)_2(OH)_8 \cdot 4H_2O$

A mineral that was originally considered to be a variety intermediate between turquoise and coeruleolactite.

ELKO COUNTY Identified by SEM and XRD as light-blue to white pulverulent coatings under variscite crystal sprays at the Rain Mine in the Carlin district. (M. Jensen, w.c., 1999)

PLATINUM

Pt

Platinum, an uncommon native metal, is generally found with palladium and other noble metals in placer accumulations and in deposits in mafic rocks. The hydrothermal Goodsprings occurrence is unusual.

CLARK COUNTY Reported in plumbojarosite-beaverite masses at the Boss and Oro Amigo Mines in the Goodsprings district. (Hewett, 1931; Jedwab, 1998b; WMK)

PLATTNERITE

PbO_2

Plattnerite is a rare mineral that occurs in the oxidation zones of lead deposits.

CLARK COUNTY At the Mobile, Whale, Ruth, and Milford No. 3 Mines in the Goodsprings district as radiating sprays of fine black needles, some with reddish-brown tips, on wulfenite plates; also as clayey coatings on, and as inclusions in, wulfenite. (Takahashi, 1957, 1960; Kepper, 1998)

EUREKA COUNTY As black, acicular crystals to 2 mm in narrow veins in gossan at the Diamond Mine in the Eureka district on stope walls about 3200 feet into the adit; crystals are typically tapered prisms. (Desautels, 1972; Jenkins, 1981)

LINCOLN COUNTY Reported in a specimen from the Dave Matthews property in the Ely Springs district. (WMK)

NYE COUNTY Black crystals to 0.5 mm on *limonite* at the Downieville Mine in the Gabbs district. (M. Jensen, w.c., 1999)

WHITE PINE COUNTY As opaque, metallic-black, acicular crystal aggregates and spherules to 1 mm with calcite; also after scrutinyite crystals on gossan with malachite at the Grand Deposit Mine in the Muncy Creek district. (UMNH; FUMT; M. Jensen, w.c., 1999; Jenkins, 1981)

PLAYFAIRITE

$Pb_{16}Sb_{18}S_{43}$

A rare sulfosalt mineral that occurs with other sulfide and sulfosalt phases.

LANDER COUNTY Reported in specimens from Austin in the Reese River district. (AMNH; RMO)

PLUMBOBETAFITE

$(Pb,U,CA)(Ti,Nb)_2O_6(OH,F)$

A rare member of the *pyrochlore* group that has been found in pegmatite in alkaline intrusive rock at two localities in Russia.

MINERAL COUNTY Intergrown with betafite and plumbopyrochlore in brown to black masses in the Zapot pegmatite in the Fitting district. (Foord et al., 1999b)

PLUMBOGUMMITE

$PbAl_3(PO_4)_2(OH)_5 \cdot H_2O$

Plumbogummite is a secondary mineral formed by the oxidation of lead ore.

EUREKA COUNTY As spherical clusters to 20 microns of colorless, leafy crystals in brecciated jasperoid with other phosphate minerals at the Gold Quarry Mine in the Maggie Creek district. (Jensen et al., 1995)

HUMBOLDT COUNTY As bluish botryoids at the Silver Coin Mine in the Iron Point district. (Thomssen, 1998; J. McGlasson, w.c., 2001)

NYE COUNTY Reported at the San Rafael Mine in the Lodi district. (Pullman and Thomssen, 1999)

PLUMBOJAROSITE

$PbFe_6^{3+}(SO_4)_4(OH)_{12}$

Plumbojarosite is a common mineral in the oxidation zones of base-metal deposits in the arid west of the United States. *Vegasite,* discovered in the Goodsprings district near Las Vegas, is a discredited species that has been shown to be identical with plumbojarosite.

CHURCHILL COUNTY Reported in oxidized portions of lead-silver veins in the Chalk Mountain district as sparse microscopic scales with descloizite and other minerals. (Schilling, 1962a; Jenkins, 1981)

An oxidized ore from an unspecified locality in the Holy Cross district. (Peters et al., 1996)

CLARK COUNTY In the Goodsprings district at the Boss, Kirby, Yellow Pine, and other mines; locally abundant enough to be an ore mineral. At the Boss Mine, bismuth-bearing plumbojarosite occurs abundantly as perfect flat pseudohexagonal (trigonal) crystals to 0.1 mm with beaverite, bismutite, and quartz; also reported in hand specimen as greenish-yellow powder in cavities and on fracture surfaces in gold-platinum-palladium ore. At the Rosella prospect, several hundred feet north of the Boss Mine, plumbojarosite was described as *vegasite*. (Knopf, 1915a; Hewett, 1931; Mount, 1964; Jedwab, 1998b; NMBM)

ELKO COUNTY With powdery cerussite in a quartz vein in gabbro and *pyroxene* diorite at the Regent Mine in the Ferber district. (Hill, 1916)

As tiny crystals in cavities in gossan with bayldonite, fornacite, carminite, and beudantite at the Killie Mine in the Spruce Mountain district, Sec. 14, T31N, R63E; also reported from the Black Forrest Mine. (Hill, 1916; Dunning and Cooper, 1987; UAMM)

Reported as a constituent of oxidized ore in the Tecoma district. (Hill, 1916)

EUREKA COUNTY In a specimen with galena, iron oxide, and cerussite at the Diamond Mine in the Diamond district. (WMK)

One of the important lead ores during early mining in the Eureka district; reported with beudantite (?) and mimetite. (Nolan, 1962)

HUMBOLDT COUNTY Reported at the Silver Coin Mine in the Iron Point district. (S. Pullman, w.c., 1999)

Reported as gold- and silver-bearing at the Silver Hill Group in the Winnemucca district. (WMK)

LANDER COUNTY With mimetite at the Nevada and Lucky Strike Mines in the Battle Mountain district. (Roberts and Arnold, 1965)

LINCOLN COUNTY With jarosite as yellow-brown, earthy masses at the Bristol Silver Mine in the Bristol district. (CSM; GCF)

As yellow powder with cerussite and opaline silica in jasperoid at the Lucky Chief Mine in the Chief district. (Callaghan, 1936)

With cerussite and *limonite* in oxidized lead-silver ore at the Comet and other veins in the Comet district. (Westgate and Knopf, 1932; Tschanz and Pampeyan, 1970)

MINERAL COUNTY With cerussite in a replacement-type orebody at the Simon Mine in the Bell district. (Knopf, 1921b)

In oxidized ore in the Broken Hills district. (Schrader, 1947)

In minor amounts with bindheimite in oxidized silver-antimony ore at the Potosi Mine in the Candelaria district. (Jenkins, 1981)

NYE COUNTY With cerussite, anglesite, wulfenite, and other minerals in oxidized base-metal ore at the San Rafael Mine in the Lodi district. (Jenkins, 1981; S. Pullman, w.c., 1999)

In strongly leached, precious-metal-rich zones of black to orange sandy quartz at the Paradise Peak Mine in the Paradise Peak district. (John et al., 1991)

Abundant as brown greasy material in lead-silver deposits in the Downieville area in the Gabbs district. (Jenkins, 1981)

PERSHING COUNTY In jasperoid in the Arabia district. (Knopf, 1918a; Lawrence, 1963)

WHITE PINE COUNTY Tentatively identified in lead-silver deposits in the Robinson district. (Spencer et al., 1917)

PLUMBOPYROCHLORE

$(Pb,Y,U,Ca)_{2-x}Nb_2O_6(OH)$

A member of the *pyrochlore* group that generally occurs in altered granite, in pegmatite in alkaline igneous rocks, and in carbonatite.

MINERAL COUNTY Intergrown with betafite and plumbobetafite in brown to black masses in the Zapot pegmatite in the Fitting district. (Foord et al., 1999b)

POKROVSKITE

$Mg_2(CO_3)(OH)_2 \cdot 0.5H_2O$

A very rare mineral that is found elsewhere in the world in only three other places.

NYE COUNTY As white rounded grains and round clusters of acicular prisms with *nakauriite*, callaghanite, and chalcopyrite at the Premier Chemicals (Gabbs Magnesite-Brucite) Mine in the Gabbs district. (M. Jensen, w.c., 1999; S. Kleine, o.c., 1999; GCF)

POLHEMUSITE

$(Zn,Hg)S$

A mineral known only from two localities: a stibnite replacement deposit in Idaho, and the Getchell Mine, Nevada.

HUMBOLDT COUNTY Identified by microprobe by S. Williams as black microgranular inclusions in silicified rock at the Getchell Mine in the

Potosi district. (Dunning, 1987; FUMT; EMP; F. Cureton, w.c., 1999)

POLYBASITE

$(Ag,Cu)_{16}Sb_2S_{11}$

Polybasite is widespread in small amounts in Nevada silver deposits, where it occurs as both a hypogene and supergene species with "ruby silver" minerals, acanthite, and base-metal sulfides.

CHURCHILL COUNTY In quartz veins at the Fairview Mine in the Fairview district. (Schrader, 1947)

As euhedral crystals to 5 mm with acanthite crystals on quartz and *adularia* at the Summit King and Summit Queen Mines in the Sand Springs district. (M. Jensen, w.c., 1999)

ELKO COUNTY Reported at the North Belle Isle Mine in the Tuscarora district. (HMM)

ESMERALDA COUNTY Questionably identified as crystals with "ruby silver" at the Florence Mine in the Goldfield district. (Ransome, 1909a)

EUREKA COUNTY With stephanite and tetrahedrite in replacement bodies and breccia in the Mineral Hill district. (Emmons, 1910; Schilling, 1979)

HUMBOLDT COUNTY With other sulfides at the Lone Tree Mine in the Buffalo Mountain district. (Bloomstein et al., 1993)

LANDER COUNTY Reported at the Garrison Mine in the Cortez district. (S. Pullman, w.c., 1999)

With stephanite, acanthite, and other minerals in quartz at the Betty O'Neal Mine in the Lewis district; also noted at the Henry Logan and Dean Mines. (Vanderburg, 1939; R. Thomssen, w.c., 1999; J. McGlasson, w.c., 2001)

Reported with other silver minerals in the Reese River district, specifically with proustite, pyrargyrite, and stephanite in the Amador Mine. (Hoffman, 1878; Melhase, 1934c; Ross, 1953; Lawrence, 1963)

LYON COUNTY Reported at the Talapoosa Mine in the Talapoosa district; also reported from the Christianson shaft dump. (Van Nieuwenhuyse, 1991; R. Thomssen, w.c., 1999)

Reported in a specimen from the Como district. (R. Thomssen, w.c., 1999)

MINERAL COUNTY Reported in gold-silver-lead ore at the Nevada Rand Mine in the Rand district. (Schrader, 1947)

NYE COUNTY With dyscrasite and galena at the Highbridge and Transylvania Mines in the Belmont district. (Jenkins, 1981)

The most important silver mineral at the Richardson Mine in the Hannapah district; associated with pyrite and quartz in a shear zone in rhyolite. (Kral, 1951)

As platy crystals, cherry-red on freshly broken surfaces, in cavities and seams in quartz with chalcopyrite and barite in the Tonopah district; specifically reported at the North Star, Tonopah Extension, Tonopah-Belmont, and Summit King Mines. (Spurr, 1903; Bastin and Laney, 1918; Melhase, 1935c; Nolan, 1935; UCB; S. Pullman, w.c., 1999; R. Thomssen, w.c., 1999)

With tetrahedrite at the Ophir and Dexter Mine in the Tybo district. (Jenkins, 1981)

Reported with linarite in a specimen from Smoky Valley. (AM)

PERSHING COUNTY Reported from the Rochester district. (Von Bargen, 1999)

STOREY COUNTY Common in bonanza ore as massive, dull, black mixtures with acanthite and stephanite at the Consolidated Virginia and other mines in the Comstock district. Reported in flat hexagonal prisms with pyrite, chalcopyrite, sphalerite, galena, *argentite, electrum,* and quartz in specimens of bonanza ore from depths of 2000 to 2500 feet in the C and C Mine; also noted as a partial replacement of acanthite at a depth of less than 350 feet in the Belcher Mine. (Becker, 1882; Bastin, 1922; Smith, 1985; LANH; HUB; WMK)

WASHOE COUNTY In a vein specimen with pyrite, acanthite, roquesite, and other minerals at the Sure Fire prospect in the Pyramid district, NW1/4 Sec. 22, T23N, R21E. (SBC)

WHITE PINE COUNTY With stephanite and "ruby silver" minerals at the Star Mine in the Cherry Creek district. (Schrader, 1931b)

POLYDYMITE

$NiNi_2S_4$

Polydymite is a rare mineral found in hydrothermal veins and magmatic deposits.

CLARK COUNTY As a violet-gray mineral replacing both pentlandite and chalcopyrite, but mainly the former, as minute veinlets to masses at the Great Eastern and Key West Mines in the Bunkerville district. (Lindgren and Davy, 1924)

POLYHALITE

$K_2Ca_2Mg(SO_4)_4 \cdot 2H_2O$

Polyhalite is a widely distributed mineral; it is commonly associated with halite and anhydrite in marine evaporite deposits.

ESMERALDA COUNTY Reported with kalinite, cinnabar, and sulfur in a specimen from the Alum Mine in the Alum district. (UCB)

POSNJAKITE

$Cu_4^{2+}(SO_4)(OH)_6 \cdot H_2O$

A rare secondary mineral that is formed from copper-bearing sulfides under acidic oxidizing conditions.

PERSHING COUNTY As blue microcrystalline laths in vuggy sulfide matrix at the Majuba Hill Mine in the Antelope district. (Jensen, 1985)

POTARITE

PdHg

A rare mineral originally described as a mercury-palladium amalgam from diamond gravels in South America.

CLARK COUNTY As micron-sized grains in cracks in siliceous rock with copper minerals, cinnabar, and palladium-bearing gold at the Boss Mine in the Goodsprings district, Sec. 27, T24S, R57E. (Jedwab, 1998a)

POTASSIUM ALUM

$KAl(SO_4)_2 \cdot 12H_2O$

A rare water-soluble mineral with nearly the same composition as kalinite but different structure.

ESMERALDA COUNTY Reported as glassy crusts to 20 cm thick associated with sulfur and minor gypsum at the Alum Mine in the Alum district; may be the mineral kalinite, which has been verified from this locality. (Spurr, 1906; UCB; LANH; M. Jensen, w.c., 1999)

Reported from Silver Peak Marsh; possibly the same occurrence as the Alum Mine. (AMNH; S. Kleine, o.c., 1999)

POTTSITE

$PbBiH(VO_4)_2 \cdot 2H_2O$

A rare Nevada type mineral that is found as a secondary phase with other bismuth- and vanadium-bearing minerals; at the type locality, it occurs in the oxidized zone of a tungsten deposit.

CHURCHILL COUNTY As sharp yellow crystals with bismutite, vanadinite, and clinobisvanite in an unnamed locality northeast of Chalk Mountain in the Clan Alpine Mountains. (R. Walstrom, w.c., 1999)

LANDER COUNTY At the type locality, sparingly as bright-yellow microcrystals with clinobisvanite, duhamelite, scheelite, bismutite, and vanadinite in thin veinlets that cut skarn in a single small outcrop at the Linka Mine in the Spencer Hot Springs district, E1/2 Sec. 12, T17N, R45-1/2E. Named for Potts, a town site to the southeast. (Williams, 1988; S. Sears, w.c., 1999)

POWELLITE

$CaMoO_4$

Powellite is a common mineral in molybdenum and tungsten skarn deposits. It is less common in hydrothermal vein deposits.

CHURCHILL COUNTY With scheelite in epidote-magnetite skarn from an unspecified locality. (Schilling, 1979)

As crystals to 0.5 inches in diameter with scheelite along fractures in skarn from the Midnight Mine in the Fairview district. (Stager and Tingley, 1988)

DOUGLAS COUNTY With scheelite in skarn at the Alpine Mine in the Gardnerville district. (Doebrich et al., 1996)

ELKO COUNTY With bismuthinite, pyrite, chalcopyrite, molybdenite, and scheelite in skarn at the Garnet (Tennessee Mountain) Mine in the Alder district. (Stager and Tingley, 1988; Smith, 1976; CSM)

ESMERALDA COUNTY As yellow crystals to 2 mm at the Black Horse Mine in the Black Horse district. (Stager and Tingley, 1988; Schilling, 1980)

Abundant with ferrimolybdite as excellent, large, distinct crystals to 5 mm in crusts and lining small vugs at the Tonopah-Divide Mine in the Divide district, Sec. 27, T2N, R42E; crystals differ in habit and color, from pyramidal to tabular, and milky brown to nearly transparent pale-yellow-brown; powellite that is reported from Tonopah in many collections is probably from this locality. (Knopf, 1921a; Pough, 1937)

As soft white flakes after molybdenite in quartz veins with ferrimolybdite at the Slate Ridge prospects in the Gold Point district. (Schilling, 1979)

HUMBOLDT COUNTY After arsenopyrite in radioactive rock at the Horse Creek uranium prospect in the Disaster district. (Castor et al., 1996; Castor and Henry, 2000)

As white crusts on quartz-molybdenite veinlets at the Molly prospect in the Gold Run district. (Schilling, 1979)

Said to occur in tungsten deposits in the Sonoma Range, possibly in reference to deposits in the Osgood Mountains. (Gianella, 1941)

LANDER COUNTY With magnetite and scheelite in epidote-quartz-*garnet* skarn at the Gold Acres tungsten prospect in the Bullion district. (Stager and Tingley, 1988)

With scheelite and sulfide minerals in *garnet*-epidote tactite at the Conquest and Linka Mines in the Spencer Hot Springs district. (Stewart et al., 1977a)

LINCOLN COUNTY Reported as a constituent of tungsten skarn deposits at the Schofield Mine and other properties in the Tem Piute district. (Stager and Tingley, 1988; Tschanz and Pampeyan, 1970)

LYON COUNTY As small golden-brown crystals in tungsten deposits in the Cambridge Hills in the Yerington district; specifically reported from the Flyboy Mine. (Jenkins, 1981; R. Thomssen, w.c., 1999)

Reported from Sweetwater in the Washington district. (Gianella, 1941)

MINERAL COUNTY In oxidized copper-zinc-antimony ore as tiny, nearly colorless and translucent, pyramidal crystals on drusy hemimorphite at the Anderson Mine in the Santa Fe district; partially replaced by hydrozincite and chrysocolla. (Goudey, 1952; S. Pullman, w.c., 1999)

Reported in contact-metamorphic deposits at unspecified localities near Luning and in the White Mountains. (Gianella, 1941)

NYE COUNTY Reported from the Heiser claims in the Oak Spring district on the Nevada Test Site. (Melhase, 1935b; Gianella, 1941; WMK)

Identified in a sample from an unspecified locality in the San Antone district. (JHS)

PERSHING COUNTY Reported with scheelite at the Star Mine in the Juniper Range district. (Johnson, 1977)

Reported with scheelite in the Mill City and Nightingale districts. (Gianella, 1941)

WHITE PINE COUNTY Reported at unspecified localities in the Bald Mountain and White Pine districts. (WMK; Gianella, 1941)

PREHNITE

$Ca_2Al_2Si_3O_{10}(OH)_2$

Prehnite is common outside Nevada in cavities with zeolites in mafic volcanic rocks and as a product of low-grade metamorphism. In Nevada it is mainly found in contact-metamorphosed rocks.

EUREKA COUNTY In white, sheaf-like aggregates to 1 mm in retrograde skarn with actinolite, titanite, and *apatite* at the Goldstrike Mine in the Lynn district. (Walck, 1989)

HUMBOLDT COUNTY Intergrown with carbonates in veinlets cutting contact-metamorphic rocks at Dutch Flat in the Dutch Flat district. (Hotz and Willden, 1964)

LANDER COUNTY With epidote, clinozoisite, and *sericite* after *plagioclase* in hydrothermally altered granodiorite in the Mount Lewis area in the Shoshone Range. (Gilluly and Gates, 1965)

MINERAL COUNTY In skarn about 7 miles north of Luning in the Fitting district. (S. Pullman, w.c., 1999)

Reported with magnesio-axinite, vesuvianite, actinolite, and epidote at a locality near Luning; this may be the same occurrence as that noted above. (Dunn et al., 1980; UCB)

NYE COUNTY As masses of white fibers associated with porphyroblasts of grossular and vesuvianite in the wall rock of gibbsite-lithiophorite veins at the Queen City Mines claims in the Queen City district. (Hewett et al., 1968)

PROSOPITE

$CaAl_2(F,OH)_8$

A mineral of fluorine-rich granites, greisens, and granite pegmatites; rarely in base-metal deposits.

MINERAL COUNTY In white to purple masses to 10 cm as alteration rims on topaz and with fluorite in pegmatite at the Zapot claims. (Foord et al., 1999b)

PROUSTITE

Ag_3AsS_3

Proustite is less common than its antimony analog pyrargyrite. Both are referred to as "ruby silver" and occur in hydrothermal silver deposits as hypogene and supergene phases. Excellent specimens have come from the Austin district.

CLARK COUNTY With tennantite and antimony oxides at the Lavina prospect in the Goodsprings district, Sec. 21, T24S, R58E. (Hewett, 1931; Melhase, 1934e)

ELKO COUNTY Identified in silver ore from the Tuscarora district. (Emmons, 1910)

ESMERALDA COUNTY Questionably identified with polybasite at the Florence Mine in the Goldfield district. (Ransome, 1909a)

With galena, sphalerite, and enargite at the Alpine Mine in the Lone Mountain district. (Phariss, 1974)

LANDER COUNTY Reported in a specimen from an unknown locality in the Battle Mountain district. (WMK)

An important component of high-grade silver ore as transparent red crystals with stephanite at the Manhattan Co. Mine in the Reese River district; also in silver ore from the Amador Mine. (Ross, 1925; Melhase, 1934c; Lawrence, 1963; UCB)

LINCOLN COUNTY Reported at the Silver Park Mine in the Atlanta district. (Hill, 1916)

In rich silver ore from the Greenwood fissure in the Pioche district. (Westgate and Knopf, 1932)

LYON COUNTY Reported in silver ore in the Como district. (Gianella, 1941)

MINERAL COUNTY Reported with other sulfides at the western base of the southern Monte Cristo Mountains. (Schrader, 1947)

NYE COUNTY As crystals in quartz vugs in veins at the Clifford Mine in the Clifford district. (Ferguson, 1916)

Proustite or arsenian pyrargyrite was noted in some ore from the Jim Butler and Mizpah Mines in the Tonopah district. (Bastin and Laney, 1918; CSM; S. Pullman, w.c., 1999)

Identified in a sample from an unspecified locality in the Union district. (JHS)

PERSHING COUNTY Reported in a specimen from an unknown locality in the Nightingale district. (WMK)

Reported in some rich gold ore in the Seven Troughs district. (Ransome, 1909b)

STOREY COUNTY In small quantities in bonanza silver ore of the Comstock district; specifically reported from the Consolidated Virginia and California Mines. (Bastin, 1922; HMM)

WHITE PINE COUNTY Reported as marvelous specimens with polybasite, stephanite, and pyrargyrite at the Star Mine in the Cherry Creek district. (Schrader, 1931b; Melhase, 1934d)

PSEUDOBROOKITE

$(Fe^{3+},Fe^{2+})_2(Ti,Fe^{3+})O_5$

Pseudobrookite is most commonly found in cavities in volcanic rock. It crystallizes from late-stage fluids related to the volcanism, along with such minerals as hematite, tridymite, and topaz.

LANDER COUNTY With topaz, fluorite, cassiterite, titanite, and other minerals in heavy mineral concentrates of samples from small tin deposits northeast of the Izenhood Ranch in the Izenhood district. (Fries, 1942)

PSEUDOMALACHITE

$Cu_5^{2+}(PO_4)_2(OH)_4$

A relatively rare mineral that occurs in the oxidation zones of some copper deposits, generally with such minerals as libethenite, cornetite, and other phosphates or arsenates. Specimens from the Yerington district are reportedly among the world's best.

EUREKA COUNTY In green spherules to 2 mm at the Gold Quarry Mine in the Maggie Creek district. (Jensen et al., 1995)

In a specimen from the Badger Mine, Hilltop claim, Lynn district. (R. Thomssen, w.c., 1999)

HUMBOLDT COUNTY As dark-green crystals to 2 mm with libethenite and pseudomalachite near Golconda Summit. (R. Thomssen, o.c., 2000)

LANDER COUNTY Tentatively identified in a specimen from the Snowstorm Mine in the North Battle Mountain district. (R. Thomssen, o.c., 2000)

LYON COUNTY As bluish-green botryoids to 0.5 mm at the Jackpot Mine in the Wellington district. (S. Pullman, w.c., 1999; S. White, w.c., 1999)

In the Yerington district with libethenite and brochantite as crystalline spherules to 2 mm in the Douglas Hill Mine; also described as drusy aggregates of bright, green to grass-green crystals with bluish *apatite* prisms on altered porphyry in the Empire-Nevada Mine, as green crystalline crusts with chrysocolla in the MacArthur Mine, and from the Blue Jay Mine. (Palache et al., 1951; UAMM; M. Jensen, w.c., 1999; R. Thomssen, w.c., 1999)

MINERAL COUNTY As emerald to blackish-green minutely botryoidal crusts at the Calavada Mine in the Bell district, Sec. 9, T8N, R37E; has no apparent crystalline structure under the binocular microscope, but under the polarizing microscope has radiating subfibrous texture and strong anisotropism. (Goudey, 1945)

NYE COUNTY As small, bright-green, finely fibrous, globular encrustations on quartz with hübnerite and rhodonite in some mines in the Tonopah district. (Melhase, 1935c)

Tentatively identified as green films on dark slate in a contact-metamorphosed zone south of Manhattan. (Ferguson, 1924)

PERSHING COUNTY With delafossite, malachite, and tyrolite at the Majuba Hill Mine in the Antelope district. (Jenkins, 1981)

PSILOMELANE

hydrous manganese oxide

Not a valid mineral species, this is a general term for massive, hard manganese oxides of secondary origin that have not been specifically identified. Romanèchite is a distinct manganese oxide species that may be present at many of the localities listed below.

CHURCHILL COUNTY In veins and wall rock at the Fairview Mine in the Fairview district. (Schrader, 1947)

CLARK COUNTY In large deposits that were mined during both world wars at the Three Kids Manganese Mine in the Las Vegas district. (Hewett and Weber, 1931)

ELKO COUNTY With pyrolusite and *wad* in a gangue of calcite, quartz, and *limonite* in replacement lenses in limestone at the Darkey Mine in the Decoy district. (Pardee and Jones, 1920)

EUREKA COUNTY As dense nodules with pyrolusite in manganese ore at the Gibellini Mine in the Gibellini district. The ore contains appreciable barium, zinc, nickel, and lesser amounts of other metals. (Roberts et al., 1967b)

As black dendrites on fractures and along bedding planes in silty limestone at the Maggie Creek Mine in the Maggie Creek district. (Rota, 1991)

HUMBOLDT COUNTY Reported as *tungomelane* from hot springs tungsten deposits that yielded more than 1.6 million pounds of tungsten oxide at the Golconda Tungsten Mine in the Golconda district, Sec. 1, T35N, R40E. (Kerr, 1940b; Willden, 1964; AMNH)

LANDER COUNTY Reported as an ore mineral with braunite and pyrolusite at the Black Rock Mine in the Buffalo Valley district. (Stewart et al., 1977a)

Abundant in lenticular bodies to 820 feet long and 35 feet thick at the Black Eagle Manganese Mine in the Jersey district. (Trengove, 1959)

LINCOLN COUNTY In a specimen with azurite, chrysocolla, and malachite in the Pioche district. (WMK)

MINERAL COUNTY With pyrolusite, *wad,* calcite, gypsum, and *chalcedony,* in veins up to 2 feet thick at the Black Jack Mine near Sodaville in the Silver Star district. (Ross, 1961)

NYE COUNTY In seams and as bunches in sheared, clay-altered rhyolite at the Defense and Victory Mines in the San Antone district. (Kral, 1951)

In botryoidal or small mammillary masses, and as impure soft velvety coatings, with jarosite crystals implanted on them in the Tonopah district. (Melhase, 1935c; WMK)

Reported with cinnabar in the Union district. (Bailey and Phoenix, 1944)

PERSHING COUNTY Abundant with pyrolusite in a calcite vein at the Victory manganese prospect in the Rose Creek district. (Iverson and Holmes, 1954)

WASHOE COUNTY Reportedly hypogene in a

shallowly dipping, 200-foot-long, 4-foot-wide, manganese-rich breccia vein in Tertiary ash-flow tuff at the Peponis prospect in the McClellan district. (Bonham and Papke, 1969)

PUMPELLYITE

$Ca_2(Al,Fe^{2+},Fe^{3+},Mg,Mn^{2+})(Al,Fe^{3+},Cr^{3+},Mn^{3+})_2$
$(SiO_4)(Si_2O_7)(OH)_2 \bullet H_2O$

Once a single species, *pumpellyite* is now subdivided on the basis of chemistry into several species, the most common of which is pumpellyite-(Mg), a widespread product of low-grade metamorphism.

ELKO COUNTY Part of the alteration assemblage at the Jerritt Canyon Mines in the Independence Mountains district. (Hofstra, 1994)

LINCOLN COUNTY Questionably identified as a constituent of calc-silicate rock adjacent to diorite in the Chief district. (Tschanz and Pampeyan, 1970)

PYRARGYRITE

Ag_3SbS_3

Pyrargyrite, one of the "ruby silver" minerals, is an economically important and widespread mineral in epithermal precious-metal deposits in Nevada. It is considered to occur both as a hypogene and as a supergene phase. Fine specimens have reportedly come from the Comstock, Austin, Tonopah, Cherry Creek, Morey, and Hannapah districts in Nevada.

CHURCHILL COUNTY With stephanite, tetrahedrite, acanthite, and other minerals in quartz veins at the Fairview Mine and the Nevada Hills deposit in the Fairview district. (Greenan, 1914; Schrader, 1947; Saunders, 1977)

CLARK COUNTY In minor amounts in ore at the Double Standard Mine in the Crescent district. (Hewett, 1956)

Identified in a specimen reported to be from the Muddy Mountains. (WMK)

ELKO COUNTY In gold-bearing quartz veins with stibnite and other minerals in the Cornucopia district. (Emmons, 1910; Lawrence, 1963)

With tetrahedrite at the Delano Mine in the Delano district. (Lawrence, 1963)

Reported with acanthite and gold-bearing pyrite in a vein in rhyolite and andesite at the

Banner Mine in the Gold Circle district. (LaPointe et al., 1991)

In quartz-*chalcedony* veins with other minerals at the Good Hope Mine in the Good Hope district. (Emmons, 1910; Lawrence, 1963)

As microcrystalline clusters with marcasite on quartz at the Meikle Mine in the Lynn district. (Jensen, 1999)

In prospects with pyrite in quartz-*chalcedony* veinlets in the Rock Creek district; also noted with pyrite or marcasite, stibnite, and barite in silicified breccia. (Emmons, 1910; Lawrence, 1963)

An important silver ore mineral with stephanite, pyrite, and quartz from mines in the Tuscarora district; reported as crystals to 2 cm at the Grand Prize Mine, and as late veinlets in rich sulfide ore from the Navajo Mine. (Emmons, 1910; Lawrence, 1963; UCB; HMM; M. Jensen, w.c., 1999; SBC)

ESMERALDA COUNTY Reported at the Mammoth Mine in the Gilbert district. (Ferguson, 1928)

EUREKA COUNTY In silver-lead ore from Mill Canyon in the Cortez district. (Emmons, 1910)

Reported in a specimen from the Yankee Mine. (WMK)

HUMBOLDT COUNTY One of the silver minerals at the Sleeper Mine in the Awakening district. (Wood, 1988)

As a secondary sulfide with stephanite, stibnite, and marcasite in joints, fractures, and quartz veins at the National Mine in the National district; also reported in ore from Auto Hill and in the Caustin-Bankers-Diffenbach area. (Lindgren, 1915; Melhase, 1934c)

Reported at the Silver Butte Mine in the Paradise Valley district. (Vanderburg, 1938a)

In a specimen with sphalerite, chalcopyrite, tetrahedrite, and quartz at the Sheeba Mine. (UCB)

LANDER COUNTY With galena, sphalerite, pyrite, and quartz in a specimen from the White Shilo Mine near the site of Galena in the Battle Mountain district. (UCB)

Reported as grains to 2 mm with pyrostilpnite in specimens from the Dean Mine in the Lewis district: also noted at the Henry Logan and Eagle Mines. (R. Thomssen, w.c., 1999; GCF)

An important ore mineral in many mines in the Reese River district as red-gray masses to several cm in quartz with polybasite and stephanite. (Melhase, 1934c; Ross, 1953; Lawrence, 1963; UCB; AMNH; SM; M. Jensen, w.c., 1999)

LINCOLN COUNTY In bonanza silver ore from the Greenwood fissure at the Prince Mine in the Pioche district. (Westgate and Knopf, 1932; Melhase, 1934d)

LYON COUNTY With proustite, tetrahedrite, and sphalerite at the Como and Wild Rose Mines in the Como district. (Gianella, 1941; WMK)

Reported at the Christenson shaft in the Talapoosa district. (WMK)

As a hypogene mineral in veins in the Wilson district. (Dircksen, 1975)

MINERAL COUNTY In ore with other silver minerals at the Broken Hills Mine in the Broken Hills district. (Schrader, 1924)

With acanthite and other sulfides at the Lucky Boy Mine in the Lucky Boy district. (Ross, 1961)

In specimens from unidentified localities in the Rawhide district. (WMK)

NYE COUNTY With acanthite in a vein in volcanic rock at the Arrowhead Mine in the Arrowhead district; also with acanthite and stibnite in veins in Paleozoic rock at the Eaton prospect. (Lincoln, 1923; Lawrence, 1963)

As crystals in quartz vugs in veins in the Clifford district. (Ferguson, 1916)

As complex prisms to 2 mm with pyrite and polybasite in the Hannapah district. (Jenkins, 1981)

Reported in the Prussian vein in the Jefferson Canyon district. (Kral, 1951)

As transparent, blood-red prisms to 3 mm at the Morey and Keyser Mines in the Morey district. (Kral, 1951; Williams, 1968; S. Pullman, w.c., 1999)

Reported with acanthite and stibnite at the Antimonial Mine in the Reveille district. (Lawrence, 1963)

An ore mineral with stephanite, chlorargyrite, and *electrum* in the Silverbow district. (Kral, 1951)

At many localities, including the Mizpah, Montana-Tonopah, Tonopah Extension, and West End Mines, in the Tonopah district, as fissure fillings in quartz and commonly intergrown with acanthite, polybasite, and wire silver; reported to contain arsenic, and in places considered to be a secondary mineral. (Spurr, 1903;

Bastin and Laney, 1918; Melhase, 1935c; Nolan, 1935; WMK)

An important source of silver at the Murphy and Teichert Mines in the Twin River district. (Hague, 1870)

PERSHING COUNTY Reported with *argentite* in a specimen from the Arizona Mine in the Buena Vista district. (WMK)

An important ore mineral at the Hecla Rosebud Mine in the Rosebud district; locally as euhedral crystals to 2 cm. (N. Prenn, w.c., 1999; M. Jensen, w.c., 1999)

In a specimen from an unknown locality in the Sacramento district. (JHS)

As euhedral crystals to 3 mm in calcite at the Florence lease in the Seven Troughs district; also reported in a specimen from the Fairview lease. (M. Jensen, w.c., 1999; WMK)

A rare late-stage mineral at the Sheba and De Soto Mines in the Star district. (Cameron, 1939)

STOREY COUNTY Noted as a minor mineral in ore specimens from the Comstock Lode in the Comstock district, but an important silver mineral in other vein systems in the district. (Bastin, 1922; Vikre, 1989)

WASHOE COUNTY Reported in the Steamboat Springs district. (White, 1980; Wilson and Thomssen, 1985)

WHITE PINE COUNTY With proustite, polybasite, stephanite, and other minerals after primary sulfides such as galena and sphalerite at the Star Mine in the Cherry Creek district. (Schrader, 1931b)

Reported in oxidized jasperoid ore in the Taylor district. (Lovering and Heyl, 1974)

PYRITE

FeS$_2$

Pyrite is the most common sulfide mineral, occurring in many geologic environments. It occurs in unoxidized rock in most of Nevada's mining districts, and the localities listed below are only a sampling of known occurrences.

CARSON CITY Gold-bearing pyrite occurs in gold-silver-copper ore from the Blue Jay lode, Eske property, in the Delaware district. (WMK)

CHURCHILL COUNTY With other sulfide minerals in quartz veins, commonly as cubes surrounded by stibnite and antimony oxides, at the Antimony King Mine in the Bernice district; also reported from the Red Bird Mine. (Lawrence, 1963; R. Thomssen, w.c., 1999)

With molybdenite and other sulfides in lead-silver veins in the Chalk Mountain district. (Schilling, 1979)

With specular hematite at the Coppereid prospect in the Copper Kettle district. (Shawe et al., 1962)

In siliceous sinter at the Senator fumaroles in Dixie Valley. (JHS)

With cinnabar, barite, gypsum, and jarosite in narrow veinlets at the Cinnabar Hill Mine in the Holy Cross district. (Vanderburg, 1940; JHS)

A modified cube about 3.5 cm across is reported to be from IXL Canyon in the IXL district. (WMK)

Noted with chalcopyrite in shear zones and breccia in the iron ore in the Mineral Basin district; also found as lustrous cubes to 0.5 cm with actinolite crystals in calcite at the Buena Vista Iron Mine. (Reeves and Kral, 1955; M. Jensen, w.c., 1999)

CLARK COUNTY With other sulfide minerals in copper-nickel ore at the Great Eastern, Key West, and other mines in the Bunkerville district. (Beal, 1965)

In quartz stockwork veinlets and adjacent rock at the Crescent Peak porphyry-molybdenum deposit in the Crescent district. (Schilling, 1979)

With gold, galena, mottramite, and numerous unidentified minerals at the Boss Mine in the Goodsprings district. (Jedwab, 1998a)

DOUGLAS COUNTY With magnetite at the Minnesota Mine in the Buckskin district, particularly near the borders of the iron deposit. (Reeves et al., 1958)

ELKO COUNTY As euhedral crystals, some zoned, in silica veinlets and disseminated in the Rossi gold deposit in the Bootstrap district. (Snyder, 1989)

With other sulfides in a quartz-calcite vein at the Mint and Mirat Mines in the Burner district. (Emmons, 1910)

As euhedral crystals and massive veins in carbonaceous sedimentary rock at the Rain Mine in the Carlin district; commonly arsenic-bearing. (D. Heitt, o.c., 1991)

Common with quartz and chalcopyrite in massive orebodies at the Prunty and Rescue Mines in the Charleston district. (Coats and Stephens, 1968)

As brilliant cubic crystals to 3 cm and larger fractured cubic crystals to 9 cm in calcite with bismuthinite at the Victoria Mine in the Dolly Varden district; also in quartz veins with cerussite and other minerals. (Melhase, 1934d; Schilling, 1979; M. Jensen, w.c., 1999)

At the Ken Snyder Mine in the Gold Circle district with calcite, *adularia,* quartz, and other minerals in veins; reportedly gold-bearing at the Yaden, Boyle, and Tayant lease. (Casteel and Bernard, 1999; WMK)

In four forms at the Jerritt Canyon and Big Springs Mines in the Independence Mountains district: framboids; pyritohedrons with rutile and quartz inclusions; subhedral to anhedral grains with sphalerite inclusions; and rims (with as much as 8 percent As) on older iron sulfides. (Youngerman, 1992; Hofstra, 1994)

The most abundant sulfide mineral at the Hollister deposit in the Ivanhoe district, comprising as much as 40 percent of rock volume as crystals to 0.3 mm with other sulfides and *electrum.* (Deng, 1991)

A major constituent of gold ore as fine-grained masses and veinlets at the Meikle Mine in the Lynn district; also as drusy crusts on post-ore barite crystals. (Jensen, 1999)

With chalcopyrite as the dominant minerals in massive sulfide copper ore at the Rio Tinto Mine in the Mountain City district; also with molybdenite and chalcopyrite in quartz stockwork veinlets at the Huber Hill molybdenum prospect. (Lawrence, 1963; Coats and Stephens, 1968; Schilling, 1979)

Finely divided with other dark sulfides, "ruby silver," and specular hematite in banded comb quartz veins in andesite at the Falcon Mines in the Rock Creek district. (Emmons, 1910; LaPointe et al., 1991)

Massive with galena and chalcopyrite at the Killie Mine in the Spruce Mountain district. (Schrader, 1931b)

Abundant in specimens with chalcopyrite, stephanite, and bornite at the Grande Prize, Navajo, North Belle Isle, and other mines in the Tuscarora district. (Melhase, 1934c; UCB; BMC; HMM; WMK)

ESMERALDA COUNTY With native gold, famatinite, bismuthinite, and other minerals in unoxidized high-grade ore in the Goldfield district;

also widespread with alunite in altered rock. (Ransome, 1909a; Ashley, 1974; WMK)

As euhedral cubic crystals to 2.5 cm from the 250-foot level at the Gold Eagle Mine in the Lone Mountain district. (M. Jensen, w.c., 1999)

Reportedly gold- and silver-bearing in a specimen from the Mary Mine in the Silver Peak district. (WMK)

With molybdenite in the large Cucomungo (Siskon) deposit in the Tule Canyon district. (JHS)

EUREKA COUNTY Reported to be widespread at the Garrison and Bullion Hill Mines and the J. DeRevierre property in the Cortez district. (Emmons, 1910; WMK)

The most common sulfide at the Carlin and Goldstrike Mines in the Lynn district, occurring in numerous forms and settings: early-formed framboids and small cubes; fine-grained arsenic-rich pyrite associated with gold; and late, bright crystalline to botryoidal and stalactitic masses. (GCF; Ferdock et al., 1997)

In masses beneath the magnetite orebody at the Barth Iron Mine in the Safford district; also found with galena, sphalerite, and chalcopyrite elsewhere in the district. (Shawe et al., 1962; WMK)

Arsenic-bearing pyrite occurs as rounded lustrous crystals to 4 cm at the Gold Quarry Mine in the Maggie Creek district. (M. Johnston, o.c., 2001)

HUMBOLDT COUNTY As framboids in breccia ore at the Sleeper Gold Mine in the Awakening district. (Nash and Trudel, 1996)

The principal sulfide at the Lone Tree Mine in the Buffalo Mountain district as arsenic-bearing pyrite in gold ore, as fine particles and veins, and as stalactitic growths to 5 cm of lustrous microcrystals. (Panhorst, 1996; M. Jensen, w.c., 1999)

A minor component in unoxidized rock as framboids and small euhedral crystals with uranium-rich zircon at the Moonlight Uranium Mine in the Disaster district. (Castor et al., 1996; Castor and Henry, 2000)

In quartz veins at the Silver Coin Mine in the Iron Point district. (R. Thomssen, w.c., 1999; J. McGlasson, w.c., 2001)

Minor in magnetite ore at the Delong (Iron King) Mine in the Jackson Mountains district. (Shawe et al., 1962)

With chalcopyrite, galena, and sphalerite in quartz-*adularia* veins in the National district. (Lindgren, 1915)

Widespread with cinnabar at the Cordero Mine in the Opalite district. (Bailey and Phoenix, 1944)

In various forms at the Getchell and Twin Creeks Mines in the Potosi district, most significantly as gold- and arsenic-bearing framboids and subhedral disseminated grains to 50 microns; also with other sulfides and scheelite in skarn at the Granite Creek Mine. (Schilling, 1962a, 1962b; Simon, et al., 1999)

As framboids in uranium-bearing petrified wood in the Virgin Valley district. (Castor et al., 1996)

Reportedly silver-bearing with acanthite from the 1910 property in the Winnemucca district. (WMK)

LANDER COUNTY In the Battle Mountain district at the Fortitude Mine in gold-bearing skarn and in the Copper Canyon mines with other sulfides; also reported attached to gold from placer deposits. (Vanderburg, 1936a; Theodore and Blake, 1978; Schilling, 1979; Doebrich, 1993)

Gold-bearing as the principal sulfide at the Buffalo Valley and Honeycomb Mines in the Buffalo Valley district. (Kizis et al., 1997)

With tetrahedrite and other minerals at the Betty O'Neal Mine in the Lewis district; also noted in a specimen from the Dean Mine. (Vanderburg, 1939; R. Thomssen, w.c., 1999)

With uranium minerals at the Apex (Rundberg) Uranium Mine in the Reese River district. (Garside, 1973)

As lustrous cubes to 1 cm with clinochlore and quartz at the McCoy Mine in the McCoy district. (M. Jensen, w.c., 1999)

In spherical clusters to 5 cm in carbonaceous shales at the Elder Creek Mine in the Bullion district. (M. Jensen, w.c., 1999)

LINCOLN COUNTY With sphalerite and galena in large, massive, silver-rich, sulfide replacement orebodies in limestone in the Pioche district, as at the Caselton and Prince Mines. (Gemmill, 1968)

With scheelite from the Lincoln, Schofield, and Tem Piute Mines in the Tem Piute district. (WMK)

LYON COUNTY A characteristic mineral in ore and altered rock in the Silver City district. (Gianella, 1936a; Vikre, 1989)

With silver and quartz in altered andesite at the Rockland Mine in the Wilson district. (WMK)

Reportedly gold-bearing as minute grains and narrow, randomly oriented, discontinuous seams at the Blue Stone Mine in the Yerington district; also in masses of crystalline calcite on the dumps at the Ludwig Mine as excellent crystals showing cube, octahedral, and pyritohedral faces. (JHS; Melhase, 1935e)

MINERAL COUNTY In quartz with tetrahedrite at the Wide West and other mines in the Aurora district. (WMK)

Encapsulating acanthite at the Candelaria Mine in the Candelaria district. (Chavez and Purusotam, 1988)

As grains and cubes replaced by *limonite* at the Lowman Mine in the Pamlico district. (Schrader, 1947)

NYE COUNTY Arsenic-bearing in gold ore at the Sterling Mine in the Bare Mountain district. (Castor, 1997)

Common in mercury-bearing quartz veins at the Senator Mine in the Belmont district. (Bailey and Phoenix, 1944; JHS; WMK)

In magnetite ore at the Phelps Stokes Mine in the Ellsworth district. (Reeves et al., 1958)

In veins with galena, cerussite, and chalcopyrite at the Star of the West Mine in the Jackson district. (Hamilton, 1993)

Reported at the White Caps Mine in the Manhattan district. (Melhase, 1935d; Gibbs, 1985)

As disseminated grains at the Northumberland Mine in the Northumberland district. (Kokinos and Prenn, 1985)

In veins and disseminated in wall rock, and commonly with microscopic gold inclusions, at the Round Mountain Mine in the Round Mountain district; also abundant at the Outlaw Mine. (Melhase, 1935d; Tingley and Berger, 1985; S. Pullman, w.c., 1999)

As fractured cubes to 2.5 cm at the Hall Mine in the San Antone district. (M. Jensen, w.c., 1999)

A principal component of altered rocks and veins in the Tonopah district; in places as inclusions in polybasite. (Spurr, 1903; Melhase, 1935c)

With other sulfides and scheelite in the Union district. (Bailey and Phoenix, 1944; Hamilton, 1993)

Common at the Golden Chariot Mine in the Jamestown district, Nellis Air Force Range. (WMK; Tingley et al., 1998a)

PERSHING COUNTY Widespread as disseminated grains and crystals in porphyry at the Majuba Hill and Antelope Mines in the Antelope district. (Lawrence, 1963; MacKenzie and Bookstrom, 1976; Jensen, 1985)

Oxidized to *limonite* in quartz veins at the Arizona and Black Warrior Mines in the Buena Vista district. (Cameron, 1939; Lawrence, 1963)

Abundant in gold-bearing quartz veins with tetrahedrite and stibnite at the Antimony Ike Mine in the Goldbanks district. (Dreyer, 1940; Lawrence, 1963)

Rare to common with scheelite, quartz, and *feldspar* at the Nevada-Massachusetts Co. Mines in the Mill City district. (JHS; WMK)

With traces of chalcopyrite at depth in iron orebodies in the Mineral Basin district; noted as lustrous, modified euhedral crystals to 1 cm at the Segerstrom-Heizer Mine. (Reeves and Kral, 1955; Johnson, 1977; M. Jensen, w.c., 1999)

Common as small grains in quartz veins and orebodies in the Rochester district. (Knopf, 1924)

As rare, lustrous, cubic crystals to 2 cm at the Hecla Rosebud Mine in the Rosebud district. (M. Jensen, w.c., 1999)

In veins with scheelite and other minerals in the Rye Patch and Sacramento districts. (Ransome, 1909b)

Reportedly gold-bearing in ore at the Dun Glen and Straub Mines in the Sierra district. (WMK)

Common in quartz-*tourmaline* veins in the Spring Valley district; partially oxidized to *limonite* and containing flakes of gold at the Pacific Matchless Mine. (Campbell, 1939; Johnson, 1977)

STOREY COUNTY In the Comstock district as tiny grains and cubes disseminated in many square miles of andesitic rock; in the Comstock Lode said to be abundant in bonanza ore and adjacent andesite wall rock, but rare elsewhere. Abundant in ore from the Occidental and Flowery lodes. (Gianella, 1936a; Schilling, 1990; JHS; WMK)

WASHOE COUNTY In quartz-calcite veins with galena and other minerals at the Union (Commonwealth) Mine in the Galena district. (Geehan, 1950)

A characteristic mineral in ore and adjacent altered rock in the Jumbo district. (Vikre, 1989; JHS)

Abundant in the Peavine district where pyrite oxidation and subsequent acid leaching has produced a white, yellow, and brown zone that extends for several miles. Present in three forms: as cubes disseminated and in veinlets in altered rock, locally with chalcopyrite; as octahedral and pyritohedral crystals in quartz veins and their wall rocks; and as anhedral grains in breccia. Associated vein minerals are enargite, galena, sphalerite, and acanthite. (Hill, 1915; Hudson, 1977)

In unoxidized parts of veins and in samples from dumps at many localities in the Pyramid district. At the Burrus Mine in enargite seams, in breccia, and disseminated in veins; also in drusy quartz vugs as euhedral octahedrons to 2 mm. (Overton, 1947; Jensen, 1994)

In cavities in sinter with chalcopyrite, stibnite, arsenopyrite, and cinnabar in the Steamboat Springs district. (Lawrence, 1963; WMK)

With quartz, galena, and sphalerite in unoxidized ore in the Wedekind district. (Overton, 1947)

As masses and crystals in breccia at the Willard Mine in the Willard district. (Jensen and Leising, 2001)

WHITE PINE COUNTY As cubes and framboids to 40 microns at the Vantage deposit on Alligator Ridge in the Bald Mountain district; arsenic-bearing pyrite (1.0 to 1.5 wt. percent As) is found in the main part of the deposit. (Ilchick, 1990)

An abundant mineral in disseminated copper ore in the Robinson district; also abundant in high-grade vein and supergene deposits. (Melhase, 1934d; Hose et al., 1976; WMK)

As cubes to 1 cm with sphalerite and other sulfides on the dump at the Ward Mine in the Ward district; also noted as cubes and crystalline masses with barite. (WMK; W. Lombardo, w.c., 2001)

PYROCHROITE
$Mn^{2+}(OH)_2$

A primary manganese mineral in some volcanogenic sulfide deposits, but also formed by the metamorphism and subsequent hydration of primary manganese ore.

WHITE PINE COUNTY Identified by XRD in black masses with hausmannite at an unspecified locality in an area of manganese deposits mined between 1917 and 1947 in the Nevada district. (S. Kleine, o.c., 1999)

PYROLUSITE
$Mn^{4+}O_2$

A common manganese oxide formed under highly oxidizing conditions in many settings, pyrolusite is an important ore mineral in large deposits elsewhere in the world, and is present in some small Nevada manganese deposits.

CHURCHILL COUNTY In precious-metal quartz veins at the Fairview Mine in the Fairview district. (Greenan, 1914; Schrader, 1947; Quade and Tingley, 1987)

ELKO COUNTY Reported in a specimen from Carlin. (WMK)

Reported in a specimen from the East pit in the Ivanhoe district. (R. Thomssen, w.c., 1999)

ESMERALDA COUNTY Reported in specimens from 7 to 8 miles south of Goldfield, probably from the Cuprite district. (WMK)

EUREKA COUNTY Reported with *wad* and *psilomelane* at the Gibellini Manganese Mine in the Gibellini district, Sec. 35, T16N, R52E. (Binyon, 1948)

As black dendrites to 10 cm long and 6 cm wide on fractures at the Goldstrike Mine in the Lynn district; also in weathered diorite as black splotches. (GCF)

As black dendrites at the Gold Quarry Mine in the Maggie Creek district. (GCF)

HUMBOLDT COUNTY As tiny patches in uranium-bearing bedded opal in the Virgin Valley district. (Castor et al., 1996)

LANDER COUNTY An ore mineral at the Black Rock Mine in the Buffalo Valley district. (Stewart et al., 1977a)

Reported with chrysocolla in a specimen from the central part of the Shoshone Range in the Bullion district. (WMK)

Reported in a specimen with silver-bearing galena in the Hilltop district. (WMK)

As botryoidal and stalactitic masses at the Black Devil Mine in the Wild Horse district. (Stewart et al., 1977a)

LINCOLN COUNTY In a pipe or chimney of silver-bearing *wad* at the Jackrabbit (Black Metals) Mine in the Bristol district. (Melhase, 1934d)

Reported in gold-silver-lead-manganese ore from the Lucky Star Mine in the Pioche district. (WMK)

Reported in a specimen from the Meadow Valley Mines. (CSM)

MINERAL COUNTY Reported in a specimen from Broken Hills. (WMK)

Locally abundant at the Candelaria Mine in the Candelaria district. (Chavez and Purusotam, 1988)

With *psilomelane* and *wad* in tungsten-bearing manganese ore at the Black Jack Mine in the Silver Star district. (Ross, 1961)

NYE COUNTY Reported in veins in the Oak Spring district. (Melhase, 1935b)

The most abundant manganese oxide in veins in metamorphosed limestone at the Queen City prospect in the Queen City district; large tabular masses of the mineral in the center of the largest vein contain abundant vugs, some lined with calcite and opal. (Hewett et al., 1968)

In cracks at the Round Mountain Mine in the Round Mountain district. (Tingley and Berger, 1985)

Reported as finely fibrous or feltlike coatings on fissure walls in the Tonopah district. (Melhase, 1935c)

Identified with hollandite and lithiophorite(?) in a quartz vein in drill core from Yucca Mountain. (Castor et al., 1999b)

PERSHING COUNTY As black dendrites at the Majuba Hill Mine in the Antelope district. (Jensen, 1985)

Abundant with *psilomelane* in a calcite vein at the Victory manganese prospect in the Rose Creek district. (Iverson and Holmes, 1954; Johnson, 1977)

As dendrites at the Willard Mine in the Willard district. (Jensen and Leising, 2001)

STOREY COUNTY Reported in a specimen from Six Mile Canyon in the Comstock district. (WMK)

WHITE PINE COUNTY Reported at the Aurora Mine in the White Pine district. (S. Pullman, w.c., 1999)

PYROMORPHITE
$Pb_5(PO_4)_3Cl$

Pyromorphite is a common mineral in oxidized lead ore; fine specimens have come from the Goodsprings district.

CHURCHILL COUNTY As arsenic-bearing, vitreous, pale-yellow-green, translucent, barrel-shaped crystals to 6 mm from the 510-foot level of the Hanlase drift of the Chalk Mountain Mine in the Chalk Mountain district. (UMNII)

CLARK COUNTY In small amounts in most of the lead-zinc mines of the Goodsprings district, commonly as small white calcian crystals. Green and yellow prisms were reported from the Root Mine. Waxy yellow masses and small yellow and orange hexagonal prisms have also been collected. (Hewett, 1931; Adams, 1998; S. Pullman, w.c., 1999; J. Kepper, w.c., 1999)

ELKO COUNTY With copper carbonates and iron oxides at the Tiger Lode Mine in the Aura district. (Emmons, 1910; Gianella, 1941; LaPointe et al., 1991)

With chlorargyrite in quartz veins at the Panther and Leopard Mines in the Cornucopia district. (Emmons, 1910; Lawrence, 1963)

As pale-green crystals to 2 mm in a specimen from Bull Run Canyon in the Edgemont district. (WMK)

Reported with silver halides at the Protection Mine in the Mountain City district. (Emmons, 1910)

Reported in a specimen with galena at the Lady of Lake Mine in the Railroad district. (WMK)

As rare, small hexagonal prisms with bayldonite in cavities in gossan at the Killie Mine in the Spruce Mountain district. (Dunning and Cooper, 1987)

Reported from various mines in the Tuscarora district. (Melhase, 1934c)

ESMERALDA COUNTY With cerussite at the Nevada-Goldfield and Courbet Mines in the Divide district. (Ransome et al., 1907)

Reported in the Tokop (Gold Mountain) district. (Hoffman, 1878)

EUREKA COUNTY Reported from an unspecified locality in the Eureka district. (Gianella, 1941)

In replacement bodies with quartz and other minerals in the Mineral Hill district on the east side of the Sulphur Springs Range. (Emmons, 1910; Schilling, 1979)

HUMBOLDT COUNTY As white prisms with hinsdalite at the Silver Coin Mine in the Iron Point district. (Thomssen, 1998; J. McGlasson, w.c., 2001)

LANDER COUNTY Reported from the Tenabo area in the Bullion district. (Gianella, 1941)

In a specimen from the Snowstorm Mine in the North Battle Mountain district. (R. Thomssen, w.c., 1999)

LINCOLN COUNTY With conichalcite as greenish-yellow cubes after galena from the Bristol Mine in the Bristol district. (T. Potucek, w.c., 1999)

Tentatively identified as minute green crystals on *limonite* with native gold in volcanic breccia in the DeLamar district. (Callaghan, 1937)

Reported with phosgenite at the Prince Mine in the Pioche district. (Westgate and Knopf, 1932)

MINERAL COUNTY As microscopic green crystals at the Endowment Mine in the Silver Star district. (R. Walstrom, w.c., 1999)

NYE COUNTY As lustrous white crystals to 2 mm at the San Rafael Mine in the Lodi district. (S. Pullman, w.c., 1999; GCF)

STOREY COUNTY Reported from an unspecified locality in the Comstock district. (Gianella, 1941)

WHITE PINE COUNTY With mimetite, wulfenite, heyite, cerussite, and other minerals 8 miles southeast of Ely at the Betty Jo claim and Vietti Mine in the Nevada mining district. (Roberts, 1943; Williams, 1973)

PYROPE

$Mg_3Al_2(SiO_4)_3$

This *garnet* group species is found in high-grade metamorphic rocks, in ultramafic and mafic intrusive rocks, and as a detrital mineral derived from such rocks.

WASHOE COUNTY Pyrope-almandine *garnet* is a major constituent of skarn in association with calcite, quartz, diopside, epidote, and tremolite in the Nightingale district. (Bonham and Papke, 1969; Johnson, 1977)

PYROPHYLLITE

$Al_2Si_4O_{10}(OH)_2$

Pyrophyllite occurs in metamorphosed aluminum-rich rocks, in epithermal deposits, and in some porphyry deposits. In Nevada, it is an important diagnostic alteration mineral in some high-sulfidation epithermal deposits.

DOUGLAS COUNTY As bold outcrops up to 50 x 100 feet in area at the Top prospect a half mile south of the Minnesota Iron Mine in the Buckskin district; also with andalusite, corundum, diaspore, and zunyite at the Blue Metal (Blue Danube) Mine. (Papke, 1975; Hudson, 1983)

With diaspore and alunite in Minnehaha Canyon. (Hudson, 1983)

ELKO COUNTY As yellow microcrystals with cerussite, azurite, and malachite at the Rip Van Winkle Mine in the Merrimac district. (M. Jensen, w.c., 1999)

ESMERALDA COUNTY Reported to occur sparingly in silicified rocks in the Goldfield district. (Ashley and Keith, 1976)

LYON COUNTY Found in a small deposit as the variety *leuchtenbergite* at the Admiral claim in the Yerington district. (Stoddard and Carpenter, 1950; M. Jensen, w.c., 1999)

MINERAL COUNTY With diaspore, andalusite, halloysite, svanbergite, and other minerals at the Dover and Green Talc Mines in the Fitting district. (Vanderburg, 1937a; F. Cureton, w.c., 1999)

In altered ash-flow tuffs in the Rand district, northern Gabbs Valley Range. (Hudson, 1983)

As rosettes to 0.5 inch in diameter at the Bismark, Deep, and Kenjay prospects in Soda Springs Valley east of Hawthorne. (JHS)

In altered andesite south of Garfield Flat. (D. Hudson, w.c., 1999)

NYE COUNTY An alteration mineral with kaolinite, diaspore, and dickite as massive replacements of wall rock at the Cactus Nevada Mine in the Cactus Springs district. (Tingley et al., 1998a)

Locally in altered rock at the Union Canyon fluorspar deposit in the Union district. (Papke, 1979)

PERSHING COUNTY Reported in a specimen from South American Canyon in the Rochester district. (WMK)

A constituent of pinite bodies along with

andalusite and *sericite* at the Pinite Mine in the Spring Valley district. (Kerr, 1940a; Johnson, 1977)

STOREY COUNTY With diaspore as an alteration mineral in the Comstock district. (D. Hudson, w.c., 1999)

WASHOE COUNTY In hydrothermal breccia as a product of advanced argillic alteration in intrusive andesite in the Peavine district with dickite, diaspore, pyrite, and *sericite;* also intergrown with quartz as pseudomorphs after *plagioclase* and mafic minerals. (Hudson, 1977)

A constituent of advanced argillic alteration adjacent to enargite-bearing veins in the Pyramid district; also as white masses associated with barite in veins at the Burrus Mine. (Wallace, 1975; GCF)

PYROSTILPNITE

Ag_3SbS_3

Pyrostilpnite, a rare "ruby silver" mineral that is dimorphous with pyrargyrite, occurs in low-temperature hydrothermal veins with other silver minerals.

LANDER COUNTY Reported as scales and microcrystals at the Dean Mine in the Lewis district; also identified as reddish-brown masses with pyrargyrite, pyrite, and quartz. (Von Bargen, 1999; J. McGlasson, o.c., 2000; GCF)

NYE COUNTY As deep-orange elongated crystals to 0.5 mm with pyrargyrite at the Morey Mine in the Morey district. (M. Jensen, w.c., 1999)

WHITE PINE COUNTY As very tiny, brilliant-red prisms on crystalline miargyrite with tetrahedrite, bournonite, enargite, and andorite at the Patriot Mine in the Taylor district. (Williams, 1965)

PYROXENE

$(Ca,Mg,Fe^{2+},Mn^{2+},Na,Li,Zn)(Al,Fe^{2+},Mn^{2+},$
$Cr^{3+},Sc,Ti,Fe^{3+},Mg,V^{3+})(Si,Al)_2O_6$ (general)

Pyroxene is a mineral group name. Occurrences for which specific species have not been identified are listed below.

LINCOLN COUNTY As splendid crystals in cavities in volcanic rocks with green *garnet,* clinochlore, and magnetite in the Bristol district. (Melhase, 1934d)

PERSHING COUNTY A major constituent of scheelite-bearing skarn in the Nightingale district. (Smith and Guild, 1944)

PYRRHOTITE

$Fe_{1-x}S$

Pyrrhotite, a high-temperature iron sulfide, occurs in magmatic, metamorphic, and hydrothermal settings; it sometimes occurs alone, but often occurs with other sulfide minerals. It resembles pyrite, but can generally be distinguished from that mineral because it is magnetic.

CLARK COUNTY With pyrite, chalcopyrite, and other sulfide minerals in copper-nickel-platinum-cobalt-gold ore at the Great Eastern and Key West Mines in the Bunkerville district. (Beal, 1965)

Reported at the Boss Mine in the Goodsprings district. (Jedwab, 1998b)

DOUGLAS COUNTY Reported in a specimen from an unknown locality in the Red Canyon district. (JHS)

ELKO COUNTY As anhedral grains to 0.15 mm enclosed by pyrite and/or marcasite in the Hollister deposit in the Ivanhoe district. (Deng, 1991)

EUREKA COUNTY With sphalerite and molybdenite in skarn at the Antelope Mine in the Fish Creek district. (Schilling, 1979)

Disseminated in diorite and rare in base-metal veins with galena, sphalerite, and chalcopyrite at the Goldstrike Mine in the Lynn district; also in gold ore as inclusions in early pyrite. (GCF)

Encountered in a drill hole southwest of the hematite-magnetite orebody at the Barth Iron Mine in the Safford district. (Shawe et al., 1962)

HUMBOLDT COUNTY As small (<100 micron) inclusions in other iron sulfides at the Lone Tree Mine in the Buffalo Mountain district; associated with chalcopyrite. (Panhorst, 1996)

Reported as a minor component in unoxidized uranium ore at the Moonlight Uranium Mine in the Disaster district. (Dayvault et al., 1985)

Reported with chalcopyrite, galena, sphalerite, and other sulfides at a locality on the northeast slope of the Sonoma Range, about 12 miles south of Golconda in the Gold Run district. (Melhase, 1934c)

As microinclusions in sphalerite and associated with chalcopyrite and pyrite at the Getchell and Twin Creeks Mines in the Potosi district. (Simon et al., 1999; D. Tretbar, o.c., 1999)

LANDER COUNTY In Copper Canyon in the Battle Mountain district as fracture fillings and small disseminated grains with other sulfides, locally after andradite; at the Fortitude Gold Mine in gold-bearing skarn as masses to 5 cm. (Theodore and Blake, 1978; Schilling, 1979; Doebrich, 1993; M. Jensen, w.c., 1999)

With pyrite at the Buffalo Valley and Honeycomb Mines in the Buffalo Valley district. (Stewart et al., 1977a)

LINCOLN COUNTY Identified in a specimen from the Lincoln Tungsten Mine in the Tem Piute district. (JHS)

LYON COUNTY With marcasite, chalcopyrite, rutile, enargite, and gold in quartz veins at the Wheeler Mine in the Wilson district. (Jackson, 1996)

NYE COUNTY Associated with magnetite ore at the Phelps Stokes Mine in the Ellsworth district; locally in massive bodies as much as 8 feet thick. (Reeves et al., 1958)

Tentatively identified in ore in the Silverbow district. (Ball, 1906)

PERSHING COUNTY As tiny grains and masses associated with arsenopyrite and chalcopyrite at the Majuba Hill Mine in the Antelope district. (Jensen, 1985)

Reportedly rare at the Stormy Day Mine in the Hooker district. (Johnson, 1977)

In quartz-pyrite veins in the Kennedy district. (JHS)

Reported to be rare at Nevada-Massachusetts Co. Mines in the Mill City district. (JHS)

In minor amounts in scheelite-bearing skarn at the Nightingale and Ranson Mines in the Nightingale district. (Smith and Guild, 1944)

Reportedly as euhedral, flattened, hexagonal crystals to 5 cm replaced by pyrite and marcasite, at the Hecla Rosebud Mine in the Rosebud district. (M. Jensen, w.c., 1999)

STOREY COUNTY Rare in drill cuttings from the Flowery lode in the Comstock district. (Schilling, 1990)

WASHOE COUNTY Disseminated in minor amounts with chalcopyrite in gabbro proximal to the Wild Horse Mine in the Cottonwood district. (Bonham and Papke, 1969)

WHITE PINE COUNTY In skarn from several sites in the Robinson district. (Spencer et al., 1917; Melhase, 1934d; James, 1976)

As massive rims to 2 cm around coarse sphalerite at the underground Ward Mine in the Ward district. (M. Jensen, w.c., 1999)

QUARTZ

SiO_2

Quartz is one of the most common minerals in the Earth's crust, and only the most notable Nevada occurrences are listed below. Many other localities for crystalline quartz and semiprecious gem varieties such as *agate* and *jasper* are probably known by collectors. *Chalcedony* is a microscopically fibrous variety commonly found in veins, and *jasper* and *agate* are similar varieties. *Amethyst, rose quartz,* and *citrine* are purple, pink, and yellow crystalline varieties, respectively.

CHURCHILL COUNTY *Agate* and *jasper* are abundant in the vicinity of old Lake Lahontan. (Mitchell, 1991; F. Cureton, w.c., 1999)

Common in the Fairview district, and reported as crystals to 20 cm long and as pseudomorphs after calcite crystals to 4 cm, especially in the Eagle vein; *chalcedony* is also a vein mineral, and *jasper* and *agate* nodules occur on the hillsides. (Greenan, 1914; Schrader, 1947; Klein, 1983; Quade and Tingley, 1987)

At the Lucky Boy quartz deposit in the Sand Springs district as massive, milky quartz in a flat-lying vein over 6 m thick; also as bladed or tabular doubly terminated crystals in vugs in large pegmatites south of Lucky Boy Canyon on the east slope of the Sand Spring Range. At an unnamed prospect in Sec. 24, T15N, R32E, as white crystals to 50 kg in a large quartz pod in granite. At the Summit King Mine as clear to milky hoppered and sceptered crystals to 10 cm, many with fluid inclusions. (Schilling et al., 1963; Walstrom, w.c., 1999; N. Prenn, w.c., 1999; M. Jensen, w.c., 1999)

As green and lace *agate* at Green Mountain east of Fallon. (Klein, 1983; Mitchell, 1991)

Agate and *jasper* occur in Bell Canyon and near Slate Mountain south of Frenchman. (Mitchell, 1991)

CLARK COUNTY As doubly terminated crystals to 2 mm in friable ore at the Boss Mine in the Goodsprings district. (Jedwab, 1998b)

In miarolitic cavities and pegmatites in the Eldorado and Newberry Mountains as clear, milky, and smoky (brown to black) crystals to 38 cm that are rarely transparent or of gem quality. Crystals with inclusions of spessartine *garnet* are much sought after, and clinochlore, hematite, and pyrite also occur as inclusions. (S. Scott, w.c., 1999; N. Prenn, w.c., 1999)

In red *jasper* and pink opal nodules along the Bitter Springs trail to the Muddy Mountains. (Klein, 1983)

Petrified wood is found alongside the road leading up to the northwest entrance to Valley of Fire State Park about 5 miles south of Overton. (Klein, 1983)

As *agate* and *jasper* near Henderson. (Mitchell, 1991)

Agate occurs along the North Shore Road near Lake Mead. (Mitchell, 1991)

DOUGLAS COUNTY As smoky crystals with *allanite* and torbernite in pegmatite dikes and also as clear crystals in the Kingsbury Grade area in the Genoa district. (Garside, 1973)

Agate and *jasper* are found in the Mount Siegel area of the Pine Nut Range. (Klein, 1983)

ELKO COUNTY At the Rossi Mine in the Bootstrap district as a peripheral alteration mineral, in crosscutting veinlets, and as euhedral crystals in gold ore. (Snyder, 1989)

As a replacement of siltstone and limestone and locally as light-gray veins at the Rain Mine in the Carlin district; also as small, doubly terminated crystals in red clay. (D. Heitt, o.c., 1991)

As fine crystals to 2.5 feet lining vugs and cavities at the Palo Alto Mine in the Contact district; also reported with rutile in a specimen from an unspecified location and as colorless crystals to 10 cm in pegmatites behind the townsite of Contact. (Bailey, 1903; MM; M. Jensen, w.c., 1999)

Amethyst occurs in gold ore veins with calcite, *adularia,* and other minerals at the Ken Snyder Mine in the Gold Circle district. (Casteel and Bernard, 1999)

Quartz crystals are found with beryl and *garnet* along Smith Creek 3 miles north of its junction with Harrison Pass Road in the Harrison Pass district. (Klein, 1983)

As gray to white, euhedral epimorphs to 5 cm after stibnite and barite crystals at the Murray Mine in the Independence Mountains district. (M. Jensen, w.c., 1999; N. Prenn, w.c., 1999)

As *agate, chalcedony,* and petrified wood near the Silver Cloud Mine in the Ivanhoe district. (Klein, 1983)

As colorless crystals to 6 cm in the Meikle Mine in the Lynn district, some overgrown with yellow barite crystals; also as epimorphs to 5 cm after stibnite crystals. (Jensen, 1999; N. Prenn, w.c., 1999)

As large doubly terminated smoky crystals in an unnamed borrow pit at Mountain City in the Mountain City district. (R. Walstrom, w.c., 1999)

Reported as *amethyst* in geodes from the Tuscarora district; also as *citrine* and *rose quartz.* (Hoffman, 1878)

As a major constituent of numerous pegmatite dikes with albite, *oligoclase,* microcline, and other pegmatite minerals in the Dawley Canyon area of the Valley View district. (Olson and Hinrichs, 1960)

Pink *jasper* and *agate,* as well as petrified wood and limb casts, come from Texas Spring Canyon, 12 miles southeast of San Jacinto. (Purington, 1903; Klein, 1983; Mitchell, 1991)

As petrified wood with crystal-lined cavities in the Hubbard Basin. (Mitchell, 1991)

As seams of *agate* along Jake Canyon, 3.5 miles north of Jarbidge. (Klein, 1983)

ESMERALDA COUNTY *Amethyst* scepters have been reported in the Coaldale district. (S. Pullman, w.c., 1999)

In petrified wood as *agate* and *jasper* in the Volcanic Hills in the Fish Lake Valley district. (Strong, 1966)

A major gangue mineral in the Goldfield district. (Ransome, 1909a)

As *agate* and *jasper* at the Gem claim 5 miles northwest of Goldfield in the Montezuma district. (Strong, 1966; Klein, 1983; Mitchell, 1991)

At the Nivloc Mine in the Silver Peak district as amethystine crystals to 1 cm; agatized wood is also reported from an unspecified locality. (M. Jensen, w.c., 1999; Strong, 1966)

Jasper and *agate* are reported from localities near Coaldale. (Mitchell, 1991)

At the Broken Toe Mine, Rock Hill district, as smoky and clear crystals to 10 cm in fractures with green muscovite; also as white crystals to 2 cm in white quartz veins. (J. Wellman, o.c., 2001)

EUREKA COUNTY As clear to white euhedral crystals to 15 mm, anhedral masses, veins, and silicified rock at the Goldstrike Mine in the Lynn district; also as sceptered late-stage crystals to 4 mm in vugs with barite crystals. (GCF)

As *agate* at the mouth of Pinto Canyon in the Pinto district. (Klein, 1983)

As amethystine crystals to 1 cm in geodes to 10 cm in the upper levels at the Barth Iron Mine in the Safford district. (M. Jensen, w.c., 1999)

HUMBOLDT COUNTY *Chalcedony* occurs with cinnabar in the Bottle Creek district. (Roberts, 1940a)

As clear to black, finely granular to coarse matrix material with pyrite, zircon, and *adularia* in radioactive breccia at the Moonlight Uranium Mine in the Disaster district. (Castor et al., 1996)

Colorless prisms with goethite and hematite at the Silver Coin Mine in the Iron Point district. (J. McGlasson, w.c., 2001)

As veins with specular hematite and as the varieties *jasper* and *chalcedony* at the Delong Mine in the Jackson Mountains district. (Shawe et al., 1962)

As drusy *amethyst* crusts to 4 cm thick in an open pit on the north side of the Eugene Mountains in the Mill City district; also as colorless to white euhedral crystals to 12 cm in lenses in Triassic argillite near the Blackbird Mine. (M. Jensen, w.c., 1999)

Common as small crystals and *chalcedony* veinlets at the Cordero Mine in the Opalite district. (Fisk, 1968)

As stubby *amethyst* crystals to 10 cm at a small prospect pit near the Getchell Mine in the Potosi district; also as smoky crystals to 6 cm with crystals in Anderson Canyon. (N. Prenn, w.c., 1999)

As *agate,* 4 miles west of Denio. (Klein, 1983)

LANDER COUNTY In Iowa Canyon in the Iowa Canyon district, as massive *agate* in veins to 10 cm with fluorite; also as spheres or nodules of radiating quartz to 5 cm with crystalline surfaces. (Fisher, 1927)

With hematite, cassiterite, and cristobalite in veins in rhyolite in the Izenhood district; also reported as *chalcedony.* (Knopf, 1916a; Fries, 1942; Anonymous, 1938; S. Pullman, w.c., 1999)

At the Mount Airy Mine in the New Pass district as massive gemmy *chalcedony* in thunder eggs to 90 cm in diameter in shades of blue to purplish blue, some with "cat's-eye" chatoyance; also reported near the New Pass Mine as colorless euhedral crystals to 8 cm. (Rose, 1997; M. Jensen, w.c., 1999)

As petrified wood in Daisy (Dacie) Creek. (Klein, 1983; Mitchell, 1991)

As *amethyst* crystals with pyrite, calcite, and sphalerite in the Fish Creek Mountains. (R. Walstrom, w.c., 1999)

Specimens of snakeskin *agate* are reported from 15 miles north of Austin. (WMK)

As *agate,* petrified wood, and *chalcedony* north of Rock Creek. (Klein, 1983)

As smoky phenocrysts in the Caetano and Fish Creek Mountains tuffs. (Stewart et al., 1977a)

LINCOLN COUNTY As clear crystals to 4 cm with reverse scepters at the April Fool Mine in the Delamar district. (N. Prenn, w.c., 1999)

Amethyst is reported as vapor-phase crystals to 3 mm with riebeckite in cavities in the peralkaline Kane Wash tuff. (Novak, 1985)

LYON COUNTY A *rose quartz* specimen is reportedly from the Silver City district. (WMK)

Crystals to 5 cm with grayish fluorapatite crystals are found in pegmatites at unspecified prospects near Sweetwater in the Washington district. (M. Jensen, w.c., 1999)

As smoky crystals to 50 cm in pegmatite dikes around the base of Quartz Hill in Nye Canyon in the Wilson district; also reported as crystals to 35 cm long with epidote and long, curving actinolite inclusions, and as rutilated quartz. (M. Rogers, w.c., 1999; N. Prenn, w.c., 1999; M. Jensen, w.c., 1999; S. Pullman, w.c., 1999)

Reported as the varieties *rose quartz, jasper,*

chalcedony, and petrified wood in the Yerington district; also at the Douglas Hill Mine as drusy *amethyst* crystals. (Melhase, 1935e; Klein, 1983; M. Jensen, w.c., 1999)

As crystals at the Topper tungsten prospect. (M. Jensen, w.c., 1999)

Reported as *agate, jasper,* and petrified wood 5 to 12 miles south and southwest of Fernley. (Klein, 1983; Mitchell, 1991)

MINERAL COUNTY As *chalcedony,* and as epimorphs after calcite and barite to 4 cm, in precious-metal veins in the Aurora district. (HMM; N. Prenn, w.c., 1999; WMK)

In crystal-lined *agate* geodes near the Kaiser (Baxter) Fluorspar Mine in the Broken Hills district, and as *agate* and *jasper* in a large area 13 miles northwest of Gabbs. (Klein, 1983; Mitchell, 1991)

As *agate, jasper,* petrified wood, and geodes near Basalt. (Strong, 1966; Klein, 1983; Mitchell, 1991)

In colorless euhedral crystals to 4 cm at the Eastside Mine in the Eastside district. (M. Jensen, w.c., 1999)

In the Fitting district as nearly black crystals in pegmatite dikes with fluorapatite, uranium minerals, and thorium minerals at the Holiday Mine, Sec. 10, T8N, R33E; also in pegmatite at the Zapot claims as lustrous smoky crystals to 20 cm with Dauphiné twinning and rare hematite inclusions. (Ross, 1961; Foord et al., 1999b; M. Jensen, w.c., 1999; N. Prenn, w.c., 1999; R. Walstrom, w.c., 1999)

As smoky crystals to 2 cm with epidote and actinolite inclusions at the Hooper No. 2 shaft in the Leonard district. (M. Jensen, w.c., 1999)

As optical-quality quartz crystals to 8 inches in quartz veins on the Malatesta claims in the Pamlico district; also as crystals to 4 cm with epidote at the Julie claim. (Ross, 1961; Smith and Benham, 1985; S. Pullman, w.c., 1999)

As smoky crystals to 20 cm in pegmatite dikes near the northwest edge of Gabbs Valley Range in the Poinsettia district. (N. Prenn, w.c., 1999)

As crystals, some with large liquid-gas inclusions, from the Grutt Brothers Mine on Burns Hill near Rawhide; petrified wood is reported 10 miles south of Rawhide. (CSM; WMK)

In thick crusts of white to pale amethystine crystals to 4 cm in large vugs at the Silver Dyke

Mine in the Silver Star district. (M. Jensen, w.c., 1999)

As *jasper* and pale-blue *chalcedony* with calcite, gypsum, iron oxides, and *psilomelane* at the Black Jack Mine in the Sodaville district. (Garside and Schilling, 1979; Klein, 1983)

As colorless to white crystals to 15 cm with rare scheelite crystals at the Kenyon claims in the Eagleville district; also as smoky crystals to 5 cm in pegmatite nearby. (M. Jensen, w.c., 1999)

NYE COUNTY As drusy amethystine crystals to 10 cm at the Original Bullfrog and Bullfrog Mines in the Bullfrog district; also reported as *jasper* in the district. (Klein, 1983; M. Jensen, w.c., 1999; GCF)

In crystal specimens from the Paradise Range near Gabbs. (WMK)

In specimens from the White Caps Mine and elsewhere in the Manhattan district, locally reported as rutilated. (Melhase, 1935d; BMC; NMBM)

In veins as crystals and *jasper* in the Oak Spring district. (Melhase, 1935b)

In a banded and vuggy vein with some crystals at the Green Top claim in the Round Mountain district; also widespread in quartz-*adularia* veins southwest of Round Mountain. (Lovering, 1954; Tingley and Berger, 1985; S. Pullman, w.c., 1999)

As white masses or crystal druses with silver minerals in the Tonopah district; also reported as pseudomorphs after analcime and as petrified wood. (Spurr, 1903; Melhase, 1935c; UCB; WMK)

As crystals from Ophir Canyon about 5 miles west of Highway 376 in the Twin River district. (WMK)

As banded *agate* in solid and hollow geodes, some lined with quartz crystals, on Agate Mountain 3 miles north of Ione. (Anonymous, 1865)

As *agate* in the Gabbs Valley area 10–12 miles west-northwest of Gabbs. (Klein, 1983)

As drusy crystalline epimorphs to 12 cm after barite crystals in an outcrop in Hot Creek Valley near Warm Springs. (N. Prenn, w.c., 1999; M. Jensen, w.c., 1999)

PERSHING COUNTY In the Antelope district as abundant well-formed smoky crystals in the Tin stope of the Majuba Hill Mine, as crystals to 1.5 cm with schorl inclusions in the Copper stope, and as crystals in cavities elsewhere in the mine. Also at the Bottomley prospect as well-formed, transparent to cloudy crystals to 5 cm with acic-

ular inclusions of stibnite, stibiconite, and cervantite to 2.5 cm. Elsewhere on Majuba Mountain, quartz occurs as amethystine and colorless crystals to 4 cm in lenses in Triassic argillite. (Jensen, 1985; UMNH; F. Cureton, w.c., 1999; N. Prenn, w.c., 1999; M. Jensen, w.c., 1999)

In the Antelope Springs district as crystals with inclusions of cinnabar at the Montgomery Mine and inclusions of stibnite at the Hollywood Mine; also as colorless crystal clusters to 10 cm from the Old Relief Canyon Mine. (R. Walstrom, w.c., 1999; M. Jensen, w.c., 1999)

Reported as pseudomorphs after arsenopyrite and pyrite crystals at the Montezuma Mine in the Arabia district. (R. Walstrom, w.c., 1999)

In hot springs sinter as *chalcedony* in the Goldbanks district. (Dreyer, 1940)

As colorless crystals to 4 cm in an outcrop near the Star Peak Mine in the Imlay district. (M. Jensen, w.c., 1999)

As colorless crystals to 2.5 cm with native silver wire inclusions at the Keystone Mine in the Mill City district. (M. Jensen, w.c., 1999)

In specimens as crystals, some with included native sulfur, from gypsum deposits 3 miles east of Lovelock in the Muttlebury district. (WMK)

As large crystals from the Nightingale Mine in the Nightingale district; also as colorless crystals to 3 cm in vugs in skarn at the Alpine Mine. (M. Rogers, w.c., 1999; M. Jensen, w.c., 1999)

As *rose quartz* that owes its color to fine inclusions of pink dumortierite in the Rochester district. (Holden, 1924; Knopf, 1924)

As *agate* geodes on the eastern slope of Star Peak about 10 miles south of Mill City. (Klein, 1983)

With inclusions of cervantite and stibnite from the Trinity district in the Trinity Mountains; also as large clear crystals with ghosts of white inclusions. (CSM; WMK; F. Cureton, o.c., 1999)

Agate is found near Trinity Peak in the Velvet district. (Klein, 1983; Mitchell, 1991)

As colorless crystals to 3 cm from an unnamed adit above the Black Warrior Mine in the Unionville district. (M. Jensen, w.c., 1999)

Crystals to 4 mm in vugs with phosphates at the Willard Mine, Willard district. (Jensen and Leising, 2001)

As colorless euhedral crystals to 3 cm, some

with stibnite inclusions, from a prospect near Imlay Canyon. (M. Jensen, w.c., 1999)

As colorless crystals to 20 cm in lenses in Triassic argillite from outcrops in Woody Canyon. (M. Jensen, w.c., 1999)

In groups of clear crystals to 20 cm, commonly with *s*-faces, from a locality at the headwaters of Corral Creek. (N. Prenn, w.c., 1999)

STOREY COUNTY In the Comstock district, massive quartz was reportedly the main constituent of the Comstock Lode; in addition, quartz was reported as pockets of milky, clear, and deep-purple to violet crystals to 3 inches in veins to 16 inches wide. Reported in the Flowery lode in massive form, as transparent crystals to 10 cm with fluid inclusions with movable bubbles, and as epimorphs after barite crystals to 2 cm. Also reported as *jasper* and *chalcedony* at American Flat. (De Quille, 1876; Schilling, 1990; Nichols, 1991; UCB; HMM; M. Jensen, w.c., 1999; R. Walstrom, w.c., 1999; WMK; GCF)

WASHOE COUNTY As subhedral amethystine crystals at the Hog Ranch Mine in the Cottonwood district. (Bussey, 1996)

As *agate, jasper,* and petrified wood in the vicinity of Leadville. (Klein, 1983; WMK)

Massive and relatively pure with minor mica and *feldspar* in a pegmatite mass about 200 feet wide and 400 feet long at the Winnemucca Lake silica deposit, Sec. 21, T24N, R24E, in the Nightingale district; crystals to 7 cm with stibiconite inclusions are also reported from this locality or nearby. Specimens of crystals to 10 cm with clinozoisite crystals came from an unspecified location in the district. (Bonham and Papke, 1969; M. Jensen, w.c., 1999; WMK)

As clear, white, and amethystine crystals to 5 cm in veins with epidote, scheelite, calcite, scolecite, *heulandite,* and gold at the Olinghouse Gold Mine in the Olinghouse district; also as crystals in cavities in a basalt flow. (Jones, 1998; GCF; WMK)

Reported as phosphorescent milky crystals in a specimen from 5 miles north of Reno; *citrine* is reported from the Votaw property in the Peavine district. (WMK)

As smoky to nearly black grains in rhyolitic volcanic ash near uranium prospects in the Pyramid district. (Brooks, 1956)

As crystals to 35 cm in pockets in a breccia pipe at the Foster Hallman claims in the Stateline Peak district on the north end of Petersen Mountain near Hallelujah Junction, including smoky quartz crystals, clear scepters with smoky stems, and smoky scepters with amethystine tips; rare doubly terminated and "dumbbell" forms have also been recovered. (N. Prenn, w.c., 1999; M. Jensen, w.c., 1999; GCF)

Silica has been mined from an area of porous quartz, cristobalite, and minor sulfur and cinnabar about 1000 feet in diameter at the Silica pit in the Steamboat Springs district; *rose quartz* is also reported in a specimen from the area. (Bonham and Papke, 1969; WMK)

As smoky crystals to 58 cm with microcline, albite, and other minerals in pegmatites at Incline Village. (Jensen, 1993a)

As *chalcedony* amygdules in basalt in a specimen from 1/2 mile southwest of Wadsworth; in a geode with calcite from the same area. (WMK)

Chalcedony, *agate,* and *jasper* are found with opalized wood in Wall Canyon. (Klein, 1983)

WHITE PINE COUNTY As colorless pseudomorphs to 1 cm after cerussite crystals at the Elijah Mine in the Robinson district. (M. Jensen, w.c., 1999)

In large, milky-white to smoky crystal clusters to 30 cm in pegmatite dikes on the west side of Wheeler Peak. (M. Jensen, w.c., 1999; N. Prenn, w.c., 1999; WMK)

RADTKEITE

Hg_2S_2ClI

A Nevada type mineral, radtkeite is a mercury species formed by reaction of cinnabar and corderoite with chlorine-rich hydrothermal solutions; named for Arthur Radtke, who did much of the pioneer work on Carlin-type gold deposits in Nevada.

HUMBOLDT COUNTY In small amounts with quartz, cinnabar, and corderoite in hydrothermally altered lake sediments as finely dispersed grains, coatings, and prismatic crystals to 30 microns at the McDermitt Mine in the Opalite district, Sec. 33, T47N, R37E; the mineral is yellow-orange when first exposed, but blackens quickly in sunlight. (McCormack et al., 1991)

RALSTONITE

$Na_xMg_xAl_{2-x}(F,OH)_6 \cdot H_2O$

Ralstonite is a rare mineral that is generally found in pegmatites and greisens rich in fluorine, and rarely in fluorspar and hydrothermal deposits.

CHURCHILL COUNTY As pseudo-octahedral colorless crystals to 1 mm associated with fluorite and quartz at the Green prospect (mopungite locality) in the Lake district. (M. Jensen, w.c., 1999; R. Walstrom, w.c., 1999)

LANDER COUNTY Detected in very small quantities by XRD methods in fluorspar ore at the Blazer prospect in the Iowa Canyon district, NW1/4 NW1/4 Sec. 26, T23N, R44E. (Papke, 1979)

MINERAL COUNTY With late stage Na-Li-alumino-fluoride minerals in *amazonite*-topaz pegmatite on the Zapot claims about 25 km northeast of Hawthorne in the Fitting district. (Foord et al., 1999b)

RAMEAUITE

$K_2CaU_6^{6+}O_{20} \cdot 9H_2O$

A secondary mineral that has been found with other uranium minerals in French deposits.

ESMERALDA COUNTY As yellow coatings on rhyolite at Rhyolite Ridge near Mina. (Weissman and Nikischer, 1999)

RAMMELSBERGITE

$NiAs_2$

A moderately rare mineral generally found in hydrothermal veins, commonly with cobalt and other nickel minerals.

CHURCHILL COUNTY Reported with annabergite and nickeline in specimens of nickel ore from the Lovelock Mine in the Table Mountain district; identified by XRD. (WMK; M. Coolbaugh, o.c., 2000)

RANCIÉITE

$(Ca,Mn^{2+})Mn_4^{4+}O_9 \cdot 3H_2O$

A manganese oxide mineral that occurs as a weathering or oxidation product, commonly in limestone.

NYE COUNTY Identified with other manganese oxides in drill core of tuff from Yucca Mountain. (Carlos et al., 1995)

RAUENTHALITE

$Ca_3(AsO_4)_2 \cdot 10H_2O$

A rare, post-mining reaction product of carbonate with arsenic-rich solutions that is found at ten localities in Europe.

HUMBOLDT COUNTY Identified by W. Wise at the Getchell Mine in the Potosi district. (M. Jensen, w.c., 1999)

REALGAR

AsS

Realgar, a common low-temperature mineral in base- and precious-metal deposits, also occurs in hot spring and fumarole deposits and in sedimentary rocks. In Nevada it is found in many Carlin-type gold deposits, and exceptional specimens have come from the Getchell and White Caps Mines.

ELKO COUNTY Generally in veins or partially replacing porous units at the Jerritt Canyon, West Generator Hill, and Murray Mines in the Independence Mountains district. (Hofstra, 1994)

EUREKA COUNTY As small crystals with orpiment in replacement deposits in limestone at the Gold Bar Mine in the Antelope district. (N. Prenn, w.c., 1999)

Reported in gold ore from Mill Canyon in the Cortez district with calcite, arsenopyrite, orpiment, stibnite, quartz, and pyrite. (WMK; Gianella, 1941)

Reported at the Windfall Mine in the Eureka district. (Gianella, 1941; M. Jensen, w.c., 1999)

With orpiment as short, prismatic crystals coating carbonaceous limestone breccia at the Carlin Mine in the Lynn district, red-orange when fresh but alters to yellow-orange when exposed to light; also as red to red-orange masses with orpiment, calcite, and fluorite, and as euhedral columnar crystals to 3 cm at the Goldstrike Mine. (Hausen and Kerr, 1968; Rota, 1989; GCF)

As massive aggregates in carbonaceous limestone with white barite at the Gold Quarry Mine in the Maggie Creek district; also as brilliant-red-orange crystals to 3 mm in small druses. (Jensen et al., 1995)

HUMBOLDT COUNTY In crusts and veinlets at the Lone Tree Mine in the Buffalo Mountain district. (Bloomstein et al., 1993)

Reportedly after arsenopyrite in quartz veins at the National Mine in the National district. (Lindgren, 1915; Melhase, 1934c)

Rare with corderoite and montmorillonite at the Cordero Mine in the Opalite district. (Jenkins, 1981)

The Getchell Mine in the Potosi district is known for attractive groups of crystals, and has produced translucent, cherry-red crystals to 12 cm (possibly the world's largest), as well as massive, red-orange aggregates in quartz-calcite rock with orpiment and stibnite; realgar is also reported in gold ore from the nearby Ogee and Pinson Mines. (Gianella, 1941; Joralemon, 1951; Stolburg and Dunning, 1985; GCF)

LANDER COUNTY With orpiment and believed to be after arsenopyrite at the Copper Canyon underground mine in the Battle Mountain district. (Roberts and Arnold, 1965)

With orpiment and cinnabar at the Lone Tree Mine in the Buffalo Mountain district. (Bloomstein et al., 1993)

In a specimen from an unknown occurrence in the Reese River district. (WMK)

NYE COUNTY At the White Caps Mine in the Manhattan district, generally as massive replacements of calcite and as small crystals in quartz vugs, but crystals to 3 cm and more were found with orpiment and stibnite in calcite, and many museum specimens came from this mine; massive and crystalline realgar also came from the nearby Manhattan Consolidated Mine. (Ferguson, 1924; Melhase, 1935d)

With orpiment, cinnabar, and other minerals in silver-gold ore at the Paradise Peak Mine in the Paradise Peak district. (Thomason, 1986)

Reported at the Round Mountain Mine in the Round Mountain district. (Ferguson, 1922; Tingley and Berger, 1985)

WASHOE COUNTY In hot spring sinter in the Steamboat Springs district. (Gianella, 1941)

WHITE PINE COUNTY Reported at the Vantage deposit on Alligator Ridge in the Bald Mountain district. (Ilchick, 1990)

RETGERSITE

$NiSO_4 \cdot 6H_2O$

Retgersite is a rare mineral that occurs as an oxidation product of primary nickel arsenides and sulfides. The Nevada occurrence is a co-type locality.

CHURCHILL COUNTY As blue-green, fibrous prisms and aggregates with granular masses of annabergite in almost completely oxidized nickeline- and millerite-bearing veins in highly altered greenstone at the Nickel and Lovelock Mines in the Table Mountain district, Sec. 34, T25N, R36E. (Ferguson, 1939; Frondel and Palache, 1949; Palache et al., 1951)

RHABDOPHANE-(CE)

(Ce,La)PO$_4$•H$_2$O

An uncommon mineral that is generally found in rare earth deposits, particularly in pegmatite and syenite; also noted from base-metal deposits.

PERSHING COUNTY As microscopic black crystals with anatase and schorl in highly altered porphyritic rhyolite in the lower and middle adits of the Majuba Hill Mine in the Antelope district. (Jensen, 1985)

RHODOCHROSITE

Mn^{2+}CO$_3$

Rhodochrosite is common as a gangue mineral in hydrothermal base- and precious-metal deposits. It also occurs in some bedded manganese deposits.

CHURCHILL COUNTY As a primary mineral in veins with quartz, calcite, and other minerals at the Fairview Silver, Eagle, and Nevada Hills Mines in the Fairview district. (Greenan, 1914; Schrader, 1947; Quade and Tingley, 1987)

Questionably identified as a gangue mineral in the Holy Cross district. (Schrader, 1947)

ELKO COUNTY With rhodonite and bementite in a bedded deposit at the Wicker Mine in the Hicks district. (Smith, 1976)

EUREKA COUNTY One of the gangue minerals in lead-silver ore in the Eureka district. (Gianella, 1941)

HUMBOLDT COUNTY As orange hemispheres to 1 cm with barite crystals and romanèchite from the Sequoia pit at the Lone Tree Mine in the Buffalo Mountain district; said to be calcium-bearing. (M. Jensen, w.c., 1999)

As spheroids to 5 mm associated with barite and marcasite at the Twin Creeks Mine in the Potosi district. (N. Prenn, w.c., 1999)

LANDER COUNTY Reported from an unspecified locality in the Hilltop district. (Gianella, 1941)

Common as a gangue mineral at the Betty O'Neal Mine in the Lewis district; also reported at the Eagle vein. (Gianella, 1941; M. Jensen, w.c., 1999; J. McGlasson, w.c., 2001)

Abundant in gangue as pale-pink cleavage masses with quartz, galena, tetrahedrite, and other minerals in the Reese River district. (Melhase, 1934c; Ross, 1953; UCB; M. Jensen, w.c., 1999)

LINCOLN COUNTY Reported from an unspecified locality in the Pioche district. (Gianella, 1941)

MINERAL COUNTY In trace amounts with sulfides, quartz, and other carbonate minerals at the Northern Belle Mine in the Candelaria district. (Moeller, 1988)

NYE COUNTY A common gangue mineral as crystals to 5 mm in silver-tin veins with sphalerite, pyrargyrite, stephanite, pyrite, and andorite at the Morey and Keyser Mines in the Morey district. (Kral, 1951; Williams, 1968; S. Pullman, w.c., 1999)

Identified as fragments in dark manganese-rich silicified rock from the Mariposa vein in the Round Mountain district. (Tingley and Berger, 1985)

Pinkish and fine-grained as a rare to major constituent of silver-bearing quartz veins in the Tonopah district; specifically reported as iron-bearing intergrowths with quartz in silver ore from the 1000-foot level of the Belmont vein, and also from the Montana-Tonopah Mine. (Bastin and Laney, 1918; Melhase, 1935c; Nolan, 1935; WMK; Pullman and Thomssen, 1999)

Widespread as a gangue mineral in silver-lead ore in the Tybo district. (Ferguson, 1933)

STOREY COUNTY In minor amounts in quartz veins in the Comstock district. (Gianella, 1941)

WASHOE COUNTY Reported in vugs in veins with quartz, pyrophyllite, diaspore, *sericite,* and dickite in Peavine district copper deposits; similar occurrences were noted in the Wedekind district. (Gianella, 1941; Hudson, 1977)

WHITE PINE COUNTY Reported with quartz and hübnerite in a specimen from the Kern Mountains in the Eagle district. (CSM)

With alabandite below the water level at the Siegel Mine in the Muncy Creek district. (Pardee and Jones, 1920)

As pale-pink crystalline masses in hydrothermally altered skarn in the Robinson district. (GCF)

Reported as a gangue mineral at the Pocotillo Mine in the White Pine district. (Smith, 1976)

RHODONITE

$(Mn^{2+},Fe^{2+},Mg,Ca)SiO_3$

Rhodonite is a widespread mineral that is found abundantly in some sedimentary and metamorphosed manganese deposits, but occurs more often as gangue in hydrothermal vein deposits.

CHURCHILL COUNTY With rhodochrosite, calcite, and quartz in banded precious-metal veins at the Fairview Mine in the Fairview district. (Greenan, 1914; Schrader, 1947; Quade and Tingley, 1987)

ELKO COUNTY With rhodochrosite and bementite in a bedded deposit at the Wicker Mine in the Hicks district. (Smith, 1976)

HUMBOLDT COUNTY Reportedly common at the Major and other mines, and at the O'Brien prospect in the Golconda district; also identified in a specimen from the Kopenhaver property. (Pardee and Jones, 1920)

Identified by XRD in pink and black replacement zones in metamorphosed shale in unnamed prospects on the southwest flank of Buffalo Mountain in the Buffalo Mountain district. (GCF)

LANDER COUNTY Identified in a specimen from the Battle Mountain district. (AMNH)

LINCOLN COUNTY Reported in brecciated volcanic rocks with stellarite, green *garnet,* and an unidentified mineral in the Bristol district. (Melhase, 1934d)

NYE COUNTY As pinkish bands in some quartz veins in the Tonopah district. (Melhase, 1935c; WMK)

PERSHING COUNTY In small amounts in massive braunite-*chalcedony* bodies at the Black Diablo Mine in the Black Diablo district. (Ferguson et al., 1951; Johnson, 1977)

WHITE PINE COUNTY The variety *fowlerite* (zinc-bearing rhodonite) is reported in a specimen from Tungstonia in the Eagle district of the Kern Mountains. (WMK)

With rhodochrosite in gangue in the White Pine district; rhodonite of flesh-red color was reported at the Pocotillo Mine. (Hague, 1870; Smith, 1976)

RHODOSTANNITE

$Cu_2FeSn_3S_8$

A rare mineral that occurs as an alteration product of stannite.

LANDER COUNTY In a 1- to 2-cm-wide sulfide vein in altered diorite at the Dean Mine in the Lewis district. (Anthony et al., 1990; J. McGlasson, w.c., 2001)

RHOMBOCLASE

$(H_5O_2)^{1+} Fe^{3+}(SO_4)_2 \cdot 2H_2O$

Rhomboclase is a secondary mineral formed by the oxidation of pyrite as an encrustation on mine walls, generally in association with other soluble sulfates.

LANDER COUNTY Reported with canfieldite in silver ore at the Dean Mine in the Lewis district. (Ream, 1990)

RICHELLITE

$Ca_3Fe_{10}^{3+}(PO_4)_8(OH,F)_{12} \cdot H_2O$ (?)

A very rare species found elsewhere with halloysite and allophane at a site in Belgium.

EUREKA COUNTY As waxy, pale-brown masses at the Gold Quarry Mine in the Maggie Creek district. (Jensen et al., 1999)

RICHELSDORFITE

$Ca_2Cu_5^{2+}Sb^{5+}(AsO_4)_4Cl(OH)_6 \cdot 6H_2O$

A rare secondary species; most other occurrences are in Europe where richelsdorfite is found with tetrahedrite-tennantite and other minerals.

WASHOE COUNTY As spongy to pulverulent crystalline aggregates that range in color from sky-blue to almost colorless at the Burrus Mine in the Pyramid district; the best specimens are stacked clusters to 1 mm of tabular, transparent blue crystals with minor malachite and chrysocolla in vugs. (Jensen, 1994)

RIEBECKITE

$\square Na_2(Fe_3^{2+}Fe_2^{3+})Si_8O_{22}(OH)_2$

Sodium-rich member of the amphibole group mainly found in alkaline rocks, granite pegmatites, and schist. The fibrous variety *crocidolite* is found in iron deposits.

HUMBOLDT COUNTY Part of the granophyric groundmass in the Double H tuff of the McDermitt caldera; gives the rock a blue-green color. (Rytuba and McKee, 1984)

LINCOLN COUNTY As vapor-phase crystals to 3 mm in pockets with *amethyst* in the peralkaline Kane Wash tuff. (Novak, 1985)

As black crystals to 3 mm in lithophysae in an outcrop near Alamo. (Jensen, w.c., 1999)

PERSHING COUNTY Blue *crocidolite* has been identified in a specimen with magnetite and *apatite* at the Segerstrom-Heizer Mine in the Mineral Basin district. (UCB)

ROBINSONITE

$Pb_4Sb_6S_{13}$

A Nevada type mineral; robinsonite is a rare primary sulfosalt mineral.

HUMBOLDT COUNTY Reported at the Silver Coin Mine in the Iron Point district. (Anthony et al., 1990)

LANDER COUNTY Reported as gray acicular crystals to 1 cm on fractures underground in the McCoy-Cove Mine in the McCoy district. (M. Johnston, o.c., 2002)

PERSHING COUNTY The type locality as a primary mineral with pyrite, sphalerite, stibnite, and boulangerite in small patches in mostly oxidized ore along faults in carbonate beds at the Red Bird Mine in the Antelope Springs district, Sec. 33, T27N, R34E; dadsonite is another mineral for which this is the type locality. (Berry et al., 1952)

ROMANÈCHITE

$(Ba,H_2O)_2(Mn^{4+},Mn^{3+})_5O_{10}$

A common mineral that is a principal component of the nonspecific materials *psilomelane, wad,* and rock varnish. Large amounts occur in weathered manganese ore deposits. Undoubtedly much more common in Nevada than the few occurrences listed below indicate; see *psilomelane* for other possible occurrences.

EUREKA COUNTY As black stalactitic masses and cauliflower-shaped bunches to 5 cm in breccia at a prospect 2 km north of the Barth Mine in the Safford district. (GCF)

HUMBOLDT COUNTY As sooty black fracture coatings associated with rhodochrosite and barite at the Lone Tree Mine in the Buffalo Mountain district. (M. Jensen, w.c., 1999)

MINERAL COUNTY Reported from Sodaville; probably from the Black Jack Mine in the Silver Star district. (Anthony et al., 1997)

ROMÉITE

$(Ca,Fe^{2+},Mn^{2+},Na)_2(Sb,Ti)_2O_6(O,OH,F)$

Roméite is a secondary mineral that generally occurs along with other antimony-bearing oxides after stibnite.

CHURCHILL COUNTY With stibiconite, senarmontite, and tripuhyite after stibnite at the Green prospect in the Lake district; encrusted by mopungite. (Williams, 1985; S. Pullman, w.c., 1999)

PERSHING COUNTY In a specimen from Oreana as yellow-brown coatings with senarmontite and other antimony minerals; verified by XRD. (UAMM; GCF)

RÖMERITE

$Fe^{2+}Fe_2^{3+}(SO_4)_4 \cdot 14H_2O$

A secondary mineral formed by the oxidation of pyrite; typically associated with copiapite and other iron sulfate minerals as a post-mining mineral.

ESMERALDA COUNTY Reported as brown cubes to 0.3 mm with other post-mining sulfate minerals in a specimen from the 250-foot level of the Nevada Eagle mine in the Goldfield district. (R. Thomssen, w.c., 1999)

ROQUESITE

$CuInS_2$

Roquesite is found with other sulfide minerals in high-temperature veins, skarn, and copper ore.

WASHOE COUNTY As 5-micron inclusions in pyrite in association with inclusions of sphalerite, galena, anglesite, polybasite, and an unnamed zinc-indium sulfide at the Sure Fire Mine in the Pyramid district, NW1/4 Sec. 22, T23N, R21E. (Tingley et al., 1998b)

ROSASITE

$(Cu^{2+},Zn)_2(CO_3)(OH)_2$

Rosasite occurs in the oxidation zones of deposits that contain both copper and zinc. It is found in similar settings to aurichalcite, but is rarer than that mineral.

CHURCHILL COUNTY Reported from an unspecified site in the Westgate district. (R. Thomssen, w.c., 1999)

CLARK COUNTY As scattered spherules to 2 mm on oxidized lead-zinc-copper ore at the Yellow Pine Mine in the Goodsprings district. (Jenkins, 1981)

ELKO COUNTY As bluish-green botryoidal crusts with fibrous to spherulitic structure in gossan with hemimorphite at the Killie Mine in the Spruce Mountain district. (Dunning and Cooper, 1987)

ESMERALDA COUNTY With fornacite and other minerals in quartz veins a mile north of the site of Columbus in the Candelaria district. (R. Walstrom, w.c., 2000)

EUREKA COUNTY In light-green fracture fillings from small prospect pits near the Copper King Mine in the Maggie Creek district. (Rota, 1991)

HUMBOLDT COUNTY Reported at the Silver Coin Mine in the Iron Point district. (Thomssen, 1998)

LANDER COUNTY Reported at the Garrison Mine in the Cortez district. (S. Pullman, w.c., 1999)

As spherules to 2 mm with scholzite at the Snowstorm Mine in the North Battle Mountain district. (J. McGlasson, w.c., 2001)

LYON COUNTY As blue-green spherules to 1 mm on gossan at the Jack Pot claim in the Wellington district. (Palache et al., 1951; Heyl and Bozion, 1962)

MINERAL COUNTY As blue crystals to 1 mm at the Simon Mine in the Bell district; also noted as 1.5-mm green spheroids on white hemimorphite crystals. (S. White, w.c., 2000; GCF)

NYE COUNTY Reported in a specimen from Grantsville in the Union district. (R. Thomssen, w.c., 1999)

Reported from the San Rafael Mine in the Lodi district. (R. Thomssen, w.c., 1999)

PERSHING COUNTY Reported at the Majuba Hill Mine in the Antelope district. (Frondel, 1949; Palache et al., 1951; Heyl and Bozion, 1962)

WHITE PINE COUNTY As green crystals to 2 mm coated with white gypsum at the Kansas Mine in the Aurum district. (GCF)

Reported from an unspecified location in the Robinson district. (Von Bargen, 1999)

ROSCOELITE

$KV_2\square AlSi_3O_{10}(OH)_2$

A vanadium-bearing, mica-group mineral that typically occurs in epithermal precious-metal deposits and sedimentary uranium-vanadium ore.

EUREKA COUNTY Reported as yellow-brown, microgranular patches to 1 cm in diameter on fracture surfaces in argillized and decalcified limestone on the 4,460-foot level of the Goldstrike Mine in the Lynn district. (GCF)

RÖSSLERITE

$MgHAsO_4 \cdot 7H_2O$

Rösslerite (*roesslerite*) is an uncommon secondary mineral in arsenic-bearing ore deposits.

NYE COUNTY As very thin white seams that are macroscopically massive, but finely crystalline under the microscope, on realgar and orpiment in limestone at the White Caps Mine in the Manhattan district; associated with minute fibers tentatively identified as pharmacolite. (Goudey, 1946; Gibbs, 1985)

ROZENITE

$Fe^{2+}SO_4 \cdot 4H_2O$

Rozenite occurs in base- and precious-metal deposits in which pyrite is oxidizing and as efflorescences on mine walls. Although found in similar settings to melanterite and other sulfates, it is relatively scarce.

ESMERALDA COUNTY With halotrichite near the Sweeney lease in the Goldfield district. (Ashley and Albers, 1975)

EUREKA COUNTY In white to pale-blue efflorescent crusts with szomolnokite in sheared and clay-altered zones at the Goldstrike Mine in the Lynn district; translucent when first exposed, but dehydrates within minutes to dull white. (GCF)

HUMBOLDT COUNTY As tiny white veinlets in arsenic-bearing ore at the Getchell Mine in the Potosi district. (Jenkins, 1981)

White efflorescence associated with marcasite, pyrite, and siderite at the Lone Tree Mine, Buffalo Mountain district; identified by XRD. (GCF)

NYE COUNTY A post-mining mineral in silver deposits as bluish crusts and white spherules to 1.0 mm in the Morey district. (Williams, 1968)

WASHOE COUNTY Reported as a white efflorescent mineral in the Peavine district. (S. Pullman, w.c., 1999)

RUCKLIDGEITE

$(Bi,Pb)_3Te_4$

A telluride mineral found in a number of types of hydrothermal metal deposits.

WASHOE COUNTY As black flakes to 5 mm on quartz from the Olinghouse open pit in the Olinghouse district; identified by ICP-MS laser ablation. (M. Hunnerlack, w.c., 2000)

RUTHERFORDINE

$UO_2(CO_3)$

A secondary mineral that has been found in oxidized uranium ore, particularly in pegmatite and sandstone-type deposits.

LANDER COUNTY As cream-colored crystalline microclusters at the Apex (Rundberg) Uranium Mine in the Reese River district. (M. Jensen, o.c., 2000)

RUTILE

TiO_2

Rutile is a common accessory mineral in many igneous and metamorphic rocks; it also occurs as detrital material in some sedimentary rocks and as a minor species in hydrothermally altered rocks.

CARSON CITY Reported as small red acicular crystals in smoky quartz with ilmenite in an unnamed barite prospect in the Carson district at the north end of Prison Hill. (R. Walstrom, w.c., 1999)

CHURCHILL COUNTY As small crystals with anatase in feldspathic masses adjacent to gold-bearing quartz veins between Table Mountain and the Dixie Comstock Mine in the Corral Canyon district in the Stillwater Mountains. (Vanderburg, 1940)

CLARK COUNTY With anatase in the Goodsprings district. (Hewett, 1931)

DOUGLAS COUNTY Identified in a sample from the north end of Smith Valley in the Buckskin district. (JHS)

ELKO COUNTY Widespread in sedimentary host rocks at the Jerritt Canyon and Big Springs Mines in the Independence Mountains district. (Youngerman, 1992; Hofstra, 1994)

As tiny grains (<50 microns) of rodlike clusters disseminated in host rock at the Hollister deposit in the Ivanhoe district, may form up to 3 percent of rock by volume; also found with quartz and zircon as sand particles. (Deng, 1991)

In small amounts with zunyite at the Zun claims in the Scraper Springs district. (Coats et al., 1979)

ESMERALDA COUNTY As microscopic crystals with alunite, quartz, and pyrite in altered dacite, probably formed by alteration of titanite or ilmenite, at the Combination and Florence Mines in the Goldfield district. (Ransome, 1909a; Ashley and Albers, 1975)

In gold-bearing placer gravels in the Tule Canyon district. (Garside, 1973)

EUREKA COUNTY As microscopic needlelike crystals resulting from the alteration of biotite in Tertiary rhyolite tuff, and as detrital grains in Paleozoic sedimentary rocks at the Goldstrike Mine in the Lynn district. (Campbell, 1994; GCF)

In silicified siltstone with zircon, xenotime, and pyrite at the Gold Quarry Mine in the Maggie Creek district. (Jensen et al., 1995)

HUMBOLDT COUNTY In the host rock accessory suite at the Lone Tree Mine in the Buffalo Mountain district. (R. Braginton, o.c., 1995)

A minor component of gangue in radioactive breccia at the Moonlight Uranium Mine in the Disaster district. (Castor et al., 1996)

LANDER COUNTY At the Cove Mine in the McCoy district. (W. Fusch, o.c., 1999)

LYON COUNTY As inclusions in quartz crystals from Nye Canyon in the Wilson district. (S. Pullman, w.c., 1999)

MINERAL COUNTY Niobian rutile occurs with ilmenite in the Zapot pegmatite in the Fitting district. (Foord et al., 1999b)

NYE COUNTY In minor amounts in a scheelite vein at the Betty O'Neal pit in the Gabbs district. (Schilling, 1968a)

PERSHING COUNTY With titanite and magnetite in placers derived from metamorphic rocks in the Scossa district. (Beal, 1963)

STOREY COUNTY With *leucoxene*, magnetite, ilmenite, and other minerals in altered volcanic rock in the Comstock district. (Whitebread, 1976)

WASHOE COUNTY In fault zones, aplite and pegmatite dikes, and in quartz veins at the Miller and Redelius prospects in the McClellan district. (Beal, 1963)

A minor constituent of highly altered rocks in the Pyramid district. (Wallace, 1980)

WHITE PINE COUNTY With mica in a specimen from McGill. (WMK)

With magnetite, pyrrhotite, and chalcocite in jasperoid at the Veteran Mine in the Robinson district. (Spencer et al., 1917)

SABUGALITE

$H_{0.5}Al_{0.5}(UO_2)_2(PO_4)_2 \cdot 8H_2O$

Sabugalite is a relatively rare secondary mineral; in the United States it is most common in sandstone-type uranium deposits.

WASHOE COUNTY With autunite and phosphuranylite and other uranium minerals along a diabase dike in welded tuff at the Red Bluff and DeLongchamps Mines in the Pyramid district. (Brooks, 1956; Garside, 1973)

Reported at the Jeannie K claims in the Stateline Peak district. (Garside, 1973)

Reported with autunite along fractures and layers in petrified wood at the Petrified Tree claims in the McClellan district. (Garside, 1973)

SAFFLORITE

$(Co,Fe)As_2$

A relatively rare mineral that occurs with cobalt and nickel minerals in mesothermal veins.

CLARK COUNTY Subhedral to anhedral silver-colored crystals to 1 cm with pyrite and other sulfides at the Blue Jay or Columbia Mine in the Goodsprings district; identified by XRD and SEM/EDX. (Vajdak, 2001)

CHURCHILL COUNTY Identified by SEM/EDX in a specimen from the Lovelock Mine, Table Mountain district. (M. Coolbaugh, o.c., 2000)

SALÉEITE

$Mg(UO_2)_2(PO_4)_2 \cdot 10H_2O$

Saléeite, a secondary uranium mineral that forms a series with nováčekite, is easily confused with autunite in hand specimen and may be more common than thought.

MINERAL COUNTY Reported with torbernite and nováčekite at the Broken Bow and Broken Bow King claims in the Santa Fe district. (Garside, 1973)

SAMARSKITE-(Y)

$(Y,Ce,U,Fe^{3+})_3(Nb,Ta,Ti)_5O_{16}$

A relatively common mineral in small quantities in rare earth–rich pegmatites with *columbite-tantalite* group minerals, *monazite,* and other phases. Reported simply as *samarskite* at all of the Nevada occurrences listed below. Samarskite-(Y) is the only known species; nevertheless, the IMA has chosen to add the chemical suffix.

CLARK COUNTY As masses with *monazite,* stibiotantalite, *allanite,* zircon, and other minerals in quartz-*feldspar* pegmatite at the Hilltop Mine in the Gold Butte district, Sec. 21, T19S, R70E. (Garside, 1973)

In a specimen reportedly from the Muddy Mountains, but probably from the Gold Butte occurrence listed above. (WMK)

MINERAL COUNTY Reported with euxenite in a pegmatite at the Lucky Susan No. 1 claim in the Buena Vista district. (Ross, 1961; Garside, 1973)

Tentatively identified in pegmatite at the Pink Lady claims in the Marietta district. (Garside, 1973)

SAMSONITE

$Ag_4MnSb_2S_6$

A rare mineral that occurs with other sulfides in hydrothermal veins.

MINERAL COUNTY identified in the Candelaria district. (Knopf, 1923)

SANIDINE

$(K,Na)(Al,Si)_4O_8$

Sanidine is a disordered alkali *feldspar,* commonly occurring instead of orthoclase in volcanic and subvolcanic rocks. It is also found in some high-temperature contact-metamorphosed rocks. Only a few locations are listed below, but the mineral occurs in many Tertiary felsic flows and tuffs in Nevada.

CHURCHILL COUNTY As a primary mineral in wall rock in the Fairview district. (Schrader, 1947)

ELKO COUNTY In a sample of quartz porphyry from 300 yards west of Alternate U.S. Highway 93, 5 miles south of Wendover. (UCB)

EUREKA COUNTY Found as phenocrysts in the host rock at the Modarelli Mine in the Modarelli-Frenchie Creek district. (Shawe et al., 1962)

LANDER COUNTY Reported as part of the skarn mineral assemblage at the West orebody in Copper Canyon, Battle Mountain district. (Theodore and Blake, 1978)

Reported with cassiterite in the Izenhood district. (Anonymous, 1938; J. McGlasson, w.c., 2001)

NYE COUNTY As clear glassy to opalescent crystals that have weathered out of a dike in the Bellehelen district just west of the summit of the Kawich Range; the crystals are both single and twinned. (Melhase, 1935d; Anonymous, 1938)

Common in the rhyolitic rocks of the southwestern Nevada volcanic field; sanidine in the Ammonia Tanks tuff typically has faint blue adularescence. (Orkild et al., 1969; SBC)

WASHOE COUNTY As abundant phenocrysts that commonly show brilliant-blue adularescence in the widespread tuff of Chimney Spring. (Castor et al., 1999a)

WHITE PINE COUNTY As white, translucent, prismatic crystals to 1 cm with dark *garnet* crystals in lithophysal cavities in rhyolite at Garnet Hill 5 miles east of Ely and 2 miles north of Ruth in the Robinson district. (GCF)

SAPONITE

$(Ca/2,Na)_{0.3}(Mg,Fe^{2+})_3(Si,Al)_4O_{10}(OH)_2 \cdot 4H_2O$

A magnesium-rich member of the *smectite* clay group generally found as a hydrothermal mineral in mafic igneous and metamorphic rocks. The commercial Ash Meadows deposits in Nevada are unusual.

DOUGLAS COUNTY Reported as mineral soap or soapstone in a specimen from the Ward & McDonald property in the Gardnerville district. (WMK)

NYE COUNTY One of the clay minerals mined commercially from large deposits near the town of Amargosa Valley in the Ash Meadows district, NE1/4 Sec. 1, T18S, R49E, where it is associated with spring-related carbonates in Tertiary lake sediments. (Papke, 1970; Hay et al., 1980)

WHITE PINE COUNTY White iron-free saponite has been identified with diopside in altered rock in the Robinson district. (James, 1976)

SARCOLITE

$NaCa_6Al_4Si_6O_{24}F(?)$

A rare mineral that is found elsewhere in the world in contact-metamorphosed volcanic ejecta at two localities.

LANDER COUNTY Reported from the McCoy Mine in the McCoy district. (Excalibur Mineral Co., 1999)

SARKINITE

$Mn_2^{2+}(AsO_4)(OH)$

A rare mineral that is found in deposits in Sweden with other manganese species such as hausmannite and bementite; may be of primary or secondary origin.

WHITE PINE COUNTY Identified in a specimen from the Kern Mountains in the Eagle district. (CSM)

SARMIENTITE

$Fe_2^{3+}(AsO_4)(SO_4)(OH) \cdot 5H_2O$

A very rare secondary mineral found elsewhere only in an occurrence with other sulfate species in gossan in Argentina.

HUMBOLDT COUNTY As brown masses from a locality near Winnemucca. (Weissman and Nikischer, 1999)

NYE COUNTY Reported at the White Caps Mine in the Manhattan district. (Pullman and Thomssen, 1999)

SASSOLITE

H_3BO_3

Sassolite (boric acid) is an uncommon mineral that occurs in hot springs deposits.

WASHOE COUNTY Naturally occurring boric acid was reported in the sinter deposits at Steamboat Hot Springs in the Steamboat Springs district. (Smith, 1985)

SCAPOLITE

$3(Na,Ca)Al_{1-2}Si_{2-3}O_8 \cdot [NaCl,(CaCO_3)]$

A mineral group name (see marialite and meionite); *scapolite* minerals generally occur in regionally metamorphosed rocks and skarn. Listings below are occurrences for which species have not been identified.

CHURCHILL COUNTY Associated with iron ore at the Buena Vista Mine in the Mineral Basin district. (Reeves and Kral, 1955)

CLARK COUNTY As large grains and aggregates in amounts up to several percent in Precambrian amphibolite in the Bunkerville district. (Beal, 1965)

LANDER COUNTY Identified in skarn at the Pipeline Mine in the Bullion district. (Foo et al., 1996)

LYON COUNTY Reported as *mizzonite* in a specimen from an unknown locality in the Yerington district. (EMP)

MINERAL COUNTY Reported in a fault zone with *uranothorite* at the Elna prospect in the Fitting district. (Garside, 1973)

As milky-white columnar crystals and radiating groups in skarn at the Julie claim in the Pamlico district. (Smith and Benham, 1985)

PERSHING COUNTY At several localities associated with iron ore in the Mineral Basin district, including the Segerstrom-Heizer and Thomas Mines, as an alteration mineral in diorite and metavolcanic rocks; also noted as well-developed crystals in veins in the district. (Reeves and Kral, 1955)

WHITE PINE COUNTY Reported from the Robinson district. (Melhase, 1934d)

SCHEELITE

$CaWO_4$

Scheelite, the most important ore mineral of tungsten, is a relatively common mineral in Nevada. It generally occurs in skarn with molybdenum, copper, and bismuth minerals, but has also been found in veins. Fine scheelite specimens reportedly came from the Snake Range of eastern White Pine County.

CARSON CITY Reported north of Carson City in the Carson district on the southwestern edge of the Virginia Range. (Moore, 1969; Stager and Tingley, 1988)

Identified in a specimen from 9 miles east of Carson City in the Delaware district; also reported from the Alex Eske and other mines. (JHS; Stager and Tingley, 1988)

In a specimen from 1 mile south of Carson City and west of U.S. Highway 395; probably from the Kings Canyon Mine in the Voltaire district. (WMK; Stager and Tingley, 1988)

CHURCHILL COUNTY As doubly terminated crystals to 2.5 cm in calcite veins with fluorite, barite, and powellite in epidote-magnetite-pyrite skarn at the Skarn No. 1 claim in the Chalk Mountain district. (Schilling, 1979; R. Walstrom, w.c., 1999)

With powellite in crystals to 1 cm in skarn at the Midnight Mine in the Fairview district. (Stager and Tingley, 1988)

In small amounts in a vein at the Gold King (Valley King) Mine in the Jessup district. (Vanderburg, 1940)

In a specimen from 25 miles southeast of Fallon in the Sand Springs district; also reported from the Red Ant and other properties. (WMK; Stager and Tingley, 1988)

With quartz and pyrite in limestone along a contact with granite at the Quick-Tung Mine in the Shady Run district. (Lawrence, 1963)

In skarn at the Toy and Hardscrabble Mines in the Toy district. (Vanderburg, 1940)

As crystals to 1 cm at the Hilltop and other mines in the Tungsten Mountain district. (Stager and Tingley, 1988)

CLARK COUNTY In quartz lenses in a sheared hornblendite dike that cuts Precambrian schist at the Silver Leaf and Tri-State Mines in the Bunkerville district. (Beal, 1965; Longwell et al., 1965)

With powellite along joints and fractures in schist at the Marron prospects in the Gold Butte district. (Longwell et al., 1965)

As bluish-white-fluorescent grains in skarn near the Iron Gold Mine in the Goodsprings district. (J. Kepper, o.c., 2001)

DOUGLAS COUNTY In skarn at the Tungsten Hills Mine, Walker and Johnson property, and Alpine Mine in the Gardnerville district. (Moore, 1969; WMK; Stager and Tingley, 1988; Doebrich et al., 1996)

In skarn in the Mountain House district in the southwestern Pine Nut Mountains. (Moore, 1969)

ELKO COUNTY In skarn at several sites in the Alder district; at the Garnet Mine in skarn and quartz veins with molybdenite and other sulfides. (Bentz et al., 1983; Stager and Tingley, 1988; Bushnell, 1967; LaPointe et al., 1991)

In skarn adjacent to granodiorite at the Tunnel prospect in the Contact district. (Stager and Tingley, 1988)

In skarn at the S. & L. Mother No. 1 claim in the Corral Creek district. (Stager and Tingley, 1988)

At the Mitchell Mine and Indian Springs prospect in the Delano district with powellite and other minerals in quartz veins in quartz-monzonite and in lesser amounts in skarn; also reported in brecciated quartz north of Delano Peak. (Stager and Tingley, 1988; LaPointe et al., 1991)

As crystals to 5 mm in seams and in a quartz vein in limestone at the Burns Scheelite Mine in the Edgemont district. (Stager and Tingley, 1988)

Sparsely scattered with molybdenite in skarn at the Pyramid Mine in the Elk Mountain district; also with molybdenite in a quartz vein at the nearby Valdez prospect. (Stager and Tingley, 1988; Lawrence, 1963)

In skarn at the Campbell, Climax, and Star Tungsten Mines in the Harrison Pass district. (Stager and Tingley, 1988; Hess and Larsen, 1920)

With stibnite at the Gribble Antimony Mine in the Island Mountain district; also in skarn with sulfides and in quartz veins at the Little Joe Mine. (Lawrence, 1963; Stager and Tingley, 1988; LaPointe et al., 1991; R. Walstrom, w.c., 1999)

With sulfides in skarn at the Coon Creek Mine in the Jarbidge district; also in skarn along granodiorite dikes at the Copper Queen prospect. (Stager and Tingley, 1988)

In skarn and alaskite with powellite and sulfides at the Phalen Mine in the Kinsley district. (Stager and Tingley, 1988)

Reported in a specimen from Tennessee Mountain in the Mountain City district. (NMBM)

With sulfides in a quartz vein at the Great Western Mine in the Proctor district; at Silver Zone Pass as crystals to 0.6 cm in a quartz vein. (Smith, 1976; Stager and Tingley, 1988)

In skarn with copper sulfides at the Aladdin and Grey Eagle Mines in the Railroad district. (Stager and Tingley, 1988; LaPointe et al., 1991)

With quartz, pyrite, and other sulfides in lenses of *chlorite* schist in granite and pegmatite at the Battle Creek Mine in the Ruby Valley district. (Stager and Tingley, 1988; Berger and Oscarson, 1998)

In skarn at the Atlantic prospect in the Spruce Mountain district. (Stager and Tingley, 1988)

As crystals to 0.5 inch in skarn at the Valley View Mine in the Valley View district. (Stager and Tingley, 1988)

Along fractures and in quartz veinlets at the Wells Tungsten Mine in the Wells district. (Stager and Tingley, 1988)

In skarn at the Kathleen claim in the White Horse district. (Stager and Tingley, 1988)

ESMERALDA COUNTY With powellite in skarn at the Black Horse Mine in the Black Horse district. (Albers and Stewart, 1972)

As small, euhedral, doubly terminated crystals with quartz, *adularia,* goethite after pyrite, and tremolite in thin veins at the Broken Toe Mine in the Rock Hill district. (R. Walstrom, w.c., 1999)

Reported at the Carrie Mine in the Gilbert district. (Ferguson, 1928)

In quartz veins at the Rock Hill Mine in the Rock Hill district; also noted in nearby placer deposits. (Schilling, 1964b, 1979)

In quartz veins and skarn at the Sylvania and other mines in the Sylvania district. (Schilling, 1979)

EUREKA COUNTY As crystals to 5 cm in a quartz-calcite vein at the Black Rock Mine in the Fish Creek district. (JHS; Stager and Tingley, 1988)

As pale-green anhedral crystals to 5 mm in calc-silicate rock at the Goldstrike Mine in the Lynn district. (GCF)

HUMBOLDT COUNTY Identified in placer gravels

at the Dutch Flat Placer Mine in the Dutch Flat district. (JHS)

Reported in specimens from the L. A. Navarine property in the Golconda district; also in skarn from the Hood Tungsten Mine. (WMK)

With molybdenite, chalcopyrite, and pyrite in quartz veinlets at the Molly prospect in the Gold Run district. (Schilling, 1979)

In the Potosi district at several localities: in skarn at the Getchell, Granite Creek, Valley View, and Riley Mines; as translucent subhedral crystals to 2 cm from the Mountain King Mine; and as crystals to 2 cm with smoky quartz and *sericite* in skarn on the west side of the Osgood Range. (Hobbs and Clabaugh, 1946; Schilling, 1962a, 1962b; Hotz and Willden, 1964; N. Prenn, w.c., 1999; S. Sears, w.c., 1999)

With molybdenite in quartz veins at the Bloody Run Mine in the Sherman district. (Vanderburg, 1938a; Stager and Tingley, 1988)

In veinlets at the Charleston Hill National Mine and elsewhere in the Shon district. (Stager and Tingley, 1988)

Fluorescent canary-yellow in skarn at the Golden Scheelite and other prospects in the Varyville district. (Stager and Tingley, 1988)

In quartz veins at the Ashdown, Defense, and Vicksburg Mines in the Vicksburg district; also in skarn at the Defense and Last Chance Mines. (Schilling, 1962a, 1962b; Willden, 1964; Stager and Tingley, 1988)

LANDER COUNTY With magnetite and powellite in epidote-quartz-*garnet* skarn at the Gold Acres Tungsten prospect in the Bullion district. (Stager and Tingley, 1988)

In skarn at the Copper King, Copper Canyon, and Carissa Mines in the Battle Mountain district. (Roberts and Arnold, 1965)

Reported in a specimen from the Yankee Blade Mine in the Reese River district. (CSM)

With bismuth minerals and molybdenite at the Linka pit and other mines in the Spencer Hot Springs district. (McKee, 1976; Williams, 1988)

LINCOLN COUNTY Disseminated in Precambrian amphibolite at the Macbruson Mine in the Gourd Springs district. (Tschanz and Pampeyan, 1970)

In a specimen from the Snowflake claims in the Highland district. (WMK)

Reported at the Pip and Ad claims in the Patterson district. (Tschanz and Pampeyan, 1970)

Mined in large quantities from skarn deposits at the Emerson (Lincoln, Tem Piute) and Schofield Mines in the Tem Piute district. (Buseck, 1967; WMK; Stager and Tingley, 1988)

LYON COUNTY In silicated limestone at the Ruth Mine in the Churchill district. (Moore, 1969)

Reported at the Pearl Harbor Mine in the Red Mountain district. (Moore, 1969)

Reported at the Cowboy Tungsten Mine in the Wilson district. (S. Pullman, w.c., 1999)

MINERAL COUNTY In skarn at the Blue Bird and Cedar Chest Mines in the Bell district. (Ross, 1961)

Reported at the Lucky Four Mine in the Buckley district. (Ross, 1961)

In skarn at the Moonlight, Queen, and Wild Rose Mines in the Buena Vista district. (Papke, 1979; Ross, 1961; Stager and Tingley, 1988)

In skarn with wollastonite at the Dry Gulch claims in the Fitting district. (Ross, 1961)

With stibnite in quartz veins at the Hartwick prospect in the Garfield district; also in skarn at the Bataan, Mabel, and Mollie Mines. (Lawrence, 1963; Ross, 1961)

In skarn at several mines in the Leonard district near Rawhide, the Nevada Scheelite Mine having been the largest producer; also noted as nearly colorless crystals and reported as *cuproscheelite* from the district. (Warner et al., 1959; Ross, 1961; Stager and Tingley, 1988; WMK; CSM; NMBM)

With sphalerite and galena in skarn at the Crystal and Lemr prospects in the Lucky Boy district; also in skarn at the Lucky Boy Mine. (Ross, 1961; Lawrence, 1963)

Reported at the Hefler Tungsten Mine in the Masonic district. (Ross, 1961)

Reported in a specimen from the George W. Hill claim in the Mountain View district near Schurz. (CSM)

In skarn with copper minerals at the Gypsy claim in the Pamlico district. (Ross, 1961)

In skarn at the Desert Scheelite, Gunmetal, and Garnet Mines in the Pilot Mountains district. (Warner et al., 1959; Ross, 1961; M. Jensen, w.c., 1999)

Identified in a specimen from the Papoose Mine in the Rand district. (WMK)

At several mines in the Santa Fe district in skarn, with copper minerals at some; specifically reported at the Kenyon claims as corroded colorless to gray crystal fragments to 2 cm in vugs with colorless quartz crystals. (Clark, 1922; Ross, 1961; M. Jensen, w.c., 1999)

With *wolframite*, ferberite, and gold in massive to banded fine-grained quartz veins at the Silver Dyke Mine in the Silver Star district, the most important tungsten mine in Nevada during World War I; minor production came from other properties in the district. (Kerr, 1936; Vanderburg, 1937a; Ross, 1961; R. Walstrom, w.c., 1999; WMK; Stager and Tingley, 1988)

With sulfides at the Qualey Mine in the Whisky Flat district. (Ross, 1961)

NYE COUNTY In skarn at the Van Ness prospect in the Barcelona district. (Kleinhampl and Ziony, 1984)

Reported with ferberite and *wolframite* in a specimen from the War Admiral claim 9 miles south of Beatty in the Bare Mountain district. (WMK)

Reported in a specimen with hübnerite in the Belmont district. (WMK)

With fluorite, vesuvianite, calcite, and *potassium feldspar* at the Roark prospect in the Currant district; also in ore from the Golden Rod Mine. (Kleinhampl and Ziony, 1984; Stager and Tingley, 1988)

In quartz veins and stringers at the Baxter-Hancock Mine in the Fairplay district; also with cinnabar and fluorite from the Scheebar Mine. (Stager and Tingley, 1988; Bailey and Phoenix, 1944)

In veins at several properties in the Gabbs district, notably in a vein in dolomite and magnesite at the Betty O'Neal Mine with *leuchtenbergite* (clinochlore) and other minerals. (Callaghan, 1933; Kerr and Callaghan, 1935; Kral, 1951; Schilling, 1968a; Stager and Tingley, 1988)

In a quartz vein at the Jett Canyon prospect in the Jett district. (Stager and Tingley, 1988)

In the Lodi district at the Kay Cooper and Victory Mines along a shear zone in granodiorite with diopside, hastingsite, fluorite, powellite, and other minerals; also in skarn and granodiorite at other properties. (Kral, 1951; Humphrey and Wyatt, 1958; Kleinhampl and Ziony, 1984; Stager and Tingley, 1988)

With beryl and fluorite in greisen in metasedimentary rocks at the Hiatt prospect in the Manhattan district. (Papke, 1979)

In the Oak Spring district on the Nevada Test Site in skarn at the Climax Mine and other properties; also reported at the Tamney Mine and the Crystal prospect as euhedral crystals in gouge with powellite and molybdenite. (Melhase, 1935b; Kral, 1951; Stager and Tingley, 1988)

Rare in placers southwest of Round Mountain in the Round Mountain district. (Vanderburg, 1936a)

Rare in skarn at the Blue Gem Mine area in the Royston district. (Stager and Tingley, 1988)

Molybdenum-bearing with molybdenite in *garnet*-epidote skarn at the Peg Leg (Greenstock) Mine in the Tonopah district. (Bonham and Garside, 1979)

In skarn at various localities in the Twin River district; specifically at the Ophir Tungsten Mine as crystals to 1/2 inch in an ore shoot in calcareous schist. (Stager and Tingley, 1988)

In replacement bodies in limestone with galena, sphalerite, and chalcopyrite at the Grantsville Mine in the Union district; also reported in veins with the same sulfide assemblage at an unknown locality in the district. (Kral, 1951; NBMG mining district file; Hamilton, 1993)

Reported in a specimen from the Wagner property in the Wagner district. (LANH)

In skarn at the Haywire prospect in the Washington district. (Stager and Tingley, 1988)

With fluorite in limestone at the Hi Grade prospect in the Willow Creek district. (Kleinhampl and Ziony, 1984; Stager and Tingley, 1988)

PERSHING COUNTY With calcite and quartz in some silver mines at Unionville in the Buena Vista district. (Johnson, 1977)

In *garnet*-epidote-magnetite skarn at the Basalt prospect in the Copper Valley district. (Shawe et al., 1962)

In skarn with powellite, molybdenite, and other sulfide minerals at the Stormy Day and Thrasher Mines and Thrabert claim in the Hooker district. (Johnson, 1977; JHS)

In quartz veins with other minerals at the Lakeview and Star Mines in the Imlay district. (Johnson, 1977; Lawrence, 1963)

As coarse crystals with stibnite in quartz veins at the Ore Drag Mine in the Iron Hat district. (Lawrence, 1963; Johnson, 1977)

Ore mineral in the Mill City district, the largest producer of tungsten in Nevada. Major mines included the Stank, Sutton, Humboldt, and Springer; by 1928 all were owned by the Nevada-Massachusetts Company. Scheelite was found as pure white crystals from a fraction of a millimeter to a few centimeters across in skarn orebodies around granodioritic intrusions with quartz, calcite, molybdenite, pyrrhotite, pyrite, *garnet,* epidote, and other minerals. (Kerr, 1934; Johnson, 1977; Stager and Tingley, 1988; WMK)

As disseminated grains, pods, and crude pseudo-octahedral crystals to 2 cm with quartz, epidote, other skarn minerals, and with pyrrhotite, chalcopyrite, and other sulfides at the Alpine, Nightingale, and other mines in the Nightingale district. (Smith and Guild, 1944; Johnson, 1977; M. Jensen, w.c., 1999)

In gold placers with cassiterite, cinnabar, and magnetite in the Rabbit Hole district. (Johnson, 1977)

Molybdenum-bearing in skarn at the Rose Creek Mine in the Rose Creek district. (Johnson, 1977)

At the Oreana Mine in the Rye Patch district, Sec. 3, T29N, R33E, in massive and crystalline form in pegmatite dikes that cut metadiorite and in lenslike masses along a contact between limestone and metadiorite; associated minerals include beryl, *oligoclase,* albite, phlogopite, fluorite, and native antimony. (Ransome, 1909b; Kerr, 1938; Olson and Hinrichs, 1960)

In quartz-calcite veins with pyrite and other sulfide minerals at the Humboldt Queen Mine in the Sacramento district. (Ransome, 1909b)

In quartz veins at the True American Mine in the Tobin and Sonoma Range district. (Johnson, 1977)

With powellite and copper minerals in iron-rich skarn at the Esther Mine in the Trinity district. (Johnson, 1977)

In skarn at the Long Tungsten Mine in the Wild Horse district. (Johnson, 1977)

In skarn in the Gold Butte, Haystack, Juniper Ridge, Ragged Top, and Seven Troughs districts. (Johnson, 1977)

WASHOE COUNTY With pyrite in skarn at the Garnet Tungsten prospect in the Cottonwood district. (Bonham and Papke, 1969)

With chalcopyrite and pyrite in skarn and crosscutting quartz veins at the Mountain View Tungsten prospect in the Deephole district. (Bonham and Papke, 1969)

Sparse in skarn near the Denver Mine in the Galena district. (Thompson, 1956; Bonham and Papke, 1969)

In epidote-rich skarn with chalcopyrite at an unnamed prospect in the Sugarloaf Peak area in the McClellan district. (Bonham and Papke, 1969)

In skarn with sulfide minerals at the Crosby and Jay Bird Mines in the Nightingale district. (Bonham and Papke, 1969)

In calc-silicate rock at the Derby Tungsten Mine in the Olinghouse district. (Rose, 1969; Overton, 1947)

Reported in a specimen of altered pegmatite with copper minerals in the Peavine district. (WMK)

In a specimen of drill core at the MGL Mine in the Pyramid district. (WMK)

WHITE PINE COUNTY In epidote-rich skarn with chalcopyrite in Water Canyon in the Bald Mountain district. (Vanderburg, 1936a; Hose et al., 1976)

In a vein at the Black Horse Mine in the Black Horse district. (Hose et al., 1976)

Reported at the Tincup Mine in the Cherry Creek district. (Hose et al., 1976; CSM; JHS)

In quartz veinlets and replacements in limestone at the Kolchek Mine in the Cleve Creek district. (Hose et al., 1976)

In veins and skarn at the Antelope and Yellow Jacket Mines in the Eagle district. (Hose et al., 1976)

With hübnerite in a quartz vein at the Deer Trail Mine in the Geyser district. (Hose et al., 1976)

With galena and calcite in breccia at the Valley View Mine in the Granite district. (Hose et al., 1976)

As crystals in calcite veins at the Bonanzy (Lexington) Mine in the Lexington district. (Hose et al., 1976)

As greasy white to light-yellow grains up to 1 cm in diameter in quartz veins with phenakite and bertrandite at the Mount Wheeler and other mines in the Lincoln district; also in greisen with fluorite, *monazite,* phenakite, and beryl. (Griffiths, 1964; Hose et al., 1976)

As coarse grains in silicified limestone with galena at the Grand View Mine in the Mount Moriah district. (Hose et al., 1976)

In stockwork veins with lead-silver ore at the Bay State Mine in the Newark district. (Hose et al., 1976)

As crystals in a quartz vein at the Dirty Shirt Mine in the Osceola district, from which a transparent crystal weighing about 2 pounds was reportedly taken; also in a breccia pipe in limestone at the Black Mule Mine. (Hess, 1917; Hose et al., 1976; WMK)

In quartz veins with pyrite and chalcopyrite at the Taylor Mine in the Robinson district. (Hose et al., 1976)

In a calcite vein at the White Horse claim in the Schellbourne district. (Hose et al., 1976)

In quartz-calcite veins at the Chief, Silver Bell, and other mines in the Shoshone district; locally with tetrahedrite, galena, powellite, *cuprodescloizite,* and silver halide minerals. (Hose et al., 1976)

Mined from the Tilford, Johnson, and Pilot Knob Mines in the Snake district; also found in a placer deposit. (Hose et al., 1976; WMK)

With more abundant hübnerite in quartz-fluorite-pyrite veins at the Hub Mine in the Tungsten district. (Hose et al., 1976; WMK)

Reported from the Sacramento and White Pine districts. (Hess, 1917)

SCHODERITE

$Al_2(PO_4)(VO_4) \cdot 8H_2O$

Schoderite, a Nevada type mineral, occurs as a secondary phase with the lower hydrate metaschoderite coating fracture surfaces in vanadium-rich oil shale.

EUREKA COUNTY The type locality is in the Gibellini district at the Van-Nav-San claim, Sec. 34, T16N, R52E, along fractures in lower Paleozoic phosphatic chert as yellowish-orange microcrystalline crusts with metaschoderite, wavellite, vashegyite, diadochite, and phosphatic materials of varying composition; also reported at the Gibellini vanadium prospect in Sec. 35. (Hausen, 1960, 1962; Zientek et al., 1979)

SCHOLZITE

$CaZn_2(PO_4)_2 \cdot 2H_2O$

A relatively uncommon secondary mineral that is found in low-grade, sedimentary rock–hosted zinc and phosphate deposits and in zinc- and phosphate-bearing pegmatite; in some deposits associated with sphalerite.

LANDER COUNTY As white microcrystalline sprays to 1 cm with rosasite at the Snowstorm Mine in the North Battle Mountain district. (Ream, 1992; Thomssen, o.c., 2000; J. McGlasson, w.c., 2001)

SCHORL

$NaFe_3^{2+}Al_6(BO_3)_3Si_6O_{18}(OH)_4$

A *tourmaline* group mineral that forms a series with dravite. Schorl occurs in granite and granite pegmatite, metamorphic rocks, and high-temperature veins.

CHURCHILL COUNTY In a specimen from the southern Trinity Range, Sec. 17, T24N, R27E. (R. Thomssen, w.c., 1999)

CLARK COUNTY As microcrystals in miarolitic cavities and pegmatites in the Eldorado and Newberry Mountains. (S. Scott, w.c., 1999)

DOUGLAS COUNTY As crystals in pegmatite dikes along Kingsbury Grade. (N. Prenn, w.c., 1999)

LANDER COUNTY In pre-Tertiary sedimentary rocks as a detrital mineral in the West orebody at Copper Canyon in the Battle Mountain district. (Theodore and Blake, 1978)

LYON COUNTY As small needlelike crystals in veins and patches in metamorphosed volcanic rock at the Adams Gypsum Mine area in the Mound House district. (SBC; M. Rogers, w.c., 1999)

MINERAL COUNTY As lustrous, elongated crystals to 4 cm in pegmatite, and as acicular inclusions in quartz crystals at the Zapot claims in the Fitting district; reportedly grades toward elbaite. (Jensen, w.c., 1999; Foord et al., 1999b)

Identified as black crystals to 1 cm in specimens from Luning. (LANH; S. Pullman, o.c., 2000)

NYE COUNTY As black crystals to 1 mm at the San Rafael Mine in the Lodi district. (B. Runner, o.c., 2000)

PERSHING COUNTY Widespread and locally abundant as black needlelike crystals to 5 mm replacing *feldspar* phenocrysts and in fist-size masses with quartz at the Majuba Hill Mine in the Antelope district; may also occur in some quartz crystals. (Jensen, 1985)

WASHOE COUNTY With epidote and chryso-colla in a fault zone cutting granodiorite in a copper prospect on a narrow ridge north of Sun Valley in the McClellan district, Sec. 5, T20N, R20E. (Bonham and Papke, 1969)

As black crystals in quartz veins with secondary copper minerals and local enargite at the Redelius claims and other properties in the Peavine district; also as matrix and euhedral needles with quartz in breccia with compositions of *ferroan-dravite* to schorl. (Overton, 1947; Bonham and Papke, 1969; Hudson, 1977)

As black needlelike crystals and stout single euhedral black crystals to 9 cm with smoky quartz, microcline, and other minerals in pegmatites at Incline Village. (Jensen, 1993a)

SCHREIBERSITE

$(Fe,Ni)_3P$

A mineral of iron meteorites.

CLARK COUNTY With kamacite, taenite, troilite, graphite, and cohenite in the Las Vegas meteorite. (Buchwald, 1975)

NYE COUNTY With kamacite, taenite, troilite, and daubréelite in the Quartz Mountain meteorite. (Buchwald, 1975)

In minor amounts as skeletal crystals intergrown with kamacite and taenite in the Quinn Canyon meteorite. (Buchwald, 1975)

SCHRÖCKINGERITE

$NaCa_3(UO_2)(CO_3)_3(SO_4)F \cdot 10H_2O$

A secondary uranium mineral found in near-surface deposits and as an efflorescence on mine walls.

HUMBOLDT COUNTY Reported with carnotite in opal of the Virgin Valley lakebeds and in fossil logs in the Virgin Valley district. (Garside, 1973)

PERSHING COUNTY As small rosettes in clay at the Four Jacks claims in the Nightingale district. (J. McGlasson, w.c., 2001)

SCHUBNELITE

$Fe^{3+}_{2-x}(V^{5+},V^{4+})_2O_4(OH)_4$

A rare mineral that is found outside Nevada in oxidized vanadium-bearing uranium deposits.

EUREKA COUNTY As brilliant, orange-brown

crystals to 0.1 mm on kazakhstanite and fervanite in strongly brecciated and fractured jasperoid at the Gold Quarry Mine in the Maggie Creek district. (Jensen et al., 1995)

SCHUETTEITE

$Hg_3(SO_4)O_2$

Schuetteite, a Nevada type mineral, was originally described from several occurrences in Nevada, California, and Texas as a post-mining oxidation product of primary mercury sulfides.

ELKO COUNTY At a co-type locality at the Silver Cloud Mine in the Ivanhoe district, Sec. 25, T37N, R47E, with dark concentrations of cinnabar on mined surfaces of opalite in bright-yellow blooms almost a foot in diameter. (Bailey et al., 1959)

ESMERALDA COUNTY At a co-type locality at the B. and B. Mine in the Fish Lake Valley district, Sec. 1, T1S, R33E, on sintery opalite mercury ore in a pit wall and on loose opalite blocks with local alunite; also noted at the Wild Rose Mine, Secs. 16 and 21, T1N, R33E. (Bailey et al., 1959)

EUREKA COUNTY As an oxidation product of cinnabar and mercury-thallium minerals at the Carlin gold deposit in the Lynn district. (Hausen and Kerr, 1968)

PERSHING COUNTY At a co-type locality at the Red Bird Mine in the Antelope Springs district, Sec. 33, T27N, R33E, coating crystals of cinnabar in a specimen of rich ore from the waste dump. (Bailey et al., 1959)

At a co-type locality at the Goldbanks Quicksilver Mine in the Goldbanks district, Sec. 14, T30N, R38E. (Bailey et al., 1959)

SCOLECITE

$Ca[Al_2Si_3O_{10}] \cdot 3H_2O$

A mineral of the zeolite group that is generally found in cavities in basalt, in alkali igneous rocks, and in metamorphic rocks.

CLARK COUNTY As white radiating crystals in volcanic rocks in the northeastern McCullough Range about 2 miles south of Railroad Pass. (Gianella and Hedquist, 1942)

STOREY COUNTY Found with calcite in the Sutro Tunnel 1 mile from the entrance and in an outcrop as fracture coatings of 1 mm acicular crystals on altered andesite in the Comstock district. (UCB; M. Jensen, w.c., 1999)

WASHOE COUNTY As white masses of radiating acicular crystals to 2 cm with calcite, *heulandite,* and quartz in gold-bearing veins at the Olinghouse Mine in the Olinghouse district; identified by XRD. (SBC; WMK)

SCORODITE
$Fe^{3+}AsO_4 \cdot 2H_2O$

Scorodite is a common oxidation product of arsenopyrite or other primary arsenic-bearing minerals. Excellent specimen material has come from the Majuba Hill and Yerington localities in Nevada.

CHURCHILL COUNTY In a specimen from the Fireball Mine in the Truckee district. (R. Thomssen, w.c., 1999)

CLARK COUNTY Reported from the Yellow Pine Mine in the Goodsprings district. (UCB)

DOUGLAS COUNTY As pale-green crusts with arsenopyrite in quartz veins in bulldozer cuts 1 mile east of Holbrook Junction in the Mountain House district. (M. Jensen, w.c., 1999)

ELKO COUNTY On fractures and in voids in breccia at the Rain Mine in the Carlin district. (D. Heitt, o.c., 1991)

ESMERALDA COUNTY Tentatively identified after arsenopyrite at the Mickspot Mine in the Gilbert district. (Lawrence, 1963)

Reported from the Weepah Mine in the Weepah district. (S. Pullman, w.c., 1999)

EUREKA COUNTY Reported on the walls of a glory hole at the Windfall Mine in the Eureka district; also reported from the Burning Moscow Mine. (Nolan, 1962; R. Thomssen, w.c., 1999)

Reported from the Carlin Mine in the Lynn district; also widespread as light-green veins to 5 cm and as fracture coatings at the Goldstrike Mine. (Hausen and Kerr, 1968; GCF)

As yellow-green, equant, platy crystals to 60 microns in siliceous siltstone with barite, kaolinite, kazakhstanite, alunite, gypsum, and quartz at the Gold Quarry Mine in the Maggie Creek district; contains significant amounts of phosphorus and vanadium. (Jensen et al., 1995)

HUMBOLDT COUNTY As gray-green microcrystals with pharmacosiderite and dufrénite at the Lone Tree Mine in the Buffalo Mountain district. (M. Jensen, w.c., 1999)

Identified in a specimen from the Getchell Mine in the Potosi district. (WMK)

LANDER COUNTY As dark-greenish-brown pods, pale-green porous aggregates, and clear to green crystals to 2 mm with arseniosiderite at the Wilson-Independence Mine in the Battle Mountain district; also noted at other properties. (Roberts and Arnold, 1965; J. McGlasson, w.c., 2001)

LINCOLN COUNTY Common as massive, earthy, greenish to brownish, radiating and granular aggregates resulting from oxidation and replacement of arsenopyrite in some veins with arseniosiderite, beudantite, jarosite, mimetite, and iron oxides at the Advance, Old Democrat, Republic, and Lucky Hobo Mines in the Chief district. (Callaghan, 1936)

Tentatively identified in gossan in the Freiburg district. (Tschanz and Pampeyan, 1970)

Reported from the Silver King Mine in the Silver King district. (R. Thomssen, w.c., 1999)

LYON COUNTY As white or pale-blue crystals to 2 mm at the Empire-Nevada Mine in the Yerington district. (Jenkins, 1981)

MINERAL COUNTY As reddish-brown microcrystals at the Simon Mine in the Bell district. (S. White, w.c., 2000)

Common as an oxidation product of arsenopyrite at the Potosi Mine 1 mile west of the Candelaria Mine in the Candelaria district. (Jenkins, 1981)

NYE COUNTY With barite, iodargyrite, iron oxide, and gold in a sample from the Gold Crater district on the Nellis Air Force Range. (Tingley et al., 1998a)

In yellow-green masses as a common oxidation product of arsenopyrite at the San Rafael Mine in the Lodi district. (Jenkins, 1981)

As greenish crusts on quartz at the Amalgamated Mine in the Manhattan district; also identified in the April Fool vein. (Jenkins, 1981; Ferguson, 1924)

As sugary crusts with pharmacosiderite on jasperoid at the Northumberland Mine in the Northumberland district. (Kokinos and Prenn, 1985)

With *limonite* on realgar at the Round Mountain Mine in the Round Mountain district. (Tingley and Berger, 1985)

PERSHING COUNTY As sky-blue crystals to 2 mm with pharmacosiderite and arthurite in crystal-lined pockets in chalcocite, and as thin green to brown coatings and tiny crystals on altered rhyolite and tin ore at the Majuba Hill Mine in the Antelope district; also noted at an unnamed arsenic prospect as massive blue-green crusts on arsenopyrite-rich vein material. (Smith and Gianella, 1942; Trites and Thurston, 1958; Jensen, 1985)

Sparse in oxidized ore in the Arabia district in the Trinity Range. (Knopf, 1918a)

Green-gray to golden microballs on quartz crystals from Wild Horse Canyon. (Excalibur Mineral Co., 2001a)

WASHOE COUNTY In brilliant clusters of transparent, colorless to very pale-blue, euhedral crystals with olivenite and barium pharmacosiderite in enargite-bearing ore at the Burrus Mine in the Pyramid district. (Jensen, 1994)

As thin coatings and stains on hot springs sinter in the Steamboat Springs district. (Palache et al., 1951)

SCRUTINYITE

α-PbO$_2$

A lead oxide mineral that is dimorphous with plattnerite and occurs in oxidized hydrothermal lead ore.

WHITE PINE COUNTY As dark-maroon-red, bladed crystal clusters to 2 mm on chrysocolla at the Grand Deposit Mine in the Muncy Creek district. (Pullman and Thomssen, 1999; M. Jensen, o.c., 2000)

SEARLESITE

NaBSi$_2$O$_5$(OH)$_2$

Searlesite is a relatively rare mineral that is found in some borate deposits, in oil shales and marls, and in vugs in volcanic rock.

ESMERALDA COUNTY As minute crystals in seams in bedded clay deposits in the Columbus Marsh district. (Rogers, 1924; LANH; UAMM)

As thin veinlets in Tertiary rocks at Cave Springs in the Fish Lake Marsh district, 2 miles south of the bedded ulexite deposits in the Fish Lake Valley. (Foshag, 1934; Melhase, 1935d)

As crystals, seldom more than 2 mm in length, and as granular masses, in banded marl, probably part of the Esmeralda Formation, in the Silver Peak Range; associated with beads of opal, small crystals of barite, yellow calcite, and fernlike splotches of manganese oxide. (Foshag, 1934)

MINERAL COUNTY Rare, replacing glass shards, in thin tuff beds in playa lake sediments in the Teels Marsh district. (JHS)

SELENIUM

Se

Native selenium is reported to occur in fumaroles, sandstone-type uranium-vanadium deposits, and in burned coal; however, in Nevada it occurs in tiny amounts in hydrothermal precious-metal deposits.

CHURCHILL COUNTY As microcrystals with mopungite at the Green prospect in the Lake district. (Williams, 1985)

ELKO COUNTY In a specimen from the Dee Mine in the Bootstrap district. (R. Thomssen, w.c., 1999)

Identified by SEM/EDX as tiny grains with sphalerite, pyrite, galena, and pyrargyrite in rich silver-gold ore at the Navajo Mine in the Tuscarora district. (Henry et al., 1999; SBC)

EUREKA COUNTY With vanadinite and descloizite as an encrustation on marcasite at the Gold Quarry Mine in the Maggie Creek district; also as dark crystalline coatings on bedding planes and joints in highly argillized sedimentary rocks at the top of the unoxidized zone. (Jensen et al., 1995)

HUMBOLDT COUNTY Tentatively reported from prospects near the summit of Buckskin Peak in the National district. (Lindgren, 1915)

NYE COUNTY Recovered in minor amounts with *electrum* by panning at the West End Mine in the Tonopah district. (Bastin and Laney, 1918)

With stibnite and barite at an unnamed prospect east of Barley Creek Summit in the Danville district. (M. Jensen, w.c., 1999)

PERSHING COUNTY In radiating clusters, to 0.1 mm, of black acicular crystals at the Willard Mine in the Willard district. (Jensen and Leising, 2001)

SELIGMANNITE

$PbCuAsS_3$

A sulfosalt mineral in series with bournonite; found in porphyry-copper and other hydrothermal deposits.

EUREKA COUNTY As rare inclusions in galena north of Ruby Hill in the Eureka district. (Vikre, 1998)

WHITE PINE COUNTY Reported with bournonite in a specimen from the Mount Hamilton area in the White Pine district. (WMK)

SENARMONTITE

Sb_2O_3

Dimorphous with valentinite; occurs in oxidized antimony-bearing hydrothermal deposits.

CHURCHILL COUNTY With stibiconite, roméite, tripuhyite, and mopungite at the Green prospect in the Lake district. (Williams, 1985)

ELKO COUNTY As crystals with stibnite in silicified breccia from the Dee Mine in the Bootstrap district. (R. Thomssen, w.c., 1999; J. McGlasson, w.c., 2001)

NYE COUNTY Reported from the White Caps Mine in the Manhattan district. (R. Thomssen, w.c., 1999)

PERSHING COUNTY With other antimony oxide minerals and stibnite in yellow-brown coatings in a specimen from Oreana. (GCF)

SEPIOLITE

$Mg_4Si_6O_{15}(OH)_2 \cdot 6H_2O$

A clay mineral that generally occurs in sedimentary rocks and *serpentine*. The Ash Meadows deposits in Nevada are the largest in the United States.

CLARK COUNTY Reported in a specimen from the Goodsprings district. (WMK)

Reported in a specimen from Boulder Canyon in the McClanahan district. (WMK)

ELKO COUNTY As tough, fibrous, cream-colored clay with silky luster, in a discordant vertical vein that averages 6 inches thick in dolomite in the Ferber district, Sec. 10, T27N, R69E; associated with calcite, opal, mica, talc, and expandable clay. (Ehlmann, 1962)

LINCOLN COUNTY Reported in a specimen from 6 miles east of Pioche. (WMK)

NYE COUNTY In extensive commercial deposits in Pliocene lake sediments in the Ash Meadows district, Sec. 21, T17S, R51E; the sepiolite occurs in a 4-foot-thick bed with dolomite and trace amounts of other minerals and is considered to have formed in a playa environment. (Papke, 1972b; SBC)

Reported in pedogenic calcite-opal veins along a fault in a trench on Yucca Mountain. (Vaniman, 1993)

SERPENTINE

$(Mg,Al,Fe,Mn,Ni,Zn)_{2-3}[(Si,Al,Fe)_2O_5](OH)_4$ (general)

A group of minerals, including chrysotile, lizardite, and antigorite, that are generally found in altered ultramafic rocks. Mineral species have not been identified for the occurrences listed below.

MINERAL COUNTY In small dikelike masses just south of Candelaria in the Candelaria district. (Ross, 1961)

NYE COUNTY In a specimen from the Cloverdale district in the northwestern part of the Royston Hills. (WMK)

SERPIERITE

$Ca(Cu^{2+},Zn)_4(SO_4)_2(OH)_6 \cdot 3H_2O$

Serpierite is a rare mineral found in the oxidation zones of base-metal deposits. It is usually associated with brochantite, langite, devilline, and other sulfate minerals.

LANDER COUNTY As blue crystals to 0.6 mm in quartz veins in shale at the Betty O'Neal Mine in the Lewis district; also noted on the Eagle Mine dump and at the Whisky Canyon adit. (S. White, w.c., 1999; R. Thomssen, w.c., 1999; J. McGlasson, w.c., 2001)

LYON COUNTY Bright turquoise-blue sprays to 1.5 mm with hemimorphite and pseudomalachite in altered limestone at the Jackpot Mine in the Wilson district. (S. White, w.c., 2000)

PERSHING COUNTY As sky-blue microcrystalline crusts in quartz vugs at an unnamed prospect near Imlay Canyon. (M. Jensen, w.c., 1999)

WHITE PINE COUNTY As pale-green crusts of

minute crystals with langite, elyite, and other minerals at the Caroline tunnel in the Ward district. (Williams, 1964, 1972)

SHADLUNITE

$(Pb,Cd)(Fe,Cu)_8S_8$

A rare sulfide found as tiny grains and veinlets in Cu-Ni ore from two mines in Siberia, Russia.

LANDER COUNTY Identified by EMP at the Cove Mine in the McCoy district. (W. Fusch, o.c., 2000)

SHATTUCKITE

$Cu_5^{2+}(SiO_3)_4(OH)_2$

Shattuckite is an uncommon mineral that occurs in the oxidation zones of copper deposits.

WHITE PINE COUNTY With mimetite, hcyite, pyromorphite, and other minerals at the Betty Jo claim 8 miles southeast of Ely in the Nevada district. (Williams, 1973)

SIDERITE

$Fe^{2+}CO_3$

A carbonate mineral found in sedimentary ironstone deposits and in hydrothermal veins. Referred to as "spathic iron" in older literature.

CHURCHILL COUNTY As a primary mineral in banded veins with quartz, calcite, and other minerals at the Fairview Mine in the Fairview district. (Schrader, 1947)

CLARK COUNTY With hematite, magnetite, and quartz replacing dolomite at an unnamed prospect in the Goodsprings district, NE1/4 SW1/4 Sec. 1, T25S, R58E. (Hewett, 1931)

ESMERALDA COUNTY The variety *manganosiderite* was identified in a specimen from the Silver Peak district. (JHS)

EUREKA COUNTY Widespread as a yellowish-brown to gray, earthy to crystalline mineral, generally with other carbonates, at the Goldstrike Mine in the Lynn district; saddle-shaped, rhombic crystals to 6 mm are locally common, and botryoidal coatings to 1 cm thick occur with barite. (GCF; Ferdock et al., 1997)

As pale-green-brown, curved crystals to 5 mm in vugs on small pyrite crystals at the Gold Quarry Mine in the Maggie Creek district. (Jensen et al., 1995)

HUMBOLDT COUNTY In the groundmass of oxidized ore at the Copper Canyon Mine in the Battle Mountain district. (Theodore and Blake, 1978; Schilling, 1979)

Identified as small, pale-brown crystals with carbonate-fluorapatite and native copper at the Lone Tree Mine in the Buffalo Mountain district. (M. Jensen, w.c., 1999)

LINCOLN COUNTY The variety *manganosiderite* is common in silver-rich orebodies in and replacing limestone in various mines in the Bristol and Pioche districts. (Melhase, 1934d; Gemmill, 1968; WMK)

NYE COUNTY Rare in veins in the Tonopah district; specifically noted in a specimen from the Liberty Mine. (Spurr, 1903; Melhase, 1935c; R. Thomssen, w.c., 1999)

WASHOE COUNTY Identified in a specimen from Great Boiling Springs. (JHS)

SIDERONATRITE

$Na_2Fe^{3+}(SO_4)_2(OH) \cdot 3H_2O$

A relatively uncommon mineral that has been found with other secondary sulfates in hydrothermal deposits; also noted as an efflorescent phase at the borate deposit at Kramer, California.

CHURCHILL COUNTY Reported in a specimen from the White Plains Mine in the Desert district. (WMK)

EUREKA COUNTY As radiating, lathlike and acicular, yellow-orange to orange crystals to 3 cm on sulfur with metacinnabar, metastibnite (?), and gypsum at the Hot Springs Point Mine in Crescent Valley, Sec. 11, T29N, R48E. (GCF)

SIDEROTIL

$Fe^{2+}SO_4 \cdot 5H_2O$

Siderotil is one of the rarer sulfate minerals that are found as efflorescences in old mine workings.

LANDER COUNTY Identified as an efflorescent mineral at the Watt Mine in the Reese River district. (Ross, 1953)

WASHOE COUNTY Identified in specimens from Steamboat Springs in the Steamboat Springs district. (Wilson and Thomssen, 1985; JHS)

SIEGENITE

$(Ni,Co)_3S_4$

A mineral found in hydrothermal veins with other sulfides.

CHURCHILL COUNTY As rims to 50 microns on cobaltite grains in calcite from an unnamed prospect near the southwest end of the Clan Alpine Mountains on the east side of Stingaree Valley. (M. Jensen, w.c., 1999)

EUREKA COUNTY As crystals less than 2 microns across with violarite on millerite crystals at the Goldstrike Mine in the Lynn district. (Ferdock et al., 1997; GCF)

SILLIMANITE

Al_2SiO_5

Trimorphous with kyanite and andalusite. Occurs in high-grade metamorphic rocks, more rarely in pegmatite and granite.

CLARK COUNTY In Precambrian schist in the Virgin Mountains in the Bunkerville district. (Beal, 1965)

In high-grade metamorphic rocks with almandine, *hypersthene,* and other minerals in the Gold Butte district. (Volborth, 1962b)

ELKO COUNTY In schist and pegmatite with andalusite, quartz, muscovite, and *garnet* 1/4 mile north of the Errington-Thiel Mine in the Valley View district. (Olson and Hinrichs, 1960)

After kyanite in gneiss in the northern East Humboldt Range. (Lush, 1982)

In gneiss with conspicuous kyanite porphyroblasts, Clover Hill area. (Snoke, 1992)

NYE COUNTY Reported from the Red Cloud claim, 16 miles south of Beatty in the Bare Mountain district. (HMM)

PERSHING COUNTY Reported as gray masses and crystals in veins with jarosite at the Montezuma Mine in the Arabia district. (R. Walstrom, w.c., 1999)

SILVER

Ag

Forms a series with native gold; intermediate compositions are often called *electrum.* In some cases a primary hydrothermal mineral, but generally a product of secondary processes in oxidized ore.

CHURCHILL COUNTY Reported in quartz veins at the Fairview Mine in the Fairview district. (Greenan, 1914; Schrader, 1947; Quade and Tingley, 1987)

In manganese-bearing calcite veinlets at the Pyramid Mine in the Holy Cross district. (Peters et al., 1996)

CLARK COUNTY With gold, galena, mottramite, pyrite, and numerous unidentified phases in a specimen from the Boss Mine in the Goodsprings district. (Jedwab, 1998a)

ELKO COUNTY As tiny isolated blebs and stringers in quartz veins at the Rossi Mine in the Bootstrap district. (Snyder, 1989)

Reported as small foliated masses in the Edgemont district. (Hoffman, 1878)

Rare in ore in the Gold Circle (Midas) district. (P. Goldstrand, w.c., 1999)

With chlorargyrite and gold in oxidized portions of fissure fillings in skarn at the Golden Ensign Mine in the Mountain City district; also with chlorargyrite in a vein in metamorphosed shaly limestone at the Mountain City Mine. (Granger et al., 1957; Stager and Tingley, 1988; Emmons, 1910)

Reported with other silver ore minerals in quartz-*adularia* veins at mines in the Tuscarora district; also reported in placer deposits; specifically found with argentojarosite in a specimen from the Dexter Mine. (Nolan, 1936; Vanderburg, 1936a; Von Bargen, 1999; SBC)

ESMERALDA COUNTY In a specimen from the Broken Toe Mine in the Rock Hill district. (R. Thomssen, w.c., 1999)

EUREKA COUNTY Reported in veins at the Nevada Star (Good Hope) Mine in the Maggie Creek district. (Rota, 1991)

Reported in a specimen with tetrahedrite at the Morning Glory Mine in a small gulch about 1 mile southeast of Barth in the Safford district. (WMK)

HUMBOLDT COUNTY Reported in quartz veins at the Neversweat claim on Buckskin Peak in the National district; also at the National Mine and Walker lease on Auto Hill. (Lindgren, 1915; Melhase, 1934c; WMK)

Reported in a specimen from the Paradise

Mine, 8 miles northwest of the town of Paradise Valley in the Paradise Valley district. (WMK)

In microscopic grains of similar size to grains of gold at the Getchell Mine in the Potosi district; megascopically visible silver has not been found. (Joralemon, 1951)

LANDER COUNTY Said to occur in delicate fibers near Galena in the Battle Mountain district; also reported in a specimen with sphalerite and pyrite. (Hoffman, 1878; UCB)

Reported in a specimen with galena, *limonite,* and quartz at the Lady of the Lake Mine in the Bullion district in the central part of the Shoshone Range. (WMK)

Formerly abundant as small wires to 2 mm in vugs in pyrite-sphalerite sulfide ore underground at the Cove Mine in the McCoy district. (M. Jensen, w.c., 1999; R. Walstrom, w.c., 1999)

As crystalline arborescent masses with sphalerite, quartz, and calcite in a specimen from near Austin in the Reese River district; also reported in specimens from the Barington, Utah, and Diana Mines. (F. Cureton, o.c., 2000; UCB)

LINCOLN COUNTY With acanthite at the Atlanta Mine in the Atlanta district; considered of supergene origin. (LaBerge, 1996)

As brilliant wires to 3 mm in vugs in fresh galena from bedded replacement ore of the Caselton ore channel in the Pioche district; also reported in specimens with gold from the Raymond Ely vein (1400-foot level) and the Meadow Valley Mine. (M. Jensen, w.c., 1999; WMK)

LYON COUNTY Reported as beautiful specimens in calcite at the Silver Hill Mine in the Silver City district. (De Quille, 1876; Gianella, 1936a)

With pyrite in quartz in a specimen of altered andesite at the Rockland Mine in the Wilson district. (WMK)

MINERAL COUNTY Identified in a specimen from the Aurora district. (JHS)

With silver-bearing galena at the Simon and Olympic Mines in the Bell district. (Knopf, 1921b)

In the Candelaria district at the Candelaria, Holmes, and Northern Belle Mines in unoxidized manganese- and iron-bearing carbonate veins with pyrite, sphalerite, galena, and jamesonite, and in oxidized rock with acanthite and chlorargyrite. Also noted with chlorargyrite at the Columbus Mine. (Ross, 1961; Moeller, 1988;

Chavez and Purusotam, 1988; HMM; F. Cureton, w.c., 1999)

Reported from the Garfield and Mabel Mines in the Garfield district. (Lincoln, 1923; Couch and Carpenter, 1943)

Reported in quartz-barite veins with tetrahedrite, acanthite, and copper carbonates at the Lucky Boy Mine in the Lucky Boy district. (Hill, 1915)

NYE COUNTY Reported in fine reticulated form from the Belmont district. (Hoffman, 1878)

Reported as a hypogene mineral that is partially replaced by chlorargyrite and acanthite at the Paradise Peak Mine in the Fairplay district. (Thomason, 1986)

Reported as wire silver from the Quartz Mountain Camp in the Lodi district. (Von Bargen, 1999)

As wires to 1 mm with pyrargyrite and rhodochrosite at the Morey Mine in the Morey district. (M. Jensen, w.c., 1999)

Reported as amalgam at the Bullion Mine in the Paradise Peak district. (Von Bargen, 1999)

As clots on a fracture surface in a quartz-*adularia* vein on the lower east flank of Stebbins Hill in the Round Mountain district. (Tingley and Berger, 1985)

In considerable quantities with arsenopyrite, acanthite, and other minerals at the Hillside, Catlin, and Blue Horse Mines in the Silverbow district. (Tingley et al., 1998a)

As a secondary mineral in wires, films, and spongy masses with acanthite, polybasite, tetrahedrite, pyrargyrite, and bromargyrite in quartz-*adularia* veins in the Tonopah district; noted as "spun wire" from the Ohio shaft. (Spurr, 1903; Melhase, 1935c; NMBM; WMK; S. Pullman, w.c., 1999)

Reported in oxidized ore at the Murphy and Teichert Mines in the Twin River district. (Hague, 1870)

Reported from the Harry Phillips, Shamrock, and Storm King Mines in the Union district. (WMK)

Reported from various mines in replacement ore in limestone in the Willow Creek district. (Hill, 1916)

Identified in a sample from an unknown locality in the Wilsons district. (JHS)

PERSHING COUNTY As tiny grains and rare

sheets to 5 mm at the Majuba Hill Mine in the Antelope district. (Jensen, 1985)

In oxidized quartz veins with acanthite and other secondary minerals at the Arizona Mine in the Buena Vista district. (Cameron, 1939)

Reported on the dumps in the Rabbit Hole district. (Melhase, 1934b)

As bright leaves to 1 cm with *sericite* at the Coeur Rochester Mine in the Rochester district. (M. Jensen, w.c., 1999)

STOREY COUNTY Reported from many mines in the Comstock district as wire silver with acanthite and other silver sulfide minerals; unrolled individual wires were said to reach lengths of a foot or more. Said to be particularly common in pockets of yellow clay in the Mexican Mine. Also noted as *electrum* with 80 percent Ag from the district, and as *electrum* from the Flowery lode that was reported to contain 1 part gold to 11 parts silver. (De Quille, 1876; Bastin, 1922; WMK; HMM; BMC; FUMT; DeMouthe, 1987; Schilling, 1990)

WHITE PINE COUNTY With chlorargyrite and acanthite in small veinlets in limestone at the Silverton Mine in the Pancake district. (Kral, 1951)

Reported as wires with chlorargyrite at the Eberhardt and Silver Plate Mines in the White Pine district. (N. Prenn, w.c., 1999; WMK)

Identified in samples from unknown localities in the Cherry Creek and Ward districts. (JHS)

SIMMONSITE

Na_2LiAlF_6

A Nevada type mineral found as one of several sodium aluminum fluoride phases in a pegmatite.

MINERAL COUNTY At the type locality in a pegmatite dike at the Zapot claims in the Fitting district; as subhedral monoclinic grains that are translucent to transparent, pale-buff to cream colored, with white streak and greasy luster, and fluoresce yellow-orange under short wave UV; associated with microcline (*amazonite*), topaz, cryolite, pachnolite, and cryolithionite. (Foord et al., 1999b)

SINCOSITE

$CaV_2^{4+}(PO_4)_2(OH)_4 \cdot 3H_2O$

A rare secondary mineral that has been found in carbonaceous shale in Peru and in siliceous gold ore from Lead, South Dakota.

EUREKA COUNTY As grass-green to pale-brown flakes to 4 mm on fractures in altered siltstone at the Gold Quarry Mine in the Maggie Creek district. (Jensen et al., 1995)

SKINNERITE

Cu_3SbS_3

A rare sulfosalt mineral originally described from zeolite veins in alkaline rock.

NYE COUNTY As tiny gray metallic grains with stetefeldtite, molybdofornacite, fornacite, and vauquelinite at the Belmont Mine in the Belmont district. (EMP; F. Cureton, w.c., 1999; Anthony et al., 1990)

SKUTTERUDITE

$CoAs_{2-3}$

Skutterudite is generally found with nickel and cobalt minerals in hydrothermal veins; it forms a series with nickel-skutterudite, with the variety *smaltite* intermediate in composition.

CHURCHILL COUNTY Reported as *smaltite* in a specimen from the Rand (1867–1902) collection from the Empress of India Mine, Pershing County, but probably from the Lovelock or Nickel Mine in the Table Mountain district in Churchill County. *Smaltite* was also identified with millerite in a specimen reported to be from Lovelock, but probably from the same locality as the Rand specimen. (BMC; UCB)

SMITHSONITE

$ZnCO_3$

Smithsonite is a common mineral in the oxidation zones of base-metal deposits. It is occasionally found in large quantities and serves as an important source of zinc.

CLARK COUNTY Reported at the Azure Ridge (Bonelli) Mine in the Gold Butte district. (Longwell et al., 1965)

Common as greenish-yellow, red, gray, or

white, massive to banded material in mines of the Goodsprings district with cerussite, hydrozincite, galena, wulfenite, dioptase, and heterogenite; in some cases a valuable zinc ore. (Hewett, 1931; Melhase, 1934e; LANH; UCB; WMK)

ELKO COUNTY With wulfenite, cerussite, and other base-metal minerals in white veins cutting partially oxidized veins rich in sphalerite at the Killie Mine in the Spruce Mountain district. (Schrader, 1931b; Schilling, 1979)

With cerussite and wulfenite in a replacement orebody at the Queen of the West Mine and Irish Boy claims in the Tecoma district. (Hill, 1916; Schilling, 1979; LaPointe et al., 1991)

ESMERALDA COUNTY In oxidized lead-zinc-copper ore in the Weepah (Lone Mountain) district; specifically reported from the 3 Metals Mine. (Phariss, 1974; R. Thomssen, w.c., 1999)

With anglesite, cerussite, and chlorargyrite at the Sylvania Mine in the Sylvania district. (Schilling, 1979)

EUREKA COUNTY Common in the mines of the Eureka district. (Nolan, 1962; JHS)

As crusts to 3 cm thick of pale-brown botryoids at the Lone Mountain Mine in the Lone Mountain district; also reported in a specimen as "drybone" smithsonite with cerussite and galena. (M. Jensen, w.c., 1999; WMK)

LANDER COUNTY As finely crystalline masses in open spaces in sphalerite ore at the Nevada and Lucky Strike Mines in the Battle Mountain district; also with iron oxides and clay minerals. (Roberts and Arnold, 1965)

With aurichalcite and other zinc minerals in the Tenabo area of the Cortez district. (Gianella, 1941)

LINCOLN COUNTY Reported with hydrozincite and other minerals in oxidized zinc ore at the Black Ledge Pioche Mine in the Pioche district. (Jenkins, 1981; WMK)

Reported from the Silver King Mine in the Silver King district. (R. Thomssen, w.c., 1999)

MINERAL COUNTY As gray-green crystals to 1 mm with adamite, mimetite, and other phases at the Simon Mine in the Bell district. (Knopf, 1921b; M. Jensen, w.c., 1999)

Questionably identified in ore from the western base of the southern Monte Cristo Mountains. (Schrader, 1947)

In minor quantities in the Candelaria district. (Page, 1959)

NYE COUNTY As red-orange, elongated microcrystals in vugs and in massive form with hydrozincite and aurichalcite in gossan at Downieville in the Gabbs district. (Jenkins, 1981; M. Jensen, w.c., 1999)

As 1-mm white spheroids at the San Rafael Mine in the Lodi district. (B. Runner, o.c., 2000)

Reported at the Alexander Mine in the Union district. (Kral, 1951)

PERSHING COUNTY In small quantities at the Last Chance Mine in the Antelope district. (MacKenzie and Bookstrom, 1976)

As orange crystals to 1 mm with quartz at the Red Bird Mine in the Antelope Springs district. (M. Jensen, w.c., 1999)

WASHOE COUNTY With cerussite, hemimorphite, malachite, and chalcanthite in oxidized ore at the Union (Commonwealth) Mine in the Galena district. (Geehan, 1950; Bonham and Papke, 1969)

Reported in oxidized quartz veins in the Pyramid district. (Overton, 1947)

WHITE PINE COUNTY Identified in a sample from an unknown site in the Cherry Creek district. (JHS)

Reported at the Grand Deposit and other mines in the Muncy Creek district. (Gianella, 1941)

Reported as sparse in the Robinson district; specifically reported from the Dauntless Mine. (Spencer et al., 1917; S. Kleine, o.c., 1999)

A minor constituent of oxidized zinc ore in the Taylor district. (Lovering and Heyl, 1974)

With galena and oxidized lead and copper minerals in the lead belt near Mount Hamilton in the White Pine district. (Humphrey, 1960; Hose et al., 1976)

In oxidized lead-silver-zinc ore from the Egan Range, possibly from the Ward district. (Hill, 1916)

SODDYITE

$(UO_2)_2SiO_4 \cdot 2H_2O$

A secondary mineral that is generally found in granite pegmatite and in sandstone-type uranium deposits.

HUMBOLDT COUNTY Tentatively identified as fracture coatings and in cavities in zeolitized and argillized tuffaceous rocks in uranium prospects near the Virgin Valley Ranch in the Virgin Valley district. (Castor et al., 1996)

SONORAITE

$Fe^{3+}Te^{4+}O_3(OH) \cdot H_2O$

A rare secondary mineral that is found only at a few localities with other tellurium-bearing oxides such as emmonsite.

ESMERALDA COUNTY Reported at the Mohawk Mine (McGinnity shaft) in the Goldfield district. (Gaines, 1972; LANH)

SORBYITE

$Pb_{19}(Sb,As)_{20}S_{49}$

Sorbyite is a very rare mineral described with other sulfide minerals in hydrothermal deposits in carbonate rock in Canada and Asia.

MINERAL COUNTY As gray to black crystalline masses and acicular crystals to 3 mm in white quartz with pyrite, arsenopyrite, jamesonite, and unidentified minerals at the Northern Belle and Lucky Strike Mines in the Candelaria district. (HUB; Jenkins, 1981; M. Jensen, w.c., 1999)

SPANGOLITE

$Cu_6^{2+}Al(SO_4)(OH)_{12}Cl \cdot 3H_2O$

Spangolite is a rare and often very beautiful mineral that occurs as small green crystals in the oxidation zones of copper deposits. It is usually associated with cyanotrichite, brochantite, and other sulfate minerals.

CHURCHILL COUNTY Locally abundant as blue-green, microcrystalline crusts with cyanotrichite and olivenite at the Lovelock Mine in the Table Mountain district. (M. Jensen, w.c., 1999; F. Cureton, w.c., 1999)

ESMERALDA COUNTY Reported from the Monte Cristo Range near Coaldale Junction. (R. Thomssen, w.c., 1999)

Reported from the 3 Metals Mine in the Weepah district. (R. Thomssen, w.c., 1999)

HUMBOLDT COUNTY Blue-green mineral at the Silver Coin Mine in the Iron Point district. (Thomssen, 1998; J. McGlasson, w.c., 2001)

LYON COUNTY As tiny, bright, greenish-blue, hexagonal plates with cyanotrichite and brochantite in prospects on Carson Hill near Mason Pass in the Yerington district. (Jenkins, 1981; M. Jensen, w.c., 1999)

PERSHING COUNTY As small (typically <1 mm), shiny, transparent, deep-blue-green crystals in thin crusts in cavities with brochantite, azurite, chalcophyllite, cyanotrichite, and parnauite at the Majuba Hill Mine in the Antelope district. (Frondel, 1949; Palache et al., 1951; Jensen, 1985)

SPERRYLITE

$PtAs_2$

The most widespread platinum mineral; occurs mainly in magmatic deposits, but is also found in other types of deposits that contain platinum.

CLARK COUNTY Reported in a specimen from the Key West Mine in the Bunkerville district. (LANH)

SPESSARTINE

$Mn_3^{2+}Al_2(SiO_4)_3$

A manganese-rich *garnet*-group mineral that forms a series with almandine. Most common in granite, granite pegmatite, and rhyolite, but also found in contact-metamorphosed rocks.

CLARK COUNTY As red-orange crystals to 1 cm in miarolitic cavities and pegmatites in the Newberry and Eldorado Mountains, in some cases as inclusions in quartz crystals; composed of about 98 percent of the spessartine molecule on the basis of EMP analyses. Also noted with quartz and *feldspar* west of Laughlin, near the California border. (S. Scott, w.c., 1999; W. Lombardo, o.c., 2001; J. Gray, o.c., 2001)

ELKO COUNTY In pegmatite as rusty-red to clear red-orange crystals to 4 mm with a composition of 54 percent spessartine, 36 percent almandine, and 10 percent andradite in the Valley View district. (Olson and Hinrichs, 1960)

MINERAL COUNTY Reported in *amazonite*- and topaz-bearing pegmatite at the Zapot claims in the Fitting district about 25 km northeast of Hawthorne. (S. Pullman, w.c., 1999; Foord et al., 1999b)

SPHALERITE

$(Zn,Fe)S$

Sphalerite is the most common zinc mineral and forms under a wide variety of hydrothermal conditions, generally with other sulfide minerals. It is common and has doubtless been the principal source of zinc in Nevada; however, only a few of the localities listed have produced well-crystallized material.

CARSON CITY Reported in a specimen from an unknown locality in the Carson City district. (UCB)

CHURCHILL COUNTY With stibnite, arsenopyrite, and pyrite in quartz veins at the Antimony King Mine in the Bernice district. (Lawrence, 1963)

In quartz veins at the Fairview Mine in the Fairview district. (Greenan, 1914; Schrader, 1947; Saunders, 1977)

Sparse in quartz-*adularia* veins with fluorite and other minerals at the Wonder Mine in the Wonder district. (Schilling, 1979)

Reported from unspecified localities in the Alpine, IXL, and White Cloud districts. (Lawrence, 1963; Vanderburg, 1940; WMK)

CLARK COUNTY Reported in copper-nickel ore at the Key West Mine in the Bunkerville district. (Lindgren and Davy, 1924; Beal, 1965)

Widespread in small amounts in the Goodsprings district; particularly noted at the Hussey's, Milford, Potosi, and Yellow Pine Mines. (Hewett, 1931; Schilling, 1979)

Reported from unspecified localities in the Eldorado, Searchlight, and Gold Butte districts. (Longwell et al., 1965; Callaghan, 1939; Hill, 1916)

DOUGLAS COUNTY Reported from an unspecified locality in the Red Canyon district. (Lawrence, 1963; Dixon, 1971)

ELKO COUNTY In clots with galena in a quartz vein in dolomitic limestone at the California Mine in the Aura district. (LaPointe et al., 1991)

With galena, pyrite, arsenopyrite, and chalcopyrite in a banded quartz-calcite vein in andesite at the Mint and Mirat Mines in the Burner district. (Emmons, 1910)

As small, translucent red veins and blebs with orpiment and native sulfur in carbonaceous sediments at the Rain Mine in the Carlin district. (D. Heitt, o.c., 1991)

Reported at the Prunty and Rescue Mines in the Charleston district. (Lawrence, 1963)

With molybdenite in quartz veins at the Indian Creek prospects in the Edgemont district. (Schilling, 1979)

Widespread at the Jerritt Canyon Mines and Big Springs Mines in the Independence Mountains district, particularly in areas of sericitic alteration where it is typically low in iron and locally has chalcopyrite inclusions; also in trace amounts as a diagenetic mineral in Paleozoic sedimentary rock. (Youngerman, 1992; Hofstra, 1994)

With galena, tennantite, and acanthite in a quartz vein at the Diamond Jim Mine in the Island Mountain district. (Bushnell, 1967)

As irregular inclusions in pyrite/marcasite and as tiny disseminated grains with galena and freibergite at the Hollister deposit in the Ivanhoe district. (Deng, 1991)

With galena in a quartz vein at the American Beauty Mine in the Lee district. (Bentz et al., 1983)

With chalcopyrite, galena, and related oxidation phases in ore at the Rip Van Winkle Mine in the Merrimac district. (Lovering and Stoll, 1943)

A common fine-grained mineral in gold ore at the Meikle Mine in the Lynn district. (Jensen, 1999)

In small amounts in massive sulfide orebodies with quartz, pyrite, and chalcopyrite at the Rio Tinto Mine in the Mountain City district. (Coats and Stephens, 1968)

With pyrite, galena, and chalcopyrite at the Killie Mine in the Spruce Mountain district; partially oxidized to smithsonite and hemimorphite in places. (Schrader, 1931b; Dunning and Cooper, 1987)

With other sulfides and *electrum* in the Tuscarora district. (SBC; WMK)

In veins in the Contact and Gold Circle districts. (Schilling, 1979; P. Goldstrand, w.c., 1999)

Reported from the Railroad and Ruby Valley districts. (Ketner and Smith, 1963; Hill, 1916)

ESMERALDA COUNTY Reported at the Carrie Mine in the Gilbert district. (Ferguson, 1928)

In a specimen with cerussite from the Hornsilver Mine in the Gold Point district. (WMK)

Very sparingly in ore from principal mines in the Goldfield district, but abundant in rich ore from the Mushette lease where it was found as botryoidal crusts over prisms of bismuthinite in places; also reported in a specimen from the Grizzly Bear Mine. (Ransome, 1909a; LANH)

Rare in quartz veinlets with scheelite and pyrite at the Red Hill (Redlich) Mine in the Rock Hill district. (Schilling, 1979)

As lustrous, transparent, red-brown crystals to 8 mm on drusy white quartz at the Sixteen-to-One Mine in the Silver Peak district; also in yellow crystals. (M. Jensen, w.c., 1999; W. Lombardo, o.c., 2001)

With quartz, pyrite, galena, and molybdenite at the Sylvania Mine in the Sylvania district. (Schilling, 1979)

Reported from the Lida, Red Mountain, Weepah (Lone Mountain), and Divide districts. (Albers and Stewart, 1972; Phariss, 1974; Bonham and Garside, 1979)

EUREKA COUNTY Reported from Mill Canyon in the Cortez district. (Emmons, 1910; Roberts et al., 1967b)

With galena, pyrite, and arsenopyrite at the Richmond-Eureka (Ruby Hill) Mine in the Eureka district. (Nolan, 1962; Schilling, 1979; S. Pullman, w.c., 1999)

With pyrrhotite and molybdenite at the Antelope Mine in the Fish Creek district; also at the Bisoni prospect. (Schilling, 1979; Papke, 1979)

As well-formed, dark-brown crystals to 6 mm with dolomite crystals at the Carlin Mine in the Lynn district; common in silicified Paleozoic sedimentary rock and in veins in diorite at the Goldstrike Mine in yellow to dark-brown, earthy, massive, and botryoidal forms. (Roberts et al., 1967b; GCF; Ferdock et al., 1997)

As brown crystals to 1 mm with pyrite and siderite at the Gold Quarry Mine in the Maggie Creek district. (M. Jensen, w.c., 1999)

In replacement bodies and breccia with galena and other sulfides at the Mineral Hill Consolidated, Live Yankee, and Mary Ann Mines in the Mineral Hill district. (Roberts et al., 1967b; Schilling, 1979; WMK)

With other sulfides in skarn at the Keystone Mine in the Roberts district. (Roberts et al., 1967a; LANH)

In barite-carbonate veins with other sulfides at the Zenoli Mine in the Safford district. (Roberts et al., 1967a; WMK)

Reported from unspecified localities in the Lone Mountain and Mount Hope districts. (Roberts et al., 1967b)

HUMBOLDT COUNTY As tiny grains in uranium-bearing silicified tuff in the Bottle Creek district. (Castor et al., 1982)

Common with pyrite and marcasite as minute grains (< 50 microns) at the Lone Tree Mine in the Buffalo Mountain district. (R. Braginton, o.c., 1995)

A minor component of zirconium- and uranium-rich breccia at the Moonlight Uranium Mine in the Disaster district. (Dayvault et al., 1985)

Reported with chalcopyrite, galena, and other sulfides in the Gold Run district. (Melhase, 1934c)

Reported at the Silver Coin Mine in the Iron Point district. (Thomssen, 1998)

Sparse with pyrite, chalcopyrite, and galena in quartz-*adularia* veins in the National district. (Lindgren, 1915)

Late yellow sphalerite occurs with jordanite at the Getchell Mine in the Potosi district; also common with other sulfides at the Riley Mine. (D. Tretbar, o.c., 1999; Hotz and Willden, 1964)

With tetrahedrite, arsenopyrite, and galena in quartz veins at the Pansy Lee Mine in the Ten Mile district. (Lawrence, 1963; WMK)

LANDER COUNTY In small amounts with chalcopyrite, galena, and other sulfides at the Copper Canyon Mines in the Battle Mountain district; also in gold-bearing skarn at the Fortitude Mine. (Roberts and Arnold, 1965; Theodore and Blake, 1978; Schilling, 1979; Doebrich, 1993)

Widespread at the Lone Tree Mine in the Buffalo Mountain district. (Bloomstein et al., 1993)

With tetrahedrite and other minerals at the Betty O'Neal Mine in the Lewis district; noted as sharp crystals to 1 mm in quartz-calcite veins. (Vanderburg, 1939; Gilluly and Gates, 1965; J. McGlasson, w.c., 2001)

As well-formed black crystals to 1 cm with pyrite and quartz at the Cove Mine in the McCoy district. (M. Jensen, w.c., 1999; M. Rogers, o.c., 2000)

Reported from the Big Creek, Hilltop, Jersey, Kingston, and Reese River districts. (Lawrence, 1963; Gilluly and Gates, 1965; Johnson, 1977; Hill, 1916; Ross, 1953)

LINCOLN COUNTY Reported at the Comet and Pyrite Mines in the Comet district. (Tschanz and Pampeyan, 1970)

Reported from the Delamar district. (Callaghan, 1937)

Common but sporadic as resinous brown (*rosinjack*) and black (*blackjack*) varieties, generally with galena, in veins and replacement bodies in limestone at the Groom and other mines in the

Groom district. (Humphrey, 1945; Tingley et al., 1998a)

Reported at the Highland Mine in the Highland district. (Klein, 1983)

Common in orebodies in the Pioche district with pyrite and galena; particularly noted in the Caselton, Ely Valley, and Prince Mines. (Westgate and Knopf, 1932; Gemmill, 1968; Tschanz and Pampeyan, 1970)

Reported at the Lincoln Tungsten Mine in the Tem Piute district. (Buseck, 1967)

LYON COUNTY Reported in a specimen from an unknown locality in the Como district. (WMK)

MINERAL COUNTY With galena, other sulfides, quartz, and calcite in a replacement-type orebody at the Simon Contact prospect in the Bell district. (Knopf, 1921b)

As traces in veins at the Gravity Barite Mine in the Buckley district; locally with wulfenite. (JHS; R. Walstrom, w.c., 1999)

Reported from the Broken Hills, Candelaria, Rand, Rawhide, Marietta, and Garfield districts. (Schrader, 1947; Knopf, 1923; Ross, 1961; Ponsler, 1977)

NYE COUNTY In quartz veins with pyrite and galena at the Antelope View Mine in the Antelope Springs district on the Nellis Air Force Range. (Tingley et al., 1998a)

With pyrite in ore at the Highbridge Mine in the Belmont district. (Kral, 1951)

As disseminated grains and stringers with galena, freibergite, and other sulfides at the Fairday Mine in the Cactus Springs district on the Nellis Air Force Range. (Tingley et al., 1998a)

With pyrite, acanthite, and other base-metal minerals in a quartz vein in welded tuff at the Iron Rail group in the Cloverdale district. (Kral, 1951; Kleinhampl and Ziony, 1984; Hamilton, 1993)

In small lenses or pipes with other sulfides at the Bluestone Lode claim in the Gabbs district; also reported from the Brown prospect. (Callaghan, 1933; Kral, 1951; WMK)

As inclusions in pyrite in a sample from the Gold Crater district on the Nellis Air Force Range. (Tingley et al., 1998a)

Reported at the San Rafael Mine in the Lodi district. (Kral, 1951; LANH)

Reported at the Morey and Keyser Mines in the Morey district. (Kral, 1951; Williams, 1968; S. Pullman, w.c., 1999)

With galena below the 900-foot level in the Mizpah shaft in the Tonopah district; also reported in a specimen from the Evergreen Mine. (Bastin and Laney, 1918; Melhase, 1935c; LANH)

With chalcopyrite, pyrite, and galena in a quartz vein at the Troy Mine in the Troy district. (Raymond, 1874; Hill, 1916; Kral, 1951)

Reported with coarse calcite and galena at the Bresnahan prospect in the Tybo district; also with other sulfide minerals in a quartz-calcite vein at the Tybo Mine. (Ferguson, 1933)

In veins with other sulfides and scheelite at the Webster Mine in the Union district. (Hamilton, 1993; WMK)

As microscopic grains with luzonite, tetrahedrite, and other sulfides at the Golden Chariot Mine in the Jamestown district on the Nellis Air Force Range. (Tingley et al., 1998a)

With pyrite and *apatite* along stylolites in Paleozoic dolomite from drill core at Yucca Mountain. (Castor et al., 1999b)

Reported from the Jefferson Canyon, Twin River, Washington, Willow Creek, Manhattan, and Oak Spring districts. (Kral, 1951; Ferguson, 1924; Melhase, 1935b)

PERSHING COUNTY Rarely with other sulfides in the Copper stope of the Majuba Hill Mine in the Antelope district; also reported from the Antelope Mine. (MacKenzie and Bookstrom, 1976; Jensen, 1993b; Lawrence, 1963)

With cinnabar in the Antelope Springs district. (Bailey and Phoenix, 1944)

With other sulfides in quartz veins at the Black Warrior Mine in the Buena Vista district. (Lawrence, 1963)

Identified in a specimen of zinc ore at the Comrade Mine in the Gerlach district. (WMK)

In quartz-*tourmaline* veins at the Bonanza King Mine in the Spring Valley district. (Campbell, 1939)

Reported from the Iron Hat, Kennedy, Muttlebury, Mill City, Rochester, Rye Patch, Sacramento, Sierra, and Star districts. (Johnson, 1977; WMK; Knopf, 1924; Ferguson et al., 1951; S. Pullman, w.c., 1999)

STOREY COUNTY A major early vein mineral, along with quartz, in specimens of bonanza ore from the Comstock Lode in the Comstock district; rare with other sulfides in quartz in the Flowery lode. (Bastin, 1922; Gianella, 1936a;

Koschmann and Bergendahl, 1968; Schilling, 1990)

WASHOE COUNTY In quartz-pyrite veins with jamesonite and tetrahedrite at the Modoc Mine and Silver Fox prospect in the Cottonwood district. (Bonham and Papke, 1969)

In quartz-calcite veins with galena, chalcopyrite, arsenopyrite, and pyrite at the Union (Commonwealth) Mine in the Galena district. (Geehan, 1950; Bonham and Papke, 1969)

With *electrum* at the Dunlop Mine in the Jumbo district. (Vikre, 1989)

In quartz-calcite veins with galena, pyrite, and chalcopyrite at the Leadville Mine in the Leadville district. (Burgess, 1926; Bonham and Papke, 1969)

Reported from the Nightingale district. (Bonham and Papke, 1969)

As trace amounts in ore at the Olinghouse Gold Mine in the Olinghouse district. (Jones, 1998)

With pyrite, galena, and enargite at the Golden Fleece Mine and other locations in the Peavine district. (Hill, 1915; Bonham and Papke, 1969)

With galena, pyrite, and other minerals at the Ruth Mine and other properties in the Pyramid district. (Bonham and Papke, 1969; Wallace, 1980; WMK)

In large masses to 20 cm of crystals to 3 cm with coarse galena and pyrite in stockworks and fracture zones at the Wedekind and other mines in the Wedekind district. (Overton, 1947; M. Jensen, w.c., 1999; WMK)

WHITE PINE COUNTY With chalcopyrite, galena, and other sulfides in gold ore at the Bald Mountain Mines in the Bald Mountain district. (Hitchborn et al., 1996)

Reported with proustite in the Cherry Creek district. (Schrader, 1931b; Melhase, 1934d)

With galena in a specimen from the Eureka claim in the Duck Creek district. (WMK)

Black to brown sphalerite is the most significant ore mineral with galena, pyrite, and chalcopyrite in zinc-lead-copper-silver skarn at the Ward Mine in the Ward district; also reported locally as masses to 10 cm and as botryoidal vug fillings. (Hill, 1916; Hasler et al., 1991; M. Jensen, w.c., 1999)

Reported from the Eagle, Robinson, Taylor, and White Pine districts. (Hill, 1916; Melhase, 1934d; Humphrey, 1960)

SPINEL

$MgAl_2O_4$

Spinel is a common accessory mineral in mafic and ultramafic igneous rocks; it also occurs in aluminum-rich schist and contact- and regionally metamorphosed limestones. *Pleonaste* is an iron-bearing variety and *picotite* is a chrome-bearing variety.

CLARK COUNTY A minor constituent of hornblendite at the Great Eastern Mine area in the Bunkerville district. (Beal, 1965)

DOUGLAS COUNTY As black crystals that make nice micromounts from an old bulldozer cut in the Red Canyon district in the southern end of the Pinenut Range, just east of Topaz Ranch Estates. (M. Rogers, o.c., 1999)

LINCOLN COUNTY With diopside in a 5-foot-wide veinlike mass in a prospect on Hill 7106 near Blind Mountain in the Highland Range. (Westgate and Knopf, 1932)

LYON COUNTY *Picotite* occurs as black octahedral crystals along a ridge 1 mile northeast of Ludwig near the Ludwig gypsum pit in the Yerington district. (HMM; R. Walstrom, w.c., 1999; S. Pullman, w.c., 1999)

MINERAL COUNTY *Pleonaste* occurs as aggregates of greenish-black octahedrons to 0.5 mm at a locality near Kinkaid siding. (WMK; Jenkins, 1981)

Pleonaste was reported as crystals to 4 mm in dark-green to nearly black granular masses with calcite and other minerals in lenticular masses to 2 feet wide and 50 feet long, 9 miles east of Hawthorne and about a half mile east of a copper skarn prospect; may be the same as the locality listed above. (Gianella, 1942)

SPIONKOPITE

$Cu_{39}S_{28}$

A rare secondary copper sulfide mineral found outside Nevada in porphyry and redbed copper deposits and in *serpentine*.

HUMBOLDT COUNTY Identified as a minor phase in the oxidation assemblage at the Lone Tree Mine in the Buffalo Mountain district. (R. Braginton, o.c., 1995)

STANNITE

Cu_2FeSnS_4

A mineral found in tin-bearing hydrothermal deposits; in some cases, as at Cornwall, England, an ore mineral.

LANDER COUNTY With sulfide minerals in stockwork veins at the Cove Mine in the McCoy district; reportedly silver- and zinc-bearing. (Emmons and Coyle, 1988; W. Fusch, o.c., 2000)

NYE COUNTY Zinc-bearing, as 10 micron grains, with scorodite in a sample from the Gold Crater district on the Nellis Air Force Range. (Tingley et al., 1998a)

PERSHING COUNTY Identified in polished sections in association with arsenopyrite, chalcopyrite, digenite, and other minerals at the Majuba Hill Mine in the Antelope district. (Jensen, 1985)

WASHOE COUNTY Identified by SEM/EDX as tiny (15 microns or less) inclusions with possible neyite in tetrahedrite at the Blue Bird Mine in the Pyramid district. (SBC)

WHITE PINE COUNTY In very minor amounts with base-metal sulfides and telluride minerals in the lower part of the RBM deposit at the Bald Mountain Mine in the Bald Mountain district. (Hitchborn et al., 1996)

STARKEYITE

$MgSO_4 \cdot 4H_2O$

A water-soluble secondary mineral that forms as an efflorescent phase in mine workings.

EUREKA COUNTY A minor component of blue efflorescent material with szomolnokite and gunningite at the Goldstrike Mine in the Lynn district. (GCF)

STAUROLITE

$(Fe^{2+},Mg,Zn)_2Al_9(Si,Al)_4O_{22}(OH)_2$

Staurolite is a common mineral in schist and gneiss; it may be found in distinctive cruciform twins.

CLARK COUNTY In metamorphic rocks in the Virgin Mountains in the Bunkerville district. (Beal, 1965)

NYE COUNTY Reported in metamorphic rocks near Beatty. (Gianella, 1941)

WHITE PINE COUNTY With almandine in schist in Hampton Creek and Hendrys Creek Canyons. (Hose et al., 1976)

STELLERITE

$Ca[Al_2Si_7O_{18}] \cdot 7H_2O$

A zeolite group mineral that typically occurs in cavities and along fractures in hydrothermally altered volcanic rocks.

LINCOLN COUNTY In joint seams in volcanic rocks with green *garnet*, red rhodonite, and an unidentified mineral in the Bristol district. (Melhase, 1934d)

NYE COUNTY In devitrified tuff at Yucca Mountain. (Carlos et al., 1995)

STEPHANITE

Ag_5SbS_4

Stephanite is a widespread late-stage sulfosalt that occurs in small amounts in many hydrothermal precious-metal deposits, generally with acanthite, "ruby silver" minerals, and other sulfides. It was particularly abundant in the Comstock Lode, which is said to have produced some fine specimens.

CHURCHILL COUNTY Reported with acanthite, pyrargyrite, and other minerals in quartz veins at the Nevada Hills and Fairview Silver Mines in the Fairview district. (Greenan, 1914; Schrader, 1947; Quade and Tingley, 1987)

Reported in oxidized ore at the Wonder Mine in the Wonder district. (Schilling, 1979)

CLARK COUNTY Reported in a specimen from the Little Eoluppus Mine in the Eldorado district. (WMK)

DOUGLAS COUNTY With acanthite in a quartz vein at the Veta Grande Mine in the Gardnerville district. (Overton, 1947)

ELKO COUNTY With "ruby silver," tetrahedrite, galena, sphalerite, and pyrite in a quartz vein at the Protection Mine in the Mountain City district. (Emmons, 1910)

Reported with pyrargyrite at the Grand Prize, Navajo, Nevada Queen, and North Belle Isle Mines in the Tuscarora district. (Emmons, 1910; Lawrence, 1963; UAMM; HMM; WMK)

ESMERALDA COUNTY Reported in a specimen from the Goldfield district. (WMK)

EUREKA COUNTY With stibnite, tetrahedrite, stromeyerite, pyrargyrite, and other minerals at the Garrison and other mines along Mill Creek and elsewhere near Mount Tenabo in the Cortez district. (Emmons, 1910)

In replacement ore with quartz, polybasite, tetrahedrite, and other minerals at the Mineral Hill Mines in the Mineral Hill district. (Emmons, 1910; Schilling, 1979)

HUMBOLDT COUNTY Reported as a secondary sulfide with pyrargyrite, stibnite, and marcasite in joints, fractures, and quartz veins at the National Mine in the National district; also reported in ore from Auto Hill and the Caustin-Bankers-Diffenbach area. (WMK)

Reported in a specimen from the Patuch Mine. (WMK)

LANDER COUNTY Reported with acanthite in high-grade silver ore in the Battle Mountain district. (WMK)

With tetrahedrite and other minerals in quartz veins at the Betty O'Neal Mine in the Lewis district; also identified by EMP in a specimen from the Dean Mine. (Vanderburg, 1939; J. McGlasson, w.c., 2001)

In ore from the Reese River district; specifically reported from the Ogden Mine dump as small crystals with proustite and other silver sulfosalts, and at the Amador Mine with pyrargyrite, polybasite, and proustite. (Melhase, 1934c; Ross, 1953; SM; WMK; Lawrence, 1963)

LINCOLN COUNTY Reported at the Rich Hill Mine in an unspecified district. (Schrader et al., 1917)

LYON COUNTY With covellite in epithermal quartz veins in the Wilson district. (Dircksen, 1975)

NYE COUNTY Reported in a specimen with acanthite at the Belmont Mine in the Belmont district. (WMK)

With pyrargyrite, proustite, pyrite, and marcasite in vugs in quartz veins at the Clifford Mine in the Clifford district. (Ferguson, 1916)

After andorite and other early sulfosalts in silver-tin veins at the Morey and Keyser Mines in the Morey district. (Shannon, 1922; Kral, 1951; Williams, 1968; S. Pullman, w.c., 1999)

Reported in a specimen from the Reveille district. (CSM)

One of the principal ore minerals with acan-

thite, chlorargyrite, and *electrum* in the Silverbow district. (Kral, 1951; Tingley et al., 1998a; JHS)

With acanthite and polybasite in ore from the Tonopah district. (Bastin and Laney, 1918; Melhase, 1935c; WMK)

PERSHING COUNTY Doubtfully reported in the Rye Patch district. (Ransome, 1909b)

In quartz-*tourmaline* veins at the Bonanza King Mine in the Spring Valley district. (Campbell, 1939)

Reported in bonanza silver ore at the Sheba Mine in the Star district. (Hague and Emmons, 1877)

STOREY COUNTY Said to have been a major component with acanthite, polybasite, and native silver, of dull black, but extremely rich, ore in the Big Bonanza of the Consolidated Virginia Mine in the Comstock district; also in specimens from the district with galena and chalcopyrite and alone on quartz. (De Quille, 1876; Smith, 1943; Palache et al., 1944; CSM; WMK)

WHITE PINE COUNTY With pyrargyrite, proustite, and polybasite after galena and sphalerite at the Star Mine in the Cherry Creek district. (Schrader, 1931b)

With andorite, miargyrite, pyrargyrite, and other minerals at the Patriot and Monitor Mines in the Taylor district. (Williams, 1965)

Intergrown with galena and hessite and rimmed by a variety of oxidation minerals in the Silver King Mine in the Ward district; also reported from the Paymaster Mine. (Jenkins, 1981; CSM)

STERNBERGITE

AgFe$_2$S$_3$

Sternbergite, which is dimorphous with argentopyrite, is a rare mineral that is found in hydrothermal veins with other silver sulfosalt minerals. The Nevada occurrences listed are suspect because they have not been verified by modern methods.

LANDER COUNTY Reported as small but fine crystals from unspecified mines in the Reese River district. (Hoffman, 1878)

STOREY COUNTY Reported as crystals with polybasite in vugs in bonanza ore from the Comstock district. (King, 1870; Smith, 1985)

STETEFELDTITE

$Ag_2Sb_2(O,OH)_7$ (?)

Stetefeldtite, a Nevada type mineral, is a relatively rare and poorly understood oxidation product that is apparently the silver analog of bindheimite and partzite, although some consider it to be a mixture. The mineral was reportedly found in large quantities in the Belmont district, the type locality, where it was the main silver ore mineral.

ESMERALDA COUNTY Identified in a specimen from the Klondyke district. (JHS)

Reported from the Black Warrior Mine in the Silver Peak district. (Anthony et al., 1997; Von Bargen, 1999)

LANDER COUNTY Reported in a specimen from an unknown locality in the Reese River district. (HUB)

LINCOLN COUNTY Reported with chlorargyrite and other secondary minerals in the Pahranagat district. (Raymond, 1869)

NYE COUNTY The principal silver mineral at the type locality on the Combination claim in the Belmont district, Sec. 36, T9N, R45E, as yellowish, black, or brown earthy coatings in association with a variety of oxidized copper and lead minerals. Additionally reported from the Highbridge, Transylvania, El Dorado South, and other properties; also identified by EMP in a pale-yellow veinlet with bindheimite and quartz in a specimen from the Green and Oder Mine. (Riotte, 1867; Hague, 1870; Raymond, 1872; Melhase, 1935d; Palache et al., 1944; Mason and Vitaliano, 1953; Anthony et al., 1997; F. Cureton, o.c., 2000)

Reported from unspecified localities in the Tybo and Union districts. (S. Pullman, w.c., 1999; WMK)

WHITE PINE COUNTY Reported at the Treasure Hill Mines in the White Pine district. (Raymond, 1869)

STEVENSITE

$(Ca/2)_{0.3}Mg_3Si_4O_{10}(OH)_2$

A clay mineral in the *smectite* group that occurs in cavities in basalt and as a hydrothermal alteration product of tuff.

WHITE PINE COUNTY Reported with magnesite,

huntite, and *serpentine* in altered tuffaceous rock in the Currant Creek district. (Hewett et al., 1936; Vitaliano, 1951; Faust, 1953)

STIBARSEN

SbAs

Stibarsen occurs in hydrothermal veins and pegmatites, commonly in intergrowths with arsenic or antimony. The discredited mineral *allemontite* has been found to be a mixture of stibarsen and native arsenic.

STOREY COUNTY Described as "antimonial arsenic" in finely crystalline, radiated, reniform masses at the Ophir Mine in the Comstock district; also reported as *allemontite*. (Hoffman, 1878; Palache et al., 1944)

STIBICONITE

$Sb^{3+}Sb_2^{5+}O_6(OH)$

Stibiconite is thought to be a relatively common mineral in Nevada where it occurs as an oxidation product of stibnite. Many localities for antimony oxides are known, but the species has been defined in only a small percentage.

CHURCHILL COUNTY With the other antimony oxide minerals senarmontite, roméite, tripuhyite, and mopungite at the Green prospect in the Lake district. (Williams, 1985)

CLARK COUNTY In radiating clusters replacing stibnite with tennantite and proustite in a quartz vein at the Lavina prospect in the Goodsprings district. (Hewett, 1931; Lawrence, 1963)

ELKO COUNTY As white pseudomorphs after stibnite crystals to 3 cm and as brown crystals to 2 mm with stibnite at the Murray Mine in the Independence district. (GCF)

ESMERALDA COUNTY With stibnite and cinnabar at the Red Rock Mine in the Fish Lake Valley district. (Lawrence, 1963)

Reported in a specimen from the Rock Hill Mine in the Rock Hill district. (R. Thomssen, w.c., 1999)

EUREKA COUNTY As yellow to yellowish-white, earthy coatings on stibnite crystals and locally as complete replacements at the Goldstrike Mine in the Lynn district; commonly associated with red antimony oxide. (GCF)

HUMBOLDT COUNTY Tentatively identified with valentinite as an oxidation product of stibnite at the W. P. Mine in the Ten Mile district. (Lawrence, 1963)

LANDER COUNTY Reported in a specimen from the Snowstorm Mine in the North Battle Mountain district. (R. Thomssen, w.c., 1999)

MINERAL COUNTY With stibnite in quartz-calcite veins at the Reservation Hill prospect in the Benway district. (Lawrence, 1963)

Reportedly ubiquitous at the Candelaria Mine in the Candelaria district. (Chavez and Purusotam, 1988)

Reported with cinnabar and stibnite at the Drew and Lost Steer Mines in the Pilot Mountains district. (Foshag, 1927; Lawrence, 1963)

With barite, sulfur, conichalcite, and jarosite in a specimen from the Old Santa Fe Mine in the Santa Fe district. (UAMM)

NYE COUNTY With finely disseminated pyrite and thin quartz veinlets at the Anvil prospect in the Jackson district. (Hamilton, 1993)

Reported at the White Caps Mine in the Manhattan district. (Gibbs, 1985)

Reported from an unspecified locality in the Tybo district. (S. Pullman, w.c., 1999)

As earthy, pinkish-white replacements of stibnite laths to 10 cm at the Titus prospect in the Squaw Hills, east of Moores Station. (Jenkins, 1981)

PERSHING COUNTY Reported in a specimen with bindheimite at the West Springs Mine in the Arabia district. (WMK)

WASHOE COUNTY Reported in a specimen from Gerlach. (HMM)

Reported in a specimen from an unknown locality in the McClellan district. (WMK)

WHITE PINE COUNTY As pseudomorphs after stibnite in quartz veinlets at the Vantage deposit on Alligator Ridge in the Bald Mountain district. (Ilchick, 1984)

Tentatively identified as a dull white to pale-yellow, partial to complete replacement of stibnite blades to 5 cm in dark-gray jasperoid at the Merrimac Mine in the Taylor district. (SBC)

As pseudomorphs after stibnite crystals to 2 cm at the Golden Butte Mine in the Butte Valley district. (M. Jensen, w.c., 1999)

After stibnite crystals at the Cherry Creek Mine in the Cherry Creek district. (W. Lombardo, w.c., 2001)

STIBIOTANTALITE

$SbTaO_4$

Stibiotantalite is a rare mineral found in some complex granite pegmatites.

CLARK COUNTY Questionably identified with *samarskite* and *monazite* in pegmatite at the Hilltop Mine in the Gold Butte district. (Longwell et al., 1965)

STIBNITE

Sb_2S_3

Stibnite is a common sulfide mineral that can be formed in a wide variety of hydrothermal environments. In Nevada it is widespread in sedimentary rock–hosted antimony deposits, Carlin-type gold deposits, and epithermal veins. Excellent specimens have come from the White Caps Mine in the Manhattan district and the Murray Mine in the Independence district. For detailed descriptions of many of the Nevada occurrences, see Lawrence (1963).

CHURCHILL COUNTY Reported in a specimen from the east slope of the Clan Alpine Mountains, probably from the Alpine district. (WMK)

In the Bernice district, from which antimony was produced during the two world wars, mainly in quartz veins at several properties including the Antimony King, Drum, IHX, and Lofthouse Mines; reported as crystals to 2 inches at the latter. (Lawrence, 1963)

In veins at the Fairview Mine in the Fairview district. (Schrader, 1947)

Reported as selenium-bearing, in bladed to fibrous form with antimony oxides in angular rubble cemented by Lake Lahontan *tufa,* at the Green prospect in the Lake district; also at the Hazel Mine in a quartz-calcite vein with jamesonite and antimony oxides. (Lincoln, 1923; Lawrence, 1963)

With quartz, scheelite, and possible valentinite along a limestone-granite contact at the Quick-Tung Mine in the Shady Run district. (Lawrence, 1963)

With antimony oxides in quartz veinlets at the Caddy Mine in the Westgate district. (Lawrence, 1963)

Reported in specimens from the Desert and Dixie Valley districts. (WMK)

CLARK COUNTY With antimony oxide at the Lavina, Yellow Pine, and Boss Mines in the Goodsprings district; also as small pods and large single crystals at an antimony prospect 3 miles southwest of Goodsprings. (Hewett, 1931; Jedwab, 1998b; Lawrence, 1963)

DOUGLAS COUNTY With antimony oxides in quartz veinlets at the Danite Mine in the Gardnerville district. (Lawrence, 1963)

With galena, pyrite, and other sulfide minerals at the Winters Mine and Iron Pot prospect in the Red Canyon district. (Lawrence, 1963)

As single crystals to 5 cm with antimony oxides in a quartz, pyrite, and calcite vein at the Fullstone prospect in the Wellington district. (Lawrence, 1963)

ELKO COUNTY At the Bootstrap Mine in the Bootstrap district in gold-bearing quartz veins, with calcite and barite, and as massive fans to 5 cm in siliceous rock; also reported from the Dee Mine. (Lawrence, 1963; R. Thomssen, w.c., 1999)

With arsenopyrite, pyrite, and other sulfides in quartz veins at the Graham, Prunty, and Rescue Mines in the Charleston district. (Lawrence, 1963)

With galena and pyrite at the Hunter prospect in the Coal Mine district. (Lawrence, 1963)

With gold, acanthite, and secondary minerals in a quartz vein near the Red Elephant incline in the Elk Mountain district; also reported at the Valdez prospect. (Schrader, 1912; Lawrence, 1963)

As crystals to 2 cm with other sulfides and antimony oxide in a quartz-*chalcedony* vein in rhyolite at the Good Hope (Buckeye and Ohio) Mine in the Good Hope district. (Lawrence, 1963; J. McGlasson, w.c., 2001)

In the Independence Mountains district at many localities including the Burns Basin, Enfield-Bell, Big Springs, Lost and Found, and Murray Mines, generally in association with barite and antimony oxide. The Murray Mine has produced some of the finest stibnite specimens from Nevada as sharp single or doubly terminated crystals to 35 cm and as radiating clusters with quartz and barite. (LaPointe et el., 1991; GCF)

With antimony oxide and quartz in faults at the Gribble Antimony Mine and Foss prospect in the Island Mountain district. (Lawrence, 1963)

Distal to gold ore at the Hollister deposit in the Ivanhoe district. (Deng, 1991)

As needles to 2 cm on fractures in the Meikle Mine in the Lynn district. (Jensen, 1999)

In gold-quartz veins at the Blue Ribbon–Boyce Mine in the Mountain City district. (Lawrence, 1963)

Reported in a specimen from the Jack Creek Mine in the Railroad district. (WMK)

In veins with pyrite, arsenopyrite, and cinnabar at the Fisher, Falcon, Red Cow, and Rock Creek prospects in the Rock Creek district. (Lawrence, 1963)

Reported in minor occurrences in the Tuscarora district. (Lawrence, 1963)

ESMERALDA COUNTY With stibiconite, cinnabar, quartz, barite, and clay in veinlets in Paleozoic rocks at the Red Rock Mine in the Fish Lake Valley district. (Bailey and Phoenix, 1944)

At the Carrie Mine in the Gilbert district with other sulfides, bindheimite, and scheelite; also at the Mickspot Mine with antimony oxide, pyrite, arsenopyrite, and copper minerals. (Ferguson, 1928; Lawrence, 1963)

Reported as silver-bearing from Columbia in the Goldfield district. (Lawrence, 1963)

Reported in a specimen with cervantite from the Silver Peak district. (WMK)

EUREKA COUNTY With antimony oxide in quartz veins at the Blue Eagle Mine in the Antelope (Baldwin) district. (Lawrence, 1963)

With other sulfide minerals, quartz, and calcite in ore in limestone and veins in granodiorite at the Garrison and other mines along Mill Creek in the Cortez district. (Emmons, 1910)

With pyrite in silicified limestone at the Dugout, Stibnite, and Young prospects in the Eureka district. (Lawrence, 1963; JHS)

In the Lynn district as a minor mineral in gold ore at the Carlin Mine. At the Goldstrike Mine as columnar to sheaflike crystals to 10 cm and as sprays with barite, quartz, and antimony oxides; also reported in a quartz-barite vein at the Morning Glory prospect. (Hausen and Kerr, 1968; GCF; Lawrence, 1963)

With quartz, calcite, barite, and other sulfides in a vein at the Zenoli Mine in the Safford district. (Emmons, 1910; Lawrence, 1963)

Rare with sulfur in calcareous and opaline hot-spring sinter at the Hot Springs Point Sulfur Mine in Crescent Valley. (Garside and Schilling, 1979)

HUMBOLDT COUNTY In a quartz vein on the east

flank of Navajo Peak in the Bottle Creek district. (Bailey and Phoenix, 1944)

A minor mineral in ore at the Lone Tree Mine in the Buffalo Mountain district. (R. Braginton, o.c., 1995)

With antimony oxide and quartz in shears at the Nevada King Mine in the Dyke district. (Lawrence, 1963)

Common in veins in the National district; specifically noted as capillary needles on quartz crystals at the Chefoo claim on Round Hill and reported with other sulfides and silver in quartz veins on Auto Hill. Found at other sites near the site of National as needles in silica veins, and as beautiful crystals from the Indian Valley vein and Radiator Hill. Also in a *chalcedony* vein at the Buckskin National group and other properties near Buckskin Mountain. (Lindgren, 1915; Melhase, 1934c; Lawrence, 1963)

Rarely with cinnabar at the Cordero Mine in the Opalite district. (JHS)

In a specimen from a road cut 10 miles north of Paradise Valley in the Paradise Valley district. (UCB)

With realgar, orpiment, ilsemannite, marcasite, magnetite, gold, and silver as hairlike crystals in restricted pockets in the ore zone at the Getchell Mine in the Potosi district; also in the Twin Creeks mine as an early mineral with inclusions of other sulfides and as crystals to 5 mm with orpiment. (Joralemon, 1951; Lawrence, 1963; Simon et al., 1999; M. Jensen, w.c., 1999)

Rare with cinnabar at the Cahill Mine in the Poverty Peak district. (Bailey and Phoenix, 1944)

With arsenopyrite, pyrite, *chloropal,* and antimony oxide in quartz on the mine dump at the Snowdrift Mine in the Red Butte district. (Melhase, 1934c; Lawrence, 1963)

In the Ten Mile district as pods to 24 by 36 feet with red and yellow antimony oxides in quartz veins and breccia in the W. P. Mine; also with other sulfides in a quartz-calcite vein at the Pansy-Lee Mine. (Lawrence, 1963)

With arsenopyrite and antimony and arsenic oxides in quartz veins at the Juanita group in the Varyville district. (Lawrence, 1963)

LANDER COUNTY In the Battle Mountain district with antimony oxides in quartz veins at the Cottonwood Canyon, Apex, and Mizpah Mines, from which antimony was produced during the

two world wars. Also with other sulfides in gouge and quartz veins at the Weber Mine. (Lawrence, 1963)

With other sulfide minerals, antimony oxides, and copper carbonates in quartz veins and breccia at the productive Antimony King, Bray-Beulah, Dry Canyon, and Hard Luck-Pradier Mines in the Big Creek district. The Bray-Beulah was the third-leading antimony producer in the state. (Lawrence, 1963)

In quartz veins at the Blue Dick, Kattenhorn, and other mines in the Hilltop district. (Lawrence, 1963; Gilluly and Gates, 1965)

With antimony oxides, arsenopyrite, and pyrite in a vein at the Blue Nose Mine in the Lewis district. (Lawrence, 1963; Gilluly and Gates, 1965)

With other sulfides in silver-rich quartz veins in the Reese River district; specifically reported from the Antimony King, San Miguel, and Silver Cliff Mines. (Ross, 1953; Lawrence, 1963; WMK)

Rare with cinnabar and pyrite in fractures at the Wildhorse Mine in the Wild Horse district. (Dane and Ross, 1943; Lawrence, 1963)

LYON COUNTY In blades to 1 inch and rosettes to 2 inches with cinnabar and antimony oxides in silicified volcanic rocks and silica stringers at the DeLongchamps prospect in the Ramsey district. (Lawrence, 1963)

Reported in a specimen from the Kernick Mine in the Silver City district. (WMK)

MINERAL COUNTY In minor amounts with stibiconite, chlorargyrite, pyrite, and malachite in a quartz-calcite vein at the Reservation Hill prospect in the Benway district. (Schrader, 1947; Lawrence, 1963)

With antimony oxide, *limonite* after pyrite, and scheelite in quartz veins at the Hartwick prospect in the Garfield district. (Lawrence, 1963)

With antimony oxides and cinnabar at the Lost Steer and Reward Mines in the Pilot Mountains district. (Foshag, 1927; Bailey and Phoenix, 1944; Lawrence, 1963)

With antimony oxides and malachite at the Lithia Mine in the Poinsettia district. (Lawrence, 1963)

In quartz veins with antimony oxide at the Happy Return Mine in the Eagleville district. (Lawrence, 1963)

With antimony oxide and pyrite in quartz-calcite veins at the Volcanic Peak Mine in the

Santa Fe district; also noted from the Old Santa Fe Mine. (Lawrence, 1963; WMK; R. Thomssen, w.c., 1999)

At several mines and prospects in the Silver Star district, the most important of which was the Lowman Mine, in veins that mainly cut Mesozoic sedimentary and volcanic rocks. (Lawrence, 1963; Ponsler, 1977)

NYE COUNTY With acanthite and pyrargyrite in veins at the Eaton prospect in the Arrowhead district. (Lawrence, 1963)

In bladed aggregates and pods to 35 cm at the Flower Mine in the Belmont district. (Lawrence, 1963)

In quartz-*chalcedony*-barite veins with yellow and white antimony oxides at the King Solomon Mine in the Danville district; locally included in white barite crystals to 2 inches in vugs. (Lawrence, 1963)

In a quartz vein at the Antimony Lode prospect in the Jefferson Canyon district. (Lawrence, 1963)

Reported in several properties in the Jett district, the most important of which, the Last Chance Mine, produced antimony during the two world wars and into the 1950s. Described from that property as pods, blebs, and individual crystals with antimony oxide, other sulfides, and copper carbonates in veins. (Lawrence, 1963; WMK)

Some of Nevada's finest stibnite specimens came from the White Caps Mine in the Manhattan district, commonly as dull gray to metallic crystals to 10 cm, radiating crystal groups, and rarely as acicular crystals; associated with cervantite, orpiment, realgar, cinnabar, and fluorite. Specimens were described by Melhase as superb clusters of radial splendent crystals, sometimes reticulated or perhaps with individual crystals capped with a crystal of nailhead calcite. Also formed as crystals to 1.3 cm in a reducing atmosphere during a mine fire in 1958. Between 1920 and 1958, antimony ore was shipped sporadically. (Ferguson, 1924; Melhase, 1935d; Lawrence, 1963; Gibbs, 1985; WMK)

Reportedly bismuth-bearing with other sulfides in silver-gold ore at the Paradise Peak Mine in the Paradise Peak district. (Thomason, 1986)

Mined during World War II from quartz veins in rhyolite with antimony oxide and pyrite from the Antimonial Mine and the Eaton prospect in the Reveille district; locally found as hairlike tufts with stubby quartz crystals in vugs. (Lawrence, 1963)

Reported in a specimen from an unknown site in the Round Mountain district. (WMK)

With antimony oxides, barite, pyrite, and quartz in brecciated rhyolite at the Blackbird prospect in the Silverton district. (Lawrence, 1963)

Reported at the Murphy and Teichert Mines in the Twin River district. (Hague, 1870)

As masses to 10 cm with antimony oxides in quartz veins at the Page Mine in the Tybo district; also reported in several prospects. (Lawrence, 1963)

Reported from the Nevada Quicksilver Mine and other properties in the Union district. (Kral, 1951; Lawrence, 1963; WMK)

As stringers in Tertiary rhyolite at the O'Toole prospect in the Washington district. (Kral, 1951)

PERSHING COUNTY In many of the antimony and mercury mines and prospects in the Antelope Springs district, generally with cinnabar in veins in Mesozoic metasedimentary rocks; some of the more notable properties were the Antimony Star, Cervantite, Hollywood, Montgomery, and Red Bird Mines. (Bailey and Phoenix, 1944; Lawrence, 1963)

Reported with cervantite in a specimen from an unspecified locality in the Arabia district. (WMK)

With antimony oxides and pyrite in quartz-calcite veins at the Sutherland Antimony Mine in the Black Knob district, Sec. 15, T27N, R33E, the largest producer of antimony in Nevada, mostly during World War I. (Lawrence, 1963)

In masses of hairlike crystals with antimony oxides, other sulfides, and copper carbonates in quartz veins at the Black Warrior Mine in the Buena Vista district; also with galena from the McCrillis prospect. (Lawrence, 1963)

With antimony oxide, pyrite, tetrahedrite, and copper carbonates in a quartz vein at the Antimony Ike Mine in the Goldbanks district. (Lawrence, 1963)

With antimony oxides and copper minerals in quartz veins in or proximal to diabase dikes at the Old Imlay and Star Mines and Motor prospect in the Imlay district. (Lawrence, 1963)

In gold-silver ore with other sulfides, gypsum, and quartz from the Kennedy district. (WMK)

With barite, antimony oxides, and cinnabar in selenium-bearing ore at the Polkinghorne prospect in the Mount Tobin district. (Lawrence, 1963)

With antimony oxides, other sulfides, and copper carbonates in a silver-rich quartz-calcite vein at the Muttlebury Canyon Mine in the Muttlebury district. (Lawrence, 1963)

As lustrous crystal sprays to 3 cm in vugs at the Hecla Rosebud Mine in the Rosebud district. (M. Jensen, w.c., 1999)

With antimony oxides and quartz in shear zones in the Rye Patch district; also as inclusions in quartz crystals at the Bradley and Panther Canyon Mines. (Lawrence, 1963)

As thin needles to 1 inch enclosed in well-formed quartz crystals in milky quartz veins in the San Jacinto district; also reported from the Poker Brown Mine, and with stibiconite pseudomorphs at the Bottomley prospect. (Lawrence, 1963; R. Thomssen, w.c., 1999; F. Cureton, w.c., 1999)

As acicular crystals in vugs in a quartz vein at the Reagan Mine in the Seven Troughs district. (Ransome, 1909b; Lawrence, 1963)

With antimony oxides, cinnabar, native mercury, calomel, and quartz in a calcite vein in limestone at the King George Mine in the Spring Valley district. (Bailey and Phoenix, 1944; WMK)

In the Star district at the Bloody Canyon Mine, Sec. 35, T31N, R34E, the second-largest antimony producer in Nevada, with antimony oxides, pyrite, and chalcopyrite in quartz veins in Triassic rhyolite; also at the De Soto, Pflum, and Sheba Mines in quartz veins. (Lawrence, 1963)

As masses and single crystals to 15 cm with antimony oxides, cinnabar, and quartz along the contact of a diorite dike and limestone at the Fencemaker Mine in the Table Mountain district. (Lawrence, 1963)

With antimony oxide, scheelite, and quartz in a shear zone at the Ore Drag Mine in the Tobin and Sonoma Range district. (Lawrence, 1963)

In quartz veins with antimony oxides at the Green Antimony Mine in the Wild Horse district; also noted in a specimen from the Sutherland Mine. (Lawrence, 1963; R. Thomssen, w.c., 1999)

In quartz veins at the Adriene, Johnson-Heizer, and Rosal Mines in the Willard district. (Lawrence, 1963)

With cervantite as inclusions in quartz crystals from the Trinity Mountains. (WMK)

STOREY COUNTY Identified in ore from the California Mine in the Comstock district. (Vikre, 1989)

WASHOE COUNTY As euhedral crystals to 10 cm and colliform to mammillary forms in a 30 m x 10 m mass at the Hog Ranch Gold Mine in the Cottonwood district; also noted at the Ross prospect and at a number of prospects in the Fox Mountains. (Bussey, 1996; Lawrence, 1963)

At the Donatelli and Sleepy Joe Mines in the McClellan district with pyrite and copper minerals in quartz veins; also at the Choates Mine and Sunset prospect with antimony oxides in quartz veins, and in a quartz crystal from the district. (Lawrence, 1963; WMK)

As masses to 6 inches and as crystals in vugs in a quartz vein and in silicified phyllite at the Angelia prospect in the Nightingale district. (Lawrence, 1963)

At a classic locality in the Steamboat Springs district as crystals to 1 mm with pyrite in stream gravels, as tiny needlelike crystals and clusters in mud and in cavities in sinter, and as scum on the surface of hot spring waters. (Lindgren, 1906; Brannock et al., 1948; White, 1955, 1980; Lawrence, 1963; Wilson and Thomssen, 1985; WMK)

With galena at the Desert King and Bell shafts in the Wedekind district. (Morris, 1903; WMK)

As platy microcrystals to 2 mm with *clinoptilolite* in veins and coatings on boulders in colluvium along West 5th Street in Reno. (R. Monnar, o.c., 2002)

WHITE PINE COUNTY In the Bald Mountain district at the Crown Point, Dees, and Gold King Mines in quartz veins in limestone. Also identified at the Vantage deposit on Alligator Ridge in quartz veins with kaolinite; commonly altered to stibiconite or entirely removed leaving casts in the clay. (Lawrence, 1963; Hill, 1916; Ilchick, 1984)

With antimony oxides, quartz, and calcite in breccia at the Lage and Nevada Antimony Mines in the Butte Valley district; noted in some cavities as needles overgrown with quartz crystals. (Lawrence, 1963)

In an oxidized specimen from the Cherry Creek district. (WMK)

As pods, veinlets, and needles to 2 inches in jasperoid at the Enterprise, Monitor, Merrimac, and Antimony Chief Mines in the Taylor district. (Lawrence, 1963; Williams, 1965; Lovering and Heyl, 1974)

STILBITE

$(Ca_{0.5},Na,K)_9[Al_9Si_{27}O_{72}] \cdot 28H_2O$ (general)

A common zeolite mineral group that generally occurs in low-grade metamorphic rocks, in cavities in mafic volcanic rocks, in hot springs deposits, and as cement in sedimentary rocks. In Nevada it has mostly been found in hydrothermal ore deposits. It has recently been subdivided into two species, stilbite-Ca and stilbite-Na; the species have not been identified for the following occurrences.

EUREKA COUNTY In skarn as fracture coatings and clear crystals to 1 mm, overgrown by calcite and barite, at the Goldstrike Mine in the Lynn district. (GCF)

HUMBOLDT COUNTY Common in silicified ore as clear tabular microcrystals lining vugs, and locally with galkhaite crystals, at the Getchell Mine, south pit and underground, in the Potosi district. (S. Sears, w.c., 1999; M. Jensen, w.c., 1999)

LYON COUNTY In volcanic rocks near the mouth of Desert Creek Canyon about 11 miles south of Smith as drusy crusts and vug fillings of weathered white to pale-pink crystals to 2.5 cm; microcrystals from vugs are clear, brilliant, and unweathered. (M. Jensen, w.c., 1999; M. Rogers, w.c., 1999)

As drusy crusts and vug fillings of white to pale-pink crystals to 1 cm with calcite and *heulandite* in Wilson Canyon a few yards east of Copper Belt Road, north side of the highway, in the Yerington district. (M. Rogers, w.c., 1999; M. Jensen, w.c., 1999; S. Pullman, w.c., 1999)

NYE COUNTY As 5-mm, pink, bladed crystals near Manhattan. (GCF)

STOREY COUNTY As white crystal druses on quartz crystals in fractures in the Flowery lode at the Golden Eagle Mine in the Comstock district; also as white fibrous material with calcite in veins in drill core from a depth of about 540 m near the Wheeler Monument. (R. Walstrom, w.c., 1999; SBC)

WASHOE COUNTY Reported as white crystalline amygdule fillings at the Olinghouse Gold Mine in the Olinghouse district. (Jones, 1998; GCF)

A rare constituent of pegmatite in Incline Village as cream-colored crystals to 5 mm. (Jensen, 1993a)

STOLZITE

$PbWO_4$

A rare secondary mineral related to scheelite that is found in oxidized lead- and tungsten-bearing hydrothermal deposits.

HUMBOLDT COUNTY As colorless to pale-yellow platy microcrystals with pyromorphite, hinsdalite, and corkite at the Silver Coin Mine in the Iron Point district. (Thomssen, 1998; M. Jensen, w.c., 1999; J. McGlasson, w.c., 2001)

STRASHIMIRITE

$Cu_8^{2+}(AsO_4)_4(OH)_4 \cdot 5H_2O$

Strashimirite is a rare mineral that occurs in the oxidation zones of arsenic-bearing copper deposits.

CHURCHILL COUNTY As pale-green, fibrous crystals to 0.5 mm with lavendulan and annabergite at the Nickel Mine in the Table Mountain district. (M. Jensen, w.c., 1999)

NYE COUNTY As pale-green to light-blue crystals to 1 mm with beudantite and bayldonite at the San Rafael Mine in the Lodi district. (S. White, w.c., 2000; B. Runner, o.c., 2000; GCF)

PERSHING COUNTY Relatively common and reportedly the finest specimens in the world at the Majuba Hill Mine in the Antelope district as lustrous, greenish-white and pale-green sheaves, and as coatings to 2 cm of felted needles. (Wise, 1978; Jensen, 1985)

WASHOE COUNTY Locally as spherules and tight clumps to 1.5 mm of pale-green, silky, sword-shaped crystals with olivenite, cornwallite, and chrysocolla in vuggy iron-stained rhyolite at the Burrus Mine in the Pyramid district. (Jensen, 1994)

STRENGITE

$Fe^{3+}(PO_4) \cdot 2H_2O$

Strengite occurs as an alteration product of primary phosphates in pegmatites and iron deposits elsewhere in the world; however, in Nevada it occurs in oxidized hydrothermal deposits.

ESMERALDA COUNTY As pink drusy crystals associated with colorless variscite crystals at an unnamed prospect in the Candelaria district, Sec. 17, T3N, R36E. (R. Walstrom, w.c., 1999)

EUREKA COUNTY As olive-green, botryoidal or radiating crystalline clusters with whitlockite, other phosphates, and iron oxides on fractures in silicified and brecciated zones at the Goldstrike Mine in the Lynn district. (GCF)

As white to gray-green, isolated spherules to 0.5 mm, and as druses, in brecciated jasperoid at the Gold Quarry Mine in the Maggie Creek district. (Jensen et al., 1995)

HUMBOLDT COUNTY As pink balls of prismatic crystals with other phosphate minerals at the Silver Coin Mine in the Iron Point district. (Thomssen, 1998; J. McGlasson, w.c., 2001)

LYON COUNTY Reported from the Yerington district at several prospects on Carson Hill, near Mason Pass. (S. Pullman, w.c., 1999)

NYE COUNTY As an aluminum-bearing variety (barrandite) in blue-black masses with vashegyite, phosphosiderite, and other minerals at the Gun and Gold Metals Mines and the Train prospect in the Manhattan district. (Wherry, 1916; Shannon, 1923; LANH; S. Pullman, w.c., 1999)

As white crusts with pharmacosiderite on jasperoid at the Northumberland Mine in the Northumberland district. (Kokinos and Prenn, 1985)

PERSHING COUNTY As white spherules to 1.5 mm and glassy radiating crystal groups to 2 mm with dufrénite, meurigite, and pharmacosiderite at the Willard Mine in the Willard district. (Jensen and Leising, 2001)

STROMEYERITE

AgCuS

Stromeyerite is thought to occur as both a primary and secondary sulfide in hydrothermal deposits; because of its similarities to chalcocite and acanthite it may go unrecognized in many deposits. Also known as "copper silver glance."

ELKO COUNTY Reported from an unspecified locality in the Gold Circle district. (Von Bargen, 1999)

Reported in silver ore at the Protection Mine in the Mountain City district. (Emmons, 1910)

Reported from the Tuscarora district. (Von Bargen, 1999)

ESMERALDA COUNTY Tentatively identified with copper carbonates and chlorargyrite in partially oxidized ore from an unspecified location in the Dyer district. (Lincoln, 1923)

Tentatively identified in primary ore in the Klondyke district. (Spurr, 1906)

Questionably identified in gold ore in the Windypah district. (Spurr, 1906)

EUREKA COUNTY Reportedly widespread at the Garrison and other mines in the Cortez district in Mill Canyon; also reported at the Cortez silver mines. (Emmons, 1910; Gilluly and Masursky, 1965)

HUMBOLDT COUNTY In sinter with acanthite, gold, tiemannite, and other minerals west of the Lewis pit in the Sulphur district. (Ebert et al., 1996)

NYE COUNTY Reported with chlorargyrite in quartz veins at the Highbridge Mine in the Belmont district. (White, 1871)

With covellite, malachite, acanthite, *electrum,* and uytenbogaardtite in quartz veins at the Lac Bullfrog and Original Bullfrog Mines in the Bullfrog district. (Castor and Sjoberg, 1993)

PERSHING COUNTY Reported from an unspecified locality in the Rochester district. (Von Bargen, 1999)

STOREY COUNTY In fine but small specimens from the Comstock Lode in the Comstock district; specifically reported with jalpaite in ore from the Savage Mine. (Hoffman, 1878; Vikre, 1989)

WHITE PINE COUNTY Reported at the Imperial Mine in the Cherry Creek district. (Schrader, 1931b)

STRONTIANITE

SrCO$_3$

A carbonate mineral that is most commonly found in masses and veins in limestone and marl; less commonly in metal-bearing veins.

CHURCHILL COUNTY As pale-tan, massive, stout, radiating crystals to 5 cm in pods in granite in an outcrop at the mouth of White Cloud Canyon in the White Cloud district. (M. Jensen, w.c., 1999)

ELKO COUNTY Reported in a specimen from an unknown locality in the Tuscarora district. (UCB)

SULFUR

S

Native sulfur is mined from large deposits in salt domes; it is also common in solfataric deposits associated with volcanism. In Nevada it is most common in hot-spring deposits and as a secondary mineral in hydrothermal sulfide deposits.

CHURCHILL COUNTY Reported at the Erway prospect in the Desert district near Fernley Hot Springs. (Bailey and Phoenix, 1944)

In a specimen from the Green prospect in the Lake district. (R. Thomssen, w.c., 1999)

In siliceous sinter at active hot springs and fumaroles at the Senator Fumaroles in Dixie Valley; cinnabar and pyrite are reported as associates. Also reported from 1 mile north of the Dixie Valley Power Plant. (Lawrence, 1971; Garside and Schilling, 1979; R. Thomssen, w.c., 1999)

CLARK COUNTY Reported as amorphous sulfur on limestone in a specimen from near the Morley Silica Plant in Moapa Valley. (WMK)

ELKO COUNTY As small, yellow-green blebs and fracture coatings with realgar and orpiment in carbonaceous sedimentary rock at the Rain Mine in the Carlin district. (D. Heitt, o.c., 1991)

As microcrystals with voltaite in oxidizing ore in workings at the Meikle Mine in the Lynn district. (D. Tretbar, o.c., 2002)

ESMERALDA COUNTY As crystals to 5 cm associated with kalinite, cinnabar, and gypsum in pipes or chimneys in rhyolite at the Alum Mine in the Alum district; also locally as coatings of cracks and crevices in rhyolite. (Spurr, 1906; UCB; M. Jensen, w.c., 1999)

Mined in small amounts from two localities in the Cuprite district, 12 miles south of Goldfield. Reported to occur in a horizontal bed with silica and alunite in rhyolitic glass or tuff; also reported as irregular seams and blebs in altered sedimentary rock and welded tuff. (Ransome, 1909a; Melhase, 1935b; Albers and Stewart, 1972)

As irregular stringers and small blebs in tuffaceous rock at the Reik (Carl Reick) prospect in the Fish Lake Valley district; also reported in mercury ore in siliceous sinter. (Albers and Stewart, 1972; WMK)

As a secondary mineral in silver-bearing veins in the Klondyke district. (Ball, 1907; JHS)

As veinlets in marble at the Silver Mountain prospect in the Tokop district. (Albers and Stewart, 1972)

EUREKA COUNTY As small brilliant crystals locally with häggite on thin marcasite coatings at the Gold Quarry Mine in the Maggie Creek district. (Jensen et al., 1995)

Common in hot-spring sinter with small amounts of cinnabar and stibnite in a fault zone at the Hot Springs Point Sulfur Mine in the Beowawe district; occurs mostly as masses with crystals to 2 mm lining fractures and cavities, but also as free-formed, single, doubly terminated, hoppered crystals to 3 cm. (Olson, 1964c; Garside and Schilling, 1979; GCF)

HUMBOLDT COUNTY Produced from deposits in Sec. 36, T35N, R29E, in the Sulphur district between the 1870s and the 1950s; total production has been estimated at 200,000 tons of ore grading 20–35 percent sulfur. The sulfur was described as crystal masses on the walls of cavities and as disseminations in highly altered siliceous rock with alunite, cinnabar, and gypsum. The deposits have been called sulfur mounds and fumaroles, and are said to be related to surficial acid-sulfate alteration. In recent years, gold has been mined at the nearby Hycroft-Lewis Mine from the same hydrothermal system. (Russell, 1882; Adams, 1904; Ebert et al., 1996)

LYON COUNTY In a specimen from Carson Hill in the Yerington district. (R. Thomssen, w.c., 1999)

MINERAL COUNTY Reported in a specimen from the Vulcan Mine in the Santa Fe district; also reported from the Old Santa Fe Mine. (WMK; R. Thomssen, w.c., 1999)

NYE COUNTY As thin crusts along seams and fractures in white, altered volcanic rock on the east flank of the hill overlooking Bailey's Hot Springs near Beatty in the Bullfrog district. (GCF)

As irregularly distributed lumps in shattered and bleached rocks in the Goldfield district 1 mile east of Tognoni Springs. (Ransome, 1909a)

Reported at the White Caps Mine in the Manhattan district. (Gibbs, 1985)

With barite and cinnabar in vugs in altered tuff, and as well-formed crystals to 4 cm, at the Paradise Peak Mine in the Fairplay district. (John et al., 1991)

Dark-orange to brown sulfur with anglesite and

bismoclite fills cavities in quartz at the Outlaw Mine in the Round Mountain district. (Dunning et al., 1991)

With cinnabar at the Titus prospect 7 miles northeast of Moores Station. (Jenkins, 1981)

PERSHING COUNTY Noted with gypsum in openings near the tops of calcareous *tufa* mounds 1/2 mile south of Humboldt. (Russell, 1885; Lee, 1915)

Reported in a specimen from an unknown locality in the Kennedy district in the East Range. (WMK)

Reported in specimens with, and included in, quartz crystals from gypsum deposits 3 miles east of Lovelock in the Muttlebury district. (WMK)

Reported on the dumps in the Rabbit Hole district. (Melhase, 1934b)

Reported in a specimen from the Willard district. (WMK)

In siliceous sinter at active hot springs and fumaroles at the Kyle Hot Springs. (JHS)

WASHOE COUNTY At the Burrus Mine in the Pyramid district as colorless to pale-yellow, equant crystals to 0.1 mm on enargite in vuggy silicified rock. Also noted with gypsum on the dumps of the Guanomi Mine and the lower adit of the Jones-Kincaid Mine as a product of recent deposition from acid mine water. (Jensen, 1994; SBC)

With gypsum, cinnabar, quartz, and opal in a 2-mile-long altered zone in Pleistocene alluvium on the southeast side of the San Emidio Desert about 20 miles south of Gerlach. (Bonham and Papke, 1969)

As yellow crystal crusts and single crystals with cinnabar in veinlets in silicified rock in and near the Silica pit in the Steamboat Springs district. The locality has been the source of some of the finest sulfur crystals in the United States; crystals to 7 cm have been collected, along with large quantities of attractive, display-quality crystal groups. (Wilson and Thomssen, 1985)

SVANBERGITE

$SrAl_3(PO_4)(SO_4)(OH)_6$

An uncommon mineral that occurs in some contact-metamorphic zones with aluminum-rich minerals such as andalusite, pyrophyllite, and diaspore.

MINERAL COUNTY A minor constituent of aluminum mineral deposits with andalusite, pyrophyllite, and other minerals at the Dover (Donnelly), Green Talc, and Champion deposits in the Fitting district. (Palache et al., 1951; MV; AMNH; HMM)

SYLVANITE

$(Au,Ag)_2Te_4$

Sylvanite is a rare mineral that occurs with other telluride minerals in some precious-metal deposits.

ESMERALDA COUNTY In telluride-rich ore with hessite and petzite at the Mohawk Mine in the Goldfield district. (Tolman and Ambrose, 1934; UCB)

LANDER COUNTY Reported with bismuth minerals at the Red Top Mine in the Hilltop district. (Vanderburg, 1939)

LINCOLN COUNTY Reported with an unidentified gold-bearing telluride mineral at the Charlie Ross Mine in the Eagle Valley district. (Higgins, 1908; Jenkins, 1981)

SYLVITE

KCl

Sylvite is a moderately common mineral in bedded evaporite deposits, as a sublimate in volcanic fumaroles, and as a daughter phase in fluid inclusions.

HUMBOLDT COUNTY With halite, gold, and other minerals in sinter west of the Lewis pit in the Sulphur district. (Ebert et al., 1996)

LANDER COUNTY As daughter crystals in fluid inclusions at the West orebody at Copper Canyon in the Battle Mountain district. (Theodore and Blake, 1978)

WHITE PINE COUNTY As a daughter mineral in fluid inclusions in jasperoid in the Robinson district. (Lovering, 1972)

SYMPLESITE

$Fe_3^{2+}(AsO_4)_2 \cdot 8H_2O$

A relatively uncommon secondary mineral in hydrothermal deposits, generally with scorodite and other arsenic minerals; also found in pegmatite.

EUREKA COUNTY With alunite veins at the Goldstrike Mine in the Lynn district. (Arehart et al., 1992)

HUMBOLDT COUNTY Reported from the Getchell Mine in the Potosi district. (Dunning, 1987)

SYNGENITE

$K_2Ca(SO_4)_2 \bullet H_2O$

An uncommon species that generally occurs in evaporite deposits; also noted as an alteration or vapor-phase mineral in volcanic rocks.

ESMERALDA COUNTY A man-made mineral as white pseudo-octahedral crystals to 2 mm that are deposited in evaporite tanks after halite crystallizes from brines pumped from the Clayton Valley playa in the Silver Peak Marsh district. (LANH; UAMM; M. Jensen, w.c., 1999)

SZAIBELYITE

$MgBO_2(OH)$

Szaibelyite is a rare mineral occurring in some contact zones in limestone with other metasomatic borate minerals.

LINCOLN COUNTY As veinlets and masses of white acicular crystals with *magnesioludwigite* and fluoborite in magnetite-*serpentine*-ludwigite rock 0.5 mile S65W from the summit of Blind Mountain in the Bristol district. (Gillson and Shannon, 1925; HMM)

LYON COUNTY As small, pinkish-white patches with other boron minerals in the Yerington district 0.5 mile east of the Ludwig gypsum pit. (Harris, 1980; S. Pullman, w.c., 1999; GCF)

SZOMOLNOKITE

$Fe^{2+}SO_4 \bullet H_2O$

A water-soluble mineral that forms as an efflorescent phase from the breakdown of pyrite in mine workings.

ELKO COUNTY As whitish crystals to 1 mm with halotrichite at the Meikle Mine in the Lynn district. (Jensen, 1999)

EUREKA COUNTY As light- to dark-blue efflorescent coatings on silicified limestone in areas of high sulfide concentration in the Betze deposit at the Goldstrike Mine in the Lynn district; also forms as a dehydration product of rozenite. (GCF)

TAENITE

(Ni,Fe)

Common nickel-rich mineral in iron meteorites; with kamacite generally forms a crosshatched lamellar texture known as the Widmanstätten pattern.

CHURCHILL COUNTY With kamacite and possible troilite in the Hot Springs meteorite. (Buchwald, 1975; Kring, 1997)

CLARK COUNTY With kamacite, troilite, schreibersite, graphite, and cohenite in the Las Vegas meteorite. (Buchwald, 1975)

NYE COUNTY With kamacite, schreibersite, troilite, and daubréelite in the Quartz Mountain meteorite. (Buchwald, 1975)

With kamacite, schreibersite, and probable troilite in the 1.5-ton, iron-nickel Quinn Canyon meteorite. (Buchwald, 1975)

TALC

$Mg_3Si_4O_{10}(OH)_2$

A mineral that is typically formed by the hydrothermal alteration of mafic and ultramafic igneous rocks or by low-grade thermal metamorphism of dolomite.

CLARK COUNTY Alteration mineral in hornblendite at the Key West Mine in the Bunkerville district. (Beal, 1965)

ELKO COUNTY As small greenish balls to 2 mm on bismuthinite crystals in calcite at the Victoria Mine in the Dolly Varden district. (M. Jensen, w.c., 1999)

ESMERALDA COUNTY In the Palmetto district in soft, white to light-gray, very fine-grained, massive bodies as much as 600 feet long and 75 feet wide with rare to abundant *chlorite* and rare quartz and calcite. Country rock is mainly Precambrian to Cambrian limestone and dolomite. About 27,000 tons of talc ore were produced from the district, mainly from the Shaw, Highway, Nevada No. 1, and Lida Mines. (Papke, 1975)

In the Sylvania district in soft, white to greenish gray, very fine-grained, massive orebodies as much as 600 feet long and 40 feet wide with some *chlorite,* quartz, and calcite, and local *sericite* and clay. *Chlorite* or *sericite* are the predominant minerals in some of the ore. The host rock is mainly Precambrian dolomite along or near contacts with Jurassic intrusive rock, and some orebodies are wholly within intrusive rock. More than 150,000 tons of ore were produced, mainly from the Oasis, Gates, White King Extension, Reed, White Top, Roseamelia, and C and C Mines. (Papke, 1975)

Identified in a sample from an unspecified locality in the Silver Peak district. (JHS)

EUREKA COUNTY Reported in a specimen from the Lanie Bull Mine in the Buckhorn district. (WMK)

LANDER COUNTY Reported in the Reese River district. (Hoffman, 1878)

LYON COUNTY Reported in a specimen from the F. V. Bovard property in the Yerington district. (WMK)

MINERAL COUNTY Reported in a specimen from Sodaville. (WMK)

NYE COUNTY Reported in a specimen of *sericite* schist in the Bellehelen district. (WMK)

Reported in a specimen from near Rhyolite in the Bullfrog district. (AMNH)

As elongate crystals to 2 mm and masses to 2 cm in magnesite and brucite bodies at the Gabbs Magnesite–Brucite Mine in the Gabbs district; thought to be an alteration product of tremolite. (Schilling, 1968a)

Reported to occur with clinochlore in an

altered zone near the border of a quartz-*feldspar* porphyry body at the Huntley Mine in the Lodi district. (Kral, 1951; Kleinhampl and Ziony, 1984)

Reported in a specimen from Tonopah. (WMK)

PERSHING COUNTY Reported in a specimen from the Mercury Mine in the Buena Vista district. (WMK)

TANGEITE

CaCuVO$_4$(OH)

Tangeite, formerly referred to as *calciovolborthite* by some mineralogists, forms a series with coni-chalcite and occurs as a secondary mineral in deposits that contain vanadium.

LYON COUNTY As yellow–green, platy crystal clusters to 0.3 mm with tyuyamunite in a specimen from the Flyboy Mine in the Flyboy district; identified by XRD. (R. Thomssen, w.c., 1999)

Reported at the Douglas Hill Mine in the Yerington district. (S. Pullman, w.c., 1999)

NYE COUNTY Reported as "calci-volborthite" in small yellowish scales with secondary copper minerals in contact-metamorphosed slate adjacent to granite at a locality south of the Manhattan district. (Ferguson, 1924)

PERSHING COUNTY As yellow films on fractures in bull quartz from an unnamed prospect pit on the north side of Sulfur Road at the southwest end of the Eugene Mountains in the Mill City district. (S. Kleine, o.c., 1999)

TANTALITE

(Fe,Mn)Ta$_2$O$_6$ (general)

Tantalite is a group mineral name that includes ferrotantalite and manganotantalite, both minerals that occur in granite pegmatite. Material identified as *columbite-tantalite*, a mix of two mineral series, at Nevada occurrences, further confuses the issue.

CLARK COUNTY Reported from the western slopes of the McCullough Range. (Volborth, 1962a)

Reported as *columbite-tantalite* in pegmatite and miarolitic cavities in the Eldorado and Newberry Mountains. (S. Scott, w.c., 1999)

ELKO COUNTY Reported as *columbite-tantalite* in pegmatites in the Gilbert and Dawley Canyon areas of the Valley View district in the southern Ruby Mountains. (Olson and Hinrichs, 1960)

TELLURITE

TeO$_2$

Tellurite, a rare mineral that occurs as an oxidation product of primary tellurides, has only been reported from a few localities in the western United States.

ESMERALDA COUNTY In the Goldfield district as a crystalline, honey-yellow mineral with sub-adamantine luster with gold telluride in the Goldfield-Belmont Mine; also reported at the Clermont (Claremont) and Mohawk Mines. (Ransome, 1909a; Palache et al., 1944; Anthony et al., 1997; EMP; HMM)

NYE COUNTY Reported at an unspecified locality in Jefferson Canyon in the Toquima Range, east of Round Mountain; this may be the same locality cited under hessite in the Jefferson Canyon district. (Palache et al., 1944)

TELLURIUM

Te

A rare native metal that occurs in precious-metal telluride deposits. It is generally found with other tellurium-bearing minerals, but never in mutual contact with native gold.

CLARK COUNTY Reported in a specimen from an unknown locality. (CSM)

ESMERALDA COUNTY In a quartz vein with fama-tinite, goldfieldite, and other tellurium-bearing minerals at the Mohawk and Clermont Mines in the Goldfield district. (Palache et al., 1944; S. Pullman, w.c. 1999)

HUMBOLDT COUNTY Reported as a microgranular native metal associated with gold-bearing arsenian pyrite at the Twin Creeks Mine in the Potosi district. (Simon et al., 1999)

LANDER COUNTY Reported from the Reward prospects in the Ravenswood district. (J. McGlasson, w.c., 2001)

LINCOLN COUNTY Reported in rich gold ore in the Delamar district. (Palache et al., 1944; Davidson, 1964)

TELLUROBISMUTHITE

Bi_2Te_3

Typically, a mineral of hydrothermal quartz veins with low sulfur content; however, the only Nevada occurrence is in a Carlin-type gold deposit.

ELKO COUNTY Reported at the Meikle Mine in the Lynn district. (Volk et al., 1996)

TENNANTITE

$(Cu,Ag,Fe,Zn)_{12}As_4S_{13}$

Tennantite, the arsenic-rich member of the tetrahedrite-tennantite series, is scarcer than its antimony analog; however, it occurs in a large number of hydrothermal and contact-metamorphic base- and precious-metal deposits.

CLARK COUNTY As sparse grains in quartz with proustite and pyrite at the Lavina prospect in the Goodsprings district. (Hewett, 1931)

ESMERALDA COUNTY With famatinite, bismuthinite, and a variety of other minerals in ore from the Goldfield district; specifically reported at Goldfield Deep Mines. (Tolman and Ambrose, 1934; WMK)

EUREKA COUNTY Locally after sphalerite in barite veins at the Carlin Mine in the Lynn district, possibly zinc-bearing; also tentatively identified in sulfide and sulfosalt remnants in oxidized ore. (Hausen and Kerr, 1968)

HUMBOLDT COUNTY As a minor mineral in unoxidized rock at the Lone Tree Mine in the Buffalo Mountain district. (R. Braginton, o.c., 1995)

LINCOLN COUNTY With questionable dufrénoysite in siliceous lead ore at the Prince Mine in the Pioche district. (Westgate and Knopf, 1932; Melhase, 1934d)

NYE COUNTY Identified in a sample from an unspecified locality in the Union district. (JHS)

WASHOE COUNTY Questionably identified as an associate of galena and pyrite at the Mountain View Mine in the Deephole district. (Bonham and Papke, 1969)

Intergrown with enargite and pyrite at the Blue Bird Mine in the Pyramid district. (SBC)

WHITE PINE COUNTY Well crystallized with enargite, bournonite, and andorite in quartz vugs at the Monitor Mine in the Taylor district. (Williams, 1965)

TENORITE

$Cu^{2+}O$

Tenorite, also known as *melaconite,* is widespread in small quantities with cuprite, copper, chrysocolla, and copper carbonate minerals in the oxidation zones of many hydrothermal deposits; it has also been identified as a volcanic sublimate.

CHURCHILL COUNTY Noted in veins and wall rock at the Fairview Mine in the Fairview district. (Warren, 1973)

CLARK COUNTY As earthy-black grains with *limonite* and as fine grains in malachite at the Yellow Pine Mine in the Goodsprings district; also identified with other copper minerals, cinnabar, potarite, and gold at the Boss Mine. (Hewett, 1931; Jedwab, 1998a)

ELKO COUNTY With cuprite and malachite in oxidized ore from the Rattler Mine in the Contact district. (Schrader, 1935b)

With chrysocolla in a copper-rich mineralized zone near the surface at the Killie (Latham) Mine in the Spruce Mountain district. (Schrader, 1931b; Dunning and Cooper, 1987)

ESMERALDA COUNTY With cuprite, chalcocite, and copper carbonates in the Cuprite district. (Ball, 1907)

Reported at some prospects in the Lone Mountain district with hemimorphite, smithsonite, cerussite, and chrysocolla. (Phariss, 1974)

EUREKA COUNTY Reported in specimens from the Eureka district. (HMM; S. Pullman, w.c., 1999)

LANDER COUNTY A pitchlike mineral in some copper deposits in the Battle Mountain district. (Roberts and Arnold, 1965)

LINCOLN COUNTY With malachite, atacamite, cuprite, and chrysocolla at the Bristol, May Day, and Monarch Mines in the Bristol district. (Tschanz and Pampeyan, 1970; CSM; S. Pullman, w.c., 1999; WMK)

As stains, coatings, small pods, and lenses with chalcocite, malachite, and chrysocolla at the Arrowhead (Southeastern) Mine in the Southeastern district on the Nellis Air Force Range. (Tingley et al., 1998a)

LYON COUNTY Reported in small quantities in oxidized ore in the Wilson district. (Dircksen, 1975)

Noted as black scales on cuprite at the Empire-

Nevada, MacArthur, Yerington pit, and other mines in the Yerington district. (Ransome, 1909e; D. Hudson, w.c., 1999)

MINERAL COUNTY In specimens of copper-silver ore from the Silver State and Lakeview lodes in the Garfield district; also reported from the Bataan and Blue Light Mines. (WMK; Ponsler, 1977)

With azurite, malachite, chrysocolla, and other minerals at the Pine Tree Mine in the Pilot Mountains district. (Ross, 1961)

Reported at the Nevada Rand Mine in the Rand district. (Schrader, 1947)

NYE COUNTY In lenses and veinlets with malachite, azurite, and chrysocolla in the Wagner district. (Tingley et al., 1998a)

PERSHING COUNTY In black coatings with clinoclase in the Copper stope of the Majuba Hill Mine in the Antelope district. (Jensen, 1993b)

As an ore mineral with cuprite in the oxidized zone of a massive sulfide deposit at the Big Mike Copper Mine in the Tobin and Sonoma Range district. (Johnson, 1977)

WASHOE COUNTY Questionably identified with copper, malachite, and cuprite in thin veins of laumontite that cut zeolitic andesitic rocks of the Alta Formation north of Interstate 80 in the Truckee River Canyon in the Olinghouse district. (Rose, 1969)

With malachite and azurite in a quartz vein in Mesozoic volcanic rock at the Antelope Mine in the Stateline Peak district. (Bonham and Papke, 1969)

WHITE PINE COUNTY In small amounts in the supergene assemblage at the Monitor Mine in the Taylor district. (Lovering and Heyl, 1974)

TERLINGUAITE

Hg_2ClO

A rare mineral that forms as an oxidation product of cinnabar or other primary mercury minerals in hydrothermal deposits.

HUMBOLDT COUNTY With eglestonite, cinnabar, and mercury in ore in tuffaceous sedimentary rocks at the McDermitt Mercury Mine in the Opalite district. (Jenkins, 1981; F. Cureton, w.c., 1999)

With cinnabar, other mercury minerals, stibnite, calcite, quartz, and gypsum in rich mercury ore along faults in sandstone and limestone at the Cahill Mine in the Poverty Peak district. (WMK; EMP; F. Cureton, w.c., 1999)

TETRADYMITE

Bi_2Te_2S

Tetradymite occurs in small amounts at numerous localities worldwide in hydrothermal gold-quartz veins and contact-metamorphic deposits.

DOUGLAS COUNTY Reported with free gold at the Gold Mint Mine in the Wellington district. (Overton, 1947; Mount, 1964)

EUREKA COUNTY Identified by EMP as blebs to rectangular grains to 40 microns along galena-boulangerite contacts in base-metal veins in diorite at the Goldstrike Mine in the Lynn district. (GCF)

LANDER COUNTY As disseminated specks in gold ore at several prospects southwest of Antler Peak (in the Telluride area) at the head of Rocky Canyon in the Battle Mountain district. (Hill, 1915; Roberts and Arnold, 1965)

Reported with montanite near the Betty O'Neal Mine in the Lewis district. (J. McGlasson, w.c., 2001)

TETRAHEDRITE

$(Cu,Fe,Ag,Zn)_{12}Sb_4S_{13}$

Tetrahedrite, a common and economically important sulfosalt mineral in many base- and precious-metal deposits, is the antimony-rich member of a series with tennantite. In Nevada, the Taylor and Tonopah districts have produced some well-crystallized specimens.

CHURCHILL COUNTY At the Lofthouse Mine in the Bernice district with stibnite, jamesonite, and antimony oxides in quartz veins; also at the Hoyt Mine with stibnite in quartz. (Lawrence, 1963)

In quartz veins with acanthite, pyrargyrite, and other sulfides at the Fairview Mine in the Fairview district. (Greenan, 1914; Schrader, 1947)

With galena, sphalerite, chalcopyrite, acanthite, pyrargyrite, and polybasite at the Pyramid Mine in the Holy Cross district. (Peters et al., 1996)

A minor mineral with erythrite and nickel minerals in Ni-Co-Cu ore at the Lovelock Mine in the Table Mountain district. (Ransome, 1909b; Ferguson, 1939)

CLARK COUNTY With pyrargyrite in ore at the Double Standard Mine in the Crescent district. (Hewett, 1956)

With antimony oxides at the New Deal (Polyanna) Mine in the Newberry district. (Lawrence, 1963)

ELKO COUNTY In minor amounts with base-metal sulfides and iron and copper oxide minerals in quartz veins in limestone at the Blue Jacket and Humboldt Mines in the Aura district. (Emmons, 1910; LaPointe et al., 1991)

Reported at the Graham and Rescue Mines in the Charleston district. (Lawrence, 1963)

In gold-bearing quartz veins with pyrargyrite, acanthite, and other minerals in the Cornucopia district near the Idaho state line. (Emmons, 1910; Lawrence, 1963)

With pyrargyrite in silver ore at the Delano (Delno) Mine in the Delano district. (Lawrence, 1963)

With stibnite at the Buckeye & Ohio Mine in the Good Hope district. (Emmons, 1910)

In minor amounts at the Big Springs Mine in the Independence Mountains district. (Youngerman, 1992)

In quartz veins at the Golden Ray Mine and Jim claims in the Loray district. (LaPointe et al., 1991)

As microcrystals with pyrite at the Meikle Mine in the Lynn district. (Jensen, 1999)

Sparsely disseminated with other copper minerals in quartz veins at a copper prospect in the Moor district about 8 miles east-southeast of Wells. (Erickson et al., 1966; LaPointe et al., 1991)

With pyrite, chalcopyrite, and galena in narrow quartz veins in granodiorite at the Resurrection Mine in the Mountain City district. (Emmons, 1910)

As small blebs in quartz veins at the Keystone Mine in the Proctor district. (LaPointe et al., 1991)

With quartz, calcite, and iron and manganese oxides in replacement orebodies in limestone at the Nora Mine in the Tecoma district. (LaPointe et al., 1991)

As euhedral crystals to 3 mm with siderite and quartz at the Grand Prize Mine in the Tuscarora district. (M. Jensen, w.c., 1999)

ESMERALDA COUNTY With galena, sphalerite, and other minerals in quartz veins a mile north of the site of Columbus in the Candelaria district. (R. Walstrom, w.c., 2000)

Questionably identified in primary ore in the Divide district. (Bonham and Garside, 1979)

Reported with sphalerite and chalcopyrite in unoxidized ore at the Combination, Mohawk, and Jumbo Extension Mines in the Goldfield district. (Collins, 1907a; JHS; S. Pullman, w.c., 1999)

Reported at some prospects southwest of Tonopah in the Lone Mountain district. (Phariss, 1974)

In a specimen from the Broken Toe Mine in the Rock Hill district. (R. Thomssen, w.c., 1999)

EUREKA COUNTY In bonanza silver ore in the Mill Canyon area of the Cortez district. (Emmons, 1910)

A relict mineral in oxidized silver-lead deposits in the Diamond district 16 miles north of Eureka. (Roberts et al., 1967b)

As small grains with pyrite at the Stibnite prospect in the Eureka district 7 miles south of Eureka. (Nolan, 1962; Lawrence, 1963)

A microgranular mineral in gold ore at the Goldstrike Mine in the Lynn district. (GCF)

As microcrystals in ore samples at the Gold Quarry Mine in the Maggie Creek district. (Jensen et al., 1995)

Reported from the east side of the Sulphur Springs Range in the Mineral Hill district. (Emmons, 1910)

Reported in a specimen with silver at the Morning Glory (Zenoli) Mine in the Safford district. (WMK)

HUMBOLDT COUNTY In gold-silver veins with *electrum* and other minerals at the Sleeper Mine in the Awakening district. (Lawrence, 1963; Wood, 1988; Saunders, 1994)

With azurite in quartz at the Low copper prospect in the Bottle Creek district. (Willden, 1964)

In high-grade sulfide veins at the Lone Tree Mine in the Buffalo Mountain district. (Bloomstein et al., 1993)

Reported in quartz veins at the Chefoo claim in the National district; also reported with *electrum,* naumannite, and other sulfides in the Birthday and National veins. (Lindgren, 1915; Vikre, 1985)

With galena, stibnite, and an unidentified lead sulfosalt in early quartz veinlets at the Twin Creeks Mine in the Potosi district. (Simon et al., 1999)

In a specimen with malachite, azurite, and galena from the Prodigal Mine in the Red Butte district. (WMK)

With other sulfides and antimony oxide in gold- and silver-bearing quartz-calcite veins at the Pansy-Lee Mine in the Ten Mile district. (Lawrence, 1963)

LANDER COUNTY Reported from a number of mines in the Battle Mountain district. (Roberts and Arnold, 1965)

With sphalerite, stibnite, malachite, azurite, and pyrite in quartz veins at the Hard Luck-Pradier and other mines in the Big Creek district. (Lawrence, 1963)

As silver-colored euhedral crystals to 2 cm with bournonite and arsenopyrite at the Little Gem Mine in the Bullion district. (S. Pullman, w.c., 1999; M. Jensen, w.c., 1999)

Reported at the Garrison Mine in the Cortez district. (S. Pullman, w.c., 1999)

Reported as silver-bearing at the Red Top and other mines in the Hilltop district. (Vanderburg, 1939; Gilluly and Gates, 1965; WMK)

With other sulfides in quartz-calcite veins at the Betty O'Neal and other mines in the Lewis district. (Vanderburg, 1939; Gilluly and Gates, 1965)

An important mineral in bonanza silver ore in the Reese River district; specifically reported at the Austin, Allsop Ledge, Plymouth, and Diana Mines. (Ross, 1925; Melhase, 1934c; Ross, 1953; UCB; BMC)

LINCOLN COUNTY With chalcocite and secondary minerals at the Jackrabbit Mine in the Bristol district. (Melhase, 1934d)

Disseminated in gold ore at the Delamar and Jumbo Mines in the Delamar district. (Callaghan, 1937)

Identified in lead-silver ore at the Groom and other mines in the Groom district on the Nellis Air Force Range. (Humphrey, 1945; Tingley et al., 1998a)

LYON COUNTY Reported in silver veins with pyrargyrite and sphalerite at the Como Consolidated Mines in the Como district. (Moore, 1969; WMK)

MINERAL COUNTY Reportedly silver-bearing with gold, pyrite, chalcopyrite, and possible gold and silver selenides in quartz-*adularia* veins at the Wide West and other mines in the Aurora district. (Hill, 1915; Lindgren, 1915; WMK)

With other sulfides in ore at the Candelaria Mine in the Candelaria district. (Chavez and Purusotam, 1988)

With chalcopyrite, galena, azurite, and malachite in quartz-barite-calcite veins at the Lucky Boy Mine in the Lucky Boy district. (Hill, 1915; Lawrence, 1963)

Reported at the Rawhide Mine in the Rawhide district. (Black, 1991)

With cerussite and galena in the Nogal vein at the Todd Mine in the Santa Fe district. (Ross, 1961)

Reported at the Endowment Mine in the Silver Star district. (Hill, 1915)

NYE COUNTY Reported with other sulfides in veins at the Antelope View Mine in the Antelope Springs district on the Nellis Air Force Range. (Tingley et al., 1998a)

With other sulfides and fluorite in vuggy quartz veins at the Perkins prospects in the Barcelona district about 8 miles northwest of Belmont. (Ervine, 1972)

With galena, azurite, malachite, stetefeldtite, and other minerals at the Transylvania Mine in the Belmont district. (Jenkins, 1981)

With *electrum* in gold ore at the Bullfrog Mine in the Bullfrog district. (Eng et al., 1996)

Reported as silver-bearing in a quartz vein at the Esta Buena Mine in the Ellsworth district; also with silver minerals and gold-bearing pyrite in quartz stringers in Tertiary rhyolite at the Return Mine. (Kral, 1951)

With galena and sphalerite in a quartz vein on a cliff south of Jett Canyon in the Jett district; also reported at the Dollar and Last Chance Mines. (Kleinhampl and Ziony, 1984; Lawrence, 1963)

In quartz-rhodochrosite veins at various locations in the Morey district. (Williams, 1968)

In hübnerite veins in the Round Mountain district. (Ferguson, 1922)

Noted as thick, tabular plates with wire silver in quartz-*feldspar* gangue at the Belmont Mine in the Tonopah district. (Melhase, 1935c)

With other sulfides and native silver at the Murphy and Teichert Mines in the Twin River district. (Hague, 1870)

With polybasite at the Rattlesnake Mine in the Tybo district. (Kral, 1951)

Reported as argentiferous *fahlerz* with galena, chalcopyrite, and pyrite in a quartz vein at the Richmond Mine in the Union district. (Kral, 1951; WMK; Hamilton, 1993)

With galena and acanthite(?) in quartzite breccia at the Werner Mine in the Washington district. (Hill, 1915; Kleinhampl and Ziony, 1984)

As microscopic grains with luzonite and other sulfides at the Golden Chariot Mine in the Jamestown district on the Nellis Air Force Range. (Tingley et al., 1998a)

PERSHING COUNTY In a silver-rich quartz vein with other sulfides at the Antelope Mine in the Antelope district. (Lawrence, 1963)

In a quartz vein with stibnite, malachite, and azurite at the Black Warrior Mine in the Buena Vista district. (Lawrence, 1963)

With stibnite and pyrite in a quartz vein at the Antimony Ike Mine in the Goldbanks district. (Lawrence, 1963)

In quartz veins with stibnite and scheelite at the Betty La Verne and Star Mines in the Imlay district. (Lawrence, 1963; WMK)

A constituent of silver ore from quartz-pyrite veins in the Kennedy district. (Johnson, 1977)

Reported at the Muttlebury Mine in the Muttlebury district. (Johnson, 1977)

Silver-bearing with jamesonite, bindheimite, sphalerite, pyrite, chalcopyrite, tetrahedrite, and *tourmaline* in quartz veins in the Rochester district. (Knopf, 1924)

With pyrite and scheelite in quartz-calcite veins in the Rye Patch district. (Ransome, 1909b)

In quartz-calcite veins with pyrite and scheelite at the Humboldt Queen Mine in the Sacramento district. (Ransome, 1909b)

Rarely with pyrite, galena, and sphalerite in gold-bearing quartz-calcite veins in the Sierra district. (Ferguson et al., 1951)

In primary silver-gold ore at the Bonanza King Mine in the Spring Valley district. (Campbell, 1939)

With owyheeite, galena, sphalerite, stibnite, pyrite, and other minerals in quartz veins in the Star district. (Cameron, 1939; Lawrence, 1963)

Reported at a copper prospect in Coal Canyon in the Willard district. (Johnson, 1977)

STOREY COUNTY With acanthite at the Califor-nia, Savage and Hall, and Norcross Mines in the Comstock district; also reported as silver-bearing with quartz in the Flowery lode. (Vikre, 1989; UCB; Schilling, 1990)

Reported as silver-bearing with pyrite at the Gooseberry Mine in the Ramsey district. (Rose, 1969)

WASHOE COUNTY In quartz-pyrite veins with jamesonite and sphalerite at the Modoc Mine and the Silver Fox prospect in the Cottonwood district. (Bonham and Papke, 1969)

With pyrite, galena, azurite, and malachite in quartz veins in altered *hornblende* gabbro in the Hungry Valley area in the McClellan district. (Bonham and Papke, 1969)

With pyrite, galena, and sphalerite in quartz veins at prospects on Black Warrior Peak in the Nightingale district. (Bonham and Papke, 1969)

With pyrite, chalcopyrite, galena, and sphalerite in copper deposits near Poeville in the Peavine district; specifically noted at the Gray Copper Standard Mine. (Hudson, 1977; WMK)

As dull-black crystals with pyrite in vugs in altered rock at the Burrus, Blue Bird, Jumbo, and other mines in the Pyramid district; also as micron-sized inclusions in pyrite. (Wallace, 1980; Jensen, 1994; SBC)

With pyrite, chalcopyrite, galena, and sphalerite in the Wedekind district. (Hudson, 1977)

WHITE PINE COUNTY With stibnite at the Crown Point Mine in the Bald Mountain district. (Hill, 1916)

With sphalerite and proustite in the Cherry Creek district. (Hill, 1916; Melhase, 1934d)

With quartz in silver-copper-gold ore at the Chihuahua and Bay State Mines in the Newark district. (Hose et al., 1976; WMK)

Sparse at the Minerva Mine in the Shoshone district. (Hose et al., 1976)

As crudely crystallized material with miar-gyrite, andorite, and pyrostilpnite at the Patriot Mine in the Taylor district. (Williams, 1965)

Questionably identified with miargyrite and andorite at the Paymaster Mine and other local-ities in the Ward district. (Hill, 1916; Jenkins, 1981)

Silver-bearing tetrahedrite was reported to be the ore mineral in exceptionally rich silver ore at the Caroline claims in the Mount Hamilton area of the White Pine district. (Humphrey, 1960)

THAUMASITE

$Ca_6Si_2(CO_3)_2(SO_4)_2(OH)_{12} \cdot 24H_2O$

A late-stage mineral in some sulfide ore deposits; also occurs in contact-metamorphic rocks and as an alteration mineral in volcanic rocks.

WHITE PINE COUNTY Reported as clear crystals to 2.5 mm in a specimen from the Liberty pit in the Robinson district. (R. Thomssen, w.c., 1999)

THENARDITE

Na_2SO_4

Thenardite is a soluble salt that is found in playa deposits and in some bedded evaporites.

CHURCHILL COUNTY Reported in a specimen from near Ragtown in the Soda Lake district. (AMNH)

CLARK COUNTY Reported in a specimen from the Saint Thomas district; probably from a deposit of sodium sulfate with glauberite and halite that is now under the waters of Lake Mead on the west side of the Virgin River about 5 miles south of Saint Thomas. (WMK; Vanderburg, 1937b)

With mirabilite and gypsum in an evaporite sequence at least 10 m thick in the Horse Spring Formation about 5 km west-northwest of Bitter Spring in the Muddy Mountains district; glauberite was also reported from this area, but not identified. (SBC)

LYON COUNTY With other sodium minerals in playa deposits in the Wabuska Marsh district. (Melhase, 1935e; Gianella, 1941)

MINERAL COUNTY As orange-brown crystals to 9 cm long and 5 cm in diameter in lenses to 150 cm thick beneath salt and silt in Rhodes Salt Marsh in the Rhodes Marsh district; about 20,000 tons of sodium sulfate was produced from this locality. (Melhase, 1935e; Heins, 1937; Vanderburg, 1937a)

Reported in a specimen from the Teels Marsh district. (AMNH)

Reported from north of Schurz in the Double Springs Marsh district. (Gianella, 1941)

NYE COUNTY With gaylussite and other minerals in salt deposits in Railroad Valley. (Gianella, 1941)

PERSHING COUNTY In an unusual occurrence as a white efflorescent phase on mine walls at the Majuba Hill Mine in the Antelope district. (Jensen, 1993b)

Reported from Buena Vista Valley on the east side of the Humboldt Range. (Gianella, 1941)

As a precipitate from steam in the main sinter terrace in the Steamboat Springs district. (D. Hudson, w.c., 1999)

WASHOE COUNTY Reported from Buffalo Salt Marsh. (Gianella, 1941)

THERMONATRITE

$Na_2CO_3 \cdot H_2O$

Thermonatrite is an uncommon, highly water-soluble mineral that occurs in playa deposits and as an efflorescence on soil and in mine workings in dry regions; it has also been reported around some volcanoes.

CHURCHILL COUNTY Thermonatrite was the final product of cold-weather preparation of sodium carbonate from brine at Soda Lake and Little Soda Lake in the Soda Lake district. (Papke, 1976)

THOMSENOLITE

$NaCaAlF_6 \cdot H_2O$

Thomsenolite occurs as an alteration product of cryolite and other alkali aluminum fluorides, most commonly in granitic pegmatites.

MINERAL COUNTY As subhedral to anhedral grains to 100 microns intergrown with fluorite as a late-stage mineral with other alkali aluminum fluoride minerals in *amazonite*-topaz pegmatite at the Zapot claims in the Fitting district about 25 km northeast of Hawthorne. (Foord et al., 1999b)

THOMSONITE

$Ca_2Na[Al_5Si_5O_{20}] \cdot 6H_2O$

A zeolite-group mineral that typically occurs in cavities in mafic volcanic rocks; also found in alkalic igneous rocks and contact-metamorphic zones.

STOREY COUNTY Reported as implanted globules with quartz in a specimen from the Comstock Lode in the Comstock district. (UCB)

THORIANITE

ThO_2

A rare accessory radioactive mineral that generally occurs in granite pegmatite, carbonatite, or *serpentine*.

EUREKA COUNTY Identified as minute grains in a polished section from the Goldstrike Mine in the Lynn district. (Berry, 1992)

THORITE

$(Th,U)SiO_4$

Thorite occurs as a minor mineral in some pegmatites and as a trace accessory in some igneous rocks. It has also been noted in placer deposits. *Uranothorite* is a variety that contains uranium.

ESMERALDA COUNTY Questionably identified as *uranothorite* with *monazite,* xenotime, and *columbite* at the Tule Canyon placers in the Tokop district. (Garside, 1973)

HUMBOLDT COUNTY Identified by SEM/EDX as a uranium-bearing trace accessory in topaz rhyolite from the top of Buff Peak in the Bottle Creek district. (Castor et al., 1996; Castor and Henry, 2000)

MINERAL COUNTY Reported at the Holiday Uranium Mine in the Fitting district with its dimorph huttonite and with the variety *uranothorite* in segregations in altered granitic rock; also reported with *scapolite* at the nearby Elna and Blue Ox prospects. (Ross, 1961; Garside, 1973; Foord et al., 1999b)

Reddish-brown to brown crystals to several mm in topaz-bearing pegmatite at the Zapot claims in the Fitting district. (Foord et al., 1999b)

PERSHING COUNTY As rare microcrystals with zircon, cassiterite, arsenopyrite, and chalcopyrite from the Tin stope of the Majuba Hill Mine in the Antelope district. (Jensen, 1993b)

THOROGUMMITE

$Th(SiO_4)_{1-x}(OH)_{4x}$

Thorogummite is a rare mineral that occurs as an oxidation product of primary thorium minerals, generally in granite pegmatite and alkaline rocks.

MINERAL COUNTY With huttonite, zircon, and thorite in altered granite at the Holiday Uranium Mine in the Fitting district. (Jenkins, 1981)

TIEMANNITE

$HgSe$

An uncommon mineral in hydrothermal veins, generally occurring in association with other selenide minerals.

EUREKA COUNTY A microgranular mineral in ore at the Goldstrike Mine in the Lynn district. (GCF)

As black, spongy crystalline masses intergrown with corderoite in association with fluellite and other phosphates on fractures in brecciated jasperoid at the Gold Quarry Mine in the Maggie Creek district. (Jensen et al., 1995)

HUMBOLDT COUNTY Reported with gold, acanthite, and other minerals in sinter west of the Lewis pit in the Sulphur district. (Ebert et al., 1996)

MINERAL COUNTY With gold in the Freedom Flat deposit at the Borealis Mine in the Borealis district. (Eng, 1991)

TIN

Sn

Native tin is rare and generally found in placer deposits, but some hard rock occurrences have also been reported.

HUMBOLDT COUNTY Reported with gold and other minerals in sinter west of the Lewis pit in the Sulphur district; may refer to analytical presence rather than to native metal. (Ebert et al., 1996)

TINCALCONITE

$Na_2B_4O_5(OH)_4 \cdot 3H_2O$

Tincalconite is found as a rare dehydration product of borax in evaporite deposits.

ESMERALDA COUNTY Reported with ulexite and borax in the Fish Lake Marsh district. (Gianella, 1941)

TINTICITE

$Fe_4^{3+}(PO_4)_3(OH)_3 \cdot 5H_2O$

A very rare inadequately described mineral that was discovered in specimens from a cave in Utah.

EUREKA COUNTY As clusters to 0.2 mm of tan to brown crystals lining fractures with fluellite, leucophosphite, and variscite in brecciated jasper-oid at the Gold Quarry Mine in the Maggie Creek district. (Jensen et al., 1995)

TINZENITE

$(Ca,Mn^{2+},Fe^{2+})_3Al_2BSi_4O_{15}(OH)$

An *axinite* group mineral that was first identified in quartz veins in a manganese deposit in Switzerland; little is known about the reported Nevada occurrence.

MINERAL COUNTY Reported in a specimen from Luning; possibly from the same occurrence de-scribed under magnesio-axinite. (WMK)

TITANITE

$CaTiSiO_5$

Titanite (*sphene*) is a very common minor acces-sory mineral in many igneous and metamorphic rocks; it may also be found in some veins and in skarn. Only a few of many Nevada localities are noted below.

CHURCHILL COUNTY With anatase in quartz veins at the mouth of Corral Canyon in the Corral Canyon district on the east slope of the Stillwater Range about 6 miles southwest of Boyer Ranch. (Ferguson, 1939)

Reported in a specimen with magnetite and calcite at the Mineral Materials Mine, Desert View claim. (UCB)

CLARK COUNTY As microcrystals in miarolitic cavities and pegmatites in the Eldorado and Newberry Mountains; also found in the granitic country rock. (S. Scott, w.c., 1999)

EUREKA COUNTY Microgranular in contact-metamorphosed rock with rutile, anatase, and *leucoxene* at the Goldstrike Mine in the Lynn dis-trict; also reported as microscopic needlelike crystals resulting from alteration of biotite. (Berry, 1992; Campbell, 1994)

LANDER COUNTY In skarn at the West orebody in Copper Canyon in the Battle Mountain dis-trict. (Theodore and Blake, 1978)

MINERAL COUNTY In granite as wedge-shaped yellow crystals to 3 mm in a large isolated road cut several miles northwest of Schurz on the south side of U.S. Alternate Highway 95. (B. Hurley, w.c., 1999)

As wedge-shaped, brilliant, honey-yellow crys-tals to 1.5 mm on quartz and epidote at the Julie claim in the Pamlico district. (Smith and Benham, 1985)

NYE COUNTY A common accessory mineral in grains to 1 mm in the ash-flow tuff of Wilson's Camp in the Antelope Springs, Mellon Mountain, and Wilsons districts on the Nellis Air Force Range. (Ekren et al., 1971)

As euhedral, fractured, cream-colored tablets to 2 mm in skarn at the Great Basin (Victory) Mine in the Lodi district. (M. Jensen, w.c., 1999)

A common accessory mineral in some tuffs and lavas erupted from the Timber Mountain caldera. (Byers et al., 1976)

PERSHING COUNTY In gangue at the American Ore Company, Beacon Hill, Buena Vista, Ford, Segerstrom-Heizer, and Thomas Mines in the Mineral Basin district; also reported as pale-golden-yellow crystals to 6 mm on clinochlore in calcite-filled vugs. (Reeves and Kral, 1955; M. Jensen, w.c., 1999)

A minor mineral in scheelite-bearing skarn in the Nightingale district. (Smith and Guild, 1944)

Abundant in titanium placers derived from the erosion of metamorphic rocks in the Scossa district. (Beal, 1963)

WASHOE COUNTY Rarely in vugs in pegmatite in the Incline Village area as golden-yellow crystals to 3 cm, some transparent, some as fourling twins. (Jensen, 1993a)

Good crystals have been found in pegmatites on Granite Peak in the San Emidio district. (M. Jensen, w.c., 1999)

WHITE PINE COUNTY As small crystals in por-phyry ore in the Robinson district. (Melhase, 1934d)

TOBERMORITE

$Ca_5Si_6(O,OH)_{18}\cdot5H_2O$

A hydrothermal alteration product in calcium carbonate rocks; also found with zeolite miner-als in cavities in basalt.

ESMERALDA COUNTY Reported from an unspec-ified mine in the Goldfield district. (M. Jensen, w.c., 1999)

TOCORNALITE

Ag,Hg)I

Tocornalite is an extremely rare and poorly defined mineral that has been found in a very rich hydrothermal silver deposit in Chile.

CLARK COUNTY Tentatively identified by J. Jedwab using EMP in a specimen from the Azurite Mine in the Goodsprings district. (J. Kepper, this volume)

TODOROKITE

$(Mn^{2+},Ca,Mg)Mn_3^{4+}O_7 \cdot H_2O$

Todorokite is derived by weathering or hydrothermal alteration of primary manganese minerals; it also occurs in marine fumarolic deposits and is the principal manganese oxide in deep-sea manganese nodules.

LANDER COUNTY Reported from Mule Canyon in the Shoshone Range. (R. Thomssen, w.c., 1999)

Associated with base-metal veins at the Cove Mine in the McCoy district. (Echo Bay Minerals Co., unpub. data, 1997)

NYE COUNTY Identified by SEM/EDX in veins in drill core from Yucca Mountain. (Carlos et al., 1995; Castor et al., 1999b)

WHITE PINE COUNTY As small acicular or bladed crystals to 70 microns with aurorite, cryptomelane, pyrolusite, birnessite (?), chlorargyrite, manganoan calcite, native silver, and other manganese minerals at the Aurora, North Aurora, and Ward Beecher Mines in the White Pine district. (Radtke et al., 1967)

TOPAZ

$Al_2SiO_4(F,OH)_2$

Topaz occurs in granite, pegmatite, and aluminum-rich high-grade metamorphic rocks. It also occurs as a vapor-phase mineral in cavities in aluminum-rich rhyolite and as an alteration mineral associated with hydrothermal ore deposits.

CLARK COUNTY As rare sherry-colored crystals to 2 cm in miarolitic cavities and pegmatites in the Newberry Mountains. (S. Scott, w.c., 1999; W. Lombardo, o.c., 2001)

DOUGLAS COUNTY With andalusite, pyrophyllite, corundum, and zunyite at the Blue Metal (Blue Danube) Mine in the Buckskin district. (M. Jensen, w.c., 1999; D. Hudson, w.c., 1999)

ELKO COUNTY Observed in a single thin section of Jarbidge rhyolite. (Coats et al., 1977)

As vapor-phase crystals to 5 mm with quartz in cavities in rhyolite in the Toano Range. (Price et al., 1992)

EUREKA COUNTY Tentatively identified as a late-magmatic mineral in the groundmass of quartz porphyry 4 km north of Mount Hope in the Mount Hope district. (Westra and Riedell, 1995)

HUMBOLDT COUNTY A common vapor-phase mineral in crystals to 3 mm in cavities in pink rhyolite flow rock on the top of Buff Peak in the Bottle Creek district. (Castor and Henry, 2000; SBC)

LANDER COUNTY Reported in veinlets and cavities in rhyolitic rocks with fluorite, cassiterite, and specular hematite in the Izenhood district. (Fries, 1942; Killeen and Newman, 1965)

LINCOLN COUNTY As vapor-phase crystals to 2 mm in devitrified lithophysal rhyolite in the western part of the Kane Springs Wash volcanic center. (Novak, 1985)

MINERAL COUNTY As pale-blue, transparent, gemmy crystals to 8 cm with *amazonite* and other pegmatite minerals at the Zapot claims in the Fitting district about 25 km northeast of Hawthorne; also reported in gemmy yellow crystals. (Foord et al., 1999)

With pyrophyllite at the Deep Mine in the Pamlico district. (D. Hudson, w.c., 1999)

WASHOE COUNTY A rare alteration mineral in veins and adjacent wall rocks in the Pyramid district. (Wallace, 1980)

TORBERNITE

$Cu^{2+}(UO_2)_2(PO_4)_2 \cdot 8\text{-}12H_2O$

Torbernite, one of the most common secondary minerals of uranium, occurs in the oxidation zones of deposits with both copper and uranium. Distinctions between it and the lower hydrate metatorbernite have not often been made, and some of the following occurrences are probably of the latter. Torbernite has the distinction of being the first mineral to be named after a person.

CHURCHILL COUNTY Questionably reported at the Martin claims in the Wonder district. (Garside, 1973)

DOUGLAS COUNTY Reported in quartz-rich pegmatite at the Kingsbury Queen prospect in the Genoa district. (Garside, 1973)

ELKO COUNTY With autunite and metatorbernite at the Autunite and October claims in the Mountain City district. (Garside, 1973)

EUREKA COUNTY As grass-green tabular crystals in clusters to 2 mm with fluellite, hewettite, leucophosphite, variscite, and anatase at the Gold Quarry Mine in the Maggie Creek district. (Jensen et al., 1995)

HUMBOLDT COUNTY Reported with quartz, fluorite, pyrite, and autunite at the Moonlight Uranium Mine in the Disaster district. (Sharp, 1955; Taylor and Powers, 1955; Garside, 1973; Dayvault et al., 1985)

Reported with secondary copper minerals at the Blue Jack property in the Varyville district. (Garside, 1973)

LANDER COUNTY With pyrite, bornite, chalcopyrite, uraninite, and other uranium minerals at the Apex (Rundberg) Uranium Mine in the Reese River district. (Garside, 1973)

LINCOLN COUNTY Reported with autunite at the Peak claims in the Eagle Valley district. (Garside, 1973)

LYON COUNTY Reported with kasolite and *pitchblende* in base-metal-bearing quartz veins at the West Willys claim group in the Washington district; also with autunite at the Silver Pick property. (Garside, 1973)

As small rosettes with autunite and phosphuranylite at the Flyboy (McCoy) Mine in the Yerington district. (Garside, 1973; WMK)

MINERAL COUNTY Questionably reported with saléeite and nováčekite at the Broken Bow and Broken Bow King claims in the Santa Fe district. (Garside, 1973)

NYE COUNTY On fractures in granite at the Nighthawk and other claims in the Belmont district. (Garside, 1973)

With autunite in Paleozoic shale at the 6-Mile claims in the Morey district. (Garside, 1973)

Reported as rare with autunite at the Hazel prospect in the Northumberland district. (Garside, 1973)

On fractures in rhyolite at the Pine claims in the Round Mountain district. (Garside, 1973)

As green, tabular, single crystals and groups in gouge at the Hall (Anaconda Moly) Mine in the San Antone district. (R. Walstrom, w.c., 1999)

PERSHING COUNTY In a brecciated zone as deep-grass-green transparent crystals to 2.5 cm that convert to metatorbernite and lose transparency when exposed to light and air in the lower adit at the Majuba Hill Mine in the Antelope district. (Smith and Gianella, 1942; Jensen, 1985)

WASHOE COUNTY Reported with autunite at uranium mines and prospects in the Pyramid district; also questionably reported at the Lost Partner claims. (Brooks, 1956; Garside, 1973)

TOSUDITE

$$[(Mg,Fe^{2+})_5Al(Si_3Al)O_{10}(OH)_8].[(Ca,Na)_{0.33}$$
$$(Al,Fe,Mg)_{2-3}(Al,Si)_4O_{10}(OH)_2 \bullet nH_2O]$$

A clay mineral in which equal proportions of *chlorite* and *smectite* are regularly interstratified; mainly occurs as a hydrothermal alteration mineral in intermediate to felsic igneous rocks.

WASHOE COUNTY With quartz at the White Mountain gold deposit in the Cottonwood district. (Bussey, 1996)

TOURMALINE

$$(Na,K,Ca)(Mg,Fe^{2+},Mn^{2+},Li,Al,Fe^{3+})_3(Al,Fe^{3+},$$
$$Cr^{3+},V^{3+})_6(BO_3)_3Si_6O_{18}(O,OH,F)_4 \text{ (general)}$$

Tourmaline is a mineral group name. Species have not been identified at the following occurrences; at many the most likely species is schorl.

CLARK COUNTY In *tourmaline*-muscovite-microcline pegmatite at the Taglo Mine in the Bunkerville district; also noted at the Virgin Mountain Chrysoberyl property. (Beal, 1965)

DOUGLAS COUNTY With quartz and pyrite in a vein at the Utopian Mine in the Delaware district. (Doebrich et al., 1996)

ELKO COUNTY With quartz, beryl, and other pegmatite minerals in Dawley Canyon in the Valley View district. (Olson and Hinrichs, 1960)

Reported in a specimen from the Youngs property in the Ruby Valley district. (WMK)

HUMBOLDT COUNTY Reported 35 miles from Winnemucca in the Jungo district. (Melhase, 1934b)

LINCOLN COUNTY Reported as crystals in milky quartz at the Bristol Mine in the Bristol district. (UCB)

As needles in veins in the Chief district at the southern tip of the Chief Range, about 8 miles north of Caliente. (Callaghan, 1936)

LYON COUNTY With calcite in a breccia zone more than 400 m long and as much as 75 m wide that cuts metamorphosed andesite in the Mound House district, Sec. 24, T16N, R20E. (Doebrich et al., 1996)

Reported at the Mason Valley Mine in the Yerington district. (Melhase, 1935e)

MINERAL COUNTY In altered intrusive rock at the Candelaria Mine in the Candelaria district. (Moeller, 1988)

Reported in a specimen with calcite and *limonite* from the Silver Star district. (WMK)

Reported in a specimen from Luning. (UCB)

NYE COUNTY Reported as greenish-brown crystals in the Morey district. (Hoffman, 1878)

PERSHING COUNTY Reported at the Montezuma Mine in the Arabia district. (Decker, 1972)

With quartz in a specimen from an unknown locality in the Gerlach district. (WMK)

As small, brownish-black crystals in quartz veins on Lincoln Hill in the Rochester district; also noted with dumortierite and in places as fibrous green veins. (Knopf, 1924; WMK; S. Pullman, o.c., 1999)

As disseminated crystals and masses in quartz veins at the Bonanza King, Pacific Matchless, and Walker Mines in the Spring Valley district. (Campbell, 1939; Johnson, 1977)

WASHOE COUNTY As tiny black crystals in altered and brecciated Tertiary igneous rock in the Jumbo district, Sec. 25, T17N, R20E. (L. Garside, o.c., 2000)

WHITE PINE COUNTY Blue *tourmaline* was reported as sparse grains in Precambrian quartzite of the McCoy Creek group. (Hose and Blake, 1976)

TREMOLITE

☐$Ca_2Mg_5Si_8O_{22}(OH)_2$

An amphibole-group mineral that occurs in regionally or contact-metamorphosed rocks. *Byssolite* is a fibrous variety.

CARSON CITY Reported in a specimen from 1 mile south of Carson City and west of U.S. Highway 395 in the Voltaire district. (WMK)

ELKO COUNTY Plentiful in contact-metamorphic zones in the Dolly Varden district. (Melhase, 1934d)

Reported in a specimen from the Pinon Range in the Railroad district. (WMK)

EUREKA COUNTY Reported as *byssolite* in dark-green, filiform to needlelike crystals with calcite and pyrite in a vug in skarn at the Goldstrike Mine in the Lynn district. (GCF)

As elongated, white, prismatic crystals on fractures in silty limestone at the Maggie Creek Mine in the Maggie Creek district. (Rota, 1991)

Reported in a specimen from Palisade in the Safford district. (WMK)

LANDER COUNTY In skarn at the West orebody in Copper Canyon in the Battle Mountain district. (Theodore and Blake, 1978)

LYON COUNTY Identified as "mountain cork" in skarn at the Douglas Hill and Mason Valley Mines in the Yerington district; also noted as acicular, radiating crystals to 6 mm in skarn. (JHS; WMK; S. Pullman, w.c., 1999; GCF)

NYE COUNTY In fibrous, radiating masses to 5 cm in metamorphosed Paleozoic carbonate near the base of Bare Mountain about 1 mile south of Beatty in the Bare Mountain district. (B. Hurley, w.c., 1999)

As radiating aggregates and crystals in the margins of magnesite bodies at the Premier Chemicals (Gabbs Magnesite-Brucite) Mine in the Gabbs district; also reported as a low-grade deposit east of Downeyville. (Schilling, 1968a; Kral, 1951)

Identified in skarn in the Oak Spring district. (JHS)

Reported as amphibole asbestos in a specimen from the Tonopah district. (WMK)

PERSHING COUNTY Part of the gangue assemblage at various iron mines in the Mineral Basin district, including the Beacon Hill, Buena Vista, Ford, and Segerstrom-Heizer Mines. (Reeves and Kral, 1955)

In minor amounts in scheelite-bearing skarn in the Nightingale district. (Smith and Guild, 1944)

WHITE PINE COUNTY Along limestone-monzonite contacts in East Ely in the Robinson district. (Melhase, 1934d)

TREVORITE

$NiFe_2^{3+}O_4$

A member of the spinel group that occurs in nickel deposits associated with ultramafic rocks. Also in minor amounts in sediments at major stratigraphic boundaries, such as the Cretaceous-Tertiary, presumably as remnants of extraterrestrial impact.

NYE COUNTY Reported in a specimen from a locality listed as Hatfield, Gabbs, Nevada. (EMP)

TRIANGULITE

$Al_3(UO_2)_4(PO_4)_4(OH)_5 \bullet 5H_2O$

A very rare species found elsewhere only at a single location in Zaire in uranium-rich pegmatite with other secondary uranium minerals.

ESMERALDA COUNTY Reported to occur as bright-yellow masses in vugs in rhyolite from Rhyolite Ridge near Mina, Mineral County; no other locality information is given, but probably from Rhyolite Ridge in the northern Silver Peak Range in Esmeralda County. (FUMT; HUB; Excalibur Mineral Co., 1991)

TRIDYMITE

SiO_2

A common silica polymorph (of quartz and cristobalite) that generally occurs as a vapor-phase mineral in volcanic rocks; less common in contact-metamorphosed sandstone and as phenocrysts in felsic volcanic rocks. Undoubtedly common in Nevada's volcanic rocks, although only a few localities are listed below.

HUMBOLDT COUNTY In opalite breccia and *chalcedony* layers at the McDermitt Mine in the Opalite district. (Roper, 1976)

LANDER COUNTY As well-crystallized hexagonal plates in radial aggregates to 1 inch in diameter in narrow veins and fissures in rhyolite in the Izenhood district. (Anonymous, 1938; S. Pullman, w.c., 1999)

MINERAL COUNTY In specimens from Montgomery Pass in the Buena Vista district. (LANH; R. Thomssen, w.c., 1999)

NYE COUNTY A vapor-phase mineral in lithophysal cavities and fractures in tuff at Yucca Mountain. (Carlos et al., 1995)

PERSHING COUNTY Reported in andesite near Lovelock. (AMNH; JHS)

WHITE PINE COUNTY A rare mineral with *garnet* in lithophysae as thick, colorless, hexagonal crystals to 1 cm at Garnet Hill in the Robinson district. (M. Jensen, w.c., 1999)

As small colorless crystals in lithophysae in rhyolite with hematite in a road cut along Highway 50 near Little Antelope Summit. (M. Jensen, w.c., 1999)

TRIPLITE

$(Mn^{2+},Fe^{2+},Mg,Ca)_2(PO_4)(F,OH)$

Triplite is an uncommon mineral that occurs as either a primary or secondary phase with other phosphates in pegmatite. The Nevada occurrence is unusual because the mineral occurs in tungsten ore.

WHITE PINE COUNTY In prospects in the Eagle district in the Kern Mountains near the Utah state line, as brownish to brownish-purple irregular masses to 2.5 cm in quartz veins with fluorite, pyrite, bismuthinite, chalcopyrite, and hübnerite; may be of pegmatitic origin. (Hess and Hunt, 1913; Hill, 1916; Melhase, 1934d)

TRIPUHYITE

$Fe^{2+}Sb_2^{5+}O_6$

Tripuhyite, which is formed where primary antimony minerals have undergone oxidation, is generally found with more common secondary antimony minerals such as stibiconite and bindheimite.

CHURCHILL COUNTY With other Fe- and Sb-oxide minerals at the mopungite type locality (Green prospect) in the Lake district. (Williams, 1985)

EUREKA COUNTY In brownish-yellow earthy crusts and veinlike masses at the Goldstrike Mine in the Lynn district. (GCF)

With stibiconite in pale-greenish-yellow to brown powdery coatings as an alteration product of stibnite at the Gold Quarry Mine in the Maggie Creek district. (Jensen et al., 1995)

HUMBOLDT COUNTY As yellow-green crystalline masses with the discredited species *mcdermittite* and other minerals at the McDermitt Mine in the Opalite district; identified by J. Wilson using XRD methods. (F. Cureton, w.c., 1999)

As crystals to 1 mm with opal in alunite veins in the Sulphur district. (Wallace, 1979; J. McGlasson, w.c., 2001)

LINCOLN COUNTY In glassy brownish masses with bindheimite and chrysocolla at a locality near Pioche. (Jenkins, 1981)

NYE COUNTY As pseudomorphs after stibnite crystals at the Paradise Peak Mine in the Fairplay district. (M. Jensen, w.c., 1999)

PERSHING COUNTY As brownish coatings with bindheimite in the Arabia district. (J. McGlasson, w.c., 2001)

TROILITE
FeS

A mineral that is found most commonly as nodules in iron meteorites; also found rarely with other sulfides in terrestrial ultramafic rocks.

CHURCHILL COUNTY Tentatively identified with kamacite, taenite, and other minerals in the Hot Springs iron meteorite. (Art Johnson, o.c., 1999)

CLARK COUNTY With kamacite, taenite, and other minerals in the Las Vegas meteorite. (Buchwald, 1975)

NYE COUNTY With kamacite, taenite, and other minerals in the Quartz Mountain meteorite. (Buchwald, 1975)

With kamacite, taenite and schreibersite in the Quinn Canyon meteorite. (Buchwald, 1975)

TRONA
$Na_3(CO_3)(HCO_3) \cdot 2H_2O$

Trona occurs in playas and bedded evaporite deposits. It is an important commercial evaporite mineral, particularly in Wyoming.

CHURCHILL COUNTY Reported as a mass more than 10 m thick in Little Soda Lake, now submerged, in the Soda Lake district; also from Big Soda Lake, but produced there by man-caused evaporation. (Gianella, 1941; Kostick, 1994)

MINERAL COUNTY Reported in the Double Springs Marsh district. (Gianella, 1941)

With halite, ulexite, glauberite, thenardite, and mirabilite in the Rhodes Marsh district. (Melhase, 1935e; Vanderburg, 1937a)

Identified in salt crust that covers about a third of the playa in the Teels Marsh district. (JHS)

NYE COUNTY One of a number of evaporite minerals in the Butterfield Marsh district. (Gianella, 1941)

TSUMEBITE
$Pb_2Cu(PO_4)(SO_4)(OH)$

A very rare secondary mineral in oxidized base-metal deposits.

WASHOE COUNTY Tentatively identified by SEM/EDX as small grains with malachite, cassiterite, and gold in a vein sample from the Wet prospect in the Pyramid district; also questionably identified in a sample with anglesite, barite, acanthite, and other minerals from an unnamed prospect in NE1/4 NE1/4 Sec. 16, T23N, R21E. (SBC)

TUNGSTENITE
WS_2

A rare mineral that generally occurs with other sulfides and other tungsten minerals in hydrothermal and skarn deposits.

EUREKA COUNTY Identified by SEM/EDX as euhedral grains to 10 microns in silicified ore from the Goldstrike Mine in the Lynn district. (GCF)

TUNGSTITE
$WO_3 \cdot H_2O$

Tungstite is an uncommon mineral that occurs as an oxidation product of scheelite, *wolframite,* and other tungsten minerals.

NYE COUNTY Reported in minor amounts in quartz-hübnerite veins in Shoshone Canyon, east of Round Mountain in the Toquima Range in the Round Mountain district. (Ferguson, 1922)

Reported in specimens from the north workings of the Victory Mine in the Lodi district. (R. Thomssen, w.c., 1999)

PERSHING COUNTY Reported with hübnerite in a specimen of tungsten ore at the Indian Peak (Royal Ann) group in the Toy district. (WMK)

TURANITE

$Cu_5^{2+}(VO_4)_2(OH)_4$ (?)

Turanite is a rare mineral that occurs as an oxidation product of primary vanadium-bearing substances.

EUREKA COUNTY At the Van-Nav-San claim in the Gibellini district as minute, yellow-green, platy aggregates, some with volborthite. (S. Pullman, w.c., 1999; Excalibur Mineral Co., 2001b)

Reported as yellow-green microcrystals and botryoidal masses with malachite from the Gold Quarry Mine in the Maggie Creek district. (S. White, w.c., 1999; J. McGlasson, w.c., 2001; GCF)

NYE COUNTY As crusts of greenish-yellow crystals to 0.25 mm and as aqua-colored coatings on black slate with volborthite, chrysocolla, and brochantite south of Manhattan; perhaps the same locality as that reported for "calcio-volborthite" by Ferguson (1924) in the Manhattan district (see tangeite). (Jenkins, 1981; FUMT; AM)

TURQUOISE

$Cu^{2+}Al_6(PO_4)_4(OH)_8 \cdot 4H_2O$

Turquoise, a common mineral in copper deposits and as a product of alteration in rocks containing small amounts of copper, has been utilized as a gemstone since prehistoric times. Nevada has many localities and has produced large amounts commercially at various times, yielding some of the best turquoise from the United States. See Morrissey (1968) for descriptions of many of the Nevada occurrences.

CHURCHILL COUNTY Reported in veins at the glory hole in the Fairview district. (Schrader, 1947)

CLARK COUNTY As excellent quality veinlets and nodules that yielded gemstones of more than 200 carats from the Crescent Peak (Wood) Mine in the Crescent district, first worked by Native Americans and rediscovered in 1889; also in narrow veinlets in quartz monzonite at the Morgan Mine. (Murphy, 1964; Morrissey, 1968; JHS)

Reported as a replacement of bone at an unspecified locality in the Goodsprings district. (J. Kepper, o.c., 2001)

Reported in sheared and altered rocks at the Sullivan Mine in the McClanahan district. (Morrissey, 1968)

In a specimen from the Toltec Mine in the Searchlight district. (AMNH)

DOUGLAS COUNTY Reported in the vicinity of the Buckskin Mine in the Buckskin district, also common in the southern part of the district. (Klein, 1983; SBC)

ELKO COUNTY As solid blue, matrix-marked, and spider-web material in small pods and veins in chert and siltstone at the Stampede Mine in the Beaver district. (Morrissey, 1968; Bentz et al., 1983; Klein, 1983)

Reported at the Carlin Black Matrix Mine in the Merrimac district. (Morrissey, 1968)

As veins to 2 cm and small masses in shear zones and on the margins of an altered dike at the Rain Mine in the Carlin district; mostly soft and chalky, but locally of gem quality. (D. Heitt, o.c., 1991)

With minor fluorite in hematitic and siliceous gossan and clay gouge at the Bald Mountain Chief Mine in the Railroad district. (Bentz et al., 1983; LaPointe et al., 1991)

As thin veins in bedded chert at the Edgar Mine in the Swales Mountain district. (Bentz et al., 1983)

Reported in a specimen from the Mt. Diablo (Wells Tungsten) Mine in the Wells district. (WMK)

ESMERALDA COUNTY Produced with variscite at the Sigmund group and Carl Riek and Miss Moffet (Los Angeles Gem Co.) Mines in the Candelaria district; also as microcrystals with variscite from an unnamed prospect in the Candelaria Hills north of Columbus. (Melhase, 1935d; Morrissey, 1968; Klein, 1983; M. Jensen, w.c., 1999)

Reported as hard with deep-sky-blue color and locally spider webbed at the Blue Bell in the Coaldale district. (Murphy, 1964; Morrissey, 1968; Klein, 1983; JHS)

Reported at several mines in the Crow Springs district, including some actually in the Royston district to the northeast; sky-blue with brown cobweb matrix at the Marguerite Mine. (Murphy, 1964; JHS)

As nodules to 2 inches and veins to 0.5 inch of hard, sky-blue to greenish and soft, pale-blue material at the Carrie Mine in the Gilbert district, which produced about $300,000 worth; also mined at the Monte Cristo and Carr-Lovejoy Mines. (Morrissey, 1968; Klein, 1983; JHS)

Reported in a specimen with famatinite in the Goldfield district. (NMBM)

As breccia fillings, seams, and veinlets in black chert at the Smith Black Matrix Mine in the Klondyke district. (Morrissey, 1968; Klein, 1983)

In high-quality stones from the Lone Mountain and Lively Mines in the Lone Mountain district; also at the Blue Silver Mine with other copper minerals and galena. (Murphy, 1964; Morrissey, 1968; Klein, 1983)

Reported from the Bonnie Blue Mine near the Columbus Salt Marsh, possibly in the Rock Hill district. (Klein, 1983)

EUREKA COUNTY Identified in a specimen from a location along Mill Creek on the east flank of Mt. Tenabo in the Cortez district. (JHS)

In the Lynn district at the Number 8 (Blue Star) Mine, which produced more than $1,400,000 worth of fine-quality gemstones, including a 150-pound nodule of gem turquoise and some of the finest spider-web turquoise found in Nevada. The turquoise, concentrated along quartz veins in chert, shale, and altered granitic rock, was mostly in nodular form but also in veins to 50 cm. Lesser quantities of greenish-blue turquoise were produced from seams and nuggets in shale and along a shale-granitic rock contact at the August Berning Mine. (Murphy, 1964; Morrissey, 1968; Klein, 1983; GCF; JHS)

Light-blue microcrystalline masses on fractures and as veins at the Copper King Mine in the Maggie Creek district, SE1/4 SE1/4 Sec. 28, T34N, R51E. (Rota, 1991)

HUMBOLDT COUNTY As lustrous, pale-blue, fluorine-bearing, euhedral crystals to 0.5 mm with wavellite, variscite, lipscombite, and crandallite at the Silver Coin Mine in the Iron Point district. (Thomssen, 1998; NMBM; M. Jensen, w.c., 1999; J. McGlasson, w.c., 2001)

LANDER COUNTY In the Copper Basin area, Battle Mountain district, as veinlets in altered granitic rock at the Turquoise King, Myron Clark, and Blue Gem Lease Mines. The latter reportedly yielded nearly $1 million worth of rough stones. (Murphy, 1964; Morrissey, 1968)

Produced in the Bullion district from many sites, including the Lander Blue, Steinlich (Gold Acres, Stone Cabin), Tom Cat, Badger, Mud Springs, Super-X, Arrowhead, Blue Eagle, Blue Matrix, and Blue Sky Mines. The most productive area was near Tenabo and Gold Acres. Turquoise nodules, veinlets, and seams were mainly found

in Paleozoic chert, shale, and quartzite. Some of the best spider-web turquoise from Nevada came from this district, particularly from the Super-X and Blue Matrix Mines. (Murphy, 1964; Morrissey, 1968; Klein, 1983; JHS)

Reported as light-blue veins in brecciated zones at the Green Tree and McGinness Mines in the Callaghan Ranch district. (Morrissey, 1968; Klein, 1983)

In a specimen from the Carico Lake Mine in the Carico Lake district. (NMBM)

At the Fox, Lone Pine, and White Horse Mines in the Cortez district. The most productive property in Nevada, the Fox mine, produced at least 500,000 pounds, much of it hard, light-colored material that was sold to German factories where it was dyed. (Morrissey, 1968; Klein, 1983; NMBM)

Reported in seams in quartzite at the Independence claim in the Hilltop district. (WMK)

Identified in a specimen from the Shoshone Mine in the New Pass district. (JHS)

As seams and nodules in shale at the Pinto (Watts) Mine in the Warm Springs district. (Morrissey, 1968)

At least $400,000 worth of mostly high-quality spider-web material came from the Godber Turquoise Mine in the Reese River district. (Morrissey, 1968)

Reported as high quality at the Jimmy Allen (Blue Goose) Mine. (Morrissey, 1968; Klein, 1983)

LINCOLN COUNTY Reported in a specimen from the Delamar Mountains. (WMK)

LYON COUNTY From the Yerington district as dark-blue to green masses and microcrystals from the Taubert Number 1 Mine, and soft, pale- to pure-blue stones at the Taubert Number 2 mine; about $50,000 worth of good quality turquoise came from these mines. Also reported from the Robin, Anaconda (Weed Heights), and Empire-Nevada Mines, and from Carson Hill. (Morrissey, 1968; Klein, 1983; WMK; S. Pullman, w.c., 1999; R. Thomssen, w.c., 1999)

MINERAL COUNTY In minor amounts with variscite as veinlets in Ordovician chert in the Candelaria Hills several miles northwest of Columbus; may be the same locality reported for the Candelaria district under Esmeralda County. (Vanderburg, 1937a)

Reported at the Basalt Mine in the Eastside district; also as nodules and veinlets in shale and limestone at the Blue Jay Gem and Blue Gem Number 1 Mines. (Murphy, 1964; Morrissey, 1968)

In the Pilot Mountains district at the Turquoise Bonanza mine as veinlets and cement in altered breccia; at the Moqui-Aztec (S. Simmons) and Montezuma (Troy Springs) Mines as veinlets and nodules in altered granitic rock. Also reported from the Pilot Mountains as pseudomorphs after orthoclase and *apatite*. (Vanderburg, 1937a; Murphy, 1964; Morrissey, 1968; Klein, 1983; S. Pullman, w.c., 1999; F. Cureton, o.c., 1999)

Reportedly mined at an unspecified locality in the Rand district. (Morrissey, 1968)

In small amounts with variscite (?) at the Halley's Comet Mine in the Silver Star district. (Morrissey, 1968)

NYE COUNTY As dark-blue material in chocolate-colored matrix at the Zabrisky Mine in the Belmont district; also at the Copper Blue and No Name Number 2 Mines. (Morrissey, 1968)

As pale-blue to dark-green nodules and veinlets in altered quartz monzonite porphyry at diggings scattered on the southwest slope of the Cactus Range in the Cactus Springs district, particularly in Sleeping Column Canyon; all on Nellis Air Force Range. (Morrissey, 1968; Tingley et al., 1998a)

Identified in a sample from an unspecified locality in the Gabbs district. (JHS)

With variscite and other phosphate minerals in Ordovician shale and slate at the Tom Molly Mine (Train prospect) in the Manhattan district. (Kral, 1951; Morrissey, 1968)

Tentatively identified in a shear zone at the Titus prospect (John Titus claim) 7 miles northeast of Moores Station. (Kleinhampl and Ziony, 1984)

As nuggets with white carbonate along bedding planes in fossiliferous shale at the Indian Blue Mine in the Northumberland district. (Morrissey, 1968; McKee, 1976)

In the Royston district, one of the better-known sources in Nevada, mainly at the Royal Blue, Bunker Hill, and Oscar Wehrend Mines. At the Royal Blue Mine, also reported to be in Esmeralda County, masses as much as 5 inches thick in altered porphyry included hard, pure-blue stones said to be the equal of turquoise from

any American mine. (Kral, 1951; Murphy, 1964; Morrissey, 1968; Kleinhampl and Ziony, 1984)

In pale-green to white masses with manganese oxides, kaolinite, and iodargyrite at the Mizpah Mine in the Tonopah district; also reported at the Little Cedars Mine. (Melhase, 1935c; Morrissey, 1968; WMK)

PERSHING COUNTY Reported with variscite from an unspecified locality in the Rye Patch district. (WMK)

STOREY COUNTY Reported in a specimen from Virginia City; the locality is questionable. (EMP)

WHITE PINE COUNTY Reported from an unspecified locality in the Robinson district. (Von Bargen, 1999)

TVALCHRELIDZEITE

$Hg_3(Sb,As)S_3$

Tvalchrelidzeite is a rare mercury sulfosalt mineral that occurs in hydrothermal deposits.

HUMBOLDT COUNTY As microscopic blebs and subhedral crystals in orpiment at the Getchell Mine in the Potosi district. (Dunning, 1987; S. Sears, w.c., 1999)

TYROLITE

$CaCu_5^{2+}(AsO_4)_2(CO_3)(OH)_4 \cdot 6H_2O$

Tyrolite, a relatively rare mineral, is generally found with other copper arsenate phases in the oxidation zones of copper deposits.

CHURCHILL COUNTY Reported at the Lovelock and Nickel Mines in the Table Mountain district. (S. Pullman, w.c., 1999; Von Bargen, 1999)

Reported in a specimen from Fireball Ridge in the Truckee district. (R. Thomssen, w.c., 1999)

ESMERALDA COUNTY Reported from the Broken Toe Mine in the Rock Hill district. (R. Thomssen, w.c., 1999)

Reported in a specimen from the 3 Metals Mine in the Weepah district. (R. Thomssen, w.c., 1999)

LANDER COUNTY Reported at the Garrison Mine in the Cortez district. (S. Pullman, w.c., 1999)

PERSHING COUNTY As pale-blue-green to green, lathlike crystals in sheaves to 3 mm, commonly partially altered to strashimirite, at the Majuba Hill Mine in the Antelope district. (MacKenzie and Bookstrom, 1976; Jensen, 1985)

WASHOE COUNTY As colorful but poorly developed bright-green crystals with chalcophyllite, clinoclase, and brochantite in oxidized enargite-bearing veins on Peavine Mountain in the Peavine district. (Jenkins, 1981)

As platy green crystals to 0.5 mm with strashimirite and chalcophyllite at the Burrus Mine in the Pyramid district. (Jensen, 1994)

TYUYAMUNITE

$Ca(UO_2)_2V_2O_8 \cdot 5\text{-}8H_2O$

The calcium analog of carnotite, tyuyamunite occurs in oxidized uranium deposits in which vanadium is present, and is particularly common in the sandstone-type deposits of the southwestern United States. It is distinguished from the commonly associated lower hydrate metatyuyamunite by X-ray diffraction.

CLARK COUNTY Reported in calcrete-gypcrete occurrences in road cuts along the Union Pacific Railroad between Erie and Arden in the Arden district. (Garside, 1973)

As yellow fracture coatings at the Blue Chip, Frank Robbin, Yellow Queen, and other claims in the Gold Butte district. (Garside, 1973)

EUREKA COUNTY In brecciated jasperoid as bright-yellow clots with kingite, thin powdery coatings associated with phosphate minerals, and relatively abundant pale-yellow dustings on kazakhstanite and fervanite at the Gold Quarry Mine in the Maggie Creek district. (Jensen et al., 1995)

LYON COUNTY As yellow rosettes and stains on fractures at the Casting Copper Mine in the Yerington district, Sec. 34, T13N, R24E; also reported from the Flyboy Mine. (GCF; R. Thomssen, w.c., 1999)

ULEXITE

$NaCaB_5O_6(OH)_6 \cdot 5H_2O$

Ulexite is one of the most common borate minerals, occurring as friable, near-surface masses (sometimes referred to as "cotton balls") in some playas and in solid bedded deposits with other borate minerals that probably represent prehistoric dry lakes. It is particularly common in Nevada playas and was an important mineral during early borate mining.

CHURCHILL COUNTY As "cotton balls" with halite and a variety of other minerals in the Dixie (Humboldt) Salt Marsh in the Dixie Marsh district. (Vanderburg, 1940)

"Cotton balls" occur in marsh deposits in the Sand Springs Marsh district. (Vanderburg, 1940)

CLARK COUNTY A little ulexite occurs with colemanite in the White Basin in the Muddy Mountains district; also reported in the Callville Wash area. (Gianella, 1941; Castor, 1993; S. Pullman, w.c., 1999)

ESMERALDA COUNTY As "cotton balls" in layers 1 to 6 inches beneath the surface in irregular areas around the margin of the Columbus Salt Marsh in the Columbus Marsh district. (Lincoln, 1923)

In veinlets in Tertiary sedimentary rocks 3 miles east of Fish Lake Marsh in the Fish Lake Valley district. (Lincoln, 1923; Albers and Stewart, 1972)

MINERAL COUNTY As "cotton balls" to several inches in mud just below the surface of the salt marsh in the Rhodes Marsh district with halite, trona, and other sodium minerals. (Melhase, 1935e; Vanderburg, 1937a; JHS)

Reported in the Teels Marsh district. (S. Pullman, w.c., 1999)

NYE COUNTY Reported in small playas in Big Smoky Valley north of Tonopah. (Gianella, 1941)

WASHOE COUNTY As encrustations on the apron of a hot spring about 1 mile northwest of Gerlach. (Cornwall, 1964; Bonham and Papke, 1969)

ULLMANNITE

NiSbS

A nickel sulfosalt that generally occurs with other nickel minerals in hydrothermal veins.

CHURCHILL COUNTY Reported in a specimen from Table Mountain, probably from the Lovelock or Nickel Mines in the Table Mountain district. (UCB)

ULVÖSPINEL

$TiFe_2^{2+}O_4$

A common mineral in titaniferous magnetite iron ore; also an accessory mineral in kimberlites and some basalts.

PERSHING COUNTY In magnetite ore in skarn at iron mines in the Buena Vista Hills in the Mineral Basin district. (M. Jensen, w.c., 1999)

URANINITE

UO_2

Uraninite is a widespread primary uranium mineral, occurring in pegmatites, hydrothermal veins, sandstone-type uranium deposits, and uranium-rich conglomerate deposits. Also called *pitchblende,* particularly if it occurs as microcrystalline banded masses or in botryoidal crusts.

CHURCHILL COUNTY Reported to occur, possibly as replacements of ferro-magnesian minerals, at the Lovelock and Nickel Mines in the Table Mountain district. (Garside, 1973)

ELKO COUNTY With chloritic gouge in fault zones that cut skarn in underground workings at the Garnet Mine in the Alder district. (Garside, 1973)

Tentatively reported in a pegmatite in the Gilbert Canyon district. (Olson and Hinrichs, 1960)

Rare in a pegmatite body at the Harrison Pass prospect in the Harrison Pass district. (Garside, 1973)

With yellow secondary uranium minerals in Tertiary bentonitic tuff that overlies quartz monzonite at the South Fork and Pixley Number 1 claims in the Mountain City district. (Garside, 1973)

EUREKA COUNTY As disseminated, micron-sized grains at the Goldstrike Mine in the Lynn district. (GCF)

As grains to 2 microns in sulfide-rich barite breccia with galena, greenockite, arsenopyrite, sphalerite, and pyrite at the Gold Quarry Mine in the Maggie Creek district. (Jensen et al., 1995)

HUMBOLDT COUNTY Reported with coffinite at the Moonlight Uranium Mine in the Disaster district. (Sharp, 1955; Williams, 1980)

Identified in a sample from an unspecified locality in the Opalite district. (JHS)

LANDER COUNTY In ore from the Apex (Rundberg) Uranium Mine in the Reese River district, the largest uranium producer in Nevada, with coffinite, pyrite, bornite, chalcopyrite, and secondary uranium minerals. (Garside, 1973)

LINCOLN COUNTY As pods of pyrite-uraninite ore in jasperoid at the Blue Bird Mine in the Atlanta district; also tentatively identified at the Nevada Rath claims. (Tschanz and Pampeyan, 1970; Garside, 1973)

LYON COUNTY As *pitchblende* in quartz veins with secondary uranium minerals and a variety of lead and copper minerals at the Teddy and West Willys claims in the Washington district. (Garside, 1973)

NYE COUNTY Questionably identified with an unknown yellow uranium mineral in an altered shear zone in granodiorite at the 66 claims in the Lodi district. (Garside, 1973)

PERSHING COUNTY As black pitchy and massive *pitchblende* with *gummite* in a mafic contact zone

at the Stalin's Present prospect in the Buena Vista district. (Lovering, 1954; Garside, 1973)

Identified in a sample from an unspecified locality in the Rye Patch district. (JHS)

With *allanite,* pyrite, and molybdenite in silicified zones in skarn at the Long Tungsten Mine in the Wild Horse district. (Garside, 1973)

WASHOE COUNTY Reported as *cleveite* from the DeLongchamps Mine in the Pyramid district; also noted in some other uranium mines and prospects. (Brooks, 1956; Bonham and Papke, 1969; Garside, 1973)

Tentatively identified with *allanite* at the Lizard claims in the Coyote Canyon area. (Garside, 1973)

URANOPHANE

$Ca(UO_2)_2[SiO_3(OH)]_2 \cdot 5H_2O$

One of the most common of the secondary minerals of uranium, occurring as an oxidation product of uraninite. Fine specimens have reportedly come from the Washington district in Lyon County.

CHURCHILL COUNTY Questionably identified with iron and copper oxide minerals in a fault zone cutting rhyolite at the K-D claims in the Copper Kettle district. (Garside, 1973)

CLARK COUNTY With dumontite and kasolite as greenish-yellow crusts of minute crystals at the Green Monster Mine in the Goodsprings district. (Jenkins, 1981)

Questionably reported in sedimentary beds of the Horse Spring Formation at the First Chance group and Long Shot Number 1 claim in the Gold Butte district. (Garside, 1973)

ELKO COUNTY Questionably reported at the Prince claims in the Contact district. (Garside, 1973)

LINCOLN COUNTY Tentatively identified in oxidized pyrite-uraninite ore at the Blue Bird Mine in the Atlanta district; also reported from the Nevada Rath claims. (Garside, 1973)

LYON COUNTY With other uranium minerals in a quartz vein at the Teddy claims in the Washington district; also reported in shear zones at the Halloween Mine and the Kateydid claim. (Garside, 1973)

Reported in a specimen from the Cambridge Hills in the Yerington district, possibly from the Flyboy Mine. (R. Thomssen, w.c., 1999)

MINERAL COUNTY With autunite in skarn iron ore at the William Johnson claims in the Fitting district. (Garside, 1973)

Reported with other secondary uranium minerals at the Blue Boy, Marietta, Silver Bell, and 4D claims in the Marietta district. (Garside, 1973)

NYE COUNTY Reported with autunite in silicified fault breccia in rhyolite at the Black Bart Extension and Black Bonanza claims in the Bullfrog district. (Garside, 1973)

PERSHING COUNTY Questionably reported at the Sage Hen Springs and Uranium Lode claims in the Nightingale district. (Garside, 1973)

WASHOE COUNTY Tentatively identified with autunite and other uranium minerals at the Lowary and Thunder Bird claims in the Pyramid district. (Brooks, 1956; Garside, 1973)

In veinlets with *gummite* and autunite at the Buckhorn Uranium Mine in the Stateline Peak district. (Hetland, 1955; Garside, 1973; Castor et al., 1996)

URANOPHANE-BETA

$Ca(UO_2)_2[SiO_3(OH)]_2 \cdot 5H_2O$

Dimorphous with uranophane; generally found with uranophane and uraninite in oxidized uranium deposits.

MINERAL COUNTY In yellow-orange groups with individual crystals to 1.5 mm in quartz veins in granitic rock at the 4D claims in the Marietta district. (Jenkins, 1981)

URANOPILITE

$(UO_2)_6(SO_4)(OH)_{10} \cdot 12H_2O$

A secondary mineral related to zippeite that mostly occurs in vein-type uranium deposits.

LANDER COUNTY As greenish-yellow crystalline clusters to 2 mm with zippeite and autunite at the Apex (Rundberg) Uranium Mine in the Reese River district. (M. Jensen, o.c., 2000; GCF)

URANOSPINITE

$Ca(UO_2)_2(AsO_4)_2 \cdot 10H_2O$

Uranospinite is a rare mineral derived from the alteration of uraninite and primary arsenic minerals in oxidized uranium deposits.

WASHOE COUNTY Reported with uraninite, autunite, phosphuranylite, sabugalite, and other minerals at the Lowary claims in the Pyramid district. (Brooks, 1956; Garside, 1973)

UVAROVITE

$Ca_3Cr_2(SiO_4)_3$

A chromium-bearing *garnet*-group mineral that occurs in serpentinite with chromite and in metamorphosed limestone and skarn.

LYON COUNTY Reported in specimens from the Ludwig Mine in the Yerington district. (UAMM; WMK)

UYTENBOGAARDTITE

Ag_3AuS_2

A recently discovered, precious-metal sulfide mineral that was simultaneously described from three worldwide localities, including the Comstock Lode. At present not known from many localities, but may prove to be more common in hydrothermal precious-metal deposits.

LANDER COUNTY Identified by EMP with *electrum* and sulfide minerals in a gravity concentrate of ore from the Dean Mine in the Lewis district. (J. McGlasson, w.c., 2001)

NYE COUNTY In quartz veins in ash-flow tuff at the Bullfrog and Original Bullfrog Mines in the Bullfrog district as irregular gray metallic grains to 2 mm that rim *electrum,* are intergrown with acanthite, and are associated with chrysocolla and *limonite.* (Castor and Weiss, 1992; Castor and Sjoberg, 1993)

With *electrum,* acanthite, and *limonite* in quartz veins at the Life Preserver Mine in the Tolicha district. (Tingley et al., 1998a)

STOREY COUNTY Co-type locality as tiny, brownish-gray, metallic grains to 100 microns with acanthite, *electrum,* and quartz in a specimen from an unknown mine in the Comstock district. (Barton et al., 1978)

VALENTINITE

Sb$_2$O$_3$

Valentinite is widespread as an oxidation product of stibnite. It occurs with stibiconite and other antimony oxide minerals, from which it is practically indistinguishable in hand specimen. There are probably many more localities for the mineral in Nevada than listed here.

CHURCHILL COUNTY Reported in a specimen of gold and silver ore in the Jessup district. (WMK)

In a specimen from the Green prospect in the Lake district. (R. Thomssen, w.c., 1999)

Questionably reported at the Quick-Tung Mine in the Shady Run district. (Lawrence, 1963)

ELKO COUNTY Reported in a specimen from the Marlboro pit in Jerritt Canyon in the Independence Mountains district. (R. Thomssen, w.c., 1999)

EUREKA COUNTY As yellow alteration coatings on stibnite crystals and as red coatings on barite crystals at the Goldstrike Mine in the Lynn district. (GCF)

HUMBOLDT COUNTY Reported in quartz veins at the National Mine and Walker lease in the National district. (Melhase, 1934c)

As yellow earthy material and stains around stibnite grains at the W. P. Mine in the Ten Mile district. (Lawrence, 1963)

MINERAL COUNTY With cinnabar and stibnite at the Drew Mine in the Pilot Mountains district. (Foshag, 1927)

NYE COUNTY Reported as resulting from a mine fire at the White Caps Mine in the Manhattan district; also reported as an alteration product of stibnite at the Sunset prospect. (Gibbs, 1985; Ferguson, 1924)

As pearly, cream-colored plates to 0.5 mm in oxidized ore from the Upper Magnolia workings of the Morey Mine in the Morey district. (Jenkins, 1981)

As yellow or white fibrous material replacing large stibnite crystals at the Titus prospect in the Squaw Hills 7 miles northeast of Moores Station. (Jenkins, 1981)

PERSHING COUNTY Reported in a sample from Bloody Canyon in the Humboldt Mountains, possibly from the Bloody Canyon Mine in the Star district. (HMM)

VANADINITE

Pb$_5$(VO$_4$)Cl

Vanadinite is a moderately common mineral in the oxidation zones of base-metal deposits in the western United States. Excellent specimens have come from the Goodsprings and Chalk Mountain districts in Nevada. *Endlichite* is an arsenic-bearing variety.

CHURCHILL COUNTY Reported with wulfenite, descloizite, and other minerals in lead-silver ore at the Chalk Mountain Mine in the Chalk Mountain district, Sec. 23, T17N, R34E; *endlichite* occurs as golden-brown, barrel-shaped prisms to 5 mm on orange descloizite in vugs at joint intersections in dolomite. (Vanderburg, 1940; Schilling, 1962a; Bryan, 1972; Jensen, 1982, 1990b)

With supergene sulfides in quartz veins at the Bimetal Mine in the Holy Cross district. (Peters et al., 1996)

CLARK COUNTY Reported at the Copper King Mine in the Bunkerville district. (WMK)

In quartz veinlets with wulfenite at the Cumberland and Budget Mines in the Crescent district. (Vanderburg, 1937b)

Noted as small, brownish, waxy crystals at several prospects east of the Silver Gem tunnels in Devil Canyon in the Goodsprings district; also reported in specimens from the Bill Nye and Whale Mines and the Hemimorphite claim. Noted as reddish- to golden-brown crystals in specimens from the Ruth Mine. (Hewett, 1931; LANH; WMK; J. Kepper, w.c., 1999)

As crystal aggregates at the Quartette Mine in the Searchlight district. (Callaghan, 1939)

With wulfenite coating quartz stockwork veinlets at the Lucy Gray Mine in the Sunset district. (Schilling, 1979)

ELKO COUNTY In trace amounts with anglesite and cerussite in a specimen from an unspecified locality in the Contact district. (WMK)

ESMERALDA COUNTY Reported in specimens from the Eureka and Exclaim Mines in the Divide district. (WMK)

EUREKA COUNTY As brownish-orange crystals to 1 mm with descloizite crystals at the Gold Quarry Mine in the Maggie Creek district. (Jensen et al, 1995)

LANDER COUNTY Reported with pottsite and other minerals at the Linka Mine in the Spencer Hot Springs district. (Williams, 1988; S. Pullman, w.c., 1999)

LINCOLN COUNTY In manganese ore from a prospect pit near the Combined Metals Mine shaft in the Pioche district; also as yellow coatings or blooms on galena at the Prince Mine. (Westgate and Knopf, 1932)

NYE COUNTY As brownish *endlichite* crystals to 3 mm in large vugs above the 350-foot level at the Downieville Lead Mine in the Gabbs district. (M. Jensen, w.c., 1999)

As brownish crystals to 2 mm from the 100-foot level of an unnamed shaft 2 km southwest of the Hasbrouck shaft at Quartz Mountain in the Lodi district. (M. Jensen, w.c., 1999)

WHITE PINE COUNTY As minute brownish crystals with cerussite and aurichalcite at the Elijah Mine in the Robinson district. (M. Jensen, w.c., 1999)

VANALITE
$NaAl_8V_{10}O_{38} \cdot 30H_2O$

A rare mineral that occurs as crusts on joints and in cavities in weathered shale at the type locality in Kazakhstan.

CARSON CITY Tentatively identified as a yellow vanadium mineral with vesuvianite, tremolite, and scheelite in a specimen from 1 mile south of Carson City and west of U.S. Highway 395 in the Voltaire district. (WMK)

VANOXITE
$V_4^{4+}V_2^{5+}O_{13} \cdot 8H_2O$ (?)

A mineral found in fossil wood and as cement in sandstone-type uranium deposits in the southwestern United States.

EUREKA COUNTY With hewettite at the Van-Nav-San claim in the Gibellini district; identified on the basis of an analysis by G. Dunning. (R. Walstrom, w.c., 1999)

VARISCITE
$AlPO_4 \cdot 2H_2O$

A common mineral that has been mined as a semiprecious gemstone. It generally occurs with turquoise, crandallite, and other phosphate minerals in oxidized hydrothermal deposits. Localities in Esmeralda and Mineral Counties in Nevada have produced considerable gem-grade variscite.

ELKO COUNTY As greenish botryoidal coatings and microcrystals at the Rain Mine in the Carlin district; also as green to brown massive veins to 10 cm thick in outcrops in Woodruff Canyon. (M. Jensen, w.c., 1999)

ESMERALDA COUNTY With and without turquoise in the matrix of breccia at the Pirate No. 3 claim in the Candelaria district; also reported as pinkish to colorless crystals to 1 mm in a specimen from a site near Columbus. (Morrissey, 1968; M. Jensen, w.c., 1999; R. Walstrom, w.c., 2000)

Identified in a specimen from the Emigrant Peak area in the Silver Peak Range in the Coaldale district; also reported with turquoise from the Holland, Wilson-Capps, and Sigmund claims east of Columbus Salt Marsh. (JHS; Murphy, 1964; UCB)

Reported in specimens at the Crow Spring and Marguerite Mines in the Crow Springs district. (WMK)

Reported in a specimen from Vindicator Mountain in the Goldfield district. (R. Thomssen, w.c., 1999)

EUREKA COUNTY *Redonuite,* an iron-rich variety of variscite, occurs as a local cement in rocks of the Ordovician Valmy Formation in the Cortez

Quadrangle, Cortez district. (Gilluly and Masursky, 1965)

At the Goldstrike Mine in the Lynn district in chert as bright- to pale-green veins to 10 cm thick and as diamond-shaped euhedral crystals and clusters to 2 mm, some translucent and gemmy; also as colorless to green crystals to 1.5 mm at the Genesis Mine. (GCF; M. Jensen, w.c., 1999)

As green masses and crystal clusters at the Gold Quarry Mine in the Maggie Creek district. (Jensen et al., 1995)

HUMBOLDT COUNTY As tiny, apple-green crystals with other phosphate minerals at the Silver Coin Mine in the Iron Point district. (Thomssen, 1998; NMBM; J. McGlasson, w.c., 2001)

With curetonite and other species in veinlets cutting barite at the Redhouse Barite prospect in the Potosi district; also as massive green pods to 2 cm at the Twin Creeks Mine. (F. Cureton, w.c., 1999; S. Sears, w.c., 1999; M. Jensen, w.c., 1999)

LANDER COUNTY As radiating crystal groups at a bulldozer cut 1 mile southwest of the Gray Eagle Mine on Granite Mountain in the Bullion district, Sec. 23, T29N, R46E. (JHS)

Mined as pale-green to sky-blue nuggets to 60 cm and veins to 5 cm in the foothills of the Shoshone Range near Crescent Valley. (Sinkankas, 1997)

As light-green veins to 9 cm thick at the Apache Variscite Mine north of Austin on the west slope of the Toiyabe Range. (Rose and Ferdock, this publication)

As light- to dark-green veins to 7.5 cm thick at the Ackerman Canyon Variscite Mine. (Rose and Ferdock, this publication)

MINERAL COUNTY At various localities, with or without turquoise, in veinlets in Ordovician chert and Triassic shale in the Candelaria district; specifically reported 100 feet west of the glory hole at the Northern Belle Mine and at the Candelaria Mine. (Page, 1959; Murphy, 1964; JHS)

Reported at the Dunwoody-Prichard group in the Silver Star district. (Murphy, 1964)

NYE COUNTY Reported as *trainite* with vashegyite and *barrandite* in Ordovician slate and along a rhyolite-limestone contact at the Tom Molly Mine (Train prospect) in the Manhattan district. (Schaller, 1918; Shannon, 1923; Ferguson, 1924; Melhase, 1935d; Morrissey, 1968; DeMouthe, 1985)

Sparse in a bedded barite deposit at the P and S Mine in the Northumberland district. (Papke, 1984)

PERSHING COUNTY Reported with turquoise in a specimen from the Rye Patch district. (WMK)

As white to pale-blue and pale-pink spherules and drusy crusts to 2 mm on quartz crystals at the Willard Mine in the Willard district; associated with other phosphate minerals. (M. Jensen, w.c., 1999; Jensen and Leising, 2001)

VASHEGYITE

$Al_{11}(PO_4)_9(OH)_6 \cdot 38H_2O$ or
$Al_6(PO_4)_5(OH)_3 \cdot 23H_2O$

Vashegyite is a moderately rare mineral found with other secondary phosphate minerals such as variscite, wavellite, and turquoise.

EUREKA COUNTY Identified with wavellite, schoderite, metahewettite, and metaschoderite at the Bisoni and Bisoni-McKay properties and on the Van-Nav-San claim in the Gibellini district. (Hausen, 1962)

NYE COUNTY Vashegyite, white or yellowish-white in color, was found in slate near a rhyolite-limestone contact at the Vashegyite Gem Mine (Tom Molly Mine, Train prospect) in the Manhattan district; the vashegyite reportedly gave way downward in the vein to variscite, which in turn graded into turquoise. (Shannon, 1923; Ferguson, 1924; Clinton, 1929; S. Pullman, w.c., 1999)

VAUQUELINITE

$Pb_2Cu^{2+}(CrO_4)(PO_4)(OH)$

Vauquelinite is a rare mineral that occurs in the oxidation zones of base-metal deposits that contain small amounts of chromium.

CLARK COUNTY As dark-yellowish-green crystals associated with chlorargyrite, wulfenite, and pyromorphite at an unnamed prospect south of Crystal Pass in the Goodsprings district. (Kepper, 1998)

ELKO COUNTY As greenish-yellow crystals to 0.25 mm on skarn fracture surfaces at the Killie (Latham) Mine in the Spruce Mountain district; also noted on gossan as yellowish-green coatings locally associated with hemimorphite. (Jenkins, 1981; Dunning and Cooper, 1987)

HUMBOLDT COUNTY As microcrystals with mottramite, carbonate-fluorapatite, and hinsdalite at the Silver Coin Mine in the Iron Point district. (Thomssen, 1998; M. Jensen, w.c., 1999)

NYE COUNTY With skinnerite, stetefeldtite, molybdofornacite, and fornacite in a specimen from the Belmont Mine in the Belmont district. (EMP)

VERMICULITE

$(Mg,Fe^{2+},Al)_3(Si,Al)_4O_{10}(OH)_2 \cdot 4H_2O$ (general)

A group of silicate minerals; listed as a species because individual minerals have not been specified. Generally formed by weathering or hydrothermal alteration of biotite or phlogopite, typically in ultramafic rocks.

CLARK COUNTY Vermiculite, hydrobiotite, and biotite occur in metamorphosed ultramafic rock as disseminations and local masses of flakes and books to 1/2 inch at the Gold Butte Vermiculite Mine in the Gold Butte district, Sec. 15, T19S, R70E. (Leighton, 1967)

LINCOLN COUNTY From a prospect in a 30- to 40-foot-thick layer of altered biotite schist between pegmatite dikes in Precambrian rocks about 3 miles southeast of Galt. (Tschanz and Pampeyan, 1970)

WASHOE COUNTY Reported as *jefferisite* in a specimen from the Steamboat Springs district. (WMK)

VERNADITE

$(\delta\text{-}MnO_2)(Mn^{4+},Fe^{3+},Ca,Na)(O,OH)_2 \cdot nH_2O$

Vernadite is found with other manganese oxides as a weathering product of primary manganese minerals; it is also an important component of ocean-floor manganese nodules.

NYE COUNTY Identified by XRD in *wad* in a vein with calcite, *chalcedony,* and quartz from unnamed shallow shafts and prospects in the northeastern Goldfield Hills, Secs. 11 and 12, T2S, R44E. (Tingley et al., 1998a)

VÉSIGNIÉITE

$BaCu_3^{2+}(VO_4)_2(OH)_2$

A rare secondary mineral that is generally found in oxidized sandstone-type uranium deposits.

EUREKA COUNTY As light-green, botryoidal and spherical clusters on hematite and goethite frac-

ture coatings with jahnsite, jarosite, whitlockite, and other minerals at the Goldstrike Mine in the Lynn district. (GCF)

With alunite, cuprite, libethenite, and malachite as druses and small rosettes of apple-green crystals that coat and partially replace barite at the Gold Quarry Mine in the Maggie Creek district. (Jensen et al., 1995)

VESUVIANITE

$Ca_{10}Mg_2Al_4(SiO_4)_5(Si_2O_7)_2(OH)_4$

Vesuvianite (*idocrase*) is a common mineral in contact-metamorphosed zones in limestones or magnesian carbonate rocks. It is frequently associated with *garnet,* epidote, diopsidic *pyroxene,* and other calc-silicate minerals.

CARSON CITY In a specimen from 1 mile south of Carson City and west of U.S. Highway 395 in the Voltaire district. (WMK)

ELKO COUNTY Reported as beryllium-bearing in skarn at the Star Mine in the Harrison Pass district, Sec. 18, T28N, R58E. (Hess and Larsen, 1920; Warner et al., 1959; Griffiths, 1964)

As olive-green prismatic crystals and crystalline masses to 5 cm at and near the contact between marble and migmatitic granite in Lamoille Canyon in the Ruby Mountains. (GCF)

Reportedly beryllium-bearing in skarn from an unspecified locality in the Wells district. (Griffiths, 1964)

EUREKA COUNTY Generally the most common calc-silicate mineral in skarn at the Goldstrike Mine in the Lynn district; occurs as brown-green masses and euhedral crystals to 5 cm with calcite. (GCF)

HUMBOLDT COUNTY Locally abundant in contact-metamorphic rocks in the Osgood Mountains. (Hotz and Willden, 1964)

LANDER COUNTY In skarn at the West orebody in Copper Canyon in the Battle Mountain district. (Theodore and Blake, 1978)

Tentatively identified in skarn with grossular, calcite, and quartz at the Linka Mine and other tungsten deposits in the Spencer Hot Springs district. (McKee, 1976)

LINCOLN COUNTY With a large variety of other minerals in skarn adjacent to igneous intrusions in the Bristol district near Blind Mountain. (Westgate and Knopf, 1932)

LYON COUNTY In garnetite skarn as rough olive-green, pale-green, and brown crystals to over 10 cm in the Yerington district; specifically noted as brilliant chrome- and grass-green crystals to 5 cm in a prospect above the Ludwig townsite. (Knopf, 1918b; Melhase, 1935e; UCB; M. Rogers, o.c., 1999; F. Cureton, w.c., 1999; M. Jensen, w.c., 1999; S. Pullman, w.c., 1999)

MINERAL COUNTY In skarn with *garnet,* epidote, and other minerals at the Queen Mine in the Buena Vista district. (Papke, 1979)

With *axinite* and other minerals in the Fitting district 7 miles north of Luning. (S. Pullman, w.c., 1999)

Reported with magnesio-axinite, prehnite, actinolite, and epidote from an unspecified locality near Luning; may be the same locality as that listed above. (Dunn et al., 1980)

Reported in a specimen from the Silver Star district. (WMK)

NYE COUNTY Reported in a specimen from Beatty. (WMK)

Reported in a specimen from Gabbs. (HMM)

As idiomorphic, zoned reddish-brown crystals in wall rock at the Queen City claims in the Queen City district. (Hewett et al., 1968)

PERSHING COUNTY Reported in a specimen from the Nightingale district. (WMK)

WHITE PINE COUNTY Reportedly beryllium-bearing on the east side of the Snake Range. (Griffiths, 1964)

VIKINGITE

$Ag_5Pb_8Bi_{13}S_{30}$

Vikingite is a rare sulfosalt mineral named from its type locality in Greenland; it occurs with other sulfides in a few hydrothermal and pyro-metasomatic deposits.

NYE COUNTY As silver-bronze clots and crystal laths to 4 mm in quartz at a prospect near Gabbs. (Jenkins, 1981; EMP)

VIOLARITE

$Fe^{2+}Ni_2^{3+}S_4$

Violarite occurs with other sulfides in nickel-bearing hydrothermal and magmatic deposits.

CHURCHILL COUNTY Reported in nickel ore with annabergite, morenosite, retgersite, and nicke-line at the Nickel Mine in the Table Mountain district. (Ransome, 1909b; Ferguson, 1939; WMK)

CLARK COUNTY Abundant after pentlandite and chalcopyrite in nickel ore with magnetite, pyrite, and pyrrhotite at the Great Eastern and Key West Mines in the Bunkerville district. (Lindgren and Davy, 1924; Beal, 1964)

EUREKA COUNTY As cubo-octahedral overgrowths to 20 microns on millerite crystals in the Betze deposit at the Goldstrike Mine in the Lynn district. (GCF; Ferdock et al., 1997)

VIVIANITE

$Fe_3^{2+}(PO_4)_2 \cdot 8H2O$

A secondary mineral in the oxidized portions of base-metal deposits and in pegmatites; has also been found in fossils in sedimentary rocks.

CARSON CITY Reported in a specimen from an unspecified locality. (UMNH)

EUREKA COUNTY As dark-blue, translucent, flattened columnar crystals to 4 cm and fracture fillings in chert at the Goldstrike Mine in the Lynn district; also as dark-greenish-blue, translucent, flattened prismatic crystals to 4 cm on fractures at the Genesis Mine. (GCF)

Reported in a specimen from the Gold Quarry Mine in the Maggie Creek district. (R. Thomssen, w.c., 1999)

LANDER COUNTY In the Fortitude pit in the Battle Mountain district as vitreous, dark-blue, transparent to translucent, bladed crystals to 2.5 cm with wavellite and marcasite; also in the Copper Canyon Mine as dark-blue-green laths to 5 cm on fractures. (UMNH; M. Jensen, w.c., 1999; B. Hurley, w.c., 1999)

STOREY COUNTY Reported in a specimen from the Cedar Ravine Tunnel in the Comstock district. (WMK)

WHITE PINE COUNTY Near the bottom of the Liberty pit in the Robinson district as nearly colorless to dark-blue crystals in seams and on joint planes, as small grains and globular masses to 1/2 inch in limestone, and as small crystals on calcite in vugs; in silvery crystals as much as 3 inches long on a specimen in the W. M. Keck Museum. Also reported from the Cedar Hill Mine. (Gianella, 1938; WMK; R. Thomssen, w.c., 1999)

VOLBORTHITE

$Cu_3^{2+}V_2^{5+}O_7(OH)_2 \bullet 2H_2O$

Volborthite, a moderately rare mineral that occurs in the oxidation zones of copper deposits containing vanadium, is generally associated with more common secondary copper minerals.

EUREKA COUNTY Reported from the Bisoni claims in the Gibellini district. (R. Thomssen, w.c., 1999)

Reported from the Carlin Mine in the Lynn district. (Anthony et al., 2000)

As bright-apple-green, platy crystals, and dark-olive-green, flat-lying, radiating spheres to 1.5 cm, on fractures at the Gold Quarry Mine in the Maggie Creek district; also as single crystals to 1 mm. (Jensen et al., 1995)

LYON COUNTY As apple-green platy crystals to 1.5 mm with brochantite at the Douglas Hill Copper Mine in the Yerington district. (M. Jensen, w.c. 1999)

NYE COUNTY As green glassy coatings with turanite, brochantite, and chrysocolla on vanadiferous slate at an unnamed prospect in the Manhattan district on the road from Manhattan to Belmont. (Jenkins, 1981; J. McGlasson, w.c., 2001)

VOLTAITE

$K_2Fe_5^{2+}Fe_4^{3+}(SO_4)_{12} \bullet 18H_2O$

A relatively common secondary mineral in oxidized portions of hydrothermal deposits; also found in solfataric deposits and as a result of burning pyrite-bearing ore.

CLARK COUNTY Reported in a specimen from Boulder City, possibly from the Alunite district. (RMO)

ELKO COUNTY Reported in oxidizing ore in workings at the Meikle Mine in the Lynn district. (D. Tretbar, o.c., 2002)

NYE COUNTY Reported from the 352-foot level of the San Rafael Mine in the Lodi district. (R. Thomssen, w.c., 1999)

VYSOTSKITE

(Pd,Ni)S

A platinum-group metal mineral that occurs in magmatic deposits in mafic and ultramafic rocks.

CLARK COUNTY Reported from near Moapa, probably from the Bunkerville district. (Anthony et al., 1990)

W

WAIRAKITE

Ca[Al$_2$Si$_4$O$_{12}$]•2H$_2$O

A zeolite-group mineral found in hydrothermal deposits; well-known occurrences are in active geothermal systems at the Geysers in California and at Wairakei, New Zealand, the type locality.

STOREY COUNTY Identified in propylitized volcanic rocks adjacent to the Comstock Lode in the Comstock district. (Vikre, 1989)

WAKABAYASHILITE

(As,Sb)$_{11}$S$_{18}$

Wakabayashilite is a rare mineral that is found with realgar and orpiment in some arsenic-containing gold ores. Specimens from the White Caps Mine in Nevada are said to be the best in the world.

HUMBOLDT COUNTY In specimens as yellow needles to 5 mm at the Getchell Mine in the Potosi district. (S. Pullman, w.c., 1999)

MINERAL COUNTY Yellow inclusions to 3 mm in barite from the Regent Mine in the Rawhide district have been identified as arsenic-antimony sulfide by SEM/EDX and may be wakabayashilite. (GCF)

NYE COUNTY As fine acicular masses of flexible, golden- to lemon-yellow hairs to 2 cm long, associated with realgar and orpiment in calcite, at the White Caps and Black Mammoth Mines in the Manhattan district, Sec. 21, T8N, R44E; identified as "hair orpiment" in many old collections. (Melhase, 1935d; Kato et al., 1970; Gibbs, 1985; AMNH)

WARDITE

NaAl$_3$(PO$_4$)$_2$(OH)$_4$•2H$_2$O

Wardite, a secondary aluminum phosphate mineral, is generally found with other phosphates such as variscite in hydrothermal deposits; it also occurs as a late-stage mineral in granite pegmatite.

HUMBOLDT COUNTY Calcian and non-calcian varieties occur as colorless octahedral crystals to 0.5 mm with lipscombite at the Silver Coin Mine in the Iron Point district. (Thomssen, 1998; M. Jensen, w.c., 1999)

NYE COUNTY Wardite, reported as *metacalciowardite,* was found to be intimately associated with variscite just north of Manhattan in the Manhattan district, where vashegyite and other phosphate minerals were also noted; this is probably the same occurrence as the Vashegyite Gem Mine where wardite was identified with strengite, fluorapatite, and vashegyite. (Melhase, 1935d; WMK)

WAVELLITE

Al$_3$(PO$_4$)$_2$(OH,F)$_3$•5H$_2$O

Wavellite is a common mineral in secondary phosphate assemblages; it generally occurs in oxidized hydrothermal deposits and in sedimentary phosphate deposits. Includes the discredited species *utahlite, lucinite,* and *peganite.*

ELKO COUNTY As green radiating aggregates to 6 mm that fill vugs to 4 cm across in carbonaceous siltstone at the Rain Mine in the Carlin district. (D. Heitt, o.c., 1991)

ESMERALDA COUNTY Reported as *peganite* at the Red Hill (Redlich) Mine in the Rock Hill district, Sec. 34, T4N, R36E. (WMK)

EUREKA COUNTY With vashegyite, schoderite, and other minerals at the Bisoni and Bisoni-

McKay properties and the Van-Nav-San claim in the Gibellini district. (Hausen, 1962; Schilling, 1962b)

As pale-green botryoidal balls and globules of radiating crystalline material to 4 mm on fractures and in open spaces in silicified breccia at the Goldstrike Mine in the Lynn district. (GCF)

In pale-green, compact masses or colorless, radiating, prismatic crystals in small clusters on fracture surfaces with strengite, fluellite, torbernite, hewettite, and opal at the Gold Quarry Mine in the Maggie Creek district. (Jensen et al., 1995)

HUMBOLDT COUNTY As radiating colorless crystals to 1 mm with turquoise and variscite at the Silver Coin Mine in the Iron Point district. (Thomssen, 1998; NMBM; M. Jensen, w.c., 1999)

As colorless or (rarely) blue microcrystals and masses at the Redhouse Barite prospect in the Potosi district. (S. Sears, w.c., 1999; S. Pullman, w.c., 1999)

LYON COUNTY In small rosettes of crystals with cacoxenite at an unnamed prospect on the south side of Carson Hill in the Yerington district. (M. Jensen, w.c., 1999)

MINERAL COUNTY Reported from the Candelaria district. (S. Pullman, w.c., 1999)

NYE COUNTY Reported on slate near Belmont. (Hoffman, 1878)

In small white spheres with concentric radial structure on vein quartz in the Tonopah district. (Melhase, 1935c)

PERSHING COUNTY In pale-yellow spherules to 1 cm at the Hecla Rosebud Mine in the Rosebud district. (M. Jensen, w.c., 1999)

Reportedly Nevada's finest specimens as colorless to pale-yellow-green, crystalline spherules to 2.5 cm and in other crystal groups in quartz-crystal-lined vugs at the Willard Mine in the Willard district; associated with variscite, fluellite, cacoxenite, crandallite, jarosite, gypsum, and pyrite; also noted as epimorphs of fluellite crystals to 6 mm. (M. Jensen, w.c., 1999; Jensen and Leising, 2001)

WEBERITE
Na_2MgAlF_7

A rare fluoride mineral that occurs with cryolite in Greenland and in pegmatite and carbonatite elsewhere.

MINERAL COUNTY With other late-stage alumino-fluoride minerals in an *amazonite*-topaz pegmatite at the Zapot claims in the Fitting district. (Foord et al., 1999b)

WEEKSITE
$K_2(UO_2)_2Si_6O_{15}\bullet 4H_2O$

A secondary uranium mineral that is found in opal veins in volcanic rocks and in sandstone-type and other types of uranium deposits.

HUMBOLDT COUNTY As bright-yellow crusts and finely drusy acicular crystals to 0.3 mm that partially fill open spaces with white opal in breccia in rhyolite in the Virgin Valley district, NE1/4 Sec. 24, T45N, R25E. (Castor et al., 1996; Castor and Henry, 2000)

MINERAL COUNTY Reported from the Teels Marsh district. (Anthony et al., 1995)

WASHOE COUNTY Tentatively identified as scaly yellow aggregates in fractures and cavities in highly radioactive rock at the Garrett prospect in the Pyramid district. (Castor et al., 1996)

WEILITE
$CaHAsO_4$

A very rare secondary mineral in oxidized arsenic-rich deposits.

HUMBOLDT COUNTY As very fine-grained white crusts on picropharmacolite in carbonaceous shale from the North Pit area at the Getchell Mine in the Potosi district, probably the result of a long-term dehydration of pharmacolite and haidingerite. (Dunning, 1988)

WEISSBERGITE
$TlSbS_2$

A Nevada type mineral, weissbergite is extremely rare; it has so far been found only in the Carlin gold deposit in Nevada and at two other localities worldwide. Discovery of the mineral resulted from deliberate search for an antimony analog to lorandite.

EUREKA COUNTY As steel-gray opaque grains to 0.5 mm in silicified rocks with quartz and stibnite from the east orebody at the Carlin Mine in the Lynn district, Sec. 13, T35N, R50E. (Dickson and Radtke, 1978)

WHITLOCKITE

$Ca_9(Mg,Fe^{2+})(PO_4)_6(PO_3OH)$

Whitlockite is an uncommon mineral that has generally been found in pegmatite with other phosphate species; it is also known from cave deposits and meteorites. In Nevada the mineral has been identified in Carlin-type gold deposits, a new environment.

EUREKA COUNTY As yellow rhombs to 5 mm in fracture coatings on oxidized silicified breccia with carbonate-fluorapatite, hydroxylapatite, jahnsite, variscite, other phosphates, and iron oxide minerals at the Goldstrike Mine in the Lynn district. (GCF)

HUMBOLDT COUNTY Verified by XRD as whitish rhombs to 2 mm with beudantite, hinsdalite, cerussite, and carbonate-fluorapatite at the Twin Creeks Mine in the Potosi district. (M. Jensen, w.c., 1999)

WICKENBURGITE

$Pb_3CaAl_2Si_{10}O_{24}(OH)_6$

A mineral that has been found with other secondary base-metal minerals at localities near Wickenburg, Arizona.

EUREKA COUNTY As a green coating on a specimen of pale-brown crystalline rock reportedly from the Fish Creek Range; formerly identified as schoderite, but new identification made by XRD. (GCF)

WILLEMITE

Zn_2SiO_4

A zinc mineral of secondary origin, especially in deposits in limestone such as at Franklin and Sterling Hill in New Jersey where it was an ore mineral; it is not uncommon in oxidized hydrothermal deposits.

CLARK COUNTY In colorless to white radiating sprays to 1 mm with galena at the Singer Mine in the Goodsprings district; also reported from the Boss and Milford Mines. (Heyl and Bozion, 1962; Adams, 1998; Jedwab, 1998b; Kepper, this volume)

Reported from a prospect 9 miles northwest of Nelson. (R. Thomssen, w.c., 1999)

ELKO COUNTY Formed by alteration of sphalerite as perfect crystals with, and on, duftite in the Railroad district. (Ketner and Smith, 1963)

LINCOLN COUNTY As millimeter-sized crystals coating oxidized lead-zinc ore at a prospect on the east side of the crest of Arizona Peak in the Pioche district. (Westgate and Knopf, 1932)

MINERAL COUNTY Reported from oxidized zinc deposits in the Mina area, presumably in the Santa Fe district. (Heyl and Bozion, 1962)

NYE COUNTY As clear complex crystals to 6.5 mm with hemimorphite in quartz vugs at the Hasbrouck Mine near Quartz Mountain in the Lodi district; also noted in a specimen from the San Rafael Mine. (Jenkins, 1981; R. Thomssen, w.c., 1999)

WHITE PINE COUNTY Reported from an unspecified locality in the Robinson district. (Heyl and Bozion, 1962)

In oxidized ore at the Willard claim in the Taylor district. (Lovering and Heyl, 1974)

WITHERITE

$BaCO_3$

A relatively rare carbonate mineral that is found in veins, generally in association with barite, galena, or fluorite.

ESMERALDA COUNTY With barite and secondary copper minerals in replacement bodies in limestone at the Mammoth No. 1 claim, NW1/4 Sec. 1, T1S, R40E. (Papke, 1984)

EUREKA COUNTY Reported in veins with barite at the Carlin Mine in the Lynn district. (Papke, 1984)

Questionably reported at the Maggie Creek Mine in the Maggie Creek district. (Papke, 1984)

LANDER COUNTY Reported in bedded barite at the Argenta Mine in the Argenta district. (Papke, 1984)

NYE COUNTY In minor amounts with celestine in bedded barite at the Ann claims in the Northumberland district. (Papke, 1984)

WOLFRAMITE

$(Mn,Fe^{2+})WO_4$ (general)

A group mineral name for a common material in hydrothermal veins, pegmatites, and some skarn deposits; end-member species are hübnerite and ferberite. Species have not been identified for localities reported below, although both *wolframite* and hübnerite are reported at some. Excellent specimens have come from the Ellsworth district.

ESMERALDA COUNTY Reported in a specimen from the Coaldale area. (WMK)

EUREKA COUNTY Occurs in association with gold at the West Archimedes deposit in the Eureka district. (Dilles et al., 1996)

HUMBOLDT COUNTY Identified in a sample of skarn from an unspecified locality in the Potosi district. (JHS)

LINCOLN COUNTY As large masses of blocky crystals with minor scheelite and abundant sulfides in gold-bearing quartz veins at the Comet Mine in the Comet district. (Tschanz and Pampeyan, 1970; Melhase, 1934d; WMK; M. Jensen, w.c., 1999)

With scheelite in light-green skarn on the western edge of a granitic stock in the Worthington Mountains in the Freiburg district. (Tschanz and Pampeyan, 1970)

In quartz veins in quartzite at the Geyser and Eagle Rock Mines in the Patterson district. (Schrader, 1931a)

MINERAL COUNTY With scheelite and bismuth minerals in a quartz vein at the Pine Crow Mine in the Silver Star (Black Mountain) district. (Ross, 1961)

NYE COUNTY Reported with scheelite 9 miles south of Beatty in the Bare Mountain district. (WMK)

In quartz veins at the Falcon prospect in the Belmont district. (Kral, 1951; Stager and Tingley, 1988)

In the Ellsworth (Mammoth) district, the type locality for hübnerite, as well-crystallized material that has been prized by mineral collectors for many years, with scheelite and hübnerite in quartz veins at the Eagle claims, Sec. 19, R13N, T38E; also reported with scheelite and hübnerite on the Commodore Mine dump. (Kral, 1951; Stager and Tingley, 1988)

With hübnerite in quartz veinlets in quartz monzonite at the N & H claims in the Round Mountain district; also reported in a specimen from the Melvin Mine. (Stager and Tingley, 1988; WMK)

In the Belmont Mine in the Tonopah district in quartz veinlets that cut earlier quartz, rhodochrosite, and silver ore veins; also in minor amounts in a quartz vein in the Valley View Mine. (Bastin and Laney, 1918)

WHITE PINE COUNTY In small quantities with more abundant scheelite in tungsten deposits in the Cherry Creek district. (Hose et al., 1976; JHS)

Reported in a specimen of tungsten ore in the Osceola district. (WMK)

In trace amounts with hübnerite and scheelite at the Hub Mine in the Tungsten district. (Hose et al., 1976)

WOLLASTONITE

$CaSiO_3$

Wollastonite is common in contact-metamorphic zones in calcareous rocks along with such minerals as vesuvianite, diopside, and grossular.

ELKO COUNTY With other skarn minerals along a limestone-granodiorite contact in the Contact district. (Schilling, 1979)

Abundant in zones of silicified limestone at the Davis tunnel in the Railroad district. (Smith, 1976)

ESMERALDA COUNTY Large amounts of rock containing about 70 percent wollastonite are present in the Gilbert district about 2 miles southeast of the site of Gilbert; associated minerals are calcite, quartz, vesuvianite, diopside, *garnet,* and epidote. (Castor, 1992; White Plains Resources Corp. pamphlet; D. Hudson, w.c., 1999)

Reported in a specimen from near Weepah in the Weepah district. (WMK)

EUREKA COUNTY As rare white lathy masses in skarn at the Goldstrike Mine in the Lynn district. (GCF)

HUMBOLDT COUNTY Very abundant in skarn in the Osgood Mountains in the Potosi district near the Getchell Gold Mine; also in large pure masses at the Pinson Mine. (Gianella, 1941; Hotz and Willden, 1964; M. Jensen, w.c., 1999)

LANDER COUNTY Identified in skarn with *garnet,* actinolite, tremolite, and other minerals at the

West orebody in Copper Canyon in the Battle Mountain district. (Roberts and Arnold, 1965; Theodore and Blake, 1978)

LINCOLN COUNTY Abundant with vesuvianite, *garnet,* and other minerals in skarn in the vicinity of Blind Mountain in the Bristol district. (Westgate and Knopf, 1932)

LYON COUNTY Wollastonite is one of the constituents of the skarns in the Yerington district, where it occurs in thick, white strata that are almost monomineralic. It was specifically noted as pure white crystals to 100 cm in skarn at the Ludwig Mine and reported with *apatite,* quartz, and epidote at the Bluestone Mine. (Knopf, 1918b; Melhase, 1935e; M. Jensen, w.c., 1999; S. Pullman, w.c., 1999)

MINERAL COUNTY As pods with scheelite-bearing skarn along a granite-limestone contact at the Dry Gulch claims in the Fitting district. (Ross, 1961)

In skarn at the Nevada Scheelite Mine in the Leonard district. (Warner et al., 1959)

PERSHING COUNTY Reportedly abundant at a locality near Gerlach, possibly in the Hooker district. (Gianella, 1941)

In relatively pure, coarsely crystalline masses to 500 pounds in hornfels with diopside and grossular at the MGL and Nightingale Tungsten Mines in the Nightingale district; reportedly has pale-orange fluorescence. (M. Rogers, o.c., 1999; WMK)

STOREY COUNTY With grossular and diopside in calcareous hornfels adjacent to granodiorite near Ramsey. (Rose, 1969)

WASHOE COUNTY With *garnet* and diopside in hornfels in the Truckee Canyon about 5 miles west of Fernley. (Rose, 1969)

WHITE PINE COUNTY Reported as a skarn mineral in the Robinson district. (James, 1976)

WOODRUFFITE

$(Zn,Mn^{2+})Mn_3^{4+}O_7 \cdot 1\text{-}2H_2O$

A manganese oxide mineral found in oxidized base-metal ore, in massive manganese ore, and in sea-floor manganese nodules.

EUREKA COUNTY Reported in a specimen from the Gibellini vanadium deposit, but probably from the Gibellini Manganese Mine in the Gibellini district. (HMM)

WROEWOLFEITE

$Cu_4^{2+}(SO_4)(OH)_6 \cdot 2H_2O$

A rare secondary mineral in copper deposits, commonly found with chalcocite and covellite.

PERSHING COUNTY As dull-greenish–blue, translucent microcrystals and masses with scorodite and pharmacosiderite from the Copper stope at the Majuba Hill Mine in the Antelope district. (UMNH)

WULFENITE

$PbMoO_4$

Wulfenite, a widespread mineral that is highly prized by collectors, at one time was the world's only source of molybdenum. No specimens from Nevada compare with those from Arizona, but good specimens have come from the Searchlight, Eureka, Goodsprings, Downeyville, Lodi, Robinson, and Chalk Mountain districts.

CHURCHILL COUNTY As dusky-brown to bright-red-orange tablets to 1.2 cm in oxidized lead-silver ore at the Chalk Mountain Silver Lead Mines in the Chalk Mountain district, most commonly in *limonite* or in small pods and veinlets of altered galena and cerussite; associated with anglesite, chlorargyrite, plumbojarosite, mimetite, and vanadinite. Also reported from a prospect on the northwest side of Chalk Mountain. (Vanderburg, 1940; Schilling, 1962a; Bryan, 1972; Jensen, 1982; WMK; R. Thomssen, w.c., 1999)

As microscopic tabular crystals at the Westgate Mine and as prismatic crystals at an unnamed prospect just south of Highway 50 at Westgate in the Westgate district; also reported as yellow tabular crystals to 3 mm in prospects on the southern end of the Clan Alpine Mountains. (R. Walstrom, w.c., 1999; M. Jensen, w.c., 1999)

Sparse in oxidized ore and quartz-*adularia* veins at the Wonder Mine in the Wonder district. (Schilling, 1979, 1980; M. Jensen, w.c., 1999)

CLARK COUNTY In quartz veinlets with vanadinite at the Budget and Cumberland Mines in the Crescent district. (Vanderburg, 1937b; Schilling, 1980)

In the Goodsprings district at many localities. Reported in the Shenandoah Mine as abundant, tabular orange to brown crystals lining cavities in lead-zinc ore; also noted as flat waxy crystals in the Smithsonite Mine. Described as translucent, paper thin, lemon-yellow plates to 15 mm in brecciated dolomite at the Milford Mine. (Hewett, 1931; Melhase, 1934e; Schilling, 1962a, 1979; Kepper, 1998)

Widespread in the northern part of the Searchlight district as tabular, yellow-orange to orange-brown crystals to 1 mm on other minerals, especially quartz and chrysocolla; specifically reported at the Quartette and Duplex Mines. (Callaghan, 1939; M. Jensen, w.c., 1999)

In a breccia pipe with vanadinite at the Lucy Gray Mine in the Sunset district. (Schilling, 1979, 1980)

ELKO COUNTY Reported in a specimen from Carlin without further notation. (UCB)

Rare in quartz veins with cerussite and other minerals in the Dolly Varden district. (Schilling, 1979, 1980)

In large masses of thin, transparent, lemon-yellow crystals with cerussite and anglesite at the Tecoma Mine, Lucin district. This notable occurrence is just east of the state line in Utah, but has been reported as a Nevada locality by some authors. (Hague and Emmons, 1877; Melhase, 1934c; Gianella, 1941; LaPointe et al., 1991; WMK)

With *limonite,* cerussite, anglesite, galena, malachite, smithsonite, and other minerals in oxidized bodies at the Killie (Latham) Mine in the Spruce Mountain district. (Schrader, 1931b; Schilling, 1979)

Reported as museum-quality crystals on calcite in the upper workings of the Jackson Mine in the Tecoma district; described as abundant, large, lemon-yellow to orange crystal aggregates containing individual crystals to 1.5 inches. (Hague and Emmons, 1877; Horton, 1916; Schilling, 1962a, 1962b)

ESMERALDA COUNTY In quartz veins with other minerals 1 mile north of the site of Columbus in the Candelaria district. (R. Walstrom, w.c., 2000)

Reported at the Stateline Mine in the Divide district. (Schilling, 1980)

Reported with massicot at the Redemption Mine in the Gold Point (Hornsilver) district. (Ransome, 1909c)

Locally abundant as crystals on galena at the McNamara Mine in the Palmetto district. (Schilling, 1979, 1980)

In a specimen from the Rock Hill Mine in the Rock Hill district. (R. Thomssen, w.c., 1999)

In oxidized ore with anglesite, cerussite, chlorargyrite, smithsonite, and hemimorphite at the Sylvania Mine in the Sylvania district; also noted as yellow-orange crystals to 1 mm with chrysocolla at a prospect near the Old Sylvania (Stateline) Mine. (Schilling, 1979, 1980; M. Jensen, w.c., 1999)

As yellow tablets to 1 mm in the Weepah Mine open pit in the Weepah district; also reported with hemimorphite, malachite, and cerussite at an unnamed prospect 1.6 miles west-northwest of the General Thomas Mine. (R. Walstrom, w.c., 1999; M. Jensen, w.c., 1999)

EUREKA COUNTY Relatively abundant and widespread in the Eureka district as orange to yellow-orange crystals to 2 cm in oxidized ore lining caverns above unoxidized ore, locally found with galena; specifically reported from the Eureka Consolidated, K. K., Lawton, Phoenix, and Richmond-Eureka Mines. (Gianella, 1941; Nolan, 1962; Schilling, 1979; UCB; M. Jensen, w.c., 1999; WMK)

HUMBOLDT COUNTY As colorless, square microcrystals at the Silver Coin Mine in the Iron Point district. (Thomssen, 1998; M. Jensen, w.c., 1999; J. McGlasson, w.c., 2001)

LANDER COUNTY Reported at the Copper Canyon Mine in the Battle Mountain district. (Schilling, 1980)

As colorless crystals to 1.1 mm from the Estella vein at the Betty O'Neal Mine in the Lewis district. (S. White, w.c., 1999; R. Thomssen, w.c., 1999)

LINCOLN COUNTY Reported in a specimen from the Shenodoah Mine in the Eagle Valley district. (WMK)

LYON COUNTY As tabular brown plates to 4 mm with linarite, caledonite, cerussite, paratacamite, and other minerals at the Old (Lost) Soldier Mine in the Churchill district. (Schilling, 1980; F. Cureton, w.c., 1999)

MINERAL COUNTY Rare as pale-yellow, thick, tabular crystals to 1.5 mm at the Simon Mine in the Bell district. (M. Jensen, w.c., 1999)

Reported in specimens from the Mount Diablo Mine in the Candelaria district; probably from the Mount Diablo pit of the Candelaria Mine. (TUB; UCB)

As yellow-orange tabular crystals to 1 cm at the

Acme Mine in the Garfield district. (M. Jensen, w.c., 1999)

Reported at an unspecified prospect near the 4-D uranium claims in the Marietta district. (Jenkins, 1981)

Reported to occur in the vicinity of Graham Spring, about 2 miles north of the Gunmetal Mine in the Pilot Mountains district. (Schilling, 1980)

As white and yellow plates in gold ore at the Nevada Rand Mine in the Rand district. (Schrader, 1947)

In skarn at the New Boston–Blue Ribbon properties in the Garfield district. (Jenkins, 1981; W. Hunt, o.c., 2000)

As bright-yellow-orange, tabular crystals to 4 mm (R. A. Walstrom specimens) in a road cut near the Gravity mine. (M. Jensen, w.c., 1999)

As unusual, white bipyramidal crystals from a locality in the Gabbs Valley Range. (Jenkins, 1981)

Reported from Egan Canyon in the Gillis Range. (R. Thomssen, w.c., 1999)

NYE COUNTY In a specimen in the Brush Collection at Yale University from the El Dorado Mine in the Belmont district. (Horton, 1916; Schilling, 1980)

As white, yellowish-green, and brilliant-yellow plates and pyramidal prisms with adamite, hemimorphite, duftite, and mimetite at the San Rafael and Hasbrouck Mines in the Lodi district. (ASDM; Jenkins, 1981)

Reported in the Round Mountain district. (Gianella, 1941)

From an unspecified locality in the Tonopah district as thin colorless plates with twinned iodargyrite crystals implanted on them; specifically reported from the Mogul Mine as yellow-orange crystals to 2 mm with hemimorphite and galena. (Bastin and Laney, 1918; Melhase, 1935c; M. Jensen, w.c., 1999)

Abundant in gossan as yellow and orange tablets to 2 cm, very thin and fragile, at an unknown locality on the west slope of the Paradise Range. (Kral, 1951)

Reported in a specimen from Downieville. (R. Thomssen, w.c., 1999)

PERSHING COUNTY Reported from the Wheeler Mine dump in the Unionville district. (R. Thomssen, w.c., 1999)

STOREY COUNTY Rare with anglesite and pyromorphite in oxidized ore of the Comstock Lode in the Comstock district. (Bastin, 1922; Koschmann and Bergendahl, 1968)

WHITE PINE COUNTY As orange crystals to 1 cm in gossan at the Success Mine in the Duck Creek district. (M. Jensen, w.c., 1999)

Reportedly tungsten-bearing in quartz with heyite, mimetite, pyromorphite, and a variety of other minerals at the Betty Jo claim in the Nevada district. (Williams, 1973)

As orange-brown, tabular crystals to 1.5 cm on jasperoid from the Ruth pit in the Robinson district; also noted as pale crystals at the Elijah Mine and in a specimen from the Kimberly pit. (M. Jensen, w.c., 1999; S. Kleine, o.c., 1999; R. Thomssen, w.c., 1999)

Reported in oxidized ore in the Taylor district. (Lovering and Heyl, 1974)

Identified in a sample from an unspecified locality in the White Pine district. (Schilling, 1980)

WURTZITE

(Zn,Fe)S

Wurtzite, a sphalerite dimorph, occurs in hydrothermal veins with other sulfides and in low-temperature concretions.

ESMERALDA COUNTY As radially fibrous crusts on and with sphalerite at the Mohawk Mine (McGinnity shaft) in the Goldfield district. (Ransome, 1909a; S. Pullman, w.c., 1999)

EUREKA COUNTY As microscopic grains in blue crusts at the Goldstrike Mine in the Lynn district. (GCF)

NYE COUNTY Noted in quartz-rhodochrosite veins in the Morey district; specifically reported as black, radially fibrous masses with sphalerite and cassiterite on bluish quartz, in the Upper Magnolia workings of the Morey Mine. (Williams, 1968; Jenkins, 1981)

WASHOE COUNTY Reported in a specimen from the Commonwealth Mine in the Galena district. (LANH)

XANTHOCONITE

Ag_3AsS_3

Xanthoconite occurs with proustite and other silver sulfosalts in hydrothermal veins.

LANDER COUNTY Reported with other silver sulfosalt minerals in the Reese River district. (Taylor, 1912)

XENOTIME-(Y)

YPO_4

Xenotime-(Y) is a moderately common mineral as a trace constituent in rare earth–bearing pegmatites, in granitic rocks, and in gneiss. It is also occasionally found in the heavy mineral fraction of some placers.

CLARK COUNTY Reported in aplite-pegmatite dikes with *allanite,* zircon, and *monazite* in the Crescent district. (Garside, 1973)

Reported in a specimen from the Boss Mine in the Goodsprings district. (Jedwab, 1998b)

ESMERALDA COUNTY With *uranothorite* and euxenite in gold-bearing placer gravels at the Tule Canyon placer deposits in the Tokop district. (Garside, 1973)

EUREKA COUNTY In trace amounts in diorite at the Goldstrike Mine in the Lynn district. (Campbell, 1994)

As wholly enclosed grains in altered siltstone with zircon, pyrite, and sphalerite at the Gold Quarry Mine in the Maggie Creek district. (Jensen et al., 1995)

NYE COUNTY Included in pyrite in sedimentary rock–hosted gold ore at the Sterling Mine in the Bare Mountain district; presumed of detrital origin. (Castor, 1997)

PERSHING COUNTY As very rare microcrystals with zircon and rhabdophane-(Ce) from the middle adit at the Majuba Hill Mine in the Antelope district. (M. Jensen, w.c., 1999)

ZAPATALITE

$Cu_3^{2+}Al_4(PO_4)_3(OH)_9 \cdot 4H_2O$

Zapatalite occurs in highly oxidized deposits with other secondary copper minerals; as the name suggests, it was first described from an occurrence in Mexico.

LANDER COUNTY As bright-blue coatings from an unspecified locality near Battle Mountain; identified by XRD. (Excalibur Mineral Co., 1989; GCF)

ZARATITE

$Ni_3(CO_3)(OH)_4 \cdot 4H_2O$

A secondary mineral that is generally found in *serpentine* and basic igneous rocks, commonly with chromite or as an alteration of primary nickel minerals; also noted in meteorites.

CHURCHILL COUNTY Identified by XRD in green earthy masses and botryoids at the Nickel Mine in the Table Mountain district. (WMK; GCF)

ZEUNERITE

$Cu^{2+}(UO_2)_2(AsO_4)_2 \cdot 10\text{-}16H_2O$

A mineral of oxidized uranium deposits. Almost all specimens of so-called zeunerite are of the lower hydrate metazeunerite.

CHURCHILL COUNTY As lustrous, green, tabular crystals to 1 mm with lavendulan at the Nickel Mine in the Table Mountain district. (M. Jensen, w.c., 1999)

LYON COUNTY Reported in milky white quartz veins with uraninite and other uranium minerals in the Washington district. (Garside, 1973)

PERSHING COUNTY Commonly in the Copper stope at the Majuba Hill Mine in the Antelope district as light-grass-green, thin, square plates to 1.5 cm; rarely as groups. (Jensen, 1985)

ZINCALUMINITE

$Zn_6Al_6(SO_4)_2(OH)_{26} \cdot 5H_2O$

An inadequately described secondary mineral reported with smithsonite from zinc mines in Greece.

LYON COUNTY Reported in a specimen from the Jackpot Mine in the Wellington district. (LANH)

ZINCITE

$(Zn,Mn^{2+})O$

Zincite is an extremely rare mineral with the exception of the remarkable Franklin and Sterling Hill localities in New Jersey, where it occurs as a primary mineral in metamorphosed zinc deposits. It has been reported elsewhere in oxidized zinc deposits and as a product of volcanism.

EUREKA COUNTY Reported with smithsonite, hydrozincite, and oxidized lead and copper minerals at the Mountain View Mine in the Lone Mountain district. (Roberts et al., 1967b)

LINCOLN COUNTY With hydrozincite and cerussite in a vein at the Lucky prospect in the Viola district. (Tschanz and Pampeyan, 1970)

WHITE PINE COUNTY Reported in a specimen with galena at the Eureka claim in the Duck Creek district. (WMK)

ZINKENITE

$Pb_9Sb_{22}S_{42}$

Zinkenite is a rare mineral with other lead sulfosalts in base-metal deposits; at many reported localities it has proved to be boulangerite.

EUREKA COUNTY Reported with pyrite from one or more unspecified localities in the Eureka district. (Palache et al., 1944; NYS; UAMM; AMNH; HMM; S. Pullman, w.c., 1999)

NYE COUNTY As minute acicular crystals in vugs in the Morey Mine in the Morey district; also described as coarse intergrowths with boulangerite and *maesonite* (?) in the Upper Magnolia workings at the same mine, and reported from the Keyser Mine. (Palache et al., 1944; Jenkins, 1981; S. Pullman, w.c., 1999; HMM)

ZINNWALDITE

$KLiFe^{2+}Al(AlSi_3)O_{10}(F,OH)_4$

A mica-group mineral that generally occurs in tin-bearing greisen, but is also found in granite, granite pegmatite, and quartz veins.

MINERAL COUNTY As anhedral to subhedral grains to 300 microns as part of the late-stage pegmatite assemblage with *feldspar,* topaz, cryolite, simmonsite, pachnolite, and cryolithionite at the Zapot claims in the Fitting district about 15 miles northeast of Hawthorne. (Foord et al., 1999b)

ZIPPEITE

$K(UO_2)_2(SO_4)(OH)_3 \cdot H_2O$

A secondary mineral that is generally found in vein- and sandstone-type uranium deposits, in many cases as an efflorescence on mine walls and dumps; variably fluorescent.

LANDER COUNTY As yellow microcrystals with other secondary uranium minerals at the Apex (Rundberg) Uranium Mine in the Reese River district. (M. Jensen, o.c., 2000)

ZIRCON

$ZrSiO_4$

A very common accessory mineral in igneous and metamorphic rocks; also a common heavy detrital mineral in sedimentary rocks and placers. Only a small percentage of Nevada occurrences are listed below.

CLARK COUNTY In aplite-pegmatite dikes with *allanite,* xenotime, and *monazite* in the Crescent district. (Garside, 1973)

In pegmatite bodies in the Gold Butte district. (Longwell et al., 1965; JHS)

In granite in the Eldorado and Newberry Mountains. (S. Scott, w.c., 1999)

ELKO COUNTY Reported with *apatite* in monzonite east of Currie in the Dolly Varden district. (Melhase, 1934d)

Widespread as an accessory in host sedimentary rocks at the Big Springs Mine in the Independence Mountains district. (Youngerman, 1992)

ESMERALDA COUNTY In gold-bearing placer gravels in the Tule Canyon district. (Garside, 1973)

EUREKA COUNTY As subhedral grains to 200 microns in diorite at the Goldstrike Mine in the Lynn district. (GCF)

As disseminated grains and subhedral crystals in altered siltstone and as euhedral, colorless crystals in malachite at the Gold Quarry Mine in the Maggie Creek district. (Jensen et al., 1995)

An accessory mineral in igneous rocks at the Barth Iron Mine in the Safford district. (Shawe et al., 1962)

HUMBOLDT COUNTY As a uranium-rich, microgranular, hydrothermal mineral in halos around clasts and as crustiform layers in radioactive breccia with quartz, *adularia,* pyrite, arsenopyrite, calcite, and other minerals at the Moonlight Uranium Mine and prospects in the Horse Creek area in the Disaster district, Sec. 9, T45N, R34E. (Dayvault et al., 1985; Castor et al., 1996; Castor and Henry, 2000)

LANDER COUNTY In pre-Tertiary sedimentary rock as a detrital mineral at the West orebody in Copper Canyon in the Battle Mountain district. (Theodore and Blake, 1978)

LYON COUNTY Reported in the Ludwig area of the Yerington district. (Melhase, 1935e)

MINERAL COUNTY Reddish-brown to grayish-brown crystals to 3 mm in topaz-bearing pegmatite at the Zapot claims in the Fitting district. (Foord et al., 1999b)

PERSHING COUNTY As rare microcrystals with anatase and rhabdophane-(Ce) in the middle adit at the Majuba Hill Mine in the Antelope district. (Jensen, 1993b)

Identified with dumortierite at the Champion Mine in the Sacramento district. (JHS)

WASHOE COUNTY In red-brown crystals to 1 mm with *feldspar,* as a very rare constituent of pegmatites at Incline Village. (M. Jensen, w.c., 1999)

WHITE PINE COUNTY As small crystals in porphyry ore in the Robinson district. (Melhase, 1934d)

ZOISITE

$Ca_2Al_3(SiO_4)_3(OH)$

Zoisite, a widespread mineral that is dimorphous with clinozoisite, mainly occurs in regionally metamorphosed schist and gneiss. *Thulite* is a pink variety that occurs in metamorphic rock and pegmatite; it has been reported in contact-metamorphosed rock from several localities in Nevada.

CHURCHILL COUNTY Reported in veins and wall rock at the Fairview Mine in the Fairview district. (Schrader, 1947)

As bright-pink terminated prisms of *thulite* with *adularia* and galena in vein material at the Westgate Mine in the Westgate district. (R. Walstrom, w.c., 1999)

As black crystals to 1 mm rimmed by clinozoisite in rock with possible hemimorphite at an unnamed prospect near the southwest end of the Clan Alpine Mountains on the east side of Stingaree Valley. (M. Jensen, w.c., 1999)

DOUGLAS COUNTY As *thulite* with *garnet,* epidote, zoisite, and other minerals about 7 miles west of Wellington in the southern Pine Nut Range. Pink to red crystals have been found, but most of the material is massive. The property has most recently been referred to as the Lapis Nevada Quarry. Rose-colored *thulite* has also been reported in a specimen from the Walker property, 10 miles south of Mountain House in the southwestern Pine Nut Mountains, possibly the same locality. (Gianella, 1936b; WMK; Sinkankas, 1959; Sinkankas, 1997)

EUREKA COUNTY An alteration product in endoskarn at the Goldstrike Mine in the Lynn district. (Walck, 1989)

LYON COUNTY As rose-pink *thulite* with *garnet* in a contact zone between limestone and granite in the Yerington district about 3/4 mile east of Ludwig. (Melhase, 1935e; Gianella, 1936b; Sinkankas, 1959)

MINERAL COUNTY Reported as *thulite* in a specimen from the northern portion of the White Mountains. (WMK)

As *thulite* in skarn with diopside and andradite in New York Canyon in the Santa Fe district. (D. Hudson, w.c., 1999)

As massive *thulite* along a limestone-granite contact in Ryan Canyon in the Fitting district 5 miles southeast of Thorne. (Gianella, 1936b; Sinkankas, 1959; NYS)

PERSHING COUNTY Reported in specimens as crystals on quartz from the Nightingale district; clinozoisite is also reported in similar specimens. (NYS; WMK)

WHITE PINE COUNTY Abundant with *garnet,* diopside, and clinozoisite in green hornfels in the Mount Hamilton area in the White Pine district; mainly as the variety *thulite.* (Humphrey, 1960)

ZUNYITE

$Al_{13}Si_5O_{20}(OH,F)_{18}Cl$

Zunyite is a moderately common mineral in hydrothermal alteration assemblages that have relatively high concentrations of fluorine; it generally occurs in association with clay minerals, topaz, and fluorite.

DOUGLAS AND LYON COUNTIES As colorless to grayish tetrahedrons and octahedrons to 5 mm with corundum, pyrophyllite, andalusite, and topaz in altered volcanic rock at the Blue Danube (Blue Metal) prospect in the Buckskin district; similar alteration extends south into Lyon County. (M. Jensen, w.c., 1999; Hudson, 1983)

ELKO COUNTY At the Zun claims in the Scraper Springs district, Sec. 3, T40N, R47E, with pyrite and rutile in massive light-colored outcrops of hydrothermally altered breccia in a zone about 2,000 feet long in andesite and tuff; also at a second occurrence in altered breccia north of Scraper Springs. (Coats et al., 1979)

WHITE PINE COUNTY With alunite and pyrite in the Robinson district. (Maher, 1996)

References

Adair, D.H., 1961, Geology of the Cherry Creek district, Nevada [M.S. thesis]: University of Utah, Salt Lake City, 125 p.

Adams, G.L., 1904, The Rabbit Hole sulphur mines, near Humboldt House, Nevada: U.S. Geological Survey Bulletin 225, p. 497–500.

Adams, P.M., 1998, On the occurrence of conichalcite and twinned fornacite from the Singer Mine, Goodsprings, Clark County, Nevada: Mineral News, v. 14, no. 5, p. 1, 6–7.

Akright, R.L., Radtke, A.S., and Grimes, D.J., 1969, Minor elements as guides to gold in the Roberts Mountains Formation, Carlin gold mine, Eureka County, Nevada, *in* Canney, F. C., ed., Proceedings of the International Geochemical Exploration Symposium, 1968: Colorado School of Mines Quarterly, v. 64, no. 1, p. 49–66.

Albers, J.P., and Cornwall, H.R., 1968, Revised interpretation of the stratigraphy and structure of the Goldfield district, Esmeralda and Nye Counties, Nevada [abs]: Geological Society of America Special Paper 101, p. 285.

Albers, J.P., and Kleinhampl, F.J., 1970, Spatial relation of mineral deposits to Tertiary volcanic centers in Nevada: U.S. Geological Survey Professional Paper 700-C, p. 1–10.

Albers, J.P., and Stewart, J.H., 1972, Geology and mineral deposits of Esmeralda County, Nevada: Nevada Bureau of Mines and Geology Bulletin 78, 80 p.

Albritton, C.C., Richards, A., Brokaw, A.L., and Reinemund, J., 1954, Geologic controls of lead and zinc deposits in Goodsprings (Yellow Pine) District, Nevada: U.S. Geological Survey Bulletin 1010, 111 p.

Altaner, S.P., Fitzpatrick, J.L., Krohn, M.D., Bethke, P.M., Hayba, D.O., Goss, J.A., and Brown, Z.A., 1988, Ammonium in alunites: American Mineralogist, v. 73, p. 145–152.

Anderson, R.E., 1978, Geologic map of the Black Canyon 15-minute quadrangle, Mohave County, Arizona, and Clark County, Nevada: U.S. Geological Survey Geologic Quadrangle Map GQ-1394.

Anonymous, 1865, Agates and diamonds: Mining and Scientific Press, v. 11, p. 369.

Anonymous, 1938, Lava minerals: The [Oregon] Mineralogist, v. 6, no. 1, p. 3–4, 22–25.

Anonymous, 1975, Industrial minerals and rocks (nonmetallics other than fuels), Lefond, S. J., ed.: American Institute of Mining, Metallurgical, and Petroleum Engineers, Inc., New York, 1360 p.

Anthony, J.W., Bideaux, R.A., Bladh, K.W., and Nichols, M.C., 1990, Handbook of Mineralogy Volume I—Elements, Sulfides, Sulfosalts: Mineral Data Publishing, Tucson, Arizona, 588 p.

Anthony, J.W., Bideaux, R.A., Bladh, K.W., and Nichols, M.C., 1995, Handbook of Mineralogy Volume II—Silica, Silicates Parts 1 and 2: Mineral Data Publishing, Tucson, Arizona, 904 p.

Anthony, J.W., Bideaux, R.A., Bladh, K.W., and Nichols, M.C., 1997, Handbook of Mineralogy Volume III—Halides, Hydroxides, Oxides: Mineral Data Publishing, Tucson, Arizona, 628 p.

Anthony, J.W., Bideaux, R.A., Bladh, K.W., and Nichols, M.C., 2000, Handbook of Mineralogy Volume IV—Arsenates, Phospahtes, Vanadates: Mineral Data Publishing, Tucson, Arizona, 680 p.

Archbold, N.L., 1966, Industrial mineral deposits of Mineral County, Nevada: Nevada Bureau of Mines and Geology Report 14, 32 p.

Arehart, G.B., Chryssoulis, S.L., and Kesler, S.E., 1993, Gold and arsenic in iron sulfides from sediment-hosted disseminated gold deposits: implications for depositional processes: Economic Geology, v. 88, p. 171–185.

Arehart, G.B., Kesler, S.E., O'Neil, J.R., and Foland, K.A., 1992, Evidence for the supergene origin of alunite in sediment-hosted micron gold deposits, Nevada: Economic Geology, v. 87, p. 263–270.

Armbrustmacher, T.J., and Wrucke, C.T., 1978, The disseminated gold deposit at Gold Acres, Lander County, Nevada, in Lovering, T.G., and McCarthy, J.H., Jr., eds., Conceptual models in exploration geochemistry; the Basin and Range Province of the western United States and northern Mexico: Journal of Geochemical Exploration, v. 9, p. 195–203.

Armstrong, A.K., Theodore, T.G., Kotlyar, B.B., Lauha, E.G., Griffin, G.L., Lorge, D.L., and Abbot, E.K., 1997, Preliminary facies analysis of Devonian autochthonous rocks that host gold along the Carlin trend, Nevada, in Vikre, P., Thompson, T.B., Bettles, K., Christensen, O., and Parratt, R., eds., Carlin-type gold deposits field conference: Society of Economic Geologists Guidebook Series, v. 28, p. 53–74.

Armstrong, R.L., Ekren, E.B., McKee, E.H., and Noble, D.C., 1969, Space-time relations of Cenozoic silicic volcanism in the Great Basin of the western United States: American Journal of Science, v. 267, p. 478–490.

Ashley, R.P., 1972, Premineralization structural history of the Goldfield volcanic center, Nevada [abs.]: Economic Geology, v. 67, p. 1002.

Ashley, R.P., 1973, Fission-track ages for premineralization volcanic and plutonic rocks of the Goldfield mining district, Esmeralda and Nye Counties, Nevada: Isochron/West, no. 8, p. 25–29.

Ashley, R.P., 1974, Goldfield mining district, in Guidebook to the geology of four Tertiary volcanic centers in central Nevada: Nevada Bureau of Mines and Geology Report 19, p. 49–66.

Ashley, R.P., 1975, Preliminary geologic map of the Goldfield mining district, Nevada: U.S. Geological Survey Miscellaneous Field Studies Map IF-682.

Ashley, R.P., and Albers, J.P., 1973, Distribution of gold and other ore-related elements near ore bodies in the oxidized zone at Goldfield, Nevada: U.S. Geological Survey Open-File Report, 126 p.

Ashley, R.P., and Albers, J.P., 1975, Distribution of gold and other ore-related elements near ore bodies in the oxidized zone at Goldfield, Nevada: U.S. Geological Survey Professional Paper 843-A, 48 p.

Ashley, R.P., and Keith, W.J., 1976, Distribution of gold and other elements in silicified rocks of the Goldfield mining district, Nevada: U.S. Geological Survey Professional Paper 843-B, 17 p.

Atkinson, W.W., Jr., Kaczmarowski, J.H., and Erickson, A.J., Jr., 1982, Geology of a skarn-breccia orebody at the Victoria Mine, Elko County, Nevada: Economic Geology, v. 77, p. 899–918.

Austin, G.T., 1991, Nevada's gem value up—thanks to jade: Colored Stone, v. 4, no. 1, p. 20–21.

Bailey, E.H., Hildebrand, F.A., Christ, C.L., and Fahey, J.J., 1959, Schuetteite, a new supergene mercury mineral: American Mineralogist, v. 44, p. 1026–1038.

Bailey, E.H., and Phoenix, D.A., 1944, Quicksilver deposits in Nevada: Nevada Bureau of Mines and Geology Bulletin 41, 206 p.

Bailey, J.T., 1903, Discussion: The ore deposits of Contact, Nevada: The Engineering and Mining Journal, v. 76, p. 612–613.

Bakken, B.M., 1990, Gold mineralization, wall-rock alteration and the geochemical evolution of the hydrothermal system in the Main Orebody, Carlin Mine, Nevada [Ph.D. diss.]: Stanford University, Stanford, Calif., 256 p.

Bakken, B.M., Hochella, M.F., Marshall, A.F., and Turner, A.M., 1989, High-resolution microscopy of gold in unoxidized ore from the Carlin Mine, Nevada: Economic Geology, v. 84, p. 171–179.

Ball, S.H., 1906, Notes on ore deposits in southwestern Nevada and eastern California: U.S. Geological Survey Bulletin 285, p. 53–73.

Ball, S.H., 1907, A geologic reconnaissance in southwestern Nevada and eastern California: U.S. Geological Survey Bulletin 308, 218 p.

Ball, S.H., 1939, Gemstones: U.S. Bureau of Mines Minerals Yearbook for 1938, p. 1385–1396.

Bancroft, H., 1910, Platinum in southern Nevada: U.S. Geological Survey Bulletin 430, p. 192–199.

Bancroft, P., 1984, Gem and crystal treasures: Fallbrook, Calif., Western Enterprises, 488 p.

Barksdale, J.D., 1939, Contact garnet from the Adelaide district, Nevada [abs.]: Geological Society of America Bulletin, v. 50, p. 1946.

Bartlett, M.W., Enders, M.S., and Hruska, D.C., 1991, Geology of the Hollister gold deposit, Ivanhoe District, Elko County, Nevada, in Raines, G.L., Lisle, R.E., Schafer, R.W., and Wilkinson, W.H., eds., Geology and ore deposits of the Great Basin; symposium proceedings: Geological Society of Nevada, Reno, p. 957–978.

Barton, M.D., Kieft, C., Burke, E.A.J., and Oen, I.S., 1978, Uytenbogaardtite, a new silver-gold sulfide: Canadian Mineralogist, v. 16, p. 651–657.

Barton, M.D., and Trim, H.E., 1991, Late Cretaceous

two-mica granites and lithophile-element mineralization in the Great Basin, *in* Raines, G.L., Lisle, R.E., Schafer, R.W., and Wilkinson, W.H., eds., Geology and ore deposits of the Great Basin; symposium proceedings: Geological Society of Nevada, Reno, p. 529–538.

Barton, P.B., 1956, Fixation of uranium in the oxidized base metal ores of the Goodsprings District, Clark County, Nevada: Economic Geology, v. 51, p. 178–191.

Bastin, E.S., 1922, Bonanza ores of the Comstock Lode, Virginia City, Nevada: U.S. Geological Survey Bulletin 735-C, p. 41–64.

Bastin, E.S., and Laney, F.B., 1918, The genesis of the ores at Tonopah, Nevada: U.S. Geological Survey Professional Paper 104, 50 p.

Bateman, A.M., 1935, The copper deposits of Ely, Nevada, *in* Copper Resources of the World: International Geological Congress, 16th, Washington, D.C., 1933, v. 1, p. 309.

Bates, R.L., and Jackson, J.A., eds., 1987, Glossary of Geology (3d ed.): American Geological Institute, Alexandria, Va., 788 p.

Bauer, H.L., Breitrick, R.A., Cooper, J.J., and Anderson, J.A., 1966, Porphyry copper deposits in the Robinson mining district, Nevada, *in* Titley, S.R., and Hicks, C.L., eds., geology of the porphyry copper deposits, southwestern North America: University of Arizona Press, Tucson, p. 233–244.

Bauer, H.L., Breitrick, R.A., Cooper, J.J., and Swinderman, J.N., 1964, Origin of the disseminated ore in metamorphosed sedimentary rocks, Robinson mining district, Nevada: Transactions of the Society of Mining Engineers, v. 229, p. 131.

Bauer, H.L., Cooper, J.J., and Breitrick, R.A., 1960, Porphyry copper deposits in the Robinson mining district, White Pine County, Nevada, *in* Guidebook to the geology of east-central Nevada: Intermountain Association of Petroleum Geologists and Eastern Nevada Geological Society, 11th Annual Field Conference, 1960, p. 220–228.

Baugh, W.M., and Kruse, F.A., 1994, Quantitative geochemical mapping of ammonium minerals using field and airborne spectrometers, Cedar Mountains, Esmeralda County, Nevada, *in* Proceedings of the Tenth Thematic Conference on Geologic Remote Sensing, Exploration, Environment, and Engineering, Environmental Research Institute of Michigan, Ann Arbor, Mich., p. II.304–II.315.

Beal, L.H., 1963, Investigation of titanium occurrences in Nevada: Nevada Bureau of Mines and Geology Report 3, 42 p.

Beal, L.H., 1964, Cobalt and nickel, *in* Mineral and water resources of Nevada: Nevada Bureau of Mines and Geology Bulletin 65, p. 78–81.

Beal, L.H., 1965, Geology and mineral deposits of the Bunkerville mining district, Clark County, Nevada: Nevada Bureau of Mines and Geology Bulletin 63, 96 p.

Beck, C.W., and Burns, J.H., 1954, Callaghanite, a new mineral: American Mineralogist, v. 39., p. 630–635.

Becker, G.F., 1882, Geology of the Comstock Lode and the Washoe district; with Atlas: U.S. Geological Survey Monograph 3, 422 p.

Becker, G.F., 1888, Geology of the quicksilver deposits of the Pacific Slope: U.S. Geological Survey Monograph 13.

Bell, J.W., and Smith, E.I., 1980, Geologic map of the Henderson Quadrangle, Nevada: Nevada Bureau of Mines and Geology Map 67.

Benoson, W.T., 1956, Investigation of mercury deposits in Nevada and in Malheur County, Oregon: U.S. Bureau of Mines Report of Investigation 5285, 54 p.

Bentz, J.L., Tingley, J.V., Smith, P.L., and Garside, L.J., 1983, A mineral inventory of the Elko resource area, Elko District, Nevada: Nevada Bureau of Mines and Geology Open-File Report 83-9, 212 p.

Berger, B.R., 1982, The geological attributes of Au-Ag-base metal epithermal deposits, *in* Erickson, R.L., comp., Characteristics of mineral deposit occurrences: U.S. Geological Survey Open-File Report 82-0795, p. 119–126.

Berger, B.R., and Tingley, J.V., 1985, History of discovery, mining, exploration of the Getchell Mine, Humboldt County, Nevada, *in* Hollister, V.F., ed., Case Histories of Mineral Discoveries, Discoveries of epithermal precious metal deposits: American Institute of Mining, Metallurgical, and Petroleum Engineers, New York, 49–51.

Berger, V.I., and Oscarson, R.L., 1998, Tungsten-, polymetallic-, and barite-mineralized rocks in the Ruby Mountains, Nevada, *in* Tosdal, R.M., ed., Contributions to the gold metallogeny of northern Nevada: U.S. Geological Survey Open-File Report 98-338, p. 151–175.

Bergman, S.C., 1982, Petrogenetic aspects of the alkali basaltic lavas and included megacrysts and nodules from the Lunar Crater volcanic field, Nevada, USA [Ph.D. diss.]: Princeton University, 447 p.

Berry, A.R., 1992, A geological study of the Betze gold deposit, Eureka County, Nevada [M.S. thesis]: University of Nevada, Reno, 173 p.

Berry, L.C., 1950, On pseudomalachite and cornetite: American Mineralogist, v. 35, p. 365–385.

Berry, L.C., Fahey, J.J., and Bailey, E.H., 1952, Robinsonite, a new lead antimony sulfide: American Mineralogist, v. 37, p. 438–446.

Berry, L.C., and Thompson, R.M., 1962, X-ray powder data for the ore minerals: The Peacock Atlas: Geological Society of America Memoir 85, 310 p.

Binyon, E.O., 1946a, Blue Metal corundum and andalusite deposit, Douglas County, Nevada: U.S. Bureau of Mines Report of Investigation 3895, 7 p.

Binyon, E.O., 1946b, Exploration of the gold, silver, lead, zinc properties, Eureka Corporation, Eureka County, Nevada: U.S. Bureau of Mines Report of Investigation 3949, 19 p.

Binyon, E.O., 1948, Gibellini manganese-zinc-nickel deposits, Eureka County, Nevada: U.S. Bureau of Mines Report of Investigation 4162, 9 p.

Binyon, E.O., Holmes, G.H., Jr., and Johnson, A.C., 1950, Investigation of the Tem Piute tungsten deposit, Lincoln County, Nevada: U.S. Bureau of Mines Report of Investigation 4626, 18 p.

Birch, W.D., Pring, A., Self, P.G., Gibbs, R.B., Keck, E., Jensen, M.C., and Foord, E.E., 1996, Meurigite, a new fibrous iron phosphate resembling kidwellite: Mineralogical Magazine, v. 60, p. 787–793.

Bird, P.H., 1932, A new occurrence and X-ray study of mosesite: American Mineralogist, v. 17, p. 541–550.

Bish, D.L., and Vaniman, D.T., 1985, Mineralogic summary of Yucca Mountain, Nevada: Los Alamos National Laboratory Report LA-10543-MS, 55 p.

Black, J.E., 1991, Geology and mineralization at the Rawhide gold-silver deposit, Mineral County, Nevada, in Buffa, R.H., and Coyner, A.R., eds., Geology and ore deposits of the Great Basin; field trip guidebook compendium: Geological Society of Nevada, Reno, p. 808–826.

Black, J.E., Mancuso, T.K., and Gant, J.L., 1991, Geology and mineralization at Aurora, Nevada, in Raines, G.L., Lisle, R.E., Schafer, R.W., and Wilkinson, W.H., eds., Geology and ore deposits of the Great Basin; symposium proceedings: Geological Society of Nevada, Reno, p. 1123–1144.

Blake, J.M., 1866, On crystals of gaylussite from Nevada Territory: American Journal of Science, v. 42, p. 221–222.

Blake, W.P., 1885, Antimony, in Mineral resources of the United States, 1883–1884: U.S. Geological Survey, p. 641–653.

Blanchard, R., 1968, Interpretation of leached outcrops: Nevada Bureau of Mines and Geology Bulletin 66, 196 p.

Bloomstein, E.I., Braginton, B., Owen, R., Pratt, R., Rabbe, K., and Thompson, W., 1993, Geology and geochemistry of the Lone Tree gold deposit, Humboldt County, Nevada: 93rd Annual Meeting, Reno, Nevada, Society for Mining, Metallurgy, and Exploration, Inc. and Society of Economic Geology, February 15–18, 1993, Preprint 93-205, 23 p.

Bohannon, R.G., 1984, Nonmarine sedimentary rocks of Tertiary age in the Lake Mead region, southeastern Nevada and northwestern Arizona: U.S. Geological Survey Professional Paper 1259, 72 p.

Bonham, H.F., Jr., 1970, Geologic map and sections of a part of the Shoshone Mountains, Lander and Nye Counties, Nevada: Nevada Bureau of Mines and Geology Map 38, 1:62,500.

Bonham, H.F., Jr., 1988, Models for volcanic-hosted epithermal precious metal deposits, in Schafer, R.W., Cooper, J.J., and Vikre, P.G., eds., Bulk mineable precious metal deposits of the western United States; symposium proceedings: Geological Society of Nevada, Reno, p. 259–271.

Bonham, H.F., Jr., and Garside, L.J., 1979, Geology of the Tonopah, Lone Mountain, Klondike, and northern Mud Lake Quadrangles, Nevada: Nevada Bureau of Mines and Geology Bulletin 92, 136 p.

Bonham, H.F., Jr., and Papke, K.G., 1969, Geology and mineral deposits of Washoe and Storey Counties, Nevada: Nevada Bureau of Mines and Geology Bulletin 70, 140 p.

Botinelly, T., Nuerburg, G.J., and Conklin, N.M., 1973, Galkhaite (Hg,Cu,Tl,Zn)(As,Sb)S_2 from the Getchell mine, Humboldt County, Nevada: U.S. Geological Survey Journal of Research, v. 1, no. 5, p. 515–517.

Brannock, W.W., Fix, P.F., Gianella, V.P., and White, D.E., 1948, Preliminary geochemical results at Steamboat Springs, Nevada: American Geophysical Union Transactions, v. 29, p. 211–226.

Brooks, H., 1956, Geology of a uranium deposit in the Virginia Mountains, Washoe County, Nevada [M.S. thesis]: University of Nevada, Reno, 50 p.

Brooks, J.W., Meinert, L.D., Kuyper, B.A., and Lane,

M.L., 1991, Petrology and geochemistry of the McCoy gold skarn, Lander County, Nevada, *in* Raines, G.L., Lisle, R.E., Schafer, R.W., and Wilkinson, W.H., eds., Geology and ore deposits of the Great Basin; symposium proceedings: Geological Society of Nevada, Reno, p. 419–442.

Brophy, C.P., Scott, E.S., and Snellgrove, R.A., 1962, Sulfate studies II—solid solution between alunite and jarosite: American Mineralogist, v. 47, p. 112–126.

Browne, J.R., 1867, Report on the mineral resources of the states and territories west of the Rocky Mountains, *in* A report upon the mineral resources of the United States: Government Printing Office, Washington, D.C., p. 7–321.

Browne, J.R., 1868, Report on the mineral resources of the states and territories west of the Rocky Mountains: Government Printing Office, Washington, D.C., 681 p.

Broxton, D.E., Bish, D.L., and Warren, R.G., 1987, Distribution and chemistry of diagenetic minerals at Yucca Mountain, Nye County, Nevada: Clays and Clay Minerals, v. 35, no. 2, p. 89–110.

Broxton, D.E., Byers, F.M., Jr., and Warren, R.G., 1989, Petrography and phenocryst chemistry of volcanic units at Yucca Mountain, Nevada: A comparison of outcrop and drill hole samples: Los Alamos National Laboratory Report LA-11503-MS, 65 p.

Brush, G.J., 1875, Manual of determinative mineralogy, with an introduction to blowpipe analysis: Wiley, New York, 104 p.

Bryan, D.P., 1972, The geology and mineralization of the Chalk Mountain and Westgate Mining Districts, Churchill County, Nevada [M.S. thesis]: University of Nevada, Reno, 78 p.

Buchwald, V.F., 1975, Handbook of iron meteorites, vols. 2 and 3, University of California Press, Berkeley, p. 247–1418.

Burchfiel, B.C., and Davis, G.A., 1988, Mesozoic thrust faults and Cenozoic low-angle normal faults, eastern Spring Mountains, Nevada and Clark Mountain thrust complex, California, *in* This extended land: Geological Society of America Field Trip Guidebook, Cordilleran section meeting, Las Vegas, Nevada, p. 87–106.

Burgess, J.A., 1909, The geology of the producing part of the Tonopah mining district: Economic Geology, v. 4, p. 681–712.

Burgess, J.A., 1911, The halogen salts of silver and associated metals at Tonopah, Nevada: Economic Geology, v. 6, p. 13–21.

Burgess, J.A., 1917, The halogen salts at Wonder, Nevada: Economic Geology, v. 12, p. 589–593.

Burgess, J.A., 1926, Leadville Mine: unpublished report for the Leadville Mining Co.

Buseck, P.R., 1967, Contact metasomatism and ore deposition: Tem Piute, Nevada: Economic Geology, v. 62, p. 331–353.

Bushnell, K., 1967, Geology of the Rowland Quadrangle, Elko County, Nevada: Nevada Bureau of Mines and Geology Bulletin 67, 44 p.

Bussey, S.D., 1996, Gold mineralization and associated rhyolitic volcanism at the Hog Ranch District, Northwest Nevada, *in* Coyner, A.R., and Fahey, P.L., eds., Geology and ore deposits of the American Cordillera; symposium proceedings: Geological Society of Nevada, Reno, p. 181–207.

Byers, F.M., Jr., Carr, W.J., Orkild, P.P., Quinlivan, W.D., and Sargent, K.A., 1976, Volcanic suites and related cauldrons of the Timber Mountain–Oasis Valley caldera complex, southern Nevada: U.S. Geological Survey Professional Paper 919, 70 p.

Calkins, F.C., 1938, Gold deposits of the Slumbering Hills, Nevada: Nevada Bureau of Mines and Geology Bulletin 30B, 23 p.

Callaghan, E., 1933, Brucite deposit, Paradise Range, Nevada—A preliminary report: Nevada Bureau of Mines and Geology Bulletin 19, 34 p.

Callaghan, E., 1936, Geology of the Chief district, Lincoln County, Nevada: Nevada Bureau of Mines and Geology Bulletin 26, 29 p.

Callaghan, E., 1937, Geology of the DeLamar district, Lincoln County, Nevada: Nevada Bureau of Mines and Geology Bulletin 30A, 69 p.

Callaghan, E., 1939, Geology of the Searchlight district, Clark County, Nevada: U.S. Geological Survey Bulletin 906-D, p. 135–188.

Callaghan, E., and Vitaliano, C.J., 1947, Type sequences of Tertiary volcanic rocks in the western part of the Great Basin: Geological Society of America Bulletin, v. 58, no. 12, pt. 2, 1171 p.

Cameron, E.N., 1939, Geology and mineralization of the northeastern Humboldt Range, Nevada: Geological Society of America Bulletin, v. 50, p. 563–634.

Campbell, D.F., 1939, Geology of the Bonanza King mine, Humboldt Range, Pershing County, Nevada: Economic Geology, v. 34, p. 96–112.

Campbell, K.B., 1994, The geology of the Post oxide deposit, Goldstrike Mine, Eureka County, Nevada: unpublished M.S. thesis, Colorado State University, Ft. Collins, Colorado, 111 p.

Carlos, B., Bish, D.L., and Chipera, S., 1991, Fracture-lining minerals in the lower Topopah Spring Tuff at Yucca Mountain: Los Alamos National Laboratory Report LA-UR 91-4354, 8 p.

Carlos, B., Bish, D.L., and Chipera, S., 1995, Distribution and chemistry of fracture-lining minerals at Yucca Mountain, Nevada: Los Alamos National Laboratory Report LA-12977-MS, 92 p.

Carpenter, A.H., 1911, Boyer copper deposits, Nevada: Mineral and Science Press, v. 103, p. 804–805.

Carpenter, J.A., 1928, The mining of dumortierite: Nevada Bureau of Mines and Geology Bulletin 8, p. 35–39.

Carr, M.D., 1983, Geometry and structural history of the Mesozoic thrust belt in the Goodsprings District, southern Spring Mountains, Clark County, Nevada: Geological Society of America Bulletin, v. 94, p. 185–198.

Carr, M.D., 1987, Geologic map of the Goodsprings District, southern Spring Mountains, Clark County, Nevada: U.S. Geological Survey Map MF-1514.

Casteel, M.V., and Bernard, L.J., 1999, Geology of the Ken Snyder mine epithermal gold and silver deposit, Midas District, Elko County, Nevada: Geological Society of Nevada Newsletter, v. 13, no. 2, p. 3–4.

Castor, S.B., 1988, Industrial minerals, in The Nevada mineral industry, 1987: Nevada Bureau of Mines and Geology Special Publication MI-1988, p. 25–28.

Castor, S.B., 1991a, Rare earth deposits in the southern Great Basin, in Raines, G.L., Lisle, R.E., Schafer, R.W., and Wilkinson, W.H., eds., Geology and ore deposits of the Great Basin; symposium proceedings: Geological Society of Nevada, Reno, p. 523–528.

Castor, S.B., 1991b, Industrial minerals, in The Nevada mineral industry, 1990: Nevada Bureau of Mines and Geology Special Publication MI-1988, p. 27–31.

Castor, S.B., 1992, Industrial minerals in Nevada, in Tooker, E.W., comp., Industrial minerals in the Basin and Range region; workshop proceedings: U.S. Geological Survey Bulletin 2013, p. 22–28.

Castor, S.B., 1993, Borates in the Muddy Mountains, Clark County, Nevada: Nevada Bureau of Mines and Geology Bulletin 107, 31 p.

Castor, S.B., 1997, Sterling Mine, Nye County, Nevada, in Vikre, P., Thompson, T.B., Bettles, K., Christensen, O., and Parratt, R., eds., Carlin-type gold deposits field conference: Society of Economic Geologists Guidebook Series, v. 28, p. 167–170.

Castor, S.B., Feldman, S.C., and Tingley, J.V., 1990, Mineral evaluation of the Yucca Mountain Addition, Nye County, Nevada: Nevada Bureau of Mines and Geology Open-File Report 90-4.

Castor, S.B., Garside, L.J., and dePolo, C.M., 1999a, Geologic map of the west half of the Moses Rock Quadrangle, Nevada: Nevada Bureau of Mines and Geology Open-File Report 99-11.

Castor, S.B., Garside, L.J., Tingley, J.V., LaPointe, D.D., Desilets, M.O., Hsu, L.C., Goldstrand, P.M., Lugaski, T.P., and Ross, H.P., 1999b, Assessment of metallic and mined energy resources in the Yucca Mountain conceptual controlled area, Nye County, Nevada: Nevada Bureau of Mines and Geology Open-File Report 99-13.

Castor, S.B., and Henry, C.D., 2000, Geology, geochemistry, and origin of volcanic rock–hosted uranium deposits in northwestern Nevada and southeastern Oregon, USA: Ore Geology Reviews, v. 16, p. 1–40.

Castor, S.B., Henry, C.D., and Shevenell, L.A., 1996, Volcanic rock–hosted uranium deposits in northwestern Nevada and southeastern Oregon—Possible sites for studies of natural analogues for the potential high-level nuclear waste repository at Yucca Mountain, Nevada: Nevada Bureau of Mines and Geology Open-File Report 96-3, 86 p.

Castor, S.B., and Hulen, J.B., 1996, Electrum and organic matter at the Gold Point Mine, Currant mining district, Nevada, in Coyner, A.R., and Fahey, P.L., eds., Geology and ore deposits of the American Cordillera; symposium proceedings: Geological Society of Nevada, Reno, p. 547–565.

Castor, S.B., Mitchell, T.P., and Quade, J.G., 1982, Vya Quadrangle, Nevada, California and Oregon: U.S. Department of Energy Open-File Report, PGJ/F135, 25 p.

Castor, S.B., and Sjoberg, J.J., 1993, Uytenbogaardtite, Ag_3AuS_2, in the Bullfrog mining district, Nevada: Canadian Mineralogist, v. 31, pt. 1, p. 89–98.

Castor, S.B., Tingley, J.V., and Bonham, H.F., 1992, Subsurface mineral resource analysis, Yucca Mountain, Nevada, preliminary report 1—lithologic logs: Nevada Bureau of Mines and Geology Open-File Report 92-4, 11 p., plus appendices.

Castor, S.B., and Weiss, S.I., 1992, Contrasting styles of epithermal precious metal mineralization in the southwestern Nevada volcanic field, USA: Ore Geology Reviews, v. 7, p. 193–223.

Chavez, W.X., and Purusotam, S., 1988, Precious metal and ore-associated mineralogy of the Candelaria Silver Mine, Mineral County, Nevada, *in* Schafer, R.W., Cooper, J.J., and Vikre, P.G., eds., Bulk mineable precious metal deposits of the western United States; symposium proceedings: Geological Society of Nevada, Reno, p. 752.

Chen, T.T., and Szymanski, J.T., 1981, The structure and chemistry of galkhaite, a mercury sulfosalt containing Cs and Tl: Canadian Mineralogist, v. 19, p. 571–581.

Chenoweth, W.L., 1998, Strategic material procurement for Manhattan: Paydirt, January issue, p. 15–16.

Chesterman, C.W., 1968, Volcanic history of the Bodie Hills, Mono County, California, *in* Coats, R.R., Haey, R.L., and Anderson, C.A., eds., Studies in Volcanology: Geological Society of America Memoir 116, p. 45–68.

Chipp, E.R., 1969, The geology of the Klondike mining district, Esmeralda County, Nevada [M.S. thesis]: University of Nevada, Reno, 52 p.

Clark, C.W., 1922, Geology and ore deposits of the Santa Fe district, Mineral County, Nevada: California University Bulletin, Department of Geological Sciences, v. 14, no.1, 74 p.

Clark, I.C., 1918, Recently recognized alunite deposits at Sulfur, Humboldt County, Nevada: Engineering and Mining Journal, v. 106, p. 159–163.

Cleveland, J.H., 1963, Paragenesis of the magnesite deposit at Gabbs, Nye County, Nevada [Ph.D. diss.]: Indiana University, Bloomington, 88 p.

Clinton, H.G., 1929, Vashegyite and barrandite in Nevada: American Mineralogist, v. 14, p. 434–436.

Coats, R.R., 1936a, Aguilarite from the Comstock Lode, Virginia City, Nevada: American Mineralogist, v. 21, p. 532–534.

Coats, R.R., 1936b, Intrusive domes of the Washoe district, Nevada: California University Bulletin, Department of Geological Sciences, v. 24, no. 4, p. 71–84.

Coats, R.R., 1940, Propylitizaton and related types of alteration on the Comstock Lode: Economic Geology, v. 35, p. 1–16.

Coats, R.R., 1964, Geology of the Jarbidge Quadrangle, Nevada-Idaho: U.S. Geological Survey Bulletin 1141-M, 24 p.

Coats, R.R., 1968, The Circle Creek Rhyolite, a volcanic dome complex in northern Elko County, Nevada, *in* Coats, R.R., Hay, R.L., and Anderson, C.A., eds., Studies in volcanology: Geological Society of America Memoir 116, p. 69–106.

Coats, R.R., Consul, J., and Neil, S.T., 1979, Massive zunyite rock from western Elko County, Nevada: U.S. Geological Survey Open-File Report 79-764, 7 p.

Coats, R.R., Green, R.C., and Cress, L.D., 1977, Mineral resources of the Jarbidge Wilderness and adjacent areas, Elko County, Nevada: U.S. Geological Survey Bulletin 1439, 79 p.

Coats, R.R., and McKee, E.H., 1972, Ages of plutons and types of mineralization, northwestern Elko County, Nevada: U.S. Geological Survey Professional Paper 800-C, p. 165–168.

Coats, R.R., and Stephens, E.C., 1968, Mountain City copper mine, Elko County, Nevada, *in* Ridge, J.D., ed., Ore deposits of the United States, 1933–1967; Graton-Sales volume 2: American Institute of Mining, Metallalurgical, and Petroleum Engineers, New York, p. 1074–1095.

Cohenour, R.E., 1963, The beryllium belt of western Utah, *in* Beryllium and uranium mineralization in western Juab County, Utah: Guidebook to the Geology of Utah 17, p. 4–7.

Collins, E.A., 1907a, The Combination mine—I. Early development and geologic structure: Mineral and Science Press, v. 91, no. 13, p. 397–399.

Collins, E.A., 1907b, The Combination mine—II. Methods of mining: Mineral and Science Press, v. 95, no. 14, p. 435–438.

Conrad, W.K., 1984, The mineralogy and petrology of compositionally zoned ash flow tuffs, and related silicic volcanic rocks, from the McDermitt caldera complex, Nevada-Oregon: Journal of Geophysical Research, v. 89, no. B10, p. 8639–8664.

Cook, D., Crackel, D., and Jensen, M., 2002, The Dee North Mine, Elko County, Nevada: Mineralogical Record, v. 33, p. 225–234.

Cook, E.F., 1965, Stratigraphy of Tertiary volcanic rocks in eastern Nevada: Nevada Bureau of Mines Report 11, 61 p.

Cook, H.E., 1968, Ignimbrite flows, plugs and dikes in the southern part of the Hot Creek Range, Nye County, Nevada, *in* Coats, R.R., Hay, R.L., and Anderson, C.A., eds., Studies in volcanology: Geological Society of America Memoir 116, p. 107–152.

Cook, R.B., 1999, Connoisseur's choice, stibnite,

Murray Mine, Elko County, Nevada: Rocks and Minerals, v. 74, p. 392–395.

Cook, R.B., 2000, Connoisseur's choice, orpiment, Twin Creeks Mine, Humboldt County, Nevada: Rocks and Minerals, v. 75, p. 112–114.

Cooke, S.R.B., and Doan, D.J., 1935, The mineralogy and X-ray analysis of stainierite from the Swansea mine, Goodsprings, Nevada: American Mineralogist, v. 20, p. 274–280.

Coombs, D.S., Alberti, A., Armbruster, T., Artioli, G., Colella, C., Galli, E., Grice, J.D., Liebau, F., Mandarino, J.A., Minato, H., Nickel, E.H., Passaglia, E., Peacor, D.R., Quartieri, S., Rinaldi, R., Ross, M., Sheppard, R.A., Tillmanns, E., Vezzalini, G., 1998, Recommended nomenclature for zeolite minerals; report of the Subcommittee on Zeolites of the International Mineralogical Association, Commission on New Minerals and Mineral Names: Mineralogical Magazine, v. 62, p. 533–571.

Coope, J.A., 1991, Carlin trend exploration history: Discovery of the Carlin deposit: Nevada Bureau of Mines and Geology Special Publication 13, 16 p.

Cooper, J.R., 1962, Bismuth in the United States: U.S. Geological Survey Mineral Inventory Resource Map MR-22.

Cornwall, H.R., 1962, Calderas and associated volcanic rocks near Beatty, Nye County, Nevada: Geological Society of America Petrologic Studies, A. F. Buddington Volume, p. 357–371.

Cornwall, H.R., 1964, Mineral and water resources of Nevada: Nevada Bureau of Mines and Geology Bulletin 65, 314 p.

Cornwall, H.R., 1966, Nickel deposits of North America: U.S. Geological Survey Bulletin 1223, 62 p.

Cornwall, H.R., 1972, Geology and mineral deposits of southern Nye County, Nevada: Nevada Bureau of Mines and Geology Bulletin 77, 49 p.

Cornwall, H.R., and Kleinhampl, F.J., 1961, Geology of the Bare Mountain Quadrangle, Nevada: U.S. Geological Survey Geologic Quadrangle Map GQ-0157.

Cornwall, H.R., and Kleinhampl, F.J., 1964, Geology of the Bullfrog Quadrangle and ore deposits related to the Bullfrog Hills caldera, Nye County, Nevada and Inyo County, California: U.S. Geological Survey Professional Paper 454-J, 25 p.

Cornwall, H.R., Lakin, H.W., Nakagawa, H.M., and Stager, H.K., 1967, Silver and mercury geochemical anomalies in the Comstock, Tonopah and Silver Reef districts, Nevada-Utah: U.S.

Geological Survey Professional Paper 575-B, p. 10–20.

Couch, B.F., and Carpenter, J.A., 1943, Nevada's metal and mineral production (1859–1940 inclusive): Nevada Bureau of Mines and Geology Bulletin 38, 159 p.

Crowley, J.A., 2000, Garnet and clinozoisite from the Nightingale mining district, Pershing County, Nevada: Rocks and Minerals, v. 75, no. 2, p. 120–125.

Curtis, J.S., 1884, Silver-lead deposits of Eureka, Nevada: U.S. Geological Survey Monograph 7, 200 p.

Cutler, H.C., 1910, National, Nevada: Mining and Scientific Press, v. 101, p. 606–607.

Cutler, H.C., 1911, Telluride, Nevada: Mining and Scientific Press, v. 102, no. 25, p. 845.

Dake, H.C., 1935, New iron meteorite found: The Mineralogist, v. 3, no. 2, p. 27.

Dana, E.S., 1898, Catalog of American localities of minerals: J. Wiley and Sons, New York, 51 p.

Dane, C.H., and Ross, C.P., 1943, The Wild Horse quicksilver district, Lander County, Nevada: U.S. Geological Survey Bulletin 931-K, p. 259–278.

Davidson, D.F., 1960, Selenium in some epithermal deposits of antimony, mercury, silver and gold: U.S. Geological Survey Bulletin 1112-A, 16 p.

Davidson, D.F., 1964, Selenium and tellurium, in Mineral and water resources of Nevada: Nevada Bureau of Mines and Geology Bulletin 65, p. 124–125.

Dayvault, R.D., Castor, S.B., and Berry, M.R., 1985, Uranium associated with volcanic rocks of the McDermitt Caldera, Nevada and Oregon, in Uranium deposits in volcanic rocks: Proceedings of a technical committee meeting: Panel Proceedings Series—International Atomic Energy Agency, STI/PUB/690, p. 379–409.

Decker, D.J., 1972, Geology of the Arabia district, Pershing County, Nevada [M.S. thesis]: University of Nevada, Reno, 44 p.

Deffeyes, K.S., 1959, Erionite from Cenozoic tuffaceous sediments, central Nevada: American Mineralogist, v. 44, p. 501–509.

Deino, A.L., 1985, Stratigraphy, chemistry, K-Ar dating, and paleomagnetism of the Nine Hill Tuff, California-Nevada, and Miocene/Oligocene ash-flow tuffs of Seven Lakes Mountain, California-Nevada [Ph.D. diss.]: University of California, Berkeley, 498 p.

Deiss, C.F., 1952, Dolomite deposit near Sloan, Nevada: U.S. Geological Survey Bulletin 973-C, p. 107–141.

DeMouthe, J.F., 1985, Type localities in Nevada: Mineralogical Record, v. 16, no. 1, p. 43–56.

DeMouthe, J.F., 1987, Gold in the California State Mineral Collection: Mineralogical Record, v. 18, no. 1, p. 81–84.

Deng, Q., 1991, Geology and trace element geochemistry of the Hollister Gold Deposit, Ivanhoe District, Elko County, Nevada [Ph.D. diss.]: University of Texas at El Paso, 287 p.

De Quille, D., 1876, History of the Big Bonanza: Alfred A. Knopf, New York, 1947, 439 p.

De Quille, D., 1889, A history of the Comstock silver lode mines: Arno Press, New York, 1973, 158 p. (Reprinted 1973.)

Desautels, P., 1972, The museum record: Mineralogical Record, v. 3, no. 6, p. 244, 275.

Desborough, G.A., Poole, F.G., Hose, R.K., and Radtke, A.S., 1979, Metals in Devonian kerogenous strata at Gibellini and Bisoni properties, southern Fish Creek Range, Eureka County, Nevada: U.S. Geological Survey Open-File Report 79-530, 31 p.

Dickson, F.W., and Radtke, A.S., 1978, Weissbergite, TlSbS$_2$, a new mineral from the Carlin gold deposit, Nevada: American Mineralogist, v. 63, p. 720–724.

Dickson, F.W., Radtke, A.S., and Peterson, J.A., 1979, Ellisite, Tl$_3$AsS$_3$, a new mineral from the Carlin gold deposit, Nevada, and associated sulfide and sulfosalt minerals: American Mineralogist, v. 64, p. 701–707.

Dickson, F.W., Radtke, A.S., and Rye, R.O., 1975, Implications of the occurrence of barium minerals and sulfur isotopic compositions of barite on late-stage processes in Carlin-type gold deposits [abs.]: Geological Society of America Abstracts with Programs, v. 7, no. 5, p. 604–605.

Dickson, F.W., and Tunnell, G., 1968, Mercury and antimony deposits associated with active hot springs in the western United States, in Ridge, J.D., ed., Ore deposits of the United States, 1933–1967; Graton-Sales volume 2: American Institute of Mining, Metallurgical, and Petroleum Engineers, New York, p. 1673–1701.

Dilles, P.A., Wright, W.A., Monteleone, S.E., Russell, K.D., Marlowe, K.E., Wood, R.A., and Margolis, J., 1996, The geology of the West Archimedes Deposit; A new gold discovery in the Eureka mining district, Eureka County, Nevada, in Coyner, A.R., and Fahey, P.L., eds., Geology and ore deposits of the American Cordillera; symposium proceedings: Geological Society of Nevada, Reno, p. 159–171.

Dircksen, P.E., 1975, Geology and mineralization of the Pine Grove-Rockland mining districts, Lyon County, Nevada [M.S. thesis]: University of Nevada, Reno, 69 p.

Divjakovic, V., and Nowacki, W., 1975, Die kristallstruktur von galchait (Hg$_{0.76}$(Cu,Zn)$_{0.24}$)$_{12}$ Tl$_{0.96}$(AsS$_3$)$_8$: Neues Jahrbuch für Mineralogie Monatshefte, no. 7, p. 291–293.

Dixon, G.L., Hedlund, D.C., and Ekren, E.B., 1973, Geologic map of the Pritchards Station Quadrangle, Nye County, Nevada: U.S. Geological Survey Miscellaneous Investigations Map I-728, 1:48,000.

Dixon, J., 1977, Geology of the Wildhorse Canyon area, Fox Range, Washoe County, Nevada [M.S. thesis]: University of Nevada, Reno, 91 p.

Dixon, R.L., 1971, The geology and ore deposits of the Red Canyon mining district, Douglas County, Nevada [M.S. thesis]: University of Nevada, Reno, 88 p.

Dobak, P.J., 1988, Alteration and paragenesis of the Paradise Peak gold/silver deposit [M.S. thesis]: Colorado State University, Fort Collins, 141 p.

Doebrich, J., 1993, The Fortitude Mine: U.S. Geological Survey Mineral Resources Newsletter, v. 4, no. 3, p. 5.

Doebrich, J.L., Garside, L.J., and Shaw, D.R., 1996, Characterization of mineral deposits in rock of the Triassic to Jurassic magmatic arc of western Nevada and eastern California: U.S. Geological Survey Open-File Report 96-9, 107 p.

Dreyer, R.M., 1940, Goldbanks mining district, Pershing County, Nevada: Nevada Bureau of Mines and Geology Bulletin 33, 36 p.

Drugman, J., 1938, On some unusual twin laws observed in the orthoclase crystals from Goodsprings, Nevada: Mineralogical Magazine, v. 25, no. 160, p. 1–14.

Dunn, P.J., Leavens, P.B., and Barnes, C., 1980, Magnesioaxinite from Luning, Nevada, and some nomenclature for the axinite group: Mineralogical Record, v. 11, p. 13–15.

Dunning, G.E., 1987, Preliminary list of mineral occurrences at the Getchell and Outlaw mines, Nevada: unpublished, 2 p.

Dunning, G.E., 1988, Calcium arsenate minerals new to the Getchell mine, Nevada: Mineralogical Record, v. 19, p. 253–257.

Dunning, G.E., and Cooper, J.F., Jr., 1987, Mineralogy of the Killie mine, Elko County, Nevada: Mineralogical Record, v. 18, p. 413–420.

Dunning, G.E., Moss, G.E., and Cooper, J.F., Jr., 1991, The Outlaw mine, Nye County, Nevada: Mineralogical Record, v. 22, p. 171–182.

Eakle, A.S., 1901, Esmeraldite, a new hydrous sesquioxide of iron from Esmeralda County, Nevada: University of California Publications in Geology, v. 2, p. 320–323.

Eakle, A.S., 1912, The minerals of Tonopah, Nevada: University of California Publications in Geological Sciences, p. 1–20.

Ebert, S.W., Groves, D.I., and Jones, J.K., 1996, Geology, alteration, and ore controls of the Crofoot/Lewis Mine, Sulphur, Nevada; a well-preserved hot-spring gold-silver deposit, *in* Coyner, A.R., and Fahey, P.L., eds., Geology and ore deposits of the American Cordillera; symposium proceedings: Geological Society of Nevada, Reno, p. 209–234.

Ehlmann, A.J., 1962, Occurrences of sepiolite in Utah and Nevada: Economic Geology, v. 57, p. 1085–1094.

Eidel, J.J., 1963, Paragenesis and geochemistry of the antimony-mercury deposits of the Antelope Springs mining district, Pershing County, Nevada [M.S. thesis]: University of California, Los Angeles, 216 p.

Ekren, E.B., Anderson, R.E., Rogers, C.L., and Noble, D.C., 1971, Geology of the northern Nellis Air Force Base, Bombing and Gunnery Range, Nye County, Nevada: U.S. Geological Survey Professional Paper 651, 91 p.

Ekren, E.B., Hinrichs, E.N., and Dixon, G.L., 1973, Geologic map of the Wall Quadrangle, Nye County, Nevada: U.S. Geological Survey Miscellaneous Investigations Map I-719, 1:48,000.

Ekren, E.B., Hinrichs, E.N., Quinlivan, W.D., and Hoover, D.L., 1974, Geologic map of the Moores Station Quadrangle, Nye County, Nevada: U.S. Geological Survey Miscellaneous Investigations Map I-756, 1:48,000.

Ekren, E.B., Rogers, C.L., and Dixon, G.L., 1974, Geologic and Bouguer gravity map of the Reveille Quadrangle, Nye County, Nevada: U.S. Geological Survey Miscellaneous Investigations Map I-806, 1:48,000.

Emmons, D.L., and Coyle, R.D., 1988, Echo Bay details exploration activities at its Cove gold deposit in Nevada: Mining Engineering, v. 40, p. 791–794.

Emmons, S.F., 1877, [Geology of the] Cortez Range [Nevada]: U.S. Geological [Survey] Exploration of the 40th Parallel (King), v. 2, p. 570–589.

Emmons, W.H., 1907, Genesis of the copper deposits of Yerington, Nevada: Engineering and Mining Journal, v. 83, p. 1143–1146.

Emmons, W.H., 1910, A reconnaissance of some mining camps in Elko, Lander and Eureka Counties, Nevada: U.S. Geological Survey Bulletin 408, 130 p.

Emmons, W.H., 1919, Geology and ore deposits of the Yerington District, Nevada: Engineering and Mining Journal, v. 107, p. 1079–1080.

Emsbo, P., Hutchinson, R.W., Hofstra, A.H., Volk, J.A., Bettles, K.H., Baschuk, G.J., Collins, T.M., Lauha, E.A., Borhauer, J.L., 1997, Newly discovered Devonian sedex-type base and precious metal mineralization, northern Carlin Trend, Nevada, *in* Vikre, P., Thompson, T.B., Bettles, K., Christensen, O., and Parratt, R., eds., Carlin-type gold deposits field conference: Society of Economic Geologists Guidebook Series, v. 28, p. 109–117.

Eng, T., 1991, Geology and mineralization of the Freedom Flats gold deposit, Borealis Mine, Mineral County, Nevada, *in* Raines, G.L., Lisle, R.E., Schafer, R.W., and Wilkinson, W.H., eds., Geology and ore deposits of the Great Basin; symposium proceedings: Geological Society of Nevada, Reno, p. 995–1019.

Eng, T., Boden, D.R., Reischman, M.R., and Biggs, J.O., 1996, Geology and mineralization of the Bullfrog Mine and vicinity, Nye County, Nevada, *in* Coyner, A.R., and Fahey, P.L., eds., Geology and ore deposits of the American Cordillera; symposium proceedings: Geological Society of Nevada, Reno, p. 353–402.

Erd, R.C., Foster, M.D., and Proctor, P.D., 1953, Faustite, a new mineral, the zinc analogue of turquoise: American Mineralogist, v. 38, p. 964–972.

Erd, R.C., White, D.E., Fahey, J.J., and Lee, D.E., 1964, Buddingtonite, an ammonium feldspar with zeolitic water: American Mineralogist, v. 49, p. 831–850.

Erickson, R.L., Marranzino, A.P., Oda, U., and James, W.W., 1964, Geochemical exploration near the Getchell mine, Humboldt County, Nevada: U.S. Geological Survey Bulletin 1198-A, 26 p.

Erickson, R.L., Van Sickle, G.H., Nakagawa, H.M., McCarthy, J.H., Jr., and Leong, K.W., 1966, Gold geochemical anomaly in the Cortez district, Nevada: U.S. Geological Survey Circular 534, 9 p.

Ervine, W.B., 1972, The geology and mineral zoning of the Spanish Belt mining district, Nye County, Nevada [Ph.D. diss.]: Stanford University, 295 p.

Everett, F.D., 1964, Reconnaissance of tellurium resources in Arizona, Colorado, New Mexico, and Utah: U.S. Bureau of Mines Report of Investigation 6350, 38 p.

Excalibur Mineral Corp., 1989, Private offering catalog no. 8–89, August, 1989: Peekskill, New York, CD-ROM.

Excalibur Mineral Corp., 1991, Private offering catalog no. 3–91, August, 1989: Peekskill, New York.

Excalibur Mineral Corp., 1999, Private offering catalog no. 9–99, September, 1999: Peekskill, New York.

Excalibur Mineral Corp., 2001a, Catalog 28, no. 1: Peekskill, New York.

Excalibur Mineral Corp., 2001b, Catalog 28, no. 2: Peekskill, New York.

Excalibur Mineral Corp., 2001c, Catalog 28, no. 3: Peekskill, New York.

Excalibur Mineral Corp., 2002a, Catalog 20205, v. 29, no. 5: Peekskill, New York.

Excalibur Mineral Corp., 2002b, Catalog 20206, v. 29, no. 6: Peekskill, New York.

Farrington, O.C., 1910, Quinn Canyon: Field Museum of Natural History—Geology, v. 3, p. 169–176.

Faust, G.T., 1953, Huntite, $Mg_3Ca(CO_3)_2$, a new mineral: American Mineralogist, v. 38, p. 4–24.

Faust, G.T., and Fahey, J.J., 1962, The serpentine group minerals: U.S. Geological Survey Professional Paper 384A, 91 p.

Felzer, B., Hauff, P., and Goetz, A.F.H., 1994, Quantitative reflection spectroscopy of buddingtonite from the Cuprite mining district, Nevada: Journal of Geophysical Research, B, Solid Earth and Planets, v. 99, no. 2, p. 2887–2895.

Ferdock, G.C., Castor, S.B., Leonardson, R.W., and Collins, T., 1997, Mineralogy and paragenesis of ore-stage mineralization in the Betze gold deposit, Goldstrike Mine, Eureka County, Nevada, in Vikre, P., Thompson, T.B., Bettles, K., Christensen, O., and Parratt, R., eds., Carlin-type gold deposits field conference: Society of Economic Geologists Guidebook Series, v. 28, p. 75–86.

Ferguson, H.G., 1916, The Golden Arrow, Clifford and Ellendale districts, Nye County, Nevada: U.S. Geological Survey Bulletin 640-F, p. 113–123.

Ferguson, H.G., 1917, Placer deposits of the Manhattan district, Nevada: U.S. Geological Survey Bulletin 640-J, p. 163–193.

Ferguson, H.G., 1917, The Golden Arrow, Clifford, and Ellendale districts, Nye County, Nevada: U.S. Geological Survey Bulletin 640-F, p. 113–123.

Ferguson, H.G., 1922, The Round Mountain district, Nevada: U.S. Geological Survey Bulletin 725-I, p. 383–406.

Ferguson, H.G., 1924, Geology and ore deposits of the Manhattan district, Nevada: U.S. Geological Survey Bulletin 723, 163 p.

Ferguson, H.G., 1928, The Gilbert district, Nevada: U.S. Geological Survey Bulletin 795-F, p. 113–146.

Ferguson, H.G., 1929, The mining districts of Nevada: Economic Geology, v. 24, p. 115–148.

Ferguson, H.G., 1933, Geology of the Tybo district, Nevada: Nevada Bureau of Mines and Geology Bulletin 20, 61 p.

Ferguson, H.G., 1939, Nickel deposits in Cottonwood Canyon, Churchill County, Nevada: Nevada Bureau of Mines and Geology Bulletin 32, 23 p.

Ferguson, H.G., 1954, Geology of the Mina Quadrangle, Nevada: U.S. Geological Survey Geologic Quadrangle Map GQ-45.

Ferguson, H.G., and Cathcart, S.H., 1954, Geology of the Round Mountain Quadrangle, Nevada: U.S. Geological Survey Geologic Quadrangle Map GQ-40.

Ferguson, H.G., and Muller, S.W., 1949, Structural geology of the Hawthorne and Tonopah Quadrangles, Nevada: U.S. Geological Survey Professional Paper 216, 55 p.

Ferguson, H.G., Muller, S.W., and Cathcart, S.H., 1953, Geology of the Coaldale Quadrangle, Nevada: U.S. Geological Survey Geologic Quadrangle Map GQ-23.

Ferguson, H.G., Muller, S.W., and Roberts, R.J., 1951, Geology of the Winnemucca Quadrangle, Nevada: U.S. Geological Survey Geologic Quadrangle Map GQ-11.

Fisher, L.W., 1927, Quartz from Iowa Canyon, Nevada: American Mineralogist, v. 12, p. 225–226.

Fisk, E.L., 1961, Cinnabar at Cordero: Mining Engineering, v. 13, p. 1228–1230.

Fisk, E.L., 1968, Cordero mine, Opalite mining district, in Ridge, J.D., ed., Ore deposits of the United States, 1933–1967; Graton-Sales volume 2: American Institute of Mining, Metallurgical, and Petroleum Engineers, New York, p. 1573–1591.

Fleck, R.J., and Reynolds, R.E., 1996, Mesozoic stratigraphic units of the eastern Mescal range, southeastern California, in Reynolds, R.E., and Reynolds, J., eds., Punctuated chaos in the northeastern Mojave Desert: San Bernardino County Museum Quarterly, v. 43, no. 1 and 2, p. 49–54.

Fleischer, M., 1961, New mineral names: Schoderite, metaschoderite: American Mineralogist, v. 46, p. 464.

Fleischer, M., 1968, New mineral names: New data on goldfieldite: American Mineralogist, v. 53, p. 2105-2106.

Fleischer, M., 1970, New mineral names: Mackayite: American Mineralogist, v. 55, p. 1072.

Fleischer, M., and Mandarino, J.A., 1995, Glossary of mineral species, 7th ed.: The Mineralogical Record Inc., Tucson, Arizona, 280 p.

Fleischer, M., Mandarino, J.A., and Pabst, A., 1980, New mineral names: Uytenbogaardtite: American Mineralogist, v. 65, p. 209.

Fontenot-Prince, R., 1998, Nevada minerals in the collections of the Museum of Geology: South Dakota School of Mines and Technology, Rapid City (unpublished data), 2 p.

Foo, S.T., Hays, R.C., and McCormack, J.K., 1996, Geology and mineralization of the Pipeline gold deposit, Lander County, Nevada, *in* Coyner, A.R., and Fahey, P.L., eds., Geology and ore deposits of the American Cordillera; symposium proceedings: Geological Society of Nevada, Reno, p. 95-109.

Foord, E.E., Berendson, P., and Storey, L.O., 1974, Corderoite, the first natural occurrence of Hg_3S2Cl_2, from the Cordero mercury deposits, Humboldt County, Nevada: American Mineralogist, v. 59, p. 652-655.

Foord, E.E., O'Conner, J.T., Hughes, J.M., Sutley, S.J., Falster, A.U., Soregaroli, A.E., Lichte, F.E., and Kile, D.E., 1999a, Simmonsite, Na_2LiAlF_6, a new mineral from the Zapot amazonite-topaz-zinnwaldite pegmatite, Hawthorne, Nevada, U.S.A.: American Mineralogist, v. 84, p. 769-772.

Foord, E.E., O'Conner, J.T., Sutley, S.J., and Soregaroli, A.E., 1998, Alumino-fluoride minerals from the Zapot pegamatite, Hawthorne, Nevada: Rocks and Minerals, v. 73, no. 3, p. 198-199.

Foord, E.E., Shawe, D.R., and Conklin, N.M., 1988, Coexisting galena, PbS_{SS}, and sulfosalts: Evidence for multiple episodes of mineralization in the Round Mountain and Manhattan gold districts, Nevada: Canadian Mineralogist, v. 26, p. 355-376.

Foord, E.E., Soregaroli, A.E., and Gordon, H.M., 1999b, The Zapot pegmatite, Mineral County, Nevada: Mineralogical Record, v. 30, p. 277-292.

Ford, W.E., 1932, A textbook of mineralogy, 4th ed.: Wiley and Sons, New York, 851 p.

Ford, W.E., and Bradley, W.M., 1916, On hydrozincite: American Journal of Science, v. 192, p. 59-62.

Foshag, W.F., 1927, Quicksilver deposits of the Pilot Mountains, Mineral County, Nevada: U.S. Geological Survey Bulletin 795-E, p. 113-124.

Foshag, W.F., 1932, Creedite from Nevada: American Mineralogist, v. 17, p. 75-78.

Foshag, W.F., 1934, Searlesite from Esmeralda County, Nevada: American Mineralogist, v. 19, p. 268-274.

Foshag, W.F., and Clinton, H.G., 1927, An occurrence of pitticite in Nevada: American Mineralogist, v. 12, p. 290-292.

Fries, C., Jr., 1942, Tin deposits of northern Lander County, Nevada: U.S. Geological Survey Bulletin 931-L, p. 279-294.

Frondel, C., 1949, Crystallography of spangolite: American Mineralogist, v. 34, p. 181-187.

Frondel, C., 1962, Polytypism in cronstedtite: American Mineralogist, v. 49, p. 781-783.

Frondel, C., and Palache, C., 1949, Retgersite, $NiSO_4•6H_2O$, a new mineral: American Mineralogist, v. 34, p. 188-194.

Frondel, C., and Pough, F.H., 1944, Two new tellurites of iron: Mackayite and blakeite, with new data on emmonsite and "durdenite": American Mineralogist, v. 29, p. 211-225.

Fulton, J.A., and Smith, A.M., 1932, Nonmetallic minerals in Nevada: Nevada Bureau of Mines and Geology Bulletin 17, 8 p.

Gaines, R.V., 1972, New data on emmonsite and a second locality for sonoraite: Mineralogical Record, v. 3, no. 2, p. 82-84.

Gaines, R.V., Skinner, H.C.W., Foord, E.E., Mason, B., Rosenzweig, A., and King, V.T., 1997, Dana's new mineralogy: The system of mineralogy of James Dwight Dana and Edward Salisbury Dana, 8th ed.: John Wiley and Sons, New York, 1819 p.

Gale, H.S., 1912, Nitrate deposits: U.S. Geological Survey Bulletin 523, 36 p.

Gale, H.S., 1921, The Callville Wash colemanite deposit: Engineering and Mining Journal, v. 112, p. 524-530.

Garside, L.J., 1973, Radioactive mineral occurrences in Nevada: Nevada Bureau of Mines and Geology Bulletin 81, 121 p.

Garside, L.J., 1979, Geologic map of the Camp Douglas Quadrangle: Nevada Bureau of Mines and Geology Map 63.

Garside, L.J., 1998, Mesozoic metavolcanic and metasedimentary rocks of the Reno–Carson City area, Nevada and adjacent California:

Nevada Bureau of Mines and Geology Report 49, 30 p.

Garside, L.J., Bonham, H.F., Tingley, J.V., and McKee, E.H., 1993, Potassium-argon ages of igneous rocks and alteration minerals associated with mineral deposits, western and southern Nevada and eastern California: Isochron/West, v. 59, p. 17–23.

Garside, L.J., and Schilling, J.H., 1979, Thermal waters of Nevada: Nevada Bureau of Mines and Geology Bulletin 91, 163 p.

Geehan, R.W., 1950, Investigation of the Union zinc-lead mine, Washoe County, Nevada: U.S. Bureau of Mines Report of Investigation 4623.

Gemmill, P., 1968, The geology of the ore deposits of the Pioche mining district, Nevada, in Ridge, J.D., ed., Ore deposits of the United States, 1933–1967; Graton-Sales volume 2, American Institute of Mining, Metallurgical, and Petroleum Engineers, New York, p. 1128–1143.

Gianella, V.P., 1936a, Geology of the Silver City district and the southern portion of the Comstock Lode, Nevada: Nevada Bureau of Mines and Geology Bulletin 29, 105 p.

Gianella, V.P., 1936b, Occurrences—thulite in Nevada: The Mineralogist, v. 4, no. 12, p. 5–6.

Gianella, V.P., 1936c, A meteorite from Quartz Mountain, Nevada: Popular Astronomy, v. 44, no. 8, p. 448–450.

Gianella, V.P., 1937, Piedmontite from Peavine Mountain, Nevada [abs.]: Geological Society of America Proceedings, 1936, p. 301.

Gianella, V.P., 1938, Vivianite from Ruth, Nevada: American Mineralogist, v. 23, p. 414.

Gianella, V.P., 1939, Mineral deposition at Steamboat Springs, Nevada [abs.]: Economic Geology, v. 34, p. 471–472.

Gianella, V.P., 1941, Nevada's common minerals, including a preliminary list of minerals found in the State: Nevada Bureau of Mines and Geology Bulletin 36, 108 p.

Gianella, V.P., 1942, Pleonaste from Mineral County, Nevada: American Mineralogist, v. 27, p. 462–463.

Gianella, V.P., 1946, Clinoclasite from Majuba Hill, Nevada: American Mineralogist, v. 31, p. 259–260.

Gianella, V.P., 1959, Period of mineralization of the Comstock Lode, Nevada [abs.]: Geological Society of America Bulletin 70, p. 1721.

Gianella, V.P., and Hedquist, W.G., 1942, Scolecite from Clark County, Nevada: Mineralogist, v. 10, p. 107–108.

Gibbs, R.B., 1985, The White Caps Mine, Manhattan, Nevada: Mineralogical Record, v. 16, p. 81–88.

Gilbert, C.M., Christiansen, M.N., Al-Rawi, Y., and Lajoie, K.L., 1968, Structural and volcanic history of Mono Basin California-Nevada, in Coats, R.R., Hay, R.L., and Anderson, C.A., eds., Studies in volcanology: Geological Society of America Memoir 116, p. 275–330.

Giles, D.L., and Schilling, J.H., 1972, Variation in rhenium content of molybdenite: International Geological Congress, Montreal 1972, Section (Volume) 10, Geochemistry, p. 145–152.

Gillson, J.L., and Shannon, E.V., 1925, Szaibelyite from Lincoln County, Nevada: American Mineralogist, v. 10, p. 137–139.

Gilluly, J., 1960, Structure of Paleozoic and early Mesozoic rocks in the northern part of the Shoshone Range, Nevada: U.S. Geological Survey Professional Paper 400-B, 265 p.

Gilluly, J., and Gates, O., 1965, Tectonic and igneous geology of the northern Shoshone Range, Nevada, with sections on gravity in Crescent Valley by D. Plouff, and economic geology by K.B. Ketner: U.S. Geological Survey Professional Paper 465, 153 p.

Gilluly, J., and Masursky, H., 1965, Geology of the Cortez Quadrangle, Nevada, with a section of gravity and aeromagnetic surveys by D. R. Mabey: U.S. Geological Survey Bulletin 1175, 117 p.

Gilmour, E.H., 1961, Orthoclase crystals from Crystal Pass, Nevada: The Compass, v. 39, no. 1, p. 9–12.

Gott, G.B., Mabey, D.R., McCarthy, H., and Oda, U., 1962, Mineralization associated with a magnetic anomaly in part of the Ely Quadrangle, Nevada: U.S. Geological Survey Professional Paper 450-E, p. 1–8.

Goudey, H., 1945, Dihydrite from Mineral County, Nevada: American Mineralogist, v. 30, p. 640.

Goudey, H., 1946, An occurrence of wapplerite in Nevada: American Mineralogist, v. 31, p. 598.

Goudey, H., 1952, Powellite and associated pseudomorphs at the Anderson mine, Mineral County, Nevada: American Mineralogist, v. 37, p. 696–697.

Granger, A.E., Bell, M.M., Simmons, G.C., and Lee, F., 1957, Geology and mineral resources of Elko County, Nevada: Nevada Bureau of Mines and Geology Bulletin 54, 190 p.

Grawe, O.R., 1928a, A table for the identification of Nevada's common minerals: University of Nevada Bulletin, v. 7, no. 1, 11 p.

Grawe, O.R., 1928b, The mineralogy of dumortierite: University of Nevada Bulletin, v. 7, no. 2, p. 7–22.

Gray, M., 1999, Transparent gemstones of Nevada: Rocks and Minerals, v. 74, p. 368–369.

Greenan, J.O., 1914, Geology of Fairview, Nevada: Engineering and Mining Journal, v. 97, p. 791–793.

Griffiths, W.R., 1964, Beryllium, *in* Mineral and water resources of Nevada: Nevada Bureau of Mines and Geology Bulletin 65, p. 70–75.

Grott, J.A., 1996, $^{40}Ar/^{39}Ar$ geochronology of gold mineralization and origin of auriferous fluids for the Getchell and Twin Creeks mines, Humboldt County, Nevada [Ph.D. diss.]: New Mexico Institute of Mining and Technology, Socorro, 291 p.

Grossman, J.N., 1997, The Meteoritical Bulletin, No. 81, 1997 July: Meteoritics and Planetary Science, v. 32, Suppl., p. A159–A166.

Gulbrandsen, R.A., and Gielow, D.G., 1960, Mineral assemblage of a pyrometasomatic deposit near Tonopah, Nevada: U.S. Geological Survey Professional Paper 400-B, p. 20–21.

Hague, A., 1870, Geology of the White Pine district, *in* King, C., ed., U.S. Geological Survey exploration of the 40th Parallel, v. 3, p. 409–421.

Hague, A., 1892, Geology of the Eureka district, Nevada: U.S. Geological Survey, Monograph 20, 419 p.

Hague, A., and Emmons, S.F., 1877, Descriptive geology, *in* King, C., ed., Report of the geological exploration of the Fortieth Parallel, v. 2, p. 469–853.

Hall, R., 1962, Sampling of Lynch Creek beryllium-tungsten prospect, Lander County, Nevada: U.S. Bureau of Mines Report of Investigation 6118, 10 p.

Hamilton, M.M., 1993, Mines, prospects and mineral occurrences in the Shoshone Range, Nevada: U.S. Bureau of Mines Open-File Report MLA 16-93.

Harder, E.C., 1909, Iron ores near Dayton, Nevada: U.S. Geological Survey Bulletin 430, p. 240–246.

Hardie, B.S., 1966, Carlin gold mine, Lynn district, Nevada: Nevada Bureau of Mines Report 13, part A, p. 73–83.

Hardy, R.A., 1938, The Getchell mine, new gold producer of Nevada: Engineering and Mining Journal, v. 139, no. 11, p. 29.

Hardy, R.A., 1940, Geology of the Getchell Mine [Nevada]: American Institute of Mining, Metallurgical, and Petroleum Engineers, Technical Publication, v. 4, no. 6, 3 p.

Harris, D.C., and Chen, T.T., 1975, Benjaminite, reinstated as a valid species: Canadian Mineralogist, v. 13, p. 402–407.

Harris, D.C., and Chen, T.T., 1976, Crystal chemistry and reexamination of nomenclature of sulfosalts in aikinite-bismuthinite series: Canadian Mineralogist, v. 14, p. 194–205.

Harris, M., 1974, Statistical treatment of selected trace elements in unoxidized gold ores of the Carlin gold deposit, Nevada [M.S. thesis]: Stanford University, 66 p.

Harris, M., and Radtke, A.S., 1976, Statistical study of selected trace elements with reference to geology and genesis of the Carlin gold deposit, Nevada: U.S. Geological Survey Professional Paper 960, 21 p.

Harris, N.B., 1980, Skarn formation near Ludwig, Yerington District, Nevada [Ph.D. diss.]: Stanford University, 218 p.

Harvey, R.D., and Vitaliano, C.J., 1964, Wall rock alteration in the Goldfield district, Nevada: Journal of Geology, v. 72, p. 564–579.

Hasler, R.W., Wilson, W.R., and Darnton, B.T., 1991, Geology and ore deposits of the Ward mining district, White Pine County, Nevada, *in* Raines, G.L., Lisle, R.E., Schafer, R.W., and Wilkinson, W.H., eds., Geology and ore deposits of the Great Basin; symposium proceedings: Geological Society of Nevada, Reno, p. 333–353.

Hausen, D.M., 1960, Schoderite, a new phosphovanadate mineral from Nevada: Geological Society of America Bulletin, v. 71, no. 12, pt. 2, p. 1883.

Hausen, D.M., 1962, Schoderite, a new phosphovanadate mineral from Nevada: American Mineralogist, v. 47, p. 637–648.

Hausen, D.M., and Kerr, P.F., 1968, Fine gold occurrence at Carlin, Nevada, *in* Ridge, J.D., ed., Ore deposits of the United States, 1933–1967; Graton-Sales volume 1: American Institute of Mining, Metallurgical, and Petroleum Engineers, New York, p. 908–940.

Hawkins, R.B., 1982, Discovery of the Bell gold mine, Jerritt Canyon District, Elko County, Nevada: American Mining Congress Journal, v. 68, no. 2, p. 28–32.

Hay, R.L., 1964, Phillipsite of saline lakes and soils: American Mineralogist, v. 49, p. 1366–1387.

Hay, R.L., Wiggins, B., and Teague, T.T., 1980, Spring-related carbonates and Mg-silicate clays in the Amargosa Basin of Nevada and California [abs.]: Geological Society of America Abstracts with Programs, v. 12, no. 7, p. 443–444.

Hays, R.C., Jr., and Foo, S.T., 1991, Geology and mineralization of the Gold Acres deposit, Lander County, Nevada, *in* Raines, G.L., Lisle, R.E., Schafer, R.W., and Wilkinson, W.H., eds., Geology and ore deposits of the Great Basin; symposium proceedings: Geological Society of Nevada, Reno, p. 677–685.

Heald, P., Foley, N.K., and Hayba, D.O., 1987, Comparative anatomy of volcanic-hosted epithermal deposits; acid-sulfate and adularia-sericite types: Economic Geology, v. 82, p. 1–26.

Heins, P.S., 1937, Thenardite crystals from Rhodes Marsh, Nevada: American Mineralogist, v. 22, p. 307–308.

Henry, C.D., Boden, D.R., and Castor, S.B., 1999, Geologic map of the Tuscarora Quadrangle, Nevada: Nevada Bureau of Mines and Geology Map 116, 20 p.

Henry, C.D., Elson, H.B., McIntosh, W.C., Heizler, M.T., and Castor, S.B., 1997, Brief duration of hydrothermal activity at Round Mountain, Nevada, determined from $^{40}Ar/^{39}Ar$ geochronology: Economic Geology, v. 92, p. 807–826.

Herting-Agthe, S., 1998, Nevada mineral holdings in the Technische Universitat Berlin, Germany: unpublished, 2 p.

Hess, F.L., 1917, Tungsten minerals and deposits: U.S. Geological Survey Bulletin 652, 85 p.

Hess, F.L., 1919, The tungsten resources of the world: Engineering and Mining Journal, v. 108, p. 715–722.

Hess, F.L., and Hunt, W.T., 1913, Triplite from eastern Nevada: American Journal of Science, v. 36, p. 51–54.

Hess, F.L., and Larsen, E.S., 1920, Contact-metamorphic tungsten deposits of the United States: U.S. Geological Survey Bulletin 725-D, p. 245–309.

Hetland, D.L., 1955, Preliminary report on the Buckhorn claims, Washoe County, Nevada and Lassen County, California: U.S. Atomic Energy Commission Report RME—2039, pt. 1, 13 p.

Hewett, D.F., 1924, Deposits of magnesia alum near Fallon, Nevada: U.S. Geological Survey Bulletin 750-E, p. 79–86.

Hewett, D.F., 1931, Geology and ore deposits of the Goodsprings Quadrangle, Nevada: U.S. Geological Survey Professional Paper 162, 172 p.

Hewett, D.F., 1956, Geology and mineral resources of the Ivanpah Quadrangle, California and Nevada: U.S. Geological Survey Professional Paper 275, 172 p.

Hewett, D.F., 1971, Coronadite—modes of occur-

rence and origin: Economic Geology, v. 66, p. 164–177.

Hewett, D.F., Callaghan, E., Moore, B.N., Nolan, T.B., Rubey, W.W., and Schaller, W.T., 1936, Mineral resources of the region around Boulder Dam: U.S. Geological Survey Bulletin 871, 197 p.

Hewett, D.F., Cornwall, H.R., and Erd, R.C., 1968, Hypogene veins of gibbsite, pyrolusite and lithiophorite in Nye County, Nevada: Economic Geology, v. 63, p. 360–371.

Hewett, D.F., and Fleischer, M., 1960, Deposits of the manganese oxides: Economic Geology, v. 55, p. 1–55.

Hewett, D.F., Fleischer, M., and Conklin, N., 1963, Deposits of the manganese oxides; supplement: Economic Geology, v. 58, p. 1–51.

Hewett, D.F., and Radtke, A.S., 1967, Silver-bearing black calcite in western mining districts: Economic Geology, v. 62, p. 1–21.

Hewett, D.F., and Rove, O.N., 1930, Occurrence and relations of alabandite: Economic Geology, v. 25, p. 36–56.

Hewett, D.F., and Webber, B.N., 1931, Bedded deposits of manganese oxides near Las Vegas, Nevada: Nevada Bureau of Mines and Geology Bulletin 13, 17 p.

Hewitt, W.P., 1968, Western Utah, eastern and central Nevada, *in* Ridge, J.D., ed., Ore deposits of the United States, 1933–1967; Graton-Sales volume 1: American Institute of Mining, Metallurgical, and Petroleum Engineers, New York, p. 857–885.

Heyl, A.V., and Bozion, C.V., 1962, Oxidized zinc deposits of the United States—Part 1: General geology: U.S. Geological Survey Bulletin 1135-A, 52 p.

Hill, J.M., 1911, Notes on the economic geology of the Ramsey, Talapoosa and White Horse mining districts, Lyon and Washoe Counties, Nevada: U.S. Geological Survey Bulletin 470, p. 99–108.

Hill, J.M., 1912, The mining districts of the western United States: U.S. Geological Survey Bulletin 507, 219 p.

Hill, J.M. 1914, The Yellow Pine Mining District, Clark County, Nevada: U. S. Geological Survey Bulletin 540, p. 223–274.

Hill, J.M., 1915, Some mining districts in northeastern California and northwestern Nevada: U.S. Geological Survey Bulletin 594, 200 p.

Hill, J.M., 1916, Notes on some mining districts in eastern Nevada: U.S. Geological Survey Bulletin 648, 214 p.

Hill, J.M., 1930, Magnesia and its compounds, *in*

Mineral resources of the U.S. 1927: U.S. Bureau of Mines, p. 167-179.

Hillebrand, W.F., and Penfield, S.L., 1902, Some additions to the alunite-jarosite group of minerals: American Journal of Science, v. 14, p. 211-220.

Hitchborn, A.D., Arbonies, D.G., Peters, S.G., Connors, K.A., Noble, D.C., Larson, L.T., Beebe, J.S., and McKee, E.H., 1996, Geology and gold deposits of the Bald Mountain mining district, White Pine County, Nevada, in Coyner, A.R., and Fahey, P.L., eds., Geology and ore deposits of the American Cordillera; symposium proceedings: Geological Society of Nevada, Reno, p. 505-546.

Hobbs, S.W., and Clabaugh, S.E., 1946, Tungsten deposits of the Osgood Range, Humboldt County, Nevada: Nevada Bureau of Mines and Geology Bulletin 44, 32 p.

Hoffman, W.J., 1878, On the mineralogy of Nevada: Bulletin of the United States Geological and Geographical Survey of the Territories, v. 4, p. 731-745.

Hofstra, A.H., 1994, Geology and genesis of Carlin-type gold deposits in the Jerritt Canyon District, Nevada [Ph.D. diss.]: University of Colorado, Boulder, 719 p.

Hofstra, A.H., Snee, L.W., Rye, R.O., Folger, H.W., Phinisey, J.D., Loranger, R.J., Dahl, A.R., Naeser, C.W., Stein, H.J., Lewchuk, M., 1999, Age constraints on Jerritt Canyon and other Carlin-type gold deposits in the western United States—relationship to mid-Tertiary extension and magmatism: Economic Geology, v. 94, no. 6, p. 769-802.

Holden, E.F., 1924, The cause of color in rose quartz: American Mineralogist, v. 9, p. 75-88.

Hollabaugh, C.L., and Purcell, V.L., 1987, Garnet Hill, White Pine County, Nevada: Mineralogical Record, v. 18, p. 195-198.

Holmes, D.A., 1994, Zeolites, in Carr, D.D., ed., Industrial minerals and rocks, 6th ed.: Society for Mining, Metallurgy, and Exploration, Inc., Littleton, Colorado, p. 1129-1158.

Holmes, G.H., Jr., 1963, Beryllium investigations in California and Nevada, 1959-62: U.S. Bureau of Mines Information Circular 7930, 19 p.

Holmes, G.H., Jr., 1965, Mercury in Nevada, in Mercury potential of the U.S.: U.S. Bureau of Mines Information Circular 8252, p. 215-300.

Holmes, P.J., 1972, Infiltration uranium deposits in ash-flow tuffs [M.S. thesis]: University of Nevada, Reno, 65 p.

Hoover, D.L., 1968, Genesis of zeolites, Nevada Test Site, in Eckel, E.B., ed., Nevada Test Site: Geological Society of America Memoir 110, p. 275-284.

Horton, F.W., 1916, Molybdenum—its ores and their concentration: U.S. Bureau of Mines and Geology Bulletin 111, 132 p.

Horton, R.C., 1961, An inventory of fluorspar occurrences in Nevada: Nevada Bureau of Mines and Geology Report 4, 37 p.

Horton, R.C., 1964a, Coal, in Mineral and water resources of Nevada: Nevada Bureau of Mines and Geology Bulletin 65, p. 49-57.

Horton, R.C., 1964b, Nonmetallic mineral resources; sodium compounds, in Mineral and water resources of Nevada: Nevada Bureau of Mines Bulletin 65, p. 247-254.

Horton, R.C., Schilling, J.H., and Slemmons, D.B., 1962, Roadlog—Reno to Fairview Peak earthquake area: Geological Society of Sacramento Annual Field Trip Guidebook, p. 69-80.

Hose, R.K., Blake, M.C., and Smith, R.M., 1976, Geology and mineral resources of White Pine County, Nevada: Nevada Bureau of Mines and Geology Bulletin 85, 105 p.

Hotz, P.E., and Willden, R., 1964, Geology and mineral deposits of the Osgood Mountains Quadrangle, Humboldt County, Nevada: U.S. Geological Survey Professional Paper 431, 128 p.

Houser, F.N., and Poole, F.G., 1960, Primary structures in pyroclastic rocks of the Oak Spring Formation (Tertiary), northeastern Nevada Test Site, Nye County, Nevada: Geological Society of America Bulletin, v. 71, no. 12, pt. 2, p. 2062-2063.

Hsu, L.C., 1991, Occurrence of cymrite in Nevada: Nevada Geology [Nevada Bureau of Mines and Geology newsletter], no. 12, p. 1-3.

Hsu, L.C., 1994, Cymrite: New occurrence and stability: Contributions to Mineralogy and Petrography, v. 118, p. 314-320.

Hudson, D.M., 1977, Geology and alteration of the Wedekind and part of the Peavine districts, Washoe County, Nevada [M.S. thesis]: University of Nevada, Reno, 164 p.

Hudson, D.M., 1983, Alteration and geochemical characteristics of the upper parts of selected porphyry systems, western Nevada [Ph.D. diss.]: University of Nevada, Reno, 229 p.

Humphrey, F.L., 1945, Geology of the Groom district, Lincoln County, Nevada: Nevada Bureau of Mines and Geology Bulletin 42, 50 p.

Humphrey, F.L., 1960, Geology of the White Pine mining district, White Pine County, Nevada:

Nevada Bureau of Mines and Geology Bulletin 57, 119 p.

Humphrey, F.L., and Wyatt, M., 1958, Scheelite in feldspathized granodiorite at the Victory mine, Gabbs, Nevada: Economic Geology, v. 53, p. 38-44.

Hurlbut, C.S., Jr., 1946, Artinite from Luning, Nevada: American Mineralogist, v. 31, p. 365-369.

Hurlbut, C.S., Jr., 1955, Wulfenite symmetry shown by crystals from Yugoslavia: American Mineralogist, v. 40, p. 857-860.

Ilchick, R.P., 1984, Hydrothermal maturation of indigenous organic matter at the Alligator Ridge Gold Deposits, Nevada [M.S. thesis]: University of California, Berkeley, 48 p.

Ilchick, R.P., 1990, Geology and genesis of the Vantage gold deposits, Alligator Ridge—Bald Mountain mining district, Nevada [Ph.D. diss.]: University of California, Los Angeles, 136 p.

Ingerson, E., and Barksdale, J.D., 1943, Iridescent garnet from the Adelaide mining district, Nevada: American Mineralogist, v. 28, p. 303-312.

Iverson, H.G., and Holmes, D.T., 1954, Concentration of oxide and silicate manganese ores from the vicinity of Winnemucca, Pershing County, Nevada: U.S. Bureau of Mines Report of Investigation 5022, 9 p.

Ivosevic, S.W., 1978, Johnnie gold district, Nevada, and implications on regional stratigraphic controls: Economic Geology, v. 73, no. 1, p. 100-106.

Jackson, P.R., 1996, Geology and gold mineralization at the Pine Grove mining distict, Lyon County, Nevada, in Coyner, A.R., and Fahey, P.L., eds., Geology and ore deposits of the American Cordillera; symposium proceedings: Geological Society of Nevada, Reno, p. 403-417.

Jackson, W.T., 1963, Treasure Hill: portrait of a silver mining camp: University of Arizona Press, Tucson, 254 p.

Jambor, J.L., 1967, New lead sulfantimonides from Modoc, Ontario, Part 2—mineral descriptions: Canadian Mineralogist, v. 9, p. 191-213.

Jambor, J.L., 1969, Dadsonite (minerals Q and QM), a new lead sulfantimonide: Mineralogical Magazine, v. 37, p. 437-441.

Jambor, J.L., and Plant, A.G., 1975, The composition of the lead sulfantimonide, robinsonite: Canadian Mineralogist, v. 13, p. 415-417.

James, L.P., 1976, Zoned alteration in limestone at porphyry copper deposits, Ely, Nevada: Economic Geology, v. 71, p. 488-512.

Jedwab, J., 1998a, Study of five pieces of copper-platinum-gold ore samples from the Boss Mine, Goodsprings Quadrangle, Clark County, Nevada: unpublished interim report, Université Libre de Bruxelles, Bruxelles, Belgique, 3 p.

Jedwab, J., 1998b, Study of a sample of platinum-gold ore from the Boss Mine, Goodsprings Quadrangle, Clark County, Nevada; Chicago Natural History Museum specimen—Economic Geology Catalogue No. E 14927: unpublished interim report, Université Libre de Bruxelles, Bruxelles, Belgique, 10 p.

Jedwab, J., Bedaut, D., and Beaunier, P., 1999, Discovery of a palladium-platinum-gold-mercury bitumen in the Boss Mine, Clark County, Nevada: Economic Geology, v. 94, p. 1163-1169.

Jenkins, R.E., 1981, Mineralogy of Nevada (Draft #6): unpublished manuscript, 280 p.

Jenney, C.P., 1935, Geology of the central Humboldt Range, Nevada: Nevada Bureau of Mines and Geology Bulletin 24, 75 p.

Jenney, W.P., 1909a, The Nevada meteorite: Mining and Scientific Press, v. 98, no. 2, p. 93-94.

Jenney, W.P., 1909b, Great Nevada meteor of 1894: American Journal of Science, v. 28, no. 167, p. 431-434.

Jennings, S., 1991, Geology of the West Sinter deposit at the Buckhorn Mine, Eureka County, Nevada, in Raines, G.L., Lisle, R.E., Schafer, R.W., and Wilkinson, W.H., eds., Geology and ore deposits of the Great Basin; symposium proceedings: Geological Society of Nevada, Reno, p. 947-956.

Jensen, M.C., 1982, Endlichite and descloizite from Chalk Mountain mine, Churchill County, Nevada: Mineralogical Record, v. 13, p. 219-221.

Jensen, M.C., 1985, The Majuba Hill mine, Pershing County, Nevada: Mineralogical Record, v. 16, p. 57-72.

Jensen, M.C., 1988a, Calcite occurrence at the Barth Iron mine, Eureka County, Nevada: Mineral News, v. 4, no. 5, p. 5-6.

Jensen, M.C., 1988b, Minerals of the San Rafael mine, Nye County, Nevada: Mineral News, v. 4, no. 9, p. 7.

Jensen, M.C., 1990a, A Nevada occurrence for glaukosphaerite at the Key West mine, Clark County: Mineral News, v. 6, no. 11, p. 1-2, 5.

Jensen, M.C., 1990b, New discoveries at the Chalk Mountain mine, Churchill County, Nevada: Mineral News, v. 6, no. 1, p. 4-5.

Jensen, M.C., 1993a, Smoky quartz crystals at Lake

Tahoe, Nevada: Rocks and Minerals, v. 68, p. 300–305.

Jensen, M.C., 1993b, Update on the Majuba Hill mine, Pershing County, Nevada: Mineralogical Record, v. 24, p. 171–180.

Jensen, M.C., 1993c, Aurichalcite and associated minerals from the Grand Deposit mine, White Pine County, Nevada: Mineral News, v. 9, no. 1, p. 1–2.

Jensen, M.C., 1993d, Kawazulite and gold from Montreal Canyon, Nevada: Mineral News, v. 9, no. 3, p. 1, 3.

Jensen, M.C., 1994, New locality for richelsdorfite and other arsenates north of Reno, Washoe County, Nevada: Mineral News, v. 10, no. 5, p. 1, 6–8.

Jensen, M.C., 1999, The Meikle mine, Elko County, Nevada: Mineralogical Record, v. 30, p. 187–196.

Jensen, M.C., and Leising, J., 2001, Wavellite, fluellite, and minyulite from the Willard Mine, Pershing County, Nevada: Mineralogical Record, v. 32, p. 297–303.

Jensen, M.C., Rota, J.C., and Foord, E.E., 1995, The Gold Quarry mine, Carlin-trend, Eureka County, Nevada: Mineralogical Record, v. 26, p. 449–469.

Jensen, M.L., Ashley, R.P., and Albers, J.P., 1971, Primary and secondary sulfates at Goldfield, Nevada: Economic Geology, v. 66, p. 618–626.

Jerome, S.E., and Cook, D.R., 1967, Relation of some metal mining districts in the western United States to regional tectonic environments and igneous activity: Nevada Bureau Mines and Geology Bulletin 69, 35 p.

John, D. A., Nash, J.T., Plouff, D., Whitebread, D.H., 1991, The Conterminous United States Mineral Appraisal Program; background information to accompany folio of geologic, geochemical, geophysical, and mineral resources maps of the Tonopah 1 degree by 2 degree Quadrangle, Nevada: U.S. Geological Survey Circular 1070, 13 p.

Johnson, M.G., 1973, Placer gold deposits of Nevada: U.S. Geological Survey Bulletin 1356, 118 p.

Johnson, M.G., 1977, Geology and mineral resources of Pershing County, Nevada: Nevada Bureau of Mines and Geology Bulletin 89, 115 p.

Johnson, R.C., 1977, Geological investigation and ore reserves estimation of the Copper Chief (Ruby Hill) mine, Douglas County, Nevada [M.S. thesis]: University of Nevada, Reno, 71 p.

Johnson, R.N., 1978, Nevada-Utah gem atlas: Cy Johnson and Son, Susanville, California, 47 p.

Jones, B.K., 1998, Geology of the Olinghouse gold mine: Geological Society of Nevada Newsletter, October 1998, p. 3–4.

Jones, C., and Jones, J., 1999, New mineral production news from Casey and Jane Jones: Mineral News, v. 15, no. 3, p. 9.

Jones, J.C., 1912, The origin of the anhydrite at the Ludwig mine, Lyon County, Nevada: Economic Geology, v. 7, p. 400–402.

Jones, J.C., 1913, The Barth iron ore deposit: Economic Geology, v. 8, p. 247–263.

Jones, J.C., 1929, Age of Lake Lahontan: Geological Society of America Bulletin, v 40, no. 3, p. 533–540.

Jones, J.C., and Grawe, O.R., 1928, The geology of the deposit of dumortierite in Humboldt Queen Canyon, Pershing County, Nevada: Nevada Bureau of Mines and Geology Bulletin 8, p. 23–34.

Jones, R.B., 1991, Metals, in The Nevada mineral industry 1990: Nevada Bureau of Mines and Geology Special Publication MI-1990, p. 11–18.

Jones, R.B., and Schilling, J.H., 1982, Metals, in The Nevada mineral industry 1981: Nevada Bureau of Mines and Geology Special Publication MI-1981, p. 3–6.

Joralemon, P., 1951, The occurrence of gold at the Getchell mine, Nevada: Economic Geology, v. 46, p. 267–310.

Jungles, G., 1974, Galkhaite, a newly discovered mineral from Siberia found at the Getchell mine, Nevada: Mineralogical Record, v. 5, p. 290.

Kamali, C., and Norman, D.I., 1996, Mineralization at the Lone Tree gold deposit [abs.]: Society for Mining, Metallurgy and Exploration, Inc., Annual Meeting, Phoenix, March 1996, p. 48.

Karup-Møeller, S., 1972, New data on pavonite, gustavite and some related sulfosalt minerals: Neues Jahrbuch Für Geologie und Palaontologie Abhandlungen, v. 117, p. 19–38.

Karup-Møeller, S., 1977, Mineralogy of some Ag-(Cu)-Pb-Bi sulphide associations: Bulletin of the Geological Society of Denmark, v. 26, p. 41–68.

Karup-Møeller, S., and Makovicky, E., 1979, On pavonite, cupropavonite, benjaminite and "oversubstituted" gustavite: Bulletin de Mineralogie, v. 102, no. 4, p. 351–367.

Kato, A., Sakurai, K.I., and Ohsumi, K., 1970, Wakabayashilite, in Introduction to Japanese minerals: Geological Survey of Japan, p. 92–93.

Keith, W.J., 1977, Geology of the Red Mountain mining district, Esmeralda County, Nevada: U.S. Geological Survey Bulletin 1423, 45 p.

Keith, W.J., Calk, L., and Ashley, R.P., 1980, Crystals of coexisting alunite and jarosite, Goldfield, Nevada: U.S. Geological Survey Professional Paper 1124C, 5 p.

Keith, W.J., Silberman, M.I., and Erd, R.C., 1976, K-Ar age of mineralization, Silver Peak (Red Mountain) mining district, Esmeralda County, Nevada: Isochron/West, v. 16, p. 13.

Kepper, J., 1995, Mineralogy of the Milford Mines, Goodsprings Quadrangle, Clark County, Nevada [abs.]: Minerals of Arizona, Third Annual Symposium, Arizona Mineral and Mining Museum Foundation and the Arizona Department of Mines and Mineral Resources, p. 24–25.

Kepper, J., 1998, Wulfenite from the Milford mine, Goodsprings Mining District, Clark County, Nevada: Mineral News, v. 14, no. 2, p. 1, 6–7.

Kepper, J., 2000, The Yellow Pine Mining District, Goodsprings, Clark County, Nevada, in Reynolds, R.E., ed., Minerals from the Mojave: San Bernardino County Museum Association, v. 47, no. 1, p. 3–14.

Kerr, P.F., 1934, Geology of the tungsten deposits near Mill City, Nevada: Nevada Bureau of Mines and Geology Bulletin 21, 46 p.

Kerr, P.F., 1936, The tungsten mineralization at Silver Dyke, Nevada: Nevada Bureau of Mines and Geology Bulletin 28, 67 p.

Kerr, P.F., 1938, Tungsten mineralization at Oreana, Nevada: Economic Geology, v. 33, p. 390–427.

Kerr, P.F., 1940a, A pinitized tuff of ceramic importance: American Ceramic Society Journal, v. 23, p. 65–70.

Kerr, P.F., 1940b, Tungsten-bearing manganese deposits at Golconda, Nevada: Geological Society of America Bulletin, v. 51, p. 1359–1389.

Kerr, P.F., 1970, Tungsten mineralization in the United States: Geological Society of America Memoir 15, 241 p.

Kerr, P.F., and Callaghan, E., 1935, Scheelite-leuchtenbergite vein in Paradise Range, Nevada: Geological Society of America Bulletin, v. 46, p. 1957–1974.

Kerr, P.F., and Jenney, P., 1935, The dumortierite-andalusite mineralization at Oreana, Nevada: Economic Geology, v. 30, p. 287–300.

Ketner, R.B., and Smith, J.F., Jr., 1963, Geology of the Railroad mining district, Elko County, Nevada: U.S. Geological Survey Bulletin 1162-B, 27 p.

Khin, B., 1970, Cornetite from Saginaw Hill, Arizona: Mineralogical Record, v. 1, no. 3, p. 117–118.

Killeen, P.L., and Newman, W.L., 1965, Tin in the United States, exclusive of Alaska and Hawaii: Mineral Resources of the United States, U.S. Geological Survey Mineral Inventory Resource Map MR-0044, 9 p.

King, C., 1870, The Comstock Lode, in King, C., U.S. Geological Survey exploration of the 40th Parallel, v. 3, p. 10–91.

King, H.O., 1942, Copper production and war demands: The Mining Journal, v. 26, no. 11, p. 5, 30.

Kirkemo, H., Anderson, C.A., and Creasey, S.C., 1965, Investigations of molybdenum deposits in the conterminous United States, 1942–60: U.S. Geological Survey Bulletin, 1182-E, p. E1–E90.

Kizis, J.A., Jr., Crist, E.M., and Bruff, S.R., 1997, Update of the Buffalo Valley intrusion-related gold project, Battle Mountain region, Nevada: unpublished oral presentation, Geological Society of Nevada, 17 October 1997.

Klein, J., 1983, Where to find gold and gems in Nevada: Gem Guides Book Company, Pico Rivera, California, 110 p.

Kleinhampl, F.J., and Ziony, J.I., 1967, Preliminary geologic map of northern Nye County, Nevada: U.S. Geological Survey Open-File Report 67-129, 1:200,000.

Kleinhampl, F.J., and Ziony, J.I., 1984, Mineral resources of northern Nye County, Nevada: Nevada Bureau of Mines and Geology Bulletin 99B, 243 p.

Kleinhampl, F.J., and Ziony, J.I., 1985, Geology of northern Nye County, Nevada: Nevada Bureau of Mines and Geology Bulletin 99A, 172 p.

Klinger, F.L., 1952, Andalusite-corundum mineralization near Hawthorne, Nevada [M.S. thesis]: University of Wisconsin-Madison, 31 p.

Knopf, A., 1915a, Plumbojarosite and other basic lead-ferric sulphates from the Yellow Pine District, Nevada: Journal of Washington Academy of Science, v. 5, no. 14, p. 497–503.

Knopf, A., 1915b, A gold-platinum-palladium lode in southern Nevada: U.S. Geological Survey Bulletin 620-A, p. 1–18.

Knopf, A., 1916a, Tin ore in northern Lander County, Nevada: U.S. Geological Survey Bulletin 640-C, p. 125–138.

Knopf, A., 1916b, Wood tin in the Tertiary rhyolites of northern Nevada: Economic Geology, v. 11, p. 652–661.

Knopf, A., 1916c, Some cinnabar deposits in western Nevada: U.S. Geological Survey Bulletin 601, p. 59–68.

Knopf, A., 1918a, The antimonial silver-lead veins of

the Arabia district, Nevada: U.S. Geological Survey Bulletin 660, p. 249–255.

Knopf, A., 1918b, Geology and ore deposits of the Yerington district, Nevada: U.S. Geological Survey Professional Paper 114, 68 p.

Knopf, A., 1921a, The Divide silver district, Nevada: U.S. Geological Survey Bulletin 715-K, p. 147–170.

Knopf, A., 1921b, Ore deposits of Cedar Mountain, Mineral County, Nevada: U.S. Geological Survey Bulletin 725-H, p. 361–382.

Knopf, A., 1923, The Candelaria silver district, Nevada: U.S. Geological Survey Bulletin 735-A, p. 1–22.

Knopf, A., 1924, Geology and ore deposits of the Rochester district, Nevada: U.S. Geological Survey Bulletin 762, 78 p.

Kokinos, M., and Prenn, N., 1985, The Northumberland Mine, Nye County, Nevada: Mineralogical Record, v. 16, p. 37–41.

Koschmann, A.H., and Bergendahl, M.H., 1968, Principal gold-producing districts of the United States: U.S. Geological Survey Professional Paper 610, 283 p.

Kostick, D.S., 1994, Soda ash, in Carr, D.D., sr. ed., Industrial minerals and rocks, 6th ed.: Society for Mining, Metallurgy, and Exploration, Inc., Littleton, Colorado, p. 929–958.

Kral, V.E., 1947, Valery fluorspar deposit, Pershing County, Nevada: U.S. Bureau of Mines Report of Investigation, 8 p.

Kral, V.E., 1951, Mineral resources of Nye County, Nevada: Nevada Bureau of Mines and Geology Bulletin 50, 22 p.

Krauskopf, K.B., and Bateman, P.C., 1977, Geologic map of the Glass Mountain Quadrangle, Mono County, California, and Mineral County, Nevada: U.S. Geological Survey Geologic Quadrangle Map GQ-1099, 1:62,500.

Kring, D.A., 1997, The Hot Springs Meteorite, Churchill County, Nevada, USA, in Grossman, J.N., The Meteoritic Bulletin, no. 81, 1997 July: Meteoritics and Planetary Science, v. 32, Suppl., p. A159–A166.

Krohn, M.D., Kendall, C., Evans, J.R., and Fries, T.L., 1993, Relations of ammonium minerals at several hydrothermal systems in the western U.S.: Journal of Volcanology and Geothermal Research, v. 56, no. 4, p. 401–413.

Kuehn, C.A., 1989, Studies of disseminated gold deposits near Carlin, Nevada: Evidence for a deep geologic setting of ore formation [Ph.D. diss.]: Pennsylvania State University, State College, Pennsylvania, 395 p.

Kunasz, I.A., 1970, Geology and geochemistry of the lithium deposit in Clayton Valley, Esmeralda County, Nevada [Ph.D. diss.]: Pennsylvania State University, University Park, Pennsylvania, 114 p.

Kunz, G.F., 1890, Gems and precious stones of North America: Scientific Publishing Company, New York, 336 p.

LaBerge, R.D., 1996, Epithermal gold mineralization related to caldera volcanism at the Atlanta District, east-central Nevada, in Coyner, A.R., and Fahey, P.L., eds., Geology and ore deposits of the American Cordillera; symposium proceedings: Geological Society of Nevada, Reno, p. 309–328.

Lang, N.R., 1918, Deposits of antimony in Nevada: Engineering and Mining Journal, v. 105, p. 797.

LaPointe, D.D., Tingley, J.V., and Jones, R.B., 1991, Mineral resources of Elko County, Nevada: Nevada Bureau of Mines and Geology Bulletin 106, 236 p.

Larsen, E.S., 1921, The microscopic determination of the nonopaque minerals: U.S. Geological Survey Bulletin 679, 294 p.

Lavkulich, R., 1993, Clay minerals and structures at the Goldstrike mine: unpublished Barrick Goldstrike Internal Memorandum, 3 p.

Lawrence, E.F., 1963, Antimony deposits of Nevada: Nevada Bureau of Mines and Geology Bulletin 61, 248 p.

Lawrence, E.F., 1971, Mercury mineralization at the Senator fumaroles, Dixie Valley, Nevada [abs.]: Geological Society of America Abstracts, v. 3, p. 147.

Lawrence, E.F., 1977, Tungsten-mercury-antimony deposits in a Jurassic carbonate reef, Stillwater Range, Churchill County, Nevada: Preprint, Pacific Southwest Mineral Industry Conference, March 1977, 11 p.

Lee, D.E., 1962, Grossularite-spessartite garnet from the Victory mine, Gabbs, Nevada: American Mineralogist, v. 47, p. 147–151.

Lee, D.E., and Bastron, H., 1962, Allanite from the Mount Wheeler area, White Pine County, Nevada: American Mineralogist, v. 47, p. 1327–1331.

Lee, D.E., and Dodge, F.C.W., 1964, Accessory minerals in some granitic rocks in California and Nevada as a function of calcium content: American Mineralogist, v. 49, p. 1660–1669.

Lee, D.E., and Erd, R.C., 1963, Phenakite from the Mount Wheeler area, Snake Range, White Pine County, Nevada: American Mineralogist, v. 48, p. 189–193.

Lee, D.E., and Van Loenen, R.E., 1979, Accessory

opaque oxides from hybrid granitoid rocks of the southern Snake Range, Nevada: U.S. Geological Survey Open-File Report 79-1608, 18 p.

Lee, W.T., 1915, Guidebook of the western United States, Part B: The Overland Route, with a side trip to Yellowstone Park: U.S. Geological Survey Bulletin 612, 244 p.

Lees, B.K., 2000, Orpiment from the Twin Creeks Mine, Humboldt County, Nevada: Mineralogical Record, v. 31, p. 311–322, 331.

Leighton, F.B., 1967, Gold Butte vermiculite deposits, Clark County, Nevada: Nevada Bureau of Mines Report 16, 18 p.

Levy, C., 1968, Contribution a la mineralogie des sulfures de cuivre du type Cu_3XS_4: Bureau de Recherches Géologiques et Minières Mémoires, v. 54, 178 p.

Lincoln, F.C., 1923, Mining districts and mineral resources of Nevada: Nevada Newsletter Publishing Company, Reno, 295 p.

Lindgren, W., 1906, The occurrence of stibnite at Steamboat Springs, Nevada: American Institute of Mining Engineers Transactions, v. 36, p. 27–31.

Lindgren, W., 1915, Geology and mineral deposits of the National mining district, Nevada: U.S. Geological Survey Bulletin 601, 58 p.

Lindgren, W., 1933, Mineral deposits, 4th ed.: McGraw-Hill Book Co., New York, 930 p.

Lindgren, W., and Davy, W.M., 1924, Nickel ores from the Key West mine, Nevada: Economic Geology, v. 19, p. 309–319.

Lipman, P.W., Christiansen, R.L., and O'Connor, J.T., 1966, A compositionally zoned ash-flow sheet in southern Nevada: U.S. Geological Survey Professional Paper 524-F, 47 p.

Longwell, C.R., Pampeyan, E.H., Bowyer, B., and Roberts, R.J., 1965, Geology and mineral deposits of Clark County, Nevada: Nevada Bureau of Mines and Geology Bulletin 62, 218 p.

Lord, E., 1883, Comstock mining and miners: U.S. Geological Survey Monograph 4, 451 p.

Louderback, G.D., 1903, Some gypsum deposits of northwestern Nevada: Journal of Geology, v. 11, p. 99.

Lovejoy, D.W., 1959, Overthrust Ordovician and the Nannie's Peak intrusive, Lone Mountain, Elko County, Nevada: Geological Society of America Bulletin 70, p. 539–563.

Lovering, T.G., 1954, Radioactive deposits of Nevada: U.S. Geological Survey Bulletin 1009-C, 106 p.

Lovering, T.G., 1972, Jasperoid in the United States—its characteristics, origin and eco-

nomic significance: U.S. Geological Survey Professional Paper 710, 164 p.

Lovering, T.G., and Heyl, A.V., 1974, Jasperoid as a guide to mineralization in the Taylor mining district and vicinity, near Ely, Nevada: Economic Geology, v. 69, p. 46–58.

Lovering, T.G., and Stoll, W.M., 1943, Preliminary report on the Rip Van Winkle Mine, Elko County, Nevada: U.S. Geological Survey Open-File Report, 9 p.

Lugaski, T.P., 2000, The W.M. Keck Museum: Rocks and Minerals, v. 75, p. 116–119.

Lumsden, W.W., Jr., 1964, Geology of the southern White Pine Range and northern Horse Range, Nye and White Pine Counties, Nevada [Ph.D. diss.]: University of California, Los Angeles, 355 p.

Lush, A.P., 1982, Geology of part of the East Humboldt Range, Elko County, Nevada [M.S. thesis]: University of South Carolina, Columbia, 138 p.

MacKenzie, W.B., and Bookstrom, A.A., 1976, Geology of the Majuba Hill area, Pershing County, Nevada: Nevada Bureau of Mines and Geology Bulletin 86, 23 p.

Maher, D.J., 1996, Spatial and temporal relationships of hydrothermal mineral assemblages at Veteran extension in the Robinson (Ely) porphyry copper system, Nevada, in Coyner, A.R., and Fahey, P.L., eds., Geology and ore deposits of the American Cordillera; a symposium: Geological Society of Nevada, Reno, p. 1595–1621.

Mallory, W., 1916, Antimony veins at Bernice, Nevada: Mining and Scientific Press, v. 112, p. 556.

Mandarino, J.A., 1997, Abstracts of new mineral descriptions: Mineralogical Record, v. 28, p. 143.

Mandarino, J.A., 1999, Fleischer's glossary of mineral species 1999: The Mineralogical Record, Inc., Tucson, Arizona, 225 p.

Martin, A.H., 1910, The Bannock mining district, Nevada: The Mining World, v. 32, no. 17, p. 835.

Martin, C., 1956, Structure and dolomitization in crystalline magnesite deposits, Paradise Range, Nye County, Nevada: Geological Society of America Bulletin, v. 69, no. 12, pt. 2, p. 1774.

Mason, B.H., and Vitaliano, C.J., 1953, The mineralogy of the antimony oxides and antimonates: Mineralogical Magazine, v. 30, p. 100–112.

Matti, J.C., Castor, S.B., Bell, J.W., Rowland, S.M.,

1993, Geologic map of the Las Vegas NE Quadrangle, Nevada: Nevada Bureau of Mines and Geology Map 3Cg, 1:24,000.

McConnell, D., 1940, Clinobarrandite and the isodimorphous series, variscite-metavariscite: American Mineralogist, v. 25, p. 719–725.

McCormack, J.K., 1986, Paragenesis and origin of sediment-hosted mercury ore at the McDermitt mine, McDermitt, Nevada [M.S. thesis]: University of Nevada, Reno, 97 p.

McCormack, J.K., and Dickson, F.W., 1998, Kenhsuite, γ-Hg$_3$S$_2$Cl$_2$, a new mineral species from the McDermitt mercury deposit, Humboldt County, Nevada: Canadian Mineralogist, v. 36, p. 201–206.

McCormack, J.K., Dickson, F.W., and Leshendok, M.P., 1991, Radtkeite, Hg$_3$S$_2$ClI, a new mineral from the McDermitt mercury deposit, Humboldt County, Nevada: American Mineralogist, v. 76, p. 1715–1721.

McJannet, G.S., 1957, Geology of the Pyramid Lake-Red Canyon area, Washoe County, Nevada [M.S. thesis]: University of California, Los Angeles, 125 p.

McKee, E.H., 1968, Geology of the Magruder Mountain area, Nevada-California: U.S. Geological Survey Bulletin 1251-H, 40 p.

McKee, E.H., 1976, Geology of the northern part of the Toquima Range, Lander, Eureka, and Nye Counties, Nevada: U.S. Geological Survey Professional Paper 931, 49 p.

McKee, E.H., Tarshis, A.L., and Marvin, R.F., 1976, Summary of radiometric ages of Tertiary volcanic and plutonic rocks in Nevada—part 5, northeastern Nevada: Isochron/West, v. 16, p. 15.

McMakin, M.R., and Prave, A.R., 1991, Stratigraphic framework of the Kingston Range, Kingston Wash, and surrounding areas, California and Nevada, in Reynolds, R., ed., Crossing the borders: Quaternary Studies in Eastern California and Southwestern Nevada: San Bernardino County Museum Special Publication, p. 189–196.

Melhase, J., 1934a, Nevada, the mineral collector's mecca—many localities described: Oregon Mineralogist, v. 2, no. 8, p. 3–4, 22–23.

Melhase, J., 1934b, Nevada, the mineral collector's mecca—many localities described: Oregon Mineralogist, v. 2, no. 9, p. 5–6, 30–31.

Melhase, J., 1934c, Numerous localities found on safari across northern Nevada: Oregon Mineralogist, v. 2, no. 10, p. 3–4, 29–31.

Melhase, J., 1934d, Nevada safari turns southward over old Pony Express route: Oregon Mineralogist, v. 2, no. 11, p. 5–6, 28–30.

Melhase, J., 1934e, Goodsprings, Nev., one of richest areas for collecting, described by engineer: Oregon Mineralogist, v. 2, no. 12, p. 5–6, 26–27.

Melhase, J., 1935a, Winter quarters of Nevada "desert rats" visited in statewide safari: [Oregon] Mineralogist, v. 3, no. 2, p. 9–10, 28.

Melhase, J., 1935b, Nevada hill with a golden lining, "ghost towns" bleaching in sun: [Oregon] Mineralogist, v. 3, no. 3, p. 9–10, 26–27.

Melhase, J., 1935c, Nevada, the mineral collector's mecca; Volcanic necks mounted in silver: [Oregon] Mineralogist, v. 3, no. 4, p. 9–10, 37–38.

Melhase, J., 1935d, The minerals of Manhattan: [Oregon] Mineralogist, v. 3, no. 5, p. 9–10, 20–21.

Melhase, J., 1935e, 3000 mile Nevada safari ends at Reno: [Oregon] Mineralogist, v. 3, no. 7, p. 3–4, 30–31.

Merriam, C.W., and Anderson, C.A., 1942, Reconnaissance survey of the Roberts Mountains, Nevada: Geological Society of America Bulletin, v. 53, p. 1675–1727.

Michell, W.D., 1945, Oxidation in a molybdenite deposit, Nye County, Nevada: Economic Geology, v. 40, p. 99–114.

Mills, B.A., 1984, Geology of the Round Mountain gold deposit, Nye County, Nevada, in Wilkins, J., Jr., ed., Gold and silver deposits of the Basin and Range Province, Western USA: Arizona Geological Society Digest, v. 15, p. 89–99.

Mitchell, J.R., 1991, Gem trails of Nevada: Gem Guides Book Co., Baldwin Park, California, 119 p.

Moeller, S.A., 1988, Geology and mineralization in the Candelaria District, Mineral County, Nevada, in Schafer, R.W., Cooper, J.J., and Vikre, P.G., eds., Bulk mineable precious metal deposits of the western United States; symposium proceedings: Geological Society of Nevada, Reno, p. 135–158.

Moiola, R.J., 1964, Authigenic mordenite in the Esmeralda Formation, Nevada: American Mineralogist, v. 49, p. 1472–1474.

Moiola, R.J., and Glover, E.D., 1965, Recent anhydrite from Clayton Playa, Nevada: American Mineralogist, v. 50, p. 2063–2069.

Monroe, S.C., 1991, Kinsley Mountain Project, Elko County, Nevada, in Buffa, R.H., and Coyner, A.R., eds., Geology and ore deposits of the Great Basin; field trip guidebook compendium: Geological Society of Nevada, Reno, p. 300–304.

Moore, J.G., 1969, Geology and mineral deposits of

Lyon, Douglas, and Ormsby Counties, Nevada: Nevada Bureau of Mines and Geology Bulletin 75, 45 p.

Moores, E.M., Scott, R.B., and Lumsden, W.W., Jr., 1968, Tertiary tectonics of the White Pine–Grant Range region, east-central Nevada, and some regional implications: Geological Society of America Bulletin, v. 79, p. 1703–1726.

Morris, H.C., 1903, Hydrothermal activity in the veins at Wedekind, Nevada: The Engineering and Mining Journal, v. 76, p. 277–278.

Morrissey, F.R., 1968, Turquoise deposits of Nevada: Nevada Bureau of Mines and Geology Report 17, 30 p.

Mount, P., 1964, Bismuth, *in* Mineral and water resources of Nevada: Nevada Bureau of Mines and Geology Bulletin 65, p. 75–78.

Muffler, L.J.P., 1964, Geology of the Frenchie Creek Quadrangle, north-central Nevada: U.S. Geological Survey Bulletin 1179, 99 p.

Muller, S.W., Ferguson, H.G., and Roberts, R.J., 1951, Geology of the Mount Tobin Quadrangle, Nevada: U.S. Geological Survey Geologic Quadrangle Map GQ-7, 1:24,000.

Murdoch, J., 1916, Microscopical determination of the opaque minerals; an aid to the study of ores: J. Wiley and Sons, New York, 165 p.

Murphy, J.B., 1964, Gems and gem materials, *in* Mineral and water resources of Nevada: Nevada Bureau of Mines and Geology Bulletin 65, p. 203–208.

Mustoe, G.E., 1999, Meteorites from the Pacific Northwest: Oregon Geology, v. 61, no. 2, p. 27–38.

Myers, G., Dennis, M.D., Wilkinson, W.H., and Wendt, C.J., 1991, Precious-metal distribution in the Mount Hamilton polymetallic skarn system, Nevada, *in* Raines, G.L., Lisle, R.E., Schafer, R.W., and Wilkinson, W.H., eds., Geology and ore deposits of the Great Basin; symposium proceedings: Geological Society of Nevada, Reno, p. 393–403.

Nash, J.T., 1972, Fluid inclusion studies of some gold deposits in Nevada: U.S. Geological Survey Professional Paper 800-C, p. 15–19.

Nash, J.T., and Trudel, W.S., 1996, Bulk mineable gold ore at the Sleeper Mine, Nevada; importance of extensional faults, breccia, framboids, and oxidation, *in* Coyner, A.R., and Fahey, P.L., eds., Geology and ore deposits of the American Cordillera; symposium proceedings: Geological Society of Nevada, Reno, p. 235–256.

Nevada Bureau of Mines, 1964, Mineral and water resources of Nevada, prepared by the U.S. Geological Survey and the Nevada Bureau of Mines: Nevada Bureau of Mines and Geology Bulletin 65, 314 p.

Nevada State Journal, 1911, Meteor falling near Doyle shakes country: 26 May, p. 1, col. 2–5.

Nichols, R.A., 1979, Opal mines of Nevada: Lapidary Journal, v. 33, p. 1638–1644.

Nickel, E.H., and Nichols, M.C., 1991, Mineral reference manual: Van Nostrand Reinhold, New York, 250 p.

Nicols, C.E., 1991, Geology and mineral potential of the Brunswick trend, Storey County, Nevada: unpublished company report, Miramar Mining Corporation, 30 p.

Nininger, R.D., 1956, Minerals for atomic energy; a guide to exploration for uranium, thorium, and beryllium: Van Nostrand, Princeton, New Jersey, 399 p.

Noble, D.C., 1968, Kane Springs Wash volcanic center, Lincoln County, Nevada, *in* Studies of geology and hydrology, Nevada Test Site: Geological Society of America Memoir 110, p. 109–116.

Noble, D.C., Sargent, K.A., Mehnert, H.H., Ekren, E.B., and Byers, F.M., Jr., 1968, Silent Canyon volcanic center, Nye County, Nevada, *in* Studies of geology and hydrology, Nevada Test Site: Geological Society of America Memoir 110, p. 65–75.

Noble, D.C., Slemmons, D.B., Korringa, M.K., Dickinson, W.R., Al-Rawi, Y., and McKee, E.H., 1974, Eureka Valley tuff, east-central California and adjacent Nevada: Geology, v. 1, p. 139–142.

Noble, J.A., 1970, Metal provinces of the western United States: Geological Society of America Bulletin, v. 81, p. 1607–1624.

Noble, L.F., 1923, Colemanite in Clark County, Nevada: U.S. Geological Survey Bulletin 735-B, p. 23–40.

Nolan, T.B., 1935, The underground geology of the Tonopah mining district, Nevada: Nevada Bureau of Mines and Geology Bulletin 23, 49 p.

Nolan, T.B., 1936, The Tuscarora mining district, Elko County, Nevada: Nevada Bureau of Mines and Geology Bulletin 25, 38 p.

Nolan, T.B., 1962, The Eureka mining district, Nevada: U.S. Geological Survey Professional Paper 406, 78 p.

Norton, O.R., 1998, Rocks from space, 2nd ed.: Mountain Press Publishing Company, Missoula, Montana, 444 p.

Novak, S.W., 1985, Geology and geochemical evolution of the Kane Springs Wash Volcanic Center, Lincoln County, Nevada (Ph.D. diss.): Stanford University, Stanford, California, 220 p.

Nuffield, E.W., 1953, Benjaminite: American Mineralogist, v. 38, p. 550–551.

Nuffield, E.W., 1975, Benjaminite—a reexamination of the type material: Canadian Mineralogist, v. 13, p. 394–401.

Odom, I.E., 1992, Hectorite deposits in the McDermitt Caldera of Nevada: Mining Engineering, v. 44, p. 586–589.

Olsen, D.R., 1960, Geology and mineralogy of the Delano mining district and vicinity, Elko County, Nevada [Ph.D. diss.]: University of Utah, Salt Lake City, 96 p.

Olsen, D.R., 1965, New variety of cassiterite from Nevada [abs.]: Geological Society of America Special Paper 82, p. 340.

Olson, J.C., and Hinrichs, E.N., 1960, Reconnaissance of beryl-bearing pegmatites in the Ruby Mountains and other areas of Nevada and northwestern Arizona: U.S. Geological Survey Bulletin 1082-D, 200 p.

Olson, R.H., 1964a, Clay, *in* Mineral and water resources of Nevada: Nevada Bureau of Mines and Geology Bulletin 65, p. 185–189.

Olson, R.H., 1964b, Magnesite, *in* Mineral and water resources of Nevada: Nevada Bureau of Mines and Geology Bulletin 65, p. 220–223.

Olson, R.H., 1964c, Sulfur, *in* Mineral and water resources of Nevada: Nevada Bureau of Mines and Geology Bulletin 65, p. 255–256.

Orkild, P.P., Sargent, K.A., and Snyder, R.P., 1969, Geologic map of Pahute Mesa, Nevada Test Site and vicinity, Nye County, Nevada: U.S. Geological Survey Map I-567, 1:48,000.

Osborne, M.A., 1985, Alteration and mineralization of the northern half of the Aurora mining district, Mineral County, Nevada [M.S. thesis]: University of Nevada, Reno, 93 p.

Osborne, M.A., 1991, Epithermal mineralization at Aurora, Nevada, *in* Raines, G.L., Lisle, R.E., Schafer, R.W., Wilkinson, W.H., eds., Geology and ore deposits of the Great Basin; symposium proceedings: Geological Society of Nevada, Reno, p. 1097–1110.

Oswald, S.G., and Crook, W.W., III, 1979, Cuprohydromagnesite and cuproartinite, two new minerals from Gabbs, Nevada: American Mineralogist, v. 64, p. 886–889.

Overton, T.D., 1947, Mineral resources of Douglas, Ormsby, and Washoe Counties, Nevada: Nevada Bureau of Mines and Geology Bulletin 46, 88 p.

Pabst, A., 1938, Garnets from vesicles in rhyolite near Ely, Nevada: American Mineralogist, v. 23, no. 2, p. 101–103.

Pabst, A., 1946, Notes on the structure of delafossite: American Mineralogist, v. 31, p. 539–546.

Page, B.M., 1959, Geology of the Candelaria mining district, Mineral County, Nevada: Nevada Bureau of Mines and Geology Bulletin 56, 67 p.

Paher, S.W., 1970, Nevada ghost towns and mining camps: Howell North Books, Berkeley, California, 492 p.

Palache, C., 1939, Antlerite: American Mineralogist, v. 24, p. 293–302.

Palache, C., Berman, H., and Frondel, C., 1944, The system of mineralogy of James Dwight Dana and Edward Salisbury Dana, 7th ed., vol. 1: Elements, sulphides, sulfosalts, oxides: J. Wiley and Sons, New York, 834 p.

Palache, C., Berman, H., and Frondel, C., 1951, The system of mineralogy of James Dwight Dana and Edward Salisbury Dana, 7th ed., vol. 2: Halides, nitrates, borates, carbonates, sulfates, phosphates, arsenates, tungstates, molybdates, etc.: J. Wiley and Sons, New York, 1124 p.

Palache, C., and Berry, L.G., 1946, Clinoclase: American Mineralogist, v. 31, p. 243–258.

Palache, C., and Modell, D., 1930, Crystallography of stibnite and orpiment from Manhattan, Nevada: American Mineralogist, v. 15, p. 365–374.

Palmer, A.R., 1981, Lower and Middle Cambrian stratigraphy of Frenchman Mountain, Nevada, *in* Taylor, M.E., and Palmer, A.R., eds., Cambrian stratigraphy and paleontology of the Great Basin and vicinity, western United States: U.S. Geological Survey Field Trip Guidebook, p. 11–13.

Panhorst, T.L., 1996, Structural control and mineralization at the Lone Tree mine, Humboldt County, Nevada, with implications on sampling protocol [Ph.D. diss.]: University of Nevada, Reno, 263 p.

Papke, K.G., 1970, Montmorillonite, bentonite, and fuller's earth deposits in Nevada: Nevada Bureau of Mines and Geology Bulletin 76, 47 p.

Papke, K.G., 1972a, Erionite and other associated zeolites in Nevada: Nevada Bureau of Mines and Geology Bulletin 79, 32 p.

Papke, K.G., 1972b, A sepiolite-rich playa deposit in southern Nevada: Clays and Clay Minerals, v. 20, p. 211–215.

Papke, K.G., 1975, Talcose minerals in Nevada—talc, chlorite and pyrophyllite: Nevada Bureau of Mines and Geology Bulletin 84, 63 p.

Papke, K.G., 1976, Evaporites and brines in Nevada playas: Nevada Bureau of Mines and Geology Bulletin 87, 35 p.

Papke, K.G., 1979, Fluorspar in Nevada: Nevada Bureau of Mines and Geology Bulletin 93, 77 p.

Papke, K.G., 1984, Barite in Nevada: Nevada Bureau of Mines and Geology Bulletin 98, 125 p.

Papke, K.G., 1987, Gypsum deposits in Nevada: Nevada Bureau of Mines and Geology Bulletin 103, 26 p.

Pardee, J.T., and Jones, E.L., Jr., 1920, Deposits of manganese ore in Nevada: U.S. Geological Survey Bulletin 710-F, p. 209–242.

Parsons, A.B., 1933, The porphyry coppers: Amercian Institute of Mining and Metallurgical Engineers, New York, 581 p.

Peacor, D.R., Simmons, W.B., Jr., Essene, E.J., and Heinrick, E.W., 1982, New data on and discreditation of "teasite," "albritonite," "cuproartinite," "cuprohydromagnesite," and "yttromicrolite": American Mineralogist, v. 67, p. 156–169.

Penrose, R.A.F., 1891, Manganese, its uses, ores, and deposits: Arkansas Geological Survey Annual Report 1890, 642 p.

Perry, J.K., 1963, Discussion of "coulsonite" by Arthur S. Radtke: American Mineralogist, v. 48, p. 948–952.

Peters, S.G., Ferdock, G.C., Woitsekhowskaya, M.B., Leonardson, R., and Rahn, J., 1998, Oreshoot zoning in the Carlin-type Betze Orebody, Goldstrike Mine, Eureka County, Nevada: U.S. Geological Survey Open-File Report 98-620, 49 p.

Peters, S.G., Hardyman, R.E., and Connors, K.H., 1996, Epithermal (silver-gold-manganese and mercury) and mesothermal (gold-quartz) deposits of the Holy Cross mining district, Churchill County, Nevada, in Coyner, A.R., and Fahey, P.L., eds., Geology and ore deposits of the American Cordillera; symposium proceedings: Geological Society of Nevada, Reno, p. 419–452.

Phariss, E.I., 1974, Geology and ore deposits of the Alpine mining district, Esmeralda County, Nevada [M.S. thesis]: University of Nevada, Reno, 114 p.

Phinisey, J.D., Hofstra, A.H., Snee, L.W., Roberts, T.T., Dahl, A.R., and Loranger, R.J., 1996, Evidence for multiple episodes of igneous and hydrothermal activity and constraints on the timing of gold mineralization, Jerritt Canyon District, Elko County, Nevada, in Coyner, A.R., and Fahey, P.L., eds., Geology and ore deposits of the American Cordillera; symposium proceedings: Geological Society of Nevada, Reno, p. 15–39.

Phoenix, D.A., and Cathcart, J.B., 1952, Quicksilver deposits in the southern Pilot Mountains, Mineral County, Nevada: U.S. Geological Survey Bulletin 973-D, p. 143–171.

Ponsler, H.E., 1977, The geology and mineral deposits of the Garfield district, Mineral County, Nevada [M.S. thesis]: University of Nevada, Reno, 133 p.

Pough, F.H., 1937, Crystallized powellite from Tonopah, Nevada: American Mineralogist, v. 22, p. 57–64.

Powers, S.L., 1978, Jasperoid and disseminated gold at the Ogee-Pinson mine, Humboldt County, Nevada [M.S. thesis]: University of Nevada, Reno, 121 p.

Price, J.G., Castor, S.B., and Miller, D.M., 1992, Highly radioactive topaz rhyolites of the Toano Range, northeastern Nevada: American Mineralogist, v. 77, p. 1067–1073.

Price, J.G., Henry, C.D., Castor, S.B., Garside, L., and Faulds, J.E., 1999, Geology of Nevada: Rocks and Minerals, v. 74, p. 357–363.

Prihar, D.W., Peters, S.G., Bourns, F.T., and McKee, E.H., 1996, Geology and gold potential of the Goat Ridge Window, Shoshone Range, Lander County, Nevada, in Coyner, A.R., and Fahey, P.L., eds., Geology and ore deposits of the American Cordillera; symposium proceedings: Geological Society of Nevada, Reno, p. 485–504.

Princehouse, D.S., and Dilles, J.H., 1996, Mesozoic porphyry gold-copper mineralization at the Wheeler Mine, Pine Grove District, Nevada, in Coyner, A.R., and Fahey, P.L., eds., Geology and ore deposits of the American Cordillera; symposium proceedings: Geological Society of Nevada, Reno, p. 1533–1566.

Pullman, S., and Thomssen, R., 1999, Nevada mineral locality index: Rocks and Minerals, v. 74, p. 370–379.

Purington, C.W., 1903, The ore deposits at Contact, Nevada: The Engineering and Mining Journal, v. 76, p. 426–427.

Purkey, B.W., and Garside, L.J., 1995, Geologic and natural history tours in the Reno area: Nevada Bureau of Mines and Geology Special Publication 19, 211 p.

Quade, J., and Tingley, J.V., 1987, Mineral resource inventory, U.S. Navy master land withdrawal area, Churchill County, Nevada: Nevada Bureau of Mines and Geology Open-File Report 87-2, 96 p.

Quade, J., Tingley, J.V., Bentz, J.L., and Smith, P.L., 1984, A mineral inventory of the Nevada Test Site and portions of the Nellis Bombing and

Gunnery Range, southern Nye County, Nevada: Nevada Bureau of Mines and Geology Open-File Report 84-2, 40 p.

Rabbitt, M.C., 1986, Minerals, lands, and geology for the common defense and general welfare, volume 3, 1904–1939: U.S. Geological Survey, U.S. Government Printing Office, Washington, D.C., 479 p.

Radtke, A.S., 1962, Coulsonite, FeV_2O_4, a spinel-type mineral from Lovelock, Nevada: American Mineralogist, v. 47, p. 1284–1291.

Radtke, A.S., 1985, Geology of the Carlin gold deposit, Nevada: U.S. Geological Survey Professional Paper 1267, 124 p.

Radtke, A.S., and Brown, C.E., 1974, Frankdicksonite, BaF_2, a new mineral from Nevada: American Mineralogist, v. 59, p. 885–888.

Radtke, A.S., and Dickson, F.W., 1975, Carlinite, Tl_2S, a new mineral from Nevada: American Mineralogist, v. 60, p. 559–565.

Radtke, A.S., Dickson, F.W., and Heropoulos, C., 1974a, Thallium-bearing orpiment, Carlin gold deposit, Nevada: U.S. Geological Survey Journal of Research, v. 2, no. 3, p. 341–392.

Radtke, A.S., Dickson, F.W., and Rytuba, J., 1974b, Genesis of disseminated gold deposits of the Carlin type: Geological Society of America Abstracts with Programs, v. 6, no. 3, p. 239–240.

Radtke, A.S., Dickson, F.W., and Slack, J.F., 1978, Occurrence and formation of avicennite, Tl_2O_3, as a secondary mineral at the Carlin gold deposit, Nevada: U.S. Geological Survey Journal of Research, v. 6, no. 2, p. 241–246.

Radtke, A.S., Dickson, F.W., Slack, J.F., and Brown, K., 1977, Christite, a new thallium mineral from the Carlin gold deposit, Nevada: American Mineralogist, v. 62, p. 421–425.

Radtke, A.S., and Taylor, C.M., 1967, A new(?) yttrium rare-earth arsenate from Hamilton, Nevada: U.S. Geological Survey Professional Paper 575-B, p. 108–109.

Radtke, A.S., Taylor, C.M., Erd, R.C., and Dickson, F.W., 1974c, Occurrence of lorandite, $TlAsS_2$, at the Carlin gold deposit, Nevada: Economic Geology, v. 69, p. 121–123.

Radtke, A.S., Taylor, C.M., and Heropoulos, C., 1973, Antimony-bearing orpiment, Carlin gold deposit, Nevada: U.S. Geological Survey Journal of Research, v. 1, no. 1, p. 85–87.

Radtke, A.S., Taylor, C.M., and Hewett, D.F., 1967, Aurorite, argentian todorokite and hydrous silver-bearing lead manganese oxide: Economic Geology, v. 62, p. 186–206.

Randolph, G.C., and Dake, H.C., 1935, Opal—Virgin Valley Field, largest deposit in U.S.A.: The Mineralogist, v. 3, no. 10, p. 9–10, 18–19.

Ransome, F.L., 1907, The association of alunite with gold in the Goldfield district, Nevada: Economic Geology, v. 2, p. 667–692.

Ransome, F.L., 1909a, Geology and ore deposits of the Goldfield district, Nevada: U.S. Geological Survey Professional Paper 66, 258 p.

Ransome, F.L., 1909b, Notes on some mining districts in Humboldt County, Nevada: U.S. Geological Survey Bulletin 414, 75 p.

Ransome, F.L., 1909c, The Hornsilver district, Nevada: U.S. Geological Survey Bulletin 380, p. 41–43.

Ransome, F.L., 1909d, Round Mountain, Nevada: U.S. Geological Survey Bulletin 380, p. 44–47.

Ransome, F.L., 1909e, The Yerington copper district, Nevada: U.S. Geological Survey Bulletin 380, p. 99–119.

Ransome, F.L., 1910a, Geology and ore deposits of the Goldfield district, Nevada. Part I—Geology: Economic Geology, v. 5, p. 301–311.

Ransome, F.L., 1910b, Geology and ore deposits of the Goldfield district, Nevada. Part II—Mines and Mining: Economic Geology, v. 5, p. 438–470.

Ransome, F.L., Emmons, W.H., and Garrey, G.H., 1910, Geology and ore deposits of the Bullfrog district, Nevada: U.S. Geological Survey Bulletin 407, 130 p.

Ransome, F.L., Garrey, G.H., and Emmons, W.H., 1907, Preliminary account of Goldfield, Bullfrog, and other mining districts in southern Nevada, with notes on the Manhattan district: U.S. Geological Survey Bulletin 303, 93 p.

Raymond, R.W., 1869, Mineral resources of the States and Territories west of the Rocky Mountains for 1868: U.S. Treasury Department 1st Annual Report, p. 84–85.

Raymond, R.W., 1872, Statistics of mines and mining in the States and Territories west of the Rocky Mountains: U.S. Treasury Department 3rd Annual Report, 566 p.

Raymond, R.W., 1874, Statistics of mines and mining in the States and Territories west of the Rocky Mountains: U.S. Treasury Department 6th Annual Report, 585 p.

Ream, L.R., 1990, The Tucson shows of the winter of 1990: Mineral News, v. 6, no. 2, p. 1–2, 5–7.

Ream, L.R., 1992, 1992 Tucson Shows: Mineral News, v. 8, no. 3, p. 1, 4–6.

Reeves, R.G., and Kral, V.E., 1955, Iron ore deposits of Nevada: Part A. Geology and iron ore deposits of the Buena Vista Hills, Churchill and

Pershing Counties, Nevada: Nevada Bureau of Mines and Geology Bulletin 53A, 37 p.

Reeves, R.G., Shawe, F.R., and Kral, V.E., 1958, Iron ore deposits of Nevada: Part B. Iron ore deposits of west-central Nevada: Nevada Bureau of Mines and Geology Bulletin 53B, 78 p.

Reno Evening Gazette, 1911, Doyle is visited by heavenly body: 26 May, p. 3, col. 2.

Rice, S.B., Papke, K.G., and Vaughan, D.E.W., 1992, Chemical controls on ferrierite crystallization during diagenesis of silicic pyroclastic rocks near Lovelock, Nevada: American Mineralogist, v. 77, p. 314–328.

Richter, D.H., Reichen, L.E., and Lemmon, D.M., 1957, New data on ferritungstite from Nevada: American Mineralogist, v. 42, p. 83–90.

Rickard, T.A., 1932, A history of American mining: McGraw-Hill, New York, 419 p.

Rickard, T.A., 1939, Ore theft—reminiscences and anecdotes from various camps: Engineering and Mining Journal, v. 140, no. 12, p. 37–39.

Rickard, T.A., 1940, Ore theft II—further reminiscences and anecdotes: Engineering and Mining Journal, v. 141, no. 1, p. 44–47.

Riotte, E.N., 1865, The new mineral: Daily Reese River Reveille, v. 4, no. 24, p. 1.

Riotte, E.N., 1867, Stetefeldtite, a new mineral from Nevada: Berg—und Huttenmannische Zeitung, v. 26, p. 253.

Roberts, A.C., Ansell, H.G., and Bonardi, M., 1980, Pararealgar, a new polymorph of AsS, from British Columbia: The Canadian Mineralogist, v. 18, p. 525–527.

Roberts, A.C., Cooper, M.A., Hawthorn, F.C., Gault, R.A., Jensen, M.C., and Foord, E.E., 2003, Goldquarryite, a new Cd-bearing phosphate mineral from the Gold Quarry Mine, Eureka County, Nevada: Mineralogical Record, v. 34, no. 3, p. 237–240.

Roberts, R.J., 1940a, Quicksilver deposits of the Bottle Creek district, Humboldt County, Nevada: U.S. Geological Survey Bulletin 922-A, p. 1–29.

Roberts, R.J., 1940b, Quicksilver deposit at Buckskin Peak, National Mining District, Humboldt County, Nevada: U.S. Geological Survey Bulletin 922-E, p. 115–133.

Roberts, R.J., 1943, Manganese deposits in the Nevada district, White Pine County, Nevada: U.S. Geological Survey Bulletin 931-M, p. 295–318.

Roberts, R.J., 1944, The Rose Creek Tungsten Mine, Pershing County, Nevada: U.S. Geological Survey Bulletin 940-A, 14 p.

Roberts, R.J., 1960, Alinement of mining districts in north-central Nevada: U.S. Geological Survey Professional Paper 400-B, p. 17–19.

Roberts, R.J., and Arnold, D.C., 1965, Ore deposits of the Antler Peak Quadrangle, Humboldt and Lander Counties, Nevada: U.S. Geological Survey Professional Paper 459-B, p. B1–B94.

Roberts, R.J., Ketner, K.B., and Radtke, A.S., 1967a, Geological environment of gold deposits in Nevada: U.S. Geological Survey Professional Paper 476-B, p. B228–B229.

Roberts, R.J., Montgomery, K.M., and Lehner, R.E., 1967b, Geology and mineral resources of Eureka County, Nevada: Nevada Bureau of Mines and Geology Bulletin 64, 152 p.

Roberts, W.L., Campbell, T.J., Rapp, G.R., Jr., and Wilson, W.E., 1990, Encyclopedia of minerals, 2nd ed: Van Nostrand Reinhold, New York, 979 p.

Rogers, A.F., 1912a, The occurrence and origin of gypsum and anhydrite at the Ludwig mine, Lyon County, Nevada: Economic Geology, v. 7, p. 185–189.

Rogers, A.F., 1912b, Dahlite (Podolite) from Tonopah, Nevada: Voelckerite, a new basic calcium phosphate; remarks on the chemical composition of apatite and phosphate rock: American Journal of Science, v. 183, p. 475–482.

Rogers, A.F., 1922, Delafossite from Kimberly, Nevada: American Mineralogist, v. 7, p. 102–103.

Rogers, A.F., 1924, The crystallography of searlesite: American Journal of Science, 5th Series, v. 7, p. 498–502.

Roper, M.W., 1976, Hot springs mercury deposition at McDermitt Mine, Humboldt County, Nevada: Society of Mining Engineers, AIME Transactions, v. 260, p. 192–195.

Rose, C., 1997, Out of the blue: Lapidary Journal, v. 51, no. 3, p. 34–36.

Rose, R.L., 1969, Geology of parts of the Wadsworth and Churchill Butte Quadrangles, Nevada: Nevada Bureau of Mines and Geology Bulletin 29, 27 p.

Ross, C.P., 1925, Quicksilver in 1924: U.S. Geological Survey Mineral Resources of the United States, p. 13–19.

Ross, C.P., 1953, The geology and ore deposits of the Reese River district, Lander County, Nevada: U.S. Geological Survey Bulletin 997, 132 p.

Ross, C.P., 1961, Geology and mineral deposits of Mineral County, Nevada: Nevada Bureau of Mines and Geology Bulletin 58, 98 p.

Rota, J.C., 1989, An occurrence of orpiment at the Carlin gold mine, Nevada: Mineralogical Record, v. 20, no. 6, p. 469–471.

Rota, J.C., 1991, Mineralogy of the Maggie Creek District, Eureka County, Nevada: unpublished Newmont Exploration Ltd. Memo, 3 p.

Russell, I.C., 1882, Sulphur deposits in Utah and Nevada: New York Academy of Science Transactions 1, p. 168–175.

Russell, I.C., 1885, Geological history of Lake Lahontan: U.S. Geological Survey Monograph 11, 288 p.

Rytuba, J.J., 1976, Volcanism and mineralization of the McDermitt Caldera, Nevada-Oregon: Geological Society of America Abstracts with Programs 8, no. 5: Rocky Mountain Section 29th annual meeting, p. 625.

Rytuba, J.J., and Glanzman, R.K., 1979, Relation of mercury, uranium and lithium deposits to the McDermitt caldera complex: Nevada Bureau of Mines and Geology Report 33, p. 109–118.

Rytuba, J.J., and Heropoulos, C., 1992, Mercury; an important byproduct in epithermal gold systems, in DeYoung, J.H., Jr., and Hammarstrom, J.M., eds., Contributions to commodity geology research: U.S. Geological Survey Bulletin 1877, p. D1–D8.

Rytuba, J.J., and McKee, E.H., 1984, Peralkaline ash flow tuffs and calderas of the McDermitt volcanic field, southeast Oregon and north central Nevada: Journal of Geophysical Research, v. 89, no. B10, p. 8616–8628.

Sainsbury, C.L., and Kleinhample, F.J., 1969, Fluorite deposits of the Quinn Canyon Range, Nevada: U.S. Geological Survey Bulletin 1272-C, p. C1–C22.

Sampson, T.R., 1993, Alteration and structural paragenetic relationships; emphasis on Mesozoic imbricate thrust faults, Goldstrike Mine area, Eureka County, Nevada [M.S. thesis]: University of Nevada, Reno, 179 p.

Sand, L.B., and Regis, A.J., 1968, Ferrierite, Pershing County, Nevada (abs.), in Abstracts for 1966: Geological Society of America Special Paper 101, p. 189.

Sanford, S., and Stone, R.W., 1914, Useful minerals of the United States: U.S. Geological Survey Bulletin 585, 250 p.

Satkoski, J.J., and Berg, A.W., 1982, Field inventory of mineral resources, Pyramid Lake Indian Reservation, Nevada: U.S. Bureau of Mines Report BIA No. 38-11, 47 p.

Saunders, F.T., 1977, Geological examination of the Fairview property, Fairview mining district, Churchill County, Nevada: unpublished Houston Oil and Minerals Co. report, 17 p.

Saunders, J.A., 1994, Silica and gold textures in Bonanza ores of the Sleeper deposit, Humboldt County, Nevada: Economic Geology, v. 89, no. 3, p. 628–638.

Saunders, J.A., Cook, R.B., and Schoenly, P.A., 1996, Electrum disequilibrium crystallization textures in volcanic-hosted bonanza epithermal gold deposits in northern Nevada, in Coyner, A.R., and Fahey, P.L., eds., Geology and ore deposits of the American Cordillera; symposium proceedings: Geological Society of Nevada, Reno, p. 173–179.

Schaller, W.T., 1918, Gems and precious stones: U.S. Geological Survey, Mineral Resources of the United States 1917, p. 163.

Schaller, W.T., 1941, Bismoclite from Goldfield, Nevada: American Mineralogist, v. 26, p. 651–656.

Schaller, W.T., and Ransome, F.L., 1910, Bismite: American Journal of Science, v. 179, p. 173–176.

Schilling, J.H., 1962a, An inventory of molybdenum occurrences in Nevada: Nevada Bureau of Mines and Geology Report 2, 48 p.

Schilling, J.H., 1962b, Molybdenum occurrences in Nevada: Nevada Bureau of Mines and Geology Map 8.

Schilling, J.H., 1963, Geology of the Sand Springs Range: U.S. Atomic Energy Commission Report VUF-1001.

Schilling, J.H., 1964a, Molybdenum, in Mineral and water resources of Nevada: Nevada Bureau of Mines and Geology Bulletin 65, p. 124–132.

Schilling, J.H., 1964b, Tungsten, in Mineral and water resources of Nevada: Nevada Bureau of Mines and Geology Bulletin 65, p. 155–161.

Schilling, J.H., 1968a, The Gabbs magnesite—brucite deposit, Nye County, Nevada, in Ridge, J.D., ed., Ore deposits of the United States, 1933–1967; Graton-Sales volume 2: American Institute of Mining, Metallurgical and Petroleum Engineers, New York, p. 1608–1621.

Schilling, J.H., 1968b, Artinite from Gabbs, Nevada: A correction in location: American Mineralogist, v. 52, p. 889.

Schilling, J.H., 1979, Molybdenum resources of Nevada: Nevada Bureau of Mines and Geology Open-File Report 79-3, 193 p.

Schilling, J.H., 1980, Molybdenum deposits and occurrences in Nevada: Nevada Bureau of Mines and Geology Map 66.

Schilling, J.H., 1988, Evaluation of the potential for deep, virgin gold placer deposits in the Hampton placer claim, Osceola mining district, White Pine County, Nevada: unpublished report on file at Nevada Bureau of Mines and Geology.

Schilling, J.H., 1990, Mineral resources of the Flowery mining district, Storey County, Nevada: unpublished company report, Miramar Mining Corporation, 49 p.

Schrader, E.L., Jr., and Furbish, W.J., 1976, Analcime and thompsonite in Humboldt County, Nevada: Rocks and Minerals, v. 51, p. 283–285.

Schrader, F.C., 1912, A reconnaissance of the Jarbidge, Contact, and Elk Mountain mining districts, Elko County, Nevada: U.S. Geological Survey Bulletin 497, 162 p.

Schrader, F.C., 1913, Notes on the Antelope District, Nevada: U.S. Geological Survey Bulletin 530-A, p. 87–98.

Schrader, F.C., 1923, The Jarbidge mining district, Nevada, with a note on the Charleston District: U.S. Geological Survey Bulletin 741, 86 p.

Schrader, F.C., 1931a, Notes on ore deposits at Cave Valley, Patterson district, Lincoln County, Nevada: Nevada Bureau of Mines and Geology Bulletin 10, 16 p.

Schrader, F.C., 1931b, Spruce Mountain District, Elko County and Cherry Creek (Egan Canyon) District, White Pine County: Nevada Bureau of Mines and Geology Bulletin 14, 39 p.

Schrader, F.C., 1935a, Epithermal antimony deposits: Economic Geology, p. 658–665.

Schrader, F.C., 1935b, The Contact mining district, Nevada: U.S. Geological Survey Bulletin B 847-A, p. 1–41.

Schrader, F.C., 1947, The Carson Sink area: unpublished report on file with Nevada Bureau of Mines and Geology in the U.S. Geological Survey Open-File section, 233 p.

Schrader, F.C., Stone, R.W., Sanford, S., 1917, Useful minerals of the United States: U.S. Geological Survey Bulletin 624, p. 190–200.

Scott, J.D., 1974, Crystalline massicot on sheet lead, waste dump, Getchell mine, Nevada: Canadian Mineralogist, v. 12, p. 286–288.

Scott, R.B., 1965, Origin of the distribution of major elements in ignimbrite cooling units, eastern Nevada: Geological Society of America Special Paper, p. 180–181.

Scull, B.J., 1951, Development of melanterite and fibroferrite from gelatinous sulfate (Nevada): Geological Society of America Bulletin, v. 62, no. 12, pt. 2, p. 1520.

Seabrook, J., 1989, A reporter at large, invisible gold: New Yorker, v. 65, April 24, p. 45–81.

Searls, F., Jr., 1948, A contribution to the published information on the geology and ore deposits of Goldfield, Nevada: Nevada Bureau of Mines and Geology Bulletin 48, 24 p.

Shamberger, H.A., 1972, The story of Seven Troughs, Nevada: Nevada Historical Press, 57 p.

Shannon, E.V., 1922, Notes on an andorite-bearing silver ore from Nevada: Proceedings of the United States National Museum, Smithsonian Institution, Washington, D.C., v. 60, 5 p.

Shannon, E.V., 1923, Barrandite from Manhattan, Nevada: American Mineralogist, v. 8, p. 182–184.

Shannon, E.V., 1925, Benjaminite, a new sulfosalt mineral of the klaprotholite group: U.S. National Museum Proceedings, v. 65, p. 1–9.

Sharp, B.J., 1955, Uranium occurrence at the Moonlight mine, Humboldt County, Nevada: U.S. Atomic Energy Commission Report RME-2032, pt. 1, 14 p.

Sharwood, W. J., 1911, Tellurium-bearing gold ores: Economic Geology, v. 6, p. 32.

Shaver, S.A., 1991, Geology, alteration, mineralization, and trace element geochemistry of the Hall (Nevada Moly) Deposit, Nye County, Nevada, in Raines, G.L., Lisle, R.E., Schafer, R.W., and Wilkinson, W.H., eds., Geology and ore deposits of the Great Basin; symposium proceedings: Geological Society of Nevada, Reno, p. 303–332.

Shawe, D.R., 1977, Preliminary generalized geologic map of the Round Mountain Quadrangle, Nye County, Nevada: U.S. Geological Survey Miscellaneous Field Studies Map MF-833, 1:24,000.

Shawe, D.R., Poole, F.G., and Brobst, D.A., 1969, Newly discovered bedded barite deposits in east Northumberland Canyon, Nye County, Nevada: Economic Geology, v. 64, p. 245–254.

Shawe, F.R., Reeves, R.G., and Kral, V.E., 1962, Iron ore deposits of Nevada: Part C. Iron ore deposits of northern Nevada: Nevada Bureau of Mines and Geology Bulletin 53C, 135 p.

Sigurdson, D.R., 1974, Fluid inclusion thermometry and paragenesis at the Silver Dyke Mine, Mineral County, Nevada [abs.]: Geological Society of America Abstracts with Programs, v. 6, no. 3, p. 252–253.

Silberman, M.L., Johnson, M.G., Koski, R.A., and Roberts, R.J., 1973, K-Ar ages of mineral deposits at Wonder, Seven Troughs, Imlay, Ten Mile, and Adelaide mining districts in central Nevada: Isochron/West, v. 8, p. 31–35.

Silliman, B., 1866, On gaylussite from Nevada Territory: American Journal of Science, v. 42, p. 220–221.

Simon, G., Kesler, S.E., and Chryssoulis, S., 1999, Geochemistry and textures of gold-bearing arsenian pyrite, Twin Creeks, Nevada: Implications for deposition of gold in Carlin-type deposits: Economic Geology, v. 94, p. 405–422.

Simonds, F.W., 1989, Geology and hydrothermal alteration in the Calico Hills, southern Nevada [M.S. thesis]: Colorado State University, Fort Collins, Colorado, 136 p.

Sinkankas, J., 1959, Gemstones of North America: Van Nostrand Reinhold, New York, 675 p.

Sinkankas, J., 1967, Mineralogy for amateurs: Van Nostrand Reinhold, New York, 585 p.

Sinkankas, J., 1997, Gemstones of North America, vol. 3: Geoscience Press, Inc., Tucson, Arizona, 526 p.

Smith, A.F., Jr., 1985, Early history of the Comstock Lode, Nevada: Mineralogical Record, v. 16, p. 5–14.

Smith, D.T., 1904, Geology of the upper region of the main Walker River, Nevada: Bulletin of the Department of Geology, University of California, v. 4, p. 1–32.

Smith, F.D., 1902, The Osceola, Nevada, tungsten deposits: Engineering and Mining Journal, v. 73, p. 304–305.

Smith, G.H., 1943, The history of the Comstock Lode 1850–1920: Nevada Bureau of Mines and Geology Bulletin 37, 279 p.

Smith, M.R., and Benham, J.A., 1985, The Julie claim, Mineral County, Nevada: Mineralogical Record, v. 16, p. 75–80.

Smith, R.M., 1970, Treasure Hill reinterpreted: Economic Geology, v. 65, no. 5, p. 538–540.

Smith, R.M., 1976a, Mineral resources, in Geology and mineral resources of White Pine County, Nevada: Nevada Bureau of Mines and Geology Bulletin 85, p. 36–99.

Smith, R.M., 1976b, Mineral resources of Elko County, Nevada: U.S. Geological Survey Open-File Report 76-56, 201 p.

Smith, W.C., and Gianella, V.P., 1942, Tin deposit at Majuba Hill, Pershing County, Nevada: U.S. Geological Survey Bulletin 931-C, p. 39–55.

Smith, W.C., and Guild, P.W., 1944, Tungsten deposits of the Nightingale district, Pershing County, Nevada: U.S. Geological Survey Bulletin 936-B, p. 39–58.

Snoke, A.W., 1992, Clover Hill, Nevada: Structural link between the Wood Hills and the East Humboldt Range, in Wilson, J.R., ed., Field guide to geologic excursions in Utah and adjacent areas of Nevada, Idaho and Wyoming: Utah Geological Survey Miscellaneous Publications 92-3, p. 107–122.

Snoke, A.W., McKee, E.H., and Stern, T.W., 1979, Plutonic, metamorphic and structural chronology in the northern Ruby Mountains, Nevada [abs.]: Geological Society of America Abstracts, v. 11, p. 520.

Snyder, K.D., 1989, Geology and mineral deposits of the Rossi Mine area, Elko County, Nevada [Ph.D. diss.]: University of Nevada, Reno, 198 p.

Speed, R.C., 1962, Humboldt gabbroic complex, Nevada [abs.]: Geological Society of America Special Paper 73, p. 66.

Spencer, A.C., Hastings, J.B., Stone, R.W., and Clapp, C.H., 1917, The geology and ore deposits of Ely, Nevada: U.S. Geological Survey Professional Paper 96, 189 p.

Spurr, J.E., 1903, The ore deposits of Tonopah, Nevada: The Mining and Engineering Journal, v. 76, p. 796–770.

Spurr, J.E., 1906, Ore deposits of the Silver Peak Quadrangle, Nevada: U.S. Geological Survey Professional Paper 55, 174 p.

Staatz, M.H., and Bauer, H.L., 1954, Virgin Valley opal district, in Lovering, T.G., ed., Radioactive deposits of Nevada: U.S. Geological Survey Bulletin 1009-C, p. 80–81.

Staatz, M.H., and Carr, W.J., 1964, Geology and mineral deposits of the Thomas and Dugway Ranges, Juab and Tooele counties, Utah: U.S. Geological Survey Professional Paper 415, 188 p.

Stager, H.K., and Tingley, J.V., 1988, Tungsten deposits in Nevada: Nevada Bureau of Mines and Geology Bulletin 105, 256 p.

Stevens, D.L., 1971, Geology and ore deposits of the Antelope mining district, Pershing County, Nevada [M.S. thesis]: University of Nevada, Reno, 89 p.

Stewart, H.J., 1913, Early knowledge of Nevada, in Davis, S.P., ed., The History of Nevada: Elms Publishing Company, Reno, Nevada, p. 214–222.

Stewart, J.H., 1980, Geology of Nevada, a discussion to accompany the geologic map of Nevada: Nevada Bureau of Mines and Geology Special Publication 4, 136 p.

Stewart, J.H., McKee, E.H., and Stager, H.K., 1977a, Geology and mineral deposits of Lander County, Nevada: Nevada Bureau of Mines and Geology Bulletin 88, p. 1–59.

Stewart, J.H., Moore, W.J., and Zietz, I., 1977b, East-west patterns of Cenozoic igneous rocks, aeromagnetic anomalies, and mineral deposits, Nevada and Utah: Geological Society of America Bulletin v. 88, no. 1, p. 67–77.

Stier, D.J., 1998, Nevada minerals in the collection of The Springfield Science Museum, Springfield, Massachusetts: unpublished, 3 p.

Stoddard, C., and Carpenter, J.A., 1950, Mineral resources of Storey and Lyon Counties, Nevada: Nevada Bureau of Mines and Geology Bulletin 49, 115 p.

Stolburg, C.S., and Dunning, G.E., 1985, The Getchell Mine, Humboldt County, Nevada: Mineralogical Record, v. 16, p. 15–23.

Stretch, R.H., 1867, Annual report of the state mineralogist of the state of Nevada for 1866: Joseph E. Eckley, State Printer, Carson City, Nevada, 151 p.

Strong, M.F., 1966, Desert gem trails; A field guide to the gem and mineral localities of the Mojave Desert, Colorado Desert and adjacent areas of Nevada and Arizona: Gembooks, Mentone, California, 81 p.

Strunz, H., 1970, Mineralogische Tabellen, 5th ed.: Akademische Verlagsgesellschaft, Leipzig, p. 621.

Swaze, G., Clark, R., and Gallagher, A., 1993, Alteration at Cuprite, Nevada: U.S. Geological Survey Mineral Resources Newsletter, v. 4, no. 3, p. 2.

Takahashi, T., 1957, Supergene alteration of zinc and lead deposits [Ph.D. diss.]: Columbia University, New York, 163 p.

Takahashi, T., 1960, Supergene alteration of zinc and lead deposits in limestone: Economic Geology, v. 55, p. 1083–1115.

Tatlock, D.B., 1964, Kyanite group aluminous minerals, in Mineral and water resources of Nevada: Nevada Bureau of Mines and Geology Bulletin 65, p. 213–217.

Taylor, A.O., and Powers, J.F., 1955, Uranium occurrences at the Moonlight mine and Granite Point claims, Humboldt County, Nevada: U.S. Geological Survey Report TEM-874A, 16 p.

Taylor, H.B., 1912, A study of ores from Austin, Nevada: School of Mines Quarterly, Columbia University, New York, p. 32–39.

Theodore, T.G., and Blake, D.W., 1978, Geology and geochemistry of the West Ore Body and associated skarns, Copper Canyon Porphyry Copper Deposits, Lander County, Nevada (Part C), in Geochemistry of the porphyry copper environment in the Battle Mountain Mining District, Nevada: U.S. Geological Survey Professional Paper 798-C, 85 p.

Theodore, T.G., and Roberts, R.J., 1971, Geochemistry and geology of deep drill holes at Iron Canyon, Lander County, Nevada: U.S. Geological Survey Bulletin 1318, 32 p.

Thole, R.H., and Prihar, D.W., 1998, Geologic map of the Eugene Mountains, northwestern Nevada: Nevada Bureau of Mines and Geology Map 115.

Thomason, R.E., 1986, Geology of the Paradise Peak gold/silver deposit, Nye County, Nevada, in Tingley, J.V., and Bonham, H.F., Jr., eds., Precious-metal mineralization in hot springs systems, Nevada-California: Nevada Bureau of Mines and Geology Report 41, p. 90–92.

Thompson, G.A., 1956, Geology of the Virginia City Quadrangle, Nevada: U.S. Geological Survey Bulletin 1042-C, p. 45–77.

Thomssen, R., 1996, What's new in minerals in western North America: Australian Journal of Mineralogy, v. 2, no. 2, p. 78–80.

Thomssen, R.W., 1998, Species list—Silver Coin Mine, Iron Point District, Edna Mountains, Humboldt County, Nevada: unpublished, 1 p.

Thurston, R.H., 1949, The Daisy Fluorspar deposit near Beatty, Nye County, Nevada: U.S. Geological Survey Strategic Minerals Inventory Preliminary Report 3-209, 10 p.

Tingley, J.V., 1992, Mining districts of Nevada: Nevada Bureau of Mines and Geology Report 47, 124 p.

Tingley, J.V., and Berger, B.R., 1985, Lode gold deposits of Round Mountain, Nevada: Nevada Bureau of Mines and Geology Bulletin 100, 62 p.

Tingley, J.V., Castor, S.B., Garside, L.J., and LaPointe, D.D., 1998b, Mineral and energy resource assessment of the Carson City urban interface area: Nevada Bureau of Mines and Geology Open-File Report 98-5, 132 p.

Tingley, J.V., Castor, S.B., Weiss, S.I., Garside, L.J., Price, J.G., LaPointe, D.D., Bonham, H.R., and Lugaski, T.P., 1998a, Mineral and energy resource assessment of the Nellis Air Force Range: Nevada Bureau of Mines and Geology Open-File Report 98-1, 2 vols., 735 p.

Tolman, C.F., and Ambrose, J.W., 1934, The rich ores of Goldfield, Nevada: Economic Geology, v. 29, no. 3, p. 255–279.

Tonopah Daily Bonanza, 1908, Fine meteorite on exhibition in this city: 17 November, p. 4, col. 3–4.

Tonopah Miner, 1908, A two ton meteor: 21 November, p. 8, col. 2.

Tonopah Sun, 1908, Two-ton meteor found in Quinn Springs Range: 16 November, p. 4, col. 1–2.

Tonopah Times-Bonanza, 1954, Last rites today for C.C. Boak, patriarch dies at 84: 6 August, p. 1, col. 5; p. 12, col. 1–2.

Trengove, R.R., 1959, Reconnaissance of Nevada

manganese deposits: U.S. Bureau of Mines Report of Investigation 5446, 40 p.

Trites, A.F., and Thurston, R.H., 1958, Geology of Majuba Hill, Pershing County, Nevada: U.S. Geological Survey Bulletin 1046-I, p. 183–203.

Tschanz, C.M., and Pampeyan, E.H., 1970, Geology and mineral deposits of Lincoln County, Nevada: Nevada Bureau of Mines and Geology Bulletin 73, 182 p.

Twain, Mark, 1872, Roughing it: American Publishing Company, Hartford, Connecticut, 591 p.

Vajdak, J., 2001, New mineral finds in the first half of 2001: Mineral News, v. 17, no. 7, p. 1–2, 4.

Vanderburg, W.O., 1936a, Placer mining in Nevada: Nevada Bureau of Mines and Geology Bulletin 27, 176 p.

Vanderburg, W.O., 1936b, Reconnaissance of mining districts in Pershing County, Nevada: U.S. Bureau of Mines Information Circular 6902, 57 p.

Vanderburg, W.O., 1937a, Reconnaissance of mining districts in Mineral County, Nevada: U.S. Bureau of Mines Information Circular 6941, 79 p.

Vanderburg, W.O., 1937b, Reconnaissance of mining districts in Clark County, Nevada: U.S. Bureau of Mines Information Circular 6964, 81 p.

Vanderburg, W.O., 1938a, Reconnaissance of mining districts in Humboldt County, Nevada: U.S. Bureau of Mines Information Circular 6995, 54 p.

Vanderburg, W.O., 1938b, Reconnaissance of mining districts in Eureka County, Nevada: U.S. Bureau of Mines Information Circular 7022, 66 p.

Vanderburg, W.O., 1939, Reconnaissance of mining districts in Lander County, Nevada: U.S. Bureau of Mines Information Circular 7043, 83 p.

Vanderburg, W.O., 1940, Reconnaissance of mining districts in Churchill County, Nevada: U.S. Bureau of Mines Information Circular 7093, 57 p.

Van Gilder, K.L., 1963, The manganese orebody at the Three Kids mine, Clark County, Nevada [M.S. thesis]: University of Nevada, Reno, 94 p.

Vaniman, D.T., 1991, Calcite, opal, sepiolite, ooids, pellets, and plant/fungal traces in laminar-fabric fault fillings at Yucca Mountain, Nevada [abs.]: Geological Society of America Abstracts with Programs, v. 23, p. A117.

Vaniman, D.T., 1993, Calcite deposits in fractures at Yucca Mountain, Nevada: High level radioac-tive waste management; proceedings of the Fourth Annual International Conference, v. 4, p. 1935–1939.

Van Nieuwenhuyse, R. 1991, Geology and ore controls of gold-silver mineralization in the Talapoosa mining district, Lyon County, Nevada, in Raines, G.L., Lisle, R.E., Schafer, R.W., and Wilkinson, W.H., eds., Geology and ore deposits of the Great Basin; symposium proceedings: Geological Society of Nevada, Reno, p. 979–993.

Vikre, P.G., 1980, Fluid inclusions in silver-anti-mony-arsenic minerals from precious metal vein deposits: Economic Geology, v. 75, no. 2, p. 338–339.

Vikre, P.G., 1985, Precious metal vein systems in the National District, Humboldt County, Nevada: Economic Geology, v. 80, no. 2, p. 360–393.

Vikre, P.G., 1989, Fluid-mineral relations in the Comstock Lode: Economic Geology, v. 84, p. 1574–1613.

Vikre, P.G., 1994, Gold mineralization and fault evolution at the Dixie Comstock Mine, Churchill County, Nevada: Economic Geology, v. 89, p. 707–719.

Vikre, P.G., 1998, Intrusion-related, polymetallic carbonate replacement deposits in the Eureka district, Eureka County, Nevada: Nevada Bureau of Mines and Geology Bulletin 110, 52 p.

Vikre, P.G., and McKee, E.H., 1994, Geology, alteration and geochronology of the Como district, Lyon County, Nevada: Economic Geology, v. 89, p. 639–646.

Vitaliano, C.J., 1944, Contact metamorphism at Rye Patch, Nevada: Geological Society of America Bulletin, v. 55, no. 8, p. 921–950.

Vitaliano, C.J., 1951, Magnesium-mineral resources of the Currant Creek District, Nevada: U.S. Geological Survey Bulletin 978-A, 25 p.

Vitaliano, C.J., and Beck, C.W., 1963, Huntite, Gabbs, Nevada: American Mineralogist, v. 48, p. 1158–1163.

Vitaliano, C.J., and Callaghan, E., 1956, Geologic map of the Gabbs magnesite and brucite deposits, Nye County, Nevada: U.S. Geological Survey Miscellaneous Field Studies Map—MF 0035.

Vitaliano, C.J., and Callaghan, E., 1963, Geology of the Paradise Peak Quadrangle, Nevada: U.S. Geological Survey Geologic Quadrangle Map GQ-0250.

Volborth, A., 1962a, Allanite pegmatites, Red Rock, Nevada, compared with allanite pegmatites in southern Nevada and California: Economic Geology, v. 57, p. 209–216.

Volborth, A., 1962b, Rapakivi-type granites in the Precambrian complex of Gold Butte, Clark County, Nevada: Geological Society of America Bulletin, v. 73, p. 813–832.

Volk, J.A., Lauha, E., Leonardson, R.W., and Rahn, J.E., 1996, Roadlog for Trip B; structural geology of Carlin Trend; structural geology of the Betze-Post and Meikle deposits, Elko and Eureka Counties, Nevada, *in* Green, S.M., and Struhscaker, E., eds. Geology and ore deposits of the American Cordillera; field trip guidebook compendium: Geological Society of Nevada, Reno, p. 180–194.

Von Bargen, D., 1999, Nevada gold, silver, and copper deposits and their minerals: Rocks and Minerals, v. 74, p. 405–414.

Walck, C.M., 1989, Petrology and petrography of the metamorphic aureole associated with the Deep Post orebody, Eureka County, Nevada [M.S. thesis]: University of Missouri, Rolla, 49 p.

Walck, C.M., 1990, Petrology and petrography of the metamorphic aureole associated with the Deep Post orebody, Eureka County, Nevada, *in* Cuffney, B. (chairperson), Geology and ore deposits of the Great Basin; programs with abstracts: Geological Society of Nevada, Reno, 119 p.

Wallace, A.B., 1975, Geology and mineral deposits of the Pyramid District, southern Washoe County, Nevada [Ph.D. diss.]: University of Nevada, Reno, 162 p.

Wallace, A.B., 1979, Possible signatures of buried porphyry-copper deposits in middle to late Tertiary volcanic rocks of western Nevada, *in* Ridge, J.D., ed., Papers on mineral deposits of western North America; proceedings of the Fifth Quadrennial Symposium of the International Association on the Genesis of Ore Deposits; vol. 2: Nevada Bureau of Mines and Geology Report 33, p. 69–75.

Wallace, A.B., 1980, Geochemistry of polymetallic veins and associated wall rock alteration, Pyramid District, Washoe County, Nevada: Mining Engineering, v. 32, no. 3, p. 314–320.

Wallace, A.B., Drexler, J.W., Grant, N.K., and Noble, D.C., 1980, Icelandite and aenigmatite-bearing pantellerite from the McDermitt caldera complex, Nevada-Oregon: Geology, v. 8, p. 380–384.

Warner, L.A., Cameron, E.N., Holser, W.T., and Wilmarth, V.R., 1959, Occurrence of non-peg-matite beryllium in the United States: U.S. Geological Survey Professional Paper 318, 198 p.

Warren, R.E., 1973, Geology of the Fairview District (Groups 2 & 20), Churchill County, Nevada: unpublished report for Summa Corp., 9 p.

Weissberg, B.G., 1965, Getchellite, $AsSbS_3$, a new mineral from Humboldt County, Nevada: American Mineralogist, v. 50, p. 1817–1826.

Weissman, J.G., and Nikischer, A.J., 1999, Photographic Guide to Mineral Species: Excalibur Mineral Company CD-ROM, Peekskill, New York.

Wells, J.D., and Silberman, M.L., 1973, K-Ar age of mineralization at Buckhorn, Eureka County, Nevada: Isochron/West, v. 8, p. 37–38.

Westgate, L.G., and Knopf, A., 1932, Geology and ore deposits of the Pioche District, Nevada: U.S. Geological Survey Professional Paper 171, 79 p.

Westra, G., and Riedell, K.B., 1996, Geology of the Mount Hope stockwork molybdenum deposit, Eureka County, Nevada, *in* Coyner, A.R., and Fahey, P.L., eds., Geology and ore deposits of the American Cordillera; symposium proceedings: Geological Society of Nevada, Reno, p. 1639–1666.

Wherry, E.T., 1916, A peculiar intergrowth of phosphate and silicate minerals: Journal of the Washington Academy of Sciences, p. 105–108.

White, A.F., 1871, Report of the state mineralogist of the State of Nevada for the years 1869 and 1870: Carson City, Nevada, 128 p.

White, D.E., 1955, Thermal springs and epithermal ore deposits, *in* Bateman, A.M., ed., Economic Geology, Fiftieth Anniversary Volume, 1905–1955, part I: Economic Geology, v. 50, p. 99–154.

White, D.E., 1980, Steamboat Springs geothermal area: Society of Economic Geologists Field Trip Guide Book, p. 44–51.

White, D.E., Thompson, G.A., and Sandberg, C.H., 1964, Rocks, structure and geologic history of Steamboat Springs thermal area, Washoe County, Nevada: U.S. Geological Survey Professional Paper 458-B, p. B1–B63.

Whitebread, D.H., 1976, Alteration and geochemistry of Tertiary volcanic rocks in parts of the Virginia City Quadrangle, Nevada: U.S. Geological Survey Professional Paper 936, 43 p.

Whitebread, D.H., and Lee, D.E., 1961, Geology of the Mount Wheeler mine area, White Pine County, Nevada: U.S. Geological Survey Professional Paper 424-C, p. 120–122.

Whitehill, H.R., 1873, Biennial report of the state mineralogist of the State of Nevada for the years 1871 and 1872: Carson City, Nevada, 191 p.

Whitehill, H.R., 1875, Biennial report of the state mineralogist of the State of Nevada for the years 1873 and 1874: Carson City, Nevada, 191 p.

Whitehill, H.R., 1877, Biennial report of the state mineralogist of the State of Nevada for the years 1875 and 1876: Carson City, Nevada, 225 p.

Willden, R., 1963, General geology of the Jackson Mountains, Humboldt County, Nevada: U.S. Geological Survey Bulletin 1141-D, p. D1–D65.

Willden, R., 1964, Geology and mineral deposits of Humboldt County, Nevada: Nevada Bureau of Mines and Geology Bulletin 59, 164 p.

Willden, R., and Speed, R.C., 1974, Geology and mineral deposits of Churchill County, Nevada: Nevada Bureau of Mines and Geology Bulletin 83, 95 p.

Williams, G.K., 1980, Amenability studies in Kings River Summary Report: unpublished Chevron Resources Company Report, p. 11.1–11.4.

Williams, S.A., 1964, A new occurrence of langite: American Mineralogist, v. 49, p. 1143–1145.

Williams, S.A., 1965, A new occurrence of andorite: American Mineralogist, v. 50: 1498–1500.

Williams, S.A., 1968, Complex silver ores from Morey, Nevada: Canadian Mineralogist, v. 9, p. 478–484.

Williams, S.A., 1972, Elyite, a new mineral from Nevada: American Mineralogist, v. 57, p. 364–367.

Williams, S.A., 1973, Heyite, a new mineral from Nevada: Mineralogical Magazine, v. 39, p. 65–68.

Williams, S.A., 1979, Curetonite, a new phosphate from Nevada: Mineralogical Record, v. 10, p. 219–221.

Williams, S.A., 1985, Mopungite, a new mineral from Nevada: Mineralogical Record, v. 16, p. 73–74.

Williams, S.A., 1988, Pottsite, a new vanadate from Lander County, Nevada: Mineralogical Magazine, v. 52, p. 389–390.

Williams, S.A., and Millett, F.B., Jr., 1970, Complex silver ores from Morey, Nevada; a correction: The Canadian Mineralogist, v. 10, p. 275–277.

Wilson, J.R., 1963, Geology of the Yerington Mine: Mining Congress Journal, v. 49, no. 6, p. 30–34.

Wilson, W.E., and Thomssen, R.W., 1985, Steamboat Springs: Mineralogical Record, v. 16, p. 25–31.

Winchell, A.N., and Winchell, H., 1909, Elements of optical mineralogy: Van Nostrand, New York, 502 p.

Wise, W.S., 1978, Parnauite and goudeyite, two new minerals from the Majuba Hill Mine, Pershing County, Nevada: American Mineralogist, v. 63, p. 704–708.

Witters, G.C., 1999, History of the Comstock Lode, Virginia City, Storey County, Nevada: Rocks and Minerals, v. 74, p. 380–390.

Woitsekhowskiya, M., and Peters, S.G., 1998, Geochemical modeling of alteration and gold deposition in the Betze deposit, in Tosdal, R.M., ed., Contributions to the gold metallogeny on Northern Nevada: U.S. Geological Survey Open-File Report 98-338, p. 211–222.

Wood, J.D., 1988, Geology of the Sleeper gold deposit, Humboldt County, Nevada, in Schafer, R.W., James, J.C., and Vikre, P.G., eds., Bulk mineable precious metal deposits of the western United States; symposium proceedings: Geological Society of Nevada, Reno, p. 293–302.

Wood, J.D., 1991, Geology of the Wind Mountain gold deposit, Washoe County, Nevada, in Raines, G.L., Lisle, R.E., Schafer, R.W., and Wilkinson, W.H., eds., Geology and ore deposits of the Great Basin; symposium proceedings: Geological Society of Nevada, Reno, p. 1051–1061.

Yates, R.G., 1942, Quicksilver deposits of the Opalite district: U.S. Geological Survey Bulletin 931-N, p. 319–348.

Youngerman, A.G., 1992, Structural control, alteration and primary mineralization at the Big Springs gold mine, Elko County, Nevada [M.S. thesis]: University of Nevada, Reno, 79 p.

Zientek, M.L., Radtke, A.S., and Oscarson, R.L., 1979, Mineralogy of oxidized vanadiferous carbonaceous siltstone, Cockalorum Wash, Nevada [abs.]: Geological Society of America Abstracts with Programs, v. 11, p. 137.

Contributors

AUTHORS

DR. STEPHEN B. CASTOR worked in the minerals industry in the 1970s and 1980s, and has been a research geologist for the Nevada Bureau of Mines and Geology (NBMG) since 1988. In addition to NBMG publications and maps such as *Borates in the Muddy Mountains, Clark County, Nevada,* and *Geologic Map of the Frenchman Mountain Quadrangle, Nevada,* he has published papers on mineralogy and economic geology in *American Mineralogist, Canadian Mineralogist, Economic Geology, Ore Geology Reviews,* and *Journal of Geochemical Exploration.*

GREGORY C. FERDOCK worked for both Newmont Gold Company and Barrick Gold Corporation in Nevada's Carlin trend during the 1980s and 1990s. He is currently a Ph.D. candidate in geology at the University of Nevada, Reno, and owns CGF Natural Resources, a company that specializes in mineral collectibles. Greg has published papers for the U.S. Geological Survey and the *Society of Economic Geologists Guidebook Series.* In 1999 he completed data compilation for the *Minerals of Nevada* catalog.

CONTRIBUTORS

FREDERICK J. BREIT JR. is a mine geologist for the Newmont Mining Corporation at its Twin Creeks Operation in Humboldt County, Nevada. He has published an article on the Brawley Peaks, California, and Aurora, Nevada, gold deposits for the Geological Society of Nevada. In 1992 he began data compilation for the *Minerals of Nevada* catalog.

FORREST E. CURETON II was owner of Cureton Mineral Company between 1950 and 1994, has published papers in the *Mineralogical Record, Canadian Mineralogist,* and *Mineralogical Magazine,* and has had a Nevada type mineral (curetonite) named after him. He is presently volunteer mineralogist at

the Empire Mine State Historic Park, Grass Valley, California, and with his wife, Barbara Cureton, has amassed a collection of rare and unusual mineral specimens.

DAVID A. DAVIS has worked as a mineral preparator for the NBMG, and is currently the information officer for the NBMG. He has contributed articles to NBMG publications and to the *Nevada State Museum Newsletter.* His interests include western Nevada mining history, metamorphic and igneous rocks in western Nevada, and meteorites.

DR. LIANG-CHI HSU was professor of geology and mineralogy at the Mackay School of Mines as well as a research mineralogist and geochemist at NBMG from 1969 to 1996. In addition to NBMG publications, he contributed papers to *American Mineralogist, Contributions to Mineralogy and Petrology, Economic Geology, Journal of Geochemical Exploration,* and *Economic Geology.*

MARTIN C. JENSEN has a master of science degree in geology/geochemistry and has worked as a research chemist for Hercules Aerospace, as an electron microscopy specialist for the Mackay School of Mines, and as an engineering geologist and inspector for a private consulting firm. He has written extensively for *The Mineralogical Record,* with comprehensive papers on the mineralogy of numerous Nevada occurrences including the Majuba Hill, Gold Quarry, and Meikle Mines. He has also contributed articles to *Canadian Mineralogist, Mineralogical Magazine, Mining Artifact Collector, Journal of Materials Science, Rocks and Minerals,* and the *Mineral News.* A mineral species new to science, jensenite, from the Tintic district, Utah, was named in his honor in 1996.

DR. JOHN C. KEPPER was Associate Professor of Geology in the University of Nevada system during the 1970s and 1980s, and subsequently worked in exploration research for several mining companies and as an environmental scientist. He is now retired, but is an active mineral collector with published articles on minerals from the Goodsprings district, Nevada. His current interest is in the application of mineralogy and ore deposit geology to the understanding of mining and ore processing at the fourth-century-B.C. lead-silver mines at Laurium, Greece.

SCOTT KLEINE is the owner of Great Basin Minerals, Reno, Nevada, a company that mines and sells collectible minerals, and is best known for his work on the rare California mineral fresnoite. He is an avid field collector, has an extensive collection of Nevada mineral specimens, and has published articles in *Rock and Gem, Rock and Mineral, The Eclectic Lapidary,* and *Mineral News.*

DAPHNE D. LAPOINTE worked in mineral exploration in the 1970s. She is now a research geologist for the NBMG, for which she has published articles and bulletins including *Mineral Resources of Elko County, Nevada.* She has also published papers in *Mining Engineering,* and is in charge of earth science education and public outreach for the NBMG.

KEITH G. PAPKE worked as a mining industry geologist in the 1950s and 1960s and served as the industrial minerals specialist for the NBMG from the 1960s to the 1990s. He is best known for his NBMG bulletins, including *Talcose Minerals in Nevada, Fluorspar in Nevada,* and *Barite in Nevada.* He also published papers in *Clays and Clay Minerals* and *American Mineralogist.*

CHRISTOPHER H. ROSE, as owner of High Desert Gems and Minerals, is a miner and purveyor of gemstones from deposits in the western United States, including *tourmaline* from California, sunstone from Oregon, and blue *chalcedony* from the Mount Airy Mine in Nevada. He has published an article on the latter in the *Lapidary Journal.*

JOHN H. SCHILLING (deceased) was Director of the NBMG during the 1960s, 1970s, and 1980s. He published an important article on the magnesium deposit at Gabbs, Nevada, for the American Institute

of Mining, Metallurgical and Petroleum Engineers. He also produced NBMG reports and maps, including inventories of Nevada molybdenum deposits and isotopic age dates of Nevada rocks. Prior to his death, he began collecting data for a book on the minerals of Nevada.

RICHARD W. THOMSSEN has worked as a professional geologist for more than 40 years, and is an avid micromineral collector with an extensive collection that includes many rare Nevada specimens. He has published articles in *Rocks and Minerals, The Mineralogical Record,* and the *Lapidary Journal,* and is known for his contribution on gold specimens in *The F. John Barlow Mineral Collection.*

JOSEPH V. TINGLEY has served as an economic geologist at the NBMG since 1978 following a career working for various mining companies in the 1960s and 1970s. He is well known for his NBMG road log and mining history publications, as well as economic geology publications, including *Mining Districts of Nevada,* and *Lode Gold Deposits of Round Mountain, Nevada.* He has also published papers in *Mining Engineering* and the *Journal of Geochemical Exploration.*

PHOTOGRAPHERS

JEFFREY SCOVIL is a world-renowned photographer of minerals, fossils, gemstones, and jewelry who regularly publishes photographs in *The Mineralogical Record, Lapidary Journal, Rocks and Minerals,* and other journals and magazines. He has also published a book entitled *Photographing Minerals, Fossils, and Lapidary Materials.*

JEFFREY G. WEISSMAN is a collector of micromineral specimens and author of *Photographic Guide to Mineral Species,* a comprehensive CD-ROM that contains images of more than 3000 mineral species.

SUGAR WHITE has been collecting micromineral specimens for many years, particularly in Nevada and the western United States. She is well known for fine mineral photomicrographs using natural light. Her work may be seen in *The Mineralogical Record, Rocks and Minerals,* and other publications.

Index

Bogwood, 77
Bonanza King Formation, 94, 95
Bonanza Opal Mine, 75
Bond, Brian, 117
Boomtowns, 4
Bootstrap district, 36
Borate minerals, 4, 11, 18, 70
Borax, 70
Borealis district, 57
Boss Mine, 92, 95, 97, 99, 105
Bottle Creek district, 27, 68
Boulder Hill Mine, 71
Breccias, 10, 56
Bristol district, 18, 30
Britton, Lee, 80
Brochantite, 111
Broken Hills district, 25
Bromargyrite, 19
Brown, T. C., 91
Browne, J. Ross, 3, 4, 18
Brucite, 8, 25, 26, 27, 71 table 11.
    See also Gabbs brucite deposits
Buckhorn district, 34
Buckskin district, 64
Buckskin Mine, 73
Buena Vista Hills, 8, 27
Buena Vista (Unionville) district,
    15
Bullfrog district, 10, 19, 38, 57, 59
Bullfrog Mine, 37, 56
Bullion district, 23, 34
Bunkerville district, 6
Burgess, J. A., 19
Burrus Mine, 57, 110
Burton, Bob, 81
Butler, Jim, 19

Cactus Springs district, 57, 59
Cahill Mine, 69
Calaverite, 53
Calcite, 38, 49, 64
Caliente, Nev., 10
Callaghan, E., 26, 28
Callaghanite, 41, 71
Callville borate deposits, 24
Cambrian Period, 7, 94
Candelaria district, 13
Candelaria Mine, 36, 110–11
Candelaria Variscite Mine, 79
Canyon Diablo meteorite, 89
Carbonate rock deposition, 7
Carico Lake Turquoise–Faustite
    Mine, 79
Carlin district, 36
Carlin Mine: and Carlin Trend,
    36; and Carlin-type gold
    deposits, 47, 48; and gold, 30;
    and mercury, 67; and Nevada
    type localities, 40
Carlin Trend: and Carlin-type

gold deposits, 47; gold produc-
    tion, 36, 37; and Gold Strike
    open pit, 4; and invisible gold,
    32, 34; and Tertiary Period, 8,
    10
Carlin-type gold deposits: and
    Cenozoic Era, 8; characteristics
    of, 47; discovery of, 31–34;
    map of, 48; and mineralogy, 8,
    37–38, 47, 48–49, 50–53 table
    3, 54, 55; and mining meth-
    ods, 36; and Nevada type local-
    ities, 40, 41, 44; production of,
    47; and Roberts, 28; specimens
    of, 54–55
Carnelian, 77
Carpenter, J. M., 25
Carson, Kit, 13
Carson and Colorado Railroad,
    70
Carson district, 65
Cassiterite, 10
Castle Peak district, 57, 68, 69
C. C. Boak's meteorite, 86 table
    12, 89
Cement, 70
Cenozoic Era, 8–11, 49
Central Pacific Railroad, 4
Chalcedony, 12, 77
Chalcopyrite, 53, 95
Chalcosiderite, 78–81
Chalk Mountain district, 64, 65
    table 9
Chalk Mountain Mine, 111
Champion Mine, 25, 71, 73
Chert, 7, 47, 77
Chimney Creek deposit, 37
Chlorargyrite, 14, 16, 19, 56
Chondrites, 85
Chromium, 7
Chrysoberyl, 73
Church, David, 76
Churchill County, 82
Cinnabar: as a gemstone, 82; and
    mercury, 22, 67, 68, 69; and
    mineral collecting, 116; and
    Native American mining, 12,
    67
Citrine, 77, 116
Clastic rocks, 7
Clay minerals, 4, 11, 70
Clinozoisite, 105
Clinton, H. G., 25, 26, 108, 117
Clipper deposit, 7
Cobalt, 6, 7, 18, 92, 94, 97
Cold Spring district, 16
Cold War: and mining, 29–31
Colemanite, 24
Colorado district, 12
Colorado Plateau province, 6

Columbia Plateau province, 6
Columbus (Candelaria) district,
    18
Computer-generated modeling,
    37
Computerized structural analy-
    ses, 37
Comstock district: and electrum,
    59; and gold, 34; and mercury,
    67; and mineral collecting,
    108; mineral species of, 18; in
    1930s, 26; and salt, 70; and sil-
    ver, 57; and volcanic rocks, 56
Comstock Lode: and develop-
    ment of Nevada, 3, 4; and elec-
    trum, 5; and epithermal
    deposits, 56; and first Nevada
    boom 1859–1900, 13–16; and
    hard-rock mining, 3, 13; and
    Mexican mining, 13; and min-
    eral collecting, 111; and min-
    ing innovations, 4; new gold
    deposits compared to, 31; pro-
    duction from, 14; and silver, 4,
    13–14, 15, 59; and Tertiary
    Period, 10
Consolidated Coppermines
    Corporation, 21, 30, 32
Consolidated Virginia Mine, 14
Contact metasomatic deposits,
    60, 62–66
Contact thrust, 93
Continental accretion, 7
Coope, Alan, 31, 34
Copper: and Bunkerville district,
    6; and Cretaceous Period, 8;
    deposits of, 3; and
    Goodsprings district, 91, 92,
    94, 95, 96, 97; and Korean War,
    33; and mineral collecting, 26,
    109; in 1930s, 26; in 1950s, 30;
    in 1960s, 30; in 1980s, 36; and
    Robinson district, 20–22, 26,
    27, 30; and Santa Fe district,
    21; skarn copper, 7, 8, 60, 62,
    63; and World War I, 23; and
    World War II, 27, 28; and
    Yerington district, 21, 60. See
    also Porphyry copper
Copper Basin, 31, 63
Copper-bearing oxide ore miner-
    als, 96–97
Copper boom, 4
Copper Canyon, 10, 36, 63
Cordero Mine, 29, 30, 40, 68, 68
    table 10
Cordero Mining Company, 29
Cordilleran Miogeocline, 6
Cornetite, 105
Cortez district, 10, 16, 34

Getchell Mine: and Carlin-type gold deposits, 47, 48, 54; and getchellite, 40; and gold, 37; and invisible gold, 32; and mineral collecting, 39, 105, 109, 112; and World War II, 28
Getchell Mining Company, 28
Getchell Trend, 4, 8, 37, 47
Gianella, V. P., 26, 28, 89
Gilbert district, 7
Gillis Range, 8
Gilmour, E. H., 96
Gneisses, 6
Godber Turquoise Mine, 81
Golconda district, 71
Golconda thrust, 7
Gold: and Comstock Lode, 4; and contact metasomatic deposits, 65; deposits of, 3; and Goodsprings district, 91–93, 92 figure 3, 94; invisible gold, 31–34, 47; and Manhattan district, 20, 26, 28; and mercury as mining by-product, 67; mining boom, 4, 8; and Mormon mining, 13; in 1930s, 26; in 1950s, 30; in 1960s, 30; in 1970s and 1980s, 35; in 1980s and 1990s, 36–38; and porphyry-gold deposits, 61–62; production of, 36; and Robinson district, 8, 34; and Round Mountain district, 19; and Tertiary Period, 8, 10; and World War II, 28, 33. *See also* Carlin-type gold deposits
Gold Acres Mine, 32, 34, 47
Goldbanks district, 22, 68, 69
Gold Bar Mine, 54
Gold-bearing pyrite, 49
Gold-bearing sulfides, 49
Gold Butte district, 6
Gold Canyon, 13, 16
Gold Circle district, 59
Gold Crater district, 57
Goldfield district: discoveries of, 19–20; and epithermal deposits, 56, 57; and gold, 34; and high-sulfidation deposits, 58; and mineral collecting, 108; minerals reported from, 3, 58, 58 table 4; and Nevada type localities, 40–41, 44; and Tertiary Period, 10; and volcanic rocks, 56
Goldfield Gem Claim, 83
Goldfieldite, 40, 44, 58, 58 table 4
Gold Hill, 13
Gold Pick pit, 54
Goldquarryite, 40

Gold Quarry Mine, 32, 40, 54, 106, 112
Goldstrike Mine, 48, 48 fig. 2, 54, 78, 112–13
Gold Strike open pit, 4
Goniometric measurements, 19
Goodsprings district: age of mineralization, 93–94; age of oxidation, 94; and contact metasomatic deposits, 66; geologic map of, 92; geology of, 93; and hard-rock metal mining, 3; and host rock minerals, 96; location of, 91; and mineral collecting, 105; minerals of, 100 table 13; and Mormon mining, 13; and Nevada type localities, 41; and 1950s, 30; primary ore mineralogy, 95–96; and railroads, 93; and secondary minerals, 96–101; structural and stratigraphic controls of ore, 94–95; and Triassic Period, 7, 93; and zinc, 20, 91–92, 93, 97–98
Gordon, Harvey, 38, 74, 113, 118
Goudey, Hatfield, 28
Goudeyite, 41, 61, 105
Granger, A. E., 26, 31
Granites, 8
Granitic porphyry, 8, 96
Granitic rocks, 6
Granodiorite, 8, 77
Graphite, 70
Grawe, O. R., 25
Gray, Edwin, 20, 21
Gray, Jerry, 74
Great Basin region, 6, 23
Great Depression, 26–27
Green, Leslie, 89
Green Monster Mine, 95, 97, 99, 101, 105
Green Talc Mine, 73
Greystone deposit, 7
Grundy, Isaac, 13
Gypsum: and Cretaceous Period, 8; as industrial mineral, 70, 71; and Mesozoic Era, 7; and 1920s, 25; and Paleozoic Era, 7; production of, 4; and Tertiary Period, 11

Hague, Arnold, 16
Halite, 4, 11, 70
Hallman, Foster, 115
Hall Mine, 8, 82, 113
Hall Molybdenum Mine, 61
Hand, Lee, 79
Hard-rock mining, 3, 13
Hausen, D. M., 34, 37

Heald, P., 38
Heavy Metals Program, 34
Hematite, 10
Hemimorphite, 111
Henderson, Nev., 3
Hess, F. L., 23
Heterogenite, 105
Hewett, D. F., 91, 93, 94, 95, 97, 98, 99, 105
Hexahedrites, 85
High Line Mine, 97
High-sulfidation deposits, 57, 58
Hill, J. M., 25
Himalaya Mining Company, 79
Hinrichs, E. N., 72
Hitchborn, A. D., 48
Hlava, P., 96, 99
Hodson family, 75
Hodson opal, 75
Hoffman, W. J., 72
Hofstra, A. H., 48, 49, 59
Hog Ranch Mine, 57, 59
Hollabaugh, C. L., 109
Holmes, G. H., 28
Holocene, 11
Honest Miner claim, 12
Horn silver, 16, 56
Horse Canyon Mine, 37
Horton, R. C., 12
Hot spring mineralization, 10
Hot Springs meteorite, 86 table 12, 90
House, Carl, 81
Howell, Eugene, 88
Hübnerite, 18, 22, 40, 44
Humboldt County, 82–83
Humboldt Iopolith, 8
Humboldt Marsh, 11
Humboldt Range, 22
Hunt, S. F., 26
Huntite, 71
Hurlbut, C. S., Jr., 28
Hydromagnesite, 25
Hydrothermal activity, 8, 49, 70, 75
Hydrothermal alteration, 19
Hydrothermal processes, 37, 49
Hydrozincite, 105

Igneous activity, 6, 7, 10
Igneous intrusions, 6
Igneous rocks, 10, 47, 49, 70
Ilchick, R. P., 48
Illite, 49
Imlay district, 32
Incline Village pegmatites, 113
Independence district, 54
Independence Mountains, 8, 47
Independence Mountains Mines, 48

McCormack, J., 99
McCormick, Harold, 90
McCoy district, 19, 37, 65
McCoy Mine, 37, 65
McDermitt caldera, 10, 30, 35, 40, 68
McDermitt Mine, 29, 40, 67, 68 table 10, 69
Megapit Mine, 4
Meikle Mine: and barite, 39, 54, 82; and gold, 36; and mineral collecting, 38, 39, 108, 110, 114; and sphalerite, 49
Melhase, John, 27, 54, 108
Mercur Mine (Utah), 31, 47
Mercury: deposits of, 3, 67–69; discoveries of, 22; and Korean War, 33; and Nevada type localities, 40; and 1920s, 24; and 1950s, 30; and 1980s, 36; and Tertiary Period, 10; and World War II, 27, 28, 29
Mesosiderites, 87
Mesozoic Era, 7, 8, 68, 93, 94
Metallic mineralization, 7, 10
Metallic-ore deposits, 7
Metamorphic rocks: and Carlin-type gold deposits, 47; and gemstones, 72, 73, 74; and mountain ranges, 6
Metaschoderite, 41
Metastibnite, 40, 44
Meteorites: appearance of, 87; composition of, 85, 87; data on, 86 table 12; examples of, 86 table 12, 88–90; location of, 87–88; and terminology, 84
Meteoroids, 84
Meteors, 84
Mexicans, and mining, 12–13
Microbeam technology, 44
Microscopic Determination of Non-Opaque Minerals (Larsen), 25
Microscopical Determination of Opaque Minerals (Murdoch), 24
Midas district, 10
Midas Mine, 37, 56, 57
Middle Proterozoic, 6
Milford #1 Mine, 97, 98, 99, 100
Milford #3 Mine, 95, 98, 99, 101
Mill City district, 8, 23, 26, 27, 64
Mineral Basin (Buena Vista) district, 18, 27, 65
Mineral collecting: and Buena Vista Hills, 8; and Majuba Hill Mine, 62; market growth in, 5, 38–39; Nevada as site for, 108–18; in 1920s, 25–26; in 1930s, 27; in 1950s and 1960s, 31; in 1980s and 1990s, 38; per-

sonal history of, 105–7; and skarn deposits, 64; and volcanic rocks, 59
Mineral County, 83
Mineralogy: and Carlin-type gold deposits, 8, 37–38, 47, 48–49, 50–53 table 3, 54, 55; and Comstock Lode, 14; and first mining boom 1859–1900, 16–19; and Goldfield district, 19–20; and Goodsprings district, 95–96; growth of field, 108; and mercury deposits, 68–69; and Nevada type localities, 41; and 1920s, 24–26; and 1950s and 1960s, 31; and 1970s and 1980s, 35; and Robinson district, 21; and Tertiary Period, 10; and Tonopah, 19; and World War I, 23–24; and World War II, 28
Minerals: importance of, 3–5; industrial, 70–71; strategic, 27–29, 33
Mineral wealth, 3, 4, 6, 8
Mining: and Cold War, 29–31; and development of Nevada, 3–4; 1859–1900, 13–19; and energy-related mining, 35–36; hard-rock, 3, 13; innovations in, 4; and invisible gold, 31–34; and Mexicans, 12–13; and mining district discovery rates, 15–16; and Mormons, 3, 13; and Native Americans, 3, 12, 67, 78; 1900–1918, 19–23; 1920s, 24–26; 1930s, 26–27; 1950s and 1960s, 29–31; 1970s and 1980s, 35–36; 1980s and 1990s, 36–38; placer mines, 13, 28; present and future, 38–39; and World War I, 23–24; and World War II, 27–29. *See also* Open-pit mining; Precious-metal mining; Underground mining
Mining camps, spread of, 15–16
Mining districts: discovery rates of, 15, 15 fig. 1, 16; ore minerals of, 18; organization of, 19; value of production in, 9 table 1. *See also specific districts*
Mining economy, and Carlin-type gold deposits, 9
Mining towns, histories of, 3–4
Minnesota Mine, 64
Miocene, 10, 11, 75, 93, 94
Miogeoclinal deposition, 6, 7
Mississippian, 7, 94
Mitchell, J. R., 77
Mobile Mine, 99

Modell, David, 26
Mohawk Mine, 40
Molybdenum: and Devonian, 7; and Goodsprings district, 92; and 1970s and 1980s, 35, 36; production of, 4; and Robinson district, 8; and Tertiary Period, 10
Monte Cristo Formation, 94, 95
Montmorillonite clay, 11
Moonlight Mine, 30
Moonstone (feldspar), 74
Moonstone (obsidian), 83
Morey district, 7
Mormons, and mining, 3, 13
Mountain City district, 16, 26, 27
Mountain district, 16
Mountain ranges, 6
Mount Airy Blue Chalcedony Mine, 77
Mount Hamilton district, 8
Mount Hope district, 10, 35, 61
Mount Wheeler, 8
Mount Wheeler Mine, 114–15
Muddy Mountains, 24
Muller, S. W., 33
Murdoch, J., 24
Murphy, L. J., 89
Murray Mine, 38, 39, 54, 110, 115
Myler Mine, 105

National district, 59
National Research Council, 23
Native Americans, and mining *See Mining, and Native Americans*
Native gold, 49, 59
Native silver, 15, 53, 59
Neutron activation analysis, 31
Nevada: economy of, 9, 16; industrial minerals of, 70–71; meteorites of, 84–90; mineralogical collection of, 17–18; as site for mineral collecting, 108–18
Nevada Bureau of Mines and Geology (NBMG), 25, 26, 28, 30–31, 35
Nevada Consolidated Copper Company, 21
Nevada district, 23
Nevada Territory, 3, 15, 16
Nevada Test Site, 13
Nevada type localities, 40–41, 42–43 table 2, 44, 61, 68, 71, 105
New Almaden Mine, 67
Newberry Mountains, 6
Newmont Mining Company, 4, 31, 32, 33, 34

Wet chemical methods, 41
Whale Mine, 99
Wheeler Peak Mine, 65, 73
White, S., 99
White Caps Mine: and Carlin-type gold deposits, 47; and invisible gold, 31–32; and mineral collecting, 105–6, 108, 117–18; in 1930s, 26; and stibnite, 54
White Pine Copper Company, 21
White Pine County, 83
White Pine district, 16, 18, 56, 63
Whitmore, Cliff, 88–89
Widmanstätten patterns, 88
Wier, Jeanne, 88
Wieting, Aaron, 116
Wieting, Jade, 116
Wildlife habitat, 38
Williams, S. A., 31, 35
Williams, Sidney, 41
Winchell, A. N., 19
Wingfield, George, 32

Winnemucca, Nev., 3
Wise, W. S., 35
Witters, G. C., 111
W. M. Keck Museum, 23, 26, 97, 105, 109, 115, 117
Woitsekhowskya, M., 49
Wollastonite, 64
Women, and mining, 27
Wonderstone, 82–83
World War I: and mining, 23–24; and tungsten, 22, 23–24
World War II, and mining, 27–29
Wulfenite, 99–101

X-ray diffraction methods, 26, 31, 44
X-ray fluorescence spectrometry, 31

Yellow Pine district. *See* Goodsprings district
Yellow Pine Mine, 94–95, 97, 98, 99, 105

Yellow Pine Mining Company, 91, 93
Yellow Pine sill, 93, 94
Yerington, Nev., 3, 7, 105
Yerington batholith, 7
Yerington district: and copper, 21, 60, 61, 62, 63, 64; discovery of, 16

Zapot claims, 8, 78, 118
Zapot Mine, 106
Zapot pegmatite, 74
Zeolite minerals, 4, 70
Zinc: and Goodsprings district, 20, 91–92, 93, 97–98; and 1950s, 30; and 1980s, 36; and Robinson district, 8; and Tertiary Period, 10; and World War I, 23; and World War II, 28
Zoisite, 82